Free Updates

with three easy steps…

1. Register today on www.pricebooks.co.uk/updates

2. We'll alert you by email when new updates are posted on our website

3. Then go to www.pricebooks.co.uk/updates
    and download the update.

All four Spon Price Books – *Architects' and Builders'*, *Civil Engineering and Highway Works*, *External Works and Landscape* and *Mechanical and Electrical Services* – are supported by an updating service. Two or three updates are loaded on our website during the year, typically in November, February and May. Each gives details of changes in prices of materials, wage rates and other significant items, with regional price level adjustments for Northern Ireland, Scotland and Wales and regions of England. The updates terminate with the publication of the next annual edition.

As a purchaser of a Spon Price Book you are entitled to this updating service for this 2014 edition – free of charge. Simply register via the website www.pricebooks.co.uk/updates and we will send you an email when each update becomes available.

If you haven't got internet access or if you've some questions about the updates please write to us at Spon Price Book Updates, Spon Press Marketing Department, 3 Park Square, Milton Park, Abingdon, Oxfordshire, OX14 4RN.

Find out more about Spon books
Visit www.pricebooks.co.uk for more details.

Spon's Architects' and Builders' Price Book

2014

Spon's Architects' and Builders' Price Book

Edited by

Davis Langdon, An AECOM Company

2014

One hundred and thirty-ninth edition

CRC Press
Taylor & Francis Group

First edition 1873
One hundred and thirty-ninth edition published 2014
by CRC Press
2 Park Square, Milton Park, Abingdon, Oxon, OX14 4RN

and by CRC Press
Taylor & Francis, Broken Sound Parkway, NW, Suite 300, Boca Raton, FL 33487

CRC Press is an imprint of the Taylor & Francis Group, an informa business

© 2014 Taylor & Francis

The right of Davis Langdon to be identified as the Author of this Work has been asserted by them in accordance with the Copyright, Designs and Patents Act 1988

Trademark notice: Product or corporate names may be trademarks or registered trademarks, and are used only for identification and explanation without intent to infringe.

The publisher makes no representation, express or implied, with regard to the accuracy of the information contained in this book and cannot accept legal responsibility or liability for any errors or omissions that may be made.

British Library Cataloguing in Publication Data
A catalogue record for this book is available from the British Library

ISBN13: 978-1-4822-0406-3
Ebook: 978-1-4822-0407-0

ISSN: 0306-3046

Typeset in Arial by Taylor & Francis Books

MIX
Paper from
responsible sources
FSC   FSC® C013056
www.fsc.org

Printed and bound in Great Britain by
TJ International Ltd, Padstow, Cornwall

Contents

PART 4: PRICES FOR MEASURED WORKS

PART 5: FEES FOR PROFESSIONAL SERVICES

PART 6: DAYWORK AND PRIME COST

PART 7: USEFUL ADDRESSES FOR FURTHER INFORMATION

PART 8: TABLES AND MEMORANDA

Preface to the One Hundred and Thirty-Ninth Edition

Recent construction cctivity trends

The economy was broadly flat in the first quarter of 2013 and reflects what has been happening to construction prices generally.

There have been increases in some construction material prices and increases in directly employed labour costs but tender prices in general remained flat as contractors continue to compete for work in a declining market.

The general feeling is that prices have nowhere further to fall, and in some sectors some prices have started to rise as the market is caught between rising input costs, both known and predicted, and the downward drag of construction workload.

Materials

The new year brought about the usual flurry of price increase announcements.

During the year ahead material prices are likely to be subdued. Steel prices are being forecast to remain flat with raw materials in surplus supply. Oil prices have recently fallen, and the exchange rate of the pound to the euro is most likely to contribute to bringing the cost of imported construction materials down.

Price adjustment formulae for construction contracts

Price Adjustment Formulae indices, compiled by the Building Cost Information Service (previously by the Department for Business, Innovation & Skills), are designed for the calculation of increased costs on fluctuating or Variation of Price contracts. They provide useful guidance on cost changes in various trades and industry sectors and on the differential movement of work sections in Spon's Price Books. Over the last 12 months between April 2012 and April 2013, the 60 Building Work categories recorded an average rise of just 1.4%.

Those works categories showing the highest increases over the last 12 months have been:	
Cladding and Covering: Lead	6.7%
Linings and Partitions: Plasterboard	5.4%
Finishes: Painting and Decorating	3.6%
Finishes: Screeds	3.6%
Finishes: Plaster	3.5%
Suspended Ceilings	3.5%
Excavation and Disposal	3.2%

Those categories showing the smallest increases or in some cases decreases are:	
Concrete: Reinforcement	−3.8%
Finishes: Bitumen, Resin and Rubber Latex Flooring	−2.7%
Insulation	−1.0%
Filling: Imported, Hardcore and Granular	0.8%
Concrete: In Situ	1.8%
Concrete: Formwork	2.0%

Labour – wage agreements

There are several increases in labour costs, most notably the 2% increase in basic wages under the national wage agreement.

Wage rates for heating and ventilating operatives rose by 1.5% in April.

Book price level

The price level of Spon's A&B 2014 has been indexed at 453. Readers of Spon's A&B are reminded that Spon's is the only known price book in which key rates are checked against current tender prices.

Measurement changes – New Rules of Measurement (NRM)

The NRM is a suite of documents to provide a standard set of measurement rules that are understandable by all those involved in a construction project.

This year's edition has been re-formatted to move towards complying with NRM1 and NRM2. There will be further refinement over the next two or three editions as NRM becomes more embedded in general practice and work-flows.

There are three parts to the NRM suite:

NRM 1:

Order of cost estimating and cost planning for capital building works provides fundamental guidance on the quantification of building works for the purpose of preparing cost estimates and cost plans. Direction on how to quantify other items forming part of the cost of a construction project, but which are not reflected in the measurable building work items, is also provided, i.e. preliminaries, overheads and profit, project team and design team fees, risk allowances, inflation, and other development and project costs.

NRM 2:

Detailed measurement for building works provides detailed rules for the measurement and description of building works for the purpose of obtaining tender prices. The rules address all aspects of bill of quantities production, including setting out the information required from the project team to enable a bill of quantities to be prepared, as well as dealing with the quantification of non-measurable work items, contractor designed works and risks. Guidance is also provided on the content, structure and format of bill of quantities, as well as the benefits and uses of bill of quantities.

NRM 3:

Order of cost estimating and cost planning for building maintenance works.

Profits and overheads

The 2014 edition includes a 2.5% mark-up for main contractor's overheads and profit.

Preliminaries

There are continuing signs that preliminaries costs are beginning to soften, but they still typically range from 10% to 13%, and sometimes, even lower. We have set our example provision for preliminaries at 11%.

Prices included within this edition do not include for VAT, which must be added if appropriate.

Part 1: General

This section contains advice on various construction specialisms: capital allowances, legislation, taxes, insurances and the building cost and tender price indices, information on regional price variations.

Part 2 Rates of Wages and Labour

Shows current industry level wage agreement rates and how we have built up labour gang rates being used in the book.

Part 3: Approximate Estimating

This section contains prices per functional unit and square metre rates for various building types, building cost models and approximate elemental estimating rates.

It also contains a section where we show typical preliminaries build-up for a £3,500,000 value project.

Part 4 Prices for Measured Works

These sections contain Prices for Measured Works organized using the NRM2 Work Sections for building works.

NOTE: *All prices in Parts 3 and 4 exclude the main contractor's preliminaries costs.*

Part 5: Fees for Professional Services

This section contains guidance on fee levels for professional services: Quantity Surveyors; Architects' and Consulting Engineers. Readers should always obtain fee proposals for their project prior to commencement.

Part 6: Daywork and Prime Cost

This section contains Daywork and Prime Cost allowances issued by the Royal Institute of Chartered Surveyors.

Part 7: Useful Addresses for Further Information

A list of useful trade associations, professional bodies' contact details.

Part 8: Tables and Memoranda

This section contains general formulae, weights and quantities of materials, other design criteria and useful memoranda associated with each trade.

While every effort is made to ensure the accuracy of the information given in this publication, neither the Editors nor Publishers in any way accept liability for loss of any kind resulting from the use of such information.

DAVIS LANGDON, An AECOM Company
MidCity Place
71 High Holborn
London WC1V 6QS

Designing Architecture

A Pressman

Designing Architecture is an indispensable tool to assist both students and young architects in formulating an idea, transforming it into a building, and making effective design decisions.

This highly focused book offers explicit guidance to students and young professionals on how to approach, analyze, and execute specific tasks; develop and refine a process to facilitate the best possible design projects; and create meaningful architectural form. Case studies augment the text and chronicle fascinating applications of the design process.

Designing Architecture will inspire readers to elevate the quality of preliminary designs and unravel some of the mystery of creating the most beautiful, responsive, and responsible architectural design possible.

December 2011: 276x219: 208pp
Hb: 978-0-415-59515-5: £105.00
Pb: 978-0-415-59516-2: £24.99

To Order: Tel: +44 (0) 1235 400524 Fax: +44 (0) 1235 400525
or Post: Taylor and Francis Customer Services,
Bookpoint Ltd, Unit T1, 200 Milton Park, Abingdon, Oxon, OX14 4TA UK
Email: book.orders@tandf.co.uk

For a complete listing of all our titles visit:
www.tandf.co.uk

Special Acknowledgements

Acodrain
ACO Business Park
Hitchin Road
Shefford
Bedfordshire
SG17 5TE
Tel: 01462 816666
buildingdrainage@aco.co.uk
www.aco.co.uk
Drainage channels

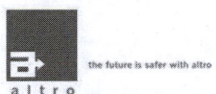

Altro Floors
Works Road
Letchworth
Hertfordshire
SG6 1NW
Tel: 0870 6065432/01462 707600
Fax: 0870 5113388/01462 707504
info@altro.co.uk
www.altro.co.uk
Floor coverings

Andrews Marble Tiles
324-330 Meanwood Road
Leeds
Yorkshire
LS7 2JE
Tel: 0113 262 4751
Fax: 0113 239 2184
sales@andrews-tiles.co.uk
www.andrews-tiles.co.uk
Floor and wall tiles

ASSA ABLOY

Assa Abloy Ltd
The Meadows
Cannock Road
Wolverhampton
West Midlands
WV10 0RR
Tel: 01902 364 648
Fax: 01902 364 666
info@assaabloy.com
www.assaabloy.co.uk
Doorsets and furniture

Building Innovation
Unit 30
Berrington Road
Sydenham Industrial Estate
Leamington Spa
Warwickshire
CV31 1NB
Tel: 01926 888808
info@building-innovation.co.uk
www.building-innovation.co.uk
Tapered insulation

Halfen Ltd
Unit 2
Humphreys Road
Woodside Estate
Dunstable
Bedfordshire
LU5 4TP
Tel: 01582 470 300
brick@halfen.co.uk
www.halfen.co.uk
Brick accessories - channels and special products

Hare Structural Engineers
Brandlesholme House
Brandlesholme Road
Bury
Lancashire
BL81JJ
Tel: 0161 609 0000
info@hare.co.uk
www.hare.co.uk
Structural steelwork

Marshalls Mono Ltd (pavings)
Landscape House
Premier Way
Lowfields Business Park
Elland
HX5 9HT
Tel: 01422 312000
marshallspaving@marshalls.co.uk
www.marshalls.co.uk
Pavings

Monier Ltd
Sussex Manor Business Park
Gatwick Road
Crawley
West Sussex
RH10 9NZ
Tel: 01293 618418
roofing.redland@monier.com
www.redland.co.uk
Redland roof tiles

N.D.M.

NDM Metal Roofing & Cladding Ltd
Mettalum House
Unit 3, 89, Manor Farm Road
Alperton
Wembley
HA0 1BA
Tel: 0208 991 7310
enquiries@ndmltd.com
www.ndmltd.com
Metal cladding and roofing

Parker & Highland Joinery Ltd
14A Chartwell Road
Lancing Business Park
Lancing
West Sussex
BN15 8TU
Tel: 01903 756 283
sales@parker-joinery.com
www.parker-joinery.com
Purpose-made joinery

Profile 22
Stafford Park 6
Telford
Shropshire
TF3 3AT
Tel: 01952 290910
mail@profile22.co.uk
www.profile22.co.uk
uPVC windows

Severfield-Rowen Structures Limited
Dalton Airfield Industrial Estate
Dalton
Thirsk
North Yorkshire
YO7 3JN
Tel: 01845 577896
www.sfrplc.com
Structural steel

Acknowledgements

Acodrain
ACO Business Park
Hitchin Road
Shefford
Bedfordshire
SG17 5TE
Tel: 01462 816666
buildingdrainage@aco.co.uk
www.aco.co.uk
Drainage channels

Allgood PLC
297, Euston Road
London
NW1 3 AQ
Tel: 0207 387 9951
info@allgood.co.uk
www.allgood.co.uk
Ironmongery

Altro Floors
Works Road
Letchworth
Hertfordshire
SG6 1NW
Tel: 0870 6065432/01462 707600
info@altro.co.uk
www.altro.co.uk
Floor coverings

Alumasc Exterior Building Products Ltd
White House Works
Bold Road
Sutton
St Helens
WA9 4JG
Tel: 01744 648400
info@alumasc-exteriors.co.uk
www.alumasc-exteriors.co.uk
Aluminium rainwater goods

Alumasc Interior Building Products Ltd
Halesfield 19
Telford
Shropshire
TF7 4QT
Tel: 01952 580590
sales@pendock.co.uk
www.pendock.co.uk
Skirting trunking and casings

Amwell Systems Ltd
Buntingford Business Park
Baldock Road
Buntingford
Hertfordshire
SG9 9ER
Tel: 01763 276200
sales@amwell-systems.com
www.amwell-systems.com
Toilet cubicles

Andrews Marble Tiles
324-330 Meanwood Road
Leeds
Yorkshire
LS7 2JE
Tel: 0113 262 4751
sales@andrews-tiles.co.uk
www.andrews-tiles.co.uk
Floor and wall tiles

Armitage Shanks Group
Armitage
Rugeley
Staffordshire
WS15 4BT
Tel: 01543 490253
arm-idealinfo@idealstandard.com
www.armitage-shanks.co.uk
Sanitary fittings

Armstrong Floor Products UK Ltd
Customer Service Centre
Fleck Way
Teeside Industrial Estate
Thornaby on Tees
Cleveland
TS17 9JT
Tel: 01642 768660
commercial_uk@armstrong.com
www.armstrong-flooring.co.uk
Flooring products

Assa Abloy Ltd
The Meadows
Cannock Road
Wolverhampton
West Midlands
WV10 0RR
Tel: 01902 364 648
Fax: 01902 364 666
info@assaabloy.com
www.assaabloy.co.uk
Ironmongery

Bison Concrete Products Ltd
Millenium Court
First Avenue Centrum 100
Burton-upon-Trent
DE14 2WR
Tel: 01283 495000
concrete@bison.co.uk
www.bison.co.uk
Precast concrete floors

Bolton Gate Co
Waterloo Street
Bolton
Lancashire
BL1 2SP
Tel: 01204 871000
sales@boltongate.co.uk
www.boltongate.co.uk
Insulated shutter doors

Bradstone Structural Sales
North End Farm Works
Ashton Keynes
Wiltshire
SN6 6QX
Tel: 01285 646 884
bradstone-structural@aggregate.com
www.bradstone-structural.com
Walling, dressings and other products

British Gypsum Ltd
Pandy Lane
Barrow-upon-Soar
Loughborough
Leicestershire
LE12 8GB
Tel: 01509 817 200
bgtechnical.enquiries@bpb.com
www.british-gypsum.com
Plasterboard and plaster products

Building Innovation
Unit 30
Berrington Road
Sydenham Industrial Estate
Leamington Spa
Warwickshire
CV31 1NB
Tel: 01926 888808
info@building-innovation.co.uk
www.building-innovation.co.uk
Tapered insulation

Burlington Slate Ltd
Cavendish House
Kirby-in-Furness
Cumbria
LA17 7UN
Tel: 01229 889 681
sales@burlingtonstone.co.uk
www.burlingtonstone.co.uk
Westmoreland slating

Buttles
Soothouse Spring
Valley Road Industrial Estate
St Albans
Hertfordshire
AL3 6NX
Tel: 01727 834 242
beeweb@buttle.co.uk
www.buttles.co.uk
Timber products

Catnic Ltd
Pontypandy Industrial Estate
Caerphilly
Mid Glamorgan
CF83 3GL
Tel: 02920 337900
www.catnic.com
Steel lintels

Cavity Trays Ltd
Administration Centre
Lufton Trading Estate
Yeovil
Somerset
BA22 8HU
Tel: 01935 474769
sales@cavitytrays.co.uk
www.cavitytrays.com
Cavity trays, closers and associated products

Cementation Foundations Skansa Ltd
Maple Cross House
Denham Way
Maple Cross
Rickmansworth
Hertfordshire
WD3 2SW
Tel: 01923 423100
cementation.foundations@skanska.co.uk
www.skanska.co.uk
Piling systems

Colour Centre
29a, Offord Road
London
N1 1AE
Tel: 0207 609 1164
www.colourcentre.com
Paints, varnishes and stains

Concrete Canvas Ltd (UK)
Unit 3
Block A22
Treforest Industrial Estate
CF37 5SP
Tel: 0845 680 1908
info@concretecanvas.co.uk
www.concretecanvas.co.uk
Concrete canvas

Cox Building Products
Unit 1
Shaw Road
Bushbury
Wolverhampton
WV10 9LA
Tel: 01902 371800
sales@coxspan.co.uk
www.coxbp.com
Rooflights

CPM-Group
Mells Road
Mells
Frome
BA11 3PD
Tel: 0117 981 2791
sales@cpm-group.co.uk
www.cpm-group.co.uk
Concrete pipes etc.

Custom Metal Fabrications
Central Way
Feltham
Middlesex
TW14 0XJ
Tel: 020 8844 0940
dgibbs@cmf.co.uk
www.cmf.co.uk
Metalwork and balustrading

Decra Roof Systems (UK) Ltd
Unit 3
Faraday Centre
Faraday Road
Crawley
West Sussex
RH10 9PX
Tel: 01293 545 058
sales@decra.co.uk
www.decra.co.uk
Roofing system and accessories

Dow Construction Products
2 Heathrow Boulevard
284 Bath Road
West Drayton
Middlesex
UB7 0DQ
Tel: 0208 917 5050
styrofoam-uk@dow.com
www.styrofoameurope.com
Insulation products

Dreadnought Clay Roof Tiles
Dreadnought Works
Pensnett
Brierley Hill
Staffordshire
DY5 4TH
Tel: 01384 77405
sales@dreadnought-tiles.co.uk
www.dreadnought-tiles.co.uk
Clay roof tiling

Dufaylite Developments Ltd
Cromwell Road
St. Neots
Cambridgeshire
PE19 1QW
Tel: 01480 215000
enquiries@dufaylite.com
www.dufaylite.com
Clayboard void former

Envirodoor Ltd
Viking Close
Great Gutter Lane East
Willerby
Hull
East Yorkshire
HU10 6BS
Tel: 01482 659375
sales@envirodoor.com
www.envirodoor.com
Sliding and folding doors

Expamet Building Products
Greatham Street
Longhill Industrial Estate (North)
Hartlepool
TS25 1PR
Tel: 01429 866611
sales@expamet.net
www.expamet.co.uk
Builder's metalwork products

Forbo Flooring UK Ltd
High Holborn Road
Ripley
Derbyshire
DE5 3NT
Tel: 01773 744 121
www.forbo-flooring.co.uk
Mats and matwells

Forticrete Ltd
Boss Avenue
off Grovebury Road
Leighton Buzzard
Bedfordshire
LU7 4SD
Tel: 01525 244 900
masonry@forticrete.com
roofing@forticrete.com
nwblocksales@forticrete.com
www.forticrete.co.uk
Blocks and roof tiles

Garador Ltd
Bunford Lane
Yeovil
Somerset
BA20 2YA
Tel: 01935 443700
www.garador.co.uk
Garage doors

Grace Construction Products
Expansion Jointing and Waterproofing Division
Ajax Avenue
Slough
Berkshire
SL1 4BH
Tel: 01753 692929
uksales@grace.com
www.uk.graceconstruction.com
Expansion joint fillers and waterbars

Gradus Wall Protection
Park Green
Macclesfield
Cheshire
SK11 7LZ
Tel: 01625 428 922
sales@gradusworld.com
www.gradusworld.com
Stair nosings

GRP Tanks
33 Rye Road
Hoddesdon
Hertfordshire
EN11 0JE
Tel: 0871 200 2082
www.grptanks.net
GRP water tanks

H+H Celcon UK Ltd
Celcon House
Ightham
Sevenoaks
Kent
TN15 9HZ
Tel: 01732 886333
info@hhcelcon.co.uk
www.hhcelcon.co.uk
Concrete blocks

Halfen Ltd
Unit 2
Humphreys Road
Woodside Estate
Dunstable
Bedfordshire
LU5 4TP
Tel: 01582 470 300
brick@halfen.co.uk
www.halfen.co.uk
Brick accessories – channels and special products

Hanson Building Products
Stewartby
Bedford
Bedfordshire
MK43 9LZ
Tel: 0870 5258258
blocks@hanson.com; info@hansonbp.com
www.heidelbergcement.com
Facing bricks

Hanson Heidelberg Cement Group
Stewartby
Bedford
Bedfordshire
MK43 9LZ
Tel: 08705 626500
blocks@hanson.com
www.heidelbergcement.com
Thermalite and Conbloc concrete blocks

Hare Structural Engineers
Brandlesholme House
Brandlesholme Road
Bury
Lancashire
BL81JJ
Tel: 0161 609 0000
info@hare.co.uk
www.hare.co.uk
Structural steelwork

Hathaway Roofing Ltd
Tindal Crescent
Bishop Auckland
County Dunham
DL14 9TL
Tel: 01388 605 636
steven.price@hathaway-roofing.co.uk
www.hathaway-roofing.co.uk
Sheet wall and roof claddings

Hillaldam Coburn Ltd
Unit 16
Merton Industrial Park
Lee Road
London
SW19 3HX
Tel: 0208 545 6680
sales@hillaldam.co.uk
www.coburn.co.uk
Sliding and folding door gear

HSS Hire Shops
Group Office
25 Willow Lane
Mitcham
Surrey
CR4 4TS
Tel: 0208 260 3100
www.hss.co.uk
Tool hire

Hudevad
Unit 5
Cyan Way
Phoenix Way (A444)
Coventry
CV2 4QP
Tel: 02476 881200
sales@hudevad.co.uk
www.hudevad.co.uk
Radiators

Hunter Plastics Ltd
Nathan Way
London
SE28 0AE
Tel: 0208 855 9851
info@hunterplastics.co.uk
www.hunterplastics.co.uks
Plastic rainwater goods

Ibstock Building Products
Leicester Road
Ibstock
Leicestershire
LE67 6HS
Tel: 01530 261999
marketing@ibstock.co.uk
www.ibstock.co.uk
Facing bricks; Tilebricks

Icopal Ltd
Barton Dock Road
Stretford
Manchester
M32 0YL
Tel: 0843 224 7400
info.uksales@icopal.com
www.icopal.co.uk
Damp proof products

James Latham
Badminton Road
Yate
Bristol
Avon
BS37 5JX
Tel: 01454 315 421
marketing@lathams.co.uk
www.lathamtimber.co.uk
Hardwood and panel products

Jeld-Wen UK Ltd
Watch House Lane
Doncaster
South Yorkshire
DN5 9LR
Tel: 0870 126 0000
marketing@jeld-wen.co.uk
www.jeld-wen.co.uk
Doors and windows

John Brash and Co Ltd
The Old Shipyard
Gainsborough
Lincolnshire
DN21 1NG
Tel: 01427 613858
info@johnbrash.co.uk
www.johnbrash.co.uk
Roofing shingles

John Guest Speedfit Ltd
Horton Road
West Drayton
Middlesex
UB7 8JL
Tel: 01895 449 233
www.speedfit.co.uk
Speedfit product range

Junkers Ltd
Unit 1a
1 Wheaton Road
Witham
Essex
CM8 3UJ
Tel: 01376 534 700
sales@junkers.co.uk
www.junkers.co.uk
Hardwood flooring

Kalzip Ltd
Haydock Lane
Haydock
St Helens
Merseyside
WA11 9TY
Tel: 01942 295500
kalzip-uk@corusgroup.com
www.kalzip.com
Kalzip roofing

KB Rebar Ltd
Unit 5, Dobson Park Industrial Estate
Dobson Park Way
Ince
Wigan
WN2 2DY
Tel: 0161 790 8635
www.kbrebar.co.uk
Reinforcement bar and mesh

Kingspan Access Floors
Burma Drive
Marfleet
Hull
HU9 5SG
Tel: 01482 781 710
enquiries@kingspanaccessfloors.co.uk
www.kingspanaccessfloors.co.uk
Raised access floors

Kingspan Environmental
College Road North
Aston Clinton
Aylesbury
Bucks
HP22 5EW
Tel: 01296 633000
pollutiongb@kingspanenv.com
www.kingspanenv.com
Interceptors and septic tanks

Kingspan Insulation Ltd
Pembridge
Leominster
Herefordshire
HR6 9LA
Tel: 0870 850 8555
info.uk@insulation.kingspan.com
www.insulation.kingspan.com
Insulation products

Kingspan Structural Products
Sheburn
Malton
YO17 8PQ
Tel: 01944 712 000
sales@kingspanstructural.com
www.kingspan.com
Multibeam purlins

Knauf Insulation Ltd
PO Box 10
Stafford Road
St Helens
WA10 3NS
Tel: 01744 766 666
sales@knaufinsulation.com
www.knaufinsulation.co.uk
Insulation products

Landpro Ltd
14 Upper Bourne Lane
Farnham
Surrey
GU10 4RQ
Tel: 01252 795030
info@landpro.co.uk
www.landpro.co.uk
Landscaping consultants

Lignacite
Meadgate Works
Nazeing
Essex
EN9 2PD
Tel: 01992 464 441
info@lignacite.co.uk
www.lignacite.co.uk
Concrete blocks

Maccaferri
7400 The Quorum
Oxford Business Park North
Garsington Road
Oxford
OX4 2JZ
Tel: 01865 770 555
marketing@maccaferri.co.uk
www.maccaferri.co.uk
Gabions

Magrini Ltd
Unit 5
Maybrook Industrial Estate
Brownhills
Walsall
West Midlands
WS8 7DG
Tel: 01543 375311
sales@magrini.co.uk
www.magrini.co.uk
Baby equipment

Manhole Covers Ltd
Airfield Industrial Estate
Cheddington Lane
Long Marston
Bucks
HP23 4QR
Tel: 01296 668850
sales@manholecovers
www.manholecovers.com
Manhole covers

Marshalls Mono Ltd (drainage)
Landscape House
Premier Way
Lowfields Business Park
Elland
HX5 9HT
Tel: 01422 312000
drainage@marshalls.co.uk
www.marshalls.co.uk
Drainage channels

Marshalls Mono Ltd (pavings)
Landscape House
Premier Way
Lowfields Business Park
Elland
HX5 9HT
Tel: 01422 312000
marshallspaving@marshalls.co.uk
www.marshalls.co.uk
Pavings

Metsec Lattice Beams Ltd
Rolls Royce Estate
Spring Road
Ettingshall
Wolverhampton
WV4 6JX
technical@metseclatticebeams.com
www.metseclatticebeams.com
Lattice beams

Monier Ltd
Sussex Manor Business Park
Gatwick Road
Crawley
West Sussex
RH10 9NZ
Tel: 01293 618418
roofing.redland@monier.com
www.redland.co.uk
Redland roof tiles

NDM Metal Roofing & Cladding Ltd
Mettalum House
Unit 3, 89 Manor Farm Road
Alperton
Wembley
HA0 1BA
Tel: 0208 991 7310
enquiries@ndmltd.com
www.ndmltd.com
Metal cladding and roofing

Parker & Highland Joinery Ltd
14 A Chartwell Road
Lancing Business Park
Lancing
West Sussex
BN15 8TU
Tel: 01903 756 283
sales@parker-joinery.com
www.parker-joinery.com
Purpose-made joinery

Pegler Yorkshire
St Catherine's Avenue
Doncaster
South Yorkshire
DN4 8DF
Tel: 0844 243 4400
info@yorkshirefittings.co.uk
www.pegleryorkshire.co.uk
Yorkshire and Kuterlite fittings

Plumb Centre
The Wolseley Centre
Harrison Way
Leamington Spa
Warwickshire
CV31 3HH
Tel: 0870 1622 557
www.plumbcentre.co.uk
Cylinders and general plumbing

Polyflor Ltd
PO Box 3
Radcliffe New Road
Whitefield
Manchester
M45 7NR
Tel: 0161 767 1111
www.polyflor.com
Polyfloor contract flooring

Polypipe Terrain
New Hythe Business Park
New Hythe Lane
Aylesford
Kent
ME20 7PJ
Tel: 01622 717811
commercialenquiries@polypipe.com
www.terraindrainage.com
Drainage goods

Premdor Ltd
Gemini House
Hargreaves Road
Groundwell Industrial Estate
Swindon
Wiltshrie
SN25 5 AJ
Tel: 01793 708200
enquiries@premdor.com
www.premdor.co.uk
Doors and windows

Premier Loft Ladders
2 Dawson Drive
Trimley
St Mary
Felixstowe
Suffolk
IP11 0YW
Tel: 0845 9000 195
sales@premierloftladders.com
www.premierloftladders.com
Loft ladders

Pressalit Care plc
100 Longwater Avenue
Green Park
Reading
Berkshire
RG2 6GP
Tel: 0844 880 6950
www.pressalitcare.com
Bathroom equipment

Profile 22
Stafford Park 6
Telford
Shropshire
TF3 3AT
Tel: 01952 290910
mail@profile22.co.uk
www.profile22.co.uk
uPVC windows

Promat UK
The Stirling Centre
Easton Road
Bracknell
Berkshire
RG12 2ST
Tel: 01344 381 301
promat@promat.co.uk
www.promat.co.uk
Fireproofing materials

Protim Solignum Ltd
Fieldhouse Lane
Marlow
Buckinghamshire
SC7 1LS
Tel: 01628 486644
info@osmose.co.uk
www.osmose.co.uk
Paints and timber treatment

Radius Systems Ltd
PO Box 1
Hillcote Plant
Blackwell
Alfreton
Derbyshire
DE55 5JD
Tel: 01773 811 112
www.radius-systems.co.uk
MDPE pipes and fittings

Rawlplug Ltd
Skibo Drive
Thronliebank Industrial Estate
Glascow
Scotland
G46 8JR
Tel: 0141 638 225
info@rawlplug.co.uk
www.rawlplug.co.uk
Anchoring and fixing systems

Richard Potter Timber Merchants
Millstone Lane
Nantwich
Cheshire
CW5 5PN
Tel: 01270 625791
richardpotter@fortimber.demon.co.uk
www.fortimber.demon.co.uk
Carcasssing softwood

Rockwool Ltd
Pencoed
Bridgend
Glamorganshire
CF35 6NY
Tel: 01656 862 261
info@rockwool.co.uk
www.rockwool.co.uk
Pipe and other insulation products

Ryton's Building Products
Design House
Orion Way
Kettering Business Park
Kettering
Northamptonshire
NN15 6NL
Tel: 01536 511874
admin@rytons.com
www.vents.co.uk
Roof ventilation products

Safeguard Europe Ltd
Redkin Close
Horsham
West Sussex
RH13 5QL
Tel: 01403 212004
www.safeguardeurope.com
Damp proofing and waterproofing

Saint Gobain PAM UK
Lows Lane
Stanton-by-Dale
Illkeston
Derbyshire
DE7 4QU
Tel: 0115 930 5000
sales.uk.pam@saint-gobain.com
www.saint-gobain-pam.co.uk
Cast iron soil, water and rainwater pipes and fittings

Sandtoft Roof Tiles Ltd
Sandtoft
Doncaster
South Yorkshire
DN8 5SY
Tel: 01427 871200
info@sandtoft.co.uk
www.sandtoft.co.uk
Clay roof tiling

Schiedel Rite-Vent
Crowther Road
Washington
Tyne and Wear
NE38 0AQ
Tel: 0191 416 1150
sales@schiedel.co.uk
www.isokern.co.uk
Flue pipes and gas blocks

Screeduct Ltd
Unit 29
Alderminster
Startford-Upon-Avon
Warwickshire
CV37 8NY
Tel: 01789 459 211
sales@screeduct.com
www.screeduct.com
Trunking systems and conduits

Severfield-Rowen Structures Limited
Dalton Airfield Industrial Estate
Dalton
Thirsk
North Yorkshire
YO7 3JN
Tel: 01845 577896
www.sfrplc.com
Structural steel

Sheet Piling UK Ltd
Oakfield House
Rough Hey Road
Grimsargh
Preston
PR2 5AR
Tel: 01772 794 141
enquiries@sheetpilinguk.com
www.sheetpilinguk.com
Sheet piling

Sheffield Insulation
Unit 1
New England Estate
Gascoigne Road
Barking
Essex
IG11 7NZ
Tel: 020 8477 9500
barking@sheffins.co.uk
www.SheffieldInsulation.co.uk
Insulation products

Siderise Insulation Ltd
Wales Office
Forge Industrial Estate
Maestag
Bridgend
CF34 0AZ
Tel: 01656 730833
sales@siderise.com
www.siderise.com
Fire barriers

Slate UK David Wallace International Ltd
Unit 6, Airfield Approach Business Park
Flookburgh
Grange-over-Sands
Lake District
Cumbria
Tel: 015395 59289
www.slate.uk.com
Spanish roof slates

Stainless UK Ltd
Newhall Road Works
Sheffield
S9 2QL
Tel: 0114 244 1333
sales@stainless-uk.co.uk
www.stainless-uk.co.uk
Stainless steel rebar

Sterling Hydraulics (Huntley & Sparks) Ltd
Building Products Division
Sterling House
Blacknell Lane
Crewkerne
Somerset
TA18 8LL
Tel: 01460 722 22
Rigifix column guards

Stirling Lloyd Polychem Ltd
Union Bank
King Street
Knutsford
Cheshire
WA16 6EF
Tel: 01565 633 111
www.stirlinglloyd.com
Integritank products

Stressline Ltd
Foxbank Industrial Estate
Stoney Stanton
Leicester
LE9 4LX
Tel: 0870 7503167
sales@stressline.ltd.uk
www.stressline.ltd.uk
Concrete lintels

Swish Building Products
Pioneer House
Lichfield Road Industrial Estate
Tamworth
Staffs
B29 7TF
Tel: 01827 317200
marketing@swishbp.co.uk
www.swishbp.co.uk
Swish Celuka

Szerelmey Ltd
369 Kennington Lane
Vauxhall
London
SE11 5QY
Tel: 0207 735 9995
info@szerelmey.com
www.szerelmey.com
Stonework

Tarkett-Marley Floors Ltd
Dickley Lane
Lenham
Maidstone
Kent
ME17 2QX
Tel: 01622 854000
uksales@tarkett.com
www.tarkett-floors.com
Sheet and tile flooring

Tarmac Ltd
Pudding Mill Lane
Bow
London
E15 2PJ
Tel: 0208 555 2415
enquiries@tarmac.co.uk
www.tarmac.co.uk
Ready-mixed concrete

Tarmac Mortar and Screeds
Tunstead House
Buxton
Derbyshire
SK17 8TG
Tel: 08701 116 116
mortar@tarmac.co.uk
www.tarmac.co.uk
Readymix screeds and mortar

Tarmac Topblock Ltd
Wergs Hall
Wergs Hall Road
Wolverhampton
West Midlands
WV8 2HZ
Tel: 01902 754 131
enquiries@tarmac.co.uk
www.topblock.co.uk
Concrete blocks

TATA Steel Ltd
PO Box 1
Brigg Road
Scunthorpe
North Lincolnshire
DN16 1BP
Tel: 01724 405 060
www.tatasteeleurope.com
Steel

Timbmet
Kemp House
Chawley Park
Cumnor Hill
Oxford
OX2 9PH
Tel: 01865 860351
marketing@timbmet.com
www.timbmet.com
Hardwood

Travis Perkins Trading Company
Lodge Way House
Lodge Way
Harlestone Road
Northampton
Northamptonshire
NN5 7UG
Tel: 01604 752484
www.travisperkins.co.uk
Builders merchant

UK Glass Force
32-34 Eldon Way Industrial Estate
Spa Road
Hockley
SS5 4AD
Tel: 0800 393 827
support@ukglassforce.co.uk
www.ukglassforce.co.uk
Glass and glazing

V.A. Hutchison Flooring Ltd
Units 1-3, Building NA
Beeding Close
Southern Cross Trading Estate
Bognor Regis
West Sussex
PO22 9TS
Tel: 01243 841 175
julia@hutchisonflooring.co.uk
www.hutchisonflooring.co.uk
Hardwood flooring

Vandex
Safeguard Europe Ltd
Redkiln Close
Horsham
Sussex
RH13 5QL
Tel: 01403 210204
info@safeguardeurope.com
www.vandex.com
Vandex Super and Premix products

Velfac Ltd
The Old Livery
Hildersham
Cambridge
CB21 6DR
Tel: 01223 897 100
post@velfac.co.uk
www.velfac.co.uk
Composite windows

Velux Company Ltd
Wellington Road
Kettering Parkway
Kettering
Northants
NN15 6XR
Tel: 0870 166 7676
enquires@velux.co.uk
www.velux.co.uk
Velux roof windows and flashings

Visqueen Building Products
Maerdy Industrial Estate
Rhymney
Tredegar
NP22 5PY
Tel: 01685 840 672
enquiries@visqueenbuilding.co.uk
www.visqueenbuilding.co.uk
Visqueen products

Wavin Plastics Ltd
Parsonage Way
Chippenham
Wiltshire
SN15 5PN
Tel: 01249 766600
info@wavin.co.uk
www.wavin.co.uk
uPVC drainage goods

Web Dynamics Ltd
Moss Lane
Station Road
Blackrod
Lancs BL6 5JB
Tel: 01204 695666
www.webdynamics.co.uk
Breather membranes

Welco
Woodgate Business Park
Kettles Wood Drive
Birmingham
B32 3GH
Tel: 0121 4219000
enquiries@welconstruct.co.uk
www.welconstruct.co.uk
Lockers and shelving systems

Welsh Slate Ltd
Penrhyn Quarry
Bethesda
Bangor
Gwynedd
LL57 4YG
Tel: 01248 600 656
enquiries@welshslate.com
www.welshslate.com
Natural Welsh slates

Wilde Contracts Ltd
Chareau House
1 Miles Street
Oldham
OL1 3NW
Tel: 0161 624 6824
www.rogerwilde.com
Glass blocks

Yeoman Aggregates Ltd
Stone Terminal
Horn Lane
Acton
London
W3 9EH
Tel: 0208 896 6800
debra.ward@yeoman-aggregates.co.uk
www.yeoman-aggregates.co.uk
Hardcore, gravels, sand

Yorkshire Copper Tube Ltd
East Lancashire Road
Kirkby
Liverpool
Merseyside
L33 7TU
Tel: 0151 546 2700
info@yct.com
www.yorkshirecopper.com
Copper tube

Architect's Pocket Book

4th Edition

Hetreed & Ross

This handy pocket book brings together a wealth of useful information that architects need on a daily basis – on site or in the studio. It provides guidance on a range of tasks, from complying with the Building Regulations, including the recent revisions to Part L, to helping with planning, use of materials and detailing.

Compact and easy to use, the Architect's Pocket Book has sold well over 65,000 copies to the nation's architects, architecture students, designers and construction professionals who do not have an architectural background but need to understand the basics, fast.

This is the famous little blue book that you can't afford to be without.

February 2011: 186x123: 268pp
Pb: 978-0-08-096959-6: £20.99

To Order: Tel: +44 (0) 1235 400524 Fax: +44 (0) 1235 400525
or Post: Taylor and Francis Customer Services,
Bookpoint Ltd, Unit T1, 200 Milton Park, Abingdon, Oxon, OX14 4TA UK
Email: book.orders@tandf.co.uk

For a complete listing of all our titles visit:
www.tandf.co.uk

How to use this Book

First time users of *Spon's Architects' and Builders' Price Book* and others who may not be familiar with the way in which prices are compiled may find it helpful to read this section before starting to calculate the costs of building work. The level of information on a scheme and availability of detailed specifications will determine which section of the book and which level of prices users should refer to.

Rates in the book include contractor's overhead and recovery margins but do not include main contractor's preliminaries except for two sections – *Building Prices per Functional Unit* and *Building Prices per Square Metre* – which are both within *Part 3: Approximating Estimating*.

Prices in the book do not necessarily reflect the lowest possible prices achievable, but are intended as a guide to expected price levels for the items described. Davis Langdon cost a series of building models using current Spon's rates to calculate a book tender index. For this edition of the book this has been set at 453.

New Rules of Measurement (NRM)

The NRM suite covers the life cycle of cost management and means that, at any point in a building's life, quantity surveyors will have a set of rules for measuring and capturing cost data. In addition, the BCIS *Standard Form of Cost Analysis (SFCA)* 4th edition has been updated so that it is aligned with the NRM suite.

The three volumes of the NRM are as follows:

- **NRM1** – Order of cost estimating and cost planning for capital building works.
- **NRM2** – Detailed measurement for building works.
- **NRM3** – Order of cost estimating and cost planning for building maintenance works.

NRM1

In May 2009 the New Rules of Measurement (NRM1) Order of cost estimating and elemental cost planning were introduced by the RICS. The NRM have been written to provide a standard set of measurement rules that are understandable by all those involved in a construction project, including the employer; thereby aiding communication between the project/design team and the employer. In addition, the NRM should assist the quantity surveyor/ cost manager in providing effective and accurate cost advice to the employer and the rest of the project/design team.

NRM2

Developed to provide a detailed set of measurement rules to support the procurement of construction works. These rules will ultimately be a direct replacement for the *Standard Method of Measurement for Building Works (SMM7)*.

This edition of Spon's is a hybrid of the NRM2 and SMM7 in order to help the user transfer between the two formats. Headings are in line with NRM2 but subheadings and content reflect SMM7 more closely. This will continue to be developed over several editions of the book.

The NRM suite is published by the RICS and readers are strongly recommended to read the documents.

APPROXIMATE ESTIMATING

For preliminary estimates/indicative costs before drawings are prepared or very little information is available, users are advised to refer to the average overall *Building Prices per Functional Unit* and multiply this by the proposed number of units to be contained within the building (i.e. number of bedrooms etc.) or *Building Prices per Square Metre* rates and multiply this by the gross internal floor area of the building (the sum of all floor areas measured within external walls) to arrive at an overall initial *Order of Cost Estimate*. These rates include preliminaries, but make no allowance for the cost of external works, VAT, or fees for provisional services.

Where preliminary drawings are available, one should be able to measure approximate quantities for all the major components of a building and multiply these by individual rates contained in the *Building Cost Models* or *Approximate Estimating Rates* sections to produce an *Elemental Cost Plan*. This should produce a more accurate estimate of cost than simply using overall prices per square metre. Labour and other incidental associated items, although normally measured separately within Bills of Quantities, are deemed included within approximate estimating rates. These rates do not include preliminaries or fees for provisional services.

MEASURED WORKS

For more detailed estimates or documents such as Bills of Quantities (Quantities of supplied and fixed components in a building, measured from drawings), use rates from Prices for Measured Works. Depending upon the overall value of the contract adjust the value as shown later. Items within the Measured Works sections are made up of many components: the cost of the material or product; any additional materials needed to carry out the work; the labour involved in unloading and fixing, etc.

Measured Works rates

These components are usually broken down into:

Prime Cost

Commonly known as the PC; Prime Cost is the actual price of the material such as bricks, blocks, tiles or paint, as sold by suppliers. Prime Cost may be given as *per square metre*, *per 100 bags* or *each* according to the way the supplier sells the product. Unless otherwise stated, prices in Spon's Architects' and Builders' Price Book (hereafter referred to as Spon's A & B), are deemed to be delivered to site (in which case transport costs will be included), and also take account of trade and quantity discounts. Part loads generally cost more than whole loads but, unless otherwise stated, Prime Cost figures are based on average prices for full loads delivered to a hypothetical site in Acton, West London. Actual prices for live projects will vary depending on the contractor, supplier, the distance from the supplier to the site, the accessibility of the site, whether the whole quantity ordered is to be supplied in one delivery or at specified dates and market conditions prevailing at the time. Prime Cost figures for commonly used alternative materials are supplied in listed form at the beginning of some work sections.

Where a PC rate is entered alongside an item rate then the cost allowed for that item is in the overall material cost. For instance, bricks need mortar; paving needs sand bedding, so the PC cost is simply for the cost of bricks or paving, thus allowing the user to simply substitute an alternative product cost if desired.

Labour

This figure covers the cost of the operation and is calculated on the gang wage rate (skilled or unskilled) and the time needed for the job. A full explanation and build-up is provided. Large regular or continuous areas of work are more economical to install than smaller complex areas.

Materials

Material prices include the cost of any ancillary materials, nails, screws, waste, etc., which may be needed in association with the main material product/s. If the material being priced varies from a standard measured rate, then identify the difference between the original PC price and the material price and add this to your alternative material price before adding to the labour cost to produce a new overall total rate. Alternative material prices, where given, are largely based upon list prices, before the deduction of quantity discounts etc., and therefore require discount adjustment before they can be substituted in place of PC figures given for Measured Work items.

Example:

Item	PC £	Labour hours	Labour £	Material £	Unit	Total rate £
100 mm Thermalite Turbo blocks	6.43	0.41	9.02	7.48	m^2	**16.50**
100 mm Hanson Conbloc standard blocks	6.70					
Calculation: (PC) £6.70 − £6.43 = £0.27 +2.5% (OHP) = £0.28 (add to materials rate of £7.48 = £7.76)						
Therefore, 100 mm Toplite block price =	6.70	0.41	9.02	7.76	m^2	**16.78**

Plant

Plant covers the use of machinery ranging from excavators and dumpers to shovels and static plant and includes running costs such as fuel, water supply, electricity and waste disposal. Items of plant are included within the Measured Works sections, but included in the Material rates.

Unit

The Unit is generally based upon measurement guidelines laid out in the New Rules of Measurement – Detailed measurement for building works (NRM2).

Total rate

Prices in the Total Rate column generally include for the supply and fix of items, unless otherwise described.

Overheads and profit

The general overheads of the Main Contractor's business – the head office overheads and any profit sought on capital and turnover employed, is usually covered under a general item of overheads and profit which is applied either to all measured rates as a percentage, or alternatively added to the tender summary or included within Preliminaries for site specific overhead costs.

Within this edition we are including an allowance of 2.5% for overheads and profit on built-up labour rates and material prices.

Preliminaries

Site specific Main Contractor's overheads on a contract, such as insurances, site accommodation, security, temporary roads and the statutory health and welfare of the labour force, are not directly assignable to individual items so they are generally added as a percentage or calculated allowances after all building component items have been costed and totalled. Preliminaries will vary from project to project according to the type of construction, difficulties of the site, labour shortage, or involvement with other contractors, etc. The overall Preliminary addition for a scheme should be adjusted to allow for these factors. For this edition we have shown a calculated typical Preliminaries cost example of 11%.

Sub/specialist-contractor's costs

For the purpose of this book, these are deemed to include all the above costs, plus 2.5% main contractor's discount.

With the exclusion of main contractor's preliminaries, the above items combine to form item rates in the Prices for Measured Works sections. It should be appreciated that a variation in any one item in any group will affect the final measured work price. Any cost variation must be weighed against the total cost of the contract, and a small variation in Prime Cost where the items are ordered in thousands may have more effect on the total cost than a large variation on a few items, while a change in design which introduces the need to use, for example, earth moving equipment, which must be brought to the site for that one task, will cause a dramatic rise in the contract cost. Similarly, a small saving on multiple items could provide a useful reserve to cover unforeseen extras.

COST PLANNING

Order of cost estimate

The purpose of an Order of Cost Estimate is to establish if the proposed building project is affordable and, if affordable, to establish a realistic cost limit for the project. The cost limit is the maximum expenditure that the employer is prepared to make in relation to the completed building project, which will be managed by the project team.

An Order of Cost Estimate is produced as an intrinsic part of RIBA Work Stages A: Appraisal and B: Design Brief or OGC Gateways 1 (Business Justification) and 2 (Delivery Strategy).

There are comprehensive guidelines within the NRM documentation and readers are recommended to read the relevant sections of the NRM where more detailed explanations and examples can be found.

At this early stage, in order for the estimate to be representative of the proposed design solution, the key variables that a designer needs to have developed to an appropriate degree of certainty are:

- The floor areas upon which the estimate is based
- Proposed elevations
- The implied level of specification

Rates will need to be updated to current estimate base date by the amount of inflation occurring from the base date of the cost data to the current estimate base date. The percentage addition can be calculated using published indices (i.e. tender price indices [TPI]).

Example 1:

New secondary school

Floor area = 15,000 m^2
Start on site January 2014, Davis Langdon Tender Price Index = 440 (*NB A&B 2014 Book TPI = 453*)
Location: North West (adjustment = −10%)
From *Building Prices per Square Metre*

Assume rate of say £1,500 per m^2

		Cost (£)
School rate, say	£1,500/m^2 × 15,000 m^2 =	2,250,000
Adjust for inflation to start date	(440/453) say −3%	−67,500
	subtotal	2,182,500
Adjust for location	say −10%	−218,250
	subtotal	1,964,250
Allow for contingencies	say 10%	196,425
Total Order of Cost Estimate		**2,160,675**

Main contractor's preliminaries, overheads and profit need not be added to the cost of building works as they are included within the Spon's building prices per square metre rates, but you will need to add on professional fees and other facilitating type costs such as demolition, external works, car parking, bringing services to site etc.

Elemental cost planning

Elemental cost plans are produced as an intrinsic part of RIBA Work Stages C: Concept, D: Design Development, E: Technical Design and F: Production Information; or when the OGC Gateway Process is used, Gateways 3A (Design Brief and Concept Approval) and 3B (Detailed Design Approval).

Cost Models can be used to quickly extract £/m^2 of GIFA.

Example 2:

Health centre

Gross Internal Floor Area = 1,000 m^2
Start on site April 2014, Davis Langdon Tender Price Index = 442 (*NB A&B 2014 Book TPI = 453*)
Location: South West (adjustment = +2%)

Element	Rate (£/m2)	Cost (£)
Substructures	106.53	106,530
Frame and Upper Floors	160.00	160,000
Roof	24.00	24,000
And so on for each element to give a total of	**1,300.56**	**1,300,560**
Contractors preliminaries, overheads and profit	say 12%	156,000
	subtotal	1,456,560
Adjust for inflation to start date	(442/453) say −2%	−29,000
	subtotal	1,427,560
Adjust for location	say 2%	28,500
	subtotal	1,456,060
Allow for contingencies	say 5%	72,800
	Total Elemental Cost Plan	**1,528,860**

Other allowances such as consultants fees, design fee, VAT, risk allowance, client costs, fixed price adjustment may need to be added to each of the examples above.

Formal cost planning stages

The NRM schedules a number of formal cost planning stages, which are comparable with the RIBA Design and Pre-Construction Work Stages and OGC Gateways 3A (Design Brief and Concept Approval) and 3B (Detailed Design Approval) for a building project. The employer is required to 'approve' the cost plan on completion of each RIBA Work Stage before authorizing commencement of the next RIBA Work Stage.

Formal Cost Plan	RIBA Work Stage
1	C: Concept
2	D: Design Development
3 {	E: Technical Design
	F: Production Information

Formal Cost Plan 1 is prepared at a point where the scope of work is fully defined and key criteria are specified but no detailed design has begun. Formal Cost Plan 1 will provide the frame of reference for Formal Cost Plan 2. Likewise, Formal Cost Plan 2 will provide the frame of reference for Formal Cost Plan 3. Neither Formal Cost Plans

2 nor 3 involve the preparation of a completely new Elemental Cost Plan; they are progressions of the previous cost plans, which are developed through the cost checking of price significant components and cost targets as more design information and further information about the site becomes available.

Cost plans can be developed from Elemental Cost Plans using both Approximate Estimating Rates and/or Prices for Measured Works depending upon the level of information available.

The cost targets within each formal cost plan approved by the employer will be used as the baseline for future cost comparisons. Each subsequent cost plan will require reconciliation with the preceding cost plan and explanations relating to changes made. In view of this, it is essential that records of any transfers made to or from the risk allowances and any adjustments made to cost targets are maintained, so that explanations concerning changes can be provided to both the employer and the project team.

Adjustment according to contract sum cost

The construction costs for a project will depend on the size, type of building, standard of finish required, location, the economic climate of the construction industry i.e. if there is a shortage of construction work available firms will reduce their tender in order to try and attract work. If the opposite is the case and there is a lot of work available, firms will increase their tenders, as they will not be too keen to obtain the contract which will stretch their resources unless it is worth their while financially. In a recession, construction firms can literally buy work in order to keep their workforce and to ensure some cash flow.

Building Costs can vary between builders/developers. This can be due to the size or purchasing abilities of a company or the discount that it receives from suppliers.

Adjustments can be made to reflect the value of a project by using the following table.

Contract Sum	% adjustment
£5,000,000	−1%
£3,500,000	0%
£2,750,000	1%
£2,000,000	2%
£1,250,000	3%
£1,000,000	4%
£850,000	5%
£700,000	6%
£600,000	7%
£500,000	8%
£400,000	9%
£300,000	10%
£250,000	11%

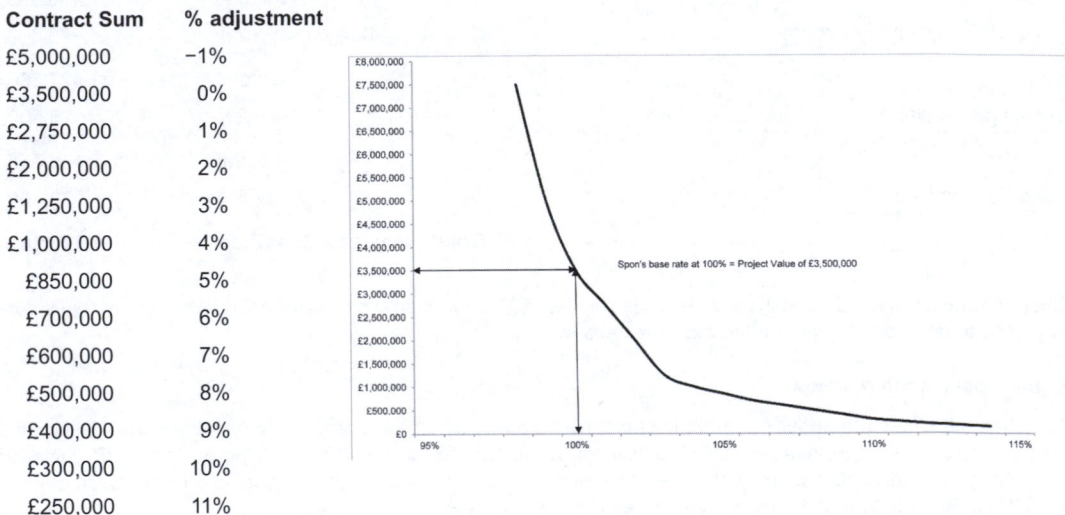

Spon's Architects' & Builders' Price Book is targeted at projects with a value of approximately £3,500,000.

In example 2 above, Health Centre contract sum is £1,528,860, so you would add between 2% and 3% to the costs to reflect how the project value affects the build rates.

Users should not simply apply percentage adjustments to any project regardless of size. We recommend project values between £250,000 and £5,000,000 for rates found in *Spon's Architects' & Builders' Price Book* could be adjusted according to the above table. This is given only as an indication and users should always remember that there are many factors that affect the overall project costs.

These numbers are only intended as a guide and are not explicit and should be applied to the total project value only, not individual rates.

Davis Langdon

An AECOM Company

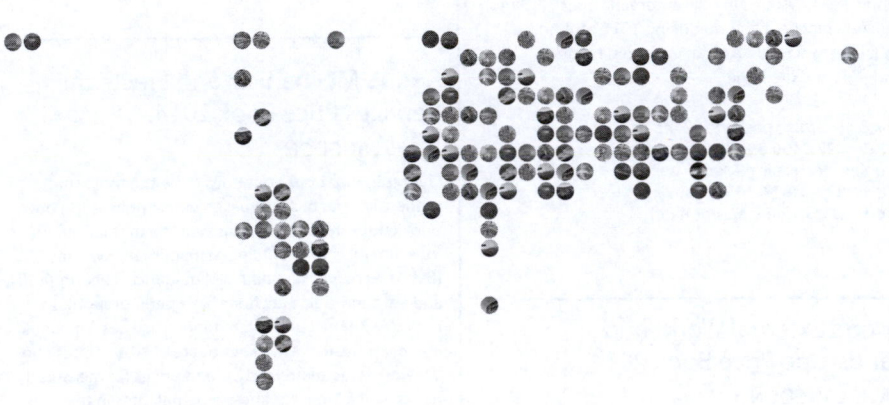

We at Davis Langdon have decided to change our name. From October 2013 we will be known as AECOM.

The time is right.

AECOM

At the heart of AECOM, we join
the dots that create, enhance and
sustain built, natural and social
environments.

Estimator's Pocket Book

Duncan Cartlidge

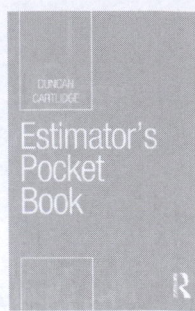

The Estimator's Pocket Book is a concise and practical reference covering the main pricing approaches, as well as useful information such as how to process sub-contractor quotations, tender settlement and adjudication. It is fully up-to-date with NRM2 throughout, features a look ahead to NRM3 and describes the implications of BIM for estimators.

It includes instructions on how to handle:

- the NRM order of cost estimate;
- unit-rate pricing for different trades;
- pro rata pricing and dayworks
- builders' quantities;
- approximate quantities.

Worked examples show how each of these techniques should be carried out in clear, easy-to-follow steps. This is the indispensible estimating reference for all quantity surveyors, cost managers, project managers and anybody else with estimating responsibilities. Particular attention is given to NRM2, but the overall focus is on the core estimating skills needed in practice.

May 2013 186x123: 310pp
Pb: 978-0-415-52711-8: £19.99

To Order: Tel: +44 (0) 1235 400524 Fax: +44 (0) 1235 400525
or Post: Taylor and Francis Customer Services,
Bookpoint Ltd, Unit T1, 200 Milton Park, Abingdon, Oxon, OX14 4TA UK
Email: book.orders@tandf.co.uk

For a complete listing of all our titles visit:
www.tandf.co.uk

Taylor & Francis
Taylor & Francis Group

PART 1

General

This part contains the following sections:

Understanding JCT Standard Building Contracts
Ninth Edition

David Chappell

This ninth edition of David Chappell's bestselling guide has been revised to take into account changes made in 2011 to payment provisions, and elsewhere. This remains the most concise guide available to the most commonly used JCT building contracts: Standard Building Contract with quantities, 2011 (SBC11), Intermediate Building Contract 2011 (IC11), Intermediate Building Contract with contractor's design 2011 (ICD11), Minor Works Building Contract 2011 (MW11), Minor Works Building Contract with contractor's design 2011 (MWD11) and Design and Build Contract 2011 (DB11).

Chappell avoids legal jargon but writes with authority and precision. Architects, quantity surveyors, contractors and students of these professions will find this a practical and affordable reference tool arranged by topic.

April 2012: 234 x 156: 160 pp
Pb: 978-0-415-50890-2: £24.99

To Order: Tel: +44 (0) 1235 400524 Fax: +44 (0) 1235 400525
or Post: Taylor and Francis Customer Services,
Bookpoint Ltd, Unit T1, 200 Milton Park, Abingdon, Oxon, OX14 4TA UK
Email: book.orders@tandf.co.uk

For a complete listing of all our titles visit:
www.tandf.co.uk

Capital Allowances

Introduction

Capital Allowances provide tax relief by prescribing a statutory rate of depreciation for tax purposes in place of that used for accounting purposes. They are utilized by government to provide an incentive to invest in capital equipment, including assets within commercial property, by allowing the majority of taxpayers a deduction from taxable profits for certain types of capital expenditure, thereby reducing or deferring tax liabilities.

The capital allowances most commonly applicable to real estate are those given for capital expenditure on existing commercial buildings in disadvantaged areas, and plant and machinery in all buildings other than residential dwellings. Relief for certain expenditure on industrial buildings and hotels was withdrawn from April 2011, although the ability to claim plant and machinery remains.

Enterprise Zone Allowances are also available for capital expenditure within designated areas only where there is a focus on high-value manufacturing. Enhanced rates of allowances are available on certain types of energy and water saving plant and machinery assets, whilst reduced rates apply to 'integral features' and items with an expected economic life of more than 25 years.

The Act

The primary legislation is contained in the Capital Allowances Act 2001. Major changes to the system were announced by the Government in 2007 and there have been further changes in subsequent Finance Acts.

The Act is arranged in 12 Parts (plus two addenda) and was published with an accompanying set of Explanatory Notes.

Plant and Machinery

The Finance Act 1994 introduced major changes to the availability of Capital Allowances on real estate. A definition was introduced which precludes expenditure on the provision of a building from qualifying for plant and machinery, with prescribed exceptions.

List A in Section 21 of the 2001 Act sets out those assets treated as parts of buildings:

- Walls, floors, ceilings, doors, gates, shutters, windows and stairs.
- Mains services, and systems, for water, electricity and gas.
- Waste disposal systems.
- Sewerage and drainage systems.
- Shafts or other structures in which lifts, hoists, escalators and moving walkways are installed.
- Fire safety systems.

Similarly, List B in Section 22 identifies excluded structures and other assets.

Both sections are, however, subject to Section 23. This section sets out expenditure, which although being part of a building, may still be expenditure on the provision of Plant and Machinery.

List C in Section 23 is reproduced below:

Sections 21 and 22 do not affect the question whether expenditure on any item in List C is expenditure on the provision of Plant or Machinery

1. Machinery (including devices for providing motive power) not within any other item in this list.
2. Gas and sewerage systems provided mainly –
    a. to meet the particular requirements of the qualifying activity, or
    b. to serve particular plant or machinery used for the purposes of the qualifying activity.
3. Omitted.
4. Manufacturing or processing equipment; storage equipment (including cold rooms); display equipment; and counters, checkouts and similar equipment.
5. Cookers, washing machines, dishwashers, refrigerators and similar equipment; washbasins, sinks, baths, showers, sanitary ware and similar equipment; and furniture and furnishings.
6. Hoists.
7. Sound insulation provided mainly to meet the particular requirements of the qualifying activity.
8. Computer, telecommunication and surveillance systems (including their wiring or other links).
9. Refrigeration or cooling equipment.
10. Fire alarm systems; sprinkler and other equipment for extinguishing or containing fires.
11. Burglar alarm systems.
12. Strong rooms in bank or building society premises; safes.
13. Partition walls, where moveable and intended to be moved in the course of the qualifying activity.
14. Decorative assets provided for the enjoyment of the public in hotel, restaurant or similar trades.
15. Advertising hoardings; signs, displays and similar assets.
16. Swimming pools (including diving boards, slides & structures on which such boards or slides are mounted).
17. Any glasshouse constructed so that the required environment (namely, air, heat, light, irrigation and temperature) for the growing of plants is provided automatically by means of devices forming an integral part of its structure.
18. Cold stores.
19. Caravans provided mainly for holiday lettings.
20. Buildings provided for testing aircraft engines run within the buildings.
21. Moveable buildings intended to be moved in the course of the qualifying activity.
22. The alteration of land for the purpose only of installing Plant or Machinery.
23. The provision of dry docks.
24. The provision of any jetty or similar structure provided mainly to carry Plant or Machinery.
25. The provision of pipelines or underground ducts or tunnels with a primary purpose of carrying utility conduits.
26. The provision of towers to support floodlights.
27. The provision of –
    a. any reservoir incorporated into a water treatment works, or
    b. any service reservoir of treated water for supply within any housing estate or other particular locality.
28. The provision of –
    a. silos provided for temporary storage, or
    b. storage tanks.
29. The provision of slurry pits or silage clamps.
30. The provision of fish tanks or fish ponds.
31. The provision of rails, sleepers and ballast for a railway or tramway.
32. The provision of structures and other assets for providing the setting for any ride at an amusement park or exhibition.
33. The provision of fixed zoo cages.

Capital Allowances on plant and machinery are given in the form of writing down allowances at the rate of 18% per annum on a reducing balance basis. For every £100 of qualifying expenditure £18 is claimable in year 1, £14.76 in year 2 and so on until either all the allowances have been claimed or the asset is sold.

Integral features

The category of qualifying expenditure on 'integral features' was introduced with effect from April 2008. The following items are integral features:

- An electrical system (including a lighting system)
- A cold water system
- A space or water heating system, a powered system of ventilation, air cooling or air purification, and any floor or ceiling comprised in such a system
- A lift, an escalator or a moving walkway
- External solar shading

A reduced writing down allowance of 8% per annum is available on integral features.

Thermal insulation

For many years the addition of thermal insulation to an existing industrial building has been treated as qualifying for plant and machinery allowances. From April 2008 this has been extended to include all commercial buildings but not residential buildings.

A reduced writing down allowance of 8% per annum is available on thermal insulation.

Long-life assets

A reduced writing down allowance of 8% per annum is available on long-life assets. Allowances were given at the rate of 6% before April 2008.

A long-life asset is defined as plant and machinery that can reasonably be expected to have a useful economic life of at least 25 years. The useful economic life is taken as the period from first use until it is likely to cease to be used as a fixed asset of any business. It is important to note that this likely to be a shorter period than an item's physical life.

Plant and machinery provided for use in a building used wholly or mainly as dwelling house, showroom, hotel, office or retail shop or similar premises, or for purposes ancillary to such use, cannot be long-life assets.

In contrast plant and machinery assets in buildings such as factories, cinemas, hospitals and so on are all potentially long-life assets.

Case law

The fact that an item appears in List C does not automatically mean that it will qualify for capital allowances. It only means that it may potentially qualify.

Guidance about the meaning of plant has to be found in case law. The cases go back a long way, beginning in 1887. The current state of the law on the meaning of plant derives from the decision in the case of *Wimpy International Ltd and Associated Restaurants Ltd v Warland* in the late 1980s.

The Judge in that case said that there were three tests to be applied when considering whether or not an item is plant.

1. Is the item stock in trade? If the answer is yes, then the item is not plant.
2. Is the item used for carrying on the business? In order to pass the business use test the item must be employed in carrying on the business; it is not enough for the asset to be simply used in the business. For example, product display lighting in a retail store may be plant but general lighting in a warehouse would fail the test.
3. Is the item the business premises or part of the business premises? An item cannot be plant if it fails the premises test, i.e. if the business use is as the premises (or part of the premises) or place on which the business is conducted. The meaning of part of the premises in this context should not be confused with the law of real property. The Inland Revenue's internal manuals suggest there are four general factors to be considered, each of which is a question of fact and degree:

- Does the item appear visually to retain a separate identity
- With what degree of permanence has it been attached to the building
- To what extent is the structure complete without it
- To what extent is it intended to be permanent or alternatively is it likely to be replaced within a short period

There is obviously a core list of items that will usually qualify in the majority of cases. However, many others still need to be looked at on a case-by-case basis. For example, decorative assets in a hotel restaurant may be plant but similar assets in an office reception area would almost certainly not be.

One of the benefits of the integral features rules, apart from simplification, is that items that did not qualify by applying these rules, such as general lighting in a warehouse or an office building, will now qualify albeit at a reduced rate.

Refurbishment schemes

Building refurbishment projects will typically be a mixture of capital costs and revenue expenses, unless the works are so extensive that they are more appropriately classified a redevelopment. A straightforward repair or a 'like for like' replacement of part of an asset would be a revenue expense, meaning that the entire amount can be deducted from taxable profits in the same year.

Where capital expenditure is incurred that is incidental to the installation of plant or machinery then Section 25 of the 2001 Act allows it to be treated as part of the expenditure on the qualifying item. Incidental expenditure will often include parts of the building that would be otherwise disallowed, as shown in the Lists reproduced above. For example, the cost of forming a lift shaft inside an existing building would be deemed to be part of the expenditure on the provision of the lift.

The extent of the application of section 25 was reviewed for the first time by the Special Commissioners in December 2007 and by the First Tier Tribunal (Tax Chamber) in December 2009, in the case of JD Wetherspoon. The key areas of expenditure considered were overheads and preliminaries where it was held that such costs could be allocated on a pro-rata basis; decorative timber panelling which was found to be part of the premises and so ineligible for allowances; toilet lighting which was considered to provide an attractive ambience and qualified for allowances; and incidental building alterations of which enclosing walls to toilets and kitchens and floor finishes did not qualify but tiled splash backs, toilet cubicles and drainage did qualify along with the related sanitary fittings and kitchen equipment.

Annual investment allowance

The annual investment allowance is available to all businesses of any size and allows a deduction for the whole of the first £250,000 from January 2013 (£25,000 before January 2013) of qualifying expenditure on plant and machinery, including integral features and long-life assets.

The Enhanced Capital Allowances Scheme

The scheme is one of a series of measures introduced to ensure that the UK meets its target for reducing green-house gases under the Kyoto Protocol. 100% first year allowances are available on products included on the Energy Technology List published on the website at www.eca.gov.uk and other technologies supported by the scheme. All businesses will be able to claim the enhanced allowances, but only investments in new and unused Machinery and Plant can qualify.

There are currently 17 technologies and 53 sub-technologies covered by the scheme.

- Air-to-air energy recovery
- Automatic monitoring and targeting (AMT)
- Boiler equipment
- Combined heat and power (CHP)
- Compact heat exchangers
- Compressed air equipment
- Heat pumps

- Heating ventilation and air conditioning equipment
- High speed hand air dryers
- Lighting
- Motors and drives
- Pipework insulation
- Radiant and warm air heaters
- Refrigeration equipment
- Solar thermal systems
- Thermal screens
- Uninterruptible power supplies

The Finance Act 2003 introduced a new category of environmentally beneficial plant and machinery qualifying for 100% first-year allowances. The Water Technology List includes 13 technologies:

- Cleaning in place equipment
- Efficient showers
- Efficient taps
- Efficient toilets
- Efficient washing machines
- Flow controllers
- Leakage detection equipment
- Meters and monitoring equipment
- Rainwater harvesting equipment.
- Small scale slurry and sludge dewatering equipment
- Vehicle wash water reclaim units
- Water efficient industrial cleaning equipment
- Water management equipment for mechanical seals

Buildings and structures and long-life assets as defined above cannot qualify under the scheme. However, following the introduction of the integral features rules, lighting in any non-residential building may potentially qualify for enhanced capital allowances if it meets the relevant criteria.

A limited payable ECA tax credit equal to 19% of the loss surrendered was also introduced for UK companies in April 2008.

From April 2012 expenditure on plant and machinery for which tariff payments are received under the renewable energy schemes introduced by the Department of Energy and Climate Change (Feed-in Tariffs or Renewable Heat Incentives) will not be entitled to enhanced capital allowances.

Enterprise zones

The creation of 11 new Enterprise Zones was announced in the 2011 Budget. Additional zones have since been added bringing the number to 24 in total. Originally introduced in the early 1980 s as a stimulus to commercial development and investment, they had virtually faded from the real estate psyche.

The original zones benefited from a 100% first year allowance on capital expenditure incurred on the construction (or the purchase within 2 years of first use) of any commercial building within a designated enterprise zone, within 10 years of the site being so designated. Like other allowances given under the industrial buildings code the building has a life of 25 years for tax purposes.

The majority of these enterprise zones had reached the end of their 10-year life by 1993. However, in certain very limited circumstances it may still be possible to claim these allowances up to 20 years after the site was first designated.

Enterprise zones benefit from a number of reliefs, including a 100% first year allowance for new and unused non-leased plant and machinery assets, where there is a focus on high-value manufacturing.

Flat conversion allowances

Tax relief is available on capital expenditure incurred on or after 11 May 2001 on the renovation or conversion of vacant or underused space above shops and other commercial premises to provide flats for rent.

In order to qualify the property must have been built before 1980 and the expenditure incurred on, or in connection with:

- Converting part of a qualifying building into a qualifying flat.
- Renovating an existing flat in a qualifying building if the flat is, or will be, a qualifying flat.
- Repairs incidental to conversion or renovation of a qualifying flat, and
- The cost of providing access to the flat(s).

The property must not have more than four storeys above the ground floor and it must appear that, when the property was constructed, the floors above the ground floor were primarily for residential use. The ground floor must be authorized for business use at the time of the conversion work and for the period during which the flat is held for letting. Each new flat must be a self-contained dwelling, with external access separate from the ground-floor premises. It must have no more than four rooms, excluding kitchen and bathroom. None of the flats can be 'high value' flats, as defined in the legislation. The new flats must be available for letting as a dwelling for a period of not more than 5 years.

An initial allowance of 100% is available or, alternatively, a lower amount may be claimed, in which case the balance may be claimed at a rate of 25% per annum in subsequent years. The allowances may be recovered if the flat is sold or ceases to be let within 7 years.

At Budget 2011 the Government announced that the relief would be abolished from April 2013.

Business Premises Renovation Allowance

The Business Premises Renovation Allowance (BPRA) was first announced in December 2003. The idea behind the scheme is to bring long-term vacant properties back into productive use by providing 100% capital allowances for the cost of renovating and converting unused premises in disadvantaged areas. The legislation was included in the Finance Act 2005 and was finally implemented on 11 April 2007 following EU state aid approval.

The legislation is identical in many respects to that for flat conversion allowances. The scheme will apply to properties within one of the areas specified in the Assisted Areas Order 2007 and Northern Ireland.

BPRA will be available to both individuals and companies who own or lease business property that has been unused for 12 months or more. Allowances will be available to a person who incurs qualifying capital expenditure on the renovation of business premises.

An announcement to extend the scheme by a further 5 years to 2017 was made within the 2011 Budget.

Other capital allowances

Other types of allowances include those available for capital expenditure on Mineral Extraction, Research and Development, Know-How, Patents, Dredging and Assured Tenancy.

Value Added Tax

Introduction

Value Added Tax (VAT) is a tax on the consumption of goods and services. The UK adopted VAT when it joined the European Community in 1973. The principal source of European law in relation to VAT is Council Directive 2006/112/EC, a recast of Directive 77/388/EEC which is currently restated and consolidated in the UK through the VAT Act 1994 and various Statutory Instruments, as amended by subsequent Finance Acts.

VAT Notice 708: Buildings and construction (June 2007) gives an interpretation of the law in connection with construction works from the point of view of HM Revenue & Customs. VAT tribunals and court decisions since the date of this publication will affect the application of the law in certain instances. The Notice is available on HM Revenue & Customs website at www.hmrc.gov.uk.

The scope of VAT

VAT is payable on:

- Supplies of goods and services made in the UK;
- By a taxable person;
- In the course or furtherance of business; and
- Which are not specifically exempted or zero-rated.

Rates of VAT

There are three rates of VAT:

- A standard rate, currently 20% since January 2011;
- A reduced rate, currently 5%; and
- A zero rate.

Additionally some supplies are exempt from VAT and others are outside the scope of VAT.

Recovery of VAT

When a taxpayer makes taxable supplies he must account for VAT at the appropriate rate of either 20% or 5%. This VAT then has to be passed to HM Revenue & Customs and will normally be charged to the taxpayer's customers.

As a VAT registered person, the taxpayer can reclaim from HM Revenue & Customs as much of the VAT incurred on their purchases as relates to the standard-rated, reduced-rated and zero-rated onward supplies they make. A person cannot however reclaim VAT that relates to any non-business activities (but see below) or to any exempt supplies they make.

At predetermined intervals the taxpayer will pay to HM Revenue & Customs the excess of VAT collected over the VAT they can reclaim. However if the VAT reclaimed is more than the VAT collected, the taxpayer can reclaim the difference from HM Revenue & Customs.

Example

X Ltd constructs a block of flats. It sells long leases to buyers for a premium. X Ltd has constructed a new building designed as a dwelling and will have granted a long lease. This sale of a long lease is VAT zero-rated. This means any VAT incurred in connection with the development which X Ltd will have paid (e.g. payments for consultants and certain preliminary services) will be reclaimable. For reasons detailed below the contractor employed by X Ltd will not have charged VAT on his construction services.

Use for business and non-business activities

Where VAT relates partly to business use and partly to non-business use then the basic rule is that it must be apportioned so that only the business element is potentially recoverable. In some cases VAT on land, buildings and certain construction services purchased for both business and non-business use could be recovered in full by applying what is known as 'Lennartz' accounting to reclaim VAT relating to the non-business use and account for VAT on the non-business use over a maximum period of 10 years. HM Revenue & Customs revised their policy in January 2010 following an ECJ case restricting the scope and its application to immovable property will be removed completely from January 2011 when the UK law is amended to comply with EU Council Directive 2009/162/EU.

Taxable persons

A taxable person is an individual, firm, company etc: who is required to be registered for VAT. A person who makes taxable supplies above certain value limits is required to be registered. The current registration limit is £77,000 for 2012–13. The threshold is exceeded if at the end of any month the value of taxable supplies in the period of 1 year then ending is over the limit, or at any time, if there are reasonable grounds for believing that the value of the taxable supplies in the period of 30 days then beginning will exceed £77,000.

A person who makes taxable supplies below these limits is entitled to be registered on a voluntary basis if they wish, for example in order to recover VAT incurred in relation to those taxable supplies.

In addition, a person who is not registered for VAT in the UK but acquires goods from another EC member state, or makes distance sales in the UK, above certain value limits may be required to register for VAT in the UK.

VAT exempt supplies

If a supply is exempt from VAT this means that no tax is payable – but equally the person making the exempt supply cannot normally recover any of the VAT on their own costs relating to that supply.

Generally property transactions such as leasing of land and buildings are exempt unless a landlord chooses to standard-rate its supplies by a process known as opting to tax. This means that VAT is added to rental income and also that VAT incurred on, say, an expensive refurbishment, is recoverable.

Supplies outside the scope of VAT

Supplies are outside the scope of VAT if they are:

- Made by someone who is not a taxable person;
- Made outside the UK; or
- Not made in the course or furtherance of business.

In course or furtherance of business

VAT must be accounted for on all taxable supplies made in the course or furtherance of business with the corresponding recovery of VAT on expenditure incurred.

If a taxpayer also carries out non-business activities then VAT incurred in relation to such supplies is generally not recoverable.

In VAT terms, business means any activity continuously performed which is mainly concerned with making supplies for a consideration. This includes:

- Anyone carrying on a trade, vocation or profession;
- The provision of membership benefits by clubs, associations and similar bodies in return for a subscription or other consideration; and
- Admission to premises for a charge.

It may also include the activities of other bodies including charities and non-profit making organizations.

Examples of non-business activities are:

- Providing free services or information;
- Maintaining some museums or particular historic sites;
- Publishing religious or political views.

Construction services

In general the provision of construction services by a contractor will be VAT standard rated at 20%, however, there are a number of exceptions for construction services provided in relation to certain residential and charitable use buildings.

The supply of building materials is VAT standard rated at 20%, however, where these materials are supplied as part of the construction services the VAT liability of those materials follows that of the construction services supplied.

Zero-rated construction services

The following construction services are VAT zero-rated including the supply of related building materials.

The construction of new dwellings

The supply of services in the course of the construction of a building designed for use as a dwelling or number of dwellings is zero-rated other than the services of an architect, surveyor or any other person acting as a consultant or in a supervisory capacity.

The following conditions must be satisfied in order for the works to qualify for zero-rating:

1. The work must not amount to the conversion, reconstruction or alteration of an existing building;
2. The work must not be an enlargement of, or extension to, an existing building except to the extent that the enlargement or extension creates an additional dwelling or dwellings;
3. The building must be designed as a dwelling or number of dwellings. Each dwelling must consist of self-contained living accommodation with no provision for direct internal access from the dwelling to any other dwelling or part of a dwelling;
4. Statutory planning consent must have been granted for the construction of the dwelling, and construction carried out in accordance with that consent;
5. Separate use or disposal of the dwelling must not be prohibited by the terms of any covenant, statutory planning consent or similar provision.

The construction of a garage at the same time as the dwelling can also be zero-rated as can the demolition of any existing building on the site of the new dwelling.

A building only ceases to be an existing building (see points 1. and 2. above) when it is:

1. Demolished completely to ground level; or when
2. The part remaining above ground level consists of no more than a single facade (or a double facade on a corner site) the retention of which is a condition or requirement of statutory planning consent or similar permission.

The construction of a new building for 'relevant residential or charitable' use

The supply of services in the course of the construction of a building designed for use as a relevant residential or charitable building is zero-rated other than the services of an architect, surveyor or any other person acting as a consultant or in a supervisory capacity.

A relevant residential use building means:

1. A home or other institution providing residential accommodation for children;
2. A home or other institution providing residential accommodation with personal care for persons in need of personal care by reason of old age, disablement, past or present dependence on alcohol or drugs or past or present mental disorder;

3.   A hospice;
4.   Residential accommodation for students or school pupils;
5.   Residential accommodation for members of any of the armed forces;
6.   A monastery, nunnery, or similar establishment; or
7.   An institution which is the sole or main residence of at least 90% of its residents.

A relevant residential purpose building does not include use as a hospital, a prison or similar institution or as a hotel, inn or similar establishment.

A relevant charitable use means use by a charity:

1.   Otherwise than in the course or furtherance of a business; or
2.   As a village hall or similarly in providing social or recreational facilities for a local community.

Non-qualifying use which is not expected to exceed 10% of the time the building is normally available for use can be ignored. The calculation of business use can be based on time, floor area or head count subject to approval being acquired from HM Revenue & Customs.

The construction services can only be zero-rated if a certificate is given by the end user to the contractor carrying out the works confirming that the building is to be used for a qualifying purpose i.e. for a 'relevant residential or charitable' purpose. It follows that such services can only be zero-rated when supplied to the end user and, unlike supplies relating to dwellings, supplies by subcontractors cannot be zero-rated.

The construction of an annex used for a 'relevant charitable' purpose

Construction services provided in the course of construction of an annexe for use entirely or partly for a 'relevant charitable' purpose can be zero-rated.

In order to qualify the annexe must:

1.   Be capable of functioning independently from the existing building;
2.   Have its own main entrance; and
3.   Be covered by a qualifying use certificate.

The conversion of a non-residential building into dwellings or the conversion of a building from non-residential use to 'relevant residential' use where the supply is to a 'relevant' housing association

The supply to a 'relevant' housing association in the course of conversion of a non-residential building or non-residential part of a building into:

1.   A building or part of a building designed as a dwelling or number of dwellings; or
2.   A building or part of a building for use solely for a relevant residential purpose, of any services related to the conversion other than the services of an architect, surveyor or any person acting as a consultant or in a supervisory capacity are zero-rated.

A relevant housing association is defined as:

1.   A private registered provider of social housing;
2.   A registered social landlord within the meaning of Part I of the Housing Act 1996 (Welsh registered social landlords);
3.   A registered social landlord within the meaning of the Housing (Scotland) Act 2001 (Scottish registered social landlords), or ;
4.   A registered housing association within the meaning of Part II of the Housing (Northern Ireland) Order 1992 (Northern Irish registered housing associations).

If the building is to be used for a 'relevant residential' purpose the housing association should issue a qualifying use certificate to the contractor completing the works.

The construction of a permanent park for residential caravans

The supply in the course of the construction of any civil engineering work 'necessary for' the development of a permanent park for residential caravans of any services related to the construction can be VAT zero-rated. This includes access roads, paths, drainage, sewerage and the installation of mains water, power and gas supplies.

Certain building alterations for disabled persons

Certain goods and services supplied to a disabled person, or a charity making these items and services available to 'disabled' persons can be zero-rated. The recipient of these goods or services needs to give the supplier an appropriate written declaration that they are entitled to benefit from zero rating.

The following services (amongst others) are zero-rated:

1. the installation of specialist lifts and hoists and their repair and maintenance;
2. the construction of ramps, widening doorways or passageways including any preparatory work and making good work;
3. the provision, extension and adaptation of a bathroom, washroom or lavatory; and
4. emergency alarm call systems.

Approved alterations to protected buildings

A supply in the course of an approved alteration to a protected building of any services other than the services of an architect, surveyor or any person acting as consultant or in a supervisory capacity can be zero-rated.

A protected building is defined as a building that is:

*   designed to remain as or become a dwelling or number of dwellings after the alterations; or
*   is intended for use for a 'relevant residential or charitable purpose' after the alterations; and which is
*   a listed building or scheduled ancient monument.

A listed building does not include buildings that are in conservation areas, but not on the statutory list, or buildings included in non-statutory local lists.

An 'approved alteration' is an alteration to a 'protected building' that requires and has obtained listed building consent or scheduled monument consent. This consent is necessary for any works that affect the character of a building of special architectural or historic interest.

It is important to note that approved alterations do not include any works of repair or maintenance or any incidental alteration to the fabric of a building that results from the carrying out of repairs or maintenance work.

A 'protected building' that is intended for use for a 'relevant residential' or charitable purpose will require the production of a qualifying use certificate by the end user to the contractor providing the alteration services.

Listed Churches are relevant charitable use buildings and where approved alterations are being carried out zero-rate VAT can be applied. Additionally since April 1 2004, listed places of worship can apply for a grant for repair and maintenance works equal to the full amount of VAT paid on eligible works carried out on or after 1 April 2004. Information relating to the scheme can be obtained from the website at www.lpwscheme.org.uk.

Budget 2012 announced plans to consult on the abolition of the zero rate for approved alterations to protected buildings from 1 October 2012.

Following the consultation period it was confirmed that projects will continue to benefit from zero rating where a contract was entered into or where listed building consent (or equivalent approval for listed places of worship) had been applied for before 21 March 2012.

Provided the application was in place before 21 March 2012, zero rating will continue under the transitional rules until 30 September 2015.

All other projects will be subject to the standard rate of VAT on or after 1 October 2012.

Sale of reconstructed buildings

Budget 2012 also announced a change in the rules that currently enable developers of substantially reconstructed listed residential properties to recover the VAT they incur on costs relating to the redevelopment through zero-rating. From 1 October 2012 a protected building shall not be regarded as substantially reconstructed unless, when the reconstruction is completed, the reconstructed building incorporates no more of the original building than the external walls, together with other external features of architectural or historical interest. Transitional arrangements protect contracts entered into before 21 March 2012 for the first grant of a major interest in the protected building made on or before 20 March 2013.

DIY builders and converters

Private individuals who decide to construct their own home are able to reclaim VAT they pay on goods they use to construct their home by use of a special refund mechanism made by way of an application to HM Revenue & Customs. This also applies to services provided in the conversion of an existing non-residential building to form a new dwelling.

The scheme is meant to ensure that private individuals do not suffer the burden of VAT if they decide to construct their own home.

Charities may also qualify for a refund on the purchase of materials incorporated into a building used for non-business purposes where they provide their own free labour for the construction of a 'relevant charitable' use building.

Reduced-rated construction services

The following construction services are subject to the reduced rate of VAT of 5%, including the supply of related building materials.

A changed number of dwellings conversion

In order to qualify for the 5% rate there must be a different number of 'single household dwellings' within a building than there were before commencement of the conversion works. A 'single household dwelling' is defined as a dwelling that is designed for occupation by a single household.

These conversions can be from relevant residential purpose buildings, non-residential buildings and houses in multiple occupation.

A house in multiple occupation conversion

This relates to construction services provided in the course of converting a 'single household dwelling', a number of single household dwellings, a non-residential building or a relevant residential purpose building into a house for multiple occupation such as bed sit accommodation.

A special residential conversion

A special residential conversion involves the conversion of a single household dwelling, a house in multiple occupation or a non-residential building into a 'relevant residential' purpose building such as student accommodation or a care home.

Renovation of derelict dwellings

The provision of renovation services in connection with a dwelling or 'relevant residential' purpose building that has been empty for 2 or more years prior to the date of commencement of construction works can be carried out at a reduced rate of VAT of 5%.

Installation of energy saving materials

The supply and installation of certain energy saving materials including insulation, draught stripping, central heating and hot water controls and solar panels in a residential building or a building used for a relevant charitable purpose.

Budget 2012 announced that buildings that are used by charities for non-business purposes, and/or as village halls, will be removed from the scope of the reduced rate for the supply of energy saving materials under legislation to be introduced in the Finance Bill 2013.

Grant-funded supply of heating equipment or connection of a gas supply

The grant-funded supply and installation of heating appliances, connection of a mains gas supply, supply, installation, maintenance and repair of central heating systems, and supply and installation of renewable source heating systems, to qualifying persons. A qualifying person is someone aged 60 or over or who is in receipt of various specified benefits.

Installation of security goods

The grant-funded supply and installation of security goods to a qualifying person.

Housing alterations for the elderly

Certain home adaptations that support the needs of elderly people were reduced rated with effect from 1 July 2007.

Building contracts

Design and build contracts

If a contractor provides a design and build service relating to works to which the reduced or zero rate of VAT is applicable then any design costs incurred by the contractor will follow the VAT liability of the principal supply of construction services.

Management contracts

A management contractor acts as a main contractor for VAT purposes and the VAT liability of his services will follow that of the construction services provided. If the management contractor only provides advice without engaging trade contractors his services will be VAT standard rated.

Construction management and project management

The project manager or construction manager is appointed by the client to plan, manage and coordinate a construction project. This will involve establishing competitive bids for all the elements of the work and the appointment of trade contractors. The trade contractors are engaged directly by the client for their services.

The VAT liability of the trade contractors will be determined by the nature of the construction services they provide and the building being constructed.

The fees of the construction manager or project manager will be VAT standard rated. If the construction manager also provides some construction services these works may be zero or reduced rated if the works qualify.

Liquidated and ascertained damages

Liquidated damages are outside of the scope of VAT as compensation. The employer should not reduce the VAT amount due on a payment under a building contract on account of a deduction of damages. In contrast an agreed reduction in the contract price will reduce the VAT amount.

Similarly, in certain circumstances HM Revenue & Customs may agree that a claim by a contractor under a JCT or other form of contract is also compensation payment and outside the scope of VAT.

Understanding the NEC3 ECC Contract

A Practical Handbook

Kelvin Hughes

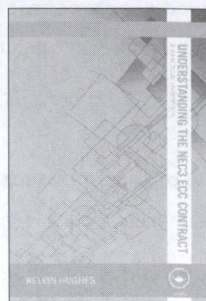

As usage of the NEC (formerly the New Engineering Contract) family of contracts continues to grow worldwide, so does the importance of understanding its clauses and nuances to everyone working in the built environment. Currently in its third edition, this set of contracts is different to others in concept as well as format, so users may well find themselves needing a helping hand along the way.

Understanding the NEC3 ECC Contract uses plain English to lead the reader through the NEC3 Engineering and Construction Contract's key features, including:

- main and secondary options
- the use of early warnings
- programme provisions
- payment
- compensation events
- preparing and assessing tenders.

Common problems experienced when using the Engineering and Construction Contract are signalled to the reader throughout, and the correct way of reading each clause explained. The way the contract effects procurement processes, dispute resolution, project management, and risk management are all addressed in order to direct the user to best practice.

Written for construction professionals, by a practicing international construction contract consultant, this handbook is the most straightforward, balanced and practical guide to the NEC3 ECC available. An ideal companion for employers, contractors, project managers, supervisors, engineers, architects, quantity surveyors, subcontractors, and anyone else interested in working successfully with the NEC3 ECC.

October 2012: 234 x156: 272 pp
Pb: 978-0-415-61496-2: £29.99

To Order: Tel: +44 (0) 1235 400524 Fax: +44 (0) 1235 400525
or Post: Taylor and Francis Customer Services,
Bookpoint Ltd, Unit T1, 200 Milton Park, Abingdon, Oxon, OX14 4TA UK
Email: book.orders@tandf.co.uk

For a complete listing of all our titles visit:
www.crcpress.com

The Aggregates Levy

The Aggregates Levy came into operation on 1 April 2002 in the UK, except for Northern Ireland where it has been phased in over 5 years from 2003.

It was introduced to ensure that the external costs associated with the exploitation of aggregates are reflected in the price of aggregate, and to encourage the use of recycled aggregate. There continues to be strong evidence that the levy is achieving its environmental objectives, with sales of primary aggregate down and production of recycled aggregate up. The Government expects that the rates of the levy will at least keep pace with inflation over time, although it accepts that the levy is still bedding in.

The rate of the levy remains at £2.00 per tonne from 1 April 2013 and is levied on anyone considered to be responsible for commercially exploiting virgin aggregates in the UK and should naturally be passed by price increase to the ultimate user.

All materials falling within the definition of Aggregates are subject to the levy unless specifically exempted.

It does not apply to clay, soil, vegetable or other organic matter.

The intention is that it will:

- Encourage the use of alternative materials that would otherwise be disposed of to landfill sites
- Promote development of new recycling processes, such as using waste tyres and glass
- Promote greater efficiency in the use of virgin aggregates
- Reduce noise and vibration, dust and other emissions to air, visual intrusion, loss of amenity and damage to wildlife habitats

Definitions

'Aggregates' means any rock, gravel or sand which is extracted or dredged in the UK for aggregates use. It includes whatever substances are for the time being incorporated in it or naturally occur mixed with it.

'Exploitation' is defined as involving any one or a combination of any of the following:

- Being removed from its original site
- Becoming subject to a contract or other agreement to supply to any person
- Being used for construction purposes
- Being mixed with any material or substance other than water, except in permitted circumstances

Incidence

It is a tax on primary aggregates production – i.e. virgin aggregates won from a source and used in a location within the UK territorial boundaries (land or sea). The tax is not levied on aggregates which are exported or on aggregates imported from outside the UK territorial boundaries.

It is levied at the point of sale.

Exemption from tax

An aggregate is exempt from the levy if it is:

- Material which has previously been used for construction purposes
- Aggregate that has already been subject to a charge to the Aggregates Levy
- Aggregate which was previously removed from its originating site before the start date of the levy
- Aggregate which is being returned to the land from which it was won
- Aggregate won from a farm land or forest where used on that farm or forest
- Rock which has not been subjected to an industrial crushing process
- Aggregate won by being removed from the ground on the site of any building or proposed building in the course of excavations carried out in connection with the modification or erection of the building and exclusively for the purpose of laying foundations or of laying any pipe or cable
- Aggregate won by being removed from the bed of any river, canal or watercourse or channel in or approach to any port or harbour (natural or artificial), in the course of carrying out any dredging exclusively for the purpose of creating, restoring, improving or maintaining that body of water
- Aggregate won by being removed from the ground along the line of any highway or proposed highway in the course of excavations for improving, maintaining or constructing the highway otherwise than purely to extract the aggregate
- Drill cuttings from petroleum operations on land and on the seabed
- Aggregate resulting from works carried out in exercise of powers under the New Road and Street Works Act 1991, the Roads (Northern Ireland) Order 1993 or the Street Works (Northern Ireland) Order 1995
- Aggregate removed for the purpose of cutting of rock to produce dimension stone, or the production of lime or cement from limestone
- Aggregate arising as a waste material during the processing of the following industrial minerals:
  - ball clay
  - barytes
  - calcite
  - china clay
  - coal, lignite, slate or shale
  - feldspar
  - flint
  - fluorspar
  - fuller's earth
  - gems and semi-precious stones
  - gypsum
  - any metal or the ore of any metal
  - muscovite
  - perlite
  - potash
  - pumice
  - rock phosphates
  - sodium chloride
  - talc
  - vermiculite

However, the levy is still chargeable on any aggregates arising as the spoil or waste from or the by-products of the above exempt processes. This includes quarry overburden.

Anything that consists 'wholly or mainly' of the following is exempt from the levy (note that 'wholly' is defined as 100% but 'mainly' as more than 50%, thus exempting any contained aggregates amounting to less than 50% of the original volumes:

- clay, soil, vegetable or other organic matter
- coal, slate or shale
- china clay waste and ball clay waste

Relief from the levy either in the form of credit or repayment is obtainable where:

- it is subsequently exported from the UK in the form of aggregate
- it is used in an exempt process
- where it is used in a prescribed industrial or agricultural process
- it is waste aggregate disposed of by dumping or otherwise, e.g. sent to landfill or returned to the originating site

The Aggregates Levy Credit Scheme (ALCS) for Northern Ireland was suspended with effect from 1 December 2010 following a ruling by the European General Court.

A new exemption for aggregate obtained as a by-product of railway, tramway and monorail improvement, maintenance and construction was introduced in 2007.

Discounts

From 1 July 2005 the standard added water percentage discounts listed below can be used. Alternatively a more exact percentage can be agreed and this must be done for dust dampening of aggregates.

- washed sand = 7%
- washed gravel = 3.5%
- washed rock/aggregate = 4%

Impact

The British Aggregates Association suggests that the additional cost imposed by quarries is more likely to be in the order of £3.40 per tonne on mainstream products, applying an above average rate on these in order that by-products and low grade waste products can be held at competitive rates, as well as making some allowance for administration and increased finance charges.

With many gravel aggregates costing in the region of £16.00 to £18.00 per tonne, there is a significant impact on construction costs.

Avoidance

An alternative to using new aggregates in filling operations is to crush and screen rubble which may become available during the process of demolition and site clearance as well as removal of obstacles during the excavation processes.

Example:

Assuming that the material would be suitable for fill material under buildings or roads, a simple cost comparison would be as follows (note that for the purpose of the exercise, the material is taken to be 1.80 tonne per m³ and the total quantity involved less than 1,000 m³):

	£/m³	£/tonne
Importing fill material:		
Cost of 'new' aggregates delivered to site	31.23	17.35
Addition for Aggregates Tax	3.78	2.00
Total cost of importing fill materials	35.01	19.35
Disposing of site material:		
Cost of removing materials from site materials	21.52	11.95
Crushing site materials:		
Transportation of material from excavations or demolition to stockpiles	3.00	1.67
Transportation of material from temporary stockpiles to the crushing plant	4.00	2.22
Establishing plant and equipment on site; removing on completion	2.00	1.11
Maintain and operate plant	9.00	5.00
Crushing hard materials on site	13.00	7.22
Screening material on site	2.00	1.11
Total cost of crushing site materials	33.00	18.33

From the above it can be seen that potentially there is a great benefit in crushing site materials for filling rather than importing fill materials.

Setting the cost of crushing against the import price would produce a saving of £2.00 per m³. If the site materials were otherwise intended to be removed from the site, then the cost benefit increases by the saved disposal cost to £23.53 per m³.

Even if there is no call for any or all of the crushed material on site, it ought to be regarded as a useful asset and either sold on in crushed form or else sold with the prospects of crushing elsewhere.

Specimen Unit rates	Unit³	£
Establishing plant and equipment on site; removing on completion:		
Crushing plant	trip	1,200.00
Screening plant	trip	600.00
Maintain and operate plant:		
Crushing plant	week	7,200.00
Screening plant	week	1,800.00
Transportation of material from excavations or demolition places to temporary stockpiles	m³	3.00
Transportation of material from temporary stockpiles to the crushing plant	m³	2.40
Breaking up material on site using impact breakers:		
mass concrete	m³	14.00
reinforced concrete	m³	16.00
brickwork	m³	6.00
Crushing material on site:		
mass concrete not exceeding 1000 m³	m³	13.00
mass concrete 1000–5000 m³	m³	12.00
mass concrete over 5000 m³	m³	11.00
reinforced concrete not exceeding 1000 m³	m³	15.00
reinforced concrete 1000–5000 m³	m³	14.00
reinforced concrete over 5000 m³	m³	13.00
brickwork not exceeding 1000 m³	m³	12.00
brickwork 1000–5000 m³	m³	11.00
brickwork over 5000 m³	m³	10.00
Screening material on site	m³	2.00

More detailed information can be found on the HMRC website (www.hmrc.gov.uk) in Notice AGL 1 Aggregates Levy published in May 2009.

Whole Life Costing
A New Approach

Peter Caplehorn

Whole life costing has been a subject waiting to come of age for many years. What was previously of mainly academic interest is now becoming a key business tool in the procurement and construction of significant projects.

With the advent of PPP and in particular of PFI, details of the project life need to be assessed and tied in to funding and operation plans. Many of these projects run to millions of pounds and are of high political or social importance, so the implications of the life of materials is crucial. A fundamental requirement of these procurement routes has been that the whole enterprise should be included within the bid, so that a company takes on not only the construction but also the running and maintenance of any building.

Additionally as sustainability has emerged and grown in importance, so has the need for a whole life time costing approach, partly driven by governmental insistence. At the heart of sustainability is an understanding of what the specification means for the future of the building and how it will affect the environment. Whole life costing considers part of this and provides an understanding of how materials may perform and what allowances are needed at the end of their life.

This book sets out the practical issues involved in the selection of materials, their performance, and the issues that need to be taken into account. The emphasis, unlike in other publications, is not to formularise or to package the issues but to leave the reader with a clear understanding and a sensible practical way of arriving at conclusions in the future.

June 2012: 246x174: 160 pp
Hb: 978-0-415-43422-5 £105.00
Pb: 978-0-415-43423-2: £34.99

To Order: Tel: +44 (0) 1235 400524 Fax: +44 (0) 1235 400525
or Post: Taylor and Francis Customer Services,
Bookpoint Ltd, Unit T1, 200 Milton Park, Abingdon, Oxon, OX14 4TA UK
Email: book.orders@tandf.co.uk

For a complete listing of all our titles visit:
www.tandf.co.uk

Taylor & Francis
Taylor & Francis Group

Land Remediation

The purpose of this section is to review the general background of ground contamination, the cost implications of current legislation and to consider the various remedial measures and to present helpful guidance on the cost of Land Remediation.

It must be emphasized that the cost advice given is an average and that costs can vary considerably from contract to contract depending on individual Contractors, site conditions, type and extent of contamination, methods of working and various other factors as diverse as difficulty of site access and distance from approved tips.

We have structured this Unit Cost section to cover as many aspects of Land Remediation works as possible.

The introduction of the Landfill Directive in July 2004 has had a considerable impact on the cost of Remediation works in general and particularly on the practice of Dig and Dump. The number of Landfill sites licensed to accept Hazardous Waste has drastically reduced and inevitably this has led to increased costs.

Market forces will determine future increases in cost resulting from the introduction of the Landfill Directive and the cost guidance given within this section will require review in light of these factors.

Statutory framework

In July 1999 new contaminated land provisions, contained in Part IIA of the Environmental Protection Act 1990, were introduced. Primary objectives of the measures included a legal definition of Contaminated Land and a framework for identifying liability, underpinned by a 'polluter pays' principle meaning that remediation should be paid for by the party (or parties) responsible for the contamination. A secondary, and indirect, objective of Part IIA is to provide the legislative context for remediation carried out as part of development activity which is controlled through the planning system. This is the domain where other related objectives, such as encouraging the recycling of brownfield land, are relevant.

Under the Act action to remediate land is required only where there are unacceptable actual or 'significant possibility of significant harm' to health, controlled waters or the environment. Only Local Authorities have the power to determine a site as Contaminated Land and enforce remediation. Sites that have been polluted from previous land use may not need remediating until the land use is changed; this is referred to as 'land affected by contamination'. This is a risk-based assessment on the site specifics in the context of future end uses. As part of planning controls, the aim is to ensure that a site is incapable of meeting the legal definition of Contaminated Land post-development activity. In addition, it may be necessary to take action only where there are appropriate, cost-effective remediation processes that take the use of the site into account.

The Environment Act 1995 amended the Environment Protection Act 1990 by introducing a new regime designed to deal with the remediation of sites which have been seriously contaminated by historic activities. The regime became operational on 1 April 2000. Local authorities and/or the Environment Agency regulate seriously contaminated sites which are known as 'special sites'. The risks involved in the purchase of potentially contaminated sites are high, particularly considering that a transaction can result in the transfer of liability for historic contamination from the vendor to the purchaser.

The contaminated land provisions of the Environmental Protection Act 1990 are only one element of a series of statutory measures dealing with pollution and land remediation that have been and are to be introduced. Others include:

- Groundwater regulations, including pollution prevention measures
- An integrated prevention and control regime for pollution
- Sections of the Water Resources Act 1991, which deals with works notices for site controls, restoration and clean up

April 2012 saw the first revision of the accompanying Part IIA Statutory Guidance. This has introduced a new categorization scheme for assessing sites under Part IIA. Category 1 is land which definitely is Contaminated Land and Category 4 is for land which definitely is not Contaminated Land. This is intended to assist prioritization of sites which pose the greatest risk. Still included in the statutory guidance are matters of inspection, definition, remediation, apportionment of liabilities and recovery of costs of remediation. The measures are to be applied in accordance with the following criteria:

- the planning system
- the standard of remediation should relate to the present use
- the costs of remediation should be reasonable in relation to the seriousness of the potential harm
- the proposals should be practical in relation to the availability of remediation technology, impact of site constraints and the effectiveness of the proposed clean-up method

Liability for the costs of remediation rests with either the party that 'caused or knowingly permitted' contamination, or with the current owners or occupiers of the land.

Apportionment of liability, where shared, is determined by the local authority. Although owners or occupiers become liable only if the polluter cannot be identified, the liability for contamination is commonly passed on when land is sold.

If neither the polluter nor owner can be found, the clean up is funded from public resources.

The ability to forecast the extent and cost of remedial measures is essential for both parties, so that they can be accurately reflected in the price of the land.

At the end of March 2012, the National Planning Policy Framework replaced relevant planning guidance relating to remediation, most significantly PPS 23 Planning and Pollution Control. This has been replaced by a need to investigate and assess land contamination, which must be carried out by a competent person.

The EU Landfill Directive

The Landfill (England and Wales) Regulations 2002 came into force on 15 June 2002 followed by Amendments in 2004 and 2005. These new regulations implement the Landfill Directive (Council Directive 1999/31/EC), which aims to prevent, or to reduce as far as possible, the negative environmental effects of landfill. These regulations have had a major impact on waste regulation and the waste management industry in the UK.

The Scottish Executive and the Northern Ireland Assembly will be bringing forward separate legislation to implement the Directive within their regions.

In summary, the Directive requires that:

- Sites are to be classified into one of three categories: hazardous, non-hazardous or inert, according to the type of waste they will receive.
- Higher engineering and operating standards will be followed.
- Biodegradable waste will be progressively diverted away from landfills.
- Certain hazardous and other wastes, including liquids, explosive waste and tyres will be prohibited from landfills.
- Pretreatment of wastes prior to landfilling will become a requirement.

On 15 July 2004 the co-disposal of hazardous and non-hazardous waste in the same landfill site ended and in July 2005 new waste acceptance criteria (WAC) were introduced which also prevents the disposal of materials contaminated by coal tar.

The effect of this Directive has been to dramatically reduce the hazardous disposal capacity post July 2004, resulting in a **SIGNIFICANT** increase in remediating costs. There are now approximately 20 commercial landfills licensed to accept hazardous waste as a direct result of the implementation of the Directive! There is one site in Scotland, none in Wales and only limited capacity in the South of England. This has significantly increased travelling distance and cost for disposal to landfill. The increase in operating expenses incurred by the landfill operators has also resulted in higher tipping costs.

However, there is now a growing number of opportunities to dispose of hazardous waste to other facilities such as soil treatment centres, often associated with registered landfills potentially eliminating landfill tax. Equally, improvements in on-site treatment technologies are helping to reduce the costs of disposal by reducing the hazardous properties of materials going off site.

All hazardous materials designated for disposal off site are subject to WAC tests. Samples of these materials are taken from site to laboratories in order to classify the nature of the contaminants. These tests, which cost approximately £200 each, have resulted in increased costs for site investigations and as the results may take up to 3 weeks this can have a detrimental effect on programme.

As from 1 July 2008 the WAC derogations which have allowed oil contaminated wastes to be disposed of in land-fills with other inert substances were withdrawn. As a result the cost of disposing of oil contaminated solids has increased.

There has been a marked slowdown in brownfield development in the UK with higher remediation costs, longer clean-up programmes and a lack of viable treatment options for some wastes.

The UK Government established the Hazardous Waste Forum in December 2002 to bring together key stake-holders to advise on the way forward on the management of hazardous waste.

Effect on disposal costs

Although most landfills are reluctant to commit to future tipping prices, tipping costs have generally stabilized. However, there are significant geographical variances, with landfill tip costs in the North of England typically being less than their counterparts in the Southern regions.

For most projects to remain viable there is an increasing need to treat soil in situ by bioremediation, soil washing or other alternative long-term remediation measures. Waste untreatable on site such as coal tar remains a problem. Development costs and programmes need to reflect this change in methodology.

Types of hazardous waste

- Sludges, acids and contaminated wastes from the oil and gas industry
- Acids and toxic chemicals from chemical and electronics industries
- Pesticides from the agrochemical industry
- Solvents, dyes and sludges from leather and textile industries
- Hazardous compounds from metal industries
- Oil, oil filters and brake fluids from vehicles and machines
- Mercury-contaminated waste from crematoria
- Explosives from old ammunition, fireworks and airbags
- Lead, nickel, cadmium and mercury from batteries
- Asbestos from the building industry
- Amalgam from dentists
- Veterinary medicines

[Source: Sepa]

Foam insulation materials containing ODP (Ozone Depletant Potential) are also considered as hazardous waste under the EC Regulation 2037/2000.

Land remediation techniques

There are two principal approaches to remediation – dealing with the contamination in situ or ex situ. The selection of the approach will be influenced by factors such as: initial and long-term cost, timeframe for remediation, types of contamination present, depth and distribution of contamination, the existing and planned topography, adjacent land uses, patterns of surface drainage, the location of existing on-site services, depth of excavation necessary for foundations and below-ground services, environmental impact and safety, interaction with geotechnical perfor-mance, prospects for future changes in land use and long-term monitoring and maintenance of in situ treatment.

On most sites, contamination can be restricted to the top couple of metres, although gasholder foundations for example can go down 8 to 10 metres. Underground structures can interfere with the normal water regime and trap water pockets.

There could be a problem if contaminants get into fissures in bedrock.

In-situ techniques

A range of in-situ techniques is available for dealing with contaminants, including:

- Clean cover – a layer of clean soil is used to segregate contamination from receptor. This technique is best suited to sites with widely dispersed contamination. Costs will vary according to the need for barrier layers to prevent migration of the contaminant.
- On-site encapsulation – the physical containment of contaminants using barriers such as slurry trench cut-off walls. The cost of on-site encapsulation varies in relation to the type and extent of barriers required, the costs of which range from £50/m² to more than £175/m².

There are also in-situ techniques for treating more specific contaminants, including:

- Bioremediation – for removal of oily, organic contaminants through natural digestion by microorganisms. Most bioremediation is ex situ, i.e. it is dug out and then treated on site in bio-piles. The process can be slow, taking up to 3 years depending upon the scale of the problem, but is particularly effective for the long-term improvement of a site, prior to a change of use.
- Phytoremediation – the use of plants that mitigate the environmental problem without the need to excavate the contaminant material and dispose of it elsewhere. Phytoremediation consists of mitigating pollutant concentrations in contaminated soils, water or air, with plants able to contain, degrade or eliminate metals, pesticides, solvents, explosives, crude oil and its derivatives and various other contaminants from the media that contain them.
- Vacuum extraction – involving the extraction of volatile organic compounds (e.g. benzene) from soil and groundwater by vacuum.
- Thermal treatment – the incineration of contaminated soils on site. Thermal processes use heat to increase the volatility to burn, decompose, destroy or melt the contaminants. Cleaning soil with thermal methods may take only a few months to several years.
- Stabilization – cement or lime, is used to physically or chemically bind oily or metal contaminants to prevent leaching or migration. Stabilization can be used in both in-situ and ex-situ conditions.
- Aeration – if the ground contamination is highly volatile, e.g. fuel oils, then the ground can be ploughed and rotovated to allow the substance to vaporize.
- Air sparging – the injection of contaminant-free air into the subsurface enabling a phase transfer of hydro-carbons from a dissolved state to a vapour phase.
- Chemical oxidization – the injection of reactive chemical oxidants directly into the soil for the rapid destruction of contaminants.
- Pumping – to remove liquid contaminants from boreholes or excavations. Contaminated water can be pumped into holding tanks and allowed to settle; testing may well prove it to be suitable for discharging into the foul sewer subject to payment of a discharge fee to the local authority. It may be necessary to process the water through an approved water treatment system to render it suitable for discharge.

Ex situ techniques

Removal for landfill disposal has, historically, been the most common and cost-effective approach to remediation in the UK, providing a broad spectrum solution by dealing with all contaminants. As mentioned above, the implementation of the Landfill Directive has resulted in other techniques becoming more competitive for the disposal of hazardous waste.

If used in combination with material-handling techniques such as soil washing, the volume of material disposed at landfill sites can be significantly reduced. The disadvantages of these techniques include the fact that the contamination is not destroyed, there are risks of pollution during excavation and transfer; road haulage may also cause a local nuisance. Ex-situ techniques include:

- Soil washing – involving the separation of a contaminated soil fraction or oily residue through a washing process. This also involves the excavation of the material for washing ex situ. The dewatered contaminant still requires disposal to landfill. In order to be cost effective, 70–90% of soil mass needs to be recovered. It will involve constructing a hard area for the washing, intercepting the now-contaminated water and taking it away in tankers.
- Thermal treatment – the incineration of contaminated soils ex situ. The uncontaminated soil residue can be recycled. By-products of incineration can create air pollution and exhaust air treatment may be necessary.

Soil treatment centres are now beginning to be established. These use a combination of treatment technologies to maximize the potential recovery of soils and aggregates and render them suitable for disposal to the landfill. The technologies include:

- Physico-chemical treatment – a method which uses the difference in grain size and density of the materials to separate the different fractions by means of screens, hydrocyclones and upstream classification.
- Bioremediation – the aerobic biodegradation of contaminants by naturally occurring microorganisms placed into stockpiles/windrows.
- Stabilization/solidification – a cement or chemical stabilization unit capable of immobilizing persistent leachable components.

Cost Considerations

Cost drivers

Cost drivers relate to the selected remediation technique, site conditions and the size and location of a project.

The wide variation of indicative costs of land remediation techniques shown below is largely because of differing site conditions.

Indicative costs of land remediation techniques for 2013 (excluding general items, testing, landfill tax and backfilling)		
Remediation technique	Unit	Rate (£/unit)
Removal – non-hazardous	disposed material (m³)	30–50
Removal – hazardous Note: excluding any pretreatment of material	disposed material (m³)	75–200
Clean cover	surface area of site (m²)	20–45
On-site encapsulation	encapsulated material (m³)	30–95
Bioremediation (in situ)	treated material (m³)	15–45
Bioremediation (ex situ)	treated material (m³)	20–50
Chemical oxidation	treated material (m³)	30–80
Stabilization/solidification	treated material (m³)	20–65
Vacuum extraction	treated material (m³)	25–65
Soil washing	treated material (m³)	35–80
Thermal treatment	treated material (m³)	100–400

Many other on-site techniques deal with the removal of the contaminant from the soil particles and not the wholesale treatment of bulk volumes. Costs for these alternative techniques are very much Engineer designed and site specific.

Factors that need to be considered include:

- Waste classification of the material
- Underground obstructions, pockets of contamination and live services
- Ground water flows and the requirement for barriers to prevent the migration of contaminants
- Health and safety requirements and environmental protection measures
- Location, ownership and land use of adjoining sites
- Distance from landfill tips, capacity of the tip to accept contaminated materials, and transport restrictions
- The cost of diesel fuel, currently approximately £1.45 per litre (at April 2013 prices)

Other project related variables include size, access to disposal sites and tipping charges; the interaction of these factors can have a substantial impact on overall unit rates.

The tables below set out the costs of remediation using dig-and-dump methods for different sizes of project, differentiated by the disposal of non-hazardous and hazardous material. Variation in site establishment and disposal cost accounts for 60–70% of the range in cost.

Variation in the costs of land remediation by removal: Non-hazardous Waste			
Item	Disposal Volume (less than 3,000 m³) (£/m³)	Disposal Volume (3,000–10,000 m³) (£/m³)	Disposal Volume (more than 10,000 m³) (£/m³)
General items and site organization costs	55–90	25–40	7–20
Site investigation and testing	5–12	2–7	2–6
Excavation and backfill	18–35	12–25	10–20
Disposal costs (including tipping charges but not landfill tax)	20–35	20–35	20–35
Haulage	15–35	15–35	15–35
Total (£/m³)	**113–207**	**74–142**	**54–116**
Allowance for site abnormals	*0–10 +*	*0–15 +*	*0–10 +*

Variation in the costs of land remediation by removal: Hazardous Waste			
Item	Disposal Volume (less than 3,000 m³) (£/m³)	Disposal Volume (3,000–10,000 m³) (£/m³)	Disposal Volume (more than 10,000 m³) (£/m³)
General items and site organization costs	55–90	25–40	7–20
Site investigation and testing	10–18	5–12	5–12
Excavation and backfill	18–35	12–25	10–20
Disposal costs (including tipping charges but not landfill tax)	80–170	80–170	80–170
Haulage	25–120	25–120	25–120
Total (£/m³)	**188–433**	**147–367**	**127–342**
Allowance for site abnormals	*0–10 +*	*0–15 +*	*0–10 +*

The strict health and safety requirements of remediation can push up the overall costs of site organization to as much as 50% of the overall project cost (see the above tables). A high proportion of these costs are fixed and, as a result, the unit costs of site organization increase disproportionally on smaller projects.

Haulage costs are largely determined by the distances to a licensed tip. Current average haulage rates, based on a return journey, range from £1.95 to £4.50 per mile. Short journeys to tips, which involve proportionally longer standing times, typically incur higher mileage rates, up to £9.00 per mile.

A further source of cost variation relates to tipping charges. The table below summarizes typical tipping charges for 2013, exclusive of landfill tax:

Typical 2013 tipping charges (excluding landfill tax)	
Waste classification	**Charges (£/tonne)**
Non-hazardous wastes	8–15
Hazardous wastes	25–85
Contaminated liquid	40–75
Contaminated sludge	55–200

Tipping charges fluctuate in relation to the grades of material a tip can accept at any point in time. This fluctuation is a further source of cost risk. Furthermore, tipping charges in the North of England are generally less than other areas of the country.

Prices at licensed tips can vary by as much as 50%. In addition, landfill tips generally charge a tip administration fee of approximately £25 per load, equivalent to £1.25 per tonne. This charge does not apply to non-hazardous wastes.

Landfill Tax, increased on 1 April 2013 to £72 a tonne for active waste, is also payable. Exemptions are no longer available for the disposal of historically contaminated material (refer also to *Landfill Tax* section).

Tax relief for remediation of contaminated land

The Finance Act 2001 included provisions that allow companies (but not individuals or partnerships) to claim tax relief on capital and revenue expenditure on the 'remediation of contaminated land' in the United Kingdom. The relief is available for expenditure incurred on or after 11 May 2001.

From 1 April 2009 there was an increase in the scope of costs that qualify for Land Remediation Relief where they are incurred on long-term derelict land. The list includes costs that the Treasury believe to be primarily responsible for causing dereliction, such as additional costs for removing building foundations and machine bases. However, while there is provision for the list to be extended, the additional condition for the site to have remained derelict since 1998 is likely to render this relief redundant in all but a handful of cases. The other positive change is the fact that Japanese Knotweed removal and treatment (on-site only) will now qualify for the relief under the existing legislation, thereby allowing companies to make retrospective claims for any costs incurred since May 2001 – provided all other entitlement conditions are met.

A company is able to claim an additional 50% deduction for 'qualifying land remediation expenditure' allowed as a deduction in computing taxable profits, and may elect for the same treatment to be applied to qualifying capital expenditure.

With Landfill Tax Exemption (LTE) now phased out, Land Remediation Relief (LRR) for contaminated and derelict land is the Government's primary tool to create incentives for brownfield development. LRR is available to companies engaged in land remediation that are not responsible for the original contamination.

Over 7 million tonnes of waste each year were being exempted from Landfill Tax in England alone, so this change could have a major impact on the remediation industry. The modified LRR scheme, which provides Corporation Tax relief on any costs incurred on qualifying land remediation expenditure, is in the long run designed to yield benefits roughly equal to those lost through the withdrawal of LTE, although there is some doubt about this stated equity in reality.

However, with much remediation undertaken by polluters or public authorities, who cannot benefit from tax relief benefits, the change could result in a net withdrawal of Treasury support to a vital sector. Lobbying and consultation continues to ensure the Treasury maintains its support for remediation.

While there are no financial penalties for not carrying out remediation, a steep escalator now affects the rate of Landfill Tax for waste material other than inert or inactive wastes, which will rise at the rate of £8/tonne per year until 2014–15. By then, the rate will be £80/tonne. This means that for schemes where there is no alternative to dig and dump and no pre-existing LTE, the cost of remediation could rise to prohibitive levels.

Looking forward, tax-relief benefits under LRR could provide a significant cash contribution to remediation. Careful planning is the key to ensure that maximum benefits are realized, with actions taken at the points of purchase, formation of JV arrangements, procurement of the works and formulation of the Final Account (including apportionment of risk premium) all influencing the final value of the claim agreed with HM Revenue & Customs.

The relief

Qualifying expenditure may be deducted at 150% of the actual amount expended in computing profits for the year in which it is incurred.

For example, a property trading company may buy contaminated land for redevelopment and incurs £250,000 on qualifying land remediation expenditure that is an allowable for tax purposes. It can claim an additional deduction of £125,000, making a total deduction of £375,000. Similarly, a company incurring qualifying capital expenditure on a fixed asset of the business is able to claim the same deduction provided it makes the relevant election within 2 years.

What is remediation?

Land remediation is defined as the doing of works including preparatory activities such as condition surveys, to the land in question, any controlled waters affected by the land, or adjoining or adjacent land for the purpose of:

- Preventing or minimizing, or remedying or mitigating the effects of, any relevant harm, or any pollution of controlled waters, by reason of which the land is in a contaminated state.

Definitions

Contaminated land is defined as land that, because of substances on or under it, is in such a condition that relevant harm is or has the significant possibility of relevant harm being caused to:

- The health of living organisms
- Ecological systems
- Quality of controlled waters
- Property

Relevant harm is defined as meaning:

- Death of living organisms or significant injury or damage to living organisms
- Significant pollution of controlled waters
- A significant adverse impact on the ecosystem, or
- Structural or other significant damage to buildings or other structures or interference with buildings or other structures that significantly compromises their use

Land includes buildings on the land, and expenditure on asbestos removal is expected to qualify for this tax relief. It should be noted that the definition is not the same as that used in the Environmental Protection Act Part IIA.

Sites with a nuclear licence are specifically excluded.

Conditions

To be entitled to claim LRR, the general conditions for all sites, which must all be met, are:

- Must be a company
- Must be land in the United Kingdom
- Must acquire an interest in the land
- Must not be the polluter or have a relevant connection to the polluter
- Must not be in receipt of a subsidy
- Must not also qualify for Capital Allowances (particular to capital expenditure only)

Additional conditions introduced since1 April 2009:

- The interest in land must be major – freehold or leasehold longer than 7 years
- Must not be obligated to carry out remediation under a statutory notice

Additional conditions particular to derelict land:

- Must not be in or have been in productive use at any time since at least 1 April 1998
- Must not be able to be in productive use without the removal of buildings or other structures

In order for expenditure to become qualifying, it must relate to substances present at the point of acquisition.

Furthermore, it must be demonstrated that the expenditure would not have been incurred had those substances not been present.

Spon's Asia-Pacific Construction Costs Handbook

Fourth Edition

Davis Langdon

Spon's Asia Pacific Construction Costs Handbook includes construction cost data for twenty countries. This new edition has been extended to include Pakistan and Cambodia. Australia, UK and America are also included, to facilitate comparison with construction costs elsewhere.

Information is presented for each country in the same way, as follows:

- key data on the main economic and construction indicators.
- an outline of the national construction industry, covering structure, tendering and contract procedures, materials cost data, regulations and standards
- labour and materials cost data
- measured rates for a range of standard construction work items
- approximate estimating costs per unit area for a range of building types
- price index data and exchange rate movements against £ sterling, $US and Japanese Yen.

The book also includes a Comparative Data section to facilitate country-to-country comparisons. Figures from the national sections are grouped in tables according to national indicators, construction output, input costs and costs per square metre for factories, offices, warehouses, hospitals, schools, theatres, sports halls, hotels and housing.

This unique handbook will be an essential reference for all construction professionals involved in work outside their own country and for all developers or multinational companies assessing comparative development costs.

April 2010: 234x156: 480pp
Hb: 978-0-415-46565-6: £140.00

To Order: Tel: +44 (0) 1235 400524 Fax: +44 (0) 1235 400525
or Post: Taylor and Francis Customer Services,
Bookpoint Ltd, Unit T1, 200 Milton Park, Abingdon, Oxon, OX14 4TA UK
Email: book.orders@tandf.co.uk

For a complete listing of all our titles visit:
www.tandf.co.uk

The Landfill Tax

The tax

The Landfill Tax came into operation on 1 October 1996. It is levied on operators of licensed landfill sites at the following rates with effect from 1 April 2013:

- Inactive or inert wastes   £2.50 per tonne   Included are soil, stones, brick, plain and reinforced concrete, plaster and glass – lower rate

- All other taxable wastes   £72 per tonne   Included are timber, paint and other organic wastes generally found in demolition work and builders skips – standard rate

The standard (higher) rate for 'all other taxable wastes' will be increased by £8 per tonne each year at least until 2014 when the rate will be £80 per tonne. There will also be a floor under the standard rate, so that the rate will not fall below £80 per tonne from 2014–15 to 2019–20. The lower rate for 'inactive or inert wastes' will be frozen at £2.50 per tonne to 31 March 2014.

The Landfill Tax (Qualifying Material) Order 2011 came into force on 1 April 2011. This has amended the qualifying criteria that are eligible for the lower rate of landfill tax. The revisions introduced arose primarily from the need to reflect changes in wider environmental policy and legislation since 1996, such as the implementation of the European Landfill Directive.

A waste will be lower rated for landfill tax from 1 April 2011 only if it is listed as a qualifying material in the Landfill Tax (Qualifying Material) Order 2011.

The principal changes to qualifying material are:

- Rocks and subsoils that are currently lower rated will remain so.
- Topsoil and peat will be removed from the lower rate, as these are natural resources that can always be recycled/reused.
- Used foundry sand, which has in practice been lower rated by extra-statutory concession since the tax's introduction in 1996, will now be included in the lower rate Order.
- Definitions of qualifying ash arising from the burning of coal and petroleum coke (including when burnt with biomass) will be clarified.
- The residue from titanium dioxide manufacture will qualify, rather than titanium dioxide itself, reflecting industry views.
- Minor changes will be made to the wording of the calcium sulphate group of wastes to reflect the implementation of the Landfill Directive since 2001.
- Water will be removed from the lower rate – water is now banned from landfill so its inclusion in the list of lower rated wastes is unnecessary; where water is used as a waste carrier the water is not waste and therefore not taxable.

Calculating the weight of waste

There are two options:

- If licensed sites have a weighbridge, tax will be levied on the actual weight of waste.
- If licensed sites do not have a weighbridge, tax will be levied on the permitted weight of the lorry based on an alternative method of calculation based on volume to weight factors for various categories of waste.

Effect on prices

The tax is paid by Landfill site operators only. Tipping charges reflect this additional cost.

As an example, Spon's A & B rates for mechanical disposal will be affected as follows:

- Inactive waste   Spon's A & B 2014 net rate   £12.09 per m³

                    Tax, 1.9 tonne per m³ (unbulked) @ £2.50   £4.75 per m³

                    Spon's rate including tax   £16.84 per m³

- Active waste   Active waste will normally be disposed of by skip and will probably be mixed with inactive waste. The tax levied will depend on the weight of materials in the skip which can vary significantly.

Exemptions

The following disposals are exempt from Landfill Tax subject to meeting certain conditions:

- dredgings which arise from the maintenance of inland waterways and harbours
- naturally occurring materials arising from mining or quarrying operations
- pet cemeteries
- material from the reclamation of contaminated land (see below)
- inert waste used to restore landfill sites and to fill working and old quarries where a planning condition or obligation is in existence
- waste from visiting NATO forces

The exemption for waste from contaminated land has been phased out completely from 1 April 2012 and no new applications for landfill tax exemption are now accepted.

For further information contact the National Advisory Service, Telephone: 0845 010 9000.

Property Insurance

The problem of adequately covering by insurance the loss and damage caused to buildings by fire and other perils has been highlighted in recent years by the increasing rate of inflation.

There are a number of schemes available to the building owner wishing to insure his property against the usual risk. Traditionally the insured value must be sufficient to cover the actual cost of reinstating the building. This means that in addition to assessing the current value an estimate has also to be made of the increases likely to occur during the period of the policy and of rebuilding which, for a moderate size building, could amount to a total of 3 years. Obviously such an estimate is difficult to make with any degree of accuracy; if it is too low the insured may be penalized under the terms of the policy and if too high will result in the payment of unnecessary premiums.

There are variations on the traditional method of insuring which aim to reduce the effects of over estimating and details of these are available from the appropriate offices. For the convenience of readers who may wish to make use of the information contained in this publication in calculating insurance cover required the following may be of interest.

1 Present cost

The current rebuilding costs may be ascertained in a number of ways:

- where the actual building cost is known this may be updated by reference to tender price changes
- by reference to average published prices per square metre of floor area. In this case it is important to understand clearly the method of measurement used to calculate the total floor area on which the rates have been based
- by professional valuation
- by comparison with the known cost of another recently built similar building

Whichever of these methods is adopted regard must be paid to any special conditions that may apply, e.g. a confined site; complexity of design; any demolition works and site clearance that may be required.

2 Allowances for inflation

The Present Cost when established will usually, under the conditions of the policy, be the rebuilding cost on the first day of the policy period. To this must be added a sum to cover future price increases over the period of the policy. For this purpose, using the historical indices as a base and taking account of the likely change in future building costs and the tender climate, an anticipated rebuild cost can be arrived at.

3 Fees

To the total of 1 and 2 above must be added an allowance for fees.

Value Added Tax (VAT)

To the total of 1 to 3 above must be added Value Added Tax.

Example:

An assessment for insurance cover is required in the fourth quarter of 2013 for a property which cost £200,000 when completed in 2000. This calculation uses Tender Price Indices.

All costs exclude demolitions and VAT.

Update to present day costs **(4Q2013)**

Known cost at end of 2000			£200,000
Calculation for present day build cost:			
Tender index fourth quarter 2000	375		
Predicted tender index fourth quarter 2013	say 438		
Increase in tender index	17%	say	£34000
Estimated present build cost (4Q2013)		say	**£234,000**

Allowance for inflation to end of policy (4Q2014)

Estimated present build cost at start of policy			£234,000
Inflation calculation to end of policy:			
Forecast tender index fourth quarter 2014	448		
Forecast tender index fourth quarter 2013	438		
Increase in tender index over policy period	2%	say	£4,600
Estimated build cost at end of policy (4Q2014)		say	**£238,600**

Assuming that total damage is suffered on the last day of the policy (end of 2014) and that planning and documentation and general project set up would require a period of 12 months before rebuilding could commence, then a further allowance must be made to cover that period.

Allowance for inflation to start reconstruction (4Q2015)

Anticipated cost at end of policy (4Q2014)			£238,600
Inflation calculation from end of policy to tender:			
Forecast tender index fourth quarter 2015	468		
Forecast tender index fourth quarter 2014	448		
Increase in tender index	4%	say	£9,500
applied to cost at expiry of policy			
Estimated build cost at tender (4Q2015)		say	**£248,100**

Assuming that reconstruction would take 1 year, allowance must be made for the increase in costs to the contractor which would directly or indirectly be met under a building contract. This is calculated using Building Cost Indices.

Anticipated cost at tender (4Q2014)			£248,100

Inflation calculation during construction period:

Predicted cost index fourth quarter 2016	922		
Predicted cost index fourth quarter 2015	895		
Increase in cost index	3%	say	£7,400

This is the total increase at the end of the 1 year period. The amount applicable to the contract would be about half

Estimated cost of reinstatement			**£255,500**

Summary

You now need to add on any other costs such as fees, demolition, VAT etc.

Estimated cost of reinstatement			£255,500
Demolitions and site clearance		say	£10,000
Add professional fees	12%	say	£30,000
			£295,500
Add for VAT	20%	say	£59,100
Total insurance cover required			**£354,600**

Portfolio and Programme Management Demystified

Managing Multiple Projects Successfully

Geoff Reiss & Paul Rayner

You're now responsible for a programme, or you've got a portfolio to manage? Where do you start? Right here!

Projects are not simply the bread and butter of an organisation. Form them into programmes or portfolios and they can be prioritised and integrated to deliver change to your organization in line with your strategic vision. You will be able to control costs and risks and bring together a complex series of themes effectively.

This overhauled second edition now combines portfolio management as a parallel theme with programme management, and it is brought in line with the current thinking of the Association for Project Management and the Project Management Institute. It is written for managers in both the public and private sectors. This new edition includes half a dozen short case studies (from Belgium's Fortis Bank, a software company, local government, and central government), along with more on cross-functional management.

Together with Project Management Demystified, also from Routledge (third edition, 2007), it provides the tools to manage your projects, your programmes and your portfolio to a very high level.

October 2012: 234x156: 322pp
Hb: 978-0-415-55834-1: £24.99

To Order: Tel: +44 (0) 1235 400524 Fax: +44 (0) 1235 400525
or Post: Taylor and Francis Customer Services,
Bookpoint Ltd, Unit T1, 200 Milton Park, Abingdon, Oxon, OX14 4TA UK
Email: book.orders@tandf.co.uk

For a complete listing of all our titles visit:
www.tandf.co.uk

Taylor & Francis
Taylor & Francis Group

Building Costs Indices, Tender Price Indices and Location Factors

The tables which follow show the changes in building costs and tender prices since 1976. To avoid confusion it is essential that the terms building costs and tender prices are clearly defined and understood.

- Building Costs are the costs incurred by the contractor in the course of his business, the principal ones being those for labour and materials, i.e. cost to contractor.
- Tender Price is the price for which the contractor offers to do the project for, i.e. cost to client.

Readers are reminded that the following adjustments for time and location should be applied to the project as a whole and not individual elements or materials.

Building cost indices

This table reflects the fluctuations since 2000 in wages and materials costs to the contractor. In compiling the table, the proportion of labour to material has been assumed to be 40:60. The wages element has been assessed from a contract wages sheet re-valued for each variation in labour costs, whilst the changes in the costs of materials have been based upon the indices prepared by the Department for Business, Innovation and Skills. No allowance has been made for changes in productivity, plus rates or hours worked which may occur in particular conditions and localities.

Year	First quarter	Second quarter	Third quarter	Fourth quarter	Annual average
2000	480	482	497	498	489
2001	498	499	516	515	507
2002	516	522	553	554	536
2003	555	560	578	577	568
2004	579	586	617	618	600
2005	621	623	660	660	641
2006	664	670	694	699	682
2007	703	707	730	730	718
2008	733	742	781	780	759
2009	773	770	771	778	773
2010	780	793	797	798	792
2011	807	818	824	827	819
2012	833	834	832	842	835
2013	858 (P)	860 (F)	861	861	860
2014	873	875	876	876	875
2015	891	894	895	895	894

Note: P = Provisional, F = Forecast thereafter

Tender price indices

Tender prices are similar to building costs but also take into account market considerations such as the availability of labour and materials, and the prevailing economic situation. This means that in boom periods, when there is a surfeit of building work to be done, tender prices may increase at a greater rate than building costs, whilst in a period when work is scarce, tender prices may actually fall when building costs are rising.

This table reflects the changes in tender prices since 2000. It indicates the level of pricing contained in the lowest competitive tenders for new work in the Greater London area with projects worth approximately £3,500,000 in value.

Year	First quarter	Second quarter	Third quarter	Fourth quarter	Annual average
2000	348	353	362	375	359
2001	375	383	388	392	384
2002	398	405	423	421	412
2003	425	424	432	434	429
2004	434	436	442	454	442
2005	459	458	466	475	465
2006	480	485	494	506	491
2007	515	520	528	538	525
2008	543	547	541	505	534
2009	500	485	460	455	475
2010	457	454	452	448	453
2011	446	447	448	444	446
2012	446	446	443	439	443
2013	438 (P)	438 (F)	438	438	438
2014	440	442	444	447	443
2015	452	457	462	467	460

Note: (P) = Provisional, (F) = Forecast thereafter

Readers will be kept abreast of tender price movements in the free Spon's *Updates* and also in the *Market Forecast* and *Cost Update* articles, published quarterly in *Building* magazine.

Davis Langdon Tender Price, Building Cost and Retail Price Indices chart

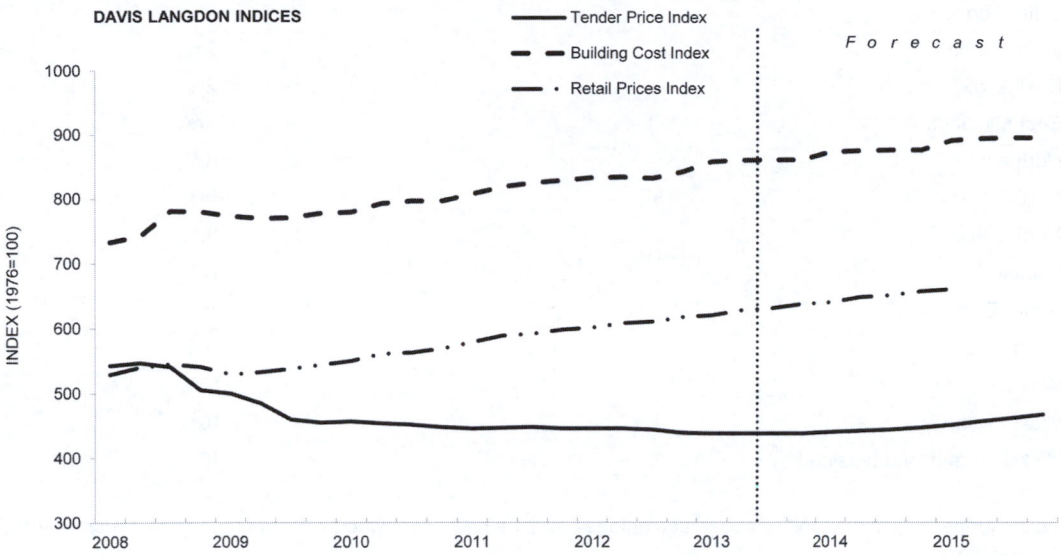

Regional variations

As well as being aware of inflationary trends when preparing an estimate, it is also important to establish the appropriate price level for the project location.

Prices throughout this book are at a price level index of 453 in Outer London. Regional variations for certain inner London boroughs can be up to 9% higher, while prices in the North and Yorkshire and Humberside can be as much as 8% lower. Broad regional adjustment factors used to assist with the preparation of initial estimates are shown in the table on the next page.

Over time, price differentials can change depending on regional workloads and local 'hot spots'. *Spon's Updates* and the *Market Forecast* and *Cost Update* featured in *Building* magazine will keep readers informed of the latest regional developments and changes as they occur.

The regional variations shown in the following table are based on our forecast of price differentials in each of the economic planning regions in the second quarter 2014 (TPI = 442). The table shows suggested adjustments required to cost plans or estimates prepared using the Spon's Price Book. Please note: these adjustments need to be applied to the total cost plan value, not to individual material items.

Region	Adjustment to Measured Works section
Outer London	−2%
Inner London	+6%
East Anglia	−2%
East Midlands	−10%
Northern	−16%
Northern Ireland	−45%
North West	−16%
Scotland	−11%
South East	0%
South West	−6%
Wales	−12%
West Midlands	−10%
Yorkshire and Humberside	−10%

Special further adjustment to the above percentages may be necessary when considering city centre or very isolated locations.

The following example illustrates the adjustment of a cost plan estimate prepared using Spon's A&B 2014 (TPI = 453) and competitive quotations, to a price level that reflects the forecast Outer London market conditions for competitive tenders in the second quarter 2014 (TPI = 442).

		£
A	Estimated value of items priced using Spon's A&B 2014	2,500,000
B	Adjustment to value of A to forecast price level for the second quarter 2014	
	(442−453) / 453 × 100 = −2% × £2,500,000 = say	−50,000
C	Value of items priced using competitive quotations for contractors / suppliers that reflect the market conditions in the second quarter 2014	725,000
	Total of A + B + C =	3,175,000
D	Allowance for preliminaries @ 11% = say	350,000
E	**Total value of estimate at second quarter 2014 price levels =**	**3,525,000**

Alternatively, for a similar estimate in North West:

		£
A	Value of items priced using Spon's A&B 2014	2,500,000
B	Adjustment to value of A to forecast price level for second quarter 2014	
	(442−453) / 453 × 100 = −2% × £2,500,000 = say	−50,000
C	Location adjustment for North West = −17% say	−425,000
D	Value of items priced using competitive quotations that reflect the market conditions in the second quarter 2014 in the North West	650,000
	Total of A + B + C + D =	2,675,000
E	Allowance for preliminaries @ 11% say	295,000
F	**Total value of estimate at second quarter 2014 price levels**	**2,970,000**

Spons Architects' and Builders' Price Book
2014 Edition
(Tender Index 453)

Scotland
−11%

Northern Ireland
−46%

Northern
−16%

Yorkshire and
Humberside
−10%

North West
−16%

East Midlands
−10%

East Anglia
−2%

West Midlands
−10%

Wales
−12%

South East
(excluding London) 0%

South West −6%

Inner London +6%
Outer London −2%

PART 2

Rates of Wages and Labour

This part contains the following sections:

Architect's Pocket Book

4th Edition

Hetreed & Ross

This handy pocket book brings together a wealth of useful information that architects need on a daily basis – on site or in the studio. It provides guidance on a range of tasks, from complying with the Building Regulations, including the recent revisions to Part L, to helping with planning, use of materials and detailing.

Compact and easy to use, the Architect's Pocket Book has sold well over 65,000 copies to the nation's architects, architecture students, designers and construction professionals who do not have an architectural background but need to understand the basics, fast.

This is the famous little blue book that you can't afford to be without.

February 2011: 186x123: 268pp
Pb: 978-0-08-096959-6: £20.99

To Order: Tel: +44 (0) 1235 400524 Fax: +44 (0) 1235 400525
or Post: Taylor and Francis Customer Services,
Bookpoint Ltd, Unit T1, 200 Milton Park, Abingdon, Oxon, OX14 4TA UK
Email: book.orders@tandf.co.uk

For a complete listing of all our titles visit:
www.tandf.co.uk

Rates of Wages

BUILDING INDUSTRY – ENGLAND, WALES AND SCOTLAND

The basic rates were agreed by CIJC in October 2012 and were applied from 7 January 2013.

The Working Rule Agreement includes a pay structure with general, additional skilled and craft operatives. Plus rates and additional payments will be consolidated into basic pay to provide the following rates (for a normal 39-hour week) which will come into effect from the following dates:

Effective from 7 January 2013

The following basic rates of pay came in to effect from 7 January 2013:	Rate per 39-hour week (£)	Rate per hour (£)
Craft Rate	416.13	10.67
Skill Rate 1	396.63	10.17
Skill Rate 2	381.81	9.79
Skill Rate 3	357.24	9.16
Skill Rate 4	337.35	8.65
General operative	313.17	8.03

Holidays with Pay and Benefits Schemes

The Building and Civil Engineering benefits scheme has unveiled a new holiday pay plan following the introduction of the Working Time Directive. From 2 August 1999 there are no fixed holiday credits, instead employers will calculate appropriate sums to fund operatives' holiday pay entitlement and make regular monthly payments into the B&CE scheme.

For full details contact B&CE on 01293 526911.

Death and accident cover is provided free. From 1 October 2012 the death benefit was increased to £25,000 and doubled to £50,000 if death occurs at work or travelling to or from work.

Young Workers

Effective from 7 January 2013

Operatives below 18 years of age will receive payment 60% of the General Operative basic rate.

At 18 years of age or over the payment is 100% of the relevant rate.

Apprentices/Trainees

The Construction Apprenticeship Scheme (CAS) operates throughout Great Britain; from 1 August 1998 it is open to all young people from the age of 16 years. For further information telephone CAS helpline – 01485 578 333.

Apprentice rates – effective from 5 September 2011	Rate per 39-hour week (£)	Rate per hour (£)
Year 1	173.16	4.44
Year 2	223.47	5.73
Year 3 without NVQ2	261.69	6.71
Year 3 With NVQ2	332.67	8.53
Year 3 With NVQ3	416.13	10.67
On Completion of Apprenticeship With NVQ2	416.13	10.67

Note: If an apprentice is 22 years and over, and in his/her second year of training, then the National Minimum Wage applies.

BUILDING AND ALLIED TRADES JOINT INDUSTRIAL COUNCIL

The Building and Allied Trades Joint Industrial Council (BATJIC), the partnership between the Federation of Master Builders (FMB) and the Transport and General Workers Union (TGWU), agreed new wage rates which became effective on 17 June 2013.

Effective from 17 June 2013

Subject to the conditions in the Working Rule Agreement the standard weekly rates of wages shall be as follows:

	Rate per 39-hour week (£)	Rate per hour (£)
Craft operative (NQV3)	426.66	10.94
Craft operative (NQV2)	366.66	9.40
Adult general operative	319.80	8.20

For the latest wage/conditions information, go to www.fmb.org.uk.

ROAD HAULAGE WORKERS EMPLOYED IN THE BUILDING INDUSTRY – effective from 7 January 2013

Authorized rates of pay for road haulage workers in the building industry recommended by the Builders Employers Confederation.

Lorry Drivers (irrespective of the gross weight of the vehicle driven) = £416.13

Employers

Construction Confederation
56–64 Leonard Street
London
EC2A 4JX
Tel: 0207 608 5039
Fax: 0207 608 5001
E-mail: enquiries@constructionconfederation.co.uk

Operatives

The Transport and General Workers Union
Transport House
128 Theobold's Road
London
WC1X 8TN
Tel: 0207 611 2500
Fax 0207 611 2555
E-mail: pgwu@tgwu.org.uk

PLUMBING AND MECHANICAL ENGINEERING SERVICES INDUSTRY

Authorized rates of wages agreed by the Joint Industry Board for the Plumbing and Mechanical Engineering Services Industry in England and Wales.

First part effective from 2 January 2012

Editors Note:
Discussions are ongoing for rates applicable from 2013.

The Joint Industry Board for Plumbing and Mechanical Engineering Services in England and Wales

Brook House
Brook Street
St Neots
Huntingdon
Cambridgeshire
PE19 2HW

Tel: 01480 476 925
Fax: 01480 403 081
E-mail: info@jib-pmes.org.uk

	Rate per hour £
Operatives	
Technical Plumber and Gas Service Technician	14.99
Advanced Plumber and Gas Service Engineer	13.50
Trained Plumber and Gas Service Fitter	11.58
Apprentices (see Note below)*	
1st year of Training	5.61
2nd year of Training	6.43
3rd year of Training	7.26
4th year of Training	8.94
Adult Trainees	
1st 6 months of Employment	9.03
2nd 6 months of Employment	9.69
3rd 6 months of Employment	10.09

Note:

*   As from 2 January 2012, overtime is payable after 37.5 hours work per week.

Authorized rates of wages agreed by the Joint Industry Board for the Plumbing Industry in Scotland and Northern Ireland.

At the time of compiling this book the parties to the Scottish & Northern Ireland Joint Industry Board for the Plumbing Industry (SNIPEF & UNITE) have failed to reach agreement on revised graded rates of wages and allowances which would have been expected to apply from Monday 4th June 2012. As a result the 2011/2012 graded rates of wages and allowances which have operated since Monday 6 June 2011(and the associated holiday credit values) **will continue from Monday 4 June 2012 (holiday credits from Monday 7 May) until further notice.**

The Joint Industry Board for the Plumbing Industry in Scotland and Northern Ireland
2 Walker Street
Edinburgh
EH3 7LB

Tel: 0131 225 2255
Fax: 0131 226 7638

	Rate per hour (£) from 4 June 2012
Operatives Plumbers & Gas Service Operatives	
Plumber and Gas Service Fitter	11.32
Advanced Plumber and Gas Service Engineer	12.89
Technician Plumber and Gas Service Technician	14.28
Plumbing Labourer	10.10
Apprentice Plumbers and Fitters	
1st Year Apprentice	3.28
2nd Year Apprentice	4.91
3rd Year Apprentice	5.94
4th Year Apprentice	7.67
Adult Trainees	8.62

Labour Rate Calculations

The format of this section is so arranged that, in the case of work normally undertaken by the Main Contractor, the constituent parts of the total rate are shown enabling the reader to make such adjustments as may be required in particular circumstances. Similar details have also been given for work normally sublet although it has not been possible to provide this in all instances.

As explained in the Preface, there is a facility available to readers, which enables a comparison to be made between the level of prices in this section and current tenders by means of a tender index. The tender index for this edition is 453.

To adjust prices for other regions and times, the reader is recommended to refer to the explanations and examples on how to apply these tender indices.

There follow explanations and definitions of the basis of costs in the Prices for Measured Work section under the following headings:

- Overhead charges and profit
- Labour hours and Labour £ column
- Material £ column
- Material/Plant £ column
- Total rate £ column

Overhead and profit charges

Rates checked against winning tenders include overhead charges and profit at current levels.

Labour hours and Labour £ columns

Labour rates are based upon typical gang costs divided by the number of primary working operatives for the trade concerned, and for general building work include an allowance for trade supervision (see below). Labour hours multiplied by Labour rate with the appropriate addition for overhead charges and profit gives Labour £. In some instances, due to variations in gangs used, Labour rate figures have not been indicated, but can be calculated by dividing Labour £ by Labour hours.

Building craft operatives and labourers

Effective from 7 January 2013 the guaranteed minimum weekly earnings rates in the London area for craft operatives and general operatives are £416.13 and £313.17 respectively; to these rates have been added allowances for the items below in accordance with the recommended procedure of the Chartered Institute of Building in its Code of Estimating Practice. The resultant hourly rates on which the Prices for Measured Work have generally been based are £14.90 and £10.46 for craft operatives and general operatives respectively.

Calculations take into account the following:

- Lost time
- Construction Industry Training Board Levy
- Holidays with pay
- Accidental injury, retirement and death benefits scheme
- Sick pay

- National Insurance
- Severance pay and sundry costs
- Employer's liability and third party insurance

NOTE: For travelling allowances and site supervision see Preliminaries section, but these have generally risen in line with inflation.

The table which follows illustrates how the all-in hourly rates have been calculated based upon CIJC Working Rule Agreement dated August 2012. Productive time has been based on a total of 1727 hours worked per year.

			Craft Operatives		General Operatives	
			£	£	£	£
Wages at standard basic rate						
Productive time	44.3	weeks	416.13	18,434.56	313.17	17,873.43
Lost time allowance	0.9	weeks	416.13	374.52	313.17	281.85
Overtime	0	weeks	0	0	0	0
				18,809.08		14,155.28
Extra payments under National Working Rules	45	weeks				
Sick Pay	1	week				
CITB Allowance (0.50% of payroll)	1	year		106.11		79.86
Holiday pay	4.20	weeks	416.13	1,747.75	313.17	1,315.31
Public Holiday pay	1.6	weeks	416.13	665.81	313.17	501.07
Employer's contribution to:						
EasyBuild Stakeholder Pension	52	weeks	5.00	260.00	5.00	260.00
National Insurance (average weekly payment)	48	weeks	32.91	1,866.80	21.96	1,141.92
				23,455.55		17,453.44
Severance pay and sundry costs	Plus		1.5%	351.83	1.5%	261.80
				23,807.38		17,715.24
Employer's Liability and Third Party Insurance	Plus		2.0%	476.15	2.0%	354.30
Total cost per annum				**£24,2836.53**		**£18,069.54**
Total cost per hour				**£14.06**		**£10.46**

Similar calculations for other CIJC labour classifications result in the following cost per hour:

The following basic rates of pay came in to effect from 7 January 2013:	Rate per hour (£)
Craft Rate	14.90
Skill Rate 1	14.06
Skill Rate 2	12.86
Skill Rate 3	12.00
Skill Rate 4	11.30
General operative	10.46

Notes:

1. Absence due to sickness has been assumed to be for periods not exceeding 3 days for which no payment is due. (Working Rule 20.7.3).
2. EasyBuild Stakeholder Pension effective from 1 July 2002. Death and accident benefit cover is provided free of charge.
    The minimum employer contribution shall be £5.00 per week. Where the operative contributes between £5.01 and £10.00 per week the employer shall increase the minimum contribution to match that of the operative up to a maximum of £10.00 per week.
3. All N.I. Payments are at not-contracted out rates applicable from April 2012.
    National Insurance is paid for 48 complete weeks (52 wks − 4.2 wks) is based on employer making regular monthly payments into the template holiday pay scheme and by doing so the employer achieves National Insurance savings on holiday wages.
    National Insurance concession on holiday pay will be withdrawn from November 2012. The revised all-in rates will be incorporated in cost update.
4. Public holiday pay includes for 9 normal public holidays.

The labour rates used in the Measured Work sections have been based on the following gang calculations which generally include an allowance for supervision by a foreman or ganger. Alternative labour rates are given showing the effect of various degrees of bonus.

Gang	Total Gang rate £/hour	Productive unit rate £/hour	Alternative labour rates £/hour			
			Normal	+10%	+20%	+30%
Groundwork Gang						
1 Ganger	1 × 11.30 =	11.30				
6 Labourers	6 × 10.46 =	62.76				
		74.06 /6.5 =	11.39	12.58	13.78	14.97
Concreting Gang						
1 Foreman	1 × 14.90 =	14.90				
4 Skilled Labourers	4 × 11.30 =	45.20				
		60.10 /4.5 =	13.36	14.75	16.14	17.53
Steelfixing Gang						
1 Foreman	1 × 14.90 =	14.90				
4 Steelfixers	4 × 11.30 =	45.20				
		60.10 /4.5 =	15.81	17.44	19.07	20.71
Formwork Gang						
1 Foreman	1 × 14.90 =	14.90				
10 Carpenters	10 × 14.06 =	140.60				
1 Labourer	1 × 10.46 =	10.46				
		165.96 /10.5 =	15.81	17.44	19.07	20.71

Gang	Total Gang rate £/hour	Productive unit rate £/hour	Alternative labour rates £/hour			
			Normal	+10%	+20%	+30%
Bricklaying/Light Blockwork Gang						
1 Foreman	1 × 14.90 =	14.90				
6 Bricklayers	6 × 14.06 =	84.36				
1 Labourer	1 × 10.46 =	31.38				
		130.64	/ 6.50 =	16.88	18.62	20.37	22.12
Dense Blockwork Gang						
1 Foreman	1 × 14.90 =	14.90				
6 Bricklayers	6 × 14.06 =	84.36				
1 Labourer	1 × 10.46=	10.46				
		109.72	/ 6.50 =	16.88	18.62	20.37	22.12
Carpentry/Joinery Gang						
1 Foreman	1 × 14.90 =	14.90				
5 Carpenters	5 × 14.06 =	70.30				
1 Labourer	1 × 10.46 =	10.46				
		95.66	/ 5.5 =	17.39	19.19	20.99	22.79
Craft Operative (Painter, Slater, etc.)		14.06	/ 1 =	14.06	15.51	16.96	18.42
1 and 1 Gang						
1 Craft Operative	1 × 14.06 =	14.06				
1 Skilled Labourer	1 × 11.30 =	11.30				
		25.36	/ 1 =	25.36	27.99	30.62	33.26
2 and 1 Gang						
2 Craft Operatives	2 × 14.06 =	28.12				
1 Skilled Labourer	1 × 11.30 =	11.30				
		39.42	/ 2 =	19.71	21.75	23.79	25.84
Small Labouring Gang (making good)						
1 Foreman	1 × 14.90 =	14.90				
4 Skilled Labourers	4 × 11.30 =	45.20				
		60.10	/ 4.5 =	13.36	14.75	16.14	17.53
Drain Laying Gang/Clayware						
2 Skilled Labourers	2 × 11.30 =	22.60	/ 2 =	11.30	12.48	13.66	14.84

Subcontractor's operatives

Similar labour rates are shown in respect of sublet trades where applicable.

Plumbing operatives

Editors Note:
Discussions are ongoing for rates applicable from 2013.

From 2 January 2012 the hourly earnings for technical and trained plumbers are £14.99 and £11.58 respectively; to these rates have been added allowances similar to those added for building operatives (see below). The resultant average hourly rate on which the Prices for Measured Work have been based is £18.41. The items referred to above for which allowance has been made are:

- Tool allowance
- Plumbers' welding supplement
- Holidays with pay
- Pension and welfare stamp
- National Insurance
- Severance pay and sundry costs
- Employer's liability and third party insurance

No allowance has been made for supervision as we have assumed the use of a team of technical or trained plumbers who are able to undertake such relatively straightforward plumbing works, e.g. on housing schemes, without supervision.

The table which follows shows how the average hourly rate referred to above has been calculated. Productive time has been based on a total of 1687.50 hours worked per year.

		Technical Plumber		Trained Plumber	
		£	£	£	£
Wages at standard basic rate					
productive time	1,687.5 hrs	14.99	25,295.63	11.58	19,541.25
Overtime (paid at standard basic rate)	0	14.99	0	0	0
Overtime	0	22.49	0	17.37	0
Plumber's welding supplement (gas and arc)	1,687.5 hrs	0.47	793.13		0.00
			26,088.76		19,541.25
Employer's contribution to:					
Holiday credit/welfare					
Stamps (to provide for 31 days)	60 credits	64.55	3,873.00	49.70	2,982.00
Pension (6.5% of earnings)	46 weeks	42.11	1,936.14	31.63	1,454.98
Holiday top-up funding	60 credits	−2.96	−177.60	−2.28	−136.80
(Provided by employer)					
National Insurance	52 weeks	58.62	3,048.24	41.35	2,150.20
			34,768.54		25,991.63
Severance pay and sundry costs	Plus	1.5%	521.53	1.5%	388.87
			35,290.07		26,381.50
Employer's Liability and Third Party Insurance	Plus	2.0%	705.80	2.0%	527.63
Total cost per annum			**35,995.87**		**26,909.13**
Total cost per hour			**21.33**		**15.95**
Average all-in rate per hour				**£18.64**	

Notes:

1. First 37.5 hours paid at normal rate (Effective from 2 January 2012).
2. 1.5 overtime rate after 37.5 hrs.
3. National Insurance is payable on Top-up funding as this is payable by the employer.
4. Absence due to sickness has been assumed to be for periods not exceeding 3 days for which no payment is due. The entitlement is payable from and including the fourth day of illness onwards.

LABOUR CATEGORIES

Schedule 1 to the WRA lists specified work establishing entitlement to the Skilled Operative Pay Rate 4, 3, 2, 1 or the Craft Rate as follows:

CIJC WORKING RULE AGREEMENT FOR THE CONSTRUCTION INDUSTRY – SCHEDULE 1	
Specified Work Establishing Entitlement to the Skilled Operative Pay Rate 4,3,2,1 or Craft Rate	
BAR BENDERS AND REINFORCEMENT FIXERS	
• Bender and Fixer of Concrete Reinforcement capable of reading and understanding drawings and bending schedules and able to set out work	Craft Rate
CONCRETE	
• Concrete Leveller or Vibrator Operator	4
• Screeder and Concrete Surface Finisher working from datum such as road-form, edge beam or wire	4
• Operative required to use trowel or float (hand or powered) to produce high quality finished concrete	4
DRILLING AND BLASTING	
• Drills, rotary or percussive: mobile rigs, operator of	3
• Operative attending drill rig	4
• Shotfirer, operative in control of and responsible for explosives including placing, connecting and detonating charges	3
• Operatives attending on shotfirer, including stemming	4
DRYLINERS	
• Operatives undergoing approved training in drylining	4
• Operatives who can produce a certificate of training achievement indicating satisfactory completion of at least one unit of approved drylining training	3
• Dryliners who have successfully completed their training in drylining fixing and finishing	Craft Rate
FORMWORK CARPENTERS	
• 1st year trainee	4
• 2nd year trainee	3
• Formwork Carpenters	Craft Rate
GANGERS AND TRADE CHARGEHANDS	
(Higher grade payments may be made at the employer's discretion)	4
GAS NETWORK OPERATIONS	
Operatives who have successfully completed approved training to the standard of:	
• GNO Trainee	4
• GNO Assistant	3
• Team Leader–Services	2
• Team Leader–Mains	1
• Team Leader–Mains and Services	1

LABOUR CATEGORIES

LINESMEN – ERECTORS

- 1st Grade   2

  Skilled in all works associated with the assembly, erection, maintenance and dismantling of Overhead Lines Transmission Lines on steel towers, concrete or wood poles, including all overhead lines construction elements.

- 2nd Grade   3

  As above but lesser degree of skill – or competent and fully skilled to carry out some of the elements of construction listed above.

- Linesmen-erector's mate   4

  Semi-skilled in works specified above and a general helper

MASON PAVIORS

- Operative assisting a Mason Pavior undertaking kerblaying, block and sett paving, flag laying, in natural stone and precast products   4
- Operative engaged in stone pitching or dry stone walling   3

MECHANICS

- **Maintenance Mechanic** capable of carrying out field service duties, maintenance activities and minor repairs   2
- **Plant Mechanic** capable of carrying out major repairs and overhauls including welding work, operating metal turning lathe or similar machine and using electronic diagnostic equipment   1
- **Maintenance/Plant Mechanics' Mate** on site or in depot   4
- **Tyre Fitter**, heavy equipment tyres   2

MECHANICAL PLANT DRIVERS AND OPERATORS

Backhoe Loaders (with rear excavator bucket and front shovel and additional equipment such as blades, hydraulic hammers and patch planers)

- Backhoe, up to and including 50 kW net engine power; driver of   4
- Backhoe, over 50 kW up to and including 100 kW net engine power; driver of   3
- Backhoe, over 100 kW net engine power; driver of   2
- **Compressors and Generators** Air compressor or generators over 10 kW; operator of   4

Concrete Mixers

- Operative responsible for operating a concrete mixer or mortar pan up to and including 400 litres drum capacity   4
- Operative responsible for operating a concrete mixer over 400 litres and up to and including 1,500 litres drum capacity   3
- Operative responsible for operating a concrete mixer over 1,500 litres drum capacity   2
- Operative responsible for operating a mobile self-loading and batching concrete mixer up to 2,500 litres drum capacity   2
- Operative responsible for operating a mechanical drag-shovel   4

LABOUR CATEGORIES

Concrete Placing Equipment	
• Trailer mounted or static concrete pumps: self-propelled concrete placers: concrete placing booms; operator of	3
• Self-propelled mobile concrete pump, with or without boom, mounted on lorry or lorry chassis; driver/operator of	2
Cranes/Mobile Cranes	
Self-propelled mobile crane on road wheels, rough terrain wheels or caterpillar tracks including lorry mounted:	
• Maximum lifting capacity at minimum radius, up to and including 5 tonne; driver of	4
• Maximum lifting capacity at minimum radius, over 5 tonne and up to and including 10 tonne; driver of	3
• Maximum lifting capacity at minimum radius, over 10 tonne	Craft Rate
Tower Cranes (including static or travelling: standard trolley or luffing jib)	
Up to and including 2 tonne max. Lifting capacity at min. radius; driver of	4
• Over 2 tonne up to and including 10 tonne max. Lifting capacity at min. radius; driver of	3
• Over 10 tonne up to and including 20 tonne max. Lifting capacity at minimum radius; driver of	2
• Over 20 tonne max. Lifting capacity at minimum radius; driver of	1
Miscellaneous Cranes and Hoists	
• Overhead bridge crane or gantry crane up to and including 10 tonne capacity; driver of	3
• Overhead bridge crane or gantry crane over 10 tonne up to and including 20 tonne capacity; driver of	2
• Power driven hoist or jib crane; operator of	4
• **Slinger/Signaller** appointed to attend crane or hoist to be responsible for fastening or slinging loads and generally to direct lifting operations	4
Dozers	
• Crawler dozer with standard operating weight up to and including 10 tonne; driver of	3
• Crawler dozer with standard operating weight over 10 tonne and up to and including 50 tonne; driver of	2
• Crawler dozer with standard operating weight over 50 tonne; driver of	1
Dumpers and Dump Trucks	
• Up to and including 10 tonne rated payload; driver of	4
• Over 10 tonne and up to and including 20 tonne rated payload; driver of	3
• Over 20 tonne and up to and including 50 tonne rated payload; driver of	2
• Over 50 tonne and up to and including 100 tonne rated payload; driver of	1
• Over 100 tonne rated payload; driver of	Craft Rate

LABOUR CATEGORIES

Excavators (360° slewing)

- Excavators with standard operating weight up to and including 10 tonne; driver of — 3
- Excavator with standard operating weight over 10 tonne and up to and including 50 tonne; driver of — 2
- Excavator with standard operating weight over 50 tonne; driver of — 1
- **Banksman** appointed to attend excavator or responsible for positioning vehicles during loading or tipping — 4

Fork-Lifts Trucks and Telehandlers

- Smooth or rough terrain fork lift trucks (including side loaders) and telehandlers up to and including 3 tonne lift capacity; driver of — 4
- Over 3 tonne lift capacity; driver of — 3

Motor Graders: driver of — 2

Motorized Scrapers: driver of — 2

Motor Vehicles (Road Licensed Vehicles) Driver and Vehicle Licensing Agency (DVLA)

- Vehicles requiring a driving licence of category C1; driver of (Goods vehicle with maximum authorized mass (mam) exceeding 3.5 tonne but not exceeding 7.5 tonne and including such a vehicle drawing a trailer with a mam not over 750 kg) — 4
- Vehicles requiring a driving licence of category C; driver of (Goods vehicle with a maximum authorized mass (mam) exceeding 3.5 tonne and including such a vehicle drawing a trailer with mam not over 750 kg) — 2
- Vehicles requiring a driving licence of category C plus E; driver of (Combination of a vehicle in category C and a trailer with maximum authorized mass over 750 kg) — 1

Power Driven Tools

- Operatives using power-driven tools such as breakers, percussive drills, picks and spades, rammers and tamping machines — 4

Power Rollers

- Roller, up to and including 4 tonne operating weight; driver of — 4
- Roller, over 4 tonne operating weight and upwards; driver of — 3

Pumps, Power-driven pump(s); attendant of — 4

Shovel Loaders, (Wheeled or tracked, including skid steer)

- Up to and including 2 cubic metre shovel capacity; driver of — 4
- Over 2 cubic metre and up to and including 5 cubic metre shovel capacity; driver of — 3
- Over 5 cubic metre shovel capacity; driver of — 2

Tractors (Wheeled or Tracked)

- Tractor, when used to tow trailer and/or with mounted compressor, up to and including 100 kW rated engine power; driver of — 4
- Tractor, ditto, over 100 kW up to and including 250 kW rated engine power; driver of — 3
- Tractor, ditto, over 250 kW rated engine power; driver of — 2

LABOUR CATEGORIES

Trenchers (Type wheel, chain or saw)	
• Trenching Machine, up to and including 50 kW gross engine power; driver of	4
• Trenching Machine, over 50 kW and up to and including 100 kW gross engine power; driver of	3
• Trenching Machine, over 100 kW gross engine power; driver of	2
Winches, Power driven winch; driver of	4
PILING	
• General Skilled Piling Operative	4
• Piling Chargehand/Ganger	3
• Pile Tripod Frame Winch Driver	3
• CFA or Rotary or Driven Mobile Piling Rig Driver	2
• Concrete Pump Operator	3
PIPE JOINTERS	
• Jointers, pipes up to and including 300 mm diameter	4
• Jointers, pipes over 300 mm diameter and up to 900 mm diameter	3
• Jointers, pipes over 900 mm diameter	2
• **except** in HDPE mains when experienced in butt fusion and/or electrofusion jointing operations	2
PIPELAYERS	
• Operative preparing the bed and laying pipes up to and including 300 mm diameter	4
• Operative preparing the bed and laying pipes over 300 mm diameter and up to and including 900 mm diameter	3
• Operative preparing the bed and laying pipes over 900 mm diameter	2
PRESTRESSING CONCRETE	
• Operative in control of and responsible for hydraulic jacks and other tensioning devices engaged in post-tensioning and/or pre-tensioning concrete elements	3
ROAD SURFACING WORK (includes rolled asphalt, dense bitumen macadam and surface dressings)	
Operatives employed in this category of work to be paid as follows:	
• Chipper	4
• Gritter Operator	4
• Raker	3
• Paver Operator	3
• Leveller on Paver	3
• Road Planer Operator	3
• Road Roller Driver, 4 tonne and upwards	3
• Spray Bar Operator	4

LABOUR CATEGORIES

TIMBERMAN	
• Timberman, installing timber supports	3
• Highly skilled timber man working on complex supports using timbers of size 250 mm by 125 mm and above	2
• Operative attending	4
WELDERS	
• Grade 4 (Fabrication Assistant)	
• Welder able to tack weld using SMAW or MIG welding processes in accordance with verbal instructions and including mechanical preparation such as cutting and grinding	3
• Grade 3 (Basic Skill Level)	
• Welder able to weld carbon and stainless steel using at least one of the following processes SMAW, GTAW, GMAW for plate-plate fillet welding in all major welding positions, including mechanical preparation and complying with fabrication drawings	2
• Grade 2 (Intermediate Skill Level)	
• Welder able to weld carbon and stainless steel using manual SMAW, GTAW, semi-automatic MIG or MAG, and FCAW welding processes including mechanical preparation, and complying with welding procedures, specifications and fabrication drawings	1
• Grade 1 (Highest Skill Level)	
• Welder able to weld carbon and stainless steel using manual SMAW, GTAW, semi-automatic GMAW or MIG or MAG, and FCA w elding processes in all modes and directions in accordance with BSEN287-1 and/or 287-2 Aluminium Fabrications including mechanical preparation and complying with welding procedures, specifications and fabrication drawings	Craft Rate
YOUNG WORKERS	
• Operatives below 18 years of age will receive payment 60% of the General Operative Basic Rate.	
• At 18 years of age or over the payment is 100% of the relevant rate.	

PART 3

Approximate Estimating

This part contains the following sections:

Designing Architecture

A Pressman

Designing Architecture is an indispensable tool to assist both students and young architects in formulating an idea, transforming it into a building, and making effective design decisions.

This highly focused book offers explicit guidance to students and young professionals on how to approach, analyze, and execute specific tasks; develop and refine a process to facilitate the best possible design projects; and create meaningful architectural form. Case studies augment the text and chronicle fascinating applications of the design process.

Designing Architecture will inspire readers to elevate the quality of preliminary designs and unravel some of the mystery of creating the most beautiful, responsive, and responsible architectural design possible.

December 2011: 276x219: 208pp
Hb: 978-0-415-59515-5: £105.00
Pb: 978-0-415-59516-2: £24.99

To Order: Tel: +44 (0) 1235 400524 Fax: +44 (0) 1235 400525
or Post: Taylor and Francis Customer Services,
Bookpoint Ltd, Unit T1, 200 Milton Park, Abingdon, Oxon, OX14 4TA UK
Email: book.orders@tandf.co.uk

For a complete listing of all our titles visit:
www.tandf.co.uk

Building Prices per Functional Unit

Prices given under this heading are average prices, on a fluctuating basis, for typical new buildings based on a tender price level index of 453 (1976 = 100). Prices includes for Preliminaries, and Overheads and Profit. Unless otherwise stated, prices do not allow for external works, furniture, loose or special equipment and are, of course, exclusive of fees for professional services.

On certain types of buildings there exists a close relationship between its cost and the number of functional units that it accommodates. During the early stages of a project therefore an approximate estimate can be derived by multiplying the proposed unit of accommodation (i.e. hotel bedrooms, car parking spaces etc.) by an appropriate cost.

The following indicative unit areas and costs have been derived from historic data. It is emphasized that the prices must be treated with reserve, as they represent the average of prices from our records and cannot provide more than a rough guide to the cost of a building. There are limitations when using this method of estimating, for example, the functional areas and costs of football stadia, hotels, restaurants etc. are strongly influenced by the extent of front and back of house facilities housed within it, and these areas can vary considerably from scheme to scheme.

The areas may also be used as a 'rule of thumb' in order to check on economy of designs. Where we have chosen not to show indicative areas, this is because either ranges are extensive or such figures may be misleading.

Costs have been expressed within a range, although this is not to suggest that figures outside this range will not be encountered, but simply that the calibre of such a type of building can itself vary significantly.

For assistance with the compilation of an *Order of Cost Estimate*, or of an *Elemental Cost Plan*, the reader is directed to the *Building Prices per Square Metre*, *Approximate Estimating Rates* and *Cost Models* sections of this book.

As elsewhere in this edition, prices do not include Value Added Tax or professional services fees.

Building type	Function	Range £		
Utilities, Civil Engineering facilities				
Car Parking				
surface car parking; 20 to 22 m²/car	car	1250.00	to	1775.00
surface car parking; landscaped 20 to 22 m²/car	car	1775.00	to	2500.00
multi-storey car parks grade & upper level; 24 to 28 m²/car	car	6400.00	to	9100.00
multi-storey car parks flat slab; 24 to 28 m²/car	car	8100.00	to	11500.00
underground car parks partially underground under buildings; natural ventilation; 27 to 30 m²/car	car	9900.00	to	14000.00
underground car parks completely underground with landscaped roof; mechanical ventilation; 28 to 32 m²/car	car	15000.00	to	21500.00
underground car parks completely underground under buildings; mechanical ventilation; 28 to 32 m²/car	car	18500.00	to	26500.00
Offices				
business park; medium quality, non air-conditioned	person	8700.00	to	12500.00
low rise; air-conditioned; high quality speculative; 4–8 storey	person	8500.00	to	12000.00
medium rise; air-conditioned; high quality speculative; 8–20 storeys	person	11000.00	to	15500.00
Health and Welfare facilities				
district hospitals; 65 to 85 m²/bed	bed	100000.00	to	145000.00
hospital teaching centres; 120+ m²/bed	bed	130000.00	to	190000.00
private hospitals; 75 to 100 m²/bed	bed	120000.00	to	170000.00
residential homes; 40 to 60 m²/bedroom	bed	38000.00	to	55000.00
nursing homes; 40 to 60 m²/bedroom	bed	61000.00	to	88000.00
Recreational facilities				
Theatres				
large – over 500 seats	seat	19000.00	to	27000.00
studio/workshop – less than 500 seats	seat	9600.00	to	14000.00
refurbishment	seat	7700.00	to	11000.00
Sports halls				
indoor bowls halls	rink	115000.00	to	160000.00
squash courts	court	65000.00	to	93000.00
Education facilities				
Schools				
nursery schools; 3 to 5 m²/pupil	pupil	4100.00	to	5900.00
secondary/middle schools; 6 to 10 m²/pupil	pupil	10000.00	to	14500.00
special schools; 18 to 20 m²/pupil	pupil	17000.00	to	24000.00
Universities				
arts buildings	student	10000.00	to	15000.00
computer buildings	student	14000.00	to	20000.00
science buildings	student	14000.00	to	20000.00
laboratories	student	27000.00	to	38000.00
Residential facilities				
Housing				
detached; four bedroom; 100 to 140 m² gifa	house	92000.00	to	130000.00
semi-detached; three bedroom; 70 to 90 m² gifa	house	58000.00	to	83000.00
terraced; 2 bedroom; 55 to 65 m² gifa	house	38000.00	to	55000.00
low rise flats; two bedroom; 55 to 65 m² gifa	flat	46000.00	to	66000.00
medium rise flats; two bedroom; 55 to 65 m² gifa	flat	56000.00	to	80000.00

BUILDING PRICES PER FUNCTIONAL UNIT

Building type	Function	Range £		
Hotels				
luxury; city centre with conference and wet leisure facilities; 70 to 120 m²/bedroom	bed	255000.00	to	365000.00
business; town centre with conference and wet leisure facilities; 70 to 100 m²/bedroom	bed	160000.00	to	230000.00
mid-range; town centre with conference and leisure facilities; 50 to 60 m²/bedroom	bed	80000.00	to	110000.00
budget; city centre with dining and bar facilities; 35 to 45 m²/bedroom	bed	51000.00	to	73000.00
budget; roadside excluding dining facilities; 28 to 35 m²/bedroom	bed	33000.00	to	47500.00
Student residences				
large budget schemes; 200+ units with en-suite accommodation; 18 to 20 m²/bedroom	bed	19500.00	to	28000.00
smaller schemes (40–100 units) with mid range specifications; some with en-suite accommodation; 19 to 24 m²/bedroom	bed	26000.00	to	37500.00
smaller high quality courtyard schemes, college scheme; 24 to 28 m²/bedroom	bed	42000.00	to	60000.00

Metric Handbook

4th Edition

D Littlefield

- Significantly updated in reference to the latest construction standards and new building types.

- Sustainable design integrated into chapters.

- Over 100,000 copies sold to successive generations of architects and designers – this book belongs on every design office desk and drawing board.

The Metric Handbook is the major handbook of planning and design data for architects and architecture students. Covering basic design data for all the major building types it is the ideal starting point for any project. For each building type, the book gives the basic design requirements and all the principal dimensional data, and succinct guidance on how to use the information and what regulations the designer needs to be aware of.

As well as buildings the Metric Handbook deals with broader aspects of design such as materials, acoustics and lighting, and general design data on human dimensions and space requirements. The Metric Handbook really is the unique reference for solving everyday planning problems.

March 2012: 297x210: 880pp
Pb: 978-1-85617-806-8: £39.99

To Order: Tel: +44 (0) 1235 400524 Fax: +44 (0) 1235 400525
or Post: Taylor and Francis Customer Services,
Bookpoint Ltd, Unit T1, 200 Milton Park, Abingdon, Oxon, OX14 4TA UK
Email: book.orders@tandf.co.uk

For a complete listing of all our titles visit:
www.tandf.co.uk

Building Prices per Square Metre

Prices given under this heading are average prices, on a *fluctuating basis*, for typical buildings based on a tender price level index of 453 (1976 = 100) adjusted for Outer London location. Prices include for Preliminaries and Overheads and Profit. Unless otherwise stated, prices do not allow for external works, furniture, loose or special equipment.

Prices are based upon the total gross internal floor area of all storeys, measured between external walls and without deduction for internal walls, columns, stairwells, lift wells and the like in accordance with the 6th edition of the RICS Code of Measuring Practice, September 2007.

As in previous editions it is emphasized that the prices must be treated with a degree of reserve as they represent the average of prices from our records and cannot provide more than an approximate guide to the cost of a building.

Prices can vary greatly between regions and even between individual sites. Professional interpretation will always need to be applied.

For assistance with the compilation of an *Order of Cost Estimate*, or of an *Elemental Cost Plan*, the reader is directed to the *Cost Models* and *Approximate Estimating Rates* sections in the book.

As elsewhere in this edition, prices do not include Value Added Tax or professional services fees.

Item	Unit	Range £		
UNICLASS D1 UTILITIES, CIVIL ENGINEERING FACILITIES				
Car parking				
surface car parking	m^2	60.00	to	75.00
surface car parking; landscaped	m^2	77.00	to	97.00
Multi-storey car parks				
grade & upper level	m^2	280.00	to	350.00
flat slab	m^2	355.00	to	445.00
Underground car parks				
partially underground under buildings; naturally ventilated	m^2	355.00	to	445.00
completely underground under buildings with mechanical ventilation	m^2	620.00	to	780.00
completely underground with landscaped roof and mechanical ventilation	m^2	760.00	to	950.00
Transport facilities				
railway stations	m^2	1925.00	to	2400.00
bus and coach stations	m^2	1825.00	to	2300.00
bus garages	m^2	690.00	to	870.00
petrol stations	m^2	1525.00	to	1925.00
Vehicle showrooms with workshops, garages etc.				
up to 2,000 m^2	m^2	900.00	to	1125.00
over 2,000 m^2	m^2	770.00	to	970.00
Vehicle showrooms without workshops, garages etc.				
up to 2,000 m^2	m^2	900.00	to	1125.00
Vehicle repair and maintenance buildings				
up to 500 m^2	m^2	1325.00	to	1650.00
over 500 m^2 up to 2000 m^2	m^2	900.00	to	1125.00
car wash buildings	m^2	770.00	to	970.00
Airport facilities (excluding aprons)				
airport terminals	m^2	2350.00	to	2950.00
airport piers/satellites	m^2	2475.00	to	3100.00
Airport campus facilities				
cargo handling bases	m^2	590.00	to	750.00
distribution centres	m^2	330.00	to	420.00
hangars (type C and D aircraft)	m^2	1025.00	to	1275.00
TV, radio and video studios	m^2	1225.00	to	1550.00
Telephone exchanges	m^2	900.00	to	1125.00
Telephone engineering centres	m^2	610.00	to	760.00
Branch post offices	m^2	900.00	to	1125.00
Postal delivery offices/sorting offices	m^2	900.00	to	1125.00
Electricity substations	m^2	1275.00	to	1600.00
Garages, domestic	m^2	550.00	to	690.00
UNICLASS D2 INDUSTRIAL FACILITIES				
B1 Light industrial/offices buildings				
economical shell, and core with heating only	m^2	540.00	to	680.00
medium shell and core with heating and ventilation	m^2	770.00	to	970.00
high quality shell and core with air conditioning	m^2	1175.00	to	1500.00
developers Category A fit-out	m^2	440.00	to	550.00
tenants Category B fit-out	m^2	290.00	to	365.00
Agricultural storage	m^2	460.00	to	580.00

Item	Unit	Range £		
Factories				
for letting (incoming services only)	m²	470.00	to	590.00
for letting (including lighting, power and heating)	m²	600.00	to	750.00
nursery units (including lighting, power and heating)	m²	600.00	to	750.00
workshops	m²	550.00	to	700.00
maintenance/motor transport workshops	m²	730.00	to	910.00
owner occupation for light industrial use	m²	770.00	to	970.00
owner occupation for heavy industrial use	m²	1275.00	to	1600.00
Factory/office buildings high technology production				
for letting (shell and core only)	m²	500.00	to	630.00
for owner occupation (controlled environment fully finished)	m²	680.00	to	860.00
High technology laboratory workshop centres, air-conditioned	m²	2175.00	to	2700.00
Warehouse and distribution centres				
high bay (10–15 m high) for owner occupation (no heating) up to 10,000 m²	m²	240.00	to	305.00
high bay (10–15 m high) for owner occupation (no heating) over 10,000 m² up to 20,000 m²	m²	180.00	to	230.00
high bay (16–24 m high) for owner occupation (no heating) over 10,000 m² up to 20,000 m²	m²	270.00	to	340.00
high bay (16–24 m high) for owner occupation (no heating) over 20,000 m²	m²	220.00	to	280.00
Fit-out cold stores, refrigerated stores inside warehouse	m²	485.00	to	610.00
Industrial buildings				
Shell with heating to office areas only				
500–1,000 m²	m²	395.00	to	500.00
1,000–2,000 m²	m²	335.00	to	420.00
greater than 2,000 m²	m²	365.00	to	460.00
Unit including services to production area				
500–1,000 m²	m²	550.00	to	690.00
1,000–2,000 m²	m²	500.00	to	630.00
greater than 2,000 m²	m²	500.00	to	630.00
UNICLASS D3 ADMINISTRATIVE, COMMERCIAL, PROTECTIVE SERVICES FACILITIES				
Embassies	m²	1625.00	to	2025.00
County courts	m²	1950.00	to	2475.00
Magistrates courts	m²	1075.00	to	1350.00
Civic offices				
non-air-conditioned	m²	1075.00	to	1350.00
fully air-conditioned	m²	1275.00	to	1600.00
Probation/registrar offices	m²	810.00	to	1025.00
Offices for owner occupation				
low rise, air-conditioned	m²	1150.00	to	1450.00
medium rise, air-conditioned	m²	1450.00	to	1825.00
high rise, air-conditioned	m²	1700.00	to	2150.00
Offices – City and West End				
To developers Cat A finish only				
high quality, speculative 8–20 storeys, air-conditioned	m²	1625.00	to	2025.00
high rise, air-conditioned, iconic speculative towers	m²	2475.00	to	3100.00

Item	Unit	Range £		
UNICLASS D3 ADMINISTRATIVE, COMMERCIAL, PROTECTIVE SERVICES FACILITIES – cont				
Business park offices				
Generally up to 3 storeys only				
functional non-air-conditioned less than 2,000 m²	m²	940.00	to	1175.00
functional non-air-conditioned more than 2,000 m²	m²	980.00	to	1225.00
medium quality non-air-conditioned less than 2,000 m²	m²	1025.00	to	1275.00
medium quality non-air-conditioned more than 2,000 m²	m²	1075.00	to	1375.00
medium quality air-conditioned less than 2,000 m²	m²	1175.00	to	1500.00
medium quality air-conditioned more than 2,000 m²	m²	1275.00	to	1600.00
good quality – naturally ventilated to meet BCO specification (exposed soffits, solar shading) less than 2,000 m²	m²	1150.00	to	1450.00
good quality – naturally ventilated to meet BCO specification (exposed soffits, solar shading) more than 2,000 m²	m²	1175.00	to	1500.00
high quality air-conditioned less than 2,000 m²	m²	1450.00	to	1825.00
high quality air-conditioned more than 2,000 m²	m²	1475.00	to	1875.00
Large trading floors in medium sized offices	m²	2350.00	to	2950.00
Two storey ancillary office accommodation to warehouses/factories	m²	940.00	to	1175.00
Fitting out offices (NIA – Net Internal Area)				
City and West End				
basic fitting out including carpets, decorations, partitions and services	m²	425.00	to	540.00
good quality fitting out including carpets, decorations, partitions and services	m²	470.00	to	590.00
high quality fitting out including carpets, decorations, partitions, ceilings, furniture, air conditioning and electrical services	m²	510.00	to	640.00
reception areas	m²	2125.00	to	2700.00
conference suites	m²	2050.00	to	2600.00
meeting areas	m²	2550.00	to	3200.00
back of house/storage	m²	470.00	to	590.00
sub-equipment room	m²	1375.00	to	1725.00
kitchen	m²	3000.00	to	3750.00
Out-of town (South East England)				
basic fitting out including carpets, decorations, partitions and services	m²	340.00	to	430.00
good quality fitting out including carpets, decorations, partitions and services	m²	380.00	to	480.00
high quality fitting out including carpets, decorations, partitions, ceilings, furniture, air conditioning and electrical services	m²	470.00	to	590.00
reception areas	m²	1275.00	to	1600.00
conference suites	m²	1750.00	to	2175.00
meeting areas	m²	1700.00	to	2150.00
back of house/storage	m²	380.00	to	480.00
sub-equipment room	m²	1375.00	to	1725.00
kitchen	m²	1950.00	to	2475.00
Office refurbishment including developers finish – (Gross Internal Floor Area – GIFA; Central London)				
minor refurbishment	m²	400.00	to	500.00
medium refurbishment	m²	850.00	to	1075.00
major refurbishment	m²	1325.00	to	1650.00

Item	Unit	Range £		
Banks and Building Societies				
banks; local	m²	1225.00	to	1550.00
banks; city centre/head office	m²	1750.00	to	2175.00
building society branches	m²	1175.00	to	1500.00
building society; refurbishment	m²	770.00	to	970.00
Shop shells				
small	m²	520.00	to	650.00
large, including department stores and supermarkets	m²	470.00	to	600.00
Fitting out shell for small shop (including shop fittings)				
simple store	m²	490.00	to	620.00
designer/fashion store	m²	980.00	to	1225.00
Fitting out shell for department store or supermarket	m²	1400.00	to	1775.00
Retail Warehouses				
shell	m²	360.00	to	450.00
fitting out, including all display and refrigeration units, check outs and IT systems	m²	210.00	to	270.00
Hypermarkets/Supermarkets				
shell	m²	430.00	to	540.00
supermarket fit-out	m²	850.00	to	1075.00
hypermarket fit-out	m²	600.00	to	760.00
Shopping Centres				
Malls, including fit-out				
comfort cooled	m²	2700.00	to	3400.00
air-conditioned	m²	3500.00	to	4400.00
food court	m²	3000.00	to	3800.00
factory outlet centre – enclosed	m²	2475.00	to	3100.00
factory outlet centre – open	m²	495.00	to	620.00
anchor tenants; capped off services	m²	810.00	to	1025.00
medium/small units; capped off services	m²	770.00	to	970.00
centre management	m²	1750.00	to	2175.00
enclosed surface level service yard	m²	1225.00	to	1550.00
landlords back of house and service corridors	m²	1225.00	to	1550.00
Refurbishment				
mall; limited scope	m²	900.00	to	1125.00
mall; comprehensive	m²	1275.00	to	1600.00
Rescue/aid facilities				
ambulance stations	m²	770.00	to	970.00
ambulance control centres	m²	1175.00	to	1500.00
fire stations	m²	1275.00	to	1600.00
Police stations	m²	1225.00	to	1550.00
Prisons	m²	1525.00	to	1925.00
UNICLASS D4 MEDICAL, HEALTH AND WELFARE FACILITIES				
District hospitals	m²	1375.00	to	1725.00
refurbishment	m²	1025.00	to	1275.00
Hospices	m²	1150.00	to	1450.00
Private hospitals	m²	1400.00	to	1775.00
Pharmacies	m²	1525.00	to	1925.00
Hospital laboratories	m²	1700.00	to	2150.00

Item	Unit	Range £		
UNICLASS D4 MEDICAL, HEALTH AND WELFARE FACILITIES – cont				
Ward blocks	m^2	1275.00	to	1600.00
refurbishment	m^2	770.00	to	970.00
Geriatric units	m^2	1275.00	to	1600.00
Psychiatric units	m^2	1175.00	to	1500.00
Psycho-geriatric units	m^2	1150.00	to	1450.00
Maternity units	m^2	1275.00	to	1600.00
Operating theatres	m^2	1700.00	to	2150.00
Outpatients/casualty units	m^2	1275.00	to	1600.00
Hospital teaching centres	m^2	1100.00	to	1375.00
Health centres	m^2	1025.00	to	1275.00
Welfare centres	m^2	1100.00	to	1375.00
Day centres	m^2	1025.00	to	1275.00
Group practice surgeries	m^2	1150.00	to	1450.00
Homes for the mentally handicapped	m^2	940.00	to	1175.00
Homes for the physically handicapped	m^2	1075.00	to	1350.00
Geriatric day hospital	m^2	1175.00	to	1500.00
Accommodation for the elderly				
residential homes	m^2	770.00	to	970.00
nursing homes	m^2	1025.00	to	1275.00
Children's homes	m^2	1275.00	to	1600.00
Homes for the aged	m^2	1175.00	to	1500.00
refurbishment	m^2	640.00	to	800.00
Observation and assessment units	m^2	850.00	to	1075.00
Primary Health Care				
doctors surgery – basic	m^2	900.00	to	1125.00
doctors surgery/medical centre	m^2	1075.00	to	1350.00
Hospitals				
diagnostic and treatment centres	m^2	2175.00	to	2700.00
acute services hospitals	m^2	2075.00	to	2600.00
radiotherapy and oncology units	m^2	2175.00	to	2700.00
community hospitals	m^2	1750.00	to	2175.00
trauma unit	m^2	1625.00	to	2025.00
UNICLASS D5 RECREATIONAL FACILITIES				
Public houses	m^2	1275.00	to	1600.00
Dining blocks and canteens	m^2	1025.00	to	1275.00
Restaurants	m^2	1150.00	to	1450.00
Community centres	m^2	1100.00	to	1375.00
General purpose halls	m^2	940.00	to	1175.00
Visitors' centres	m^2	1525.00	to	1925.00
Youth clubs	m^2	1275.00	to	1600.00
Arts and drama centres	m^2	1375.00	to	1725.00
Theatres, including seating and stage equipment				
large – over 500 seats	m^2	2900.00	to	3650.00
studio/workshop – less than 500 seats	m^2	1925.00	to	2400.00
refurbishment	m^2	1525.00	to	1925.00
Concert halls, including seats and stage equipment	m^2	2175.00	to	2700.00

Item	Unit	Range £		
Cinema				
shell	m²	680.00	to	860.00
multiplex; shell only	m²	1375.00	to	1725.00
fitting out, including all equipment, air-conditioned	m²	850.00	to	1075.00
Exhibition centres	m²	1275.00	to	1600.00
Swimming pools				
international standard	m²	2900.00	to	3650.00
local authority standard	m²	1925.00	to	2400.00
school standard	m²	900.00	to	1125.00
leisure pools, including wave making equipment	m²	2250.00	to	2850.00
Ice rinks	m²	1025.00	to	1275.00
Rifle ranges	m²	850.00	to	1075.00
Leisure centres				
dry	m²	1175.00	to	1500.00
extension to hotels; shell and fit-out, including small pool	m²	1700.00	to	2150.00
wet and dry	m²	1775.00	to	2250.00
Sports halls including changing rooms	m²	810.00	to	1025.00
School gymnasiums	m²	770.00	to	970.00
Squash courts	m²	810.00	to	1025.00
Indoor bowls halls	m²	570.00	to	720.00
Bowls pavilions	m²	700.00	to	880.00
Health and Fitness Clubs	m²	1175.00	to	1500.00
Sports pavilions	m²	850.00	to	1075.00
changing only	m²	980.00	to	1225.00
social and changing	m²	980.00	to	1225.00
Clubhouses	m²	810.00	to	1025.00
UNICLASS D6 RELIGIOUS FACILITIES				
Temples, mosques, synagogues	m²	1100.00	to	1375.00
Churches	m²	1025.00	to	1275.00
Mission halls, meeting houses	m²	1100.00	to	1375.00
Convents	m²	1175.00	to	1500.00
Crematoria	m²	1375.00	to	1725.00
Mortuaries	m²	1925.00	to	2350.00
UNICLASS D7 EDUCATION, SCIENTIFIC AND INFORMATION FACILITIES				
Nursery schools	m²	1025.00	to	1275.00
Primary/junior schools	m²	1075.00	to	1350.00
Secondary/middle schools	m²	1275.00	to	1600.00
Secondary schools and further education colleges				
classrooms	m²	850.00	to	1075.00
laboratories	m²	940.00	to	1175.00
craft design and technology	m²	940.00	to	1175.00
music	m²	1075.00	to	1350.00
Extensions to schools				
classrooms	m²	1025.00	to	1275.00
laboratories	m²	1075.00	to	1350.00
Sixth form colleges	m²	1025.00	to	1275.00
Special schools	m²	850.00	to	1075.00
Training colleges	m²	850.00	to	1075.00
Management training centres	m²	1100.00	to	1375.00

Item	Unit	Range £		
UNICLASS D7 EDUCATION, SCIENTIFIC AND INFORMATION FACILITIES – cont				
Universities				
arts buildings	m^2	940.00	to	1175.00
science buildings	m^2	1150.00	to	1450.00
College/University Libraries	m^2	940.00	to	1175.00
Laboratories and offices, low-level servicing	m^2	1775.00	to	2250.00
Computer buildings	m^2	1525.00	to	1925.00
Museums and Art Galleries				
national standard museum	m^2	4100.00	to	5200.00
national standard independent specialist museum, excluding fit-out	m^2	2900.00	to	3650.00
regional, including full air conditioning	m^2	2475.00	to	3100.00
local, including full air conditioning	m^2	1825.00	to	2300.00
conversion of existing warehouse to regional standard museum	m^2	1150.00	to	1450.00
conversion of existing warehouse to local standard museum	m^2	980.00	to	1225.00
Galleries				
international standard art gallery	m^2	2500.00	to	3150.00
national standard art gallery	m^2	1925.00	to	2400.00
independent commercial art gallery	m^2	1075.00	to	1350.00
Arts and drama centre	m^2	1025.00	to	1275.00
Learning resource centre				
economical	m^2	940.00	to	1175.00
high quality	m^2	1175.00	to	1500.00
Libraries				
regional; 5,000 m^2	m^2	2175.00	to	2700.00
national; 15,000 m^2	m^2	2850.00	to	3500.00
international; 20,000 m^2	m^2	3700.00	to	4550.00
Conference centres	m^2	1475.00	to	1875.00
UNICLASS D8 RESIDENTIAL FACILITIES				
Local Authority and Housing Association schemes				
Housing Asociation Developments (Code for Sustainable Homes Level 3)				
Bungalows				
semi-detached	m^2	770.00	to	970.00
terraced	m^2	730.00	to	910.00
Two storey housing				
detached	m^2	770.00	to	970.00
semi-detached	m^2	730.00	to	910.00
terraced	m^2	640.00	to	800.00
Three storey housing				
semi-detached	m^2	770.00	to	970.00
terraced	m^2	640.00	to	800.00
Apartments/flats				
low rise	m^2	770.00	to	970.00
medium rise	m^2	940.00	to	1175.00
Sheltered housing with wardens accommodation	m^2	730.00	to	910.00

Item	Unit	Range £		
Private developments				
single detached houses	m²	1025.00	to	1275.00
two and three storey houses	m²	1075.00	to	1350.00
Apartments/flats generally				
standard quality; 3–5 storeys	m²	1275.00	to	1600.00
warehouse/office conversion to apartments	m²	1400.00	to	1775.00
high quality apartments in residential tower – Inner London	m²	1700.00	to	2150.00
Hotels (including fittings, furniture and equipment)				
luxury city-centre with conference and wet leisure facilities	m²	2550.00	to	3200.00
business town centre with conference and wet leisure facilities	m²	1875.00	to	2350.00
mid-range with conference and leisure facilities	m²	1450.00	to	1825.00
budget city-centre with dining and bar facilities	m²	1275.00	to	1600.00
budget roadside excluding dining facilities	m²	1100.00	to	1375.00
Hotel accommodation facilities (excluding fittings, furniture and equipment)				
bedroom areas	m²	770.00	to	970.00
front of house and reception	m²	980.00	to	1225.00
restaurant areas	m²	1150.00	to	1450.00
bar areas	m²	1025.00	to	1275.00
function rooms/conference facilities	m²	980.00	to	1225.00
Students residences				
large budget schemes with en-suite accommodation	m²	980.00	to	1225.00
smaller schemes (40–100 units) with mid range specifications	m²	1175.00	to	1500.00
smaller high quality courtyard schemes, college style	m²	1625.00	to	2025.00

Quantity Surveyor's Pocket Book

2nd Edition

D Cartlidge

This second edition of the Quantity Surveyor's Pocket Book is fully updated in line with NRM1, NRM2 and JCT(11), and remains a must-have guide for students and qualified practitioners. Its focussed coverage of the data, techniques, and skills essential to the quantity surveying role make it an invaluable companion for everything from initial cost advice to the final account stage.

Key features include:

- the structure of the construction industry

- cost forecasting and feasibility studies

- measurement and quantification, with NRM2 and SMM7 examples

- estimating and bidding

- whole life costs

- contract selection

- final account procedure.

June 2012: 186x123: 440pp
Pb: 978-0-415-50110-1: £18.99

To Order: Tel: +44 (0) 1235 400524 Fax: +44 (0) 1235 400525
or Post: Taylor and Francis Customer Services,
Bookpoint Ltd, Unit T1, 200 Milton Park, Abingdon, Oxon, OX14 4TA UK
Email: book.orders@tandf.co.uk

For a complete listing of all our titles visit:
www.tandf.co.uk

Building Cost Models

Davis Langdon has been producing cost models for publication in *Building* magazine since 1993.

During the intervening period, over 90 have been published examining most building types as well as providing detailed coverage of broader issues including sustainability, infrastructure and off-site manufacturing.

Trends continue to change. Sustainability, mixed-use developments, conversions and the increasing size and complexity of schemes have become more evident recently.

Although the scope and coverage of the cost models have expanded considerably, the objectives have always remained constant. They are:

- To provide detailed elemental cost information derived from a generic building that can be applied to other projects
- To provide a commentary on cost drivers and other design and specification issues
- To compare suitable procurement routes that secure the clients objectives

For this edition of Spon's, Davis Langdon have published updated elemental cost data for 18 building types. All models have been updated to reflect a tender index of 453. Locations do vary for each model, so please make a note of the location and location factor for any adjustments that may need to be made.

A summary of UK location factors can be found in *Building Costs Indices, Tender Price Indices and Location Factors* section of this book, together with examples showing how to apply those factors.

Readers should note that these models are based around actual projects and are intended as an indication to build costs only. There are many factors which need to be considered when putting together an early cost plan or estimate and all need to be considered when arriving at a Cost Plan for any particular project.

Readers may refer to the *Approximate Estimating Rates* section of this book to make adjustments to the models for alternative specifications within any of the elements.

As elsewhere in this edition prices do not include Value Added Tax or professional services fees.

LAND REMEDIATION

This cost model features the remediation of a 2 hectare brownfield site, using a combination of bioremediation, stabilization and solidification, on site screening and some off-site disposal. Eighty per cent of excavated material is reused on site. Factors that need to be considered include: Waste classification; Underground obstructions and pockets of contamination; Ground water flows and barriers; Distance to landfill tips able to accept the contaminated materials.

Gross site area   20,000 m²

Model location is Outer London   TPI = 453; LF = 1.00

This updated cost model is copyright of Davis Langdon and was originally published in Building Magazine in Nov−09

Land Remediation	Quantity	Unit	Cost (£)	Total (£)	Rate (£/m²)
General items				**353,900**	**17.70**
Performance Bond and Insurances		item	25,000		
Site mobilization		item	8,900		
Site running costs	32	weeks	10,000		
Ground Investigation				**54,900**	**2.75**
Trial pits and boreholes		item	4,900		
Laboratory analyses and sampling		item	50,000		
Geotechnical and specialist services				**304,900**	**15.25**
Bioremediation – mobilization		item	10,000		
Bioremediation – treatment	8000	m³	20		
Stabilization and solidification – mobilization		item	4,900		
Stabilization and solidification – treatment	2000	m³	65		
Demolition				**10,000**	**0.50**
General site clearance	20000	m²	0.50		
Earthworks				**2,284,200**	**114.21**
General excavations	50000	m³	1.5		
Extra over for hand excavation	200	m³	50		
Breaking out tarmacadam surfaces	300	m³	4.9		
Breaking out mass concrete	1000	m³	7.4		
Breaking out reinforced concrete	800	m³	12.30		
On-site material management	51800	m³	3.9		
Trimming and preparation of excavated surfaces	20000	m²	0.5		
Screening of excavated material	50000	m³	1		
Crushing and screening of hard material	1800	m³	7.4		
Disposal of tarmacadam off site including haulage	300	m³	39		
Disposal of non-hazardous material off site including haulage	6000	m³	50		

LAND REMEDIATION

Land Remediation	Quantity	Unit	Cost (£)	Total (£)	Rate (£/m²)
Disposal of hazardous material off site including haulage	3000	m³	120		
Disposal of WAC-failing hazardous material off site including haulage	1000	m³	150		
Landfill Tax on disposed material, based on 1.9 tonnes/m³	19570	tonne	39		
Filling of material arising from site screening and crushing	31800	m³	1.5		
Filling of material arising from on-site treatment	10000	m³	1.5		
Imported granular material to make up levels to existing	10300	m³	25		
Miscellaneous works				**10,000**	**0.50**
Repairs to boundary fencing, walls etc.		item	10,000		
Contractor overheads and profits				**434,100**	**21.71**
Allowance contractor's overheads and profit @	4%				
Design reserve @	10%				
Construction cost (rate based on gross site area)				**3,452,000**	**172.62**

DISTRIBUTION CENTRE

This cost model features a detailed cost breakdown of a new build high bay distribution centre with a 15 m haunch height. The costs are based on a generic solution with a gross internal floor area of 70,000 m², which includes 5% office and ancillary accommodation. Costs of enhancements including the warehouse and office area fit-out and ancillary buildings, together with costs of external works are detailed. Costs of racking and materials handling installations are not included. The model has been prepared on the assumption that ground conditions are good and that minimal site preparation is required.

Warehouse: Gross internal floor area	71,700 m²
Office Shell: Net internal floor area	3,500 m²
Model location is based on UK average	TPI = 435; LF = 1.00

This updated cost model is copyright of Davis Langdon and was originally published in Building Magazine in Aug−04

Warehouse Shell and Core	Quantity	Unit	Cost (£)	Total (£)	Rate (£/m²)
Substructure				**3,642,000**	**50.79**
225 mm reinforced concrete ground slab; laser levelled; surface hardener; subbase; perimeter ground beam; lift pits	70,000	m²	45		
In situ concrete pad foundations and ground beams	260	nr	1,200		
Allowance for foundations and retaining walls to dock levellers	100	nr	1,800		
Frame				**2,940,000**	**41.00**
Steel propped portal frame, cold rolled purlin sections, surface treatment, including decorations	70,000	m²	44		
Roof				**3,206,000**	**44.71**
Composite roof panels; powder coated galvanized steel	70,000	m²	30		
Extra over for 10% rooflights	7,000	m²	33		
Roof drainage generally; syphonic system	70,000	m²	6		
Eaves/valley gutter; galvanized steel; insulation; stop ends	3,100	m²	130		
Allowance for mansafe system, hatches and access ladders	1	item	80,000		
External walls, windows and doors				**983,500**	**13.72**
Wall cladding system, composite panels and built-up cladding systems; mineral fibre insulation; polyester powder coating	17,750	m²	50		
Allowance for personnel escape doors	25	m²	875		
Cladding and details to inside face of parapet walls	1,150	m²	44		
Level access doors; insulated sectional overhead dock doors	10	nr	2,350		

DISTRIBUTION CENTRE

Warehouse Shell and Core	Quantity	Unit	Cost (£)	Total (£)	Rate (£/m²)
Dock leveller installations				**1,089,500**	**15.20**
Insulated sectional overhead dock doors	100	nr	2,800		
Dock leveller; precast concrete dock pits; wheel guides	100	nr	5,900		
Dock shelter; heavy duty scissors retracting frame	100	nr	1,650		
Protection, bollards to door tracks; heavy duty rubber dock buffers	1	item	32,5000		
Traffic control lights	100	nr	220		
Services installations				**222,200**	**3.10**
Water installations; hot and cold water services	1	item	32,500		
Mechanical installations; gas and water connections	1	item	22,000		
Electrical installation; general sub-mains and distribution	1	item	75,000		
Electrical installation; to dock levellers and access doors	1	item	85,000		
Allowance for lightning protection	1	item	7,700		
Preliminaries and contingency				**1,268,800**	**17.70**
Overheads and profit, site establishment and supervision @	8.5%				
Contingency @	2%				
Construction cost (Warehouse shell and core only; rate based on GIFA)				**13,352,000**	**186.22**

Office Shell and Fit-Out	Quantity	Unit	Cost (£)	Total (£)	Rate (£/m²)
Substructure				**73,500**	**21.00**
Extra for foundations to offices; in situ concrete pad foundations; ground beams; lift pit (upper floor footprint)	1,750	m²	42		
Frame, upper floors and stairs				**302,000**	**86.29**
Steel frame, universal sections; surface treatment, fire protection and decoration	3,500	m²	45		
Upper floors; 200 mm thick precast concrete plank and structural screed	1,750	m²	42		
Allowance for fire stopping to perimeter		item	16,000		
Precast concrete stairs, mild steel balustrades and handrails; polyester powder coated	2	nr	27,500		

DISTRIBUTION CENTRE

Office Shell and Fit-Out	Quantity	Unit	Cost (£)	Total (£)	Rate (£/m²)
External walls, windows and doors				404,200	115.49
Extra over wall cladding for double glazed ribbon windows	750	m²	380		
Extra over wall cladding for louvres		item	70,000		
Allowance for glazed screen		item	27,500		
Glazed entrance doors; to match glazed screen (per leaf)	14	nr	1,550		
Internal partitions and doors				46,200	13.20
140 mm thick blockwork; head restraint; fire stopping	700	m²	55		
Doors; ironmongery (cost per leaf)	14	nr	550		
Finishes				287,500	82.14
Allowance for wall finishes generally, emulsion paint and ceramic tile (allowance based on floor area)	3,500	m²	12		
Raised floor to office areas only; 150 cavity; fire barriers	3,200	m²	33		
Ceramic tiles to reception and WCs	650	m²	44		
Vinyl sheet and skirtings to corridor areas	750	m²	36		
Suspended ceiling; mineral fibre tile in exposed lay in grid	3,500	m²	23		
Extra for moisture-resistant tiles	300	m²	10		
Fittings				587,300	167.80
Allowance for open plan office fit-out to category B	3,500	m²	150		
Allowance for kitchen fittings	1	item	9,000		
Allowance for reception fittings and features	1	item	42,500		
Allowance for matwells and frames	1	item	5,600		
Allowance for WC fittings	1	item	5,200		
Services installations				89,100	254.57
Sanitary fittings generally	75	nr	420		
Hot and cold water services. Disposal installations		item	55,000		
Low temperature hot water heating	3,500	m²	27		
Mechanical ventilation and comfort cooling	3,500	m²	110		
Allowance for toilet and Lift Motor Room ventilation		item	27,500		
Gas and Electrical installation		item	100,000		
Lighting and emergency lighting and small power	3,500	m²	38		
Lift installation; 8 person electro-hydraulic		item	22,500		

DISTRIBUTION CENTRE

Office Shell and Fit-Out	Quantity	Unit	Cost (£)	Total (£)	Rate (£/m²)
Allowance for builder's work in connection say	5.0%				
Preliminaries and contingency				272,300	77.80
Overheads and profit, site establishment and supervision @	8.5%				
Contingency @	2.0%				
Construction cost (Office shell and fit-out only; rate based on Office NIA)				2,864,000	818.29

Warehouse Fit-Out	Quantity	Unit	Cost (£)	Total (£)	Rate (£/m²)
Fixtures and fittings				335,000	4.67
Allowance for general fixtures and fittings		item	120,000		
Protection; secondary steelwork, armco barriers, bollards		item	85,000		
Allowance for internal and external signage		item	45,000		
Allowance for jockey wheel strips and wheel stops		item	85,000		
Services and communication installations				5,254,000	73.28
Gas fired heating to warehouse areas; high level nozzle system	68,200	m²	9		
Roof smoke ventilation system	68,200	m²	7		
Increased incoming power supply		item	220,000		
Mains power to mechanical installations; high level lighting	68,200	m²	33		
Electrical installations; standby generator		item	250,000		
Roof level sprinklers and storage tanks (category 3)	68,200	m²	13		
Communications; fire detection and alarm; CCTV; PA	68,200	m²	7		
Allowance for builder's work in connection		item	85,000		
Preliminaries and contingency				754,000	10.52
Overheads and profit, site establishment and supervision @	8.5%				
Contingency @	5%				
Construction cost (Warehouse fit-out only; rate based on GIFA)				6,343,000	88.47

Building Cost Models

DISTRIBUTION CENTRE

External Works	Quantity	Unit	Cost (£)	Total (£)	Rate (£/m²)
Site works				**4,159,200**	**58.01**
Allowance for site preparation	175,000	m²	4		
Excavation to form ramp to dock levellers		item	55,000		
Heavy duty access road and service yard	41,700	m²	33		
Extra for ramped vehicle access		item	55,000		
Car parking; tarmacadam on subbase	29,500	m²	27		
Paved areas for pedestrian and maintenance access	3800	m²	27		
Allowance for soft landscaping, including reuse of topsoil	30,000	m²	8		
Boundary fencing; 2.4 m high; gates and entrance barriers	1,900	m	70		
Signage		item	50,000		
Hardstanding drainage	77,000	m²	8		
External services				**720,000**	**10.04**
Gas, water, electricity and telecommunications connections		item	190,000		
External lighting installations including BWIC		item	230,000		
Fire hydrant main; 12 nr hydrants		item	150,000		
Allowance for builder's works in connection with utilities		item	150,000		
Ancillary buildings				**990,000**	**13.81**
Vehicle wash; steam clean facility; fuel pump and canopy		item	490,000		
Sprinkler tank base and housing		item	55,000		
Gatehouse; transport office; axle weigher; cycle storage		item	490,000		
Preliminaries and contingency				**616,800**	**8.60**
Overheads and profit, site establishment and supervision @	8.5%				
Contingency @	2%				
Construction cost (External works only; rate based on GIFA)				**6,486,000**	**90.46**
TOTAL DISTRIBUTION CENTRE CONSTRUCTION COST (rate based on GIFA)				**29,045,000**	**386.24**

SMALL INDUSTRIAL UNIT

A single-storey new building with a gross internal floor area of 900 m², subdivided into five industrial units. Reinforced concrete ground bearing slab and pads to receive a steel portal frame. Wall and roof cladding is aluminium built up system, with internal blockwork division walls. Each of the five units has a separate entrance door and one roller shutter door, together with a single WC. Units vary in size from 150 m² to 360 m².

Gross internal floor area		900 m²
Model location is South East England	TPI = 462; LF = 1.02	

This updated cost model is copyright of Davis Langdon and was originally published in Building Magazine in Oct–10

Small Industrial Unit	Quantity	Unit	Cost (£)	Total (£)	Rate (£/m²)
Substructure				66,070	73.41
Excavation and disposal off site	190	m³	23		
Reinforced concrete ground slab, including ground beams and column bases	900	m²	50		
Power floated and hardener	900	m²	7		
Strip foundations for party walls	80	m	130		
Frame and upper floors				55,400	61.56
Steel propped portal frame, cold rolled purlins, surface treatments (@ 40 kg/m²)	36	tonne	1,250		
Intumescent paint fire protection to steelwork	36	tonne	240		
Allowance for miscellaneous works, protecting columns		item	1,800		
Roof				50,950	56.61
Built up aluminium roof cladding with 180 thick insulation, including all labours	950	m²	35		
Extra over for Rooflights (10% of total roof area)	95	m²	60		
Mansafe system	80	m	75		
Rainwater drainage, aluminium gutters and downpipes	120	m	50		
External wall, windows and doors				113,290	125.88
Built up aluminium wall cladding with 130 thick insulation	520	m²	38		
2.5 m high inner leaf of 140 thick fairface blockwork	380	m²	31		
3000 × 4600 high steel sectional overhead doors	5	nr	3150		
Aluminium single entrance doors	5	nr	1100		
Coated aluminium double glazed window system	150	m²	370		
Polycarbonate canopy entrance – approx 1500 × 1000	5	nr	1000		

SMALL INDUSTRIAL UNIT

Small Industrial Unit	Quantity	Unit	Cost (£)	Total (£)	Rate (£/m²)
Internal walls and partitions and doors				36,900	41.00
2 hour fire-resistant blockwork party walls	450	m²	65		
Fireproofing between blockwork and roof		item	1,800		
Metal stud partitions	50	m²	50		
Laminated faced internal doorset with softwood frames and ironmongery	5	nr	675		
Wall finishes				4,300	4.78
Emulsion paint to blockwork wall surfaces generally	1,370	m²	3		
Ceramic wall tiles splashbacks to WC area	5	m²	45		
Floor finishes				700	0.78
Screed and non-slip vinyl sheeting to WC areas	15	m²	45		
Ceiling finishes				500	0.56
Moisture-resistant plasterboard to WC with ceiling grid and paint finish	15	m²	36		
Sanitary appliances				5,500	6.11
Disabled WC Suite including all sanitary and fittings	5	nr	1,100		
Disposal installations				2,000	2.22
Waste, soil and vent installation; uPVC pipework and fittings	900	m²	2		
Hot and cold water installations				3,100	3.44
Hot and cold water supplies to WCs	5	nr	625		
Electrical installations				25,900	28.78
Small power, basic and emergency lighting	900	m²	20		
Supply to WC for ventilation, heater etc.	5	nr	1,450		
External lighting generally item		item	680		
Incoming services				13,500	15.00
Allowance for incoming, electrical, gas and water services		item	13,500		
Protective installations				900	1.00
Lightning protection	900	m²	1		
Communication installations				9,100	10.11
Fire and intruder alarm	900	m²	10		
Builder's work in connection				600	0.67
Forming holes and chases etc.		item	600		
Preliminaries and contingency				72,290	80.32
Overheads; profit, site establishment and site supervision @	13%				
Contingency @	5%				
Construction cost (Shell only; rate based on GIFA)				461,000	512.23

CENTRAL LONDON OFFICES

This cost model features a high quality City office scheme arranged over 13 floors and one basement with a gross internal area of 21,300 m². The scheme is steel framed and incorporates an internally ventilated double-wall facade. The wall–floor ratio is 0.46. Air treatment is by a four-pipe fan-coil unit. Costs are based on construction management procurement. Demolitions, site preparation, external works and services beyond Category A, tenant enhancement are excluded.

Gross internal floor area	21,300 m²
Net internal floor area	14,600 m²
Model location is the City of London	TPI = 498; LF = 1.10

This updated cost model is copyright of Davis Langdon and was originally published in Building Magazine in Dec−04

Shell and Core Works	Quantity	Unit	Cost (£)	Total (£)	Rate (£/m²)
Substructure				**3,355,200**	**157.52**
Dewatering		item	250,000		
Break out existing slabs, piles, obstructions and allowance for probing/testing; dewatering		item	510,000		
Foundations; bored piles with under-ream; ground beams; pile caps	1,940	m²	360		
Piling platform; mini piles and other works to boundary walls		item	310,000		
RC basement slab 300 mm thick, including waterproofing, excavation and disposal	1,940	m²	160		
RC mat slab 1200 mm thick, including waterproofing, excavation and disposal	200	m²	470		
Reinforced concrete retaining walls 300 mm thick	600	m²	280		
Reinforced concrete ground floor slab 130 mm thick on profiled metal sheet decking	1,760	m²	65		
Allowance for car park ramp, slab thickenings to stair foundations, lift/escalator pits, drainage channels, concrete transfer walls etc.		item	310,000		
Allowance for crane base including base piles		item	30,000		
Attendance on archaeologists and movement monitoring		item	100,000		
Below slab drainage; other items and sundries		item	460,000		
Frame and upper floors				**6,393,000**	**300.14**
Structural steel frame including fittings	1,704	tonne	1,550		
Extra for built up beams	440	tonne	230		
Secondary steelwork, based on an extra 5 kg/m²	110	tonne	2,100		
Extra for concrete encased beams at ground floor		item	65,000		
Fire protection to steel frame (90 mins intumescent paint)	1,350	tonne	550		

CENTRAL LONDON OFFICES

Shell and Core Works	Quantity	Unit	Cost (£)	Total (£)	Rate (£/m²)
Reinforced concrete core walls average	3,300	m²	210		
Allowance for other structures (e.g. within plant rooms etc.)		item	100,000		
Allowance for expansion joints and other sundries		item	50,000		
Lightweight reinforced concrete 130 mm thick on profiled steel decking; upstands plinths; walkways etc.	17,430	m²	80		
Roof				648,000	30.42
Profiled steel decking with 200 mm lightweight concrete inc. mesh reinforcement; Insulation and acoustics to soffit	1,760	m²	150		
Proprietary roof; paving slabs; insulation and ballast	1,760	m²	150		
Insulation to exposed soffits and acoustic treatment		item	60,000		
Upstands/plinths, hatches/ladders, safety hooks and latchways		item	60,000		
Stairs				560,000	26.29
Steel pan staircases; concrete in-fills to stair treads; painted mild steel balustrades and handrails (basement to roof; 26 flights)	2	nr	200,000		
Ditto, basement to ground: 2 flights	2	nr	1,5000		
Feature entrance stairs		item	100,000		
Allowance for stairs/cat ladders and safety rails to plant rooms		item	30,000		
External walls				7,867,000	369.34
Unitized curtain walling; solid spandrel panels; selective high performance glass and some solar control	8,600	m²	825		
Aluminium screening to plant enclosures	450	m²	480		
Glass entrance canopies; cantilevered from building	250	m²	1000		
Extra for louvres		item	50,000		
Blockwork walls at roof level, including wind posts	60	m²	100		
Allowance for visual mock-ups and performance tests		item	250,000		
External windows and doors				256,000	12.02
Single and double doors, including disabled pass doors		item	55,000		
Extra over cladding for revolving doors	2	nr	60,000		
Extra over screen enclosures for single and double doors		item	15,000		

CENTRAL LONDON OFFICES

Shell and Core Works	Quantity	Unit	Cost (£)	Total (£)	Rate (£/m²)
Steel roller shutter to loading bay and car park	2	nr	18,000		
Metal doors in service areas		item	30,000		
Internal walls and partitions				**1,733,400**	**81.38**
In situ concrete walls in basement, etc.	540	m²	160		
Fairfaced blockwork walls at basement, ground and roof levels	3,500	m²	80		
Curved blockwork entrance feature wall	300	m²	180		
Drylined core walls	6,950	m²	90		
Extra for double thickness drylined core walls	1,000	m²	90		
Other walls/partitions to plant areas, additional walls		item	160,000		
Glazed screen to shopfronts	70	m²	850		
Veneer faced WC cubicles/doors; access panelling	90	nr	4,200		
Internal doors				**420,800**	**19.76**
Single timber doors	140	nr	1,700		
Double timber doors	30	nr	2,800		
Profilex riser doors	35	nr	1,250		
Other doors: plantrooms; additional access door hatches		item	55,000		
Wall finishes				**1,065,400**	**50.02**
Stone cladding to main entrance lobby	880	m²	380		
Back-lit glass panelling on steel frame in main entrance lobby	150	m²	1,100		
Paint to fair face block walls	2,150	m²	7		
Plaster and paint to blockwork/concrete	3,820	m²	15		
Skim coat and paint to drylined walls	1,700	m²	8		
Stone cladding to toilets	450	m²	310		
Granite cladding to lift lobbies	800	m²	330		
Lift architraves		item	75,000		
Floor finishes				**821,700**	**38.58**
Granite/stone tiles to main entrance lobby and lift lobbies	1,250	m²	330		
Stone tiles to toilets including, waterproofing, screed; skirtings	440	m²	360		
Lightweight screed to circulation and core areas	1,280	m²	36		
Sealant/hardener to car park, loading bay and plant rooms	1,140	m²	85		
Vinyl flooring to security areas		item	7,800		
Entrance mats and matwells		item	45,000		

CENTRAL LONDON OFFICES

Shell and Core Works	Quantity	Unit	Cost (£)	Total (£)	Rate (£/m²)
Allowance for white lining to carpark and loading bay		item	25,000		
Allowance for other floor finishes		item	30,000		
Ceiling finishes				**653,800**	**30.69**
GRG feature ceiling to main entrance lobby	870	m²	360		
Feature drylined ceiling to lift lobbies	380	m²	200		
Metal tile suspended ceilings to toilets	440	m²	80		
Painted plasterboard on metal framing to corridors etc.	840	m²	70		
Insulation to car park/loading bay soffits	1,030	m²	20		
Access panels, bulkheads; detailing; sundry ceiling finishes		item	150,000		
Fittings/fitting out (excludes loose furniture)				**573,000**	**26.90**
Main entrance reception desk and security desks		item	100,000		
Stone vanity tops in toilets for basins/taps with mirrors behind	70	m	1,900		
Soap dispensers/tanks, roll holders, paper towels etc.	90	nr	550		
Extra for fittings to disabled toilets	10	nr	1,550		
Rubbish compactor		item	25,000		
Column guards, bollards/crash rails to loading bay/car park, cycle racks, traffic management, statutory signage		item	250,000		
Sanitary appliances				**110,000**	**5.16**
WCs, basins, cleaners sinks, urinals (average rate per point)	200	nr	500		
Extra for disabled toilets	10	nr	1,000		
Disposal installations				**2,65,600**	**12.47**
Rainwater disposal system	21,300	m²	3		
Soil waste and vent installation	21,300	m²	8		
Extra for drainage to retail areas		item	10,000		
Condensate drainage	21,300	m²	1		
Water installations				**372,200**	**17.47**
Cold water services: incoming, storage, pumps, distribution	21,300	m²	9		
Hot water heaters and distribution	21,300	m²	2		
Supplies to mechanical systems, basement/plant	21,300	m²	3		
Water services for vending area	21,300	m²	2		
Supply to retail areas		item	25,000		

CENTRAL LONDON OFFICES

Shell and Core Works	Quantity	Unit	Cost (£)	Total (£)	Rate (£/m²)
Space heating and air treatment				**2,159,700**	**101.39**
Gas installation		item	20,000		
Boilers		item	80,000		
Air handling units	21,300	m²	10		
Chillers	21,300	m²	10		
Heat rejection plant	21,300	m²	8		
LTHW heating installation including pumps and boiler flues	21,300	m²	18		
Air conditioning installation including fans and ductwork	21,300	m²	16		
CHW installation including pumps and riser pipework	21,300	m²	17		
Condenser water installation	21,300	m²	12		
Metering LTHW/CHW installations	21,300	m²	4		
Ventilation installation				**631,100**	**29.63**
Toilet and smoke extract ventilation	21,300	m²	14		
Ventilation to plant room, lift motor rooms, refuse area, etc.		item	47,500		
Car park and basement ventilation	21,300	m²	7		
Stair and lobby pressurization	21,300	m²	6		
Electrical installation				**1,705,500**	**80.07**
HV Switchgear and transformer	21,300	m²	8		
LV distribution; busbars	21,300	m²	25		
Power to mechanical plant	21,300	m²	3		
Small power installation	21300	m²	5		
Lighting, emergency lighting, including basement and car park	21,300	m²	18		
Earthing and bonding	21,300	m²	3		
Enhanced lighting in lobby and other areas		item	50,000		
External building lighting		item	170,000		
Standby power installation, including oil system		item	150,000		
Lifts				**1,700,000**	**79.81**
Passenger lifts, 21 person serving 10 floors with enhanced finishes	6	nr	200,000		
Goods lift serving 14 floors	1	nr	180,000		
Car park lift	1	nr	160,000		
Fire fighting lift	1	nr	160,000		

CENTRAL LONDON OFFICES

Shell and Core Works	Quantity	Unit	Cost (£)	Total (£)	Rate (£/m²)
Protective installations				**511200**	**24.00**
Sprinkler Installations; tanks, pumps, risers etc.	21,300	m²	20		
Dry riser installation	21,300	m²	2		
Lightning protection	21,300	m²	2		
Communication installations				**480,800**	**22.57**
Fire alarm installations	21,300	m²	15		
Containment for BMS, security, data, etc.	21,300	m²	2		
Landlord security provisions	21,300	m²	4		
Disabled alarms		item	25,000		
Special installations				**827,300**	**38.84**
Building management system	21,300	m²	20		
Leak detection system	21,300	m²	1		
Allowance for facade cleaning equipment		item	380,000		
Builder's work				**405,000**	**19.01**
Builder's work in connection with services installations, including machine bases,	21,300	m²	19		
Preliminaries and contingency				**7,307,000**	**343.05**
Contractor's overheads and profit, site establishment and supervision @	16%				
Contingency @	5%				
Construction cost (Shell and core works only; rate based on GIFA)				**40,827,000**	**1,916.73**

Category A Works	Quantity	Unit	Cost (£)	Total (£)	Rate (£/m²)
Wall finishes				**109,300**	**5.13**
Emulsion paint finish to office side of core walls	1,770	m²	5		
Column casings, including paint, subframe, etc.	1,180	m²	85		
Floor finishes				**598,300**	**28.09**
Dust sealer to concrete slabs	15,340	m²	1		
Medium grade fully accessible raised floor, metal faced plycore; 150 nominal depth; including fire barriers	14,600	m²	38		
Ceiling finishes				**721,000**	**33.85**
Concealed grid metal tray suspended ceiling to office areas; acoustic quilt and fire breaks	15,340	m²	47		
Fittings/fitting out				**15,300**	**0.72**
Statutory signage	15,340	m²	1		
Space heating and air treatment				**2,089,300**	**98.09**
Four pipe fancoil units	15,340	m²	20		

CENTRAL LONDON OFFICES

Category A Works	Quantity	Unit	Cost (£)	Total (£)	Rate (£/m²)
Distribution ductwork, grilles etc.	15,340	m²	50		
CHW installation; insulation	15,340	m²	31		
LTHW installation; insulation	15,340	m²	25		
Condensate installation; insulation	15,340	m²	10		
Electrical installations				**1,233,300**	**57.90**
Lighting and emergency lighting installation	15,340	m²	60		
Distribution boards	15,340	m²	5		
Earthing and bonding	15,340	m²	3		
Lighting control	15,340	m²	12		
Protective installations				**306,800**	**14.40**
Sprinkler protection to offices	15,340	m²	20		
Communications installations				**202,500**	**9.51**
Fire alarm installation	15,340	m²	13		
Special installations				**306,800**	**14.40**
Building management system	15,340	m²	20		
Builder's work in connection				**78,200**	**3.67**
Builder's work in connection with Category A services	15,340	m²	5		
Preliminaries and contingency				**928,200**	**43.58**
Contractor's overheads and profit, site establishment and supervision @	13%				
Contingency @	3%				
Construction cost (Category A only; rate based on GIFA)				**6,589,000**	**309.34**

OFFICE FIT-OUT

The cost model is based on a notional scheme that is typical for a West End based professional firm. The design is of a high specification, with a mixture of open-planned and cellular office use. It should be noted that the cost of internal finishes can vary considerably and is dependent on a client's aspirations. The cost model assumes that the floor space has been taken from shell and core with a negotiated landlord contribution for the Category A installation.

The office fit-out model is presented as a cost breakdown by functional area rather than in a traditional elemental format. The fit-out is designed to provide the following areas:

Open plan/circulation areas	1,300 m²	Tea points	40 m²
Client meeting rooms	280 m²	Staff dining	150 m²
Client reception	80 m²	Kitchen	60 m²
Cellular offices	1,000 m²	Communications/data rooms	60 m²
Staff meeting rooms	500 m²	Ancillary areas	175 m²
Total usable area	3,645 m²		

Model location is London   TPI = 480; LF = 1.06

This updated cost model is copyright of Davis Langdon and was originally published in Building Magazine in Nov-11

Office Fit-Out	Quantity	Unit	Rate	Total (£)	Cost (£/m²)
Open plan/circulation				**766,000**	**589.23**
Medium grade raised access floor 150 mm high	1,300	m²	40		
Carpet tiles	1,300	m²	25		
Suspended perforated metal ceiling	1,300	m²	70		
Emulsion paint to core walls		item	10,000		
Mechanical and electrical services; including heating, air treatment, ventilation, power, lighting, protective installations, communications, special installations	1,300	m²	380		
Structured cabling	1,300	m²	35		
Roller blinds to perimeter windows	1,300	m²	20		
Statutory signage		item	2,000		
Builders work in connection with services	1,300	m²	10		
Client meeting rooms				**801,800**	**2,863.57**
Acoustic rated metal stud partitions, 60 minute fire-rated, hardwood skirting and ply strengthening	170	m	270		
Double glazed partitions, manifestation	50	m	900		
Folding wall; acoustic finish	15	nr	2,000		
American Walnut doors	15	nr	2,250		
Medium grade raised access floor 150 mm high	280	m²	40		
Carpet tiles	280	m²	40		
Suspended perforated metal ceiling with feature raft	280	m²	100		

OFFICE FIT-OUT

Office Fit-Out	Quantity	Unit	Rate	Total (£)	Cost (£/m²)
Fabric wall coverings	900	m²	270		
Mechanical and electrical services; including heating, air treatment, ventilation, power, lighting, protective installations, communications, special installations	280	m²	600		
Structured cabling	280	m²	30		
Blackout blinds to perimeter windows	280	m²	50		
Statutory signage		item	2,250		
Branding and corporate signage		item	10,000		
Specialist joinery including credenzas		item	150,000		
Builders work in connection with services	280	m²	10		
Client reception				**212,400**	**2,655.00**
Double glazed door	1	nr	4000		
Natural stone flooring on boarded substrate	80	nr	270		
Painted plasterboard suspended ceilings including bulkheads, access panels and ceiling feature	80	m²	120		
Timber panel and polished plaster wall finishes		item	3,0000		
Mechanical and electrical services; including heating, air treatment, ventilation, power, lighting, protective installations, communications, special installations	80	m²	600		
Structured cabling	80	m²	30		
Statutory signage		item	1,000		
Branding and corporate signage		item	20,000		
Specialist joinery including reception desk		item	75,000		
Builders work in connection with services	80	m²	10		
Cellular offices				**1,174,000**	**1,174.00**
Acoustic rated metal stud partitions, 60 minute fire-rated, MDF skirting and ply strengthening	600	m	250		
Single glazed partitions, manifestation, blinds	150	m	700		
Solid core timber veneer doors	120	nr	2,000		
Medium grade raised access floor 150 mm high	1,000	m²	40		
Carpet tiles	1,000	m²	25		
Suspended perforated metal tiled ceiling	1,000	m²	70		
Emulsion paint wall finish	3,000	m²	6		
Mechanical and electrical services; including heating, air treatment, ventilation, power, lighting, protective installations, communications, special installations	1,000	m²	450		
Structured cabling	1,000	m²	35		
Roller blinds to perimeter windows	1,000	m²	25		

OFFICE FIT-OUT

Office Fit-Out	Quantity	Unit	Rate	Total (£)	Cost (£/m²)
Statutory and corporate signage		item	6,000		
Builders work in connection with services	1,000	m²	10		
Staff meeting rooms				**564,500**	**1,129.00**
Acoustic rated metal stud partitions, 60 minute fire-rated, MDF skirting and ply strengthening	320	m	250		
Double glazed partitions, manifestation	80	m	900		
Solid core timber veneer doors	15	nr	2,000		
Medium grade raised access floor 150 mm high	500	m²	40		
Carpet tiles	500	m²	25		
Suspended perforated metal tiled ceiling	500	m²	70		
Emulsion paint wall finish	1,750	m²	6		
Mechanical and electrical services; including heating, air treatment, ventilation, power, lighting, protective installations, communications, special installations	500	m²	450		
Structured cabling	500	m²	30		
Roller blinds to perimeter windows	500	m²	25		
Statutory and corporate signage		item	2,000		
Specialist joinery		item	45,000		
Builders work in connection with services	500	m²	10		
Tea points				**79,600**	**1,990.00**
Standard metal stud partitions, 60 minute fire-rated, MDF skirting and ply strengthening	40	m	200		
Medium grade raised access floor 150 mm high including tanking to slab	40	m²	60		
Slip-resistant floor finish on two layers of boarding	40	m²	60		
Suspended perforated metal tiled ceiling	40	m²	70		
Emulsion paint wall finish and ceramic splashbacks		item	3,500		
Mechanical and electrical services; including disposal, water installations, zip taps, heating, air treatment, transfer ventilation, power, lighting, protective installations, communications, leak detection, special installations	40	m²	500		
Statutory and signage		item	100		
Joinery units, work surfaces and equipment		item	40,000		
Builders work in connection with services	40	m²	10		
Staff dining				**189,500**	**1,263.33**
Standard metal stud partitions, 60 minute fire-rated, MDF skirting and ply strengthening	40	m	200		
Medium grade raised access floor 150 mm high including tanking to slab	150	m²	40		

OFFICE FIT-OUT

Office Fit-Out	Quantity	Unit	Rate	Total (£)	Cost (£/m²)
Porcelain floor tiles on two layers of boarding and ditra matting	150	m²	150		
Suspended perforated metal tiled ceiling	150	m²	100		
Emulsion paint wall finish and ceramic splash backs		item	12,000		
Mechanical and electrical services; including disposal, water installations, zip taps, heating, air treatment, transfer ventilation, power, lighting, protective installations, communications, leak detection, special installations	150	m²	470		
Statutory and signage		item	4,000		
Servery and specialist joinery item		item	50,000		
Builders work in connection with services	150	m²	10		
Kitchen				**199,200**	**3,320.00**
Standard metal stud partitions, 60 minute fire-rated, MDF skirting and ply strengthening	50	m	200		
Metal pivot door	1	nr	3000		
Medium grade raised access floor 150 mm high including tanking to slab	60	m²	60		
Slip-resistant floor finish on two layers of boarding and ditra matting	60	m²	60		
Hygienic finished ceiling	60	m²	100		
Altro Whiterock wall finish	150	m²	130		
Mechanical and electrical services; including disposal, water installations, zip taps, heating, air treatment, ventilation, dedicated extract, power, lighting, protective installations, communications, leak detection, special installations	60	m²	875		
Statutory signage		item	1,000		
Stainless steel kitchen units, worktops and equipment		item	100,000		
Builders work in connection with services	60	m²	25		
Communications/data rooms				**172,200**	**2,870.00**
Standard metal stud partitions, 90 minute fire-rated, MDF skirting and ply strengthening	40	m	230		
Solid timber painted doors; 90 mins. fire-rated	1	nr	1,750		
Heavy duty raised access floor 150 mm high with anti-static floor finish	60	m²	120		
Dust sealant to concrete soffit	60	m²	15		
Emulsion paint wall finish	120	m²	6		

OFFICE FIT-OUT

Office Fit-Out	Quantity	Unit	Rate	Total (£)	Cost (£/m²)
Mechanical and electrical services; including air treatment, ventilation, stand alone air conditioning units, electrical, power, lighting, protective installations, communications, special installations, leak detection	60	m²	2,500		
Structured cabling	60	m²	35		
Statutory signage		item	100		
Builders work in connection with services	60	m²	10		
Ancillary areas				**197,200**	**1,126.86**
(including stores, reprographics, mail room etc.)					
Standard metal stud partitions, 60 minute fire-rated, MDF skirting and ply strengthening	150	m	200		
Solid timber painted doors	10	nr	1,500		
Medium grade raised access floor 150 mm high	175	m²	35		
Sheet vinyl flooring including boarding substrate	175	m²	40		
Suspended perforated metal tiled ceiling	175	m²	70		
Emulsion paint wall finish	800	m²	6		
Mechanical and electrical services; including disposal, water installations, heating, air treatment, ventilation, power, lighting, protective installations, communications, special installations	175	m²	400		
Statutory signage		item	250		
Fixed fittings and shelving		item	50,000		
Builders work in connection with services	175	m²	10		
Preliminaries and contingencies				**584,600**	**160.38**
Main Contractor's management, site establishment and site supervision @	6%				
Main Contractor's overhead & profit @	2%				
Design reserve @	5%				
TOTAL				**4,941,000**	**1,355.56**

SUSTAINABLE OFFICE REFURBISHMENT

The cost breakdown is based on the major refurbishment of a 10,000 m², 4 storey city centre scheme, involving the stripping back of a 1960's building to its original frame.

The scope of the refurbishment includes structural alterations, new facades which facilitate a cross-ventilation strategy and a complete reconstruction of the interior fit-out and building services. Other aspects of the work include the construction of a roof level plant platforms and green roofs.

Demolitions are included in the costs. External works, furniture and fittings, professional fees and VAT are excluded.

Gross internal floor area	10,000 m²
Model location is South East	TPI = 462; LF = 1.02

This updated cost model is copyright of Davis Langdon and was originally published in Building Magazine in Oct−09

Sustainable Office Refurbishment	Quantity	Unit	Cost (£)	Total (£)	Rate (£/m²)
Demolitions and alterations				2,024,000	202.40
Cladding removal	5,800	m²	65		
Allowance for soft strip	10,000	m²	32		
Allowance for asbestos removal	10,000	m²	65		
Structural alterations including alterations to lift/ door openings, works to new and existing risers and upstands, new steps on stairs to accommodate new raised floor level etc.	10,000	m²	55		
Repairs to existing concrete frame, exposed soffit and columns	10,000	m²	13		
Roof				454,200	45.42
Replace existing mastic asphalt covering on existing concrete slab including insulation, vapour barrier, gravel topping, paviors, solar reflective paint, flashings, roof access hatches, pigeon nets and spikes, rainwater outlets and accessories	2,800	m²	150		
Extra over for green roof; substrate and planting	950	m²	37		
Stairs				125,400	12.54
New balustrade and handrails to existing core stairs; repairs to stairs	18	nr	6,300		
Sundry steps and ramps		item	12,000		
External walls, windows and doors				3,107,200	310.72
Brickwork; single skin on brickwork angle support system; flashing; trims; weather board and insulation	580	m²	210		
Solid cladding, aluminium panels, composite system with aluminium cladding on weather board within timber frame; insulation: flashings; accessories	3,200	m²	480		

SUSTAINABLE OFFICE REFURBISHMENT

Sustainable Office Refurbishment	Quantity	Unit	Cost (£)	Total (£)	Rate (£/m²)
Glazing, composite system with aluminium sealed double glazed units within timber frame; opening lights; flashings; accessories	1,650	m²	525		
Extra over for actuator control to opening lights		item	270,000		
Solar shading; vertical and horizontal; aluminium aerofoil sections		item	90,000		
Aluminium plant screening including framing	425	m²	525		
Internal walls and partitions				509,700	50.97
Internal plasterboard walls	7,000	m²	60		
Extra for make up acoustic ventilation grilles detail to cellular spaces		item	55,000		
Skirtings; MDF; primed	5,500	m	6.3		
Internal Doors				453,000	45.30
Flush doorsets; solid; softwood frames; ironmongery	300	nr	1,150		
Extra over for motorized door devices	40	nr	2,600		
Extra over for glazed side screens to meeting rooms	25	nr	160		
Wall finishes				172,100	17.21
Independent wall lining; insulation; plasterboard lining to inner face of external walls	3,200	m²	42		
Painting	10,200	m²	3.7		
Floor finishes				740,400	74.04
Sand cement screed; average 75 mm thick, steel fabric reinforcement, 100 thick insulation and separating layer	2,650	m²	20		
Raised access floor	7,200	m²	39		
Extra for ramps		item	11,000		
Carpet	8,850	m²	32		
Vinyl floor; plywood	600	m²	90		
Reconstituted stone tiling to reception	400	m²	140		
Floor paint to plant rooms	150	m²	15.9		
Ceiling finishes				297,200	29.72
Suspended ceilings; plasterboard on MF; including access panels	2,000	m²	85		
Decoration and making good of exposed soffits	8,000	m²	15.9		
Furniture and fittings				519,100	51.91
Allowance for general joinery and fittings		item	55,000		
Kitchenette fit-out	14	nr	6,900		
Canteen fit-out, including servery and catering equipment		item	330,000		

SUSTAINABLE OFFICE REFURBISHMENT

Sustainable Office Refurbishment	Quantity	Unit	Cost (£)	Total (£)	Rate (£/m²)
Allowance for furniture and fit-out mock ups		item	37,500		
Sanitary fittings and disposal installations				**571,500**	**57.15**
Toilet core fit-out, WCs, urinals, wash hand basins; vanity units; handryers and mirrors, showers and shower cabinets; 190 fittings; overall rate for each WC block	14	nr	19000		
Allowance for cubicles and back panels	70	nr	1650		
Soil, waste and disposal: rainwater disposal; cast iron downpipes and fittings	10,000	m²	19		
Water installations				**240,000**	**24.00**
Hot and cold water service; hot and cold water storage; distribution	10,000	m²	24		
Heat source				**137000**	**13.70**
Gas fired boilers, flue, pumps, heat exchanger	10,000	m²	13.7		
Space heating and air treatment				**2,279,000**	**227.90**
LTHW installation to plant and risers, on floor distribution and radiators, insulation	10,000	m²	105		
Hot and cold water supply to plant	10,000	m²	8.9		
Dedicated cooling systems, VRV/DX installation	10,000	m²	34		
Ventilation; air handling plant and ductwork; floor grilles and diffusers; toilet and kitchen extract	10,000	m²	80		
Electrical installations				**2,030,000**	**203**
LV Switchgear and panels, distribution boards, power to main plant and lifts	10,000	m²	33		
Small power installation, Busbar and floor grommets. Power supply to actuators. Allowance for containment to security and power	10,000	m²	95		
Office lighting, high level exposed luminaires including emergency lighting, lighting control, PIR sensors and daylight sensors	10,000	m²	75		
Gas installations				**21,000**	**2.10**
Revised gas distribution including new supply to kitchen	10,000	m²	2.1		
Lift installations				**375,000**	**37.50**
13 person, electro-hydraulic lift serving 4 nr floors; 1.6 m/s	5	nr	75,000		
Protective installations				**21,000**	**2.10**
Earthing and bonding, lightning protection	10,000	m²	2.1		
Communications installations				**804,000**	**80.40**
Fire and smoke detection and alarm system, security installation	10,000	m²	48		

SUSTAINABLE OFFICE REFURBISHMENT

Sustainable Office Refurbishment	Quantity	Unit	Cost (£)	Total (£)	Rate (£/m²)
Disabled refuge alarm, disabled toilet alarm, induction loop	10,000	m²	5.4		
Data cabling	10,000	m²	27		
Special installations				**700,000**	**70.00**
Building management system	10,000	m²	70		
Builder's work in connection				**358,900**	**35.89**
Forming holes etc. @	5%				
Preliminaries and contingencies				**2,805,350**	**280.54**
Contractor's overheads, site establishment and site supervision @	12%				
Design reserve @	5%				
Construction cost (rate based on GIFA)				**18,746,000**	**1,874.61**

OUT OF TOWN RETAIL UNIT

A 6,500 m² (about 70,000 sq ft) non-food retail unit on an existing retail park in the South East. The unit is comprised of two full height levels which will give the operator the flexibility to trade over both floors without having to compromise on a reduced head height that is often encountered on mezzanine levels of retail park units. The cost model is based on a shell only specification and allows for the construction of the retail unit whilst maintaining continuous trade for all other tenants within the retail park. The overall programme duration is based on 18 weeks.

Gross internal floor area   6,500 m²

Model location is South East England   TPI = 462; LF = 1.02

This updated cost model is copyright of Davis Langdon and was originally published in Building Magazine in Jul-12

Shell and Core	Quantity	Unit	Cost (£)	Total (£)	Rate (£/m²)
Substructure				**511,000**	**78.62**
Excavation, disposal, underground drainage, backfill		item	60,000		
Concrete pad foundations	64	nr	1,500		
Reinforced concrete ground floor slab, 250 mm thick including gas and vapour barrier, edge beams	3,250	m²	100		
Lift and escalator pits	4	nr	7,500		
Frame and upper floors				**827,500**	**127.31**
Steel propped portal frame, cold rolled purlins, surface treatments (@ 60 kg/m²)	390	tn	1,500		
Secondary steelwork and signage framework		item	20,000		
Holorib and concrete upper floor slab 150 mm thick	3,250	m²	50		
Precast concrete stairs and metal balustrades to cores	4	nr	15,000		
Roof				**443,100**	**68.17**
Standing seam metal roof, insulated roof	2,844	m²	65		
Feature rooflight, clerestory glazed		item	60,000		
Concrete plant deck, inverted roof including waterproofing and ballast	540	m²	80		
Sundry roof works, flashings, forming openings etc.		item	90,000		
Roof parapet screens		item	30,000		
Handrails and GRP designated maintenance walkways		item	35,000		
External walls, windows and doors				**1,353,600**	**208.25**
Structural glazing system to front and side elevations, planar type system, low 'G' value, low iron glass, glazed fins and capless mastic joints, stainless steel trims to perimeter framing	830	m²	650		
Acrylic render system, including insulation and blockwork, to stair cores and soffits of colonnade	740	m²	150		
Punched windows to office areas	9	nr	1,000		
Framed curtain walling to stair cores	130	m²	300		

OUT OF TOWN RETAIL UNIT

Shell and Core	Quantity	Unit	Cost (£)	Total (£)	Rate (£/m²)
Half round profiled composite cladding, including block work inner skin	1,270	m²	110		
Granite cladding system to columns	336	m²	400		
Granite cladding system to feature entrance columns	166	m²	575		
Brise soleil, aluminium fins to glazed facades		item	150,000		
Entrance lobby, glazed sides and 2 nr sets of double plus single door and security screens		item	70,000		
Engineering brick base to facades	105	m²	100		
Screens to sprinkler tanks and plant		item	30,000		
Roller shutter doors to service yard		item	10,000		
Fire escape doors		item	15,000		
Internal walls and partitions				**30,000**	**4.62**
Sound reducing board and metal stud partitions; generally 2 layers of board each side; various levels of fire and sound insulations; part glazed as appropriate	500	m²	60		
Internal doors				**12,000**	**1.85**
Internal doorsets to stair cores; solid core; including vision panels and overpanels; painted finish; ironmongery		item	5500		
Disposal installations				**30,000**	**4.62**
Rainwater disposal; downpipes and fittings		item	30,000		
Water installations				**30,000**	**4.62**
Hot and cold water supply to WC, kitchen		item	30,000		
Electrical and gas installations				**85,000**	**13.08**
Incoming electrical services including new substation and termination cubicle		item	80,000		
Incoming gas supply		item	5,000		
Protective, communications and special installations				**115,000**	**17.69**
Lightning protection, earthing and bonding		item	5,000		
Sprinkler tanks and pump set (internal pipework and sprinkler heads part of fit-out)		item	110,000		
Builder's work				**25,000**	**3.85**
Builder's work in connection with services		item	25,000		
Preliminaries and contingency				**608,800**	**93.66**
Management costs; site establishment; site supervision @		12%	3,462,110		
Contingency @		5%			
Construction cost (rate based on GIFA)				**4,071,000**	**626.34**

CAR SHOWROOM

A 1,200 m² building of which 580 m² is workshop, 420 m² is showroom and 200 m² is office space. The level of specification is targeted at a mid-level car dealership built to achieve a BREEAM Good rating. There is a reinforced concrete ground bearing slab, full height glazing to the showroom, microrib wall cladding generally and a standing seam roof.

Gross internal floor area	1,200 m²
Model location is South East England	TPI = 462; LF = 1.02

This updated cost model is copyright of Davis Langdon and was originally published in Building Magazine in Mar–11

Car Showroom	Quantity	Unit	Cost (£)	Total (£)	Rate (£/m²)
Substructure				**104,300**	**86.92**
Excavation and disposal off site	1,200	m³	16		
Reinforced concrete ground slab, hardcore, dpm, ground beams and column bases for steel frame (0.25 W/m².K)	1,000	m²	80		
Power floated and hardener to workshop	580	m²	9		
Frame, upper floors and stairs				**132,800**	**110.67**
Steel propped portal frame, cold rolled purlins, surface treatments (@ 50 kg/m²)	60	tonne	1,650		
Intumescent paint to give 30 minute fire protection to steelwork	60	tonne	330		
Staircase handrails and balustrades	2	nr	5,900		
Allowance for miscellaneous works, protecting columns		item	2,150		
Roof				**128,700**	**107.25**
Built up aluminium insulated roof cladding including all barge boards, trim etc. (0.20 W/m².K)	1,260	m²	85		
Extra for rooflights (10% of total roof area)	95	m²	70		
Mansafe system	80	m	90		
Rainwater drainage, aluminium gutters and downpipes	70	m	80		
External wall, windows and doors				**310,700**	**258.92**
Built up aluminium insulated wall cladding (0.25 W/m².K)	610	m²	85		
2.5 m high inner leaf of 140 thick fairface blockwork	400	m²	37		
3000 × 4600 high steel insulated sectional overhead doors (1.5 W/m².K)	2	nr	3,800		
Aluminium single entrance doors	5	nr	1,300		
Coated aluminium double glazed window system to showroom	300	m²	650		

Building Cost Models

CAR SHOWROOM

Car Showroom	Quantity	Unit	Cost (£)	Total (£)	Rate (£/m²)
Aluminium louvres to glass frontage 1500 projection; including support and brackets:	90	m²	320		
Steel fire escape doors	3	nr	975		
Canopy over entrance	1	nr	3,250		
Internal walls and partitions and doors				**66,700**	**55.58**
Plasterboard and metal stud partitions; generally 2 layers of plasterboard each side; various levels of fire and sound insulations; part glazed as appropriate	450	m²	80		
140 thick fairface blockwork to workshop/ showroom wall	245	m²	37		
Fireproofing between partitions and roof		item	2,150		
Laminated faced internal doorset with softwood frames and ironmongery; including fire-resisting where necessary	15	nr	1300		
Wall finishes				**4,700**	**3.92**
Emulsion paint to blockwork wall surfaces generally	1,370	m²	3		
Ceramic wall tiles splashbacks to WC area	5	m²	55		
Floor finishes				**42,400**	**35.33**
Screed and non-slip vinyl sheeting to WC areas	15	m²	55		
Screed and ceramic floor tiles to showroom areas, including skirtings	15	m²	80		
Screed and carpet tiling to office and circulation areas	180	m²	37		
Entrance matwell	5	m²	270		
Ceiling finishes				**7,700**	**6.42**
Plasterboard and skim to offices, WC etc. with ceiling grid and paint finish	180	m²	42		
Sanitary appliances				**3,200**	**2.67**
WC suite, including disabled facility including all sanitary and fittings	3	nr	1050		
Disposal installations				**700**	**0.58**
Waste, soil and vent installation; uPVC pipework and fittings	3	nr	220		
Water installations				**1,400**	**1.17**
Hot and cold water supplies to WCs and kitchen	5	nr	270		
Space heating and air treatment and ventilation				**74,500**	**62.08**
Gas fired VRV comfort cooling system, condensers, air handling unit, ductwork, ceiling void mounted fan coil units	620	m²	80		
Workshop heating; ambirad type heaters	580	m²	43		

CAR SHOWROOM

Car Showroom	Quantity	Unit	Cost (£)	Total (£)	Rate (£/m²)
Electrical and gas installations				**102,400**	**85.33**
Small power, basic and emergency lighting	1,200	m²	27		
Showroom lighting, including all supply and installation of luminaires	420	m²	55		
Supply to WC for ventilation, heater etc.	5	nr	1750		
Task lighting to workshop to 300 Lux	580	m²	27		
External building lighting generally	1,200	m²	5		
Allowance for incoming, electrical, gas and water services		item	16,000		
Protective, communications and special installations				**31,800**	**26.50**
Lightning protection, earthing and bonding	900	m²	2		
Sprinkler installation	1,200	m²	16		
Fire and intruder alarms, panic alarm buttons	900	m²	12		
Lift installation				**16,000**	**13.33**
4 person hydraulic lift		item	16,000		
Builder's work in connection				**1,600**	**1.33**
Forming holes and chases etc.		item	1,550		
Preliminaries and contingency				**192,400**	**160.33**
Overheads, profit, site establishment and site supervision @	13%				
Contingency @	5%				
Construction cost (Shell only; rate based on GIFA)				**1,222,000**	**1,018.33**

HEALTH CENTRE

This cost model details a large joint service centre scheme housing GP, dentistry and social services in a single building located on a tight urban site. The accommodation included 60 consulting rooms. Internal circulation is a key design aspect and the building design is based on two wings of largely cellular space arranged around an enclosed 'street', providing space for reception, cafes and other public facilities. There is extensive service instal-lation which included data infrastructure.

Gross internal floor area, including tiers	8,435 m²
Model location is Outer London	TPI = 453; LF = 1.00

This updated cost model is copyright of Davis Langdon and was originally published in Building Magazine in Oct–05

Health Centre	Quantity	Unit	Cost (£)	Total (£)	Rate (£/m²)
Substructure				**845,500**	**100.24**
450 mm dia. reinforced concrete piling	1,955	m²	150		
Excavating basement; disposal off site; breaking out obstructions; dewatering	1,900	m³	45		
Reinforced concrete slab to basement area (565 m²) and ground floor with vapour barrier, insulation; underslab ducts and drainage	1,955	m²	150		
Extra for formation of lift pits		item	16,000		
Reinforced concrete retaining walls to basement area, including temporary sheet piling; blockwork lining wall	450	m²	350		
Frame and upper floors				**1,349,600**	**160.00**
Reinforced concrete frame, 7.2 × 7.2 m grid; 200 mm thick flat slab; in situ concrete shear walls	8,435	m²	160		
Roof				**201,400**	**23.88**
Inverted roof coverings, polymeric roof coverings, insulation; flashings and copings	1,955	m²	75		
Mansafe system		item	14,000		
Extra for polycarbonate rooflights; 800 dia.; including all flashings etc.	15	nr	1,050		
Allowance for entrance canopies		item	25,000		
Stairs				**325,600**	**38.60**
Precast reinforced concrete stair cases; including fins smooth finish to exposed surfaces	6	nr	9,000		
Balustrades; 1.1 m average high; glazed with profiled timber handrails; average rate used	665	m	370		
Profiled timber handrails	150	m	110		
Miscellaneous metalwork; cat ladders; open mesh flooring in risers etc.		item	9,000		

HEALTH CENTRE

Health Centre	Quantity	Unit	Cost (£)	Total (£)	Rate (£/m²)
External walls, windows and doors				**1,437,300**	**170.40**
Multi-coloured curtain walling; double-glazed units; composite cladding panels; secondary steel and internal liner panel	2,520	m²	300		
Coloured render system on mesh, insulation, dpm and metsec backing wall	1,260	m	200		
Aluminium framed windows	95	m²	340		
Perforated clay tile rain screen cladding complete including vertical and horizontal support rails	1,300	m²	180		
Entrance door unit; aluminium framed glazed doors; automatic operation	2	nr	15000		
External doors; aluminium framed; polyester coated; to match windows	10	nr	2900		
Freestanding screens to roof plant; louvre panels as required	650	m	160		
Internal walls and partitions				**550,500**	**65.26**
Concrete blockwork; head restraints, movement joints	2,000	m²	42		
Plasterboard partitions; 2 layers of 12.5 mm wall board; including 25 mm insulation and skim coat finish	7,350	m²	50		
Internal glazed screens; blinds	225	m²	440		
Internal doors				**397,500**	**47.13**
Flush doors; beech veneer; fire-rated; solid core; veneered frame; vision panel and ironmongery	325	nr	800		
Glazed doors and screens; fire-rated; frames, fittings and ironmongery	22	nr	2800		
Riser cupboard doors; beech veneer; fire-rated, fittings and ironmongery	66	nr	1150		
Wall finishes				**160,900**	**19.08**
Emulsion paint to wall surfaces generally	17,250	m²	3		
Single layer plasterboard dry lining 12.5 mm thick	5,150	m²	14		
Ceramic wall tiling 150 × 150 mm	900	m²	45		
Floor finishes				**459,600**	**54.49**
Screed; latex self levelling screed 15 to 25 mm thick	8,435	m²	11		
Ceramic floor tiling 600 × 600 mm to ground floor atrium	435	m²	90		
Non-slip vinyl sheet flooring; skirtings	315	m²	34		
Linoleum sheet; skirtings	1,100	m²	38		
Carpet; softwood skirtings	5,500	m²	39		

HEALTH CENTRE

Health Centre	Quantity	Unit	Cost (£)	Total (£)	Rate (£/m²)
Carpet and nosing to stair cases	4	m²	10,000		
Allowance for entrance barrier matting		item	19,000		
Ceiling finishes				256,900	30.46
Suspended ceilings; 600 mm × 600 mm clip in metal tiles	4,170	m²	37		
Plasterboard MF suspended ceilings and bulkheads	1,280	m²	44		
Plaster to soffit of reinforced concrete slabs	1,900	m²	15		
Painting to plasterboard ceilings and bulkheads	3,180	m²	6		
Furniture and fittings				525,000	62.24
Supply and install Group 1 medical equipment (fixed furniture); fit only Group 2 and 3 medical equipment (loose furniture)		item	440,000		
Allowance for additional fittings and furniture		item	85,000		
Sanitary fittings and disposal installations				247,300	29.32
Sanitary fittings generally	200	nr	775		
Waste, soil and vent pipework	200	nr	310		
Extra for lift sump pumps		item	6,700		
Rainwater installation	8,435	m²	3		
Water installations				313,300	37.14
Cold water plant room installation; storage and booster unit	8,435	m²	6		
Hot water plant room installation; gas fired water heaters, pumps etc.	8,435	m²	10		
Hot and cold water distribution pipework, insulation	8,435	m²	17		
Allowance for water treatment		item	37,500		
Space heating and air treatment				1,424,100	168.83
Chilled water plant room installation; air cooled chiller; pumps, pipework distribution in plant rooms and risers		item	200,000		
LTHW plant room installation; boiler and flue; pumps, pipework distribution in plant rooms and risers		item	75,000		
Air handling units; total combined capacity 10 m³/s; duct mounted heating and cooling batteries		item	80,000		
Chilled water distribution; pipework; valves; insulation, trace heating	3,900	m²	50		
LTHW heating; LST panels and trench heaters; pipework; valves; insulation	8,435	m²	29		
Air curtains and door heaters; generally		item	10,000		

HEALTH CENTRE

Health Centre	Quantity	Unit	Cost (£)	Total (£)	Rate (£/m²)
Supply and extract ductwork installation serving active chilled beams	3900	m²	40		
Extra for enhanced attenuation measures in ductwork		item	15,000		
Active chilled beam installation	3,900	m²	55		
Thermal insulation to ductwork	8,435	m²	17		
Allowance for packaged cooling systems to IT rooms		item	9,000		
Allowance for dedicated ventilation systems (7 nr systems)		item	20,000		
Dirty extract system	8,435	m²	8		
Electrical and gas installations				**902,000**	**106.94**
Incoming supply and Main LV panel		item	45,000		
Submains distribution, busbars in risers, distribution boards and isolators; including electrical supplies to main plant	8,435	m²	19		
Lighting: standard luminaires, wiring, containment, accessories	8,435	m²	36		
Allowance for emergency lighting	8,435	m²	6		
Allowance for external lighting		item	25,000		
General small power, including cabling, trunking and socket outlets	8,435	m²	33		
Allowance for standby power and UPS		item	32,500		
Incoming gas supply, including steel pipework, valves etc.		item	10,000		
Lift installations				**125,000**	**14.82**
Electric traction lift; 10 person; serving 4 nr floors	2	nr	27,500		
Electric traction lift; 10 person; serving 6 nr floors	2	nr	35,000		
Protective, communications and special installations				**954,500**	**113.16**
Lightning protection; earthing installations:		item	1,9000		
Data and telephone pathway and cable infrastructure	8,435	m²	29		
Fire detection and alarm system	8,435	m²	11		
Security system; access control and intruder detection; wiring and equipment	8,434	m²	13		
Building management and automatic control systems	8,435	m²	39		
CCTV system		item	30,000		
Patient call system; LED signboards		item	65,000		
Public Address system		item	9,000		

HEALTH CENTRE

Health Centre	Quantity	Unit	Cost (£)	Total (£)	Rate (£/m²)
Induction loop system		item	18,000		
Video/display monitor system in public areas		item	13,000		
Gas scavenging systems to dental rooms		item	10,000		
Compressor and vacuum installation to dental rooms		item	12,000		
Builder's work				**99,200**	**11.76**
Builder's work in connection with services @	2.5%				
Preliminaries and contingency				**1,800,800**	**213.49**
Testing and commissioning of building services installations		item	62,500		
Contractor's preliminaries, overheads and profit @	13%				
Contractor's contingency @	3%				
Construction cost (rate based on GIFA)				**12,376,000**	**1467.24**

CARE HOME

The cost model is for a 50 bed unit housed in a two storey building of 3,000 m² gross floor area, including kitchen and communal areas. It includes traditional masonry construction, however timber frame solutions are also frequently employed at generally similar cost. The model allows for achieving Lifetime Homes, Building for Life, Design and Quality Standards and Code for Sustainable Homes Level 3 standards.

Gross internal floor area, including tiers	3,000 m²
Model location is Outer London	TPI = 453; LF = 1.00

This updated cost model is copyright of Davis Langdon and was originally published in Building Magazine in Nov–11

Care Home	Quantity	Unit	Cost (£)	Total (£)	Rate (£/m²)
Substructure				217,800	72.60
Trench fill concrete foundation, beam and block floor, insulation and screed (0.25 W/m².K)	1,556	m²	140		
Upper floors				106,800	35.60
Precast concrete beam and block upper floors, screed and insulation	1,525	m²	70		
Roof				188,500	62.83
Pitched roof with timber trusses, battens, felt and concrete tiles	1,450	m²	130		
Stairs				16,800	5.60
Main feature staircase including balustrade and finish	1	nr	1,800		
Staircase including balustrades, handrails and half landings	2	nr	7,500		
External walls, windows and doors				170,400	56.80
Traditional external wall construction of medium quality facing bricks, cavity, inner blockwork skin and insulation	1,420	m²	120		
Windows and external doors				123,200	41.07
uPVC windows and doors	365	m²	230		
Aluminium curtain walling to residents lounges	35	m²	525		
Main entrance external door, polyester powder coated aluminium	1	nr	4,000		
Powder coated aluminium external doorsets			-		
double doors	2	nr	2,700		
single doors	6	nr	1,750		
Aluminium louvre double plant room doors	1	nr	1,000		
Internal walls and partitions				119,800	39.93
Cavity wall, 2 skins of 100 mm blockwork, acoustic insulation	675	m²	55		

CARE HOME

Care Home	Quantity	Unit	Cost (£)	Total (£)	Rate (£/m²)
100 mm blockwork	1180	m²	25		
140 mm blockwork	1380	m²	31		
215 mm blockwork in lift shaft	95	m²	41		
Internal glazed screens, softwood	10	m²	650		
Internal doors				83,500	**27.83**
Hardwood veneered solid core doors					
single doors	105	nr	220		
leaf and a half doors	55	nr	625		
1 hour fire-rated hardwood veneered doors with vision panels					
single doors	40	nr	500		
double doors	8	nr	750		
Wall finishes				121,300	**40.43**
Plasterboard, skim and emulsion	6,300	m²	15		
Ceramic wall tiles to kitchens and bathrooms	725	m²	32		
Wall protector rails	325	m	13		
Floor finishes				76,700	**25.57**
Carpet to floors and staircases	1,650	m²	18		
Safety vinyl flooring with coved skirting	450	m²	37		
Laminate flooring	325	m²	26		
MDF skirting, painted	1,825	m	8		
Sundries		item	7,500		
Ceiling finishes				74,800	**24.93**
Plasterboard and skim suspended ceiling to apartments and communal areas with emulsion finish	2100	m²	27		
Moisture-resistant plasterboard and skim suspended ceiling to wet areas with emulsion finish	525	m²	29		
Access hatches	22	nr	140		
Fittings and furnishings				273,500	**91.17**
Commercial kitchen installation		item	110,000		
Kitchen fittings and appliance to kitchenettes	4	nr	3,000		
Kitchen fittings and appliances to extra care flats	8	nr	1,500		
Laundry equipment		item	45,000		
Bathroom fittings and ironmongery					
apartments	50	nr	400		
elsewhere		item	1,500		
Wardrobes	50	nr	370		

CARE HOME

Care Home	Quantity	Unit	Cost (£)	Total (£)	Rate (£/m²)
Reception desk		item	20,000		
Boxing to bathrooms	50	nr	250		
Sluice room fittings		item	7,000		
Sundry items		item	15,000		
Sanitary fittings				**218,500**	**72.83**
Apartment sanitary ware consisting of walk-in shower, shower seat, WC, wash hand basin and grab rails	50	nr	2,500		
Sanitary ware to assisted bathrooms, including bath, WC and basin	4	nr	10,000		
Sanitary ware to communal WCs	10	nr	1,250		
Sluice room equipment	6	nr	6,000		
Additional items		item	5,000		
Mechanical installations				**552,000**	**184.00**
Central heating, hot water, domestic cold water, gas distribution, ventilation, waste disposal	3,000	m²	180		
Builder's work in connection	3,000	m²	4		
Electrical				**480,000**	**160.00**
Mains supply and distribution, power and lighting, smoke detectors, door bell, immersion heater, telephone wiring, TV aerial	3,000	m²	160		
Lift installations				**63,000**	**21.00**
Lift installation including pit	1	nr	35,000		
Platform lift	2	nr	14,000		
Builder's work				**9,000**	**3.00**
Builder's work in connection	3,000	m²	3		
Preliminaries and contingency				**485,400**	**161.80**
Management costs; site establishment; site supervision @	10%		0		
Main contractor's overheads and profit @	3%		3,200,00		
Contingency @	3%		3,300,		
Construction cost (rate based on GIFA)				**3,381,000**	**1126.99**

PRIMARY SCHOOL EXTENSION

A single storey, three-classroom extension to a primary school. Constructed using traditional masonry cavity walls on concrete strip foundations, with a pitched tiled roof over. Individual classrooms are formed by load bearing blockwork partitions.

     Gross internal floor area including tiers   310 m²

     Model location is South East England   TPI = 462; LF = 1.02

This updated cost model is copyright of Davis Langdon and was originally published in Building Magazine in Mar–08

Three Classroom Extension	Quantity	Unit	Cost (£)	Total (£)	Rate (£/m²)
Substructure				**36,400**	**117.42**
Excavation and disposal	140	m³	26		
Concrete strip foundations, masonry work below DPC; blockwork and facing brickwork	90	m	140		
Reinforced in situ concrete ground slab, including service trench; vapour barrier; hardcore, excavation and disposal	310	m²	65		
Roof				**50,600**	**163.23**
Softwood roof trusses	380	m²	40		
Board insulation to roof	310	m²	18		
Cement slate roofing including all eaves, ridge tiles and labours; measured on plan	380	m²	55		
Aluminium rainwater down pipes	40	m	70		
Aluminium gutters	90	m	60		
Fire barriers	40	m	18		
External walls				**30,800**	**99.35**
Brick cavity wall, facing brick outer skin with cavity, 140 mm inner blockwork leaf	220	m²	140		
Windows and external doors				**29,400**	**94.84**
Proprietary aluminium framed, double glazed windows, doors and solid aluminium faced panels; powder coated finish	55	m²	370		
Double door steel security doors, including all ironmongery	1	nr	1500		
Double door aluminium framed glazed doors and screens to paved areas, including all ironmongery	15	m²	500		
Internal walls and partitions				**11,900**	**38.39**
Partitions; 100/140 blockwork	285	m²	35		
WC cubicle partitions; laminated plastics, including all ironmongery	4	nr	500		

PRIMARY SCHOOL EXTENSION

Three Classroom Extension	Quantity	Unit	Cost (£)	Total (£)	Rate (£/m²)
Internal doors				**9,700**	**31.29**
Internal fire doors, Georgian wired glass vision panel, stainless steel ironmongery to classrooms	5	nr	900		
Double fire door, Georgian wired glass, stainless steel ironmongery to corridor	1	nr	1,250		
Wrot softwood storage cupboard door, stainless steel ironmongery	5	nr	450		
Non-fire-rated doors, stainless steel ironmongery to WC	3	nr	575		
Wall finishes				**14,200**	**45.81**
Plaster and 1 mist coat, 3 coats of emulsion paint	760	m²	16		
Ceramic wall tiles, full height in toilets and selected classroom areas	40	m²	55		
Floor finishes				**15,400**	**49.68**
Sand cement screed	310	m²	18		
Carpet tiles	150	m²	22		
Safety vinyl to WC areas and practical areas, including skirtings	90	m²	35		
Heavy duty vinyl to circulation areas, including skirtings	68	m²	40		
Entrance matting with aluminium matwell	2	m²	310		
Ceiling finishes				**9,700**	**31.29**
Plasterboard ceiling, plaster skim and emulsion paint finish	280	m²	31		
Moisture resistance plasterboard ceiling, plaster skim and emulsion paint finish	30	m²	35		
Furniture and fittings				**18,700**	**60.32**
Storage trays containers	3	nr	2350		
Storage units – allowance of 1 double cupboard per classroom	3	nr	475		
Worktops, including cut out for sink, 3000 × 600	3	nr	120		
Coat hooks, fixed to masonry	100	nr	26		
Pinboards, 1000 × 2000	6	nr	70		
Pinboards, 1000 × 1200	2	nr	50		
Whiteboards: Interactive	3	nr	1400		
Whiteboards: Magnetic	3	nr	90		
Signage	1	item	1950		
Mirrors 640 × 460	7	nr	40		

PRIMARY SCHOOL EXTENSION

Three Classroom Extension	Quantity	Unit	Cost (£)	Total (£)	Rate (£/m²)
Sanitary fittings				**6,200**	**20.00**
WCs	6	nr	300		
Urinals, including side panel	2	nr	200		
Hand basins	8	nr	180		
Disabled toilet, including WC, wash hand basin, grab rails and other fittings	1	nr	1150		
Single stainless steel sinks to classrooms, 1200 × 600	3	nr	220		
Cleaners sink, 510 × 380	1	nr	700		
Disposal installations				**5,600**	**18.06**
Waste, soil and vent installation; uPVC pipework and fittings	310	m²	18		
Water installations				**9,500**	**30.65**
Cold water points	21	nr	260		
Hot water points	13	nr	310		
Space heating and air treatment				**34,100**	**110.00**
Space heating, all costs associated with the supply and installation of the heating system, temperature control and distribution pipework	310	m²	110		
Ventilation installations				**4,400**	**14.19**
Ventilation extraction to toilet areas		item	4400		
Electrical and gas installations				**35,100**	**113.23**
Mains and sub-mains installation	310	m²	22		
Small power installation	310	m²	31		
Lighting and general luminaires; emergency lighting	310	m²	55		
Gas installations, all costs associated with the supply and installation of gas	310	m²	5		
Protective, communications and special installations				**13,200**	**42.58**
Lightning protection		item	1,050		
Fire alarm installation; smoke detectors; call points		item	3,950		
Telephone and data wireways; internal telephone system		item	2,200		
Security installation; intruder detection, CCTV to existing control unit etc.		item	6,000		
Builder's work				**5,300**	**17.10**
Forming holes, chases etc.		item	5,300		

PRIMARY SCHOOL EXTENSION

Three Classroom Extension	Quantity	Unit	Cost (£)	Total (£)	Rate (£/m²)
Preliminaries and contingency				**59,800**	**192.90**
Management costs; site establishment; site supervision @	12%				
Contractor's contingency @	5%				
Construction cost (rate based on GIFA)				**400,000**	**1,290.33**

SECONDARY SCHOOL BLOCK EXTENSION

A small secondary school extension to provide a new science block with two teaching rooms attached to an existing building.

Gross internal floor area including tiers 200 m²

Model location is South East England TPI = 462; LF = 1.02

This updated cost model is copyright of Davis Langdon and was originally published in Building Magazine in Mar–11

Science Block and Classrooms	Quantity	Unit	Cost (£)	Total (£)	Rate (£/m²)
Substructure				**14,200**	**71.00**
Excavation and disposal	45	m³	30		
Concrete strip foundations, masonry work below DPC; blockwork and facing brickwork	45	m	140		
Reinforced in situ concrete ground slab, vapour barrier; granular fill, insulation (0.25 W/m².K)	100	m²	65		
Upper floors and stairs				**11,800**	**59**
Prestressed precast concrete plank floor 200 mm thick	100	m²	55		
Precast concrete staircase with half landing	1	nr	4800		
Nylon coated steel handrail	20	m	75		
Roof				**15,600**	**78.00**
Steel profiled sheet roof decking with single layer waterproof membrane, insulation and vapour control layer	100	m²	150		
Aluminium rainwater pipework and hopper heads	14	m	43		
External walls				**44,300**	**221.50**
Facing brickwork, insulated cavity, 100 mm concrete blockwork inner leaf	85	m²	120		
Timber cladding on battens with breather membrane on block cavity wall	170	m²	180		
Aluminium flashings to timber boarding	115	m	30		
Windows and external doors				**26,500**	**132.50**
Polyester powder coated double glazed aluminium windows, steel lintels, painted window boards	44	m²	410		
Polyester powder coated double glazed aluminium doors, steel lintels					
Single leaf external doors	3	nr	1,800		
Leaf and a half entrance doors	1	nr	3,100		

SECONDARY SCHOOL BLOCK EXTENSION

Science Block and Classrooms	Quantity	Unit	Cost (£)	Total (£)	Rate (£/m²)
Internal walls and partitions				**4,800**	**24.00**
Lightweight aerated concrete blockwork 140 mm thick	130	m²	37		
Internal doors				**4,100**	**20.50**
Painted solid core timber doors with vision panels, linings and ironmongery	6	nr	675		
Wall finishes				**9,700**	**48.50**
Plaster and eggshell paint	600	m²	16		
Floor finishes				**14,300**	**71.50**
Screed, latex and slip-resistant vinyl sheet flooring and skirtings to toilets	9	m²	60		
Screed, latex and vinyl safety flooring and skirtings to classrooms	170	m²	60		
Screed, latex and heavy contract polypropylene carpet to lobbies and landings	50	m²	43		
Aluminium stair edging	30	m	32		
Painted timber skirting	50	m	9		
Ceiling finishes				**5,400**	**27.00**
Demountable suspended ceiling, sound-absorbing tiles	200	m²	27		
Furniture and fittings				**15,200**	**76.00**
Blackout blinds	9	nr	120		
Science laboratory furniture	2	nr	6800		
Mirrors	3	nr	50		
Sanitary fittings				**4,400**	**22.00**
Plastic laminate wall linings	3	nr	350		
WC suite	4	nr	370		
Wash hand basin	3	nr	270		
Warm air hand dryers	4	nr	270		
Disposal installations				**4,000**	**20.00**
Waste, soil and vent installation; uPVC pipework and fittings	200	m²	20		
Water installations				**4,600**	**24.00**
Hot and cold water supply	200	m²	24		
Space heating and air treatment				**28,000**	**140.00**
Space heating, all costs associated with the supply and installation of the heating system, temperature control and distribution pipework	200	m²	140		
Ventilation installations				**7,200**	**36.00**
Ventilation extraction to toilet areas	200	m²	36		

SECONDARY SCHOOL BLOCK EXTENSION

Science Block and Classrooms	Quantity	Unit	Cost (£)	Total (£)	Rate (£/m²)
Electrical and gas installations				**36,600**	**183.00**
Distribution boards and sub-main	200	m²	16		
General power installation	200	m²	48		
General lighting installation	200	m²	95		
Gas installation	200	m²	24		
Protective, communications and special installations				**21,700**	**108.50**
Lightning protection	200	m²	9		
Data installation	200	m²	18		
Fire alarm installation	200	m²	18		
Intruder alarm installation	200	m²	17		
CCTV installation	200	m²	19		
Door access installation	200	m²	27		
Builder's work				**14,600**	**73.00**
Forming holes, chases etc.	200	m²	13		
Alterations in connecting to existing building		item	12000		
Preliminaries and contingency				**58,800**	**294.00**
Management costs; site establishment; site supervision @	12%				
Contingency @	7.5%				
Construction cost (rate based on GIFA)				**346,000**	**1,730.00**

SCHOOL REFURBISHMENT

The model is based on the updating of a group of Victorian and sixties schools buildings with a mix of general classrooms and specialist science and music departments. The refurbishment includes repairs to the existing fabric of the 60s building, including replacement glazing, comprehensive updating of finishes and a complete overhaul of the building services, external works and new FF&E and ICT installation. Professional fees and VAT are excluded.

Gross internal floor	9,850 m²
Model location is Outer London	TPI = 453; LF = 1.0

This updated cost model is copyright of Davis Langdon and was originally published in Building Magazine in Feb–10

School Refurbishment	Quantity	Unit	Cost (£)	Total (£)	Rate (£/m²)
Demolitions and alterations				370,000	37.56
Demolition, structural alterations, removal of windows, soft strip out and temporary works, disposal		item	37,000		
Minor and temporary building works				1,970,000	200.00
Allowance for temporary accommodation and decant costs		item	1,700,000		
Asbestos removal		item	130,000		
Temporary services supply		item	140,000		
Substructure				70,000	7.11
Concrete cleaning and repairs		item	40,000		
Structural steelwork, universal sections, erection, surface finishes	20	tonne	1,400		
Allowance for fire-rating of existing steel beams		item	1,950		
Roof				38,500	3.91
Replacement roofing asphalt complete, additional insulation, removal and disposal of existing	165	m²	150		
Mansafe; including access ladder and fall arrest system	125	m	110		
External walls, windows and doors				670,300	68.05
Replacement window wall panels; stick system curtain walling; insulated aluminium faced spandrel panels; concealed vents; powder-coated aluminium frame; double glazed units; opening lights with actuators; making good to existing openings	600	m²	800		
Aluminium rainscreen cladding to stairs	160	m²	440		
Entrance screens, full height double glazing in aluminium frames	30	m²	1,100		
Aluminium solar louvres; integrated with curtain walling	25	m²	390		

SCHOOL REFURBISHMENT

School Refurbishment	Quantity	Unit	Cost (£)	Total (£)	Rate (£/m²)
External render; insulated proprietary self-coloured render system to replace existing	685	m²	110		
Glazing and window film; 2 mm thick		item	1,750		
Internal walls and partitions				**149,200**	**15.15**
Independent wall lining and metal stud partition; boarded both sides; including access panel, cavity barriers and sound barriers	1,700	m²	55		
Toilet cubicles and back panels; Formica laminated WC cubicle partitions, wall linings and doors; including all relevant ducting, fixing, fittings accessories		item	50,000		
Allowance for upgrading fire-rating of walls		item	3,050		
Allowance for mounting plates and patressing in existing buildings		liem	2,650		
Internal doors				**230,600**	**23.41**
Internal doorsets; solid core; including vision panels and over panels; painted finish; ironmongery	150	nr	825		
Allowance for refurbishing existing doorsets, vision panels, fire-rating, ironmongery, painting	225	nr	475		
Wall, floor and ceiling finishes				**508,100**	**51.58**
Suspended ceiling; 2 layers plasterboard on MF; including access panels	132	m²	50		
Plasterwork, removal of existing, 2 coat plasterwork	4,000	m²	21		
Sheet carpet	7,350	m²	23		
Vinyl safety flooring, latex screed	1,500	m²	37		
Ceramic tiling to walls	250	m²	50		
Painting, generally, including preparing and making good existing surfaces	18,500	m²	9		
Allowance for skirting and other joinery		item	12,000		
Allowances for specialist acoustic ceiling finishes generally		item	17,000		
Furniture and fittings				**287,000**	**29.14**
Internal roller blind system; average size 6000 × 1800 mm	180	nr	160		
Internal fabric blackout blinds; average size 1500 × 1800 mm	22	nr	420		
Fixed FF&E to art and design		item	45,000		
Fixed FF&E to food technology		item	60,000		
Fixed FF&E to science laboratories and prep rooms		item	130,000		

SCHOOL REFURBISHMENT

School Refurbishment	Quantity	Unit	Cost (£)	Total (£)	Rate (£/m²)
Allowance for additional fixed FF&E		item	14,000		
Sanitary fittings				**46,000**	**4.67**
Toilet replacement; removal of existing, replacement WCs, urinals, wash hand basins, hand dryers, water points, average rate per item	100	nr	460		
Disposal installations				**111,300**	**11.30**
Soil, waste and disposal: rainwater disposal; cast iron down pipes and fittings	9,850	m²	11		
Water Installations				**285,700**	**29.01**
Hot and cold water service; hot and cold water storage; distribution	9,850	m²	29		
Space heating and air treatment				**1,274,000**	**129.34**
Heat source, gas-fired boilers, flues, removing existing, complete	9,850	m²	7		
LTHW system, on floor distribution and radiators, insulation	9,850	m²	70		
Mechanical ventilation installation to high heat gain areas	4,500	m²	75		
Allowance for package cooling to server rooms		item	5,700		
Allowance for toilet and laboratory extract systems	9,850	m²	18		
Electrical and gas installations				**136,900**	**138.98**
LV switchgear and panels, distribution boards, power to main plant and lifts	9,850	m²	22		
Small power installation, perimeter trunking	9,850	m²	34		
Lighting and emergency lighting, lighting control, PIR sensors and daylight sensors	9,850	m²	70		
External lighting		item	55,000		
Gas supply to boilers and science laboratories	9,850	m²	7		
Lift installations				**135,000**	**13.71**
8 person, electro-hydraulic lifts	3	nr	45,000		
Protective, communications and special installations				**790,000**	**80.20**
Earthing and bonding, lightning protection	9,850	m²	5		
Fire and smoke detection and alarm system, security installation	9,850	m²	23		
Disabled refuge alarm, disabled toilet alarm, induction loop	9,850	m²	6		
Data cabling included in main building contract	9,850	m²	19		
Building management system, sensors, valves and interfaces with window actuators	9,850	m²	28		

SCHOOL REFURBISHMENT

School Refurbishment	Quantity	Unit	Cost (£)	Total (£)	Rate (£/m²)
Builder's work				200,400	20.35
Forming holes, chases etc. @	5%				
Preliminaries and contingency				2,031,900	206.28
Management costs; site establishment; site supervision @	18%				
Design reserve @	5%				
Construction cost (rate based on GIFA)				**10,535,000**	**1,069.55**

AFFORDABLE HOUSING

AFFORDABLE HOUSING

This cost model is based on a new build scheme of a mix of two and three bedroom houses. The buildings comply with Code for Sustainable Homes Level 3, are of a timber frame construction with a reinforced in situ concrete ground slab, facing brick external wall and concrete tiles on timber roof trusses. The softwood doors and windows are painted and internal subdivision is by load-bearing timber stud partition. Features for the Lifetimes Homes standard, National Housing Federation Standards and Housing Quality Indicators are also included.

Gross internal floor area	820 m²
Model location is South East England	TPI = 462; LF = 1.02

This updated cost model is copyright of Davis Langdon and was originally published in Building Magazine in Mar–11

Affordable Housing	Quantity	Unit	Rate	Total (£)	Cost (£/m²)
Substructure				**57,400**	**70.00**
Excavation and disposal	144	m²	23		
Strip foundation with cavity masonry wall to DPC level	244	m	120		
Ground floor slab with thickenings, including insulation (0.13 W/m².K)	414	m²	60		
Frame, upper floors and stairs				**88,400**	**107.80**
Structural timber frame	820	m²	80		
Timber cassette upper floor, including any lintels required	402	m²	38		
Wooden, straight flight stairs	10	nr	750		
Roof				**55,800**	**68.05**
Timber roof trusses and associated timbers	432	m²	60		
Insulation to roof at rafter level (0.13 W/m².K)	402	m²	25		
Concrete interlocking roof tiles	432	m²	29		
Entrance canopy, approx 3 m²	10	nr	270		
uPVC rainwater down pipes	108	m	12		
uPVC gutters	220	m	15		
External walls, windows and doors				**133,800**	**163.17**
Facing brick outer skin to structural timber frame	950	m²	55		
Forming cavity and fit insulation (0.22 W/m².K)	950	m²	13		
Application of enhanced construction details to reduce air tightness to 8.0 m³/(h.m²)	950	m²	8		
Party wall, 240 mm thick timber stud frame with Isowool quilt, enhanced insulation and plasterboard plank (0.2 W/m².K)	300	m²	60		
Softwood double glazed windows to meet secured by design requirements (0.15 W/m².K)	88	m²	340		

AFFORDABLE HOUSING

Affordable Housing	Quantity	Unit	Rate	Total (£)	Cost (£/m²)
Softwood window boards	55	nr	36		
Single external doors (2.0 W/m².K)	20	nr	575		
Internal walls and partitions				**24,100**	**29.39**
Loadbearing timber stud partition faced with plasterboard, skim finish	250	m²	50		
Non-load-bearing timber stud partition, faced with plasterboard, skim finish	270	m²	43		
Internal doors				**26,700**	**32.56**
Internal doors, wood veneer, ironmongery	70	nr	250		
Wrot softwood storage cupboard door, ironmongery	20	nr	180		
Softwood door linings, architrave, painted	800	m	7		
Wall finishes				**14,700**	**17.93**
Plaster skim with emulsion paint finish	2,200	m²	6		
Ceramic wall tiles	50	m²	29		
Floor finishes				**25,600**	**31.22**
Screed	432	m²	14		
Carpet	520	m²	18		
Vinyl tiling including skirting	150	m²	31		
Softwood skirtings with gloss paint finish	670	m	8		
Ceiling finishes				**17,100**	**20.85**
Plasterboard ceiling, plaster skim and emulsion paint finish	656	m²	20		
Moisture resistance plasterboard ceiling, plaster skim and emulsion paint finish	164	m²	24		
Fittings and furnishings				**15,000**	**18.29**
General fittings – kitchen units, worktops, sink	10	nr	1,500		
Sanitary fittings				**13,000**	**15.85**
Sanitary fittings, full fitted bathroom, wash hand basin, bath with shower over	10	nr	1,300		
Disposal installations				**4,900**	**5.98**
Waste, soil and vent installation; uPVC pipework and fittings	820	m²	6		
Water installations				**35,400**	**43.17**
Hot water	60	nr	320		
Cold water	60	nr	270		
Space heating and ventilation				**30,300**	**36.95**
Electric High efficiency condensing boiler, distribution pipework and panels with temperature control zones	820	m²	37		

AFFORDABLE HOUSING

Affordable Housing	Quantity	Unit	Rate	Total (£)	Cost (£/m²)
Electrical and gas installations				**23,500**	**28.66**
Mains and sub-mains connection, small power distribution, cooker point, lighting distribution	820	m²	25		
Gas Installations, all costs associated with the supply and installation of gas supply	10	nr	300		
Protective, communications and special installations				**5000**	**6.10**
Earthing and bonding	860	m²	1		
Smoke detectors; telephone points and TV aerial, including additional data and power sockets and wiring	820	m²	5		
Special installations				**30,800**	**37.56**
Solar thermal hot water installation	10	nr	2,250		
Mechanical ventilation with heat recovery	820	m²	10		
Builder's Work				**11,500**	**14.02**
Builder's work in connection with services installation	10	nr	700		
Additional works to comply with Lifetime Homes requirements	10	nr	450		
Preliminaries and contingency				**94,000**	**114.63**
Management costs; site establishment; site supervision @	10%				
Contingency @	5%				
Construction cost (rate based on GIFA)				**707,000**	**862.18**

APARTMENTS

This cost model is based on a mixed-tenure apartment building in a south-east location, featuring 65 open-market apartments and 35 flats for the affordable sector, in a mix of one and two-bedroom configurations. The scheme also features a 50-place semi-basement car park, providing secure spaces for the open-market element of the scheme. Demolition and site preparation, and external works are excluded.

Apartment block: Gross internal floor area	7,000 m²
Open Market Apartments: Net internal floor area	3,660 m²
Affordable Apartments: Net internal floor area	1,930 m²
Car Park: Gross internal floor area	1,750 m²
Model location is South East England	TPI = 462; LF = 1.02

This updated cost model is copyright of Davis Langdon and was originally published in Building Magazine in Feb–04

Apartment Shell and Core	Quantity	Unit	Cost (£)	Total (£)	Rate (£/m²)
Substructure				596,200	85.17
Substructure, piled foundations, pile caps, ground slab	1,000	m²	550		
Allowance for drainage	1,000	m²	36		
Allowance for lift pits etc.	2	nr	5,100		
Frame and upper floors				1,329,000	189.86
In situ reinforced concrete frame and upper floors	6,650	m²	150		
Balconies, primary and secondary frame, decking, balustrade	65	nr	5,100		
Roof				121,300	17.33
Flat roof coverings, single ply membrane, insulation, ballast; allowance for details to upstands	1,000	m²	75		
Extra for roof terraces and paving to terraces	350	m²	46		
Allowance for roof drainage, roof sundries	1,000	m²	15		
Roof access equipment, latchways, cat ladder, access hatch, safety balustrade		item	15,000		
Stairs				145,600	20.80
RC concrete stairs, mild steel balustrades and handrails	14	m²	7,100		
Extra over for enhanced finishes to entrance level staircases	2	m²	5,100		
Balustrade and parapet to terraces; polyester powder coated	120	m	300		
External walls, windows and doors				1,906,100	272.30
Unitized curtain walling; powder coated insulated aluminium spandrel panels; double glazed tilt and turn windows	4,000	m²	460		
Extra for doors to balconies, ironmongery	65	nr	800		

APARTMENTS

Apartment Shell and Core	Quantity	Unit	Cost (£)	Total (£)	Rate (£/m²)
Entrance doors, aluminium framed glazed door and screen	1	nr	10,000		
External fire escape doors, metal, polyester powder coated	2	nr	2,050		
Internal walls and partitions				**287,000**	**41.00**
Core walls, in situ concrete, 225 thick	420	m²	100		
Party walls to apartments and corridors; dense concrete block; head restraint, fire stopping	4,900	m²	50		
Internal doors				**135,000**	**19.29**
Fire doors to cores and corridors; hardwood architraves/frames, including basic ironmongery	60	nr	750		
Fire doors to risers; hardwood architraves/frames, including basic ironmongery	20	nr	500		
Apartment entrance doors; solid core doors, hardwood architraves/frames, including basic quality ironmongery	100	nr	800		
Wall finishes				**90,000**	**12.86**
Plasterboard to concrete and blockwork, with specialist painted finish; entrance hall	500	nr	50		
Plasterboard to concrete and blockwork, with emulsion paint finish; lift lobbies and corridors	2,600	nr	25		
Floor finishes				**94,300**	**13.47**
Feature ceramic floor tiles, sand cement screed; entrance hall	50	m²	120		
Heavy duty carpet, sand cement screed; corridors	1,050	m²	60		
Ceramic tile, sand cement screed; lift lobbies	170	m²	100		
Skirtings, surface fixed skirting, painted MDF	810	m	10		
Ceiling finishes				**41,600**	**5.94**
Painted plasterboard with feature bulkheads; reception	50	m²	100		
Painted plasterboard on battens; lift lobbies and corridors	1,220	m²	30		
Furniture and fittings				**15,000**	**2.14**
Allowance for reception area fittings; mailboxes, signage		item	15,000		
Sanitary fittings and disposal installations				**106,400**	**15.20**
Allowance for cleaners sinks	14	nr	500		
Rainwater disposal	7,000	m²	3		
Soil, waste and overflow installations; stacks and connections to below ground drainage	7,000	m²	11		

APARTMENTS

Apartment Shell and Core	Quantity	Unit	Cost (£)	Total (£)	Rate (£/m²)
Water installations				127,500	18.21
Cold water storage tanks, booster pumps, mains distribution pipework, trace heating, water softener/conditioner etc.	7,000	m²	15		
Hot and cold water services to landlord's areas, including local water storage heaters	1,220	m²	17		
Space heating and ventilation				161,300	23.04
Electric panel heaters; landlord's areas	1,220	m²	3		
Central extract system for bathrooms; ductwork, extract fans	7,000	m²	20		
Reception area air treatment		item	7,600		
Supply and extract; plantroom areas		item	10,000		
Electrical installations				180,100	25.73
Mains switchgear, cabling, containment and landlord's distribution boards	7,000	m²	10		
Small power; landlord's areas	1,220	m²	6		
Power supply to mechanical services	7,000	m²	5		
Lighting and emergency lighting to landlord's areas	1,220	m²	30		
Feature lighting to entrances		item	1,500		
Earthing and bonding	7,000	m²	2		
Lift installation				180,000	25.71
Lift installation; 13 person fire fighting lifts serving 7 storeys	2	nr	90,000		
Protective, communications and special installations				205,600	29.37
Allowance for dry riser inlets	1,220	m²	30		
Lightning protection	7,000	m²	2		
Fire alarm system to landlord's areas	1,220	m²	36		
Telephone containment only	7,000	m²	4		
TV/Satellite system; central aerial and distribution	7,000	m²	4		
Localized controls for cold water system	7,000	m²	4		
CCTV and access control to perimeter		item	25,000		
Builder's work				48,000	6.86
Forming holes and chases; firestopping @	5%				
Preliminaries and contingency				1,198,000	171.14
Testing and commissioning of building services @	2.5%				
Contractor's overheads and profit, site establishment and supervision @	15%				
Contingency @	5%				
Construction cost (Apartment shell and core only, rate based on GIFA)				**6,968,000**	**995.42**

APARTMENTS

Open Market Apartment Fit-Out	Quantity	Unit	Cost (£)	Total (£)	Rate (£/m²)
Internal walls and partitions and doors				**384,800**	**105.14**
Metal stud partitions; 1 layer wall board each side; insulation; skim coat	4,575	m²	50		
Flush doors; non-fire-rated; single leaf; solid core hardwood veneered; softwood frames; decorations; ironmongery	260	nr	600		
Wall finishes				**266,800**	**72.90**
Plasterboard dry lining; MF framing; to external facade; emulsion paint finish	650	m²	46		
Plasterboard; to concrete and blockwork walls; emulsion paint	3,730	m²	25		
Ceramic tiles to kitchens	290	m²	85		
Ceramic tiles to bathrooms	1,400	m²	85		
Floor finishes				**314,700**	**85.98**
Suspended floor construction; ply on timber battens	3,080	m²	25		
Edge fixed carpet; PC sum £20/m²; underlay	3,080	m²	36		
Screed; ceramic tiling; to kitchens and bathrooms	730	m²	110		
Skirtings; surface fixed skirting, painted MDF	3,700	m	10		
Skirting; ceramic to match tiling	580	m	15		
Ceiling finishes				**115,400**	**31.53**
Plasterboard suspended ceiling on battens; painting	3,660	m²	28		
Feature bulkhead to junction with external wall	420	m	15		
Plasterboard bulkhead for bathroom extract ductwork	65	nr	100		
Furniture and fittings				**550,700**	**150.46**
Fully fitted kitchen to developer's specification with quality laminate worktops; appliances	65	nr	5,100		
Additional fittings to kitchens to 2 bed apartments	30	nr	2,050		
Built-in furniture to bedrooms; MDF, softwood frame and doors	95	nr	650		
Allowance for built-in cloak, meter and airing cupboards	95	nr	300		
Bathroom furniture, cistern enclosure; shelving	95	nr	410		
Bathroom accessories, mirrors etc.	95	nr	300		
Sanitary fittings and disposal installations				**310,000**	**84.70**
Fully fitted bathroom; WC, bidet, wash hand basin, pressed steel bath with power shower and screen	65	nr	2,650		
Fully fitted en-suite shower room; WC, wash hand basin, power shower, tray and screen; including all fixtures and fittings	30	nr	2,200		
Kitchen sink; including all fixtures and fittings	65	nr	320		

APARTMENTS

Open Market Apartment Fit-Out	Quantity	Unit	Cost (£)	Total (£)	Rate (£/m²)
Soil waste and vent installation within apartments; connections to stacks	545	nr	80		
Allowance for overflow pipework	3,660	nr	2		
Water installations				**174,900**	**47.79**
Cold water supply; connection, meter	65	nr	160		
Cold water distribution within apartments; final connections with sanitary fittings and appliances	545	nr	110		
Domestic electric water heaters	65	nr	500		
Hot water distribution within apartments; final connections with sanitary fittings and appliances	450	nr	160		
Space heating, air treatment and ventilation				**195,500**	**53.42**
Electrical panel heaters; local thermostatic control; power supply measured separately	320	nr	220		
Electric heated towel rails	95	nr	390		
Kitchen and bathroom extract, centralized bathroom system; localized kitchen extract with vent to facade, extract fans	160	nr	550		
Electrical installation				**230,600**	**63.01**
Mains and sub-mains; connection; LV distribution boards to apartments; meters	65	nr	390		
Small power distribution; sockets and fused connection points; wiring	1,685	nr	32		
Cooker point; wiring	65	nr	100		
Lighting; pendants, ceiling roses and bulkhead connections, wiring; to general areas	520	nr	27		
Lighting; low energy fluorescent and low voltage fittings, wiring; to kitchens and bathrooms	580	nr	100		
Shaving outlet; wiring	95	nr	70		
Lighting; 5 amp lighting sockets; wiring	390	nr	32		
Lighting distribution; switches and wiring	640	nr	30		
Extra for; kitchen pelmet lighting	65	nr	180		
Extra for; bathroom mirror lighting	95	nr	130		
Allowance for earthing and bonding	65	nr	160		
Communication installation				**114,400**	**31.26**
Fire alarm; combined detector/sounder; mains supply	130	nr	220		
Phone points and wiring; 2 nr points	65	nr	110		
TV sockets and wiring; 2 nr sockets	65	nr	110		
Video entry phone system	65	nr	1,100		

APARTMENTS

Open Market Apartment Fit-Out	Quantity	Unit	Cost (£)	Total (£)	Rate (£/m²)
Builder's work				**51,300**	**14.02**
Forming holes and chases; firestopping @	5%				
Preliminaries and contingency				**592,800**	**161.97**
Testing and commissioning of building services @	2.5%				
Contractor's overheads and profit, site establishment and supervision @	16%				
Contingency @	5%				
Construction cost (Open market apartment fit-out only; rate based on NIA area)				**3,301,900**	**902.18**

Affordable Market Apartment Fit-Out	Quantity	Unit	Cost (£)	Total (£)	Rate (£/m²)
Internal walls and partitions and doors				**175,000**	**90.67**
Metal stud partitions; 1 layer wall board each side; insulation; skim coat	2,100	m²	50		
Flush doors; non fire-rated; single leaf; solid core hardwood veneered; softwood frames; decorations; ironmongery	140	nr	500		
Wall finishes				**108,900**	**56.42**
Plasterboard dry lining; MF framing; to external facade; emulsion paint finish	380	m²	46		
Plasterboard; to concrete and blockwork walls; emulsion paint	2,510	m²	25		
Ceramic tiles to kitchens	70	m²	70		
Ceramic tiles to bathrooms	340	m²	70		
Floor finishes				**116,600**	**60.41**
Suspended floor construction; ply on timber battens	1,450	m²	25		
Edge fixed carpet; PC sum £20/m²; underlay	1,450	m²	25		
Sand cement screed; ceramic tiling; to kitchens and bathrooms	455	m²	55		
Skirtings; surface fixed skirting, painted MDF	1,410	m	10		
Skirting; ceramic to match tiling	460	m	10		
Ceiling finishes				**53,300**	**27.62**
Plasterboard suspended ceiling on battens; painting	1,905	m²	28		
Furniture and fittings				**141,000**	**73.06**
Kitchen fittings to housing association specifications	35	nr	2,550		
Additional fittings to kitchens to 2 bed apartments	15	nr	500		
Allowance for built-in furniture to bedrooms; MDF, softwood frame and doors	50	nr	500		
Allowance for built-in cloak, meter and airing cupboards	35	nr	250		

APARTMENTS

Affordable Market Apartment Fit-Out	Quantity	Unit	Cost (£)	Total (£)	Rate (£/m²)
Allowance for bathroom furniture, cistern enclosure; shelving	35	nr	150		
Allowance for bathroom accessories, mirrors etc.	35	nr	150		
Sanitary fittings and disposal installations				**75,600**	**39.17**
Fully fitted bathroom; WC, bidet, wash hand basin, pressed steel bath with power shower and screen; including all fixtures and fittings	35	nr	1,300		
Kitchen sink; including all fixtures and fittings	35	nr	270		
Soil waste and vent installation within apartments; connections to stacks	210	nr	80		
Allowance for overflow pipework	1,930	nr	2		
Water installations				**78,100**	**40.47**
Cold water supply; connection, meter	35	nr	160		
Cold water distribution within apartments; final connections with sanitary fittings and appliances	245	nr	110		
Domestic electric water heaters	35	nr	500		
Hot water distribution within apartments; final connections with sanitary fittings and appliances	175	nr	160		
Space heating, air treatment and ventilation				**85,700**	**44.40**
Electrical panel heaters; local thermostatic control; power supply measured separately	170	nr	220		
Electric heated towel rails; power supply measured separately	35	nr	280		
Kitchen and bathroom extract, centralized bathroom system; localized kitchen extract with vent to facade, extract fans	70	nr	550		
Electrical installation				**69,100**	**35.80**
Mains and sub-mains; connection; LV distribution boards to apartments; meters	35	nr	390		
Small power distribution; sockets and fused connection points; wiring	905	nr	32		
Cooker point; wiring	35	nr	100		
Lighting; pendants, ceiling roses and bulkhead connections, wiring; to general areas	280	nr	27		
Shaving outlet; wiring	35	nr	70		
Lighting; 5 amp lighting sockets; wiring	245	nr	30		
Allowance for earthing and bonding	35	nr	160		
Communication installation				**17,200**	**8.91**
Fire alarm; combined detector/sounder; mains supply	35	nr	220		
Phone points and wiring; 2 nr points	35	nr	60		

APARTMENTS

Affordable Market Apartment Fit-Out	Quantity	Unit	Cost (£)	Total (£)	Rate (£/m²)
TV sockets and wiring; 2 nr sockets	35	nr	60		
Audio entry phone system	35	nr	150		
Builder's work				**16,300**	**8.45**
Forming holes and chases; firestopping @	5%				
Preliminaries and contingency				**204,000**	**105.70**
Testing and commissioning of building services @	2.5%				
Contractor's overheads and profit, site establishment and supervision @	16%				
Contingency @	5%				
Construction cost (Affordable market apartment fit-out only; rate based on NIA)				**1,148,800**	**591.08**

Semi-Basement Car Park	Quantity	Unit	Cost (£)	Total (£)	Rate (£/m²)
Substructure				**400,100**	**228.63**
Concrete retaining wall; temporary propping	510	m²	230		
Excavation and disposal, including dewatering	5,250	m²	49		
Tie in slab edge to retaining wall	170	m	150		
Frame and upper floors				**404,000**	**230.86**
Reinforced in situ concrete columns and suspended slab to ground floor	1,750	m²	180		
Extra for vehicle ramp		item	40,000		
Allowance for louvres for natural ventilation	175	m²	280		
Stairs				**16,200**	**9.26**
In situ concrete stairs and half landings; mild steel, polyester coated handrails and balustrades; finishes	2	nr	8,100		
Internal walls and partitions				**12,500**	**7.14**
Blockwork partitions; facework; 215 average thickness; emulsion paint finish	60	m²	50		
Reinforced concrete core walls; emulsion paint finish	100	m²	95		
Internal doors				**8,600**	**4.91**
Flush doors; fire-rated; double leaf; solid core ply faced; softwood frames; decorations; ironmongery; complete	2	nr	1,000		
Flush doors; fire-rated; single leaf; solid core ply faced; softwood frames; decorations; ironmongery; complete	4	nr	800		
Fire shutters; 120 minutes fire resistance; frame and subframe; electric operation	2	nr	1700		

APARTMENTS

Semi-Basement Car Park	Quantity	Unit	Cost (£)	Total (£)	Rate (£/m²)
Finishes				**49,400**	**28.23**
Emulsion paint to concrete and blockwork	320	m²	3		
Allowance for painted floor finish, with car parking demarcation	1,750	m²	7		
Allowance for insulation to underside of building footprint	1,000	m²	36		
Fittings				**20,100**	**11.49**
Car park barriers and operating system		item	15,000		
Protective bollards; kerbs; barriers; column guards etc.		item	5,100		
Electrical installations				**52,900**	**30.23**
Mains and sub-mains installation	1,750	m²	5		
Lighting and luminaires to car park areas	1,750	m²	20		
Emergency lighting and luminaires	1,750	m²	5		
Protective and communications iInstallations				**80,900**	**46.23**
Sprinkler installation; ordinary hazard group 1	1,750	m²	36		
Fire, smoke detection and alarm system	1,750	m²	10		
Builder's work				**4,000**	**2.29**
Forming holes and chases; firestopping @	3%				
Preliminaries and contingency				**232,400**	**132.80**
Testing and commissioning of building services @	2.5%				
Overheads and profit, site establishment and supervision @	16%				
Contingency @	5%				
Construction cost (Semi-basement car park only; rate based on GIFA)				**1,281,100**	**732.07**

HOTEL

This cost model is for a new build business hotel located in an urban location in Manchester. The hotel floor area is 8,400 m² and utilizes a proportion of MMC including bathroom pods and precast structural concrete beams, slabs, crosswalls and external wall panels. Amenities include meeting rooms, bar and restaurants. The costs cover all areas – front of house, back of house and guestrooms. The costs of site preparation, external works and incoming services are excluded.

Gross internal floor area including tiers	8,400 m²
Model location is Manchester	TPI = 417; LF = 0.92

This updated cost model is copyright of Davis Langdon and was originally published in Building Magazine in Dec–06

Hotel	Quantity	Unit	Cost (£)	Total (£)	Rate (£/m²)
Substructure				**583,500**	**69.46**
Excavation, ground beams, filling to levels, lift pits, ground slab	1,500	m²	250		
Rotary bored piles	1,500	m³	110		
Underslab drainage	1,500	m²	29		
Frame and upper floors				**122,500**	**145.83**
In situ concrete frame and flat slab acting as transfer structure; 400 thick slab	1100	m²	150		
Precast concrete floor slab, cross walls and stairs (quantity based on floor slab area); self finish quality to cross walls	5,300	m²	200		
Roof				**392,900**	**46.77**
Precast concrete roof slab	1,500	m²	120		
Extra for forming upstands and copings		item	12,000		
Single ply roof membrane; insulation; rainwater outlets	1,500	m²	100		
Roof plant room; louvre screens and cladding	100	m²	320		
Mansafe system		item	6,900		
Allowance for roof level ancillaries; walkways, plant bases etc.		item	12,000		
Stairs				**124,200**	**14.79**
In situ concrete; ground to first floor (other stairs included in frame and upper floors package)	3	nr	7,400		
Handrails and balustrades; stainless steel	21	m	2,950		
Floor and soffit finishes to stairs; nosings		item	40,000		
External walls, windows and doors				**1,210,600**	**144.12**
Precast concrete wall panels; insulation and self-coloured render	2,850	m²	230		

HOTEL

Hotel	Quantity	Unit	Cost (£)	Total (£)	Rate (£/m²)
Extra for coated aluminium-framed double glazed windows to guest room floors	670	m²	250		
Full height glazed window wall; ground floor elevations	500	m²	400		
Extra over above for double doorsets	9	nr	1,950		
Extra for glazed entrance lobby		item	80,000		
Allowance for additional masonry works		item	90,000		
Internal walls and partitions				**220,700**	**26.27**
Blockwork; 100 mm and 140 m thick	1,800	m²	33		
Acoustic metal stud partitions	2,500	m²	40		
Hardwood glazed partitions; fire-rated glazing	135	m²	400		
WC cubicles	10	nr	725		
Internal doors				**385,800**	**45.93**
(Bathroom doors included in pod costs)					
Hardwood doors and frames; vision panels; stainless steel ironmongery	80	nr	900		
Bedroom doors; card access control; ironmongery	200	nr	1,100		
Melamine faced doors and hardwood frames; ironmongery; back of house areas	50	nr	725		
Riser access doors; ironmongery	100	nr	575		
Wall finishes				**346,700**	**41.27**
(Finishes to bathroom interiors included in bathroom pod costs; some bedroom finishes included in FF&E)					
Drylining to bathroom pod	1800	m²	29		
Specialist finish to public areas; decorative panels	400	m²	160		
Applied finish to bedrooms and corridors; vinyl	10,500	m²	16		
Whiterock cladding to kitchen areas	200	m²	50		
Ceramic tiling	200	m²	50		
Timber window boards	400	m	25		
Corridor corner guards		item	32,500		
Floor finishes				**359,000**	**42.74**
(Finishes to bathrooms included in bathroom pod cost)					
Guest room flooring; edge fixed carpet	6,400	m²	16		
Screeds generally	7,500	m²	10		
Wood flooring – front of house	750	m²	120		
Ceramic tiling	100	m²	80		

HOTEL

Hotel	Quantity	Unit	Cost (£)	Total (£)	Rate (£/m²)
Specialist flooring	225	m²	110		
Allowance for skirtings and joints generally		Item	55,000		
Entrance matting		Item	5,300		
Ceiling finishes				**186,400**	**22.19**
(Finishes to bathrooms included in bathroom pod cost)					
Plasterboard ceiling and bulkheads	6,400	m²	25		
Extra for acoustic treatment in public areas	900	m²	16		
Allowance for feature ceilings in front of house areas		item	12,000		
Furniture and fittings					
(Furniture and fittings to front of house and guestrooms generally included in FF&E budget)					
Sanitary fittings				**20,000**	**2.38**
(Sanitary ware to guestrooms in bathroom pod cost)					
Sanitary ware and fittings to front and back of house only		item	20,000		
Services equipment				**410,000**	**48.81**
Installation of kitchen and servery complete; including catering equipment		item	410,000		
Disposal installations				**440,00**	**5.24**
Waste, soil and vent pipework to guestrooms; stub connections to pods	200	nr	180		
Rainwater installation		item	8,000		
Hot and cold water installations				**168,000**	**20.00**
Hot and cold water installation; incoming main, storage, distribution; valves and accessories in front and back of house; stub connections to pods only	8,400	m²	20		
Space heating, air treatment and ventilation				**811,000**	**96.55**
Air conditioning to public areas; main plant; ductwork, pipework, insulation, terminal units, grilles and diffusers	1,000	m²	90		
Extract ventilation and heating/cooling to guest rooms; main plant; ductwork, pipework, insulation, wall mounted units	200	nr	2,550		
Extra for supplementary supply ventilation to inboard guest rooms	30	nr	650		
Staircase pressurization		item	70,000		
Toilet extract ventilation; public areas only		item	6,500		

HOTEL

Hotel	Quantity	Unit	Cost (£)	Total (£)	Rate (£/m²)
Supplementary supply and extract ventilation to front and back of house areas; dedicated systems serving restaurant, meeting rooms, lobby etc.		item	70,000		
Kitchen supply and extract		item	40,000		
Electrical and gas installations				**1,070,600**	**127.45**
(Lighting to guestroom bathrooms included in bathroom pods, decorative lighting included in FF&E)					
Mains and sub-mains distribution	8,400	m²	16		
Lighting installation to front and back of house; luminaires; emergency lighting	1,500	m²	80		
Small power to front and back of house	1,500	m²	16		
Lighting installation to guestrooms; luminaires; emergency lighting	6,400	m²	65		
Small power to guestrooms	6,400	m²	33		
Fix only allowance for connecting lighting and appliances included in FF&E	200	nr	400		
Electrical supplies to mechanical plant, including guestroom ventilation units		item	32,500		
External building lighting		item	27,500		
Incoming gas supply, including steel pipework, valves etc.		item	25,000		
Lift installations				**200,000**	**23.81**
Public lifts	2	nr	70,000		
Service lift	1	nr	40,000		
Platform lift	1	nr	20,000		
Protective installations				**27,900**	**3.32**
Lightning protection		item	6,500		
Dry riser		item	8,000		
Earthing and bonding	8,400	m²	2		
Communication installations				**250,800**	**29.86**
Fire alarm and smoke detection	8,400	m²	12		
Disabled WC alarm system		item	6,500		
Allowance for containment		item	16,000		
Telephone and data cabling		item	45,000		
Audio and TV distribution network		item	40,000		
Security and CCTV systems		item	40,000		

HOTEL

Hotel	Quantity	Unit	Cost (£)	Total (£)	Rate (£/m²)
Specialist installations				**923,300**	**109.92**
Bathroom pods including protection, sealing duct, door handles and locks	200	nr	4,000		
BMS Controls; Installations and PC	8,400	m²	12		
Builder's work				**27,500**	**3.27**
Builder's work in connection with services		item	27,500		
Preliminaries and contingency				**1,489,100**	**177.27**
Testing and commissioning of building services installations		item	16,000		
Management costs, site establishment and site supervision. Contractor's preliminaries, overheads and profit @	13%				
Design reserve @	3%				
Construction cost (Hotel only; rate based on GIFA)				**10,477,000**	**1,247.25**
FF&E items				**1,972,000**	**234.76**
Front-of-house and back-of-house items	8,400	m²	80		
Guestroom fit-out, casework, fixed and loose furniture, lighting fittings and appliances	200	nr	6,500		

OFFICE TO RESIDENTIAL CONVERSION

The cost model considers the potential costs for converting an existing office building into a residential building that provides a good quality level of accommodation. The existing building is arranged over 15 floors, with a single level of basement.

Gross internal floor area including tiers   17,600 m²

Model location is Outer London   TPI = 453; LF = 1.00

This updated cost model is copyright of Davis Langdon and was originally published in Building Magazine in Sep–11

Office to Residential Conversion	Quantity	Unit	Cost (£)	Total (£)	Rate (£/m²)
Demolitions and alterations				1,870,500	106.28
Soft strip	17,600	m²	30		
Asbestos removal		item			
Remove existing glazing	3,500	m²	140		
Structural alterations including repairs to existing concrete frame (exposed soffit and columns), alterations to lift/door openings, works to new and existing risers and cores	17,600	m²	65		
Substructures				250,000	14.20
Sundry remedial work including extension to new entrance, ground slab and structural walls including foundations and lift pit bases		item	250,000		
Roof				185,500	10.37
Replacement roofing asphalt complete, additional insulation, removal and disposal of existing	1,100	m²	140		
Stairs				104,400	5.93
New balustrades and handrails to existing stairs including repairs to stairs and finishes	32	nr	2,950		
Cat ladders and plant access		item	10,000		
External walls, windows and external doors				5,338,000	303.30
Allowance for replacing facade with aluminium rainscreen cladding system	5,100	m²	400		
Windows, aluminium framed, double glazed units	3,500	m²	550		
Entrance screens, aluminium framed, double glazing	100	m²	600		
Stainless steel balustrade with glass infill panels	190	m²	500		
Glazed manual revolving door, aluminium framed, 2,200 mm dia.	1	nr	30,000		
Allowance for facade cleaning system, entrance canopy and sundry doors		item	150,000		

OFFICE TO RESIDENTIAL CONVERSION

Office to Residential Conversion	Quantity	Unit	Cost (£)	Total (£)	Rate (£/m²)
New projecting balconies, steel structure with timber decking and glazed balustrade	150	nr	5,000		
Timber decking to roof terrace,	500	m²	120		
Single doors to balconies and roof terrace	152	nr	1,500		
Internal walls and partitions				**1,382,000**	**78.52**
Single timber core/apartment entrance doors, including frame and ironmongery	200	nr	800		
Double timber core doors, including frame and ironmongery	2	nr	1,500		
Single timber internal apartment doors, including frame and ironmongery	1,200	nr	600		
Riser doors and ironmongery	85	nr	500		
Internal doors				**925,500**	**52.59**
Single timber core/apartment entrance doors, including frame and ironmongery	200	nr	800		
Double timber core doors, including frame and ironmongery	2	nr	1,500		
Single timber internal apartment doors, including frame and ironmongery	1,200	nr	600		
Riser doors and ironmongery	85	nr	500		
Wall finishes				**1,886,800**	**107.20**
Plasterboard and paint to inside face of external walls	19,350	m²	40		
Paint to drylined/plasterboard internal partitions	36,050	m²	5		
Paint to fair faced blockwork walls	1,000	m²	8		
Ceramic wall tiling to bathrooms	6,900	m²	75		
Allowance for enhanced wall finishes to entrance		item	25,000		
Painted MDF skirting	25,650	m²	15		
Floor finishes				**1,142,400**	**64.91**
Sand and cement screed including acoustic underlay	14,800	m²	20		
Carpet floor finishes	5,850	m²	30		
Ceramic floor tiling to kitchen and bathrooms	2,700	m²	75		
Timber strip flooring to other areas	6,150	m²	65		
Epoxy paint to concrete slabs in plant areas	600	m²	25		
Enhanced floor finishes to entrance		item	25,000		
Surface hardener/anti-skid finish to car park	1,100	m²	8		
Lining/markings to car park		item	20,000		

OFFICE TO RESIDENTIAL CONVERSION

Office to Residential Conversion	Quantity	Unit	Cost (£)	Total (£)	Rate (£/m²)
Ceiling finishes				**546,500**	**31.05**
Painted plasterboard suspended ceiling generally	14,800	m²	35		
Access panels	34	nr	150		
Painted finish to underside of concrete soffit	1,700	m²	8		
Sundry ceiling finishes		item	10,000		
Furniture and fittings				**1,723,000**	**97.90**
Main entrance reception desk		item	15,000		
Post boxes	150	nr	120		
Kitchens, including fitted base/wall units, worktops, white goods and ceramic tiled splashback	150	nr	7,400		
Vanity units with cistern enclosure shelf	300	nr	500		
Bathroom fittings including mirrors, toilet roll holder etc.	300	nr	350		
Bespoke timber wardrobe to master bedrooms	150	nr	1,500		
Column guards, protective bollards, barriers to car park, cycle rack, car park traffic management system to basement		item	50,000		
Statutory signage and other fittings		item	50,000		
Sanitary fittings				**671,000**	**38.13**
Sanitaryware fittings to apartments based upon 2 nr bathrooms per unit	150	nr	4,450		
Cleaners' sink on every other floor		item	3,450		
Disposal installations				**295,700**	**16.80**
Waste, soil and vent installation; cast iron downpipes	17,600	nr	17		
Water installations				**598,400**	**34.00**
Hot and cold water supply	17,600	m²	34		
Space heating and air treatment and ventilation				**1,298,900**	**73.80**
Full mechanical ventilation to basement car park	1,100	m²	110		
Heat source, gas-fired boilers, flues, removing existing	16,500	m²	4		
Landlords heating, including distribution	16,500	m²	9		
Mechanical ventilation to plantrooms and staircase pressurization installations via Automatic Openable Vents (AOVs)		item	65,000		
Heat plate exchangers incuding heat meter	150	nr	2,000		

OFFICE TO RESIDENTIAL CONVERSION

Office to Residential Conversion	Quantity	Unit	Cost (£)	Total (£)	Rate (£/m²)
LTHW system including radiators, bathroom heated towel rails, associated pipework and fittings	150	nr	2,000		
Whole house ventilation system with heat recovery, including fan, ductwork and grilles	150	nr	2,000		
Electrical and gas installations				2,106,800	119.70
LV Switchgear and panels, distribution boards, power to main plant and lifts	17,600	m²	17		
Standby generation and associated flue		item	65,000		
Power installation, including distribution board, meter, small power points and fused connection units	17,600	m²	34		
Lighting and emergency lighting to landlord areas (including basement), lighting control, PIR sensors and daylight sensors	17,600	m²	27		
Lighting installation with apartments, including low energy downlighters, kitchen pelmet lighting, bathroom mirror lighting, balcony lighting and basic controls	150	nr	4,250		
Gas supply to boilers		item	35,000		
Lift installations				14,600	73.00
13 person, 1,100 kg, 16 stops, 1.6 m/s monospace passenger lift, including lighting and power. 1 nr having fire fighting capability	4	nr	180,000		
Protective, communications and special installations				1,639,000	93.13
Earthing and bonding	17,600	m²	4		
Wet riser, sprinkler installation, including lightning and surge protection	17,600	m²	45		
Fire and smoke detection and alarm system, security installation	17,600	m²	6		
Telephone/data/satellite installation, including containment, wiring and dishes	17,600	m²	15		
Security installation, including CCTV, door entry, access control, intruder alarms and car park barrier	17,600	m²	7		
Containment and wiring only for future security, sound system and home automation systems	150	nr	1,000		
Building management system, sensors, valves and interfaces with window actuators	17,600	m²	8		
Builder's work				334,400	19.00
Builder's work in connection with services installations, including machine bases, fire stopping, forming holes, chases in the existing building fabric	17,600	m²	19		

OFFICE TO RESIDENTIAL CONVERSION

Office to Residential Conversion	Quantity	Unit	Cost (£)	Total (£)	Rate (£/m²)
Preliminaries and contingency				**5,016,900**	**285.05**
Contractor's management, site establishment and site supervision, including overheads and profit @	16%				
Design reserve and construction contingency @	5%				
Construction cost (rate based on GIFA)				**28,033,000**	**1592.77**

COMMUNITY CENTRE

An 860 m² building built to achieve BREEAM Very Good level of performance. Main structure is a concrete ground bearing slab and steel frame on concrete pads. External walls are a mixture of facing bricks at ground level and render above. Roof of single ply warm roof construction with a small amount of patent glazing.

Gross internal floor area	860 m²
Model location is South East England	TPI = 462; LF = 1.02

This updated cost model is copyright of Davis Langdon and was originally published in Building Magazine in Mar−11

Community Centre	Quantity	Unit	Rate	Total (£)	Cost (£/m²)
Substructure				81,300	94.53
Excavation and disposal off site	430	m³	19		
Reinforced concrete ground slab, hardcore, dpm, ground beams and column bases for steel frame (0.25 W/m².K)	860	m²	85		
Frame				101,700	118.26
Steel propped portal frame, cold rolled purlins, surface treatments (@ 60 kg/m²)	52	tonne	1,600		
Intumescent paint to give 30 minute fire protection to steelwork	52	tonne	320		
Allowance for miscellaneous works, protecting columns		item	2,650		
Roof				12,8100	148.95
Ply roofing with rigid board insulation and single ply polymeric roof cladding (0.18 W/m².K)	887	m²	130		
Roof glazing, aluminium, double glazed (2.2 W/m².K)	16	m²	625		
Rainwater drainage, aluminium gutters and downpipes	50	m	55		
External walls, windows and doors				178,600	207.67
Light gauge steel framing; board; natural insulation; clay facing brick externally (0.20 W/m².K)	47	m²	190		
Light gauge steel framing; board; natural insulation; rendered finish on cement profile substrate (0.20 W/m².K)	497	m²	160		
Curtain walling with powder coated aluminium frame	125	m²	625		
Aluminium triple glazed windows	25	m²	320		
Steel fire escape doors	2	nr	950		
Canopy over entrance	1	nr	2,100		

COMMUNITY CENTRE

Community Centre	Quantity	Unit	Rate	Total (£)	Cost (£/m²)
Internal walls and partitions				**54,300**	**63.14**
Sound reducing board and metal stud partitions; generally 2 layers of board each side; various levels of fire and sound insulations; part glazed as appropriate	700	m²	70		
Fireproofing between partitions and roof		item	5,300		
Internal doors				**34,800**	**40.47**
Timber internal doorset with softwood frames and ironmongery; vision panels; including fire-resisting where necessary	38	nr	750		
Timber acoustic sliding stacking panel partition; 6 m × 2.6 m high	1	nr	6,300		
Wall finishes				**22,200**	**25.81**
Plaster blocks and board; emulsion paint generally	1,100	m²	14		
Ceramic wall tiles splashbacks to WC areas	130	m²	55		
Floor finishes				**48,400**	**56.28**
Demountable stage	20	m²	160		
Vinyl flooring	300	m²	80		
Carpet tiling	275	m²	37		
Laminate timber	250	m²	42		
Entrance matwell	2	m²	260		
Ceiling finishes				**28,400**	**33.02**
Concealed grid plasterboard suspended ceiling	500	m²	42		
Plasterboard fixed directly to roof purlins	350	m²	21		
Fittings and furnishings				**27,500**	**31.98**
Kitchen fitting; worktops; shelving		item	27,500		
Sanitary fittings				**20,000**	**23.26**
WC suite; urinals, including disabled facility including all sanitary and fittings	15	nr	1,050		
Equipment for shower rooms; including disabled person equipment	2	nr	2,100		
Disposal installations				**3,900**	**4.52**
Waste, soil and vent installation; uPVC pipework and fittings	15	nr	260		
Water installations				**31,800**	**36.98**
Hot and cold water supply to WC, kitchen	860	m²	37		
Space heating and air treatment and ventilation				**70,400**	**81.86**
Boiler		item	27,500		
Underfloor heating system	860	m²	48		
Toilet and kitchen extract		item	1,600		

COMMUNITY CENTRE

Community Centre	Quantity	Unit	Rate	Total (£)	Cost (£/m²)
Electrical and gas installations				52,900	61.51
Small power, basic and emergency lighting	860	m²	16		
General lighting	860	m²	21		
External building lighting generally		item	5,300		
Incoming services		item	16,000		
Protective, communications and special installations				44,700	51.98
Lightning protection, earthing and bonding	860	m²	4		
Fire and intruder alarms, panic alarm buttons	860	m²	13		
Data installations and containment	860	m²	11		
Basic audio visual installation, projector, screen, satellite TV installation		item	21,000		
Builder's work				7,900	9.19
Forming holes, chases etc.		item	7,900		
Preliminaries and contingency				194,100	225.70
Management costs; site establishment; site supervision @	15%				
Contingency @	5%				
Construction cost (rate based on GIFA)				1,131,000	1,315.11

MUSEUM FIT-OUT

The cost model is for a standalone new pavilion as an addition to an existing museum with high quality internal and external finishes. The building contains 500 m² gross floor area and is sited in the South East of England.

Costs exclude exhibition fit-out costs. Costs are at third quarter 2012 price levels. The costs exclude external works, professional fees, VAT and site-specific abnormals.

Gross internal floor area		500 m²
Model location is South East England		TPI = 462; LF = 1.02

This updated cost model is copyright of Davis Langdon and was originally published in Building Magazine in Jul–12

Museum Fit-Out	Quantity	Unit	Rate	Total (£)	Cost (£/m²)
Facilitating works				**1,000**	**2.00**
Sundry demolitions and removals		item	1,000		
Substructures				**375,400**	**750.80**
Retaining walls		item	100,000		
Excavation, disposal and filling	1,900	m³	36		
Concrete strip footings, ground floor slab, walls	900	m³	230		
Frame				**221,500**	**443.00**
Structural steel columns	40	tonne	2,650		
Structural steel roof beams and trusses	26	tonne	3,000		
Secondary steelwork to cladding, roof deck, windows and services		item	37,500		
Roof				**358,800**	**717.60**
Clerestory glazing	3	nr	40,000		
Green roof	600	m²	350		
Aluminium louvres	30	m²	625		
Fall restraint system		item	10,000		
External walls, windows and doors				**610,000**	**1,220.00**
Stone cladding		item	360,000		
Capless silicone glazed curtain walling and main entrance doors	110	m²	2,000		
Loading bay doors	1	nr	30,000		
Internal walls and partitions				**98,000**	**196.00**
Jumbo metal stud partitions, plasterboard both sides					
up to 5 m high	50	m	600		
up to 7 m high	55	m	800		
up to 9 m high	20	m	1,200		
Internal doors				**15,000**	**30.00**
Frameless doors 3 m wide	2	nr	7,500		

MUSEUM FIT-OUT

Museum Fit-Out	Quantity	Unit	Rate	Total (£)	Cost (£/m²)
Wall finishes				**5,500**	**11.00**
Plaster blocks and board; emulsion paint generally	325	m²	17		
Floor finishes				**72.100**	**144.20**
Raised access flooring	325	m²	120		
Oak composite wood flooring	325	m²	75		
Polished concrete finish	120	m²	60		
Resin flooring	25	m²	60		
Ceiling finishes				**111,600**	**223.20**
Concealed grid plasterboard suspended ceiling	465	m²	240		
Fittings and furnishings				**28,100**	**56.20**
Blackout blinds to roof lights	45	m²	625		
Space heating and air conditioning				**265,000**	**530.00**
HVAC assuming DX cooling system, full humidity control	500	m²	450		
Automatic controls	500	m²	80		
Electrical installation				**118,000**	**236.00**
Mains electric and switchgear	500	m²	35		
Small power distribution	500	m²	21		
Lighting installation and controls	500	m²	60		
Luminaires	500	m²	120		
Protective, communications and special installations				**51,500**	**103.00**
Lightning protection, earthing and bonding	500	m²	6		
Communications and data	500	m²	15		
Fire detection system	500	m²	26		
Disabled facilities/induction loops	500	m²	16		
Security	500	m²	40		
Builder's work				**16,500**	**33.00**
Builder's work in connection with M&E services	500	m²	33		
Preliminaries and contingency				**607,00**	**1,214.00**
Management costs; site establishment; site supervision @	17%				
Main contractor's overheads and profit @	3%				
Contingency @	5%				
Construction cost (rate based on GIFA)				**2,955,000**	**5,910.00**

Spon's First Stage Estimating Handbook

Third Edition

Bryan Spain

Have you ever had to provide accurate costs for a new supermarket or a pub "just an idea...a ballpark figure..."?

The earlier a pricing decision has to be made, the more difficult it is to estimate the cost and the more likely the design and the specs are to change. And yet a rough-and-ready estimate is more likely to get set in stone.

Spon's First Stage Estimating Handbook is the only comprehensive and reliable source of first stage estimating costs. Covering the whole spectrum of building costs and a wide range of related M&E work and landscaping work, vital cost data is presented as:

*   costs per square metre
*   elemental cost analyses
*   principal rates
*   composite rates.

Compact and clear, Spon's First Stage Estimating Handbook is ideal for those key early meetings with clients. And with additional sections on whole life costing and general information, this is an essential reference for all construction professionals and clients making early judgements on the viability of new projects.

January 2010: 216x138: 244 pp
Pb: 978-0-415-54715-4: £46.99

To Order: Tel: +44 (0) 1235 400524 Fax: +44 (0) 1235 400525
or Post: Taylor and Francis Customer Services,
Bookpoint Ltd, Unit T1, 200 Milton Park, Abingdon, Oxon, OX14 4TA UK
Email: book.orders@tandf.co.uk

For a complete listing of all our titles visit:
www.tandf.co.uk

Preliminaries Build-up Example

The number of items priced in the preliminaries section of Bills of Quantities and the manner in which they are priced vary considerably between contractors. Some contractors, by modifying their percentage factor for over-heads and profit, attempt to cover the costs of preliminary items in their Prices for Measured Work. However, the cost of Preliminaries will vary widely according to job size and complexity, site location, accessibility, degree of mechanization practicable, position of the contractor's head office and relationships with local labour/domestic subcontractors. It is therefore usually far safer to price preliminary items separately on their merits according to the project.

It is not possible for the quantity surveyor/cost manager to quantify the main contractor's preliminaries and it is left for the contractor to interpret the tender documentation and ascertain his resources and method of working he requires to complete the works.

The preliminaries bill is therefore a relatively simple pricing schedule of headings under which the contractor prices his preliminaries items.

An example in pricing preliminaries follows, and this assumes the form of contract used is the JCT Standard Building Contract With Quantities (SBC/Q) and the value, including preliminaries, is approximately £3,500,000. The contract is estimated to take 52 weeks to complete and the value is built up as follows:

	£
Labour value	1,200,000
Material value	950,000
Provisional sums and all subcontractors	1,000,000
Preliminaries, say	£350,000
Project value	**£3,500,000**

At the end of the section the data is summarized to give a total value of preliminaries for the project example.

NOTE: The term 'Not priced', where used throughout this section means either that the cost implication is negligible or that it is usually included elsewhere in the tender.

This section has been amended in accordance with the Tabulated Work Sections defined in NRM2.

1 Preliminaries

Part B Pricing Schedule

1.1   Employer's requirements

1.1.1   Site accommodation – not priced

1.1.2   Site records – not priced

1.1.3   Completion and post-completion requirements

Defects after completion	Based on 0.20% of the contract sum	£3,500,000 × 0.20%, say = **£7,000**

1.2   Main Contractor's cost items

1.2.1   Management and staff – main contractor's project-specific management and staff

Management and staff	Based on 4.5% of the tender (excluding preliminaries)	£3,200,000 × 4.5%, say = **£140,000**

1.2.2   Site establishment – main contractor's and common user temporary site accommodation

Site accommodation	Based on 0.35% of the tender (excluding preliminaries)	£3,200,000 × 0.35%, say = **£11,200**

1.2.3   Temporary services

Lighting and power	Based on 1.0% of the tender (excluding preliminaries)	£3,200,000 × 1.0%, say = **£32,000**
Water for the works (provided from mains)	Based on 0.1% of the tender (excluding preliminaries)	£3,200,000 × 0.10%, say = **£3,200**
Temporary telephones etc.	Based on 0.15% of the tender (excluding preliminaries)	£3,200,000 × 0.15%, say = **£4,800**

1.2.4   Security – allow for staff and security equipment

Security	Based on 0.10% of the tender (excluding preliminaries)	£3,200,000 × 0.15%, say = **£4,800**
Temporary hoarding, fencing etc.	2.4 m high fencing of OSB boarding one side; pair of gates	100 m @ £100, say = **£10,000**

1.2.5   Safety and environmental protection – compliance with all welfare facilities, first aid etc.

Safety, health and welfare	Based on 0.1% of the tender (excluding preliminaries)	£3,200,000 × 0.20%, say = **£6,400**

1.2.6   Control and protection – allowance for setting out, protection of the works, sampling

Control and protection of the works	Based on 0.10% of the tender (excluding preliminaries)	£3,200,000 × 0.10%, say = **£3,200**
Drying the works	Based on 0.03% of the tender (excluding preliminaries)	£3,200,000 × 0.03%, say = **£960**

1.2.7   Mechanical plant – common user mechanical plant and equipment

Plant/Transport	Based on 2.00% of the tender (excluding preliminaries)	£3,000,000 × 2.00%, say = **£64,000**
Small plant and tools	Based on 0.20% of the tender (excluding preliminaries)	£3,200,000 × 0.2%, say = **£6,400**

1.2.8   Temporary works – common user access scaffolding

Temporary roads and walkways	Say 250 × 3.5 m hardcore/recycled concrete	875 m² @ £9.00/m², say = **£8,000**
Access scaffolding	Based on 0.75% of the tender (excluding preliminaries)	£3,500,000 × 0.75%, say = **£24,000**

1.2.9   Site records – photographs, progress reporting etc. – not priced

1.2.10 Completion and post-completion requirements – testing and commissioning, handover plan etc. – not priced

1.2.11 Cleaning – the cost would normally represent an allowance for final clearing of the works on completion with the residue for cleaning throughout the contract period. This amount being reduced with the introduction of waste management plans

Removing rubbish and cleaning	Based on 0.15% of the tender (excluding preliminaries)	£3,200,000 × 0.15%, say = **£4,800**

1.2.12 Fees and charges – any miscellaneous fees charges (rates etc.) – not priced

1.2.13 Site services – temporary works not specific to an element

Traffic regulations	Based on 0.05% of the tender (excluding preliminaries)	£3,200,000 × 0.05%, say = **£1,600**
Additional temporary items	Based on 0.25% of the tender (excluding preliminaries)	£3,200,000 × 0.25%, say = **£8,000**

1.2.14 Insurances, bonds, guarantees and warranties

Contract value (including preliminaries), say	£3,500,000
Estimated increased costs during contract period, say 2%	£70,000
	£3,570,000
Estimated increased costs incurred during period of reinstatement, say 3%	£107,100
	£3,677,100
Professional fees, say 12%	£441,300
	£4,118,400

If at the Contractor's risk, the insurance cover must be sufficient to include the full cost of reinstatement, all increases in cost, professional fees and any consequential costs such as demolition. The average provision for fire risk is 0.15% of the value of the work after adding for increased costs and professional fees.

Insurance of the works	Allow £4,118,400 × 0.15%, say = **£6,200**

NOTE: Insurance premiums are liable to considerable variation, depending on the contractor, the nature of the work and the market in which the insurance is placed.

Summary of Preliminaries Costs Example	
Defects after completion	£7,000
Management and staff	£140,000
Site accommodation	£11,200
Lighting and power for the works	£32,000
Water for the works	£3,200
Temporary telephones	£4,800
Security	£4,800
Hoardings, fans, fencing, etc.	£10,000
Safety, health and welfare	£6,400
Protection of the works	£3,200
Drying the works	£960
Large plant/transport	£64,000
Small plant and tools	£6,400
Temporary roads and walkways	£8,000
Access scaffolding	£24,000
Removing rubbish, etc., and cleaning	£4,800
Traffic regulations	£1,600
Additional temporary works	£8,000
Insurance	£6,200
TOTAL PRELIMINARIES	**£346,560**

It is emphasized that the above is an example only of the way in which Preliminaries may be priced and it is essential that for any particular contract or project the items set out in Preliminaries should be assessed on their respective values. The value of the Preliminaries items in recent tenders received by the editors varies from a 10% to 13% addition to all other costs. The above example represents approximately an 11% addition to the value of measured work.

Building Systems

Moe & Smith

We can no longer view building components as artifacts (a brick or a boiler) or as autonomous systems (air conditioning or prefabrication). Rather these components and systems are part of much larger systems of which architects are one agent. This book will help architects more broadly envision these networks including :

- canonical texts as well as contemporary thinking from well known theorists and practitioners, each contribution frames a specific range of technology in relation to society such as building process, products, economies and ecologies
- clearly structured, the book is divided into three parts; each accompanied by a comprehensive introduction by the editors
- an annotated bibliography provides a glossary of further reading
- illustrated throughout with over 100 illustrations.

The book calls for integration, a convergence and confluence of social and technical factors, discovering the capability and culpability of such; for architects to finally realize that the term building systems is best grasped as a verb, not a set of nouns.

This reader presents students, faculty and practicing architects with an expanded view of technology in architecture that transcends naive determinisms and technocratic applications; forming a more pithy intellectual context for the complex and contingent roles of technology in twenty-first century architecture.

March 2012: 234x156: 272pp
Hb: 978-0-415-61793-2: £105.00
Pb: 978-0-415-61794-9: £24.99

To Order: Tel: +44 (0) 1235 400524 Fax: +44 (0) 1235 400525
or Post: Taylor and Francis Customer Services,
Bookpoint Ltd, Unit T1, 200 Milton Park, Abingdon, Oxon, OX14 4TA UK
Email: book.orders@tandf.co.uk

For a complete listing of all our titles visit:
www.tandf.co.uk

Green Buildings Pay
Third Edition

Design, Productivity and Ecology

Brian W Edwards & Emanuele Naboni

This third edition of Green Buildings Pay presents new evidence and new arguments concerning the institutional and business case that can be made for green design. The green argument has moved a long way forward since the previous edition, and this fully updated book addresses the key issues faced by architect, engineer and client today.

Green Buildings Pay: Design, Productivity and Ecology examines, through a range of detailed case studies, how different approaches to green design can produce more sustainable patterns of development. These cases are examined from three main perspectives: that of the architect, the client and the user. Completely revised with all new chapters, cases, sections and introductory material the third edition presents:

- over 20 new researched case studies drawn from the UK, Europe and the USA, written in collaboration with the architects, engineers, clients and user groups
- examples of office and educational buildings of high sustainable and high architectural quality
- an exploration of the architectural innovations that have been driven by environmental thinking, such as the new approaches to the design of building facades, roofs, and atria
- cases which demonstrate current practice in the area of energy/eco-retrofits of existing buildings
- documentation of the benefit impact assessment schemes such as LEED and BREEAM have had upon client expectations and on design approaches over the past decade
- beautiful full color illustrations throughout.

January 2013 276x219: 296 pp
Hb: 978-0-415-68534-4: £105.00
Pb: 978-0-415-68535-1: £35.00

To Order: Tel: +44 (0) 1235 400524 Fax: +44 (0) 1235 400525
or Post: Taylor and Francis Customer Services,
Bookpoint Ltd, Unit T1, 200 Milton Park, Abingdon, Oxon, OX14 4TA UK
Email: book.orders@tandf.co.uk

For a complete listing of all our titles visit:
www.tandf.co.uk

Taylor & Francis
Taylor & Francis Group

Approximate Estimating Rates

Estimating by means of priced approximate quantities is always more accurate than by using overall building prices per square metre. Prices given in this section, which is arranged in elemental order, are derived from the *Prices from Measured Works* section, but also include for all the incidental items and labours which are normally measured separately in Bills of Quantities. They have been established with a tender price level of 453. They include overheads and profit but do not include for main contractors preliminaries, an example of which is given in an earlier section, and which in the current tendering climate amount to approximately 11% of the value of the measured works.

Whilst every effort is made to ensure the accuracy of these figures, they have been prepared for approximate estimating purposes and guidance only and on no account should they be used for the preparation of tenders.

Unless otherwise described units denoted as m² refer to appropriate unit areas (rather than gross internal floor areas).

As elsewhere in this edition prices do not include Value Added Tax or professional services fees.

Cost plans can be developed using from Elemental Cost Plans using both *Approximate Estimating Rates* and/or *Prices for Measured Works* depending upon the level of information available.

The cost targets within each formal cost plan approved by the employer will be used as the baseline for future cost comparisons. Each subsequent cost plan will require reconciliation with the preceding cost plan and explanations relating to changes made. In view of this, it is essential that records of any transfers made to or from the risk allowances and any adjustments made to cost targets are maintained, so that explanations concerning changes can be provided to both the employer and the project team.

PRELIMINARIES – TYPICAL PLANT HIRE RATES

Item	Unit	Range £		
PRELIMINARIES – TYPICAL PLANT HIRE RATES				
Preliminaries items – temporary items				
Minimum hire periods generally apply				
Typical rates for site accommodation				
Office cabins – 24 ft × 9 ft (20 m²)	week	25.00	to	30.00
Office cabins – 32 ft × 10 ft (30 m²)	week	31.00	to	38.00
Fire-rated cabins – 24 ft × 9 ft (20 m²)	week	52.00	to	63.00
Fire-rated cabins – 32 ft × 10 ft (30 m²)	week	71.00	to	86.00
Meeting room – 24 ft × 9 ft (20 m²)	week	25.00	to	30.00
Meeting room – 32 ft × 9 ft (30 m²)	week	31.00	to	38.00
Mess cabins (incl Furniture)	week	29.50	to	35.50
Drying rooms (incl Furniture)	week	33.00	to	40.00
Safestore – 20ft × 8ft (15 m²)	week	13.70	to	16.60
Container with padlock – 20ft × 8ft (15 m²)	week	9.50	to	11.50
Toilets – 13 ft × 9 ft	week	38.00	to	46.00
Signs and notices	week	47.50	to	58.00
Fire extinguishers	week	7.60	to	9.20
Kitchen with cooker, fridge, sink, water heater; 32 ft × 10 ft	week	140.00	to	170.00
Mess room with wash basin, water heater, seating; 16 ft × 7 ft	week	86.00	to	100.00
Toilets – fitted to mains; three pan unit	week	150.00	to	180.00
Toilets – fitted to mains; four pan unit	week	180.00	to	220.00
Haulage to and from site; Site offices, Storage sheds, Toilets	item	4300.00	to	5200.00
Power and lighting				
3 kVA transformer	week	7.80	to	9.45
5 kVA transformer	week	13.00	to	15.70
10 kVA transformer	week	30.00	to	36.00
4 Way distribution Box; 110 v	week	7.60	to	9.20
14 m extension cable	week	3.65	to	4.40
25 m extension cable	week	9.60	to	11.60
2 × 500 W floodlights	week	11.20	to	13.50
2 × 500 W floodlights on stands	week	14.00	to	16.90
Generators				
1 kVA 240 v portable (low noise)	week	35.00	to	42.00
2 kVA dual volt, portable petrol	week	24.00	to	29.50
5 kVA silenced	week	68.00	to	83.00
7.5 kVA silenced	week	100.00	to	120.00
15 kVA silenced	week	145.00	to	175.00
40 kVA silenced	week	220.00	to	265.00
60 kVA silenced	week	230.00	to	275.00
108 kVA silenced	week	280.00	to	340.00
165 kVA silenced	week	350.00	to	420.00
Mixers and small plant				
Small mixer; 5–3 1/2 diesel	week	29.00	to	35.50
Mixer; 7–5 diesel	week	36.50	to	44.00
Compresser	week	110.00	to	130.00
Small dehumidifers	week	19.90	to	24.00
Large dehumidifers	week	53.00	to	64.00
Turbo dryer	week	33.00	to	40.00
Gas jet air heater; 260,000 Btu	week	40.00	to	48.00

1.1 SUBSTRUCTURE

Item	Unit	Range £		
1 SUBSTRUCTURE				
1.1 SUBSTRUCTURE				
All-in rates for simple foundations and ground floor slab. Ground floor area				
Strip or trenchfill foundations with masonry up to 150 mm above floor level only; blinded hardcore bed; slab insulation; reinforced ground bearing slab 200 mm thick. To suit residential and small commercial developments with good ground bearing capacity				
shallow foundations up to 1.00 m deep	m²	130.00	to	165.00
shallow foundations up to 1.50 m deep	m²	145.00	to	190.00
Foundations in poor ground; mini piles; typically 300 mm dia.; 15 m long; 175 mm thick reinforced concrete slab; for single storey commercial type development				
minipiles to building columns only; 1 per column	m²	110.00	to	140.00
minipiles to columns and perimeter ground floor beam; piles at 2.00 m centres to ground beams	m²	215.00	to	275.00
minipiles to entire building at 2.00 m × 2.00 m grid	m²	240.00	to	305.00
Raft foundations				
simple reinforced concrete raft on poorer ground for development up to two storey high	m²	130.00	to	165.00
Basements				
Reinforced concrete basement floors; NOTE: excluding bulk excavation costs and hardcore fill				
200 mm thick waterproof concrete with 2 × layers of A252 mesh reinforcement	m²	52.00	to	67.00
Reinforced concrete basement walls; NOTE: excluding bulk excavation costs and hardcore fill				
200 mm thick waterproof reinforced concrete	m²	290.00	to	370.00
Trench fill foundations				
Machine excavation, disposal, plain in situ concrete 20.00 N/mm² – 20 mm aggregate (1:2:4) trench fill, 300 mm high cavity masonry in cement mortar (1:3), pitch polymer damp roof course				
With 3 courses of common brick outer skin up to DPC level (PC bricks @ £240/1000)				
600 mm × 1000 mm deep	m	85.00	to	100.00
600 mm × 1500 mm deep	m	115.00	to	140.00
With 3 courses of facing bricks brick outer skin up to DPC level (PC bricks @ £350/1000)				
600 mm × 1000 mm deep	m	87.00	to	105.00
600 mm × 1500 mm deep	m	120.00	to	145.00
With 3 courses of facing bricks brick outer skin up to DPC level (PC bricks @ £500/1000)				
600 mm × 1000 mm deep	m	89.00	to	110.00
600 mm × 1500 mm deep	m	120.00	to	150.00

1.1 SUBSTRUCTURE

Item	Unit	Range £		

1.1 SUBSTRUCTURE – cont

Strip foundations

Excavate trench 600 mm wide, partial backfill, partial disposal, earthwork support (risk item), compact base of trench, plain in situ concrete 20.00 N/mm² – 20 mm aggregate (1:2:4), 250 mm thick, 3 courses of common brick up to DPC level, cavity brickwork/blockwork in cement mortar (1:3), pitch polymer damp-proof course machine excavation

With 3 courses of common brick outer skin up to DPC level (PC bricks @ £240/1000)

Item	Unit	Range £		
600 mm × 1000 mm deep	m	89.00	to	110.00
600 mm × 1500 mm deep	m	115.00	to	140.00

With 3 courses of common brick outer skin up to DPC level (PC bricks @ £350/1000)

Item	Unit	Range £		
600 mm × 1000 mm deep	m	93.00	to	115.00
600 mm × 1500 mm deep	m	120.00	to	150.00

With 3 courses of common brick outer skin up to DPC level (PC bricks @ £500/1000)

Item	Unit	Range £		
600 mm × 1000 mm deep	m	97.00	to	120.00
600 mm × 1500 mm deep	m	125.00	to	150.00

Column bases

Excavate pit in firm ground by machine, partial backfill, partial disposal, support, compact base of pit; in situ concrete 25.00 N/mm²; reinforced at 50 kg/m³

rates per base for base size

Item	Unit	Range £		
600 mm × 600 mm × 300 mm; 1000 mm deep pit	nr	97.00	to	120.00
900 mm × 900 mm × 450 mm; 1250 mm deep pit	nr	150.00	to	180.00
1500 mm × 1500 mm × 600 mm; 1500 mm deep pit	nr	360.00	to	440.00
2700 mm × 2700 mm × 1000 mm; 1500 mm deep pit	nr	1450.00	to	1775.00

rates per m³ for base size

Item	Unit	Range £		
600 mm × 600 mm × 300 mm; 1000 mm deep pit	m³	780.00	to	950.00
900 mm × 900 mm × 450 mm; 1250 mm deep pit	m³	420.00	to	510.00
1500 mm × 1500 mm × 600 mm; 1500 mm deep pit	m³	280.00	to	345.00
2700 mm × 2700 mm × 1000 mm; 1500 mm deep pit	m³	210.00	to	250.00

extra for

Item	Unit	Range £		
reinforcement at 75 kg/m³ concrete, base size	m³	70.00	to	86.00
reinforcement at 100 kg/m³ concrete, base size	m³	87.00	to	105.00

Piling

Enabling Works

Item	Unit	Range £		
excavate to form piling mat; supply and lay imported hardcore – recycled brick and similar to form piling mat	m³	6.80	to	8.30
provision of plant (1 nr rig); including bringing to and removing from site; maintenance, erection and dismantling at each pile position	item	14500.00	to	18000.00

Supply and install concrete Continuous Flight Auger (CFA) piles; cart away inactive spoil; measure total length of pile

Item	Unit	Range £		
450 mm dia. reinforced concrete CFA piles	m	41.50	to	51.00
600 mm dia. reinforced concrete CFA piles	m	66.00	to	81.00
750 mm dia. reinforced concrete CFA piles	m	100.00	to	125.00
900 mm dia. reinforced concrete CFA piles	m	145.00	to	180.00

1.1 SUBSTRUCTURE

Item	Unit	Range £		
Pile testing				
using tension piles as reaction; typical loading test 1000 kN to 2000 kN	item	6700.00	to	9700.00
integrity testing; minimum 20 per visit	nr	13.10	to	14.40
Secant wall piling, 750 mm dia. piles including site mobilization and demobilization; Measure total area of piling (total length × total depth)				
supply and install secant wall piling	m²	210.00	to	265.00
Steel sheet piling; Measure total area of piling (total length × total depth)				
interlocking steel sheet piling to excavation perimeter; Corus LX or similar; extraction on completion	m²	140.00	to	170.00
Pile caps				
Excavate pit in firm ground by machine, partial backfill, partial disposal, support, compaction, cut off pile and prepare reinforcement, formwork; reinforced in situ concrete cap 25 N/mm²; reinforcement at 50 kg/m³; rates per cap for cap size				
900 mm × 900 mm × 1000 mm; 1–2 piles	nr	290.00	to	355.00
2100 mm × 2100 mm × 1000 mm; 2–3 piles	nr	930.00	to	1125.00
2700 mm × 2700 mm × 1500 mm; 3–5 piles	nr	2075.00	to	2550.00
Excavate pit in firm ground by machine, partial backfill, partial disposal, support, compaction, cut off pile and prepare reinforcement, formwork; reinforced in situ concrete cap 25 N/mm²; reinforcement at 50 kg/m³; rates per m³ for typical cap size				
up to 900 mm × 900 mm × 1000 mm; 1–2 piles	m³	335.00	to	410.00
up to 2100 mm × 2100 mm × 1000 mm; 2–3 piles	m³	215.00	to	260.00
up to 2700 mm × 2700 mm × 1500 mm; 3–5 piles	m³	190.00	to	235.00
extra for				
reinforcement at 75 kg/m³ concrete	m³	69.00	to	85.00
reinforcement at 100 kg/m³ concrete	m³	87.00	to	105.00
alternative strength concrete C30 (30.00 N/mm²)	m³	1.15	to	1.40
alternative strength concrete C40 (40.00 N/mm²)	m³	5.50	to	6.70
Concrete ground beams				
Reinforced in situ concrete ground beams; bar reinforcement; formwork				
300 mm × 300 mm, reinforcement at 180 kg/m³	m³	37.50	to	46.00
450 mm × 450 mm, reinforcement at 200 kg/m³	m³	73.00	to	89.00
450 mm × 600 mm, reinforcement at 270 kg/m³	m³	110.00	to	130.00
Precast concrete ground beams; square ends and dowelled. Supplied and installed in lengths to span between stranchion bases; concrete strength C50; maximum UDL of 4.5 kN/m; in situ work at beam ends not included				
350 mm × 425 mm	m	72.00	to	92.00
350 mm × 875 mm	m	170.00	to	220.00
350 mm × 1075 mm	m	230.00	to	300.00
450 mm × 400 mm	m	85.00	to	110.00
Concrete lift pits				
Excavate and disposal; reinforced concrete floor and walls; bitumen tanking as necessary				
1.65 m × 1.81 m × 1.60 m deep pit – 8 person lift/630 kg	nr	1650.00	to	2025.00
1.80 m × 2.50 m × 1.60 m deep pit –13 person lift/1000 kg	nr	2075.00	to	2500.00
3.00 m × 2.50 m × 1.60 m deep pit – 21 person lift/1700 kg	nr	2425.00	to	2950.00

1.1 SUBSTRUCTURE

Item	Unit	Range £		
1.1 SUBSTRUCTURE – cont				
Ground floor				
Mechanical excavation to reduce levels, disposal, level and compact, hardcore bed blinded with sand, 1200 gauge polythene damp-proof membrane, in situ concrete 20.00 N/mm² – 20 mm aggregate (1:2:4)				
150 mm thick concrete slab with 1 layer of A195 fabric reinforcement	m²	59.00	to	72.00
200 mm thick concrete slab with 1 layer of A252 fabric reinforcement	m²	65.00	to	79.00
250 mm thick concrete slab with 1 layer of A393 fabric reinforcement	m²	66.00	to	81.00
extra for				
every additional 50 mm thick concrete	m²	5.35	to	6.50
alternative strength concrete C30 (30.00 N/mm²)	m³	1.15	to	1.40
alternative strength concrete C40 (40.00 N/mm²)	m³	5.50	to	6.70
reinforcement at 25 kg/m³ concrete	m³	26.00	to	32.00
reinforcement at 50 kg/m³ concrete	m³	43.00	to	53.00
reinforcement at 75 kg/m³ concrete	m³	69.00	to	85.00
reinforcement at 100 kg/m³ concrete	m³	87.00	to	105.00
Warehouse ground floor				
Steel fibre reinforced floor slab placed using large pour construction techniques providing a finish floor flatness complying with FM2 special +/- 15 mm from datum. NOTE: Excavation, subbase and damp-proof membrane not included				
nominal 200 mm thick in situ concrete floor slab, concrete grade C40, reinforced with steel fibres, surface power floated and cured with a spray application of curing and hardening agent	m²	24.50	to	30.00
Suspended ground floor				
Beam and block flooring				
suspended floor with 150 mm deep precast concrete beams and infill blocks	m²	19.80	to	25.00
suspended floor with 225 mm deep precast concrete beams and infill blocks	m²	24.00	to	30.00
Board or slab insulation				
Kingspan Thermafloor TF70 (Thermal conductivity 0.022 W/mK) rigid urethane floor insulation for solid concrete and suspended ground floors				
50 mm thick	m²	10.60	to	11.80
75 mm thick	m²	14.50	to	16.10
100 mm thick	m²	18.00	to	20.00
115 mm thick	m²	10.00	to	11.10
125 mm thick	m²	21.00	to	23.50
150 mm thick	m²	25.50	to	28.50
Styrofoam Floormate 500 (Thermal conductivity 0.033 W/mK) extruded polystyrene foam or other equal and approved				
50 mm thick	m²	11.40	to	12.70
80 mm thick	m²	13.90	to	15.40
120 mm thick	m²	16.90	to	18.80

2.1 FRAME

Item	Unit	Range £		

Underpinning
In stages not exceeding 1500 mm long from one side of existing wall and foundation, excavate preliminary trench by machine and underpinning pit by hand, partial backfill, partial disposal, earthwork support (open boarded), cutting away projecting foundations, prepare underside of existing, compact base of pit, plain in situ concrete 20.00 N/mm² – 20 mm aggregate (1:2:4), formwork, brickwork in cement mortar (1:3), pitch polymer damp-proof course, wedge and pin to underside of existing with slates.
Commencing at 1.00 m below ground level with common bricks, depth of underpinning

Item	Unit	Range £		
900 mm high, one brick wall	m²	275.00	to	340.00
1500 mm high, one brick wall	m²	415.00	to	510.00
extra for excavating commencing				
2.00 m below ground level	m²	55.00	to	68.00
3.00 m below ground level	m²	115.00	to	140.00
4.00 m below ground level	m²	160.00	to	200.00

Temporary works
Roadways
formation of temporary roads to building perimeter comprising of geoxtile membrane and 300 mm MOT type 1; reduce level; spoil to heap on site

Item	Unit	Range £		
within 25 m of excavation	m²	15.00	to	18.30
installation of wheel wash facility and maintenance	nr	2300.00	to	2800.00

2 SUPERSTRUCTURE

2.1 FRAME
Comparative frame and upper floors; upper floor area (unless otherwise described)

Concrete frame; flat slab reinforced concrete floors up to 250 mm thick
Suspended slab; no coverings or finishes

Item	Unit	Range £		
up to six storeys	m²	110.00	to	140.00
six to twelve storeys	m²	125.00	to	155.00
thirteen to eighteen storeys	m²	145.00	to	180.00

Steel frame; composite beam and slab floors
Suspended slab; permanent steel shuttering with 130 mm thick concrete; no coverings or finishes

Item	Unit	Range £		
up to six storeys	m²	110.00	to	140.00
seven to twelve storeys	m²	130.00	to	160.00
thirteen to eighteen storeys	m²	140.00	to	180.00

Frame only; reinforced in situ concrete columns, bar reinforcement, formwork. Generally all formwork assumes four uses
Reinforcement rate 180 kg/m³; column size

Item	Unit	Range £		
225 mm × 225 mm	m	58.00	to	75.00
300 mm × 600 mm	m	130.00	to	165.00
450 mm × 900 mm	m	235.00	to	305.00

2.1 FRAME

Item	Unit	Range £		
2.1 FRAME – cont				
Frame only – cont				
Reinforcement rate 240 kg/m³; column size				
225 mm × 225 mm	m	65.00	to	85.00
300 mm × 600 mm	m	140.00	to	180.00
450 mm × 900 mm	m	265.00	to	340.00
In situ concrete casing to steel column, formwork; column size				
225 mm × 225 mm	m	51.00	to	66.00
300 mm × 600 mm	m	120.00	to	155.00
450 mm × 900 mm	m	205.00	to	270.00
Frame only; reinforced in situ concrete beams, bar reinforcement, formwork. Generally all formwork assumes four uses				
Reinforcement rate 200 kg/m³; beam size				
225 mm × 450 mm	m	74.00	to	96.00
300 mm × 600 mm	m	120.00	to	160.00
450 mm × 600 mm	m	160.00	to	205.00
600 mm × 600 mm	m	195.00	to	250.00
Reinforcement rate 240 kg/m³; beam size				
225 mm × 450 mm	m	81.00	to	105.00
300 mm × 600 mm	m	130.00	to	170.00
450 mm × 600 mm	m	165.00	to	215.00
600 mm × 600 mm	m	210.00	to	270.00
In situ concrete casing to steel beams, formwork; beam size				
225 mm × 450 mm	m	67.00	to	86.00
300 mm × 600 mm	m	100.00	to	130.00
450 mm × 600 mm	m	135.00	to	180.00
600 mm × 600 mm	m	165.00	to	210.00
Other floor and frame constructions				
reinforced concrete cantilevered balcony; up to 1.50 m wide × 1.20 deep	nr	2075.00	to	2700.00
reinforced concrete cantilevered walkways; up to 1.00 m wide	m²	140.00	to	180.00
reinforced concrete walkways and supporting frame; up to 1.00 m wide × 2.50 m high	m²	180.00	to	235.00
reinforced concrete core with steel umbrella frame, twelve to twenty four storeys	m²	340.00	to	440.00
Frame only; steel frame				
Fabricated steelwork erected on site with bolted connections, primed				
smaller sections ne 40 kg/m	tonne	1625.00	to	1975.00
universal beams; grade S275	tonne	1450.00	to	1750.00
universal beams; grade S355	tonne	1500.00	to	1850.00
universal columns; grade S275	tonne	1475.00	to	1775.00
universal columns; grade S355	tonne	1525.00	to	1875.00
composite columns	tonne	1350.00	to	1650.00
hollow section circular	tonne	1775.00	to	2175.00
hollow section square or rectangular	tonne	1700.00	to	2075.00
cellular beams (FABSEC)	tonne	1775.00	to	2175.00
lattice beams	tonne	2075.00	to	2500.00
roof trusses	tonne	2175.00	to	2650.00

2.1 FRAME

Item	Unit	Range £		
Steel finishes				
grit blast and one coat zinc chromate primer	m²	6.65	to	8.60
touch up primer and one coat of two pack epoxy zinc phosphate primer	m²	4.60	to	5.95
blast cleaning	m²	2.30	to	3.00
galvanizing	m²	10.30	to	13.40
galvanizing	tonne	150.00	to	200.00
Other floor and frame constructions				
space deck on steel frame, unprotected	m²	240.00	to	315.00
exposed steel frame for tent/mast structures	m²	225.00	to	290.00
columns and beams to 18.00 m high bay warehouse unprotected	m²	135.00	to	175.00
columns and beams to mansard protected	m²	105.00	to	135.00
feature columns and beams to glazed atrium roof unprotected	m²	125.00	to	160.00
extra for				
wrought formwork	m²	5.90	to	7.65
sound reducing quilt in screed	m²	5.10	to	6.65
insulation to avoid cold bridging	m²	6.40	to	8.30
Fire protection to steelwork				
Sprayed mineral fibre; gross surface area				
60 minute protection	m²	8.55	to	11.10
90 minute protection	m²	15.70	to	20.00
Sprayed vermiculite cement; gross surface area				
60 minute protection	m²	9.50	to	12.30
90 minute protection	m²	14.30	to	18.50
Supply and fit fire-resistant boarding to steel columns and beams; noggins, brackets and angles, intumescent paste. Beamclad or similar; measure board area				
30 minute protection for concealed applications	m²	29.00	to	37.00
60 minute protection for concealed applications	m²	41.00	to	53.00
30 minute protection left exposed for decoration	m²	35.00	to	45.50
60 minute protection left exposed for decoration	m²	48.00	to	62.00
Intumescent fire protection coating/decoration to exposed steelwork; Gross surface area (m²) or per tonne; On site application, spray applied				
30 minute protection per m²	m²	8.70	to	11.10
30 minute protection per tonne	tonne	220.00	to	280.00
60 minute protection per m²	m²	11.00	to	14.10
60 minute protection per m²	tonne	280.00	to	360.00
Intumescent fire protection coating/decoration to exposed steelwork; Gross surface area (m²) or per tonne; Off site application, spray applied				
60 minute protection per m²	m²	19.20	to	23.50
60 minute protection per tonne	tonne	490.00	to	600.00
90 minute protection per m²	m²	32.00	to	39.00
90 minute protection per tonne	tonne	810.00	to	990.00
120 minute protection per m²	m²	48.00	to	59.00
120 minute protection per tonne	tonne	1225.00	to	1475.00

Approximate Estimating Rates

2.3 ROOF

Item	Unit	Range £		
2.2 UPPER FLOORS				
Rates for area of flooring				
Composite steel and concrete upper floors (Note: all floor thicknesses are nominal); A142 mesh reinforcement				
TATA Slimdek SD225 steel decking 1.25 mm thick; 130 mm reinforced concrete topping	m²	78.00	to	94.00
Re-entrant type steel deck 0.90 mm thick; 150 mm reinforced concrete	m²	48.00	to	58.00
Re-entrant type steel deck 60 mm deep, 1.20 mm thick; 200 mm reinforced concrete	m²	53.00	to	64.00
Trapezoidal steel decking 0.90 mm thick; 150 mm reinforced concrete topping	m²	46.00	to	56.00
Trapezoidal steel decking 0.90 mm thick; 150 mm reinforced concrete topping	m²	49.00	to	59.00
Trapezoidal steel decking 1.20 mm thick; 150 mm reinforced concrete topping	m²	48.00	to	58.00
Composite precast concrete and in situ concrete upper floors				
Omnidec (Hanson Building Products) flooring system comprising of 50 mm thick precast concrete deck (as permanent shuttering); 210 mm polysterene void formers; concrete topping and reinforcement	m²	78.00	to	94.00
Post-tensioned concrete upper floors				
reinforced post-tensioned suspended concrete slab 150–225 mm thick, 40 N/mm², reinforcement 60 kg/m³, formwork	m²	88.00	to	105.00
Precast concrete suspended floors				
1200 mm wide suspended slab; 75 mm thick screed; no coverings or finishes				
3.00 m span; 150 mm thick planks; 5.00 kN/m² loading	m²	61.00	to	70.00
6.00 m span; 150 mm thick planks; 5.00 kN/m² loading	m²	62.00	to	71.00
7.50 m span; 200 mm thick planks; 5.00 kN/m² loading	m²	65.00	to	75.00
9.00 m span; 250 mm thick planks; 5.00 kN/m² loading	m²	68.00	to	79.00
12.00 m span; 350 mm thick planks; 5.00 kN/m² loading	m²	72.00	to	83.00
3.00 m span; 150 mm thick planks; 8.50 kN/m² loading	m²	59.00	to	68.00
6.00 m span; 200 mm thick planks; 8.50 kN/m² loading	m²	61.00	to	71.00
7.50 m span; 250 mm thick planks; 8.50 kN/m² loading	m²	68.00	to	79.00
3.00 m span; 150 mm thick planks; 12.50 kN/m² loading	m²	62.00	to	72.00
6.00 m span; 250 mm thick planks; 12.50 kN/m² loading	m²	68.00	to	79.00
Softwood floors				
joisted floor; plasterboard ceiling; skim; emulsion; t&g chipboard, sheet vinyl flooring and painted softwood skirtings	m²	52.00	to	68.00
2.3 ROOF				
Timber; roof plan area (unless otherwise decribed)				
Timber roof trusses; insulation; roof coverings; PVC rainwater goods; plasterboard; skim and emulsion to ceilings (U-value = 0.25 W/m² K)				
concrete interlocking tile coverings	m²	110.00	to	140.00
clay pan tile coverings	m²	120.00	to	160.00
plain clay tile coverings	m²	140.00	to	180.00
natural slate coverings	m²	150.00	to	200.00
composite slate coverings	m²	125.00	to	160.00
reconstructed stone coverings	m²	140.00	to	180.00

2.3 ROOF

Item	Unit	Range £		
Timber dormer roof trusses; insulation; roof coverings; PVC rainwater goods; plasterboard; skim and emulsion to ceilings (U-value = 0.25 W/m² K)				
concrete interlocking tile coverings	m²	150.00	to	190.00
clay pantile coverings	m²	150.00	to	200.00
plain clay tile coverings	m²	170.00	to	220.00
natural slate coverings	m²	175.00	to	225.00
composite slate coverings	m²	155.00	to	200.00
reconstructed stone coverings	m²	165.00	to	215.00
extra for				
end of terrace semi/detached configuration	m²	30.50	to	39.50
hipped roof configuration	m²	33.00	to	43.00
Flat roofing systems				
Includes insulation and vapour control barrier; excludes decking (U-value = 0.25 W/m² K)				
single layer polymer roofing membrane	m²	68.00	to	88.00
single layer polymer roofing membrane with tapered insulation	m²	120.00	to	150.00
20 mm thick polymer modified asphalt roofing including felt underlay	m²	70.00	to	90.00
high performance bitumen felt roofing system	m²	87.00	to	110.00
high performance polymer modified bitumen membrane	m²	91.00	to	120.00
Edges to felt flat roofs; softwood splayed fillet				
280 mm × 25 mm painted softwood fascia; no gutter aluminium edge trim	m	32.50	to	42.00
Edges to flat roofs; code 4 lead drip dresses into gutter; 230 mm × 25 mm painted softwood fascia				
100 mm uPVC gutter	m	34.50	to	44.50
100 mm cast iron gutter; decorated	m	57.00	to	73.00
Roof walkways				
600 mm × 600 mm × 600 mm precast concrete slabs on support system; heavy pedestrian access	m²	49.50	to	64.00
extra for				
solar reflective paint	m²	2.20	to	2.85
limestone chipping finish	m²	4.10	to	5.30
grip tiles in hot bitumen	m²	28.00	to	36.50
Softwood trussed pitched roofs				
Structure only				
comprising 75 mm × 50 mm Fink roof trusses at 600 mm centres (measured on plan)	m²	26.00	to	33.00
comprising 100 mm × 38 mm Fink roof trusses at 600 mm centres (measured on plan)	m²	28.00	to	36.50
Mansard type roof comprising 100 mm × 50 mm roof trusses at 600 mm centres; 70° pitch	m²	29.00	to	38.00
forming dormers	m²	495.00	to	640.00
Steel trusses with metal sheet cladding				
Steel roof trusses and beams; thermal and acoustic insulation (U-value = 0.25 W/m² K)				
aluminium profiled composite cladding	m²	205.00	to	265.00
copper roofing on boarding	m²	220.00	to	280.00

2.3 ROOF

Item	Unit	Range £	
2.3 ROOF – cont			
Flat roof decking			
Softwood flat roofs; structure only			
comprising roof joists; 100 mm × 50 mm wall plates; herringbone strutting;			
no coverings or finishes	m²	41.00 to	54.00
Metal decking			
galvanized steel roof decking; insulation; three layer felt roofing and			
chippings; 0.70 mm thick steel decking (U-value = 0.25 W/m² K)	m²	58.00 to	74.00
aluminium roof decking; three layer felt roofing and chippings; 0.90 mm			
thick aluminium decking (U-value = 0.25 W/m² K)	m²	66.00 to	86.00
Concrete decking flat roofs; structure only			
precast concrete suspended slab with sand: cement screed over	m²	58.00 to	75.00
reinforced concrete slabs; on steel permanent steel shuttering; 150 mm			
reinforced concrete topping	m²	49.00 to	59.00
Screeds/decks to receive roof coverings			
18 mm thick external quality plywood boarding	m²	18.30 to	23.50
50 mm thick cement and sand screed	m²	12.20 to	15.40
75 mm thick lightweight bituminous screed and vapour barrier	m²	17.90 to	23.00
Roof claddings			
Fibre cement sheet profiled cladding			
Profile 6; single skin; natural grey finish	m²	17.40 to	22.50
P61 Insulated System; natural grey finish; metal inner lining panel			
(U-value = 0.25 W/m² K)	m²	32.00 to	42.00
extra for			
coloured fibre cement sheeting	m²	2.80 to	3.65
single skin GRP translucent roof sheets	m²	39.00 to	50.00
double skin GRP translucent roof sheets	m²	54.00 to	70.00
Steel PVF2 coated galvanized trapezoidal profile cladding on steel purlins			
single skin trapezoidal	m²	17.90 to	23.00
built up system; insulation; metal inner lining panel (U-value =			
0.25 W/m² K)	m²	33.50 to	43.50
composite insulated roofing system; 80 mm overall panel thickness			
(U-value = 0.25 W/m² K)	m²	48.00 to	62.00
standing seam joints composite insulated roofing system; 80 mm overall			
panel thickness (U-value = 0.25 W/m² K)	m²	81.00 to	105.00
Aluminum roofing; standing seam			
Kalzip standard natural aluminium; 0.9 mm thick; 180 mm glassfibre			
insulation; vapour control layer; liner sheets; (U-value = 0.25 W/m² K)	m²	54.00 to	66.00
Copper roofing			
copper roofing with standing seam joints; insulation breather membrane			
or vapour barrier (U-value = 0.25 W/m² K)	m²	87.00 to	110.00
Comparative tiling and slating finishes			
Including underfelt, battening, eaves courses and ridges			
concrete troughed or bold roll interlocking tiles; sloping	m²	28.00 to	36.00
Tudor clay pantiles; sloping	m²	31.00 to	40.00
natural red pantiles; sloping	m²	36.00 to	46.00

2.3 ROOF

Item	Unit	Range £		
blue composition (cement fibre) slates; sloping	m^2	32.00	to	42.00
machine-made clay plain tiles; sloping	m^2	48.00	to	62.00
Welsh natural slates; sloping	m^2	100.00	to	130.00
Spanish slates; sloping	m^2	59.00	to	77.00
man-made slates; sloping	m^2	54.00	to	70.00
reconstructed stone slates; random slates; sloping	m^2	43.50	to	56.00
handmade sandfaced plain tiles; sloping	m^2	72.00	to	94.00
Comparative cladding finishes (including boarding, underfelt, labourers, etc.)				
0.91 mm thick aluminium roofing; commercial grade; fixed to boarding				
flat	m^2	62.00	to	80.00
sloping	m^2	64.00	to	83.00
0.81 mm thick zinc roofing; fixed to boarding				
flat	m^2	76.00	to	98.00
sloping	m^2	85.00	to	110.00
Copper roofing; fixed to boarding				
0.56 mm thick; flat	m^2	72.00	to	94.00
0.56 mm thick; sloping	m^2	81.00	to	105.00
Stainless steel sheeting				
0.40 mm thick; sloping	m^2	83.00	to	110.00
Lead roofing				
code 5 sheeting; sloping	m^2	120.00	to	160.00
code 5 sheeting; vertical to mansard; including insulation	m^2	165.00	to	215.00
Landscaped roofs				
Polyester-based elastomeric bitumen waterproofing and vapour equalization layer, copper lined bitumen membrane root barrier and waterproofing layer, separation and slip layers, protection layer, 50 mm thick drainage board, filter fleece, insulation board, Sedum vegetation blanket				
intensive (high maintenance – may include trees and shrubs, require deeper substrate layers, are generally limited to flat roofs)	m^2	105.00	to	135.00
extensive (low maintenance – herbs, grasses, mosses and drought tolerant succulents such as Sedum)	m^2	100.00	to	130.00
Ethyletetrafluoroethylene (ETFE) systems				
Multiple layered ETFE inflated cushions supported by a lightweight aluminium or steel structure	m^2	630.00	to	840.00
Insulation				
Glass fibre roll; Crown Loft Roll 44 (Thermal conductivity 0.044 W/mK) or other equal; laid loose				
100 mm thick	m^2	2.00	to	2.45
150 mm thick	m^2	2.60	to	3.15
200 mm thick	m^2	3.15	to	3.80
Glass fibre quilt; Isover Modular roll (Thermal conductivity 0.043 W/mK) or other equal and approved; laid loose				
100 mm thick	m^2	2.55	to	3.10
150 mm thick	m^2	3.30	to	4.05
170 mm thick	m^2	4.10	to	4.95
200 mm thick	m^2	5.00	to	6.00

2.4 STAIRS AND RAMPS

Item	Unit	Range £	
2.3 ROOF – cont			
Insulation – cont			
Crown Rafter Roll 32 (Thermal conductivity 0.032 W/mK) glass fibre flanged building roll; pinned vertically or to slope between timber framing			
50 mm thick	m²	4.40 to	5.30
75 mm thick	m²	5.70 to	6.90
100 mm thick	m²	6.90 to	8.40
Thermafleec EcoRoll (0.039W/mK)			
50 mm thick	m²	5.20 to	6.30
75 mm thick	m²	6.60 to	7.95
100 mm thick	m²	8.25 to	10.00
140 mm thick	m²	11.00 to	13.40
Rooflights/patent glazing and glazed roofs			
Rooflights			
individual polycarbonate rooflights; rectangular	m²	310.00 to	400.00
individual polycarbonate rooflights; circular	m²	510.00 to	660.00
feature/ventilating	m²	305.00 to	395.00
Velux style rooflights to traditional roof construction (tiles/slates)	m²	540.00 to	700.00
Patent glazing; including flashings, standard aluminium Georgian wired			
single glazed	m²	305.00 to	395.00
double glazed	m²	360.00 to	460.00
Glazed roof; purpose-made polyester powder coated aluminium;			
double glazed low emissivity glass	m²	290.00 to	380.00
feature; to covered walkways	m²	335.00 to	435.00
Rainwater disposal			
Gutters			
100 mm uPVC gutter	m	22.50 to	29.00
150 mm uPVC gutter	m	28.00 to	36.50
100 mm cast iron gutter; decorated	m	36.00 to	47.00
150 mm cast iron gutter; decorated	m	43.50 to	56.00
Rainwater downpipes pipes; fixed to backgrounds; including offsets and shoes			
68 mm dia. uPVC	m	8.10 to	10.50
110 mm dia. uPVC	m	11.50 to	14.90
75 mm dia. cast iron; decorated	m	29.00 to	37.00
100 mm dia. cats iron; decorated	m	34.00 to	44.00
2.4 STAIRS AND RAMPS			
Reinforced concrete construction			
Escape staircase; granolithic finish; mild steel balustrades and handrails			
3.00 m rise; dogleg	nr	4800.00 to	6200.00
plus or minus for each 300 mm variation in storey height	nr	465.00 to	600.00
Staircase; terrazzo finish; mild steel balustrades and handrails; plastered soffit; balustrades and staircase soffit decorated			
3.00 m rise; dogleg	nr	7200.00 to	9400.00
plus or minus for each 300 mm variation in storey height	nr	700.00 to	910.00

2.4 STAIRS AND RAMPS

Item	Unit	Range £		
Staircase; terrazzo finish; stainless steel balustrades and handrails; plastered and decorated soffit				
3.00 m rise; dogleg	nr	8600.00	to	11000.00
plus or minus for each 300 mm variation in storey height	nr	850.00	to	1075.00
Staircase; high quality finishes; stainless steel and glass balustrades; plastered and decorated soffit				
3.00 m rise; dogleg	nr	14000.00	to	18000.00
plus or minus for each 300 mm variation in storey height	nr	1625.00	to	2125.00
Metal construction				
Steel access/fire ladder				
3.00 m high	nr	570.00	to	740.00
4.00 m high; epoxide finished	nr	890.00	to	1150.00
Light duty metal staircase; galvanized finish; perforated treads; no risers; balustrades and handrails; decorated				
3.00 m rise; straight; 900 mm wide	nr	3250.00	to	4200.00
plus or minus for each 300 mm variation in storey height	nr	260.00	to	330.00
Light duty circular metal staircase; galvanized finish; perforated treads; no risers; balustrades and handrails; decorated				
3.00 m rise; straight; 1548 mm dia.	nr	3650.00	to	4700.00
plus or minus for each 300 mm variation in storey height	nr	320.00	to	410.00
Heavy duty cast iron staircase; perforated treads; no risers; balustrades and hand rails; decorated				
3.00 m rise; straight	nr	4100.00	to	5400.00
plus or minus for each 300 mm variation in storey height	nr	425.00	to	550.00
3.00 m rise; spiral; 1548 mm dia.	nr	4700.00	to	6100.00
plus or minus for each 300 mm variation in storey height	nr	470.00	to	610.00
Feature metal staircase; galvanized finish perforated treads; no risers; decorated				
3.00 m rise; spiral balustrades and handrails	nr	5400.00	to	7000.00
3.00 m rise; dogleg; hardwood balustrades and handrails	nr	6400.00	to	8300.00
3.00 m rise; dogleg; stainless steel balustrades and handrails	nr	8400.00	to	11000.00
plus or minus for each 300 mm variation in storey height	nr	550.00	to	710.00
galvanized steel catwalk; nylon coated balustrading 450 mm wide	m	305.00	to	395 00
Timber construction				
Softwood staircase; softwood balustrades and hardwood handrail; plasterboard; skim and emulsion to soffit				
2.60 m rise; standard; straight flight	nr	800.00	to	1025.00
2.60 m rise; standard; top three treads winding	nr	910.00	to	1175.00
2.60 m rise; standard; dogleg	nr	1000.00	to	1275.00
Oak staircase; balustrades and handrails; plasterboard; skim and emulsion to soffit				
2.60 m rise; purpose-made; dogleg	nr	6800.00	to	8800.00
plus or minus for each 300 mm variation in storey height	nr	910.00	to	1175.00
Comparative finishes/balustrading				
Wall handrails				
Softwood handrail and brackets	m	60.00	to	78.00
Hardwood handrail and brackets	m	85.00	to	110.00
Mild steel handrail and brackets	m	120.00	to	160.00
Stainless steel handrail and brackets	m	140.00	to	185.00

2.5 EXTERNAL WALLS

Item	Unit	Range £		
2.4 STAIRS AND RAMPS – cont				
Comparative finishes/balustrading – cont				
Balustrading and handrails				
Mild steel balustrade and steel or timber handrail	m	235.00	to	305.00
Balustrade and handrail with metal infill panels	m	295.00	to	380.00
Balustrade and handrail with glass infill panels	m	325.00	to	420.00
Stainless steel balustrade and handrail	m	375.00	to	485.00
Stainless steel and structural glass balustrade	m	650.00	to	850.00
Finishes to treads and risers; including nosings etc.				
vinyl or rubber	m	38.50	to	50.00
carpet (PC sum £25/m²)	m	55.00	to	71.00
2.5 EXTERNAL WALLS				
Wall area (unless otherwise described)				
Brick/block walling				
Common brick solid walls; bricks PC £240.00/1000				
half brick thick	m²	34.00	to	44.00
one brick brick	m²	63.00	to	81.00
one and a half brick thick	m²	91.00	to	120.00
two brick thick	m²	115.00	to	150.00
extra for				
fair face one side	m²	1.80	to	2.30
Engineering brick walls; class B; bricks PC £375.00/1000				
half brick thick	m²	38.00	to	49.50
one brick thick	m²	70.00	to	91.00
one and a half brick thick	m²	100.00	to	130.00
two brick thick	m²	130.00	to	170.00
Facing brick walls; sand faced facings; bricks PC £350.00/1000				
half brick thick; pointed one side	m²	46.00	to	59.00
one brick thick; pointed both sides	m²	89.00	to	115.00
Facing bricks solid walls; handmade facings; bricks PC £500.00/1000				
half brick thick; fair face one side	m²	56.00	to	73.00
one brick thick; fair face both sides	m²	110.00	to	140.00
ADD or DEDUCT for each variation of £10.00/1000 in PC value				
half brick thick	m²	0.65	to	0.85
one brick thick	m²	1.30	to	1.70
one and a half brick thick	m²	1.95	to	2.50
two brick thick	m²	2.60	to	3.40
Cavity wall; facing brick outer skin; insulation; plasterboard on stud inner skin; emulsion (U-value = 0.30 W/m² K)				
machine-made facings; PC £350.00/1000	m²	70.00	to	91.00
machine-made facings; PC £500.00/1000	m²	80.00	to	105.00
Cavity wall; facing brick outer skin; insulation; with plaster on standard weight block inner skin; emulsion (U-value = 0.30 W/m² K)				
machine-made facings; PC £350.00/1000	m²	84.00	to	110.00
machine-made facings; PC £500.00/1000	m²	94.00	to	120.00
add or deduct for				
each variation of £10.00/1000 in PC value	m²	0.65	to	0.85

2.5 EXTERNAL WALLS

Item	Unit	Range £		
Aerated lightweight block walls				
100 mm thick	m²	15.80	to	20.50
140 mm thick	m²	19.90	to	26.00
215 mm thick	m²	27.00	to	35.00
Dense aggregate block walls				
100 mm thick	m²	19.20	to	25.00
140 mm thick	m²	25.00	to	33.00
Coloured dense aggregate masonry block walls; Lignacite or similar				
100 mm thick	m²	41.00	to	53.00
140 mm thick	m²	49.50	to	64.00
Cavity wall; block; insulated to U-value = 0.30 W/m² K				
coloured masonry block outer skin and paint grade block inner skins; fair faced both sides	m²	66.00	to	86.00
block outer skin; insulation; lightweight block inner skin outer block rendered	m²	61.00	to	79.00
Reinforced concrete walling				
In situ reinforced concrete 25.00 N/mm²; 15 kg/m² reinforcement; formwork both sides				
150 mm thick	m²	84.00	to	110.00
225 mm thick	m²	96.00	to	125.00
300 mm thick	m²	105.00	to	140.00
Panelled walling				
Precast concrete panels; including insulation; lining and fixings generally 7.5 m × 0.15 thick × storey height (U-value = 0.30 W/m² K)				
standard panels	m²	190.00	to	250.00
standard panels; exposed aggregate finish	m²	210.00	to	275.00
reconstructed stone faced panels	m²	240.00	to	310.00
brick clad panels (P.C £350.00/1000 for bricks)	m²	320.00	to	420.00
natural stone faced panels (Portland Stone or similar)	m²	410.00	to	540.00
marble or granite faced panels	m²	540.00	to	700.00
Kingspan TEK cladding panel; 142 mm thick (15 mm thick OSB board and 112 mm thick rigid eurathane core) fixed to frame; metal edge trim flashings	m²	240.00	to	310.00
Sheet claddings				
Non-asbestos profiled cladding				
Profile 6; single skin; natural grey finish	m²	21.00	to	27.00
P61 Insulated System; natural grey finish; metal inner lining panel (U-value = 0.30 W/m² K)	m²	40.50	to	52.00
extra for				
coloured fibre cement sheeting	m²	2.20	to	2.85
insulated; with 2.80 m high block inner skin; emulsion	m²	26.00	to	34.00
insulated; with 2.80 m high block inner skin plasterboard lining on metal tees; emulsion	m²	40.00	to	52.00
Metal profiled cladding (U-value = 0.30 W/m² K)				
coated steel profiled cladding on steel rails; insulated built up system	m²	41.00	to	53.00
coated steel micro-rib profiled cladding on steel rails; composite sandwich panel system	m²	75.00	to	97.00
coated aluminium profiled cladding on steel rails; insulated built up system	m²	43.50	to	56.00
coated aluminium flat panel cladding on steel rails; insulated built up system	m²	110.00	to	145.00

2.5 EXTERNAL WALLS

Item	Unit	Range £		
2.5 EXTERNAL WALLS – cont				
Glazed walling				
Curtain walling				
stick curtain walling with double glazed units, aluminium structural framing and spandrel rails. Standard colour powder coated capped	m^2	380.00	to	465.00
unitized curtain walling system with double glazed units, aluminium structural framing and spandrel rails. Standard colour powder coated	m^2	790.00	to	960.00
unitized curtain walling bespoke project specific system with double glazed units, aluminium structural framing and spandrel rails. Standard colour powder coated	m^2	790.00	to	960.00
visual mock-ups for project specific curtain walling solutions	item	25000.00	to	31000.00
Solar shading				
fixed aluminium Brise Soleil including uni-strut supports; 300 mm deep	m	125.00	to	155.00
Other systems				
lift surround of double glazed or laminated glass with aluminium or stainless steel framing	m^2	720.00	to	880.00
Other cladding systems				
Tiles (clay/slate/glass/ceramic)				
machine-made clay tiles; including battens	m^2	39.50	to	51.00
best handmade sand faced tiles; including battens	m^2	51.00	to	67.00
concrete plain tiles; including battens	m^2	44.50	to	57.00
natural slates; including battens	m^2	65.00	to	85.00
20 mm × 20 mm thick mosaic glass or ceramic; in common colours; fixed on prepared surface	m^2	91.00	to	120.00
Steel				
vitreous enamelled insulated steel sandwich panel system; with insulation board on inner face	m^2	150.00	to	190.00
Formalux sandwich panel system; with coloured lining tray; on steel cladding rails	m^2	170.00	to	220.00
Rainscreen				
25 mm thick tongued and grooved tanalized softwood boarding; including timber battens	m^2	35.00	to	45.00
timber shingles, Western Red Cedar, preservative treated in random widths; not including any subframe or battens	m^2	44.50	to	57.00
25 mm thick tongued and grooved Western Red Cedar boarding including timber battens	m^2	39.50	to	51.00
25 mm thick Western Red Cedar tongued and grooved wall cladding on and including treated softwood battens on breather mambrane, 10 mm Eternit Blueclad board and 50 mm insulation board; the whole fixed to Metsec frame system; including sealing all joints etc.	m^2	86.00	to	110.00
Trespa single skin rainscreen cladding, 8 mm thick panel, with secondary support/frame system; adhesive fixed panels; open joints; 100 mm Kingspan K15 insulation; aluminium subframe fixed to masonry/concrete	m^2	185.00	to	235.00
Terracotta rainscreen cladding; aluminium support rails; anti-graffiti coating	m^2	280.00	to	375.00
Corium brick tiles in metal tray system; standard colour; including all necessary angle trim on main support system; comprising of polythene vapour check; 75 mm × 50 mm timber studs; 50 mm thick rigid insulation with taped joints; 38 mm × 50 mm timber counterbattens	m^2	260.00	to	340.00

2.5 EXTERNAL WALLS

Item	Unit	Range £		
Comparative external finishes				
Comparative concrete wall finishes				
wrought formwork one side including rubbing down	m²	4.30	to	5.60
shotblasting to expose aggregate	m²	5.35	to	6.90
bush hammering to expose aggregate	m²	13.80	to	17.80
Comparative in situ finishes				
two coats Sandtex Matt cement paint to render	m²	5.05	to	6.10
13 mm thick cement and sand plain face rendering	m²	17.00	to	22.00
ready-mixed self-coloured acrylic resin render on blockwork	m²	42.00	to	57.00
three coat Tyrolean rendering; including backing	m²	31.00	to	40.00
Stotherm Lamella; 6 mm thick work in 2 coats	m²	59.00	to	77.00
Insulation				
Crown Dritherm Cavity Slab 37 (Thermal conductivity 0.037 W/mK) glass fibre batt or other equal; as full or partial cavity fill; including cutting and fitting around wall ties and retaining discs				
50 mm thick	m²	4.30	to	5.15
75 mm thick	m²	4.70	to	5.65
100 mm thick	m²	4.70	to	5.65
Crown Dritherm Cavity Slab 34 (Thermal conductivity 0.034 W/mK) glass fibre batt or other equal; as full or partial cavity fill; including cutting and fitting around wall ties and retaining discs				
65 mm thick	m²	3.60	to	4.35
75 mm thick	m²	4.00	to	4.85
85 mm thick	m²	5.10	to	6.10
100 mm thick	m²	5.20	to	6.30
Crown Dritherm Cavity Slab 32 (Thermal conductivity 0.032 W/mK) glass fibre batt or other equal; as full or partial cavity fill; including cutting and fitting around wall ties and retaining discs				
65 mm thick	m²	4.10	to	5.00
75 mm thick	m²	4.65	to	5.60
85 mm thick	m²	5.00	to	6.05
100 mm thick	m²	5.60	to	6.80
Crown Frametherm Roll 40 (Thermal conductivity 0.040 W/mK) glass fibre semi-rigid or rigid batt or other equal; pinned vertically in timber frame construction				
90 mm thick	m²	4.70	to	5.70
140 mm thick	m²	6.30	to	7.60
Kay-Cel (Thermal conductivity 0.033 W/mK) expanded polystyrene board standard grade SD/N or other equal; fixed with adhesive				
20 mm thick	m²	3.45	to	4.20
25 mm thick	m²	3.55	to	4.30
30 mm thick	m²	3.65	to	4.40
40 mm thick	m²	4.00	to	4.90
50 mm thick	m²	4.40	to	5.30
60 mm thick	m²	4.75	to	5.75
75 mm thick	m²	5.20	to	6.30
100 mm thick	m²	5.85	to	7.10

2.6 WINDOWS AND EXTERNAL DOORS

Item	Unit	Range £		

2.5 EXTERNAL WALLS – cont

Insulation – cont
KIngspan Kooltherm K8 (Thermal conductivity 0.022 W/mK) zero ODP rigid urethene insulation board or other equal; as partial cavity fill; including cutting and fitting around wall ties and retaining discs

Item	Unit		Range £	
40 mm thick	m²	7.40	to	8.95
50 mm thick	m²	8.60	to	10.40
60 mm thick	m²	9.70	to	11.80
75 mm thick	m²	11.40	to	13.80

KIngspan Thermawall TW50 (Thermal conductivity 0.022 W/mK) zero ODP rigid urethene insulation board or other equal and approved; as partial cavity fill; including cutting and fitting around wall ties and retaining discs

Item	Unit		Range £	
25 mm thick	m²	7.40	to	9.00
50 mm thick	m²	10.50	to	12.70
75 mm thick	m²	14.50	to	17.60
100 mm thick	m²	18.10	to	22.00

Thermafleece TF35 high density wool insulating batts (0.035 W/mK); 60% British wool, 30% recycled polyester and 10% polyester binder with a high recycled content

Item	Unit		Range £	
50 mm thick	m²	13.10	to	15.90
75 mm thick	m²	16.30	to	19.70

2.6 WINDOWS AND EXTERNAL DOORS
Window and external door area (unless otherwise described)

Softwood windows (U-value = 1.6 W/m² K)
Standard windows

Item	Unit		Range £	
painted; double glazed; up to 1.50 m²	m²	230.00	to	300.00
painted; double glazed; over 1.50 m², up to 3.20 m²	m²	190.00	to	240.00

Purpose-made windows

Item	Unit		Range £	
painted; double glazed; up to 1.50 m²	m²	440.00	to	570.00
painted; double glazed; over 1.50 m²	m²	390.00	to	510.00

Hardwood windows (U-value = 1.6 W/m² K)
Standard windows; stained

Item	Unit		Range £	
double glazed	m²	460.00	to	600.00

Purpose-made windows; stained

Item	Unit		Range £	
double glazed	m²	520.00	to	670.00

Steel windows (U-value = 1.6 W/m² K)
Standard windows

Item	Unit		Range £	
double glazed; powder coated	m²	280.00	to	365.00

Purpose-made windows

Item	Unit		Range £	
double glazed; powder coated	m²	350.00	to	450.00

2.6 WINDOWS AND EXTERNAL DOORS

Item	Unit	Range £		
uPVC windows				
Windows; standard ironmongery; cills and factory glazed with low E 24 mm double glazing				
standard with low E 24 mm double glazing	m²	155.00	to	200.00
WER A rating	m²	170.00	to	220.00
WER C rating	m²	155.00	to	200.00
Secured by Design accreditation	m²	160.00	to	210.00
extra for				
colour finish to uPVC	m²	26.00	to	33.50
Softwood external doors				
Standard external softwood doors and hardwood frames; doors painted; including ironmongery				
Matchboarded, framed, ledged and braced door, 838 mm × 1981 mm	nr	355.00	to	460.00
flush door; cellular core; plywood faced; 838 mm × 1981 mm	nr	370.00	to	475.00
Heavy duty solid flush door				
single leaf	nr	810.00	to	1050.00
double leaf	nr	1375.00	to	1775.00
single leaf; emergency fire exit	nr	1075.00	to	1400.00
double leaf; emergency fire exit	nr	1625.00	to	2125.00
Steel external doors				
Standard doors				
single external steel door, including frame, ironmongery, powder coated finish	nr	620.00	to	800.00
single external steel security door, including frame, ironmongery, powder coated finish	nr	1350.00	to	1725.00
double external steel security door, including frame, ironmongery, powder coated finish	nr	1975.00	to	2600.00
Steel roller shutter doors				
single skin; manual	m²	210.00	to	275.00
single skin; electric	m²	280.00	to	360.00
insulated; manual; 1.4 W/m² K	m²	210.00	to	275.00
insulated; electric; 1.4 W/m² K; 1 hour fire-resistant	m²	275.00	to	355.00
uPVC external doors				
Entrance doors; residential standard; uPVC frame; brass furniture (spyhole/ security chain/letter plate/draught excluder/multipoint locking)				
overall 1480 × 2100 mm with glazed side panel	nr	550.00	to	690.00
overall 1800 × 2100 mm with glazed side panel	nr	520.00	to	650.00
overall 2430 × 2100 mm with glazed side panel each side	nr	610.00	to	770.00
Composite aluminium/timber windows, entrance screens and doors; U value = 1.5 W/m² K				
Purpose-made windows; stainless steel ironmongery; Velfac System 200 or similar				
fixed windows up to 1.50 m²	m²	205.00	to	265.00
fixed windows over 1.50 m² up to 4.00 m²	m²	185.00	to	240.00
outward opening pivot windows up to 1.50 m²	m²	510.00	to	660.00
outward opening pivot windows over 1.50 m² up to 4.00 m²	m²	220.00	to	290.00
round porthole window 1.40 m dia.	nr	640.00	to	820.00
round porthole window 1.80 m dia.	nr	960.00	to	1250.00
round porthole window 2.60 m dia.	nr	1825.00	to	2375.00
purpose-made entrance screens and doors double glazed	m²	810.00	to	1050.00

2.7 INTERNAL WALLS AND PARTITIONS

Item	Unit	Range £		

2.6 WINDOWS AND EXTERNAL DOORS – cont

Composite aluminium/timber windows, entrance screens and doors – cont

Purpose-made doors

glazed single personnel door; stainless steel ironmongery; Velfac System100 or similar	nr	1425.00	to	1850.00
glazed double personnel door; stainless steel ironmongery; Velfac System 100 or similar	pair	2350.00	to	3050.00
revolving door; 2000 mm dia.; clear laminated glazing; 4 nr wings; glazed curved walls	nr	3900.00	to	5000.00
automatic sliding door; bi-parting	m²	2125.00	to	2750.00

Stainless steel entrance screens and doors

Purpose-made screen; double glazed

with manual doors	m²	1450.00	to	1875.00
with automatic doors	m²	1775.00	to	2275.00
purpose-made revolving door 2000 mm dia.; clear laminated glazing; 4 nr wings; glazed curved walls	m²	3900.00	to	5100.00
automatic sliding door; bi-parting	m²	2125.00	to	2750.00

Shop fronts, shutters and grilles

Purpose-made screen

temporary timber shop fronts	m²	61.00	to	79.00
hardwood and glass; including high enclosed window beds	m²	650.00	to	850.00
flat facade; glass in aluminium framing; manual centre doors only	m²	570.00	to	730.00
grilles or shutters	m²	250.00	to	325.00
fire shutters; power operated	m²	275.00	to	355.00

2.7 INTERNAL WALLS AND PARTITIONS

Internal partition area (unless otherwise described)

Frame and panel partitions

Timber stud partitions

structure only comprising 100 mm × 38 mm softwood studs at 600 mm centres; head and sole plates	m²	17.70	to	21.50
softwood stud comprising 100 mm × 38 mm softwood studs at 600 mm centres; head and sole plates; 12.5 mm thick plasterboard each side; tape and fill joints; emulsion finish	m²	46.00	to	55.00

Metal stud and board partitions; height range from 2.40 m to 3.30 m

73 mm partition; 48 mm studs and channels; one layer of 12.5 mm Gyproc Wallboard each side; joints filled with joint filler and joint tape; emulsion paint finish; softwood skirtings with gloss finish	m²	31.00	to	38.00
102 mm partition; 70 mm studs and channels; one layer of 15 mm Gyproc Wallboard each side; joints filled with joint filler and joint tape; emulsion paint finish; softwood skirtings with gloss finish	m²	33.00	to	40.00
102 mm thick partition; 70 mm steel studs at 600 mm centres generally; 1 layer 15 mm Fireline board each side; joints filled with joint filler and joint tape; emulsion paint finish; softwood skirtings with gloss finish	m²	37.50	to	45.00
130 mm thick partition; 70 mm steel studs at 600 mm centres generally; 2 layers 15 mm Fireline board each side; joints filled with joint filler and joint tape; emulsion paint finish; softwood skirtings with gloss finish	m²	44.00	to	53.00

2.7 INTERNAL WALLS AND PARTITIONS

Item	Unit	Range £		
102 mm thick partition; 70 mm steel studs at 600 mm centres generally; 1 layer 15 mm Soundbloc board each side; joints filled with joint filler and joint tape; emulsion paint finish; softwood skirtings with gloss finish	m²	39.00	to	47.50
130 mm thick partition; 70 mm steel studs at 600 mm centres generally; 2 layers 15 mm Soundbloc board each side; joints filled with joint filler and joint tape; emulsion paint finish; softwood skirtings with gloss finish	m²	53.00	to	64.00
Alternative board finishes				
12 mm plywood boarding	m²	9.40	to	11.40
one coat Thistle board finish 3 mm thick work to walls; plasterboard base	m²	5.90	to	7.15
extra for				
curved work	%	15.00	to	20.00
Acoustic insulation to partitions				
Mineral fibre quilt; Isover Acoustic Partition Roll (APR 1200) or other equal; pinned vertically to timber or plasterboard				
25 mm thick	m²	2.10	to	2.10
50 mm thick	m²	2.70	to	2.70
Brick/block masonry partitions				
Brick				
Common brick half brick thick wall; bricks PC £240.00/1000	m²	32.50	to	42.00
Block				
aerated/lightweight block partitions				
100 mm thick	m²	20.00	to	26.00
140 mm thick	m²	27.00	to	34.50
200/215 mm thick	m²	33.50	to	43.50
dense aggregate block walls				
100 mm thick	m²	19.20	to	25.00
140 mm thick	m²	25.00	to	33.00
extra for				
fair face (rate per side)	m²	0.80	to	1.00
plaster and emulsion (rate per side)	m²	16.60	to	21.00
curved work	%	15.00	to	20.00
Concrete partitions				
Reinforced concrete walls; C25 strength; standard finish; reinforced at 100 kg/m³				
150 mm thick	m²	120.00	to	150.00
225 mm thick	m²	135.00	to	170.00
300 mm thick	m²	155.00	to	195.00
extra for				
plaster and emulsion (rate per side)	m²	16.60	to	21.00
curved work	%	20.00	to	30.00
Glass				
Hollow glass block walling; Pittsburgh Corning sealed Thinline or other equal and approved; in cement mortar joints; reinforced with 6 mm dia. stainless steel rods; pointed both sides with mastic or other equal and approved				
240 mm × 240 mm × 80 mm Flemish; cross reeded or clear blocks	m²	300.00	to	400.00
190 mm × 190 mm × 100 mm glass blocks; 30 minute fire-rated	m²	460.00	to	610.00
190 mm × 190 mm × 100 mm glass blocks; 60 minute fire-rated	m²	830.00	to	1100.00

2.7 INTERNAL WALLS AND PARTITIONS

Item	Unit	Range £		
2.7 INTERNAL WALLS AND PARTITIONS – cont				
Demountable/folding partitions				
Demountable partitioning; aluminium framing; veneer finish doors				
medium quality; 46 mm thick panels factory finish vinyl faced	m²	120.00	to	150.00
high quality; 46 mm thick panels factory finish vinyl faced	m²	155.00	to	200.00
Demountable aluminium/steel partitioning and doors				
high quality	m²	36.00	to	46.00
high quality; sliding	m²	690.00	to	900.00
Demountable fire partitions				
enamelled steel; half hour	m²	500.00	to	650.00
stainless steel; half hour	m²	810.00	to	1050.00
soundproof partitions; hardwood doors luxury veneered	m²	235.00	to	305.00
Aluminium internal patent glazing				
single glazed laminated	m²	110.00	to	145.00
double glazed; 1 layer toughened and 1 layer laminated glass	m²	180.00	to	230.00
Stainless steel glazed manual doors and screens				
high quality; to inner lobby of malls	m²	680.00	to	890.00
Acoustic folding partition; headtrack suspension and bracing; aluminium framed with high density particle board panel with additional acoustic insulation; melamine laminate finish; acoustic seals. Nominal weight approximately 55 kg per m²				
sound reduction 48 db (Rw)	m²	485.00	to	640.00
sound reduction 55 db (Rw)	m²	580.00	to	770.00
Framed panel cubicles				
Changing and WC cubicles; high pressure laminate faced MDF; proprietary system				
standard quality WC cubicle partition sets; aluminium framing; melamine face chipboard dividing panels and doors; ironmongery; small range (up to 5 cubicles); standard cubicle set; (rate per cubicle)	nr	370.00	to	445.00
medium quality WC cubicle partition sets; stainless steel framing; real wood veneer face chipboard dividing panels and doors; ironmongery; small range (up to 6 cubicles); standard cubicle set; (rate per cubicle)	nr	750.00	to	910.00
high end specification flush fronted system floor to ceiling 44 mm doors, no visible fixings; real wood veneer or HPL or high Gloss paint and lacquer; cubicle set; 800 mm × 1500 mm × 2400 mm high per cubicle, with satin finished stainless steel ironmongery; 30 mm high pressure laminated (HPL) chipboard divisions and 44 mm solid cored real wood veneered doors and pilasters; small range (up to 6 cubicles); standard cubicle set; (rate per cubicle)	nr	2075.00	to	2500.00
IPS duct panel systems; melamine finish chipboard; softwood timber subframe				
2.70 m high IPS back panelling system; to accomodate wash hand basins or urinals; access hatch; frame support	m²	105.00	to	130.00

2.8 INTERNAL DOORS

Item	Unit	Range £		
2.8 INTERNAL DOORS The following rates include for the supply and hang of doors, complete with all frames, architrave, typical ironmongery set and appropriate finish.				
Standard doors Standard doors; cellular core; softwood; softwood architrave; aluminuim ironmongery (latch only)				
single leaf; hardboard face; gloss paint finish	nr	245.00	to	300.00
single leaf; moulded panel; gloss paint finish	nr	250.00	to	300.00
single leaf; Sapele veneered finish	nr	265.00	to	320.00
Purpose-made doors Softwood panelled; softwood lining; softwood architrave; aluminium ironmongery (latch only); brass or stainless ironmongery (latch only); painting and polishing				
single leaf; four panels; mouldings	nr	330.00	to	445.00
double leaf; four panels; mouldings	nr	650.00	to	880.00
Hardwood panelled; hardwood lining; hardwood architrave; aluminium ironmongery (latch only); brass or stainless ironmongery (latch only); painting and polishing				
single leaf; four panelled doors; mouldings	nr	740.00	to	890.00
double leaf; four panelled doors; mouldings	nr	1425.00	to	1750.00
Fire doors Standard fire doors; cellular core; softwood lining; softwood architrave; aluminuim ironmongery (lockable, self-closure); painting or polishing;				
single leaf; flush hardboard faced; 30 min fire resistance; painted	nr	495.00	to	600.00
single leaf; Oak veneered; 30 min fire resistance; polished	nr	410.00	to	500.00
double leaf; Oak veneered; 30 min fire resistance; polished	nr	980.00	to	1175.00
single leaf; flush hardboard faced; 60 min fire resistance; painted	nr	600.00	to	730.00
single leaf; Oak veneered; 60 min fire resistance; polished	nr	650.00	to	780.00
double leaf; Oak veneered; 60 min fire resistance; polished	nr	1200.00	to	1450.00
Ironmongery sets Stainless steel ironmongery; euro locks; push plates; kick plates; signage; closures; standard sets				
office door; non-locking; fire-rated	nr	210.00	to	250.00
office/store; lockable; fire-rated	nr	235.00	to	285.00
classroom door; lockable; fire-rated	nr	310.00	to	370.00
maintenance/plant room door; lockable; fire-rated	nr	220.00	to	270.00
standard bathroom door (unisex)	nr	185.00	to	220.00
accessible toilet door	nr	110.00	to	135.00
fire escape door	nr	1125.00	to	1350.00

3.1 WALL FINISHES

Item	Unit	Range £		
3 INTERNAL FINISHES				
3.1 WALL FINISHES				
Internal wall area (unless otherwise described)				
In situ wall finishes				
Comparative finishes				
one mist and two coats emulsion paint	m²	3.05	to	3.85
two coats of lightweight plaster	m²	12.00	to	15.20
two coats of lightweight plaster with emulsion finish	m²	15.10	to	19.00
two coat sand cement render and emulsion finish	m²	18.30	to	25.00
plaster and vinyl wallpaper coverings	m²	17.30	to	22.00
polished plaster system; Armourcoat or similar; 11 mm thick first coat and 2 mm thick finishing coat with a polished finish	m²	87.00	to	120.00
Rigid tile/panel/board finishes				
Timber boarding/panelling; on and including battens; plugged to wall				
12 mm thick softwood boarding	m²	29.50	to	40.00
hardwood panelling; t&g & v-jointed	m²	87.00	to	120.00
Ceramic wall tiles; including backing				
economical quality	m²	29.00	to	36.50
medium to high quality	m²	38.00	to	48.00
Porcelain; including backing				
Porcelain mosaic tiling; walls and floors	m²	70.00	to	88.00
Marble				
Roman Travertine marble wall linings 20 mm thick; polished	m²	250.00	to	320.00
Granite				
Dakota mahogany granite cladding 20 mm thick; polished finish; jointed and pointed in coloured mortar	m²	330.00	to	420.00
Dakota mahogany granite cladding 40 mm thick; polished finish; jointed and pointed in coloured mortar	m²	550.00	to	690.00
Comparative woodwork finishes				
Knot; one coat primer; two undercoats; gloss on wood surfaces; number of coats:				
two coats gloss	m²	7.45	to	10.10
three coats gloss	m²	8.20	to	11.10
three coats gloss; small girth n.e. 300 mm	m	5.40	to	7.30
Polyurethane lacquer				
two coats	m²	5.35	to	7.20
three coats	m²	5.85	to	7.95
Flame-retardant paint				
Unitherm or similar; two coats	m²	13.60	to	18.40
Polish				
wax polish; seal	m²	9.60	to	13.00
wax polish; stain and body in	m²	12.90	to	17.40
French polish; stain and body in	m²	18.80	to	25.00

3.2 FLOOR FINISHES

Item	Unit	Range £		
Other wall finishes				
Wall coverings				
decorated paper backed vinyl wall paper	m²	5.30	to	6.40
PVC wall linings – Altro or similar; standard satins finish	m²	59.00	to	71.00
3.2 FLOOR FINISHES				
Internal floor area (unless otherwise described)				
In situ screed and floor finishes; laid level; over 300 mm wide				
Cement and sand (1:3) screeds; steel trowelled				
50 mm thick	m²	12.20	to	15.40
75 mm thick	m²	16.10	to	20.00
100 mm thick	m²	20.00	to	25.00
Flowing screeds				
Latex screeds; self colour; 3 mm to 5 mm thick	m²	5.60	to	7.05
Lafarge Gyvlon flowing screed 50 mm thick	m²	10.70	to	13.00
Treads, steps and the like (small areas)	m²	29.00	to	35.00
Granolithic; laid on green concrete				
20 mm thick	m²	25.50	to	32.00
38 mm thick	m²	28.00	to	36.00
Resin floor finish				
Altrotect; 2 coat application nominally 350–500 micron thick	m²	7.70	to	9.70
AltroFlow EP; 3 part solvent-system; up to 3 mm thick	m²	27.00	to	34.00
AltroSafe; slip-resistant; 3 part solvent-free system; 5 mm to 6 mm thick	m²	62.00	to	78.00
Sheet/board flooring				
Chipboard				
18 mm to 22 mm thick chipboard flooring; t&g joints	m²	14.50	to	18.30
Softwood				
22 mm thick wrought softwood flooring; 150 mm wide; t&g joints;	m²	22.00	to	27.50
softwood skirting, gloss paint finish	m	11.30	to	15.30
MDF skirting, gloss paint finish	m	10.90	to	14.80
Hardwood				
Wrought hardwood t&g strip flooring; polished; including fillets	m²	80.00	to	110.00
Hardwood skirting, stained finish	m	13.90	to	18.80
Sprung floors				
Taraflex Combisport 85 System with Taraflex Sports M Plus; t&g plywood 22 mm thick on softwood battens and crumb rubber cradles	m²	125.00	to	170.00
Sprung composition block flooring (sports), court markings, sanding and sealing	m²	83.00	to	110.00
Rigid Tile/slab finishes (includes skirtings; excludes screeds)				
Quarry tile flooring	m²	51.00	to	65.00
Glazed ceramic tiled flooring				
standard plain tiles	m²	34.50	to	46.50
anti-slip tiles	m²	37.50	to	51.00
designer tiles	m²	78.00	to	105.00
Terrazzo tile flooring 28 mm thick polished	m²	37.50	to	51.00
York stone 50 mm thick paving	m²	125.00	to	170.00
Slate tiles, smooth; straight cut	m²	49.50	to	67.00
Portland stone paving	m²	205.00	to	280.00

3.3 CEILING FINISHES

Item	Unit	Range £		
3.2 FLOOR FINISHES – cont				
Rigid Tile/slab finishes (includes skirtings; excludes screeds) – cont				
Roman Travertine marble paving; polished	m²	200.00	to	270.00
Granite paving 20 mm thick paving	m²	305.00	to	410.00
Parquet/wood block wrought hardwood block floorings; 25 mm thick;				
polished; t&g joints	m²	105.00	to	140.00
Flexible tiling; welded sheet or butt joint tiles; adhesive fixing				
vinyl floor tiling; 2.00 mm thick	m²	12.80	to	16.20
vinyl safety flooring; 2.00–2.50 mm thick	m²	33.50	to	42.50
vinyl safety flooring; 3.5 mm thick heavy duty	m²	42.50	to	54.00
linoleum tile flooring; 333 mm × 333 mm × 3.20 mm tiles	m²	35.00	to	44.00
linoleum sheet flooring; 2.00 mm thick	m²	28.50	to	36.00
rubber studded tile flooring; 500 mm × 500 mm × 2.50 mm thick	m²	32.00	to	40.50
Carpet tiles; including underlay, edge grippers				
heavy domestic duty; to floors; PC Sum £24/m²	m²	40.00	to	50.00
heavy domestic duty; to treads and risers; PC Sum £24/m²	m	33.00	to	42.00
heavy contract duty; Forbo Flotex HD	m²	41.00	to	52.00
Entrance matting				
door entrance and circulation matting; soil and water removal	m²	63.00	to	79.00
Entrance matting and matwell				
Gradus Topguard barrier matting with aluminium frame	m²	310.00	to	390.00
Nuway Tuftiguard barrier matting with aluminium frame	m²	345.00	to	435.00
Gradus Topguard barrier matting with stainless steel frame	m²	305.00	to	390.00
Nuway Tuftiguard barrier matting with stainless steel frame	m²	340.00	to	430.00
Gradus Topguard barrier matting with polished brass frame	m²	455.00	to	580.00
Nuway Tuftiguard barrier matting with polished brass frame	m²	490.00	to	620.00
Access floors and finishes				
Raised access floors: including 600 mm × 600 mm steel encased particle				
boards on height adjustable pedestals < 300 mm				
light grade duty	m²	33.00	to	45.00
medium grade duty	m²	33.00	to	45.00
heavy grade duty	m²	44.50	to	60.00
battened raft chipboard floor with sound insulation fixed to battens;				
medium quality carpeting	m²	61.00	to	83.00
Common floor coverings bonded to access floor panels				
anti-static vinyl	m²	21.00	to	28.00
heavy duty fully flexible vinyl tiles	m²	18.50	to	25.00
needlepunch carpet	m²	12.70	to	17.20
3.3 CEILING FINISHES				
Internal ceiling area (unless otherwise described)				
In situ finishes				
Decoration only to soffits; one mist and two coats emulsion paint				
to exposed steelwork (surface area)	m²	4.40	to	6.00
to concrete soffits (surface area)	m²	3.10	to	4.20
to plaster/plasterboard	m²	2.80	to	3.80

3.3 CEILING FINISHES

Item	Unit	Range £		
Plaster to soffits				
plaster skim coat to plasterboard ceilings	m²	5.40	to	7.30
lightweight plaster to concrete	m²	12.40	to	16.70
Plasterboard to soffits				
9 mm Gyproc board and skim coat	m²	21.00	to	28.50
12.50 mm Gyproc board and skim coat	m²	21.50	to	29.00
Other board finishes; with fire-resisting properties; excluding decoration				
12.50 mm thick Gyproc Fireline board	m²	15.60	to	21.00
15 mm thick Gyproc Fireline board	m²	16.00	to	21.50
12 mm thick Supalux	m²	34.00	to	46.00
Specialist plasters; to soffits				
sprayed acoustic plaster; self-finished	m²	29.50	to	40.00
rendering; Tyrolean finish	m²	31.00	to	42.50
Other ceiling finishes				
timber boarding	m²	19.60	to	26.50
Suspended and integrated ceilings				
Armstrong suspended ceiling; assume large rooms over 250 m²				
mineral fibre; basic range; Cortega, Tatra, Academy; exposed grid	m²	15.70	to	22.00
mineral fibre; medium quality; Corline; exposed grid	m²	21.00	to	29.50
mineral fibre; medium quality; Corline; concealed grid	m²	37.00	to	52.00
mineral fibre; medium quality; Corline; silhoutte grid	m²	46.00	to	64.00
mineral fibre; Specific; Bioguard; exposed grid	m²	18.40	to	25.50
mineral fibre; Specific; Bioguard; clean room grid	m²	30.00	to	41.50
mineral fibre; Specific; Hygeine, Cleanroom; exposed grid	m²	37.00	to	52.00
metal open cell; Cellio Global White; exposed grid	m²	37.00	to	52.00
metal open cell; Cellio Black; exposed grid	m²	46.00	to	64.00
wood – Maderal laminates; plain; exposed grid	m²	75.00	to	100.00
wood – Maderal laminates; perforated; exposed grid	m²	120.00	to	165.00
wood – Maderal laminates; plain; concealed grid grid	m²	65.00	to	91.00
wood – Maderal laminates; perforated; concealed grid	m²	110.00	to	150.00
wood – Maderal veneers; plain; exposed grid	m²	93.00	to	130.00
wood – Maderal veneers; perforated; exposed grid	m²	140.00	to	190.00
wood – Maderal veneers; plain; concealed grid grid	m²	84.00	to	115.00
wood – Maderal veneers; perforated; concealed grid	m²	130.00	to	180.00
Other suspended ceilings				
perforated aluminium ceiling tiles 600 mm × 600 mm	m²	26.00	to	35.00
perforated aluminium ceiling tiles 1200 mm × 600 mm	m²	32.00	to	43.00
metal linear strip; Dampa/Luxalon	m²	39.50	to	54.00
metal linear strip microperforated acoustic ceiling with Rockwool acoustic infill	m²	49.00	to	66.00
metal tray	m²	41.00	to	56.00
egg-crate	m²	64.00	to	86.00
open grid; Formalux/Dimension	m²	84.00	to	110.00
Integrated ceilings				
coffered; with steel surfaces	m²	110.00	to	150.00
acoustic suspended ceilings on anti-vibration mountings	m²	51.00	to	68.00

4.1 FITTINGS, FURNISHINGS AND EQUIPMENT

Item	Unit	Range £		

3.3 CEILING FINISHES – cont

Comparative ceiling finishes

Emulsion paint

two coats	m²	1.90	to	2.60
one mist and two coats	m²	3.10	to	4.20

Gloss paint

primer and two coats	m²	5.35	to	7.25
primer and three coats	m²	7.50	to	10.10

Other coatings

Artex plastic compound one coat; textured	m²	10.10	to	13.60

4 FITTINGS, FURNISHINGS AND EQUIPMENT

4.1 FITTINGS, FURNISHINGS AND EQUIPMENT

Residential fittings (volume housing)

Kitchen fittings for residential units (not including white goods). NB: quality and quantity of units can varies enormously. Always obtain costs from your preferrred supplier. The following assume medium standard units from a large manufacturer and includes all units, worktops, stainless steel sink and taps, not including white goods

one person flat	nr	1575.00	to	2025.00
two person flat/house	nr	1825.00	to	2375.00
three person house	nr	3050.00	to	3950.00
four person house; includes utility area	nr	5800.00	to	7400.00
five person house; includes utility area	nr	7100.00	to	9200.00

Office furniture and equipment

There is a large quality variation for office furniture and we have assumed a medium level, even so prices will vary between suppliers

Reception desk

straight counter; 3500 mm long; 2 person	nr	1575.00	to	2025.00
curved counter; 3500 mm long; 2 person	nr	3900.00	to	5000.00
curved counter; 3500 mm long; 2 person; real wood veneer finish	nr	7100.00	to	9200.00

Furniture and equipment to general office area; standard off the shelf specification

workstation; 2000 mm long desk; drawer unit; task chair	nr	570.00	to	730.00

Hotel bathroom pods

Fully fitted out, finished and furnished bathroom pods; installed; suitable for business class hotels

standard pod (4.50 m plan area)	nr	3900.00	to	5100.00
accessible pod (4.50 m plan area)	nr	4800.00	to	6200.00

Window blinds

Louvre blind

89 mm louvres, manual chain operation, fixed to masonry	m²	38.50	to	49.50
127 mm louvres, manual chain operation, fixed to masonry	m²	34.00	to	43.00

5.1 SANITARY INSTALLATIONS

Item	Unit	Range £		
Roller blind				
fabric blinds, roller type, manual chain operation, fixed to masonry	m²	30.00	to	39.00
solar blackout blinds, roller type, manual chain operation, fixed to masonry	m²	51.00	to	65.00
Fire curtains				
Electrically operated automatic fire curtains to form a virtually continuous barrier against both fire and smoke	m²	1225.00	to	1475.00
5 SERVICES				
5.1 SANITARY INSTALLATIONS				
Rates are for gross internal floor area – GIFA unless otherwise described				
shopping malls and the like (not tenant fit-out works)	m²	0.75	to	1.00
office building – multi-storey	m²	6.00	to	8.30
office building – business park	m²	6.00	to	8.30
performing arts building (medium specification)	m²	10.60	to	14.70
sports hall	m²	8.80	to	12.20
hotel	m²	29.50	to	41.00
hospital; private	m²	17.50	to	24.00
school; secondary	m²	8.80	to	12.20
residential; multi-storey tower	m²	0.45	to	0.60
supermarket	m²	1.10	to	1.50
Comparative sanitary fittings/sundries				
Note: Material prices vary considerably, the following composite rates are based on average prices for mid priced fittings:				
Individual sanitary appliances (including fittings)				
low level WCs; vitreous china pan and cistern; black plastic seat; low pressure ball valve; plastic flush pipe; fixing brackets	nr	240.00	to	290.00
bowl type wall urinal; white glazed vitreous china flushing cistern; chromium plated flush pipes and spreaders; fixing brackets	nr	200.00	to	240.00
sink; glazed fireclay; chromium plated waste; plug and chain	nr	455.00	to	560.00
sink; stainless steel; chromium plated waste; plug and chain				
single drainer; double bowl (bowl and half)	nr	260.00	to	320.00
double drainer; double bowl (bowl and half)	nr	280.00	to	340.00
bath; reinforced acrylic; chromium plated taps; overflow; waste; chain and plug; 'P' trap and overflow connections	nr	410.00	to	500.00
bath; enamelled steel; chromium plated taps; overflow; waste; chain and plug; 'P' trap and overflow connections	nr	500.00	to	610.00
shower tray; glazed fireclay; chromium plated waste; riser pipe; rose and mixing valve	nr	370.00	to	455.00
Soil waste stacks; 3.15 m storey height; branch and connection to drain				
110 mm dia. PVC	nr	350.00	to	430.00
extra for				
additional floors	nr	185.00	to	225.00
100 mm dia. cast iron; decorated	nr	475.00	to	580.00
extra for				
additional floors	nr	350.00	to	430.00

5.7 VENTILATING SYSTEMS

Item	Unit	Range £		

5.4 WATER INSTALLATIONS

Hot and cold water installations; mains supply; hot and cold water distribution. Gross internal floor area (unless described otherwise)

Item	Unit		Range £	
shopping malls and the like (not tenant fit-out works)	m²	10.10	to	14.00
office building – multi-storey	m²	14.80	to	20.50
office building – business park	m²	11.00	to	15.30
performing arts building (medium specification)	m²	17.30	to	24.00
sports hall	m²	13.10	to	18.20
hotel	m²	37.00	to	51.00
hospital; private	m²	44.00	to	62.00
school; secondary; potable and non potable to labs, art rooms	m²	34.00	to	47.50
residential; multi-storey tower	m²	41.50	to	58.00
supermarket	m²	23.00	to	32.00
distribution centre	m²	1.60	to	2.20

5.6 SPACE HEATING AND AIR CONDITIONING

Gross internal floor area (unless described otherwise)

Item	Unit		Range £	
shopping malls and the like (not tenant fit-out works); LTHW, air conditioning, ventilation	m²	65.00	to	90.00
office building – multi-storey; LTHW, ductwork, chilled water, ductwork	m²	92.00	to	130.00
office building – business park; LTHW, ductwork, chilled water, ductwork	m²	100.00	to	140.00
performing arts building (medium specification); LTHW, ductwork, chilled water, ductwork	m²	185.00	to	255.00
sports hall; warm air to main hall, LTHW radiators to ancillary	m²	32.00	to	45.00
hotel; air conditioning	m²	120.00	to	165.00
hospital; private; LPHW	m²	185.00	to	255.00
school; secondary; LTHW, cooling to ICT server room	m²	100.00	to	140.00
residential; multi-storey tower; LTHW to each apartment	m²	55.00	to	77.00
supermarket	m²	13.80	to	19.20
distribution centre; LTHW to offices, displacement system to warehouse	m²	18.40	to	25.50

5.7 VENTILATING SYSTEMS

Gross internal area (unless otherwise described)

Item	Unit		Range £	
shopping malls and the like (not tenant fit-out works)	m²	32.00	to	45.00
office building – multi-storey; toilet and kitchen areas only	m²	5.55	to	7.70
office building – business park; toilet and kitchen areas only	m²	6.45	to	8.95
performing arts building (medium specification); toilet and kitchen areas, workshop	m²	12.90	to	17.90
sports hall	m²	12.00	to	16.70
hotel; general toilet extraction, bathrooms, kitchens	m²	37.00	to	51.00
hospital; private; toilet and kitchen areas only	m²	6.90	to	9.60
school; secondary; toilet and kitchen areas, science labs	m²	15.70	to	22.00
residential; multi-storey tower	m²	32.00	to	45.00
supermarket	m²	4.60	to	6.40
distribution centre; smoke extract system	m²	4.60	to	6.40

5.9 FUEL INSTALLATIONS

Item	Unit	Range £		
5.8 ELECTRICAL INSTALLATIONS				
Electrical installations				
Including LV distribution, HV distribution, lighting, small power. Gross internal floor area (unless described otherwise)				
shopping malls and the like (not tenant fit-out works)	m²	100.00	to	140.00
office building – multi-storey	m²	88.00	to	120.00
office building – business park	m²	46.00	to	64.00
performing arts building (medium specification)	m²	165.00	to	230.00
sports hall	m²	51.00	to	70.00
hotel	m²	155.00	to	220.00
hospital; private	m²	185.00	to	255.00
school; secondary	m²	110.00	to	150.00
residential; multi-storey tower	m²	74.00	to	100.00
supermarket	m²	60.00	to	83.00
distribution centre	m²	55.00	to	77.00
Comparative fittings/rates per point				
Consumer control unit; 63–100 Amp 230 volt; switched and insulated; RCDB protection. Gross internal floor area (unless described otherwise)	nr	270.00	to	325.00
Fittings; excluding lamps or light fittings				
lighting point; PVC cables	nr	39.00	to	48.00
lighting point; PVC cables in screwed conduits	nr	45.00	to	55.00
lighting point; MICC cables	nr	60.00	to	73.00
Switch socket outlet; PVC cables				
single	nr	5.55	to	6.75
double	nr	65.00	to	79.00
Switch socket outlet; PVC cables in screwed conduit				
single	nr	74.00	to	90.00
double	nr	85.00	to	100.00
Switch socket outlet; MICC cables				
single	nr	71.00	to	87.00
double	nr	82.00	to	100.00
Other power outlets				
immersion heater point (excluding heater)	nr	91.00	to	110.00
cooker point; including control unit	nr	150.00	to	185.00
5.9 FUEL INSTALLATIONS				
Gas mains service to plantroom. Gross internal floor area (unless described otherwise)				
shopping mall/supermarket	m²	2.50	to	3.10
warehouse/distribution centre	m²	0.75	to	0.90
office/hotel	m²	1.30	to	1.45

5.10 LIFT AND CONVEYOR INSTALLATIONS

Item	Unit	Range £		
5.10 LIFT AND CONVEYOR INSTALLATIONS				
Passenger lifts				
Passenger lifts (standard brushed stainless steel finish; 2 panel centre opening doors)				
8-person; 4 stops; 1.0 m/s speed	nr	65000.00	to	79000.00
8-person; 4 stops; 1.6 m/s speed	nr	68000.00	to	83000.00
13-person; 6 stops; 1.0 m/s speed	nr	81000.00	to	99000.00
13-person; 6 stops; 1.6 m/s speed	nr	87000.00	to	105000.00
13-person; 6 stops; 2.0 m/s speed	nr	88000.00	to	110000.00
21-person; 8 stops; 1.0 m/s speed	nr	110000.00	to	135000.00
21-person; 8 stops; 1.6 m/s speed	nr	115000.00	to	140000.00
21-person; 8 stops; 2.0 m/s speed	nr	120000.00	to	140000.00
extra for				
enhanced finishes; mirror; carpet	nr	3200.00	to	3950.00
lift car LCD TV	nr	6900.00	to	8500.00
intelligent group control; 5 cars; 11 stops	nr	33000.00	to	40000.00
Special installations				
wall climber lift; 10-person; 0.50 m/sec; 10 levels	nr	320000.00	to	395000.00
disabled platform lift single wheelchair; 400 kg; 4 stops; 0.16 m/s	nr	7900.00	to	9700.00
Non-passenger lifts				
Goods lifts; prime coated internal finish				
2000 kg load; 4 stops; 1.0 m/s speed	nr	81000.00	to	90000.00
2000 kg load; 4 stops; 1.6 m/s speed	nr	88000.00	to	97000.00
2500 kg load; 4 stops; 1.0 m/s speed	nr	87000.00	to	96000.00
2000 kg load; 8 stops; 1.0 m/s speed	nr	110000.00	to	120000.00
2000 kg load; 8 stops; 1.6 m/s speed	nr	115000.00	to	130000.00
Other goods lifts				
hoist	nr	22000.00	to	27000.00
kitchen service hoist 50 kg; 2 levels	nr	9400.00	to	11500.00
Dock levellers				
dock levellers	nr	20000.00	to	24000.00
dock leveller and canopy	nr	27500.00	to	34000.00
Escalators				
30° escalator; 0.50 m/sec; enamelled steel glass balustrades				
3.50 m rise; 800 mm step width	nr	70000.00	to	86000.00
4.60 m rise; 800 mm step width	nr	76000.00	to	92000.00
5.20 m rise; 800 mm step width	nr	78000.00	to	96000.00
6.00 m rise; 800 mm step width	nr	87000.00	to	105000.00
extra for				
enhanced finish; enamelled finish; glass balustrade	nr	9200.00	to	11000.00

5.13 SPECIAL INSTALLATIONS

Item	Unit	Range £		

5.11 FIRE AND LIGHTNING PROTECTION

Gross internal floor area (unless described otherwise)
Including lightning protection, sprinklers unless stated otherwise

Item	Unit		Range £	
shopping malls and the like (not tenant fit-out works)	m²	12.90	to	17.90
office building – multi-storey	m²	27.50	to	38.50
office building – business park	m²	4.60	to	6.40
performing arts building (medium specification); lightning protection only	m²	1.40	to	1.90
sports hall; lightning protection only	m²	2.80	to	3.85
hospital; private; lightning protection only	m²	1.80	to	2.55
school; secondary; lightning protection only	m²	1.70	to	2.40
hotel	m²	33.00	to	46.00
residential; multi-storey tower	m²	46.00	to	64.00
supermarket; lightning protection only	m²	0.90	to	1.30
distribution centre	m²	37.00	to	51.00

5.12 COMMUNICATION AND SECURITY INSTALLATIONS

Gross internal floor area (unless described otherwise)
Including fire alarm, public address system, security installation unless stated otherwise

Item	Unit		Range £	
shopping malls and the like (not tenant fit-out works)	m²	32.00	to	45.00
office building – multi-storey	m²	23.00	to	32.00
office building – business park; fire alarm and wireways only	m²	11.10	to	15.40
performing arts building (medium specification)	m²	71.00	to	99.00
sports hall	m²	46.00	to	64.00
hotel	m²	78.00	to	110.00
hospital; private	m²	60.00	to	83.00
school; secondary	m²	46.00	to	64.00
residential; multi-storey tower	m²	46.00	to	64.00
supermarket	m²	13.80	to	19.20
distribution centre	m²	16.60	to	23.00

5.13 SPECIAL INSTALLATIONS

Rates for area of material or per item
Solar

Item	Unit		Range £	
Residential solar water heating including collectors; dual coil cylinders; pump; controller (NB excludes any grant allowance)	m²	890.00	to	1125.00
Solar power including 2.2 kWp monocrystalline solar modules (12 × 185 Wp) on roof mounting kit; certified inverter; DC and AC iolation switches and connection to the grid and certification	nr	9700.00	to	12000.00
Window cleaning equipment				
twin track	m	130.00	to	165.00
manual trolley/cradle	nr	8900.00	to	11000.00
automatic trolley/cradle	nr	21000.00	to	26500.00
Sauna				
2.20 m × 2.20 m internal Finnish sauna; benching; heater; made of Sauna grade Aspen or similar	nr	6900.00	to	8800.00
Jacuzzi Installation				
2.20 m × 2.20 m × 0.95 m; 5 adults; 99 jets; lights; pump	nr	9200.00	to	13000.00

8.2 ROADS, PATHS, PAVINGS AND SURFACINGS

Item	Unit	Range £		

5.14 BUILDER'S WORK IN CONNECTION WITH SERVICES

Gross internal floor area (unless described otherwise)
Warehouses, sports halls and shopping malls.

Item	Unit	Range £		
main supplies, lighting and power to landlord areas	m²	2.80	to	3.60
central heating and electrical installation	m²	8.50	to	10.80
central heating, electrical and lift installation	m²	9.60	to	12.30
air conditioning, electrical and ventilation installations	m²	21.50	to	28.00
Offices and hotels.				
main supplies, lighting and power to landlord areas	m²	8.15	to	10.40
central heating and electrical installation	m²	11.90	to	15.20
central heating, electrical and lift installation	m²	14.10	to	18.00
air conditioning, electrical and ventilation installations	m²	26.00	to	33.50

8 EXTERNAL WORKS

8.2 ROADS, PATHS, PAVINGS AND SURFACINGS

Paved areas
Gravel paving rolled to falls and chambers paving on subbase; including

Item	Unit	Range £		
excavation	m²	9.75	to	12.60
Resin bound paving				
16 mm–24 mm deep of natural gravel	m²	62.00	to	76.00
16 mm–24 mm deep of crushed rock	m²	63.00	to	77.00
16 mm–24 mm deep of marble chips	m²	67.00	to	82.00
Tarmacadam paving				
two layers; limestone or igneous chipping finish paving on subbase;				
including excavation	m²	43.50	to	56.00
Slab paving				
precast concrete paving slabs on subbase; including excavation	m²	32.00	to	42.00
precast concrete tactile paving slabs on subbase; including excavation	m²	43.50	to	56.00
York stone slab paving on subbase; including excavation	m²	130.00	to	170.00
imitation York stone slab paving on subbase; including excavation	m²	65.00	to	85.00
Brick/Block/Setts paving				
brick paviors on subbase; including excavation	m²	58.00	to	76.00
precast concrete block paviors to footways including excavation; subbase	m²	48.00	to	62.00
granite setts on subbase; including excavation	m²	105.00	to	135.00
cobblestone paving cobblestones on subbase; including excavation	m²	110.00	to	145.00
Reinforced grass construction				
Grasscrete or similar on 200 mm type 1 subbase; topsoil spread across				
units and grass seeded upon completion	m²	44.00	to	53.00
Car parking alternatives				
Surface parking; include drains, kerbs, lighting				
surface level parking	m²	73.00	to	89.00
surface car parking with landscape areas	m²	94.00	to	110.00
Rates per car				
surface level parking	car	1250.00	to	1825.00
surface car parking with landscape areas	car	1750.00	to	2600.00

8.3 SOFT LANDSCAPING, PLANTING AND IRRIGATION SYSTEMS

Item	Unit	Range £		
Other surfacing options				
permable concrete block paving; 80 mm thick Formpave Aquaflow or equal; max gradient 5%; laid on 50 mm sand/gravel; subbase 365 mm thick clean crushed angular non-plastic aggregate, geotextiles filter layer to top and bottom of subbases	m²	44.00	to	53.00
All purpose roads				
Tarmacadam or reinforced concrete roads, including all earthworks, drainage, pavements, lighting, signs, fencing and safety barriers				
single 7.30 m wide carriageway	m	960.00	to	1275.00
wide single 10.00 m wide carriageway	m	3600.00	to	4800.00
dual two lane road 7.30 m wide carriageway	m	2700.00	to	3650.00
dual three lane road 11.00 m wide carriageway	m	3500.00	to	4700.00
Road crossings				
NOTE: Costs include road markings, beacons, lights, signs, advance danger signs etc.				
Zebra crossing	nr	17500.00	to	23500.00
Pelican crossing	nr	30500.00	to	41000.00
Underpass				
Provision of underpasses to new roads, constructed as part of a road building programme				
Precast concrete pedestrian underpass				
3.00 m wide × 2.50 m high	m	4350.00	to	5900.00
Precast concrete vehicle underpass				
7.00 m wide × 5.00 m high	m	19000.00	to	26000.00
14.00 m wide × 5.00 m high	m	44500.00	to	60000.00
8.3 SOFT LANDSCAPING, PLANTING AND IRRIGATION SYSTEMS				
Preparatory excavation and subbases				
Surface treatment				
spread and lightly consolidate top soil form spoil 150 mm thick; by machine	m²	1.60	to	2.00
spread and lightly consolidate top soil form spoil 150 mm thick; by hand	m²	6.55	to	8.40
Seeded and planted areas				
Plant supply, planting, maintenance and 12 months guarantee				
seeded areas	m²	4.15	to	5.10
turfed areas	m²	5.60	to	6.85
Planted areas (per m² of planted area)				
herbaceous plants	m²	4.05	to	5.25
climbing plants	m²	6.40	to	8.30
general planting	m²	15.50	to	20.00
woodland	m²	23.00	to	30.00
shrubbed planting	m²	39.00	to	51.00
dense planting	m²	39.00	to	50.00
shrubbed area including allowance for small trees	m²	51.00	to	67.00

8.4 FENCING, RAILINGS AND WALLS

Item	Unit	Range £		

8.3 SOFT LANDSCAPING, PLANTING AND IRRIGATION SYSTEMS – cont

Seeded and planted areas – cont

Trees

light standard bare root tree (PC £9.75)	nr	43.50	to	56.00
standard root balled tree (PC 23.50)	nr	51.00	to	67.00
heavy standard root ball tree (PC £39.25)	nr	96.00	to	125.00
semi-mature root balled tree (PC £125.00)	nr	305.00	to	395.00

Parklands

NOTE: Work on parklands will involve different techniques of earth shifting and cultivation. The following rates include for normal surface excavation, they include for the provision of any land drainage

parklands, including cultivating ground, applying fertilizer, etc. and seeding with parks type grass	ha	15000.00	to	20000.00

Lakes including excavation average 1.0 m deep, laying 1000 micron sheet with welded joints; spreading top soil evenly on top 200 mm deep

regular shaped lake	1000 m²	18500.00	to	23500.00

8.4 FENCING, RAILINGS AND WALLS

Crib retaining walls

Permacrib timber crib retaining walls on concrete foundation; granular infill material; measure face area m². Excludes backfill material to the rear of the wall

wall heights up to 2.0 m	m²	150.00	to	190.00
wall heights up to 3.0 m	m²	155.00	to	200.00
wall heights up to 4.0 m	m²	165.00	to	210.00
wall heights up to 5.0 m	m²	175.00	to	220.00
wall heights up to 6.0 m	m²	185.00	to	235.00

Andacrib concrete crib retaining walls on concrete foundation; granular infill material; measure face area m². Excludes backfill material to the rear of the wall

wall heights up to 2.0 m	m²	155.00	to	200.00
wall heights up to 3.0 m	m²	165.00	to	210.00
wall heights up to 4.0 m	m²	175.00	to	220.00
wall heights up to 5.0 m	m²	185.00	to	235.00
wall heights up to 6.0 m	m²	190.00	to	250.00

Textomur green faced reinforced soil system; geogrid/geotextile reinforcement and compaction of reinforced soil mass

wall heights up to 3.0 m	m²	100.00	to	130.00
wall heights up to 4.0 m	m²	105.00	to	135.00
wall heights up to 5.0 m	m²	110.00	to	140.00
wall heights up to 6.0 m	m²	115.00	to	150.00

Titan/Geolock vertical modular block reinforced soil system; steel ladder; geogrid reinforcement and soil fill material

wall heights up to 3.0 m	m²	165.00	to	210.00
wall heights up to 4.0 m	m²	175.00	to	220.00
wall heights up to 5.0 m	m²	185.00	to	235.00
wall heights up to 6.0 m	m²	190.00	to	250.00

8.5 EXTERNAL FIXTURES

Item	Unit	Range £		
Guard rails and parking bollards etc.				
Post and rail fencing				
open metal post and rail fencing 1.00 m high	m	115.00	to	150.00
galvanized steel post and rail fencing 2.00 m high	m	130.00	to	170.00
Guard rails				
steel guard rails and vehicle barriers	m	47.00	to	60.00
Bollards and barriers				
parking bollards precast concrete or steel	nr	130.00	to	165.00
vehicle control barrier; manual pole	nr	730.00	to	930.00
Chain link fencing; plastic coated				
1.20 m high	m	17.00	to	21.50
1.80 m high	m	22.50	to	29.00
Timber fencing				
1.20 m high chestnut pale facing	m	17.10	to	22.00
1.80 m high cross-boarded fencing	m	22.50	to	29.00
Screen walls; one brick thick; including foundations etc.				
1.80 m high facing brick screen wall	m	250.00	to	320.00
1.80 m high coloured masonry block boundary wall	m	270.00	to	350.00
8.5 EXTERNAL FIXTURES				
Street furniture				
Roadsigns				
reflected traffic signs 0.25 m² area on steel post	nr	110.00	to	140.00
internally illuminated traffic signs; dependent on area	nr	180.00	to	230.00
externally illuminated traffic signs; dependent on area	nr	690.00	to	880.00
Lighting				
lighting to pedestrian areas an estate roads on 4.00 m–6.00 m columns				
with up to 70 W lamps	nr	210.00	to	265.00
lighting to main roads				
10.00 m–12.00 m columns with 250 W lamps	nr	430.00	to	550.00
12.00 m–15.00 m columns with 400 W high pressure sodium lighting	nr	550.00	to	710.00
Benches; bolted to ground				
benches – hardwood and precast concrete	nr	880.00	to	1125.00
Litter bins; bolted to ground				
precast concrete	nr	155.00	to	200.00
hardwood slatted	nr	160.00	to	205.00
cast iron	nr	320.00	to	405.00
large aluminium	nr	470.00	to	600.00
Bus stops	nr	450.00	to	570.00
Bus stops including basic shelter	nr	2250.00	to	2900.00
Pillar box	nr	450.00	to	570.00
Galvanized steel cycle stand	nr	36.00	to	46.00
Galvanized steel flag staff	nr	950.00	to	1200.00
Playground equipment				
Modern swings with flat rubber safety seats: four seats; two bays	nr	1175.00	to	1525.00
Stainless steel slide, 3.40 m long	nr	1400.00	to	1775.00
Climbing frame – igloo type 3.20 m × 3.75 m on plan × 2.00 m high	nr	950.00	to	1200.00

8.6 EXTERNAL DRAINAGE

Item	Unit	Range £		
8.5 EXTERNAL FIXTURES – cont				
Playground equipment – cont				
See-saw comprising timber plank on sealed ball bearings				
3960 mm × 230 mm × 70 mm thick	nr	78.00	to	100.00
Wickstead Tumbleguard type safety surfacing around play equipment	m²	11.00	to	14.00
Bark particles type safety surfacing 150 mm thick on hardcore bed	nr	11.30	to	14.40
8.6 EXTERNAL DRAINAGE				
Overall £/m² of drained area allowances				
site drainage (per m² of paved area)	m²	13.10	to	18.50
building drainage (per m² of gross internal floor area)	m²	9.60	to	13.50
Machine excavation, grade bottom, earthwork support, laying and jointing pipes and accessories, backfill and compact, disposal of surplus soil. Vitrified clay pipes and fittings, Hepseal socketted, with push fit flexible joints; shingle bed and surround				
up to 1.00 m deep; nominal size				
150 mm dia. pipe	m	48.00	to	60.00
225 mm dia. pipe	m	67.00	to	85.00
300 mm dia. pipe	m	88.00	to	110.00
over 1.50 m not exceeding 3.00 m deep; nominal size				
150 mm dia. pipe	m	78.00	to	98.00
225 mm dia. pipe	m	97.00	to	120.00
300 mm dia. pipe	m	115.00	to	150.00
Machine excavation, grade bottom, earthwork support, laying and jointing pipes and accessories, backfill and compact, disposal of surplus soil. Class M tested concrete centrifugally spun pipes and fittings, flexible joints; concrete bed and surround				
up to 3.00 m deep; nominal size				
300 mm dia. pipe	m	130.00	to	160.00
525 mm dia. pipe	m	200.00	to	250.00
900 mm dia. pipe	m	340.00	to	430.00
1200 mm dia. pipe	m	485.00	to	610.00
Machine excavation, grade bottom, earthwork support, laying and jointing pipes and accessories, backfill and compact, disposal of surplus soil. Cast iron Timesaver drain pipes and fittings, mechanical coupling joints				
up to 1.50 m deep; nominal size				
100 mm dia. pipe	m	73.00	to	92.00
150 mm dia. pipe	m	100.00	to	130.00
over 1.50 m not exceeding 3.00 m deep; nominal size				
100 mm dia. pipe	m	100.00	to	130.00
150 mm dia. pipe	m	130.00	to	165.00

8.6 EXTERNAL DRAINAGE

Item	Unit	Range £		

Machine excavation, grade bottom, earthwork support, laying and jointing pipes and accessories, backfill and compact, disposal of surplus soil. uPVC pipes and fittings, lip seal coupling joints

up to 1.50 m deep; nominal size

Item	Unit	Range £		
100 mm dia. pipe	m	38.00	to	48.00
160 mm dia. pipe	m	45.00	to	57.00

over 1.50 m not exceeding 3.00 m deep; nominal size

Item	Unit	Range £		
100 mm dia. pipe	m	68.00	to	86.00
160 mm dia. pipe	m	76.00	to	95.00

Machine excavation, grade bottom, earthwork support, laying and jointing pipes and accessories, backfill and compact, disposal of surplus soil. uPVC Ultra-Rib ribbed pipes and fittings, sealed ring push fit joints

up to 1.50 m deep; nominal size

Item	Unit	Range £		
150 mm dia. pipe	m	38.00	to	48.00
225 mm dia. pipe	m	46.00	to	58.00
300 mm dia. pipe	m	53.00	to	67.00

over 1.50 m not exceeding 3.00 m deep; nominal size

Item	Unit	Range £		
150 mm dia. pipe	m	68.00	to	86.00
225 mm dia. pipe	m	80.00	to	100.00
300 mm dia. pipe	m	90.00	to	110.00

Brick manholes

Excavate pit in firm ground, partial backfill, partial disposal, earthwork support, compact base of pit, 150 mm plain in situ concrete 20.00 N/mm² – 20 mm aggregate (1:2:4) base, formwork, one brick wall of engineering bricks in cement mortar (1:3) finished fair face, vitrified clay channels, plain in situ concrete 25.00 N/mm² – 20 mm aggregate (1:2:4) cover and reducing slabs, fabric reinforcement, formwork step irons, medium duty cover and frame; Internal size of manhole

600 mm × 450 mm; cover to invert

Item	Unit	Range £		
not exceeding 1.00 m deep	nr	325.00	to	395.00
over 1.00 m not exceeding 1.50 m deep	nr	425.00	to	510.00
over 1.50 m not exceeding 2.00 m deep	nr	530.00	to	640.00

900 mm × 600 mm; cover to invert

Item	Unit	Range £		
not exceeding 1.00 m deep	nr	425.00	to	510.00
over 1.00 m not exceeding 1.50 m	nr	580.00	to	700.00
over 1.50 m not exceeding 2.00 m	nr	730.00	to	880.00

900 mm × 900 mm; cover to invert

Item	Unit	Range £		
not exceeding 1.00 m deep	nr	510.00	to	620.00
over 1.00 m not exceeding 1.50 m	nr	690.00	to	830.00
over 1.50 m not exceeding 2.00 m	nr	870.00	to	1050.00

1200 × 1800 mm; cover to invert

Item	Unit	Range £		
not exceeding 1.00 m deep	nr	1025.00	to	1250.00
over 1.00 m not exceeding 1.50 m deep	nr	1350.00	to	1625.00
over 1.50 m not exceeding 2.00 m deep	nr	1650.00	to	1975.00

8.6 EXTERNAL DRAINAGE

Item	Unit	Range £	
8.6 EXTERNAL DRAINAGE – cont			
Concrete manholes			
Excavate pit in firm ground, disposal, earthwork support, compact base of pit, plain in situ concrete 20.00 N/mm² – 20 mm aggregate (1:2:4) base, formwork, reinforced precast concrete chamber and shaft rings, taper pieces and cover slabs bedded jointed and pointed in cement: mortar (1:3) weak mix concrete filling to working space, vitrified clay channels, plain in-situ concrete 25.00 N/mm² – 20 mm aggregate (1:1:5:3) benchings, step irons, medium duty cover and frame; depth from cover to invert; Internal diameter of manhole			
900 mm dia.; cover to invert			
over 1.00 m not exceeding 1.50 m deep	nr	485.00 to	590.00
1050 mm dia.; cover to invert			
over 1.00 m not exceeding 1.50 m deep	nr	540.00 to	650.00
over 1.50 m not exceeding 2.00 m deep	nr	610.00 to	740.00
over 2.00 m not exceeding 3.00 m deep	nr	760.00 to	920.00
1500 mm dia.; cover to invert			
over 1.50 m not exceeding 2.00 m deep	nr	1000.00 to	1200.00
over 2.00 m not exceeding 3.00 m deep	nr	1275.00 to	1550.00
over 3.00 m not exceeding 4.00 m deep	nr	1575.00 to	1875.00
1800 mm dia.; cover to invert			
over 2.00 m not exceeding 3.00 m deep	nr	1725.00 to	2075.00
over 3.00 m not exceeding 4.00 m deep	nr	2125.00 to	2550.00
2100 mm dia.; cover to invert			
over 2.00 m not exceeding 3.00 m deep	nr	2500.00 to	3050.00
over 3.00 m not exceeding 4.00 m deep	nr	3100.00 to	3750.00
Polypropylene inspection chambers			
475 mm dia. PPIC inspection chamber including all excavations; earthwork support; cart away surplus spoil; concrete bed and surround; lightweight cover and frame			
600 mm deep	nr	230.00 to	280.00
900 mm deep	nr	280.00 to	340.00
Stormwater management			
Aquacell storm water system; lightweight polypropylene high-void box units; including excavation and backfilling as required	m³	260.00 to	330.00
Land drainage			
NOTE: If land drainage is required on a project, the propensity of the land to flood will decide the spacing of the land drains. Costs include for excavation and backfilling of trenches and laying agricultural clay drain pipes with 75 mm dia. lateral runs average 600 mm deep, and 100 mm dia. mains runs average 750 mm deep			
land drainage to parkland with laterals at 30 m centres and main runs at 100 m centres	ha	5800.00 to	7600.00

8.8 ANCILLARY BUILDINGS AND STRUCTURES

Item	Unit	Range £		

8.7 EXTERNAL SERVICES

Service runs
Water main; all laid in trenches including excavation and backfill with excavated material

Item	Unit	Range £		
up to 75 mm dia. uPVC main	m	44.00	to	57.00

Electric main; all laid in trenches including excavation and backfill with excavated material

600/1000 volt cables. Two core 25 mm dia. cable including 100 mm dia. clayware duct	m	28.00	to	36.00

Gas main; all laid in trenches including excavation and backfill with excavated material

150 mm dia. gas pipe	m	48.00	to	61.00

Telephone duct; all laid in trenches including excavation and backfill with excavated material

100 mm dia. uPVC duct	m	22.50	to	29.00

Service connection charges
The privatization of telephone, water, gas and electricity has complicated the assessment of service connection charges. Typically, service connection charges will include the actual cost of the direct connection plus an assessment of distribution costs from the main. The latter cost is difficult to estimate as it depends on the type of scheme and the distance from the mains. In addition, service charges are complicated by discounts that may be offered. For instance, the electricity boards will charge less for housing connections if the house is all electric. However, typical charges for a reasonably sized housing estate might be as follows:

Item	Unit	Range £		
Water and sewerage connections				
water connections; water main up to 2 m from property	house	450.00	to	550.00
water infrastructure charges for new properties	house	360.00	to	440.00
sewerage infrastructure charges for new properties	house	360.00	to	440.00
Electric				
all electric	house	920.00	to	1175.00
pre-packaged substation housing	nr	23000.00	to	29500.00
Gas				
gas connection to house	house	230.00	to	295.00
governing station	nr	14000.00	to	17500.00
Telephone	house	92.00	to	120.00

8.8 ANCILLARY BUILDINGS AND STRUCTURES

Footbridges
Footbridge of either precast concrete or steel construction up to 6.00 m wide, 6.00 m high including deck, access stairs and ramp, parapets etc.

Item	Unit	Range £		
5 m span between piers or abutments	m²	1025.00	to	1325.00
20 m span between piers or abutments	m²	1450.00	to	1850.00
Footbridge of timber (stress graded with concrete piers)				
12 m span between piers or abutments	m²	770.00	to	980.00

8.8 ANCILLARY BUILDINGS AND STRUCTURES

Item	Unit	Range £		
8.8 ANCILLARY BUILDINGS AND STRUCTURES – cont				
Roadbridges				
Reinforced concrete bridge with precast beams; including all excavation, reinforcement, formwork, concrete, bearings, expansion joints, deck waterproofing and finishings, parapets etc. deck area				
10.00 m span	m²	1050.00	to	1350.00
15.00 m span	m²	1450.00	to	1850.00
Reinforced concrete bridge with prefabricated steel beams; including all excavation, reinforcement, formwork, concrete, bearings, expansion joints, deck waterproofing and finishings, parapets etc. deck area				
20.00 m span	m²	970.00	to	1225.00
30.00 m span	m²	920.00	to	1175.00
Multiparking systems/stack parkers				
Fully automatic systems				
integrated robotic parking system using robotic car transporter to store vehicles	car	16000.00	to	21000.00
Semi-automatic systems				
integrated parking system, transerve and vertical positioning; semi-automatic parking achieving 17 spaces in a 6 car width × 3 car height grid	car	8800.00	to	11000.00
Integrated stacker systems				
integrated parking system, vertical positioning only; double width, double height pit stacker achieving 4 spaces with each car stacker	car	4400.00	to	5600.00
integrated parking system, vertical positioning only; triple stacker achieving 3 spaces with each car stacker, generally 1 below ground and 2 above ground	car	9200.00	to	12000.00
integrated parking system, vertical positioning only; triple height double width stacker, achieving 6 spaces with each car stacker, generally 2 below ground and 4 above ground	car	6700.00	to	8500.00

The World's Greenest Buildings Promise Versus Performance in Sustainable Design

Jerry Yudelson & Ulf Meyer

The World's Greenest Buildings tackles an audacious task. Among the thousands of green buildings out there, which are the best, and how do we know?

Authors Jerry Yudelson and Ulf Meyer examined hundreds of the highest-rated large green buildings from around the world and asked their owners to supply one simple thing: actual performance data, to demonstrate their claims to sustainable operations.

- an overview of the rating systems and shows "best in class" building performance in North America, Europe, the Middle East, India, China, Australia and the Asia-Pacific region
- practical examples of best practices for greening both new and existing buildings
- a practical reference for how green buildings actually perform at the highest level, one that takes you step-by-step through many different design solutions
- a wealth of exemplary case studies of successful green building projects using actual performance data from which to learn
- interviews with architects, engineers, building owners and developers and industry experts, to provide added insight into the greening process

This guide uncovers some of the pitfalls that lie ahead for sustainable design, and points the way toward much faster progress in the decade ahead.

January 2013: 276 x219: 264 pp
Pb: 978-0-415-60629-5: £29.99

The Structural Basis of Architecture
Second Edition

Bjorn N. Sandaker, Arne P. Eggen & Mark R. Cruvellier

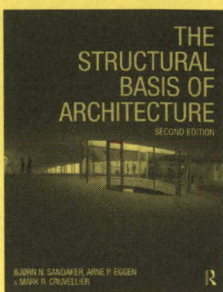

This is a book about structures that shows students how to "see" structures as integral to architecture, and how knowledge of structures is the basis for understanding both the mechanical and conceptual aspects inherent to the art of building.

Analyzing the structural principles behind many of the best known works of architecture from past and present alike, this book places the subject within a contemporary context.

The subject matter is approached in a qualitative and discursive manner, and is illustrated by many photographs of architectural projects and structural behaviour diagrams. This new edition is revised and updated throughout, includes worked-out examples, and is perfect as either an introductory structures course text or as a designer's sourcebook for inspiration.

April 2011: 276x219: 414 pp
Pb: 978-0-415-41547-7: £24.99

To Order: Tel: +44 (0) 1235 400524 Fax: +44 (0) 1235 400525
or Post: Taylor and Francis Customer Services,
Bookpoint Ltd, Unit T1, 200 Milton Park, Abingdon, Oxon, OX14 4TA UK
Email: book.orders@tandf.co.uk

For a complete listing of all our titles visit:
www.tandf.co.uk

Taylor & Francis
Taylor & Francis Group

Dwelling with Architecture

Platt & Kemsley

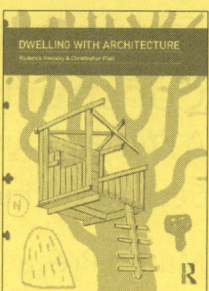

The dwelling is the most fundamental building type, nowhere more so than in the open landscape.

This book can be read in a number of ways. It is first a book about houses and particularly the theme 'dwelling and the land'. It examines the poetic and prosaic issues inherent in claiming a piece of the landscape to live on. It could also be seen as a kind of road map, full of both warnings and encouragements for all those involved with, or just interested in, the making of houses.

That the domestic realm and the landscape can be vehicles for significant architectural insights is hardly an original observation. However this book seeks to bring the two topics together in a unique way. In exploring a building type that lies on the cusp of what is commonly understood as 'building' and 'architecture', it asks fundamental questions about what the very nature of architecture is. Who indeed is the architect and what is their role in the process of creating meaningful buildings?

March 2012: 216x156: 248pp
Hb: 978-0-415-56903-3: £105.00
Pb: 978-0-415-56904-0: £29.99

Taylor & Francis
Taylor & Francis Group

Groundwater Lowering in Construction
A Practical Guide to Dewatering

Second Edition

Martin Preene

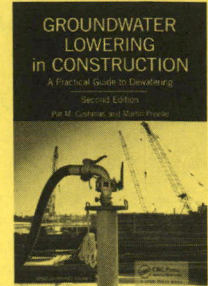

This new edition of a popular reference covers the design, construction, and environmental management of groundwater control and dewatering works for construction projects. Written for practicing construction professionals—including engineers, geologists, and hydrogeologists—and postgraduate students, this book focuses on practical aspects and is well illustrated with case studies. This edition has been updated to reflect the considerable developments in environmental management practice and procedures, as well as groundwater control regulations. The book contains brand new chapters on permanent dewatering systems and groundwater cutoff walls, as well as heavily revised coverage of other dewatering systems, environmental impacts, and the design of groundwater lowering systems.

August 2012: 234 x 156: 673pp
Hb: 978-0-415-66837-8: £100.00

To Order: Tel: +44 (0) 1235 400524 Fax: +44 (0) 1235 400525
or Post: Taylor and Francis Customer Services,
Bookpoint Ltd, Unit T1, 200 Milton Park, Abingdon, Oxon, OX14 4TA UK
Email: book.orders@tandf.co.uk

For a complete listing of all our titles visit:
www.tandf.co.uk

PART 4

Prices for Measured Works

This part contains the following NRM2 work sections:

INTRODUCTION

The rates contained in Prices for Measured Works are intended to apply to a project in the Outer London area costing about £3,500,000 including Preliminaries and assume that reasonable quantities of all types of work are required. Similarly it has been necessary to assume that the size of the project warrants the subletting of all types of work normally sublet. Adjustments should be made to standard rates for time, location, local conditions, site constraints and all the other factors likely to affect costs of any individual project.

The distinction between builders' work and work normally sublet is stressed because prices for work which can be sublet may well be inadequate for the contractor who is called upon to carry out relatively small quantities of such work themselves.

Measured Works prices are generally based upon wage rates and known material costs from May 2013. Built up prices and subcontractor rates include an allowance of 2½% for overheads and profit.

As elsewhere in this edition, prices do not include Value Added Tax or professional services fees.

This year we have restructured the *Prices for Measured Works* sections to reflect the recent introduction of NRM2.

To aid readers with the transition we have scheduled the SMM7 Work Sections and shown the new NRM2 Work Sections alongside:

SMM7 – Work Section	NRM – 2 Work Section
	2 Off-site manufactured materials, components and buildings
C Demolition/Alteration/Renovation	3 Demolitions
	4 Alterations, repairs and conservation
D Groundwork	5 Excavating and filling
D20 Excavation and Filling	6 Ground remediation and soil stabilization
D30 Cast In Place Piling	7 Piling
D32 Steel Piling	8 Underpinning
D40 Embedded Retaining Walling	9 Diaphragm walls and embedded retaining walls
D41 Crib Walls/Gabions/Reinforced Earthworks	10 Crib walls, gabions and reinforced earth
D50 Underpinning	
E In Situ Concrete/Large Precast Concrete	11 In situ concrete works
E10 In Situ Concrete Construction	12 Precast/composite concrete
E20 Formwork For In Situ Concrete	13 Precast concrete
E30 Reinforcement For In Situ Concrete	
E40 Designed Joints In Situ Concrete	
E41 Worked Finishes/Cutting To In Situ Concrete	
E42 Accessories Cast Into In Situ Concrete	
E50 Precast Concrete Large Units	
E60 Precast/Composite Concrete Decking	
F Masonry	14 Masonry
F10 Brick/Block Walling	
F11 Glass Block Walling	
F20 Natural Stone Rubble Walling	
F22 Cast Stone Walling/Dressings	

INTRODUCTION

SMM7 – Work Section	NRM – 2 Work Section
F30 Accessories/Sundry Items For Brick/Block/Stone Walling F31 Precast Concrete Sills/Lintels/Coping Features	
G Structural/Carcassing Metal/Timber G10 Structural Steel Framing G12 Isolated Structural Metal Members G20 Carpentry/Timber Framing/First Fixing	15 Structural metalwork 16 Carpentry
H Cladding/Covering H10 Patent Glazing H11 Curtain Walling H20 Rigid Sheet Cladding H30 Fibre Cement Profile Sheet Cladding H31 Metal Profiled/Flat Sheet Cladding/Covering/Siding H32 Plastic Profiled Sheet Cladding/Covering/Siding H41 Glass Reinforced Plastic Panel Cladding Features H51 Natural Stone Cladding Features H53 Clay Slab/Cladding/Features H60 Plain Roof Tiling H61 Fibre Cement Slating H62 Natural Slating H63 Reconstructed Stone Slating/Tiling H64 Timber Shingling H71 Lead Sheet Coverings/Flashes H72 Aluminium Sheet Coverings/Flashings H73 Copper Strip Sheet Coverings/Flashings H74 Zinc Strip Sheet Coverings/Flashings H75 Stainless Steel Sheet Coverings/Flashings H76 Fibre Bitumen Thermoplastic Sheet Coverings/Flashings H92 Rainscreen Cladding	17 Sheet roof coverings 18 Tile and slate roof and wall coverings 21 Cladding and covering 26 Metalwork
J Waterproofing J10 Specialist Waterproof Rendering J20 Mastic Asphalt Tanking/Damp-Proof Membranes J21 Mastic Asphalt Roofing/Insulation/Finishes J30 Liquid Applied Tanking/Damp-Proof Membranes J40 Flexible Sheet Tanking/Damp-Proof Membranes J41 Built Up Felt Roof Coverings J42 Single Layer Plastic Roof Coverings	19 Waterproofing

INTRODUCTION

SMM7 – Work Section	NRM – 2 Work Section
J43 Proprietary Roof Decking With Felt Finish	
K Linings/Sheathing/Dry partitioning	20 Proprietary linings and partitions
K10 Plasterboard Dry Lining/Partitions/Ceilings	30 Suspended ceilings
K11 Rigid Sheet Flooring/Sheathing/Linings/Casings	
K13 Rigid Sheet Fine Linings/Panelling	
K14 Glass Reinforced Gypsum Linings/Panelling	
K20 Timber Board Flooring/Sheathing/Linings/Casings	
K30 Demountable Partitions	
K32 Framed Panel Cubicle Partitions	
K33 Concrete/Terrazzo Partitions	
K40 Demountable Suspended Ceilings	
K41 Raised Access Floors	
L Windows/Doors/Stairs	23 Windows, screens and lights
L10 Windows/Rooflights/Screens/Louvres	24 Doors, shutters and hatches
L20 Doors/Shutters/Hatches	25 Stairs, walkways and balustrades
L30 Stairs/Walkways/Balustrades	27 Glazing
L40 General Glazing	
M Surface Finishes	28 Floor, wall, ceiling and roof finishings
M10 Cement: Sand/Concrete Screeds/Granolithic Screeds/Topping	29 Decoration
M11 Mastic Asphalt Flooring/Floor Underlays	
M12 Trowelled Bitumen/Resin/Rubber Latex	
M20 Plastered/Rendered/Roughcast Coating	
M21 Insulation With Rendered Finish	
M22 Sprayed Mineral Fibre Coatings	
M30 Metal Mesh Lathing/Anchored Reinforcement For Plastered Ceilings	
M31 Fibrous Plaster	
M40 Stone/Concrete/Quarry/Ceramic Tiling	
M41 Terrazzo Tiling/In Situ Terrazzo	
M42 Wood Block/Composition Block/Parquet	
M50 Rubber/Plastic/Cork/Lino/Carpet Tiling/Sheeting	
M51 Edge Fixed Carpeting	
M52 Decorative Papers/Fabrics	
M60 Painting/Clear Finishing	
N Furniture/Equipment	32 Furniture, fittings and equipment
N10/11 General Fixtures/Kitchen Fittings	
N13 Sanitary Appliances/Fittings	

INTRODUCTION

SMM7 – Work Section	NRM – 2 Work Section
N15 Signs/Notices	
P Building fabric sundries	22 General joinery
P10 Sundry Insulation/Proofing Work/Fire Stops	31 Insulation, fire stopping and fire protection
P20 Unframed Isolated Trims/Skirtings/Sundry Items	
P21 Ironmongery	
P30 Trenches/Pipeways/Pits For Buried Engineering Services	
P31 Holes/Chases/Covers/Supports for Services	
Q Paving/Planting/Fencing/Site furniture	35 Site works
Q10 Kerbs/Edgings/Channels/Paving Access	36 Fencing
Q20 Hardcore/Granular/Cement Bound Bases	37 Soft landscaping
Q21 In Situ Concrete Roads/Pavings	
Q22 Coated Macadam/Asphalt Roads/Pavings	
Q23 Gravel/Hoggin/Woodchip Roads/Pavings	
Q25 Slab/Brick/Block/Sett/Cobble Pavings	
Q26 Special Surfacings/Pavings For Sport	
Q30 Seeding/Turfing	
Q31 Planting	
Q40 Fencing	
R Disposal systems	33 Drainage above ground
R10 Rainwater Pipework/Gutters	34 Drainage below ground
R11 Foul Drainage Above Ground	
R12 Drainage Below Ground	
R13 Land Drainage	
S Piped supply systems	
S10 / S11 Hot And Cold Water	
S13 Pressurized Water	
T Mechanical heating/Cooling/Refrigeration Systems	38 Mechanical services
T10 Gas/Oil Fired Boilers	
T31 Low Temperature Hot Water	
V Electrical systems	39 Electrical services
V21/V22 General Lighting and Low Voltage Power	
W Security systems	
W20 Lightning Protection	
X Transport Systems	40 Transportation
	41 Builder's work in connection with mechanical, electrical and transportation

Prices for Measured Works

3 DEMOLITIONS

Item	PC £	Labour hours	Labour £	Material £	Unit	Total rate £
DEMOLITIONS						
NOTE: Demolition rates vary enormously from project to project, depending upon access, type of construction, method of demolition, redundant or recycleable materials etc. Always obtain specific quotations for each project under consideration. The following rates for simple demolition works may be of some assistance for comparative purposes. Generally scaffold and other access equipment is not included in the rates.						
Demolishing structures; by machine; note disposal not included						
Demolishing to ground level; single-storey brick outbuilding; timber flat roof; grub up shallow foundations; volume						
single storey outbuilding approximately 50 m³	–	–	–	–	m³	54.02
single storey outbuilding approximately 200 m³	–	–	–	–	m³	36.74
single storey outbuilding approximately 500 m³	–	–	–	–	m³	17.83
Demolishing to ground level; light steel framed and sheet roofed cycle shelter; grub up shallow foundations; volume cycle shelter approximately 120 m³	–	–	–	–	m³	41.06
Concrete framed multi-storey carpark	–	–	–	–	m³	27.02
Demolishing to ground level; multi-storey masonry building; flat roof; 1–4 storeys; volume	–	–	–	–	m³	17.02
Warehouse to slab level; 7 m eaves	–	–	–	–	m²	11.34
Warehouse to below slab; 7 m eaves; grub up shallow foundations (pads, ground beams and ground slab)	–	–	–	–	m²	19.99
Reinforced concrete frame building; easy access	–	–	–	–	m³	32.41
Reinforced concrete frame building; restricted access	–	–	–	–	m³	59.42
Concrete encased steel frame building; easy access	–	–	–	–	m³	21.61
Concrete encased steel frame building; restricted access	–	–	–	–	m³	54.02
Demolishing parts of structures; by hand; cost of skip included						
Breaking up plain concrete bed; load into skip						
100 mm thick	–	0.45	8.61	6.21	m²	14.82
150 mm thick	–	0.67	12.75	11.39	m²	24.14
200 mm thick	–	0.90	17.22	12.39	m²	29.61
300 mm thick	–	1.33	25.49	18.16	m²	43.65
Breaking up reinforced concrete bed; load into skip						
100 mm thick	–	0.50	9.65	6.99	m²	16.64
150 mm thick	–	0.75	14.30	12.17	m²	26.47
200 mm thick	–	1.00	19.12	14.01	m²	33.13
300 mm thick	–	1.50	28.76	20.88	m²	49.64

3 DEMOLITIONS

Item	PC £	Labour hours	Labour £	Material £	Unit	Total rate £
Demolishing reinforced concrete column or cutting away concrete casing to steel column; load into skip	–	9.99	191.19	87.20	m³	**278.39**
Demolishing reinforced concrete beam or cutting away concrete casing to steel beam; load into skip	–	11.47	219.62	92.75	m³	**312.37**
Demolishing reinforced concrete wall; load into skip						
100 mm thick	–	0.99	18.95	8.59	m²	**27.54**
150 mm thick	–	1.50	28.76	12.77	m²	**41.53**
225 mm thick	–	2.25	43.06	19.19	m²	**62.25**
300 mm thick	–	3.00	57.36	25.85	m²	**83.21**
Demolishing reinforced concrete suspended slabs; load into skip						
100 mm thick	–	0.84	16.02	8.17	m²	**24.19**
150 mm thick	–	1.25	23.95	11.95	m²	**35.90**
225 mm thick	–	1.87	35.84	17.91	m²	**53.75**
300 mm thick	–	2.50	47.88	24.28	m²	**72.16**
Breaking up small concrete plinth; make good structures; load into skip	–	3.83	73.38	51.23	m³	**124.61**
Breaking up small precast concrete kerb; make good structures; load into skip	–	0.41	7.93	8.22	m	**16.15**
Remove precast concrete window sill; set aside for reuse	–	1.33	25.49	–	m	**25.49**
Remove brick on-edge-coping; prepare walls for raising; load into skip						
one brick thick	–	0.38	9.13	0.79	m	**9.92**
one and a half brick thick	–	0.50	12.18	1.17	m	**13.35**
Demolishing brick chimney to 300 mm below roof level; sealing off flues with slates; piecing in treated sawn softwood rafters and making good roof coverings; load into skip (excluding scaffold access)						
680 mm × 680 mm × 900 mm high above roof	–	11.40	331.33	61.51	nr	**392.84**
add for each additional 300 mm height	–	2.08	58.13	5.94	nr	**64.07**
680 mm × 1030 mm × 900 mm high above roof	–	16.86	486.36	78.79	nr	**565.15**
add for each additional 300 mm height	–	3.11	87.09	8.90	nr	**95.99**
1030 mm × 1030 mm × 900 mm high above roof	–	25.52	732.21	112.00	nr	**844.21**
add for each additional 300 mm height	–	4.71	131.58	11.88	nr	**143.46**
Demolishing defective brick chimney to roof level; rebuild using 25% new facing bricks; new lead flashings; core flues; re-set chimney pot; load into skip (excluding scaffold access)						
680 mm × 680 mm × 900 mm high above roof	–	10.73	298.76	107.88	nr	**406.64**
add for each additional 300 mm height	–	2.08	58.13	12.02	nr	**70.15**
680 mm × 1030 mm × 900 mm high above roof	–	15.66	437.85	127.82	nr	**565.67**
add for each additional 300 mm height	–	3.11	87.09	16.40	nr	**103.49**
1030 mm × 1030 mm × 900 mm high above roof	–	24.02	671.57	159.22	nr	**830.79**
add for each additional 300 mm height	–	4.71	131.58	21.46	nr	**153.04**
Demolishing external brick walls; load into skip						
half brick thick	–	0.87	24.64	5.82	m²	**30.46**
two half brick thick skins if cavity wall	–	1.46	41.15	11.29	m²	**52.44**
one brick thick	–	1.50	42.42	11.29	m²	**53.71**
one and a half brick thick	–	2.04	57.66	16.93	m²	**74.59**
two brick thick	–	2.62	73.91	22.56	m²	**96.47**
extra for plaster, render or pebbledash finish per side	–	0.08	2.28	1.17	m²	**3.45**

Prices for Measured Works

3 DEMOLITIONS

Item	PC £	Labour hours	Labour £	Material £	Unit	Total rate £
DEMOLITIONS – cont						
Demolishing parts of structures – cont						
Demolish external stone walls; load into skip						
300 mm thick	–	1.00	28.22	11.88	m²	**40.10**
400 mm thick	–	1.33	37.60	15.45	m²	**53.05**
600 mm thick	–	2.00	56.45	23.31	m²	**79.76**
Demolish external stone walls; clean off and set aside for reuse						
300 mm thick	–	1.50	42.34	2.96	m²	**45.30**
400 mm thick	–	2.00	56.45	4.45	m²	**60.90**
600 mm thick	–	3.00	84.67	5.94	m²	**90.61**
Remove fireplace surround and hearth						
breaking up 500 mm wide concrete hearth; load into skip	–	1.50	42.42	3.89	m	**46.31**
remove fire surround; tiled interior; load into skip	–	1.54	43.44	8.90	nr	**52.34**
cast iron surround; set aside for reuse	–	2.58	72.90	–	nr	**72.90**
stone surround; set aside for reuse	–	6.74	190.26	–	nr	**190.26**
Fill in opening with common bricks; air brick; 2 coat plaster; timber skirting; re-screed floor	–	0.95	22.96	47.65	m²	**70.61**
Removing roof timbers complete (NB not coverings); including rafters, purlins, ceiling joists, plates etc.; load into skip; measured flat on plan	–	0.28	7.64	4.58	m²	**12.22**
Remove softwood floor structure; load into skip						
joists at ground floor level	–	0.20	5.78	1.55	m²	**7.33**
joists at first floor level	–	0.41	11.69	1.55	m²	**13.24**
joists at second floor	–	0.58	16.52	1.55	m²	**18.07**
individual timber members	–	0.23	6.35	0.05	m	**6.40**
Remove boarding; withdraw nails; set aside for reuse						
softwood flooring at ground floor level	–	0.30	8.46	–	m²	**8.46**
softwood flooring at first floor level	–	0.51	14.46	–	m²	**14.46**
softwood flooring at second floor level	–	0.61	17.27	–	m²	**17.27**
fascia/barge board/gutter board second floor level; load into skip	–	0.52	14.53	0.05	m	**14.58**
Remove doors and windows; set aside for reuse						
solid single door only (frame left in place)	–	0.33	9.40	1.17	nr	**10.57**
solid single door; frame to skip	–	0.55	15.52	1.94	nr	**17.46**
solid double door only (frame left in place)	–	0.66	18.63	2.33	nr	**20.96**
solid double door; frame to skip	–	1.00	28.19	3.89	nr	**32.08**
glazed screen and doors	–	1.00	28.22	5.82	m²	**34.04**
casement window frame; set aside frame; glass into skip; up to 1.00 m²	–	0.75	21.17	0.29	m²	**21.46**
casement window frame; set aside frame; glass into skip; 1.00 to 2.00 m²	–	1.00	28.22	0.29	m²	**28.51**
casement window frame; set aside frame; glass into skip; 2.00 to 3.00 m²	–	1.10	31.05	0.29	m²	**31.34**
pair French windows and frame; up to 1200 mm × 1200 mm	–	3.33	93.99	5.82	nr	**99.81**
removing double hung sash window and frame; store for reuse	–	2.50	70.49	–	nr	**70.49**

3 DEMOLITIONS

Item	PC £	Labour hours	Labour £	Material £	Unit	Total rate £
Remove doors and windows; load into skip						
remove solid timber door	–	0.30	8.46	1.94	nr	**10.40**
remove door frame	–	0.10	2.82	1.94	nr	**4.76**
casement window frame; up to 1.00 m²	–	0.60	16.94	1.94	m²	**18.88**
casement window frame; 1.00 to 2.00 m²	–	0.80	22.57	1.94	m²	**24.51**
casement window frame; 2.00 to 3.00 m²	–	0.90	25.40	1.94	m²	**27.34**
Demolishing internal partitions; load into skip						
half brick thick brickwork	–	0.87	24.64	5.82	m²	**30.46**
one brick thick brickwork	–	1.50	42.42	11.29	m²	**53.71**
one and a half brick thick brickwork	–	2.04	57.66	16.93	m²	**74.59**
75 mm blockwork	–	0.58	16.52	4.27	m²	**20.79**
90 mm blockwork	–	0.62	17.53	5.05	m²	**22.58**
100 mm blockwork	–	0.67	18.80	5.82	m²	**24.62**
115 mm blockwork	–	0.71	20.07	5.82	m²	**25.89**
125 mm blockwork	–	0.75	21.08	6.22	m²	**27.30**
140 mm blockwork	–	0.79	22.36	6.60	m²	**28.96**
150 mm blockwork	–	0.84	23.63	7.37	m²	**31.00**
190 mm blockwork	–	0.98	27.69	9.32	m²	**37.01**
215 mm blockwork	–	1.08	30.48	10.10	m²	**40.58**
255 mm blockwork	–	1.25	35.28	12.04	m²	**47.32**
extra for plaster finish per side	–	0.08	2.28	1.17	m²	**3.45**
breaking up brick plinths	–	3.33	93.99	38.84	m³	**132.83**
stud walls (metal or timber); solid board each side	–	0.38	10.67	3.89	m²	**14.56**
stud walls (metal or timber); glazed panels	–	0.50	14.23	3.89	m²	**18.12**
lightweight steel mesh security screen	–	0.41	11.69	1.93	m²	**13.62**
solid steel demountable partition	–	0.62	17.53	2.72	m²	**20.25**
glazed demountable partition including removal of glass	–	0.84	23.63	3.89	m²	**27.52**
glazed screen including any doors incorporated	–	1.00	28.22	5.82	m²	**34.04**
remove large folding partition and track	–	9.90	279.40	59.09	m²	**338.49**
remove timber skirting, architrave, dadi rail; load into skip	–	0.10	2.82	0.05	m	**2.87**
Removing timber staircase and balustrades						
single straight flight	–	2.92	82.30	38.84	nr	**121.14**
single dogleg flight	–	4.17	106.67	52.85	nr	**159.52**
Perimeter hoarding						
security fencing; 2400 mm high painted plywood hoarding	–	–	–	–	m	**97.24**

3 DEMOLITIONS

Item	PC £	Labour hours	Labour £	Material £	Unit	Total rate £
TEMPORARY SUPPORT OF STRUCTURES, ROADS AND THE LIKE						
Temporary support						
NOTE: The requirement for shoring and strutting for the formation of large openings are dependent upon a number of factors; for example the weight of the superimposed structure to be supported; the number of windows; number of floors; roof type; whether raking shores are required; depth to load bearing surface; duration support is to be left in place. Prices therefore are best built-up by assessing the use and waste of materials and the labours involved. The following guide needs to be tested against each particular project.						
Support of structures not to be demolished						
strutting to window openings over proposed new openings	–	0.56	13.94	3.55	nr	**17.49**
plates, struts, braces and hardwood wedges in supports to floors and roof of openings	–	1.11	27.64	10.38	nr	**38.02**
dead shore and needle using die square timber with sole plates, braces, hardwood wedges and steel dogs	–	27.75	691.01	51.74	nr	**742.75**
set of two raking shores using die square timber with 50 mm thick wall pieces, hardwood wedges and steel dogs, including forming holes for needles and making good	–	33.30	829.22	52.97	nr	**882.19**
cut holes through one brick wall for die square needle and make good on completion of works, including facings externally and plaster internally	–	5.36	117.49	16.98	nr	**134.47**

4 ALTERATIONS, REPAIRS AND CONSERVATION

Item	PC £	Labour hours	Labour £	Material £	Unit	Total rate £
REMOVING						
Remove materials from existing buildings						
NOTE: It is highly unlikely for exactly the same composite items of alteration works will be encountered on different projects. The following spot items have been included to allow the reader to build up the composite rates for estimating purposes. Rates include for the removal of debris from site, but do not include for any temporary works, shoring, scaffolding etc. or for any redecorations.						
Removing coverings load into skip						
Roof coverings						
slates	–	0.45	7.75	0.71	m²	**8.46**
slates; set aside for reuse	–	0.54	9.43	–	m²	**9.43**
nibbed tiles	–	0.36	6.23	0.71	m²	**6.94**
nibbed tiles; set aside for reuse	–	0.45	7.75	–	m²	**7.75**
underfelt and nails	–	0.04	0.67	0.35	m²	**1.02**
felt or polymer membrane	–	0.22	3.87	0.71	m²	**4.58**
profiled metal sheeting; insulation; steel liner	–	0.75	13.05	1.35	m²	**14.40**
sheet metal coverings	–	0.45	7.75	0.71	m²	**8.46**
remove tiling battens	–	0.07	1.18	0.35	m²	**1.53**
Remove roof coverings; select and refix including providing 25% new tiles including nails						
natural slates; Welsh Blue	–	1.08	39.48	13.83	m²	**53.31**
clay plain tiles	–	0.99	36.28	6.74	m²	**43.02**
asbestos-free artificial blue/black slates	–	0.99	36.28	5.95	m²	**42.23**
concrete interlocking tiles	–	0.63	23.12	2.68	m²	**25.80**
removing ridge or hip tile; provide and fit new	–	0.45	16.36	8.01	m	**24.37**
Fixtures and fittings; remove and load into skip						
Handrails and balustrades						
timber or metal handrail and brackets	–	0.27	7.66	0.39	m	**8.05**
timber balustrades	–	0.30	8.49	1.17	m	**9.66**
metal balustrades	–	0.45	12.59	1.17	m	**13.76**
Kitchen fittings						
wall units (up to 500 mm wide)	–	0.41	11.50	5.82	nr	**17.32**
floor units units (up to 500 mm wide)	–	0.27	7.66	8.55	nr	**16.21**
larder units units (up to 500 mm wide)	–	0.36	10.13	19.43	nr	**29.56**
Bathroom fittings						
bath panels and bearers	–	0.36	10.13	0.39	nr	**10.52**
toilet roll holders or soap dispensers; 2 or 3 screw fixings	–	0.27	5.61	0.05	nr	**5.66**
towel rails; 4 or 6 screw fixings	–	0.54	10.40	0.05	nr	**10.45**
mirror; up to 600 mm × 600 mm; 4 screw fixings	–	0.58	12.03	0.07	nr	**12.10**
pipe casings	–	0.27	7.66	1.55	m	**9.21**

4 ALTERATIONS, REPAIRS AND CONSERVATION

Item	PC £	Labour hours	Labour £	Material £	Unit	Total rate £
REMOVING – cont						
Fixtures and fittings – cont						
Other fixtures and fittings						
shelves, window boards and the like; screw fixings	–	0.30	6.22	0.39	m	**6.61**
curtain track, plastic; screw fixing at 600 mm centres	–	0.10	2.06	–	m	**2.06**
small nameplates or individual numerals/letters; screw fixings	–	0.25	5.17	–	nr	**5.17**
small notice board and frame from walls; up to 600 mm × 900 mm; screw fixings	–	0.50	10.33	–	nr	**10.33**
Finishes; removing and load into skip unless stated otherwise						
Breaking up floor screeds						
asphalt paving	–	0.54	15.34	0.69	m²	**16.03**
floor screed up to 100 mm thick unreinforced	–	0.33	9.31	0.69	m²	**10.00**
granolithic floor screed up to 100 mm thick unreinforced	–	0.58	16.43	0.69	m²	**17.12**
Floor finishes						
carpet and underfelt	–	0.11	3.01	1.49	m²	**4.50**
carpetgrip edge fixing strip	–	0.01	0.21	–	m	**0.21**
vinyl or similar sheet flooring	–	0.09	2.46	1.02	m²	**3.48**
vinyl or similar tile flooring with tile remover	–	0.17	4.65	1.13	m²	**5.78**
ceramic floor tiles with tile remover	–	0.40	7.65	1.05	m²	**8.70**
woodblock flooring; set aside for reuse	–	0.67	12.81	–	m²	**12.81**
Hack off wall finishes with chipping hammer						
plaster to walls	–	0.33	6.32	0.97	m²	**7.29**
cement rendering, pebbledash or similar	–	0.25	4.79	0.70	m²	**5.49**
ceramic wall tiles	–	0.40	7.65	0.97	m²	**8.62**
Strip off wallpaper using steam paper stripper						
1 layer	–	0.10	1.92	0.11	m²	**2.03**
2 layers	–	0.14	2.68	0.11	m²	**2.79**
3 layers	–	0.17	3.35	0.11	m²	**3.46**
Remove wall linings including battening behind						
plain sheet boarding	–	0.27	7.66	1.55	m²	**9.21**
matchboarding	–	0.36	10.13	2.33	m²	**12.46**
Ceiling finishes						
plasterboard and skim; removes nails or screws	–	0.20	3.47	1.06	m²	**4.53**
wood lathe and plaster; removes nails or screws	–	0.22	3.82	1.76	m²	**5.58**
suspended ceilings; lay in grid; hangers and tee sections	–	0.45	7.75	1.76	m²	**9.51**
plain sheet boarding; including battening	–	0.41	7.07	1.41	m²	**8.48**
matchboarding; including battening	–	0.54	9.43	2.11	m²	**11.54**
Wall panelling						
remove timber wall panelling; clean off and set aside for reuse	–	0.58	16.43	–	m²	**16.43**

4 ALTERATIONS, REPAIRS AND CONSERVATION

Item	PC £	Labour hours	Labour £	Material £	Unit	Total rate £
CUTTING OR FORMING OPENINGS						
Cutting openings or recesses; spoil into adjacent skip						
Through reinforced concrete walls; make good; not including any props or supports						
150 mm thick wall	–	5.02	94.27	11.03	m²	**105.30**
225 mm thick wall	–	6.95	133.68	15.67	m²	**149.35**
300 mm thick wall	–	8.75	161.85	16.29	m²	**178.14**
Through reinforced concrete suspended slab; make good; not including any props or supports						
150 mm thick wall	–	3.81	70.77	14.17	m²	**84.94**
225 mm thick wall	–	5.66	105.07	15.67	m²	**120.74**
300 mm thick wall	–	7.04	130.07	16.29	m²	**146.36**
Through brick or block walls or partitions						
half brick thick	–	2.38	60.84	5.99	m²	**66.83**
one brick thick	–	3.95	101.06	11.98	m²	**113.04**
one and a half brick thick	–	5.52	141.29	17.97	m²	**159.26**
two brick thick	–	7.09	181.52	23.96	m²	**205.48**
100 mm thick blockwork	–	1.75	44.70	5.28	m²	**49.98**
140 mm thick blockwork	–	2.10	53.88	7.40	m²	**61.28**
215 mm thick blockwork	–	2.60	66.55	11.28	m²	**77.83**
Cut opening through wall for new window (not included); 1200 mm × 1200 mm; quoining up jambs; cut and pin galvanized steel lintel; new soldier course in facing bricks						
one brick thick wall or two half brick thick skins	5.60	7.18	169.26	82.08	nr	**251.34**
one and a half brick thick wall	5.60	9.44	227.19	90.71	nr	**317.90**
two brick thick wall	5.60	11.70	285.12	99.33	nr	**384.45**
Cut opening through wall for new door (not included); 1200 mm × 2100 mm; quoining up jambs; cut and pin galvanized steel lintel; new soldier course in facing bricks						
one brick thick wall or two half brick thick skins	5.60	17.11	402.51	131.64	nr	**534.15**
one and a half brick thick wall	5.60	22.75	540.51	165.05	nr	**705.56**
two brick thick wall	5.60	28.39	679.48	198.46	nr	**877.94**
Diamond cutting						
Cutting to create openings in block wall						
up to 1.00 m girth; up to 150 mm thick	–	–	–	–	m	**18.83**
up to 1.00 m girth; 200 mm thick	–	–	–	–	m	**22.46**
up to 1.00 m girth; 250 mm thick	–	–	–	–	m	**26.10**
up to 1.00 m girth; 300 mm thick	–	–	–	–	m	**29.73**
3 m–3.5 m girth; up to 150 mm thick	–	–	–	–	m	**84.75**
3 m–3.5 m girth; 200 mm thick	–	–	–	–	m	**101.09**
3 m–3.5 m girth; 250 mm thick	–	–	–	–	m	**117.45**
3 m–3.5 m girth; 300 mm thick	–	–	–	–	m	**133.79**
9 m–10 m girth; up to 150 mm thick	–	–	–	–	m	**281.25**
9 m–10 m girth; 200 mm thick	–	–	–	–	m	**375.00**
9 m–10 m girth; 250 mm thick	–	–	–	–	m	**468.75**
9 m–10 m girth; 300 mm thick	–	–	–	–	m	**562.50**

4 ALTERATIONS, REPAIRS AND CONSERVATION

Item	PC £	Labour hours	Labour £	Material £	Unit	Total rate £
CUTTING OR FORMING OPENINGS – cont						
Diamond cutting – cont						
Cutting to create openings in brick walll						
up to 1.00 m girth; up to 150 mm thick	–	–	–	–	m	23.54
up to 1.00 m girth; 200 mm thick	–	–	–	–	m	28.08
up to 1.00 m girth; 250 mm thick	–	–	–	–	m	32.63
up to 1.00 m girth; 300 mm thick	–	–	–	–	m	37.16
3 m–3.5 m girth; up to 150 mm thick	–	–	–	–	m	169.51
3 m–3.5 m girth; 200 mm thick	–	–	–	–	m	202.18
3 m–3.5 m girth; 250 mm thick	–	–	–	–	m	234.90
3 m–3.5 m girth; 300 mm thick	–	–	–	–	m	267.57
9 m–10 m girth; up to 150 mm thick	–	–	–	–	m	562.50
9 m–10 m girth; 200 mm thick	–	–	–	–	m	750.00
9 m–10 m girth; 250 mm thick	–	–	–	–	m	937.50
9 m–10 m girth; 300 mm thick	–	–	–	–	m	1125.00
Cutting to create openings in concrete walll						
up to 1.00 m girth; up to 150 mm thick	–	–	–	–	m	94.17
up to 1.00 m girth; 200 mm thick	–	–	–	–	m	112.32
up to 1.00 m girth; 250 mm thick	–	–	–	–	m	130.50
up to 1.00 m girth; 300 mm thick	–	–	–	–	m	148.65
3 m–3.5 m girth; up to 150 mm thick	–	–	–	–	m	847.53
3 m–3.5 m girth; 200 mm thick	–	–	–	–	m	1010.88
3 m–3.5 m girth; 250 mm thick	–	–	–	–	m	1174.50
3 m–3.5 m girth; 300 mm thick	–	–	–	–	m	1337.85
9 m–10 m girth; up to 150 mm thick	–	–	–	–	m	2812.50
9 m–10 m girth; 200 mm thick	–	–	–	–	m	3750.00
9 m–10 m girth; 250 mm thick	–	–	–	–	m	4687.50
9 m–10 m girth; 300 mm thick	–	–	–	–	m	5625.00
Cutting to create openings in timber walll						
up to 1.00 m girth; up to 150 mm thick	–	–	–	–	m	28.25
up to 1.00 m girth; 200 mm thick	–	–	–	–	m	33.70
up to 1.00 m girth; 250 mm thick	–	–	–	–	m	39.15
up to 1.00 m girth; 300 mm thick	–	–	–	–	m	44.60
3 m–3.5 m girth; up to 150 mm thick	–	–	–	–	m	254.26
3 m–3.5 m girth; 200 mm thick	–	–	–	–	m	303.26
3 m–3.5 m girth; 250 mm thick	–	–	–	–	m	352.35
3 m–3.5 m girth; 300 mm thick	–	–	–	–	m	401.36
9 m–10 m girth; up to 150 mm thick	–	–	–	–	m	843.75
9 m–10 m girth; 200 mm thick	–	–	–	–	m	1125.00
9 m–10 m girth; 250 mm thick	–	–	–	–	m	1406.25
9 m–10 m girth; 300 mm thick	–	–	–	–	m	1687.50
Quoining up jambs to openings						
Common bricks						
half brick thick wall	–	0.90	19.77	4.36	m	24.13
one brick thick wall	–	1.35	29.55	8.72	m	38.27
one and a half brick thick wall	–	1.75	38.27	13.08	m	51.35
two brick thick wall	–	2.15	47.22	17.44	m	64.66
100 mm blockwork wall	–	0.63	13.82	4.13	m	17.95
140 mm blockwork wall	–	0.78	17.01	6.19	m	23.20
215 mm blockwork wall	–	0.97	21.26	9.34	m	30.60

4 ALTERATIONS, REPAIRS AND CONSERVATION

Item	PC £	Labour hours	Labour £	Material £	Unit	Total rate £
Cutting back projections; spoil into skip						
Brick projection flush with adjacent wall						
225 mm × 112 mm	–	0.27	5.95	0.35	m	**6.30**
225 mm × 225 mm	–	0.45	9.78	0.71	m	**10.49**
337 mm × 112 mm	–	0.63	13.82	1.06	m	**14.88**
450 mm × 225 mm	–	0.81	17.65	1.41	m	**19.06**
chimney breast	–	1.57	34.44	11.28	m²	**45.72**
Temporary screens						
Providing and erecting; maintaining temporary dust proof screens; 50 mm × 75 mm softwood framing and 12 mm thick plywood covering to one side; single layer of polythene sheet to the other; clear away on completion	–	0.90	20.37	12.09	m²	**32.46**
FILLING IN OPENINGS						
Fill in small holes						
Make good hole where small (up to 12 mm dia.) pipe removed; cement mortar	–	0.19	4.59	0.14	nr	**4.73**
Larger holes fill in with common brickwork or blockwork						
half brick thick	–	1.66	36.36	17.32	m²	**53.68**
one brick thick	–	2.74	59.96	36.13	m²	**96.09**
one and a half brick thick	–	3.77	82.71	54.19	m²	**136.90**
two brick thick	–	4.71	103.33	72.26	m²	**175.59**
100 mm blockwork	–	0.94	20.62	8.55	m²	**29.17**
140 mm blockwork	–	1.13	24.66	11.70	m²	**36.36**
215 mm blockwork	–	1.44	31.47	17.50	m²	**48.97**
cavity wall; facing bricks; 100 mm cavity fill insulation; 140 mm block	–	4.13	90.58	41.64	m²	**132.22**
REMOVE EXISTING AND REPLACING						
Remove existing material for replacement						
NB Access equipment is not included						
Cutting out decayed, defective or cracked work and replacing with new common bricks; gauged mortar						
small areas; half brick thick walling	–	4.56	110.20	26.03	m²	**136.23**
small areas; one brick thick walling	–	8.88	214.60	51.81	m²	**266.41**
small areas; one and a half brick thick walling	–	12.58	304.02	79.20	m²	**383.22**
small areas; two brick thick walling	–	16.10	389.07	105.10	m²	**494.17**
individual bricks; half brick thick walling	–	0.28	6.77	0.46	nr	**7.23**

4 ALTERATIONS, REPAIRS AND CONSERVATION

Item	PC £	Labour hours	Labour £	Material £	Unit	Total rate £
REMOVE EXISTING AND REPLACING – cont						
Remove existing material for replacement – cont						
Cutting out decayed, defective or cracked work and replacing with new facing bricks; gauged mortar; facing to one side						
small areas; half brick thick walling (PC £ per 1000)	350.00	6.75	163.13	35.12	m²	**198.25**
small areas; half brick thick walling (PC £ per 1000)	500.00	6.75	163.13	47.53	m²	**210.66**
individual bricks; half brick thick walling (PC £ per 1000)	350.00	0.42	10.15	0.61	m²	**10.76**
Cutting out decayed, defective or cracked soldier course arch and replace with new						
facing bricks (PC £ per 1000)	350.00	1.80	43.50	8.85	m²	**52.35**
facing bricks (PC £ per 1000)	500.00	1.80	43.50	12.16	m²	**55.66**
Cutting out raking cracks in brickwork; stitching in new common bricks						
half brick thick (PC £ per 1000)	240.00	2.96	71.53	13.27	m²	**84.80**
one brick thick (PC £ per 1000)	240.00	5.41	130.74	27.39	m²	**158.13**
one and a half brick thick (PC £ per 1000)	240.00	8.09	195.51	40.66	m²	**236.17**
Cutting out raking cracks in brickwork; stitching in new facing bricks; half brick thick; facing to one side						
facing bricks (PC £ per 1000)	350.00	4.40	106.33	18.81	m²	**125.14**
facing bricks (PC £ per 1000)	500.00	4.44	107.30	25.01	m²	**132.31**
Cutting out raking cracks in cavity brickwork; stitching in new common bricks one side; facing bricks the other side; both skins half brick thick; facing to one side						
facing bricks (PC £ per 1000)	350.00	7.59	183.42	31.09	m²	**214.51**
facing bricks (PC £ per 1000)	500.00	7.59	183.42	39.28	m²	**222.70**
Cutting away old angle fillets and replacing with new; cement mortar; 50 mm face width	–	0.23	5.55	3.71	m	**9.26**
Cutting out ends of joists and plates from walls; making good in common bricks in cement mortar						
175 mm joists; 400 mm centres	–	0.60	14.50	8.52	m	**23.02**
225 mm joists; 400 mm centres	–	0.74	17.88	9.84	m	**27.72**
Cutting and pinning to existing brickwork ends of joists	–	0.37	8.95	–	nr	**8.95**
Remove defective wall and rebuild with new materials						
defective parapet wall; 600 mm high; with two courses of tiles and brick-on-edge coping over; rebuilding with new facing bricks, tiles and coping	–	6.16	148.86	65.99	m	**214.85**
defective capping stones and haunching; replace stones and re-haunch in cement mortar	–	1.39	33.59	7.41	m	**41.00**

4 ALTERATIONS, REPAIRS AND CONSERVATION

Item	PC £	Labour hours	Labour £	Material £	Unit	Total rate £
PREPARING EXISTING STRUCTURES FOR CONNECTION						
Prepare existing structures for connection						
Making good where intersecting wall has been removed						
half brick thick	–	0.28	6.14	1.12	m	**7.26**
one brick thick	–	0.37	8.11	2.24	m	**10.35**
100 mm blockwork	–	0.23	5.04	1.12	m	**6.16**
150 mm blockwork	–	0.27	5.92	1.12	m	**7.04**
215 mm blockwork	–	0.32	7.01	2.24	m	**9.25**
225 mm blockwork	–	0.36	7.89	2.24	m	**10.13**
REPAIRING						
Repairing concrete						
Reinstating plain concrete bed with site mixed in situ concrete; mix 20.00 N, where opening no longer required						
100 mm thick	–	0.44	9.05	11.49	m²	**20.54**
150 mm thick	–	0.72	14.41	17.23	m²	**31.64**
Reinstating reinforced concrete bed with site mixed in situ concrete; mix 20.00 N; mesh reinforcement, where opening no longer required						
100 mm thick	–	0.66	13.26	15.50	m²	**28.76**
150 mm thick	–	0.91	18.05	21.24	m²	**39.29**
Reinstating reinforced concrete suspended floor with site mixed in situ concrete; mix 25.00 N; mesh reinforcement; formwork, where opening no longer required						
150 mm thick	–	2.96	60.48	24.45	m²	**84.93**
225 mm thick	–	3.12	63.55	33.34	m²	**96.89**
300 mm thick	–	3.60	72.73	42.21	m²	**114.94**
Reinstating small hole through concrete suspended slab with site mixed in situ concrete; mix 25.00 N, where opening no longer required						
50 mm dia.	–	0.39	7.47	0.06	nr	**7.53**
100 mm dia.	–	0.51	9.75	0.15	nr	**9.90**
150 mm dia.	–	0.65	12.43	0.34	nr	**12.77**
Clean out minor crack and fill with cement: mortar mixed with bonding agent	–	0.31	8.34	1.12	m	**9.46**
Clean out crack to form a 20 mm × 20 mm groove and fill with cement: mortar mixed with bonding agent	–	0.61	17.22	4.58	m	**21.80**
Repairing metal						
Overhauling and repairing metal casement windows; adjust and oil ironmongery; prepare afffected parts for redecoration	–	1.39	34.61	–	nr	**34.61**

4 ALTERATIONS, REPAIRS AND CONSERVATION

Item	PC £	Labour hours	Labour £	Material £	Unit	Total rate £
REPAIRING – cont						
Repairing timber						
Removing or punching projecting nails; refixing timber flooring						
loose boards	–	0.14	3.16	–	m²	**3.16**
refix floorboards previously set aside	–	0.74	16.71	0.64	m²	**17.35**
Remove damaged softwood flooring and fix new plain edge softwood flooring						
small areas	–	1.06	23.94	23.36	m²	**47.30**
individual boards 150 mm wide	–	0.28	6.32	2.37	m	**8.69**
Sanding down and resurfacing existing flooring; preparing; body in with shellac and wax polish						
softwood	–	–	–	–	m²	**13.43**
softwood	–	–	–	–	m²	**16.20**
Fitting existing softwood skirting or architrave to new frames						
75 mm high	–	0.09	2.03	–	m	**2.03**
150 mm high	–	0.12	2.71	–	m	**2.71**
225 mm high	–	0.15	3.39	–	m	**3.39**
Piecing in new 25 mm × 150 mm moulded softwood skirtings to match existing where old has been removed; prepare for decoration	–	0.35	7.25	2.85	m	**10.10**
Repair doors						
Easing and adjusting softwood doors; oil ironmongery	–	0.59	13.33	–	nr	**13.33**
Remove softwood doors; easing and adjust; oil ironmongery; rehang	–	0.75	16.94	–	nr	**16.94**
Remove softwood door; plane 12 mm from bottom edge; rehang	–	1.11	25.07	–	nr	**25.07**
Take off existing softwood doorstops; supply and fit new 25 mm × 38 mm doorstop	–	0.10	2.26	0.91	m	**3.17**
Cutting out infected or decayed structural members; supply and fix new treated sawn softwood members						
Floors and flat roofs						
50 mm × 125 mm	–	0.37	8.36	4.04	m	**12.40**
50 mm × 150 mm	–	0.41	9.26	4.85	m	**14.11**
50 mm × 175 mm	–	0.44	9.94	4.52	m	**14.46**
Pitched roofs						
38 mm × 100 mm	–	0.33	7.45	2.58	m	**10.03**
50 mm × 100 mm	–	0.42	9.49	3.23	m	**12.72**
50 mm × 125 mm	–	0.46	10.39	4.04	m	**14.43**
50 mm × 150 mm	–	0.51	11.52	4.85	m	**16.37**
Kerb bearers						
50 mm × 75 mm	–	0.42	9.49	2.01	m	**11.50**
50 mm × 100 mm	–	0.52	11.74	3.23	m	**14.97**
75 mm × 100 mm	–	0.63	14.23	4.61	m	**18.84**

4 ALTERATIONS, REPAIRS AND CONSERVATION

Item	PC £	Labour hours	Labour £	Material £	Unit	Total rate £
REPOINTING						
Repointing masonry						
Raking out decayed masonry joints and repointing in cement mortar						
brickwork walls generally	–	0.69	16.67	1.24	m²	**17.91**
cutting out staggered cracks and repointing to match existing along brick joints	–	0.37	8.95	–	m	**8.95**
brickwork and re-wedge horizontal flashing	–	0.23	5.55	0.62	m	**6.17**
brickwork and re-wedge stepped flashing	–	0.34	8.21	0.62	m	**8.83**
brickwork in chimney stacks	–	1.11	26.83	1.24	m²	**28.07**
uncoursed stonework	–	1.11	26.83	0.62	m	**27.45**
CLEANING SURFACES						
Cleaning surfaces by hand using hand held manual tools only (brushes, scrapers etc.)						
Masonry/concrete						
Cleaning surface of moss and lichen from walls	–	0.28	7.91	–	m²	**7.91**
Cleaning bricks of mortar, sort and stack for reuse	–	9.25	261.06	–	1000	**261.06**
Clean surface of concrete to receive new damp-proof membrane	–	0.14	3.95	–	m²	**3.95**
DECONTAMINATION						
Insecticide or fungicidal treatments						
Treating individual timbers with two coats of proprietary insecticide and fungicide by brush or spray application as appropriate						
general boarding	–	–	–	–	m²	**9.59**
structural members	–	–	–	–	m²	**8.27**
skirtings, architraves etc.	–	–	–	–	m	**8.55**
remove cobwebs, dust and roof insulation; treat exposed timbers with proprietary insecticide and fungicide by spray application	–	–	–	–	m²	**15.93**
lifting necessary floorboards; treating flooring with two coats of proprietary insecticide and fungicide by spray application; refix boards	–	–	–	–	m²	**14.89**
treating surfaces of adjoining masonry or concrete with two coats of proprietary dry rot by spray application	–	–	–	–	m²	**9.37**

5 EXCAVATING AND FILLING

Item	PC £	Labour hours	Labour £	Material £	Unit	Total rate £
SITE CLEARANCE AND PREPARATION						
Site clearance						
Prices are applicable to excavation in firm soil						
Removing trees						
girth 500–1500 mm	–	18.50	215.98	–	nr	**215.98**
girth 1500–3000 mm	–	32.50	379.42	–	nr	**379.42**
girth exceeding 3000 mm	–	46.50	542.87	–	nr	**542.87**
Removing tree stumps						
girth 500 mm–1.50 m	–	0.93	10.85	31.97	nr	**42.82**
girth 1.50 – 3.00 m	–	0.93	10.85	46.29	nr	**57.14**
girth exceeding 3.00 m	–	0.93	10.85	63.16	nr	**74.01**
Clearing site vegetation						
bushes, scrub, undergrowth, hedges and trees and tree stumps not exceeding 500 mm girth	–	0.03	0.35	–	m²	**0.35**
Lifting turf for preservation						
stacking	–	0.32	3.73	–	m²	**3.73**
Topsoil for preservation; to spoil heap less than 50 m from excavations						
average depth 150 mm	–	0.02	0.24	0.58	m²	**0.82**
add or deduct for each 25 mm variation in average depth	–	0.01	0.12	0.14	m²	**0.26**
EXCAVATION						
Excavating by machine						
To reduce levels						
maximum depth not exceeding 0.25 m	–	0.03	0.39	0.45	m³	**0.84**
maximum depth not exceeding 1.00 m	–	0.03	0.39	0.45	m³	**0.84**
maximum depth not exceeding 2.00 m	–	0.04	0.43	0.49	m³	**0.92**
maximum depth not exceeding 4.00 m	–	0.04	0.47	0.54	m³	**1.01**
Basements and the like; commencing level exceeding 0.25 m below existing ground level						
maximum depth not exceeding 1.00 m	–	0.06	0.70	0.83	m³	**1.53**
maximum depth not exceeding 2.00 m	–	0.07	0.82	0.83	m³	**1.65**
maximum depth not exceeding 4.00 m	–	0.08	0.93	1.01	m³	**1.94**
maximum depth not exceeding 6.00 m	–	0.09	1.06	1.27	m³	**2.33**
maximum depth not exceeding 8.00 m	–	0.12	1.40	1.47	m³	**2.87**
Pits						
maximum depth not exceeding 0.25 m	–	0.31	3.62	2.98	m³	**6.60**
maximum depth not exceeding 1.00 m	–	0.33	3.85	2.98	m³	**6.83**
maximum depth not exceeding 2.00 m	–	0.39	4.55	3.36	m³	**7.91**
maximum depth not exceeding 4.00 m	–	0.47	5.48	3.81	m³	**9.29**
maximum depth not exceeding 6.00 m	–	0.49	5.72	4.00	m³	**9.72**
Extra over pit excavating for commencing level exceeding 0.25 m below existing ground level						
1.00 m below	–	0.03	0.35	0.44	m³	**0.79**
2.00 m below	–	0.05	0.58	0.64	m³	**1.22**
3.00 m below	–	0.06	0.70	0.83	m³	**1.53**
4.00 m below	–	0.09	1.06	1.08	m³	**2.14**

5 EXCAVATING AND FILLING

Item	PC £	Labour hours	Labour £	Material £	Unit	Total rate £
Trenches; width not exceeding 0.30 m						
maximum depth not exceeding 0.25 m	–	0.26	3.03	2.35	m³	**5.38**
maximum depth not exceeding 1.00 m	–	0.28	3.27	2.35	m³	**5.62**
maximum depth not exceeding 2.00 m	–	0.33	3.85	2.73	m³	**6.58**
maximum depth not exceeding 4.00 m	–	0.40	4.67	3.36	m³	**8.03**
maximum depth not exceeding 6.00 m	–	0.46	5.37	4.00	m³	**9.37**
Trenches; width exceeding 0.30 m						
maximum depth not exceeding 0.25 m	–	0.23	2.69	2.09	m³	**4.78**
maximum depth not exceeding 1.00 m	–	0.25	2.92	2.09	m³	**5.01**
maximum depth not exceeding 2.00 m	–	0.30	3.51	2.54	m³	**6.05**
maximum depth not exceeding 4.00 m	–	0.35	4.09	2.98	m³	**7.07**
maximum depth not exceeding 6.00 m	–	0.43	5.02	3.81	m³	**8.83**
Extra over trench excavating for commencing level exceeding 0.25 m below existing ground level						
1.00 m below	–	0.03	0.35	0.44	m³	**0.79**
2.00 m below	–	0.05	0.58	0.64	m³	**1.22**
3.00 m below	–	0.06	0.70	0.83	m³	**1.53**
4.00 m below	–	0.09	1.06	1.08	m³	**2.14**
For pile caps and ground beams between piles						
maximum depth not exceeding 0.25 m	–	0.35	4.09	4.07	m³	**8.16**
maximum depth not exceeding 1.00 m	–	0.39	4.55	4.00	m³	**8.55**
maximum depth not exceeding 2.00 m	–	0.39	4.55	4.45	m³	**9.00**
To bench sloping ground to receive filling						
maximum depth not exceeding 0.25 m		0.07	0.82	1.08	m³	**1.90**
maximum depth not exceeding 1.00 m	–	0.09	1.06	1.08	m³	**2.14**
maximum depth not exceeding 2.00 m	–	0.09	1.06	1.27	m³	**2.33**
Extra over any types of excavating irrespective of depth						
excavating below ground water level	–	0.13	1.52	1.47	m³	**2.99**
next to existing services	–	0.35	4.09	0.83	m³	**4.92**
around existing services crossing excavation	–	0.60	7.00	2.35	m³	**9.35**
Extra over any types of excavating irrespective of depth for breaking out existing materials						
rock	–	2.95	34.44	13.19	m³	**47.63**
concrete	–	2.55	29.77	10.39	m³	**40.16**
reinforced concrete	–	3.60	42.02	15.16	m³	**57.18**
brickwork, blockwork or stonework	–	1.85	21.60	7.66	m³	**29.26**
Extra over any types of excavating irrespective of depth for breaking out existing hard pavings, 75 mm thick						
coated macadam or asphalt	–	0.19	2.21	0.66	m²	**2.87**
Extra over any types of excavating irrespective of depth for breaking out existing hard pavings, 150 mm thick						
concrete	–	0.39	4.55	1.58	m²	**6.13**
reinforced concrete	–	0.58	6.78	2.15	m²	**8.93**
coated macadam or asphalt and hardcore	–	0.26	3.03	0.74	m²	**3.77**

5 EXCAVATING AND FILLING

Item	PC £	Labour hours	Labour £	Material £	Unit	Total rate £
EXCAVATION – cont						
Excavating by machine – cont						
Working space allowance to excavations 600 mm wide						
reduce levels, basements and the like	–	0.07	0.82	0.83	m²	**1.65**
pits	–	0.19	2.21	2.35	m²	**4.56**
trenches	–	0.18	2.10	2.09	m²	**4.19**
pile caps and ground beams between piles	–	0.20	2.34	2.35	m²	**4.69**
Extra over excavating for working space for backfilling in with special materials						
hardcore	–	0.13	1.52	13.21	m²	**14.73**
sand	–	0.13	1.52	20.82	m²	**22.34**
40 mm–20 mm gravel	–	0.13	1.52	26.98	m²	**28.50**
plain in situ ready mixed designated concrete C7.5–40 mm aggregate	–	0.93	12.73	49.66	m²	**62.39**
Excavating by hand						
Topsoil for preservation						
average depth 150 mm	–	0.23	2.69	–	m²	**2.69**
add or deduct for each 25 mm variation in average depth	–	0.03	0.35	–	m²	**0.35**
To reduce levels						
maximum depth not exceeding 0.25 m	–	1.44	16.81	–	m³	**16.81**
maximum depth not exceeding 1.00 m	–	1.63	19.03	–	m³	**19.03**
maximum depth not exceeding 2.00 m	–	1.80	21.01	–	m³	**21.01**
maximum depth not exceeding 4.00 m	–	1.99	23.24	–	m³	**23.24**
Basements and the like; commencing level exceeding 0.25 m below existing ground level						
maximum depth not exceeding 1.00 m	–	1.90	22.18	–	m³	**22.18**
maximum depth not exceeding 2.00 m	–	2.04	23.82	–	m³	**23.82**
maximum depth not exceeding 4.00 m	–	2.73	31.87	–	m³	**31.87**
maximum depth not exceeding 6.00 m	–	3.33	38.88	–	m³	**38.88**
maximum depth not exceeding 8.00 m	–	4.02	46.93	–	m³	**46.93**
Pits						
maximum depth not exceeding 0.25 m	–	2.13	24.87	–	m³	**24.87**
maximum depth not exceeding 1.00 m	–	2.75	32.10	–	m³	**32.10**
maximum depth not exceeding 2.00 m	–	3.30	38.53	–	m³	**38.53**
maximum depth not exceeding 4.00 m	–	4.18	48.80	–	m³	**48.80**
maximum depth not exceeding 6.00 m	–	5.17	60.36	–	m³	**60.36**
Extra over pit excavating for commencing level exceeding 0.25 m below existing ground level						
1.00 m below	–	0.42	4.90	–	m³	**4.90**
2.00 m below	–	0.88	10.27	–	m³	**10.27**
3.00 m below	–	1.30	15.18	–	m³	**15.18**
4.00 m below	–	1.71	19.97	–	m³	**19.97**
Trenches; width not exceeding 0.30 m						
maximum depth not exceeding 0.25 m	–	1.85	21.60	–	m³	**21.60**
maximum depth not exceeding 1.00 m	–	2.76	32.23	–	m³	**32.23**

5 EXCAVATING AND FILLING

Item	PC £	Labour hours	Labour £	Material £	Unit	Total rate £
Trenches; width exceeding 0.30 m						
maximum depth not exceeding 0.25 m	–	1.80	21.01	–	m³	**21.01**
maximum depth not exceeding 1.00 m	–	2.46	28.72	–	m³	**28.72**
maximum depth not exceeding 2.00 m	–	2.88	33.62	–	m³	**33.62**
maximum depth not exceeding 4.00 m	–	3.66	42.73	–	m³	**42.73**
maximum depth not exceeding 6.00 m	–	4.68	54.63	–	m³	**54.63**
Extra over trench excavating for commencing level exceeding 0.25 m below existing ground level						
1.00 m below	–	0.42	4.90	–	m³	**4.90**
2.00 m below	–	0.88	10.27	–	m³	**10.27**
3.00 m below	–	1.30	15.18	–	m³	**15.18**
4.00 m below	–	1.71	19.97	–	m³	**19.97**
For pile caps and ground beams between piles						
maximum depth not exceeding 0.25 m	–	2.78	32.45	–	m³	**32.45**
maximum depth not exceeding 1.00 m	–	2.96	34.55	–	m³	**34.55**
maximum depth not exceeding 2.00 m	–	3.52	41.09	–	m³	**41.09**
To bench sloping ground to receive filling						
maximum depth not exceeding 0.25 m	–	1.30	15.18	–	m³	**15.18**
maximum depth not exceeding 1.00 m	–	1.48	17.28	–	m³	**17.28**
maximum depth not exceeding 2.00 m	–	1.67	19.50	–	m³	**19.50**
Extra over any types of excavating irrespective of depth						
excavating below ground water level	–	0.32	3.73	–	m³	**3.73**
next existing services	–	0.93	10.85	–	m³	**10.85**
around existing services crossing excavation	–	1.85	21.60	–	m³	**21.60**
Extra over any types of excavating irrespective of depth for breaking out existing materials						
rock	–	4.63	54.05	11.86	m³	**65.91**
concrete	–	4.16	48.56	9.88	m³	**58.44**
reinforced concrete	–	5.55	64.79	13.84	m³	**78.63**
brickwork, blockwork or stonework	–	2.78	32.45	5.92	m³	**38.37**
Extra over any types of excavating irrespective of depth for breaking out existing hard pavings, 60 mm thick						
precast concrete paving slabs	–	0.28	3.27	–	m²	**3.27**
Extra over any types of excavating irrespective of depth for breaking out existing hard pavings, 75 mm thick						
coated macadam or asphalt	–	0.37	4.32	0.80	m²	**5.12**
Extra over any types of excavating irrespective of depth for breaking out existing hard pavings, 150 mm thick						
concrete	–	0.65	7.59	1.38	m²	**8.97**
reinforced concrete	–	0.83	9.69	1.98	m²	**11.67**
coated macadam or asphalt and hardcore	–	0.46	5.37	0.99	m²	**6.36**
Working space allowance to excavations						
reduce levels, basements and the like	–	2.13	24.87	–	m²	**24.87**
pits	–	2.22	25.92	–	m²	**25.92**
trenches	–	1.94	22.65	–	m²	**22.65**
pile caps and ground beams between piles	–	2.31	26.97	–	m²	**26.97**

5 EXCAVATING AND FILLING

Item	PC £	Labour hours	Labour £	Material £	Unit	Total rate £
EXCAVATION – cont						
Excavating by hand – cont						
Extra over excavation for working space for backfilling with special materials						
hardcore	–	0.74	8.64	13.04	m²	**21.68**
sand	–	0.74	8.64	22.85	m²	**31.49**
40 mm–20 mm gravel	–	0.74	8.64	25.94	m²	**34.58**
plain in situ concrete ready mixed designated concrete; C7.5–40 mm aggregate	–	1.02	13.96	48.57	m²	**62.53**
SUPPORT TO EXCAVATIONS						
Earthwork support (average risk prices)						
Maximum depth not exceeding 1.00 m						
distance between opposing faces not exceeding 2.00 m	–	0.10	1.11	0.15	m²	**1.26**
distance between opposing faces 2.00–4.00 m	–	0.10	1.22	0.18	m²	**1.40**
distance between opposing faces exceeding 4.00 m	–	0.11	1.33	0.23	m²	**1.56**
Maximum depth not exceeding 2.00 m						
distance between opposing faces not exceeding 2.00 m	–	0.11	1.33	0.18	m²	**1.51**
distance between opposing faces 2.00–4.00 m	–	0.12	1.45	0.23	m²	**1.68**
distance between opposing faces exceeding 4.00 m	–	0.13	1.55	0.28	m²	**1.83**
Maximum depth not exceeding 4.00 m						
distance between opposing faces not exceeding 2.00 m	–	0.15	1.77	0.23	m²	**2.00**
distance between opposing faces 2.00–4.00 m	–	0.15	1.77	0.28	m²	**2.05**
distance between opposing faces exceeding 4.00 m	–	0.17	2.00	0.35	m²	**2.35**
Maximum depth not exceeding 6.00 m						
distance between opposing faces not exceeding 2.00 m	–	0.17	2.00	0.27	m²	**2.27**
distance between opposing faces 2.00–4.00 m	–	0.18	2.11	0.35	m²	**2.46**
distance between opposing faces exceeding 4.00 m	–	0.21	2.44	0.43	m²	**2.87**
Maximum depth not exceeding 8.00 m						
distance between opposing faces not exceeding 2.00 m	–	0.22	2.55	0.35	m²	**2.90**
distance between opposing faces 2.00–4.00 m	–	0.27	3.11	0.43	m²	**3.54**
distance between opposing faces exceeding 4.00 m	–	0.31	3.66	0.52	m²	**4.18**
Earthwork support (open boarded)						
Maximum depth not exceeding 1.00 m						
distance between opposing faces not exceeding 2.00 m	–	0.27	3.11	0.31	m²	**3.42**
distance between opposing faces 2.00–4.00 m	–	0.29	3.43	0.35	m²	**3.78**
distance between opposing faces exceeding 4.00 m	–	0.33	3.88	0.43	m²	**4.31**

5 EXCAVATING AND FILLING

Item	PC £	Labour hours	Labour £	Material £	Unit	Total rate £
Maximum depth not exceeding 2.00 m						
distance between opposing faces not exceeding 2.00 m	–	0.33	3.88	0.35	m²	**4.23**
distance between opposing faces 2.00–4.00 m	–	0.37	4.33	0.42	m²	**4.75**
distance between opposing faces exceeding 4.00 m	–	0.42	4.88	0.52	m²	**5.40**
Maximum depth not exceeding 4.00 m						
distance between opposing faces not exceeding 2.00 m	–	0.42	4.88	0.40	m²	**5.28**
distance between opposing faces 2.00–4.00 m	–	0.47	5.55	0.48	m²	**6.03**
distance between opposing faces exceeding 4.00 m	–	0.53	6.21	0.60	m²	**6.81**
Maximum depth not exceeding 6.00 m						
distance between opposing faces not exceeding 2.00 m	–	0.53	6.21	0.43	m²	**6.64**
distance between opposing faces 2.00–4.00 m	–	0.58	6.76	0.55	m²	**7.31**
distance between opposing faces exceeding 4.00 m	–	0.67	7.76	0.70	m²	**8.46**
Maximum depth not exceeding 8.00 m						
distance between opposing faces not exceeding 2.00 m	–	0.70	8.21	0.57	m²	**8.78**
distance between opposing faces 2.00–4.00 m	–	0.79	9.20	0.66	m²	**9.86**
distance between opposing faces exceeding 4.00 m	–	0.92	10.76	0.87	m²	**11.63**
Earthwork support (close boarded)						
Maximum depth not exceeding 1.00 m						
distance between opposing faces not exceeding 2.00 m	–	0.70	8.21	0.60	m²	**8.81**
distance between opposing faces 2.00–4.00 m	–	0.77	8.98	0.70	m²	**9.68**
distance between opposing faces exceeding 4.00 m	–	0.85	9.98	0.87	m²	**10.85**
Maximum depth not exceeding 2.00 m						
distance between opposing faces not exceeding 2.00 m	–	0.88	10.31	0.70	m²	**11.01**
distance between opposing faces 2.00–4.00 m	–	0.97	11.32	0.83	m²	**12.15**
distance between opposing faces exceeding 4.00 m	–	1.05	12.31	1.05	m²	**13.36**
Maximum depth not exceeding 4.00 m						
distance between opposing faces not exceeding 2.00 m	–	1.10	12.86	0.79	m²	**13.65**
distance between opposing faces 2.00–4.00 m	–	1.24	14.42	0.97	m²	**15.39**
distance between opposing faces exceeding 4.00 m	–	1.36	15.86	1.22	m²	**17.08**
Maximum depth not exceeding 6.00 m						
distance between opposing faces not exceeding 2.00 m	–	1.37	15.97	0.87	m²	**16.84**
distance between opposing faces 2.00–4.00 m	–	1.49	17.41	1.10	m²	**18.51**
distance between opposing faces exceeding 4.00 m	–	1.67	19.52	1.39	m²	**20.91**

5 EXCAVATING AND FILLING

Item	PC £	Labour hours	Labour £	Material £	Unit	Total rate £
SUPPORT TO EXCAVATIONS – cont						
Earthwork support (close boarded) – cont						
Maximum depth not exceeding 8.00 m						
distance between opposing faces not exceeding 2.00 m	–	1.67	19.52	1.13	m²	**20.65**
distance between opposing faces 2.00–4.00 m	–	1.84	21.51	1.30	m²	**22.81**
distance between opposing faces exceeding 4.00 m	–	2.11	24.62	1.57	m²	**26.19**
Extra over earthwork support for						
Curved	–	0.02	0.23	0.15	m²	**0.38**
Below ground water level	–	0.27	3.11	0.14	m²	**3.25**
Unstable ground	–	0.44	5.10	0.27	m²	**5.37**
Next to roadways	–	0.35	4.10	0.23	m²	**4.33**
Left in	–	0.57	6.65	6.10	m²	**12.75**
Earthwork support (average risk prices – inside existing buildings)						
Maximum depth not exceeding 1.00 m						
distance between opposing faces not exceeding 2.00 m	–	0.18	2.10	0.23	m²	**2.33**
distance between opposing faces 2.00–4.00 m	–	0.19	2.21	0.26	m²	**2.47**
distance between opposing faces exceeding 4.00 m	–	0.22	2.57	0.31	m²	**2.88**
Maximum depth not exceeding 2.00 m						
distance between opposing faces not exceeding 2.00 m	–	0.22	2.57	0.26	m²	**2.83**
distance between opposing faces 2.00–4.00 m	–	0.24	2.80	0.34	m²	**3.14**
distance between opposing faces exceeding 4.00 m	–	0.32	3.73	0.37	m²	**4.10**
Maximum depth not exceeding 4.00 m						
distance between opposing faces not exceeding 2.00 m	–	0.28	3.27	0.34	m²	**3.61**
distance between opposing faces 2.00–4.00 m	–	0.31	3.62	0.40	m²	**4.02**
distance between opposing faces exceeding 4.00 m	–	0.34	3.97	0.46	m²	**4.43**
Maximum depth not exceeding 6.00 m						
distance between opposing faces not exceeding 2.00 m	–	0.34	3.97	0.37	m²	**4.34**
distance between opposing faces 2.00–4.00 m	–	0.38	4.44	0.46	m²	**4.90**
distance between opposing faces exceeding 4.00 m	–	0.43	5.02	0.55	m²	**5.57**
DISPOSAL						
Disposal of excavated material						
Basic Landfill Tax rates						
inactive waste	–	–	–	2.50	tonne	**2.50**
all other taxable wastes	–	–	–	72.00	tonne	**72.00**

5 EXCAVATING AND FILLING

Item	PC £	Labour hours	Labour £	Material £	Unit	Total rate £
Excavated material deposited off site						
inactive waste off site; to tip not exceeding 13 km (using lorries); including Landfill Tax	–	–	–	17.00	m³	**17.00**
active non-hazardous waste off site; to tip not exceeding 13 km (using lorries); including Landfill Tax	–	–	–	145.46	m³	**145.46**
RETAINING EXCAVATED MATERIAL ON SITE						
Excavated material deposited on site using dumpers to transport material						
on site; spreading; average 25 m distance	–	0.20	2.34	0.32	m³	**2.66**
on site; depositing in spoil heaps; average 50 m distance	–	–	–	0.77	m³	**0.77**
on site; spreading; average 50 m distance	–	0.20	2.34	0.58	m³	**2.92**
on site; depositing in spoil heaps; average 100 m distance	–	–	–	1.35	m³	**1.35**
on site; spreading; average 100 m distance	–	0.20	2.34	0.90	m³	**3.24**
on site; depositing in spoil heaps; average 200 m distance	–	–	–	1.71	m³	**1.71**
on site; spreading; average 200 m distance	–	0.20	2.34	1.22	m³	**3.56**
FILLING OBTAINED FROM EXCAVATED MATERIAL						
Filling to make up levels						
By machine						
average thickness over 50 mm not exceeding 500 mm	–	0.22	2.52	1.46	m³	**3.98**
average thickness exceeding 500 mm	–	0.18	2.10	1.06	m³	**3.16**
By hand						
average thickness over 50 mm not exceeding 500 mm	–	1.25	14.60	2.86	m³	**17.46**
average thickness exceeding 500 mm	–	1.02	11.91	2.33	m³	**14.24**
Surface treatments						
Surface packing to filling						
To vertical or battered faces	–	0.17	1.99	0.14	m²	**2.13**
Compacting						
bottoms of excavations	–	0.04	0.47	0.03	m²	**0.50**
Trimming						
sloping surfaces	–	0.17	1.99	–	m²	**1.99**

5 EXCAVATING AND FILLING

Item	PC £	Labour hours	Labour £	Material £	Unit	Total rate £
IMPORTED FILLING						
Basic material prices delivered to site						
Supply only in full loads						
D.O.T. type 1	–	–	–	13.61	tonne	**13.61**
D.O.T. type 2	–	–	–	12.92	tonne	**12.92**
10 mm single size aggregate	–	–	–	17.44	tonne	**17.44**
20 mm single size aggregate	–	–	–	17.44	tonne	**17.44**
20 mm all-in aggregate	–	–	–	17.44	tonne	**17.44**
40 mm all-in aggregate	–	–	–	17.44	tonne	**17.44**
40 mm scalpings	–	–	–	12.20	tonne	**12.20**
hardcore	–	–	–	10.78	tonne	**10.78**
soft/building sand	–	–	–	16.73	tonne	**16.73**
recycled type 1	–	–	–	11.07	tonne	**11.07**
E-blend (50% type 1 and 50% recycled type 1)	–	–	–	12.46	tonne	**12.46**
Filling to make up levels; by machine						
Average thickness over 50 mm not exceeding 500 mm						
obtained off site; imported topsoil	17.00	0.22	2.52	18.64	m³	**21.16**
obtained off site; hardcore	21.71	0.25	2.94	23.66	m³	**26.60**
obtained off site; granular fill type one	31.41	0.25	2.94	34.63	m³	**37.57**
obtained off site; granular fill type two	29.80	0.25	2.94	32.98	m³	**35.92**
obtained off site; sand	38.60	0.25	2.94	42.00	m³	**44.94**
Average thickness over 500 mm						
obtained off site; imported topsoil	17.00	0.18	2.10	18.79	m³	**20.89**
obtained off site; hardcore	21.49	0.22	2.52	22.29	m³	**24.81**
obtained off site; granular fill type one	31.41	0.22	2.52	34.10	m³	**36.62**
obtained off site; granular fill type two	29.80	0.22	2.52	32.45	m³	**34.97**
obtained off site; sand	38.60	0.22	2.52	41.47	m³	**43.99**
Filling to make up levels; by hand						
Average thickness over 50 mm not exceeding 500 mm						
obtained off site; imported topsoil	17.00	1.25	14.60	20.28	m³	**34.88**
obtained off site; hardcore	21.71	1.39	16.23	25.43	m³	**41.66**
obtained off site; granular fill type one	31.41	1.54	17.98	35.70	m³	**53.68**
obtained off site; granular fill type two	29.80	1.54	17.98	34.06	m³	**52.04**
obtained off site; sand	38.60	1.54	17.98	43.07	m³	**61.05**
Average thickness over 500 mm						
obtained off site; imported topsoil	17.00	1.02	11.91	19.75	m³	**31.66**
obtained off site; hardcore	21.71	1.34	15.64	25.33	m³	**40.97**
obtained off site; granular fill type one	31.41	1.43	16.70	35.48	m³	**52.18**
obtained off site; granular fill type two	29.80	1.43	16.70	33.82	m³	**50.52**
obtained off site; sand	38.60	1.43	16.70	42.84	m³	**59.54**

5 EXCAVATING AND FILLING

Item	PC £	Labour hours	Labour £	Material £	Unit	Total rate £
Surface treatments						
Surface packing to filling						
To vertical or battered faces	–	0.17	1.99	0.14	m²	**2.13**
Compacting						
bottoms of excavations	–	0.04	0.47	0.03	m²	**0.50**
filling; blinding with sand 50 mm thick	–	0.04	0.47	1.63	m²	**2.10**
Trimming						
sloping surfaces	–	0.17	1.99	–	m²	**1.99**
sloping surfaces; in rock	–	0.93	10.77	2.77	m²	**13.54**
GEOTEXTILE FABRIC						
Erosion mats						
Filter membrane; one layer; laid on earth to receive granular material; over 500 mm wide						
Terram 500 filter membrane or other equal and approved; one layer; laid on earth	0.23	0.04	0.47	0.24	m²	**0.71**
Terram 700 filter membrane or other equal and approved; one layer; laid on earth	0.29	0.04	0.47	0.30	m²	**0.77**
Terram 1000; filter membrane or other equal and approved; one layer; laid on earth	0.37	0.04	0.47	0.38	m²	**0.85**
Terram 2000; filter membrane or other equal and approved; one layer; laid on earth	0.43	0.04	0.47	0.44	m²	**0.91**
MEMBRANES						
Sheeting to prevent moisture loss/gas barrier						
Building paper; lapped joints						
subsoil grade 410; horizontal on foundations	–	0.02	0.28	0.59	m²	**0.87**
standard grade 420; horizontal on slabs	–	0.04	0.54	0.88	m²	**1.42**
Polythene sheeting; lapped joints; horizontal on slabs						
250 microns; 0.25 mm thick	–	0.04	0.54	0.46	m²	**1.00**
Visqueen sheeting or other equal and approved; lapped joints; horizontal on slabs						
250 microns; 0.25 mm thick	–	0.04	0.54	0.39	m²	**0.93**
300 microns; 0.30 mm thick	–	0.05	0.69	0.43	m²	**1.12**
gas-resistant damp-proof membrane taped joints	–	0.05	0.69	4.88	m²	**5.57**
sealing at perimeter of slab	–	0.10	1.37	1.25	m	**2.62**
sealing around penetrations; drain pipes etc.	–	1.00	13.69	3.83	nr	**17.52**
CUTTING OFF TOPS OF PILES						
Cut off and prepare reinforcement to receive reinforced concrete pile cap; not including disposal						
450 mm dia. piles	–	1.00	19.71	–	m	**19.71**
600 mm dia. piles	–	1.20	23.65	–	m	**23.65**
750 mm dia. piles	–	1.50	29.57	–	m	**29.57**
900 mm dia. piles	–	1.80	35.48	–	m	**35.48**

7 PILING

Item	PC £	Labour hours	Labour £	Material £	Unit	Total rate £
INTERLOCKING SHEET PILES						
Arcelor 600 mm wide 'U' shaped steel sheeting piling; or other equal approved; pitched and driven						
Provision of all plant for sheet pile installation; including bringing to and removing from site; maintenance, erection and dismantling; assuming one rig for 1500 m² of piling						
leader rig with vibratory hammer	–	–	–	–	item	3450.00
conventional rig	–	–	–	–	item	4500.00
silent vibrationless rig	–	–	–	–	item	5000.00
Supply only of standard sheet pile sections						
PU12	–	–	–	–	m²	88.92
PU18–1	–	–	–	–	m²	97.71
PU22–1	–	–	–	–	m²	110.20
PU25	–	–	–	–	m²	125.97
PU32	–	–	–	–	m²	153.61
Pitching and driving of sheet piles; using the following plant						
leader rig with vibratory hammer	–	–	–	–	m²	19.00
conventional rig	–	–	–	–	m²	26.60
silent vibrationless rig	–	–	–	–	m²	39.90
Provision of all plant for sheet pile extraction; including bringing to and removing from site; maintenance, erection and dismantling; assuming one rig for 1500 m² of piling						
leader rig with vibratory hammer	–	–	–	–	item	3500.00
conventional rig	–	–	–	–	item	4500.00
silent vibrationless rig	–	–	–	–	item	4500.00
Extraction of sheet piles; using the following plant						
leader rig with vibratory hammer	–	–	–	–	m²	14.25
conventional rig	–	–	–	–	m²	16.63
silent vibrationless rig	–	–	–	–	m²	24.70
Credit on extracted piles; recovered in reusable lengths; for sheet pile sections						
PU12	–	–	–	–	m²	62.70
PU18–1	–	–	–	–	m²	68.97
PU22–1	–	–	–	–	m²	77.81
PU25	–	–	–	–	m²	88.92
PU32	–	–	–	–	m²	108.39

7 PILING

Item	PC £	Labour hours	Labour £	Material £	Unit	Total rate £
BORED PILES						
NOTE: The following approximate prices, for the quantities of piling quoted, are for work on clear open sites with reasonable access. They are based on normal concrete mix 35.00 N/mm^2; reinforced and include up to 0.15 m of projecting reinforcement at top of pile. The prices do not allow for removal of spoil.						
Minipile cast-in-place concrete piles						
300 mm nominal dia.						
Establish equipment on site						
Provision of all plant (1 nr rig) ; including bringing to and removing from site; maintenance, erection and dismantling at each pile position for 50 nr piles	–	–	–	–	item	4700.00
Bored piles						
300 mm dia. piles; nominally reinforced; 15 m long	–	–	–	–	nr	203.00
set up at each pile position	–	–	–	–	nr	20.00
add for additional piles length	–	–	–	–	m	13.50
deduct for reduction in pile length	–	–	–	–	m	5.00
Delays						
rig standing time	–	–	–	–	hour	175.00
Pile tests						
integrity tests (minimum 20 per visit)	–	–	–	–	nr	14.00
300 mm nominal dia.; small area with difficult access and limited headroom						
Establish equipment on site						
Provision of all plant (1 nr rig) ; including bringing to and removing from site; maintenance, erection and dismantling at each pile position for 50 nr piles	–	–	–	–	item	2500.00
Bored piles						
300 mm dia. piles; nominally reinforced; 15 m long	–	–	–	–	nr	203.00
set up at each pile position	–	–	–	–	nr	20.00
add for additional piles length	–	–	–	–	m	44.00
deduct for reduction in pile length	–	–	–	–	m	18.00
Delays						
rig standing time	–	–	–	–	hour	170.00
Pile tests						
integrity tests (minimum 20 per visit)	–	–	–	–	nr	14.00
450 mm nominal dia.						
Establish equipment on site						
Provision of all plant (2 nr rigs) ; including bringing to and removing from site; maintenance, erection and dismantling at each pile position for 100 nr piles	–	–	–	–	item	20500.00

7 PILING

Item	PC £	Labour hours	Labour £	Material £	Unit	Total rate £
BORED PILES – cont						
450 mm nominal dia. – cont						
Bored piles						
450 mm dia. piles; nominally reinforced; 20 m long	–	–	–	–	nr	2900.00
add for additional piles length	–	–	–	–	m	145.00
deduct for reduction in pile length	–	–	–	–	m	20.00
Blind bored piles						
450 mm dia.	–	–	–	–	m	125.00
Delays						
rig standing time	–	–	–	–	hour	375.00
Extra over piling						
breaking through obstructions	–	–	–	–	hour	425.00
Pile tests						
working to 600 kN; using tension piles as reaction; first pile	–	–	–	–	nr	10000.00
working to 600 kN; using tension piles as reaction; subsequent piles	–	–	–	–	nr	10000.00
integrity tests (minimum 20 per visit)	–	–	–	–	nr	12.75
Rotary CFA bored cast-in-place concrete piles						
450 mm dia. piles						
Provision of plant						
Provision of all plant (1 nr rig); including bringing to and removing from site; maintenance, erection and dismantling at each pile position for 100 nr piles 20 m long	–	–	–	–	item	12000.00
Bored piles						
450 mm dia. piles; reinforced; 20 m long	–	–	–	–	nr	820.00
add for additional piles length	–	–	–	–	m	40.00
deduct for reduction in pile length	–	–	–	–	m	16.50
Delays						
rig standing time	–	–	–	–	hour	400.00
Pile tests						
working to 600 kN; using tension piles as reaction; first pile	–	–	–	–	nr	5800.00
working to 600 kN; using tension piles as reaction; subsequent piles	–	–	–	–	nr	5800.00
integrity tests (minimum 20 per visit)	–	–	–	–	nr	12.75
600 mm dia. piles						
Provision of all plant (1 nr rig); including bringing to and removing from site; maintenance, erection and dismantling at each pile position for 100 nr piles 20 m long	–	–	–	–	item	12000.00
Bored piles						
600 mm dia. piles; reinforced; 20 m long	–	–	–	–	nr	1300.00
add for additional piles length	–	–	–	–	m	65.00
deduct for reduction in pile length	–	–	–	–	m	28.50

7 PILING

Item	PC £	Labour hours	Labour £	Material £	Unit	Total rate £
Delays						
rig standing time	–	–	–	–	hour	400.00
Pile tests						
working to 1000 kN; using tension piles as reaction; first pile	–	–	–	–	nr	7250.00
working to 1000 kN; using tension piles as reaction; subsequent piles	–	–	–	–	nr	7250.00
integrity test 600 mm dia. piles; min 20 per visit	–	–	–	–	nr	12.75
750 mm dia. piles						
Provision of all plant (1 nr rig); including bringing to and removing from site; maintenance, erection and dismantling at each pile position for 100 nr piles 20 m long	–	–	–	–	item	15750.00
Bored piles						
750 mm dia. piles; reinforced; 20 m long	–	–	–	–	nr	2025.00
add for additional piles length	–	–	–	–	m	100.00
deduct for reduction in pile length	–	–	–	–	m	43.50
Delays						
rig standing time	–	–	–	–	hour	475.00
Pile tests						
working to 1500 kN; using tension piles as reaction; first pile	–	–	–	–	nr	8725.00
working to 1500 kN; using tension piles as reaction; subsequent piles	–	–	–	–	nr	8725.00
integrity test 600 mm dia. piles; min 20 per visit	–	–	–	–	nr	12.75
900 mm dia. piles						
Provision of all plant (1 nr rig); including bringing to and removing from site; maintenance, erection and dismantling at each pile position for 100 nr piles 20 m long	–	–	–	–	item	16000.00
Bored piles						
900 mm dia. piles; reinforced; 20 m long	–	–	–	–	nr	2900.00
add for additional piles length	–	–	–	–	m	145.00
deduct for reduction in pile length	–	–	–	–	m	60.00
Delays						
rig standing time	–	–	–	–	hour	475.00
Pile tests						
working to 2000 kN; using tension piles as reaction; first pile	–	–	–	–	nr	11000.00
working to 2000 kN; using tension piles as reaction; subsequent piles	–	–	–	–	nr	11000.00
integrity test 900 mm dia. piles; min 20 per visit	–	–	–	–	nr	12.75

8 UNDERPINNING

Item	PC £	Labour hours	Labour £	Material £	Unit	Total rate £
UNDERPINNING						
Excavating; by machine						
Preliminary trenches						
maximum depth not exceeding 1.00 m	–	0.23	2.69	4.46	m³	**7.15**
maximum depth not exceeding 2.00 m	–	0.28	3.27	5.37	m³	**8.64**
maximum depth not exceeding 4.00 m	–	0.32	3.73	6.28	m³	**10.01**
Extra over preliminary trench excavating for breaking out existing hard pavings, 150 mm thick						
concrete	–	0.65	7.59	1.38	m²	**8.97**
Excavating; by hand						
Preliminary trenches						
maximum depth not exceeding 1.00 m	–	2.68	31.28	–	m³	**31.28**
maximum depth not exceeding 2.00 m	–	3.05	35.61	–	m³	**35.61**
maximum depth not exceeding 4.00 m	–	3.93	45.88	–	m³	**45.88**
Extra over preliminary trench excavating for breaking out existing hard pavings, 150 mm thick						
concrete	–	0.28	3.27	1.91	m²	**5.18**
Underpinning pits; commencing from 1.00 m below existing ground level						
maximum depth not exceeding 0.25 m	–	4.07	47.52	–	m³	**47.52**
maximum depth not exceeding 1.00 m	–	4.44	51.83	–	m³	**51.83**
maximum depth not exceeding 2.00 m	–	5.32	62.10	–	m³	**62.10**
Underpinning pits; commencing from 2.00 m below existing ground level						
maximum depth not exceeding 0.25 m	–	5.00	58.37	–	m³	**58.37**
maximum depth not exceeding 1.00 m	–	5.37	62.69	–	m³	**62.69**
maximum depth not exceeding 2.00 m	–	6.24	72.85	–	m³	**72.85**
Underpinning pits; commencing from 4.00 m below existing ground level						
maximum depth not exceeding 0.25 m	–	5.92	69.12	–	m³	**69.12**
maximum depth not exceeding 1.00 m	–	6.29	73.43	–	m³	**73.43**
maximum depth not exceeding 2.00 m	–	7.17	83.70	–	m³	**83.70**
Extra over any types of excavating irrespective of depth						
excavating below ground water level	–	0.32	3.73	–	m³	**3.73**
Earthwork support to preliminary trenches (open boarded – in 3.00 m lengths)						
Maximum depth not exceeding 1.00 m						
distance between opposing faces not exceeding 2.00 m	–	0.37	4.32	0.57	m²	**4.89**
Maximum depth not exceeding 2.00 m						
distance between opposing faces not exceeding 2.00 m	–	0.46	5.37	0.70	m²	**6.07**
Maximum depth not exceeding 4.00 m						
distance between opposing faces not exceeding 2.00 m	–	0.59	6.89	0.87	m²	**7.76**

8 UNDERPINNING

Item	PC £	Labour hours	Labour £	Material £	Unit	Total rate £
Earthwork support to underpinning pits (open boarded – in 3.00 m lengths)						
Maximum depth not exceeding 1.00 m						
distance between opposing faces not exceeding 2.00 m	–	0.41	4.79	0.60	m²	**5.39**
Maximum depth not exceeding 2.00 m						
distance between opposing faces not exceeding 2.00 m	–	0.51	5.96	0.79	m²	**6.75**
Maximum depth not exceeding 4.00 m						
distance between opposing faces not exceeding 2.00 m	–	0.65	7.59	0.96	m²	**8.55**
Earthwork support to preliminary trenches (closed boarded – in 3.00 m lengths)						
Maximum depth not exceeding 1.00 m						
1.00 m deep	–	0.93	10.85	0.96	m²	**11.81**
Maximum depth not exceeding 2.00 m						
distance between opposing faces not exceeding 2.00 m	–	1.16	13.54	1.22	m²	**14.76**
Maximum depth not exceeding 4.00 m						
distance between opposing faces not exceeding 2.00 m	–	1.43	16.70	1.49	m²	**18.19**
Earthwork support to underpinning pits (closed boarded – in 3.00 m lengths)						
Maximum depth not exceeding 1.00 m						
distance between opposing faces not exceeding 2.00 m	–	1.02	11.91	1.05	m²	**12.96**
Maximum depth not exceeding 2.00 m						
distance between opposing faces not exceeding 2.00 m	–	1.28	14.94	1.30	m²	**16.24**
Maximum depth not exceeding 4.00 m						
distance between opposing faces not exceeding 2.00 m	–	1.57	18.33	1.65	m²	**19.98**
Extra over earthwork support for						
Left in	–	0.69	8.06	6.10	m²	**14.16**
Cutting away existing projecting foundations						
Concrete						
maximum width 150 mm; maximum depth 150 mm	–	0.15	1.75	0.15	m	**1.90**
maximum width 150 mm; maximum depth 225 mm	–	0.22	2.57	0.23	m	**2.80**
maximum width 150 mm; maximum depth 300 mm	–	0.30	3.51	0.30	m	**3.81**
maximum width 300 mm; maximum depth 300 mm	–	0.58	6.78	0.57	m	**7.35**
Masonry						
maximum width one brick thick; maximum depth one course high	–	0.04	0.47	0.06	m	**0.53**
maximum width one brick thick; maximum depth two courses high	–	0.13	1.52	0.14	m	**1.66**
maximum width one brick thick; maximum depth three courses high	–	0.25	2.92	0.25	m	**3.17**
maximum width one brick thick; maximum depth four courses high	–	0.42	4.90	0.41	m	**5.31**

8 UNDERPINNING

Item	PC £	Labour hours	Labour £	Material £	Unit	Total rate £
UNDERPINNING – cont						
Preparing the underside of existing work to receive the pinning up of the new work						
Width of existing work						
380 mm wide	–	0.56	6.54	–	m	**6.54**
600 mm wide	–	0.74	8.64	–	m	**8.64**
900 mm wide	–	0.93	10.85	–	m	**10.85**
1200 mm wide	–	1.11	12.96	–	m	**12.96**
Disposal; by hand						
Excavated material						
off site; to tip not exceeding 13 km (using lorries); including Landfill Tax based on inactive waste	–	0.75	8.75	16.75	m³	**25.50**
Filling to excavations; by hand						
Average thickness exceeding 0.25 m						
arising from the excavations	–	0.93	10.85	–	m³	**10.85**
Surface treatments						
Compacting						
bottoms of excavations	–	0.04	0.47	0.03	m²	**0.50**
Plain in situ ready mixed designated concrete; C10 – 40 mm aggregate; poured against faces of excavation						
Underpinning						
thickness not exceeding 150 mm	–	3.42	46.82	89.43	m³	**136.25**
thickness 150–450 mm	–	2.87	39.29	89.43	m³	**128.72**
thickness exceeding 450 mm	–	2.50	34.22	89.43	m³	**123.65**
Plain in situ ready mixed designated concrete; C20 – 20 mm aggregate; poured against faces of excavation						
Underpinning						
thickness not exceeding 150 mm	–	3.42	46.82	91.35	m³	**138.17**
thickness 150–450 mm	–	2.87	39.29	91.35	m³	**130.64**
thickness exceeding 450 mm	–	2.50	34.22	91.35	m³	**125.57**
Extra for working around reinforcement	–	0.28	3.83	–	m³	**3.83**
Sawn formwork; sides of foundations in underpinning						
Plain vertical						
height exceeding 1.00 m	–	1.48	23.97	2.18	m²	**26.15**
height not exceeding 250 mm	–	0.51	8.26	0.63	m²	**8.89**
height 250–500 mm	–	0.79	12.80	1.17	m²	**13.97**
height 500 mm–1.00 m	–	1.20	19.44	2.18	m²	**21.62**

8 UNDERPINNING

Item	PC £	Labour hours	Labour £	Material £	Unit	Total rate £
Reinforcement bar; BS 4449 hot rolled deformed square high yield steel bars						
20 mm dia. nominal size						
straight	381.60	22.80	364.67	577.00	tonne	**941.67**
bent	417.60	22.80	364.67	614.83	tonne	**979.50**
16 mm dia. nominal size						
straight	381.60	24.70	395.46	599.01	tonne	**994.47**
bent	417.60	24.70	395.46	636.83	tonne	**1032.29**
12 mm dia. nominal size						
straight	381.60	26.60	426.25	621.02	tonne	**1047.27**
bent	417.60	26.60	426.25	658.84	tonne	**1085.09**
10 mm dia. nominal size						
straight	381.60	28.50	457.04	658.79	tonne	**1115.83**
bent	417.60	28.50	457.04	696.61	tonne	**1153.65**
8 mm dia. nominal size						
straight	417.60	30.40	485.43	718.63	tonne	**1204.06**
straight	417.60	30.40	485.43	718.63	tonne	**1204.06**
Extra over for cutting and bending to shape codes 67 to 99	–	–	–	51.25	tonne	**51.25**
Common bricks; in cement: mortar (1:3)						
Walls in underpinning						
one brick thick (PC £ per 1000)	240.00	2.22	38.41	35.32	m²	**73.73**
one and a half brick thick	–	3.05	52.77	52.64	m²	**105.41**
two brick thick	–	3.79	65.57	72.81	m²	**138.38**
Class A engineering bricks; in cement: mortar (1:3)						
Walls in underpinning						
one brick thick (PC £ per 1000)	450.00	2.22	38.41	47.43	m²	**85.84**
one and a half brick thick	–	3.05	52.77	70.80	m²	**123.57**
two brick thick	–	3.79	65.57	97.00	m²	**162.57**
Class B engineering bricks; in cement: mortar (1:3)						
Walls in underpinning						
one brick thick (PC £ per 1000)	375.00	2.22	38.41	40.49	m²	**78.90**
one and a half brick thick	–	3.05	52.77	60.38	m²	**113.15**
two brick thick	–	3.79	65.57	83.12	m²	**148.69**
ADD or DEDUCT for variation of £10.00/1000 in PC of bricks						
one brick thick	–	–	–	1.23	m²	**1.23**
one and a half bricks thick	–	–	–	1.85	m²	**1.85**
two bricks thick	–	–	–	2.46	m²	**2.46**
Zedex CPT (Co-Polymer Thermoplastic) damp-proof course or other equal and approved; 200 mm laps; in gauged mortar (1:1:6)						
Horizontal						
width exceeding 225 mm	–	0.23	3.98	5.74	m²	**9.72**
width not exceeding 225 mm	–	0.46	7.95	5.74	m²	**13.69**

8 UNDERPINNING

Item	PC £	Labour hours	Labour £	Material £	Unit	Total rate £
UNDERPINNING – cont						
Hyload (pitch polymer) damp-proof course or similar; 150 mm laps; in cement: mortar (1:3)						
Horizontal						
width exceeding 225 mm	–	0.23	3.98	4.36	m²	**8.34**
width not exceeding 225 mm	–	0.46	7.95	4.46	m²	**12.41**
Alumite aluminium cored bitumen gas retardant damp-proof course or other equal and approved; 200 mm laps; in gauged mortar (1:1:6)						
Horizontal						
width exceeding 225 mm	–	0.31	5.36	6.06	m²	**11.42**
width not exceeding 225 mm	–	0.60	10.38	6.06	m²	**16.44**
Two courses of slates in cement: mortar (1:3)						
Horizontal						
width exceeding 225 mm	–	1.39	24.05	31.44	m²	**55.49**
width not exceeding 225 mm	–	2.31	39.96	32.15	m²	**72.11**
Wedging and pinning						
To underside of existing construction with slates in cement:mortar (1:3)						
width of wall – half brick thick	–	1.02	17.65	7.10	m	**24.75**
width of wall – one brick thick	–	1.20	20.76	14.22	m	**34.98**
width of wall – one and a half brick thick	–	1.39	24.05	21.32	m	**45.37**

9 DIAPHRAGM WALLS AND EMBEDDED RETAINING WALLS

Item	PC £	Labour hours	Labour £	Material £	Unit	Total rate £
EMBEDDED RETAINING WALLS						
Diaphragm walls; contiguous panel construction; panel lengths not exceeding 5 m						
Provision of all plant; including bringing to and removing from site; maintenance, erection and dismantling; assuming one rig for 1000 m² of walling	–	–	–	–	item	220000.00
Excavation for diaphragm wall; excavated material removed from site; Bentonite slurry supplied and disposed of						
600 mm thick walls	–	–	–	–	m³	600.00
1000 mm thick walls	–	–	–	–	m³	372.50
Ready mixed reinforced in situ concrete; normal Portland cement; C30 – 10 mm aggregate in walls	–	–	–	–	m³	120.00
Reinforcement bar; BS 4449 cold rolled deformed square high yield steel bars; straight or bent						
25 mm–40 mm dia.	–	–	–	–	tonne	1225.00
20 mm dia.	–	–	–	–	tonne	1225.00
16 mm dia.	–	–	–	–	tonne	1225.00
Formwork 75 mm thick to form chases	–	–	–	–	m²	75.00
Construct twin guide walls in reinforced concrete; together with reinforcement and formwork along the axis of the diaphragm wall	–	–	–	–	m	460.00
Delays						
rig standing	–	–	–	–	hour	1450.00

10 CRIB WALLS, GABIONS AND REINFORCED EARTH

Item	PC £	Labour hours	Labour £	Material £	Unit	Total rate £
GABION BASKET WALLS						
Gabion baskets						
Wire mesh gabion baskets; Maccaferri Ltd or other equal and approved; galvanized mesh						
80 mm × 100 mm; filling with broken stones 125 mm–200 mm size						
2.00 × 1.00 × 0.50	16.54	1.00	20.20	77.72	nr	**97.92**
2.00 × 1.00 × 0.50 PVC coated	20.91	1.00	20.20	82.85	nr	**103.05**
2.00 × 1.00 × 1.00	23.18	2.00	40.41	145.04	nr	**185.45**
2.00 × 1.00 × 1.00 PVC coated	29.34	2.00	40.41	151.52	nr	**191.93**
Reno mattress gabion baskets or other equal and approved; Maccaferri Ltd; filling with broken stones 125 mm–200 mm size						
6.00 × 2.00 × 0.17	73.98	2.00	40.41	196.13	nr	**236.54**
6.00 × 2.00 × 0.23	80.08	2.50	50.50	244.08	nr	**294.58**
6.00 × 2.00 × 0.30	93.94	3.00	60.61	304.21	nr	**364.82**

11 IN SITU CONCRETE WORKS

Item	PC £	Labour hours	Labour £	Material £	Unit	Total rate £
IN SITU CONCRETE						
Basic mixed concrete prices						
NOTE: The following prices are for concrete ready for placing excluding any allowance for waste, discount or overheads and profit. Prices are based upon delivery to site within a 5 mile (8 km) radius of concrete mixing plant, using full loads.						
Designed mixes						
grade C7.5; cement to BS12; 10 mm aggregate	–	–	–	72.89	m³	**72.89**
grade C7.5; cement to BS12; 20 mm aggregate	–	–	–	71.46	m³	**71.46**
grade C7.5; cement to BS12; 40 mm aggregate	–	–	–	71.02	m³	**71.02**
grade C7.5; sulphate-resistant cement; 10 mm aggregate	–	–	–	79.57	m³	**79.57**
grade C7.5; sulphate-resistant cement; 20 mm aggregate	–	–	–	78.98	m³	**78.98**
grade C7.5; sulphate-resistant cement; 40 mm aggregate	–	–	–	77.70	m³	**77.70**
grade C10; cement to BS12; 10 mm aggregate	–	–	–	73.57	m³	**73.57**
grade C10; cement to BS12; 20 mm aggregate	–	–	–	72.15	m³	**72.15**
grade C10; cement to BS12; 40 mm aggregate	–	–	–	71.35	m³	**71.35**
grade C10; sulphate-resistant cement; 10 mm aggregate	–	–	–	80.29	m³	**80.29**
grade C10; sulphate-resistant cement; 20 mm aggregate	–	–	–	78.84	m³	**78.84**
grade C10; sulphate-resistant cement; 40 mm aggregate	–	–	–	78.04	m³	**78.04**
grade C15; cement to BS12; 10 mm aggregate	–	–	–	73.96	m³	**73.96**
grade C15; cement to BS12; 20 mm aggregate	–	–	–	72.51	m³	**72.51**
grade C15; cement to BS12; 40 mm aggregate	–	–	–	71.69	m³	**71.69**
grade C15; sulphate-resistant cement; 10 mm aggregate	–	–	–	80.65	m³	**80.65**
grade C15; sulphate-resistant cement; 20 mm aggregate	–	–	–	79.20	m³	**79.20**
grade C15; sulphate-resistant cement; 40 mm aggregate	–	–	–	78.37	m³	**78.37**
grade C20; cement to BS12; 10 mm aggregate	–	–	–	74.31	m³	**74.31**
grade C20; cement to BS12; 20 mm aggregate	–	–	–	72.86	m³	**72.86**
grade C20; cement to BS12; 40 mm aggregate	–	–	–	72.05	m³	**72.05**
grade C20; sulphate-resistant cement; 10 mm aggregate	–	–	–	81.00	m³	**81.00**
grade C20; sulphate-resistant cement; 20 mm aggregate	–	–	–	79.55	m³	**79.55**
grade C20; sulphate-resistant cement; 40 mm aggregate	–	–	–	78.73	m³	**78.73**
grade C25; cement to BS12; 10 mm aggregate	–	–	–	76.34	m³	**76.34**
grade C25; cement to BS12; 20 mm aggregate	–	–	–	74.88	m³	**74.88**
grade C25; cement to BS12; 40 mm aggregate	–	–	–	74.05	m³	**74.05**
grade C25; sulphate-resistant cement; 10 mm aggregate	–	–	–	83.87	m³	**83.87**

11 IN SITU CONCRETE WORKS

Item	PC £	Labour hours	Labour £	Material £	Unit	Total rate £
IN SITU CONCRETE – cont						
Designed mixes – cont						
grade C25; sulphate-resistant cement; 20 mm aggregate	–	–	–	82.41	m³	**82.41**
grade C25; sulphate-resistant cement; 40 mm aggregate	–	–	–	81.58	m³	**81.58**
grade C30; cement to BS12; 10 mm aggregate	–	–	–	71.68	m³	**71.68**
grade C30; cement to BS12; 20 mm aggregate	–	–	–	75.22	m³	**75.22**
grade C30; cement to BS12; 40 mm aggregate	–	–	–	74.40	m³	**74.40**
grade C30; sulphate-resistant cement; 10 mm aggregate	–	–	–	84.23	m³	**84.23**
grade C30; sulphate-resistant cement; 20 mm aggregate	–	–	–	82.75	m³	**82.75**
grade C30; sulphate-resistant cement; 40 mm aggregate	–	–	–	81.93	m³	**81.93**
grade C40; cement to BS12; 10 mm aggregate	–	–	–	82.26	m³	**82.26**
grade C40; cement to BS12; 20 mm aggregate	–	–	–	80.86	m³	**80.86**
grade C40; sulphate-resistant cement; 10 mm aggregate	–	–	–	90.62	m³	**90.62**
grade C40; sulphate-resistant cement; 20 mm aggregate	–	–	–	89.24	m³	**89.24**
grade C50; cement to BS12; 10 mm aggregate	–	–	–	82.48	m³	**82.48**
grade C50; cement to BS12; 20 mm aggregate	–	–	–	80.88	m³	**80.88**
grade C50; sulphate-resistant cement; 10 mm aggregate	–	–	–	90.84	m³	**90.84**
grade C50; sulphate-resistant cement; 20 mm aggregate	–	–	–	89.25	m³	**89.25**
Standard mixes						
NOTE: The following prices are for concrete ready for placing excluding any allowance for waste, discount or overheads and profit. Prices are based upon delivery to site within a 5 mile (8 km) radius of concrete mixing plant, using full loads						
designated concrete mix; GEN0	–	–	–	70.57	m³	**70.57**
designated concrete mix; GEN1	–	–	–	71.44	m³	**71.44**
designated concrete mix; GEN2	–	–	–	72.31	m³	**72.31**
designated concrete mix; GEN3	–	–	–	73.19	m³	**73.19**
designated concrete mix; RC20/25	–	–	–	74.06	m³	**74.06**
designated concrete mix; RC25/30	–	–	–	75.80	m³	**75.80**
designated concrete mix; RC30/37	–	–	–	77.10	m³	**77.10**
designated concrete mix; RC35/45	–	–	–	81.03	m³	**81.03**
designated concrete mix; RC40/50	–	–	–	81.90	m³	**81.90**
designated concrete mix; FND3	–	–	–	77.54	m³	**77.54**
designated concrete mix; FND4	–	–	–	79.28	m³	**79.28**
designed concrete mix; ST 1	–	–	–	67.22	m³	**67.22**
designed concrete mix; ST 2	–	–	–	68.36	m³	**68.36**
designed concrete mix; ST 3	–	–	–	69.49	m³	**69.49**
designed concrete mix; ST 4	–	–	–	70.63	m³	**70.63**
designed concrete mix; ST 5	–	–	–	72.15	m³	**72.15**

11 IN SITU CONCRETE WORKS

Item	PC £	Labour hours	Labour £	Material £	Unit	Total rate £
Lightweight concrete; pumped						
grade 25; pumped; Lytag medium and natural sand	–	–	–	113.09	m^3	**113.09**
grade 30; pumped; Lytag medium and natural sand	–	–	–	117.25	m^3	**117.25**
grade 35; pumped; Lytag medium and natural sand	–	–	–	121.41	m^3	**121.41**
Site mixed concrete (on site batching plant)						
mix 7.50 N/mm^2; cement (1:8); 40 mm aggregate	–	–	–	80.21	m^3	**80.21**
mix 7.50 N/mm^2; sulphate-resisting cement (1:8); 40 mm aggregate	–	–	–	88.92	m^3	**88.92**
mix 10.00 N/mm^2; cement (1:8): 40 mm aggregate	–	–	–	81.95	m^3	**81.95**
mix 10.00 N/mm^2; sulphate-resisting cement (1:8); 40 mm aggregate	–	–	–	90.66	m^3	**90.66**
mix 20.00 N/mm^2; cement (1:2:4); 20 mm aggregate	–	–	–	85.43	m^3	**85.43**
mix 20.00 N/mm^2; sulphate-resisting cement (1:2:4); 20 mm aggregate	–	–	–	94.15	rn^3	**94.15**
mix 25.00 N/mm^2; cement to (1:1:5:3); 20 mm aggregate	–	–	–	88.06	m^3	**88.06**
mix 25.00 N/mm^2; sulphate-resisting cement (1:1:5:3); 20 mm aggregate	–	–	–	95.90	m^3	**95.90**
Add for						
rapid-hardening cement	–	–	–	9.61	m^3	**9.61**
polypropylene fibre additive	–	–	–	5.34	m^3	**5.34**
air entrained concrete	–	–	–	4.69	m^3	**4.69**
water repellent additive	–	–	–	5.02	m^3	**5.02**
foamed concrete	–	–	–	–	%	**25.00**
distance per mile in excess of 5 miles (8 km)	–	–	–	0.56	m^3	**0.56**
Other material prices						
Cements						
Ordinary Portland to BS12	–	–	–	108.91	tonne	**108.91**
high alumina	–	–	–	492.26	tonne	**492.26**
Sulfacrete sulphate-resisting	–	–	–	133.30	tonne	**133.30**
Ferrocrete rapid hardening	–	–	–	253.53	tonne	**253.53**
Snowcrete white cement	–	–	–	173.38	tonne	**173.38**
Cement admixtures						
Febtone colorant – red, marigold, yellow, brown, black	–	–	–	5.02	kg	**5.02**
Febproof waterproof	–	–	–	1.62	litre	**1.62**
Febond PVA bonding agent	–	–	–	1.38	litre	**1.38**
Febspeed frostproofer and hardener	–	–	–	1.03	litre	**1.03**

11 IN SITU CONCRETE WORKS

Item	PC £	Labour hours	Labour £	Material £	Unit	Total rate £
IN SITU CONCRETE – cont						
SUPPLY AND FIX PRICES						
NOTE: The following concrete material prices include an allowance for shrinkage and waste. PC Sums are concrete supply only prices						
Plain in situ ready mixed designated concrete; C7.5 – 40 mm aggregate						
Foundations	73.47	1.02	14.03	79.08	m³	**93.11**
Isolated foundations	73.47	1.13	15.40	79.08	m³	**94.48**
Beds						
thickness not exceeding 300 mm	73.47	1.02	14.03	79.08	m³	**93.11**
thickness exceeding 300 mm	73.47	0.93	12.73	79.08	m³	**91.81**
Screeded beds; protection to compressible formwork exceeding 600 mm wide						
50 mm thick	73.47	0.10	1.37	3.96	m²	**5.33**
75 mm thick	73.47	0.15	2.05	5.93	m²	**7.98**
100 mm thick	73.47	0.20	2.74	7.90	m²	**10.64**
Filling hollow walls						
thickness not exceeding 150 mm	73.47	3.15	43.12	79.08	m³	**122.20**
Column casings						
stub columns beneath suspended ground slabs	73.47	4.50	61.60	79.08	m³	**140.68**
Plain in situ ready mixed designated concrete; C10 – 40 mm aggregate						
Foundations	73.79	1.02	14.03	75.63	m³	**89.66**
Isolated foundations	73.79	1.13	15.44	75.63	m³	**91.07**
Beds						
thickness not exceeding 300 mm	73.79	1.02	14.03	75.63	m³	**89.66**
thickness exceeding 300 mm	73.79	0.93	12.66	75.63	m³	**88.29**
Filling hollow walls						
thickness not exceeding 300 mm	73.79	3.15	43.12	79.42	m³	**122.54**
Plain in situ ready mixed designated concrete; C10 – 40 mm aggregate; poured on or against earth or unblinded hardcore						
Foundations	73.79	1.02	14.03	79.42	m³	**93.45**
Isolated foundations	73.79	1.13	15.44	79.42	m³	**94.86**
Beds						
thickness not exceeding 300 mm	73.79	1.02	14.03	79.42	m³	**93.45**
thickness exceeding 300 mm	73.79	0.93	12.66	79.42	m³	**92.08**
Plain in situ ready mixed designated concrete; C20 – 20 mm aggregate						
Foundations	75.37	0.98	13.42	77.25	m³	**90.67**
Isolated foundations	75.37	1.13	15.44	77.25	m³	**92.69**
Beds						
thickness not exceeding 300 mm	75.37	1.02	13.96	77.25	m³	**91.21**
thickness exceeding 300 mm	75.37	0.93	12.73	77.25	m³	**89.98**
Filling hollow walls						
thickness not exceeding 300 mm	75.37	3.00	41.07	81.12	m³	**122.19**

11 IN SITU CONCRETE WORKS

Item	PC £	Labour hours	Labour £	Material £	Unit	Total rate £
Plain in situ ready mixed designated concrete; C20 – 20 mm aggregate; poured on or against earth or unblinded hardcore						
Foundations	75.37	1.02	14.03	81.12	m³	**95.15**
Isolated foundations	75.37	1.13	15.44	81.12	m³	**96.56**
Beds						
thickness not exceeding 300 mm	75.37	1.02	14.03	82.74	m³	**96.77**
thickness exceeding 300 mm	75.37	0.93	12.73	77.25	m³	**89.98**
Reinforced in situ ready mixed designated concrete; C25 – 20 mm aggregate						
Foundations	77.47	1.02	14.03	83.37	m³	**97.40**
Ground beams	77.47	2.59	35.45	79.41	m³	**114.86**
Isolated foundations	77.47	1.13	15.40	83.37	m³	**98.77**
Beds						
thickness not exceeding 300 mm	77.47	1.02	13.96	83.37	m³	**97.33**
thickness exceeding 300 mm	77.47	0.93	12.73	79.41	m³	**92.14**
Slabs						
thickness not exceeding 300 mm	77.47	1.25	17.11	83.37	m³	**100.48**
thickness exceeding 300 mm	77.47	1.15	15.74	83.37	m³	**99.11**
Coffered and troughed slabs						
thickness not exceeding 300 mm	77.47	2.96	40.52	83.37	m³	**123.89**
thickness exceeding 300 mm	77.47	2.59	35.45	83.37	m³	**118.82**
Extra over for sloping work						
not exceeding 15°	–	0.23	3.15	–	m³	**3.15**
exceeding 15°	–	0.46	6.29	–	m³	**6.29**
Walls						
thickness not exceeding 300 mm	77.47	3.15	43.12	83.37	m³	**126.49**
thickness exceeding 300 mm	77.47	2.40	32.92	83.37	m³	**116.29**
Beams						
isolated	77.47	3.70	50.65	83.37	m³	**134.02**
isolated deep	77.47	4.07	55.72	83.37	m³	**139.09**
attached deep	77.47	3.70	50.65	83.37	m³	**134.02**
Beam casings						
isolated	77.47	4.07	55.72	85.36	m³	**141.08**
isolated deep	77.47	4.44	60.78	83.37	m³	**144.15**
attached deep	77.47	4.07	55.72	83.37	m³	**139.09**
Columns	77.47	4.20	57.49	83.37	m³	**140.86**
Column casings	77.47	4.90	67.08	79.41	m³	**146.49**
Staircases	77.47	5.25	71.87	83.37	m³	**155.24**
Upstands	77.47	3.30	45.17	83.37	m³	**128.54**
Reinforced in situ ready mixed designated concrete; C35 – 20 mm aggregate						
Foundations	81.12	1.02	14.03	87.31	m³	**101.34**
Ground beams	81.12	2.59	35.45	87.31	m³	**122.76**
Isolated foundations	81.12	1.13	15.47	87.31	m³	**102.78**
Beds						
thickness not exceeding 300 mm	81.12	1.02	13.96	87.31	m³	**101.27**
thickness exceeding 300 mm	81.12	0.95	13.01	87.31	m³	**100.32**

11 IN SITU CONCRETE WORKS

Item	PC £	Labour hours	Labour £	Material £	Unit	Total rate £
IN SITU CONCRETE – cont						
Reinforced in situ ready mixed designated concrete – cont						
Slabs						
thickness not exceeding 300 mm	81.12	1.25	17.11	87.31	m³	**104.42**
thickness exceeding 300 mm	81.12	1.15	15.74	87.31	m³	**103.05**
Coffered and troughed slabs						
thickness not exceeding 300 mm	81.12	2.96	40.52	87.31	m³	**127.83**
thickness exceeding 300 mm	81.12	2.59	35.45	87.31	m³	**122.76**
Extra over for sloping						
not exceeding 15°	–	0.23	3.15	–	m³	**3.15**
exceeding 15°	–	0.46	6.29	–	m³	**6.29**
Walls						
thickness not exceeding 300 mm	81.12	2.67	36.62	87.31	m³	**123.93**
thickness exceeding 300 mm	81.12	2.41	32.95	87.31	m³	**120.26**
Beams						
isolated	81.12	3.70	50.65	87.31	m³	**137.96**
isolated deep	81.12	4.07	55.72	87.31	m³	**143.03**
attached deep	81.12	3.70	50.65	87.31	m³	**137.96**
Beam casings						
isolated	81.12	4.07	55.72	87.31	m³	**143.03**
isolated deep	81.12	4.44	60.78	87.31	m³	**148.09**
attached deep	81.12	4.07	55.72	87.31	m³	**143.03**
Columns	81.12	4.44	60.78	87.31	m³	**148.09**
Column casings	81.12	4.90	67.08	87.31	m³	**154.39**
Staircases	81.12	5.55	75.97	87.31	m³	**163.28**
Upstands	81.12	3.56	48.74	87.31	m³	**136.05**
Reinforced in situ ready mixed designated concrete; C40 – 20 mm aggregate						
Foundations	83.29	1.02	14.03	89.65	m³	**103.68**
Isolated foundations	83.29	1.13	15.44	89.65	m³	**105.09**
Ground beams	83.29	2.59	35.45	89.65	m³	**125.10**
Beds						
thickness not exceeding 300 mm	83.29	0.95	13.01	89.65	m³	**102.66**
thickness exceeding 300 mm	83.29	0.93	12.73	89.65	m³	**102.38**
Slabs						
thickness not exceeding 300 mm	83.29	1.20	16.43	89.65	m³	**106.08**
thickness exceeding 300 mm	83.29	1.15	15.74	89.65	m³	**105.39**
Coffered and troughed slabs						
thickness not exceeding 300 mm	83.29	2.96	40.52	89.65	m³	**130.17**
thickness exceeding 300 mm	83.29	2.59	35.45	89.65	m³	**125.10**
Extra over for sloping						
not exceeding 15°	–	0.23	3.15	–	m³	**3.15**
exceeding 15°	–	0.46	6.29	–	m³	**6.29**
Walls						
thickness not exceeding 300 mm	83.29	2.73	37.37	89.65	m³	**127.02**
thickness exceeding 300 mm	83.29	2.41	32.99	89.65	m³	**122.64**

11 IN SITU CONCRETE WORKS

Item	PC £	Labour hours	Labour £	Material £	Unit	Total rate £
Beams						
isolated	83.29	3.70	50.65	89.65	m³	**140.30**
isolated deep	83.29	4.07	55.72	89.65	m³	**145.37**
attached deep	83.29	3.70	50.65	89.65	m³	**140.30**
Beam casings						
isolated	83.29	4.07	55.72	89.65	m³	**145.37**
isolated deep	83.29	4.44	60.78	89.65	m³	**150.43**
attached deep	83.29	4.07	55.72	89.65	m³	**145.37**
Columns	83.29	4.44	60.78	89.65	m³	**150.43**
Column casings	83.29	4.90	67.08	89.65	m³	**156.73**
Staircases	83.29	5.55	75.97	89.65	m³	**165.62**
Upstands	83.29	3.56	48.74	89.65	m³	**138.39**
Sundry in situ concrete work						
Filling; plain in situ concrete; mixed on site						
mortices	–	0.09	1.23	0.62	nr	**1.85**
holes	–	0.23	3.15	91.98	m³	**95.13**
chases not exceeding 300 mm wide or thick	–	0.14	1.92	1.19	m	**3.11**
chases exceeding 300 mm wide or thick	–	0.19	2.60	91.91	m³	**94.51**
Proprietary voided Bubbledeck, Cobiax or other equal and approved slab; concrete mix RC35; to achieve design loadings of 5.0 kN/m² live and 3.0 kN/m² dead; with trowelled finish						
Beds						
360 mm overall thickness	–	–	–	–	m²	**96.66**
additional concrete up to 600 mm wide at edges where formers omitted at junctions with walls etc.	–	–	–	–	m	**39.54**
Extra over vibrated concrete for						
Reinforcement content over 5%	–	0.51	6.98	–	m³	**6.98**
Grouting with cement mortar (1:1)						
Stanchion bases						
10 mm thick	–	0.93	12.73	0.12	nr	**12.85**
25 mm thick	–	1.16	15.88	0.30	nr	**16.18**
Grouting with epoxy resin						
Stanchion bases						
10 mm thick	–	1.16	15.88	7.74	nr	**23.62**
25 mm thick	–	1.39	19.02	19.78	nr	**38.80**
Grouting with Conbextra GP cementitious grout						
Stanchion bases						
10 mm thick	–	1.16	15.88	1.30	nr	**17.18**
25 mm thick	–	1.39	19.02	3.33	nr	**22.35**
Grouting with Conbextra HF flowable cementitious grout						
Stanchion bases						
10 mm thick	–	1.16	15.88	1.61	nr	**17.49**
25 mm thick	–	1.39	19.02	4.11	nr	**23.13**

11 IN SITU CONCRETE WORKS

Item	PC £	Labour hours	Labour £	Material £	Unit	Total rate £
SURFACE FINISHES						
Worked finishes						
Tamping by mechanical means	–	0.02	0.28	0.09	m²	**0.37**
Power floating	–	0.16	2.19	0.26	m²	**2.45**
Trowelling	–	0.31	4.24	–	m²	**4.24**
Hacking						
by mechanical means	–	0.31	4.24	0.31	m²	**4.55**
by hand	–	0.65	8.90	–	m²	**8.90**
Lightly shot blasting surface of concrete	–	0.37	5.06	–	m²	**5.06**
Blasting surface of concrete to produce textured finish	–	0.65	8.90	0.65	m²	**9.55**
Sand blasting (blast and vac method)	–	–	–	–	m²	**34.94**
Wood float finish	–	0.12	1.64	–	m²	**1.64**
Tamped finish						
level or to falls	–	0.06	0.82	–	m²	**0.82**
to falls	–	0.09	1.23	–	m²	**1.23**
Spade finish	–	0.14	1.92	–	m²	**1.92**
Diamond drilling/cutting						
Cutting holes in block wall						
up to 52 mm dia.; up to 150 mm thick	–	–	–	–	nr	**14.00**
up to 52 mm dia.; 200 mm thick	–	–	–	–	nr	**17.45**
up to 52 mm dia.; 250 mm thick	–	–	–	–	nr	**20.91**
up to 52 mm dia.; 300 mm thick	–	–	–	–	nr	**24.37**
79 mm–107 mm dia.; up to 150 mm thick	–	–	–	–	nr	**21.97**
79 mm–107 mm dia.; 200 mm thick	–	–	–	–	nr	**26.21**
79 mm–107 mm dia.; 250 mm thick	–	–	–	–	nr	**30.45**
79 mm–107 mm dia.; 300 mm thick	–	–	–	–	nr	**34.69**
251 mm–300 mm dia.; ne 150 mm thick	–	–	–	–	nr	**54.23**
251 mm–300 mm dia.; 200 mm thick	–	–	–	–	nr	**68.68**
251 mm–300 mm dia.; 250 mm thick	–	–	–	–	nr	**83.15**
251 mm–300 mm dia.; 300 mm thick	–	–	–	–	nr	**97.60**
Cutting holes in brick wall						
up to 52 mm dia.; up to 150 mm thick	–	–	–	–	nr	**16.00**
up to 52 mm dia.; 200 mm thick	–	–	–	–	nr	**19.94**
up to 52 mm dia.; 250 mm thick	–	–	–	–	nr	**23.90**
up to 52 mm dia.; 300 mm thick	–	–	–	–	nr	**27.85**
79 mm–107 mm dia.; up to 150 mm thick	–	–	–	–	nr	**25.11**
79 mm–107 mm dia.; 200 mm thick	–	–	–	–	nr	**29.95**
79 mm–107 mm dia.; 250 mm thick	–	–	–	–	nr	**34.80**
79 mm–107 mm dia.; 300 mm thick	–	–	–	–	nr	**39.64**
251 mm–300 mm dia.; ne 150 mm thick	–	–	–	–	nr	**61.98**
251 mm–300 mm dia.; 200 mm thick	–	–	–	–	nr	**78.50**
251 mm–300 mm dia.; 250 mm thick	–	–	–	–	nr	**95.02**
251 mm–300 mm dia.; 300 mm thick	–	–	–	–	nr	**111.54**
Cutting holes in concrete wall						
up to 52 mm dia.; up to 150 mm thick	–	–	–	–	nr	**20.00**
up to 52 mm dia.; 200 mm thick	–	–	–	–	nr	**24.93**
up to 52 mm dia.; 250 mm thick	–	–	–	–	nr	**29.87**

11 IN SITU CONCRETE WORKS

Item	PC £	Labour hours	Labour £	Material £	Unit	Total rate £
up to 52 mm dia.; 300 mm thick	–	–	–	–	nr	34.81
79 mm–107 mm dia.; up to150 mm thick	–	–	–	–	nr	31.39
79 mm–107 mm dia.; 200 mm thick	–	–	–	–	nr	37.44
79 mm–107 mm dia.; 250 mm thick	–	–	–	–	nr	43.50
79 mm–107 mm dia.; 300 mm thick	–	–	–	–	nr	49.55
251 mm–300 mm dia.; ne 150 mm thick	–	–	–	–	nr	77.47
251 mm–300 mm dia.; 200 mm thick	–	–	–	–	nr	98.12
251 mm–300 mm dia.; 250 mm thick	–	–	–	–	nr	118.78
251 mm–300 mm dia.; 300 mm thick	–	–	–	–	nr	139.43
Cutting holes in timber wall						
up to 52 mm dia.; up to 150 mm thick	–	–	–	–	nr	19.00
up to 52 mm dia.; 200 mm thick	–	–	–	–	nr	23.68
up to 52 mm dia.; 250 mm thick	–	–	–	–	nr	28.38
up to 52 mm dia.; 300 mm thick	–	–	–	–	nr	33.07
79 mm–107 mm dia.; up to 150 mm thick	–	–	–	–	nr	29.82
79 mm–107 mm dia.; 200 mm thick	–	–	–	–	nr	35.57
79 mm–107 mm dia.; 250 mm thick	–	–	–	–	nr	41.33
79 mm–107 mm dia.; 300 mm thick	–	–	–	–	nr	47.07
251 mm–300 mm dia.; ne 150 mm thick	–	–	–	–	nr	73.60
251 mm–300 mm dia.; 200 mm thick	–	–	–	–	nr	93.21
251 mm–300 mm dia.; 250 mm thick	–	–	–	–	nr	112.84
251 mm–300 mm dia.; 300 mm thick	–	–	–	–	nr	132.46
Chasing for concealed conduits						
up to 35 mm deep, up to 50 mm wide in block wall	–	–	–	–	m	6.50
up to 35 mm deep, up to 50 mm wide in brick wall	–	–	–	–	m	7.50
up to 35 mm deep, up to 50 mm wide in concrete wall	–	–	–	–	m	25.00
up to 35 mm deep, up to 50 mm wide in concrete floor	–	–	–	–	m	30.00
up to 35 mm deep, 75 mm–100 mm wide in block wall	–	–	–	–	m	10.16
up to 35 mm deep, 75 mm–100 mm wide in brick wall	–	–	–	–	m	11.72
up to 35 mm deep, 75 mm–100 mm wide in concrete wall	–	–	–	–	m	39.06
up to 35 mm deep, 75 mm–100 mm wide in concrete floor	–	–	–	–	m	46.88
recess 100 mm × 100 mm × 35 mm deep in block wall	–	–	–	–	nr	3.50
recess 100 mm × 100 mm × 35 mm deep in brick wall	–	–	–	–	nr	4.04
recess 100 mm × 100 mm × 35 mm deep in concrete wall	–	–	–	–	nr	16.16
recess 100 mm × 100 mm × 35 mm deep in concrete floor	–	–	–	–	nr	13.46
Floor sawing						
diamond floor sawing; 25 mm depth	–	–	–	–	m	5.00
Ring sawing						
up to 25 mm deep in block wall	–	–	–	–	m	6.50
up to 25 mm deep in brick wall	–	–	–	–	m	7.50
up to 25 mm deep in concrete floor	–	–	–	–	m	10.00
up to 25 mm deep in concrete wall	–	–	–	–	m	12.00

11 IN SITU CONCRETE WORKS

Item	PC £	Labour hours	Labour £	Material £	Unit	Total rate £
FORMWORK						
Sides of foundations; 18 mm thick external quality plywood; basic finish; four uses						
Plain vertical						
height not exceeding 250 mm	–	0.38	6.12	1.48	m	**7.60**
height not exceeding 250 mm; left in	–	0.38	6.12	2.62	m	**8.74**
height 250–500 mm	–	0.71	11.52	2.83	m	**14.35**
height 250–500 mm; left in	–	0.62	10.06	5.76	m	**15.82**
height 500 mm–1.00 m	–	1.00	16.18	3.61	m	**19.79**
height 500 mm–1.00 m; left in	–	0.95	15.46	8.95	m	**24.41**
height exceeding 1.00 m	–	1.33	21.58	3.61	m²	**25.19**
height exceeding 1.00 m; left in	–	1.17	18.95	8.95	m²	**27.90**
Sides of foundations; polystyrene sheet formwork; Cordek Claymaster or equal						
50 mm thick; plain vertical						
height not exceeding 250 mm; left in	–	0.09	1.46	1.98	m	**3.44**
height 250–500 mm; left in	–	0.14	2.34	3.96	m	**6.30**
height 500 mm–1.00 m; left in	–	0.22	3.50	7.91	m	**11.41**
height exceeding 1.00 m; left in	–	0.27	4.38	7.55	m²	**11.93**
75 mm thick; plain vertical						
height not exceeding 250 mm; left in	–	0.09	1.46	2.83	m	**4.29**
height 250–500 mm; left in	–	0.14	2.34	5.67	m	**8.01**
height 500 mm–1.00 m; left in	–	0.22	3.50	11.88	m	**15.38**
height exceeding 1.00 m; left in	–	0.27	4.38	11.34	m²	**15.72**
100 mm thick; plain vertical						
height not exceeding 250 mm; left in	–	0.09	1.46	3.96	m	**5.42**
height 250–500 mm; left in	–	0.14	2.34	7.91	m	**10.25**
height 500 mm–1.00 m; left in	–	0.22	3.50	15.84	m	**19.34**
height exceeding 1.00 m; left in	–	0.27	4.38	15.12	m²	**19.50**
150 mm thick; plain vertical						
height not exceeding 250 mm; left in	–	0.09	1.46	5.93	m	**7.39**
height 250–500 mm; left in	–	0.14	2.34	11.88	m	**14.22**
height 500 mm–1.00 m; left in	–	0.22	3.50	23.75	m	**27.25**
height exceeding 1.00 m; left in	–	0.28	4.46	22.67	m²	**27.13**
175 mm thick; plain vertical						
height not exceeding 250 mm; left in	–	0.10	1.62	6.93	m	**8.55**
height 250–500 mm; left in	–	0.15	2.43	13.86	m	**16.29**
height 500 mm–1.00 m; left in	–	0.22	3.57	27.72	m	**31.29**
height exceeding 1.00 m; left in	–	0.30	4.86	26.46	m²	**31.32**
200 mm thick; plain vertical						
height not exceeding 250 mm; left in	–	0.10	1.62	7.92	m	**9.54**
height 250–500 mm; left in	–	0.15	2.43	15.84	m	**18.27**
height 500 mm–1.00 m; left in	–	0.22	3.57	31.67	m	**35.24**
height exceeding 1.00 m; left in	–	0.30	4.86	30.24	m²	**35.10**
250 mm thick; plain vertical						
height not exceeding 250 mm; left in	–	0.15	2.43	9.90	m	**12.33**
height 250–500 mm; left in	–	0.17	2.84	19.79	m	**22.63**
height 500 mm–1.00 m; left in	–	0.25	4.05	39.60	m	**43.65**
height exceeding 1.00 m; left in	–	0.35	5.67	37.79	m²	**43.46**

11 IN SITU CONCRETE WORKS

Item	PC £	Labour hours	Labour £	Material £	Unit	Total rate £
Combined heave pressure relief insulation and compressible board substructure formwork; Cordeck Cellcore CP or other equal and approved; butt joints; securely fixed in place						
Plain horizontal						
200 mm thick; beneath slabs; left in	–	0.54	8.74	10.23	m^2	**18.97**
250 mm thick; beneath slabs; left in	–	0.58	9.48	11.38	m^2	**20.86**
300 mm thick; beneath slabs; left in	–	0.63	10.21	12.26	m^2	**22.47**
Dufaylite Clayboard void former; butt joints						
KN30; compressive strength of 30 kN/m^2; board thickness						
60 mm	–	0.05	0.58	15.54	m^2	**16.12**
90 mm	–	0.05	0.58	17.62	m^2	**18.20**
110 mm	–	0.05	0.58	18.59	m^2	**19.17**
160 mm	–	0.05	0.58	21.00	m^2	**21.58**
KN90; compressive strength of 90 kN/m^2; board thickness						
60 mm	–	0.06	0.70	18.20	m^2	**18.90**
90 mm	–	0.06	0.70	19.94	m^2	**20.64**
110 mm	–	0.06	0.70	21.61	m^2	**22.31**
160 mm	–	0.06	0.70	22.87	m^2	**23.57**
600 mm voidpack pipe						
36 mm dia.	–	0.05	0.58	5.33	nr	**5.91**
Sides of ground beams and edges of beds; basic finish; 18 mm thick external quality plywood; basic finish; four uses						
Plain vertical						
height not exceeding 250 mm	–	0.41	6.70	1.46	m	**8.16**
height 250–500 mm	–	0.75	12.11	2.80	m	**14.91**
height 500 mm–1.00 m	–	1.04	16.91	3.58	m	**20.49**
height exceeding 1.00 m	–	1.38	22.30	3.58	m^2	**25.88**
Edges of suspended slabs; basic finish; 18 mm thick external quality plywood; basic finish; four uses						
Plain vertical						
height not exceeding 250 mm	–	0.62	10.06	1.51	m	**11.57**
height 250–500 mm	–	0.92	14.87	2.37	m	**17.24**
height 500 mm–1.00 m	–	1.46	23.62	3.63	m	**27.25**
Sides of upstands; basic finish; 18 mm thick external quality plywood; basic finish; four uses						
Plain vertical						
height not exceeding 250 mm	–	0.52	8.46	1.56	m	**10.02**
height 250–500 mm	–	0.84	13.56	2.90	m	**16.46**
height 500 mm–1.00 m	–	1.46	23.62	4.33	m	**27.95**
height exceeding 1.00 m	–	1.67	26.98	4.33	m^2	**31.31**

11 IN SITU CONCRETE WORKS

Item	PC £	Labour hours	Labour £	Material £	Unit	Total rate £
FORMWORK – cont						
Steps in top surfaces; basic finish; 18 mm thick external quality plywood; basic finish; four uses						
Plain vertical						
height not exceeding 250 mm	–	0.41	6.70	1.58	m	**8.28**
height 250–500 mm	–	0.67	10.79	2.93	m	**13.72**
Steps in soffits; basic finish; 18 mm thick external quality plywood; basic finish; four uses						
Plain vertical						
height not exceeding 250 mm	–	0.46	7.43	1.22	m	**8.65**
height 250–500 mm	–	0.73	11.81	2.18	m	**13.99**
Machine bases and plinths; basic finish; 18 mm thick external quality plywood; basic finish; four uses						
Plain vertical						
height not exceeding 250 mm	–	0.41	6.70	1.46	m	**8.16**
height 250–500 mm	–	0.71	11.52	2.80	m	**14.32**
height 500 mm–1.00 m	–	1.04	16.91	3.58	m	**20.49**
height exceeding 1.00 m	–	1.33	21.58	3.58	m²	**25.16**
Soffits of slabs; basic finish; 18 mm thick external quality plywood; basic finish; four uses						
Slab thickness not exceeding 200 mm						
horizontal; height to soffit not exceeding 1.50 m	–	1.50	24.34	3.30	m²	**27.64**
horizontal; height to soffit 1.50–2.40 m	–	1.46	23.62	3.35	m²	**26.97**
horizontal; height to soffit 2.40–2.70 m	–	1.38	22.30	2.78	m²	**25.08**
horizontal; height to soffit 2.70–3.00 m	–	1.33	21.58	2.40	m²	**23.98**
horizontal; height to soffit 3.00–4.50 m	–	1.41	22.89	3.47	m²	**26.36**
horizontal; height to soffit 4.50–6.00 m	–	1.50	24.34	3.61	m²	**27.95**
Slab thickness 200–300 mm						
horizontal; height to soffit 1.50–3.00 m	–	1.50	24.34	4.24	m²	**28.58**
Slab thickness 300–400 mm						
horizontal; height to soffit 1.50–3.00 m	–	1.54	24.93	4.69	m²	**29.62**
Slab thickness 400–500 mm						
horizontal; height to soffit 1.50–3.00 m	–	1.62	26.24	5.15	m²	**31.39**
Slab thickness 500–600 mm						
horizontal; height to soffit 1.50–3.00 m	–	1.75	28.29	5.15	m²	**33.44**
Extra over soffits of slabs for						
sloping not exceeding 15°	–	0.17	2.77	–	m²	**2.77**
sloping exceeding 15°	–	0.33	5.39	–	m²	**5.39**
Soffits of landings; basic finish; 18 mm thick external quality plywood; basic finish; four uses						
Slab thickness not exceeding 200 mm						
horizontal; height to soffit 1.50–3.00 m	–	1.50	24.34	3.53	m²	**27.87**
Slab thickness 200–300 mm						
horizontal; height to soffit 1.50–3.00 m	–	1.58	25.67	4.51	m²	**30.18**
Slab thickness 300–400 mm						
horizontal; height to soffit 1.50–3.00 m	–	1.62	26.24	5.01	m²	**31.25**

11 IN SITU CONCRETE WORKS

Item	PC £	Labour hours	Labour £	Material £	Unit	Total rate £
Slab thickness 400–500 mm						
horizontal; height to soffit 1.50–3.00 m	–	1.71	27.71	5.50	m²	**33.21**
Slab thickness 500–600 mm						
horizontal; height to soffit 1.50–3.00 m	–	1.84	29.75	5.50	m²	**35.25**
Extra over soffits of landings for						
sloping not exceeding 15°	–	0.17	2.77	–	m²	**2.77**
sloping exceeding 15°	–	0.33	5.39	–	m²	**5.39**
Top formwork; basic finish; 18 mm thick external quality plywood; basic finish; four uses						
Sloping exceeding 15°	–	1.25	20.26	2.58	m²	**22.84**
Walls; basic finish; 18 mm thick external quality plywood; basic finish; four uses						
Vertical	–	1.50	24.34	4.24	m²	**28.58**
Vertical; height exceeding 3.00 m above floor level	–	1.84	29.75	4.38	m²	**34.13**
Vertical; interrupted	–	1.75	28.29	4.38	m²	**32.67**
Vertical; to one side only	–	2.92	47.24	5.40	m²	**52.64**
Battered	–	2.33	37.76	4.60	m²	**42.36**
Beams; basic finish; 18 mm thick external quality plywood; basic finish; four uses						
Attached to slabs						
regular shaped; square or rectangular; height to soffit 1.50–3.00 m	–	1.84	29.75	4.09	m²	**33.84**
regular shaped; square or rectangular; height to soffit 3.00–4.50 m	–	1.92	31.06	4.24	m²	**35.30**
regular shaped; square or rectangular; height to soffit 4.50–6.00 m	–	2.00	32.37	4.38	m²	**36.75**
Attached to walls						
regular shaped; square or rectangular; height to soffit 1.50–3.00 m	–	1.92	31.06	4.09	m²	**35.15**
Isolated						
regular shaped; square or rectangular; height to soffit 1.50–3.00 m	–	2.00	32.37	4.09	m²	**36.46**
regular shaped; square or rectangular; height to soffit 3.00–4.50 m	–	2.08	33.68	4.24	m²	**37.92**
regular shaped; square or rectangular; height to soffit 4.50–6.00 m	–	2.17	35.14	4.38	m²	**39.52**
Extra over beams for						
regular shaped; sloping not exceeding 15°	–	0.25	4.08	0.42	m²	**4.50**
regular shaped; sloping exceeding 15°	–	0.50	8.17	0.84	m²	**9.01**
Beam casings; basic finish; 18 mm thick external quality plywood; basic finish; four uses						
Attached to slabs						
regular shaped; square or rectangular; height to soffit 1.50–3.00 m	–	1.92	31.06	4.09	m²	**35.15**
regular shaped; square or rectangular; height to soffit 3.00–4.50 m	–	2.00	32.37	4.24	m²	**36.61**

11 IN SITU CONCRETE WORKS

Item	PC £	Labour hours	Labour £	Material £	Unit	Total rate £
FORMWORK – cont						
Beam casings – cont						
Attached to walls						
regular shaped; square or rectangular; height to soffit 1.50–3.00 m	–	2.00	32.37	4.09	m²	**36.46**
Isolated						
regular shaped; square or rectangular; height to soffit 1.50–3.00 m	–	2.08	33.68	4.09	m²	**37.77**
regular shaped; square or rectangular; height to soffit 3.00–4.50 m	–	2.17	35.14	4.24	m²	**39.38**
Extra over beam casings for						
regular shaped; sloping not exceeding 15°	–	0.25	4.08	0.42	m²	**4.50**
regular shaped; sloping exceeding 15°	–	0.50	8.17	0.84	m²	**9.01**
Columns; basic finish; 18 mm thick external quality plywood; basic finish; four uses						
Attached to walls						
regular shaped; square or rectangular; height to soffit 1.50–3.00 m	–	1.84	29.75	3.61	m²	**33.36**
Isolated						
regular shaped; square or rectangular; height to soffit 1.50–3.00 m	–	1.92	31.06	3.61	m²	**34.67**
regular shaped; circular; not exceeding 300 mm dia.; height to soffit 1.50–3.00 m	–	3.33	53.95	5.78	m²	**59.73**
regular shaped; circular; 300–600 mm dia.; height to soffit 1.50–3.00 m	–	3.12	50.59	5.15	m²	**55.74**
regular shaped; circular; 600–900 mm dia.; height to soffit 1.50–3.00 m	–	2.92	47.24	5.01	m²	**52.25**
Column casings; basic finish; 18 mm thick external quality plywood; basic finish; four uses						
Attached to walls						
regular shaped; square or rectangular; height to soffit 1.50–3.00 m	–	1.92	31.06	3.61	m²	**34.67**
Isolated						
regular shaped; square or rectangular; height to soffit 1.50–3.00 m	–	2.00	32.37	3.61	m²	**35.98**
Special shapes and finishes						
Recesses or rebates						
12 × 12 mm	–	0.05	0.87	0.27	m	**1.14**
25 × 25 mm	–	0.05	0.87	0.55	m	**1.42**
25 × 50 mm	–	0.05	0.87	0.72	m	**1.59**
50 × 50 mm	–	0.05	0.87	0.69	m	**1.56**
Nibs						
50 × 50 mm	–	0.46	7.43	0.47	m	**7.90**
100 × 100 mm	–	0.65	10.50	0.62	m	**11.12**
100 × 200 mm	–	0.86	14.00	9.36	m	**23.36**

11 IN SITU CONCRETE WORKS

Item	PC £	Labour hours	Labour £	Material £	Unit	Total rate £
Extra over a basic finish for fine formed finishes						
slabs	–	0.27	4.38	–	m^2	4.38
walls	–	0.27	4.38	–	m^2	4.38
beams	–	0.27	4.38	–	m^2	4.38
columns	–	0.27	4.38	–	m^2	4.38
Add to prices for basic formwork for						
curved radius 6.00 m	–	–	–	–	%	50.00
curved radius 2.00 m	–	–	–	–	%	100.00
coating with retardant agent	–	0.01	0.15	0.21	m^2	0.36
Wall kickers; basic finish						
height 150 mm	–	0.41	6.70	1.05	m	7.75
height 225 mm	–	0.54	8.74	1.28	m	10.02
Suspended wall kickers; basic finish						
height 150 mm	–	0.52	8.46	0.98	m	9.44
Wall ends, soffits and steps in walls; basic finish						
Plain						
width exceeding 1.00 m	–	1.58	25.67	4.24	m^2	29.91
width not exceeding 250 mm	–	0.50	8.17	1.10	m	9.27
width 250–500 mm	–	0.79	12.83	2.38	m	15.21
width 500 mm–1.00 m	–	1.25	20.26	4.24	m	24.50
Openings in walls						
Plain						
width exceeding 1.00 m	–	1.75	28.29	4.24	m^2	32.53
width not exceeding 250 mm	–	0.54	8.74	1.10	m	9.84
width 250–500 mm	–	0.92	14.87	2.38	m	17.25
width 500 mm–1.00 m	–	1.41	22.89	4.24	m	27.13
Stairflights						
Width 1.00 m; 150 mm waist; 150 mm undercut risers						
string, width 300 mm	–	4.17	67.51	9.47	m	76.98
Width 2.00 m; 200 mm waist; 150 mm undercut risers						
string, width 350 mm	–	7.50	121.45	49.57	m	171.02
Mortices						
Girth not exceeding 500 mm						
depth not exceeding 250 mm; circular	–	0.13	2.04	0.35	nr	2.39
Holes						
Girth not exceeding 500 mm						
depth not exceeding 250 mm; circular	–	0.17	2.77	0.47	nr	3.24
depth 250–500 mm; circular	–	0.25	4.08	1.11	nr	5.19
Girth 500 mm–1.00 m						
depth not exceeding 250 mm; circular	–	0.21	3.35	0.78	nr	4.13
depth 250–500 mm; circular	–	0.32	5.10	2.05	nr	7.15
Girth 1.00–2.00 m						
depth not exceeding 250 mm; circular	–	0.38	6.12	2.05	nr	8.17
depth 250–500 mm; circular	–	0.56	9.04	4.33	nr	13.37
Girth 2.00–3.00 m						
depth not exceeding 250 mm; circular	–	0.50	8.17	4.16	nr	12.33
depth 250–500 mm; circular	–	0.75	12.11	65.38	nr	77.49

11 IN SITU CONCRETE WORKS

Item	PC £	Labour hours	Labour £	Material £	Unit	Total rate £
REINFORCEMENT						
Bars; BS 4449; hot rolled deformed high steel bars; grade 500C						
40 mm dia. nominal size						
straight	–	14.44	233.99	587.00	tonne	**820.99**
bent	–	17.50	283.57	624.83	tonne	**908.40**
32 mm dia. nominal size						
straight	–	15.34	248.57	592.41	tonne	**840.98**
bent	–	18.95	307.10	630.24	tonne	**937.34**
25 mm dia. nominal size						
straight	–	16.25	263.23	600.98	tonne	**864.21**
bent	–	18.95	307.10	638.80	tonne	**945.90**
20 mm dia. nominal size						
straight	–	16.25	263.23	619.00	tonne	**882.23**
bent	–	18.95	307.10	656.82	tonne	**963.92**
16 mm dia. nominal size						
straight	–	19.86	321.73	640.17	tonne	**961.90**
bent	–	22.56	365.60	678.00	tonne	**1043.60**
12 mm dia. nominal size						
straight	–	21.66	350.98	666.75	tonne	**1017.73**
bent	–	24.37	394.85	704.57	tonne	**1099.42**
10 mm dia. nominal size						
straight	–	23.46	380.22	709.09	tonne	**1089.31**
bent	–	26.17	424.09	746.92	tonne	**1171.01**
8 mm dia. nominal size						
straight	–	24.37	394.85	735.68	tonne	**1130.53**
links	–	27.07	438.72	795.15	tonne	**1233.87**
bent	–	27.07	438.72	773.50	tonne	**1212.22**
Extra over for cutting and bending to shape codes 67 to 99	–	–	–	51.25	tonne	**51.25**
Bars; stainless steel; to EN 1.4301						
40 mm dia. nominal size						
straight	–	17.00	275.47	2928.72	tonne	**3204.19**
bent	–	21.00	335.26	3107.33	tonne	**3442.59**
32 mm dia. nominal size						
straight	–	17.00	275.47	2928.72	tonne	**3204.19**
bent	–	21.00	335.26	3107.33	tonne	**3442.59**
25 mm dia. nominal size						
straight	–	18.00	291.67	2928.74	tonne	**3220.41**
bent	–	18.00	291.67	3107.35	tonne	**3399.02**
20 mm dia. nominal size						
straight	–	20.00	324.07	2944.48	tonne	**3268.55**
bent	–	20.00	324.07	3123.08	tonne	**3447.15**
16 mm dia. nominal size						
straight	–	22.00	356.48	2966.49	tonne	**3322.97**
bent	–	22.00	356.48	3145.10	tonne	**3501.58**

11 IN SITU CONCRETE WORKS

Item	PC £	Labour hours	Labour £	Material £	Unit	Total rate £
12 mm dia. nominal size						
straight	–	24.00	388.90	2988.50	tonne	**3377.40**
bent	–	24.00	388.90	3167.11	tonne	**3556.01**
10 mm dia. nominal size						
straight	–	26.00	421.31	3026.27	tonne	**3447.58**
bent	–	26.00	421.31	3204.88	tonne	**3626.19**
8 mm dia. nominal size						
straight	–	28.00	451.19	3048.28	tonne	**3499.47**
bent	–	28.00	451.19	3226.88	tonne	**3678.07**
Bars; stainless steel; to EN 1.4462						
40 mm dia. nominal size						
straight	–	17.00	275.47	3955.70	tonne	**4231.17**
bent	–	21.00	335.26	4178.97	tonne	**4514.23**
32 mm dia. nominal size						
straight	–	17.00	275.47	3955.70	tonne	**4231.17**
bent	–	21.00	335.26	4178.97	tonne	**4514.23**
25 mm dia. nominal size						
straight	–	18.00	291.67	3821.77	tonne	**4113.44**
bent	–	18.00	291.67	4045.04	tonne	**4336.71**
20 mm dia. nominal size						
straight	–	20.00	324.07	3837.51	tonne	**4161.58**
bent	–	20.00	324.07	4060.76	tonne	**4384.83**
16 mm dia. nominal size						
straight	–	22.00	356.48	3859.52	tonne	**4216.00**
bent	–	22.00	356.48	4082.78	tonne	**4439.26**
12 mm dia. nominal size						
straight	–	24.00	388.90	3881.53	tonne	**4270.43**
bent	–	24.00	388.90	4060.14	tonne	**4449.04**
10 mm dia. nominal size						
straight	–	26.00	421.31	4142.56	tonne	**4563.87**
bent	–	26.00	421.31	4276.52	tonne	**4697.83**
8 mm dia. nominal size						
straight	–	28.00	451.19	4164.56	tonne	**4615.75**
bent	–	28.00	451.19	4298.52	tonne	**4749.71**
Bars; stainless steel; to EN 1.4362 (low nickel alloys or Lean Duplexes)						
40 mm dia. nominal size						
straight	–	17.00	275.47	2794.76	tonne	**3070.23**
bent	–	21.00	335.26	2973.37	tonne	**3308.63**
32 mm dia. nominal size						
straight	–	17.00	275.47	2794.76	tonne	**3070.23**
bent	–	21.00	335.26	2973.37	tonne	**3308.63**
25 mm dia. nominal size						
straight	–	18.00	291.67	2794.80	tonne	**3086.47**
bent	–	18.00	291.67	2973.40	tonne	**3265.07**
20 mm dia. nominal size						
straight	–	20.00	324.07	2810.52	tonne	**3134.59**
bent	–	20.00	324.07	2989.13	tonne	**3313.20**

11 IN SITU CONCRETE WORKS

Item	PC £	Labour hours	Labour £	Material £	Unit	Total rate £
REINFORCEMENT – cont						
Bars – cont						
16 mm dia. nominal size						
straight	–	22.00	356.48	2832.54	tonne	**3189.02**
bent	–	22.00	356.48	3011.14	tonne	**3367.62**
12 mm dia. nominal size						
straight	–	24.00	388.90	2854.54	tonne	**3243.44**
bent	–	24.00	388.90	3033.15	tonne	**3422.05**
10 mm dia. nominal size						
straight	–	26.00	421.31	2892.31	tonne	**3313.62**
bent	–	26.00	421.31	3070.92	tonne	**3492.23**
8 mm dia. nominal size						
straight	–	28.00	451.19	2914.32	tonne	**3365.51**
bent	–	28.00	451.19	3092.93	tonne	**3544.12**
Bars; stainless steel; to EN 1.4162 (low nickel alloys or Lean Duplexes)						
40 mm dia. nominal size						
straight	–	17.00	275.47	2660.81	tonne	**2936.28**
bent	–	21.00	335.26	2794.76	tonne	**3130.02**
32 mm dia. nominal size						
straight	–	17.00	275.47	2660.81	tonne	**2936.28**
bent	–	21.00	335.26	2794.76	tonne	**3130.02**
25 mm dia. nominal size						
straight	–	18.00	291.67	2660.84	tonne	**2952.51**
bent	–	18.00	291.67	2794.80	tonne	**3086.47**
20 mm dia. nominal size						
straight	–	20.00	324.07	2676.57	tonne	**3000.64**
bent	–	20.00	324.07	2810.52	tonne	**3134.59**
16 mm dia. nominal size						
straight	–	22.00	356.48	2698.58	tonne	**3055.06**
bent	–	22.00	356.48	2832.54	tonne	**3189.02**
12 mm dia. nominal size						
straight	–	24.00	388.90	2720.59	tonne	**3109.49**
bent	–	24.00	388.90	2854.54	tonne	**3243.44**
10 mm dia. nominal size						
straight	–	26.00	421.31	2758.36	tonne	**3179.67**
bent	–	26.00	421.31	2892.31	tonne	**3313.62**
8 mm dia. nominal size						
straight	–	28.00	451.19	2780.37	tonne	**3231.56**
bent	–	28.00	451.19	2914.32	tonne	**3365.51**
Fabric; to BS4483 in standard 4.8 m × 2.4 m sheets						
Ref D98 (1.54 kg/m^2)						
400 mm minimum laps	–	0.12	1.95	1.12	m^2	**3.07**
strips in one width; 600 mm width	–	0.15	2.43	1.12	m^2	**3.55**
strips in one width; 900 mm width	–	0.14	2.27	1.12	m^2	**3.39**
strips in one width; 1200 mm width	–	0.13	2.11	1.12	m^2	**3.23**

11 IN SITU CONCRETE WORKS

Item	PC £	Labour hours	Labour £	Material £	Unit	Total rate £
Ref A142 (2.22 kg/m^2)						
400 mm minimum laps	–	0.12	1.95	1.42	m^2	**3.37**
strips in one width; 600 mm width	–	0.15	2.43	1.42	m^2	**3.85**
strips in one width; 900 mm width	–	0.14	2.27	1.42	m^2	**3.69**
strips in one width; 1200 mm width	–	0.13	2.11	1.42	m^2	**3.53**
Ref A193 (3.02 kg/m^2)						
400 mm minimum laps	–	0.12	1.95	1.93	m^2	**3.88**
strips in one width; 600 mm width	–	0.15	2.43	1.93	m^2	**4.36**
strips in one width; 900 mm width	–	0.14	2.27	1.93	m^2	**4.20**
strips in one width; 1200 mm width	–	0.13	2.11	1.93	m^2	**4.04**
Ref A252 (3.95 kg/m^2)						
400 mm minimum laps	–	0.13	2.11	2.49	m^2	**4.60**
strips in one width; 600 mm width	–	0.16	2.59	2.49	m^2	**5.08**
strips in one width; 900 mm width	–	0.15	2.43	2.49	m^2	**4.92**
strips in one width; 1200 mm width	–	0.14	2.27	2.49	m^2	**4.76**
Ref A393 (6.16 kg/m^2)						
400 mm minimum laps	–	0.15	2.43	3.87	m^2	**6.30**
strips in one width; 600 mm width	–	0.18	2.92	3.87	m^2	**6.79**
strips in one width; 900 mm width	–	0.17	2.76	3.87	m^2	**6.63**
strips in one width; 1200 mm width	–	0.16	2.59	3.87	m^2	**6.46**
Ref B196 (3.05 kg/m^2)						
400 mm minimum laps	–	0.12	1.95	3.61	m^2	**5.56**
strips in one width; 600 mm width	–	0.15	2.43	3.61	m^2	**6.04**
strips in one width; 900 mm width	–	0.14	2.27	3.61	m^2	**5.88**
strips in one width; 1200 mm width	–	0.13	2.11	3.61	m^2	**5.72**
Ref B283 (3.73 kg/m^2)						
400 mm minimum laps	–	0.12	1.95	2.46	m^2	**4.41**
strips in one width; 600 mm width	–	0.15	2.43	2.46	m^2	**4.89**
strips in one width; 900 mm width	–	0.14	2.27	2.46	m^2	**4.73**
strips in one width; 1200 mm width	–	0.13	2.11	2.46	m^2	**4.57**
Ref B385 (4.53 kg/m^2)						
400 mm minimum laps	–	0.13	2.11	2.98	m^2	**5.09**
strips in one width; 600 mm width	–	0.16	2.59	2.98	m^2	**5.57**
strips in one width; 900 mm width	–	0.15	2.43	2.98	m^2	**5.41**
strips in one width; 1200 mm width	–	0.14	2.27	2.98	m^2	**5.25**
Ref B503 (5.93 kg/m^2)						
400 mm minimum laps	–	0.15	2.43	3.85	m^2	**6.28**
strips in one width; 600 mm width	–	0.18	2.92	3.85	m^2	**6.77**
strips in one width; 900 mm width	–	0.17	2.76	3.85	m^2	**6.61**
strips in one width; 1200 mm width	–	0.16	2.59	3.85	m^2	**6.44**
Ref B785 (8.14 kg/m^2)						
400 mm minimum laps	–	0.17	2.76	5.29	m^2	**8.05**
strips in one width; 600 mm width	–	0.20	3.24	5.29	m^2	**8.53**
strips in one width; 900 mm width	–	0.19	3.08	5.29	m^2	**8.37**
strips in one width; 1200 mm width	–	0.18	2.92	5.29	m^2	**8.21**
Ref B1131 (10.90 kg/m^2)						
400 mm minimum laps	–	0.18	2.92	7.08	m^2	**10.00**
strips in one width; 600 mm width	–	0.24	3.88	7.08	m^2	**10.96**
strips in one width; 900 mm width	–	0.22	3.57	7.08	m^2	**10.65**
strips in one width; 1200 mm width	–	0.20	3.24	7.08	m^2	**10.32**
Ref D49 (0.77 kg/m^2)						
100 mm minimum laps; bent	–	0.24	3.88	1.53	m^2	**5.41**

11 IN SITU CONCRETE WORKS

Item	PC £	Labour hours	Labour £	Material £	Unit	Total rate £
DESIGNED JOINTS						
Formed; Fosroc impregnated fibreboard joint filler or other equal						
Width not exceeding 150 mm						
12.50 mm thick	–	0.14	2.27	1.44	m	**3.71**
20 mm thick	–	0.19	3.08	3.82	m	**6.90**
25 mm thick	–	0.23	3.73	2.48	m	**6.21**
Width 150–300 mm						
12.50 mm thick	–	0.23	3.73	2.11	m	**5.84**
20 mm thick	–	0.23	3.73	3.78	m	**7.51**
25 mm thick	–	0.23	3.73	4.51	m	**8.24**
Width 300–450 mm						
12.50 mm thick	–	0.28	4.54	2.95	m	**7.49**
20 mm thick	–	0.28	4.54	5.24	m	**9.78**
25 mm thick	–	0.28	4.54	6.24	m	**10.78**
Formed; Grace Servicised Kork-pak waterproof bonded cork joint filler board or other equal						
Width not exceeding 150 mm						
10 mm thick	–	0.14	2.27	2.81	m	**5.08**
13 mm thick	–	0.14	2.27	2.86	m	**5.13**
19 mm thick	–	0.14	2.27	3.71	m	**5.98**
25 mm thick	–	0.14	2.27	4.24	m	**6.51**
Width 150–300 mm						
10 mm thick	–	0.19	3.08	5.07	m	**8.15**
13 mm thick	–	0.19	3.08	5.16	m	**8.24**
19 mm thick	–	0.19	3.08	6.88	m	**9.96**
25 mm thick	–	0.19	3.08	7.94	m	**11.02**
Width 300–450 mm						
10 mm thick	–	0.23	3.73	7.78	m	**11.51**
13 mm thick	–	0.23	3.73	7.91	m	**11.64**
19 mm thick	–	0.23	3.73	10.49	m	**14.22**
25 mm thick	–	0.23	3.73	12.08	m	**15.81**
Sealants; Fosroc Pliastic 77 hot poured rubberized bituminous compound or other equal						
Width 10 mm						
25 mm depth	–	0.17	2.76	0.74	m	**3.50**
Width 12.50 mm						
25 mm depth	–	0.18	2.91	0.91	m	**3.82**
Width 20 mm						
25 mm depth	–	0.19	3.08	1.48	m	**4.56**
Width 25 mm						
25 mm depth	–	0.20	3.24	1.81	m	**5.05**

11 IN SITU CONCRETE WORKS

Item	PC £	Labour hours	Labour £	Material £	Unit	Total rate £
Sealants; Fosroc Thioflex 600 gun grade two part polysulphide or other equal						
Width 10 mm						
25 mm depth	–	0.05	0.81	3.54	m	**4.35**
Width 12.50 mm						
25 mm depth	–	0.06	0.97	4.42	m	**5.39**
Width 20 mm						
25 mm depth	–	0.07	1.14	7.06	m	**8.20**
Width 25 mm						
25 mm depth	–	0.08	1.29	8.84	m	**10.13**
Sealants; Grace Servicised Paraseal polysulphide compound or other equal; priming with Grace Servicised Primer P						
Width 10 mm						
25 mm depth	–	0.19	2.60	3.13	m	**5.73**
Width 13 mm						
25 mm depth	–	0.19	2.60	4.00	m	**6.60**
Width 19 mm						
25 mm depth	–	0.23	3.15	5.76	m	**8.91**
Width 25 mm						
25 mm depth	–	0.23	3.15	7.51	m	**10.66**
Waterstops; Grace Servicised or other equal						
Hydrophilic strip water stop; lapped joints; cast into concrete						
50 × 20 mm Adcor 500S	5.50	0.30	4.11	7.35	m	**11.46**
Servitite Internal 10 mm thick PVC water stop; flat dumbbell type; heat welded joints; cast into concrete						
Servitite 150; 150 mm wide	–	0.23	3.73	9.85	m	**13.58**
flat angle	–	0.28	4.54	23.26	nr	**27.80**
vertical angle	–	0.28	4.54	23.11	nr	**27.65**
flat three way intersection	–	0.37	6.00	33.75	nr	**39.75**
vertical three way intersection	–	0.37	6.00	38.16	nr	**44.16**
four way intersection	–	0.46	7.45	41.98	nr	**49.43**
Servitite 230; 230 mm wide	–	0.23	3.73	14.13	m	**17.86**
flat angle	–	0.28	4.54	28.52	nr	**33.06**
vertical angle	–	0.28	4.54	33.62	nr	**38.16**
flat three way intersection	–	0.37	6.00	41.67	nr	**47.67**
vertical three way intersection	–	0.37	6.00	71.68	nr	**77.68**
four way intersection	–	0.46	7.45	52.51	nr	**59.96**
Servitite AT200; 200 mm wide	–	0.23	3.73	15.60	m	**19.33**
flat angle	–	0.28	4.54	28.08	nr	**32.62**
vertical angle	–	0.28	4.54	30.37	nr	**34.91**
flat three way intersection	–	0.37	6.00	48.47	nr	**54.47**
vertical three way intersection	–	0.37	6.00	38.33	nr	**44.33**
four way intersection	–	0.46	7.45	57.70	nr	**65.15**
Servitite K305; 305 mm wide	–	0.28	4.54	22.95	m	**27.49**
flat angle	–	0.32	5.19	48.80	nr	**53.99**
vertical angle	–	0.32	5.19	52.52	nr	**57.71**
flat three way intersection	–	0.42	6.81	69.68	nr	**76.49**
vertical three way intersection	–	0.42	6.81	81.31	nr	**88.12**
four way intersection	–	0.51	8.26	94.35	nr	**102.61**

11 IN SITU CONCRETE WORKS

Item	PC £	Labour hours	Labour £	Material £	Unit	Total rate £
DESIGNED JOINTS – cont						
Waterstops – cont						
Serviseal External PVC water stop; PVC water stop; centre bulb type; heat welded joints; cast into concrete						
Serviseal 195; 195 mm wide	–	0.23	3.73	6.57	m	**10.30**
flat angle	–	0.28	4.54	14.61	nr	**19.15**
vertical angle	–	0.28	4.54	24.09	nr	**28.63**
flat three way intersection	–	0.37	6.00	24.50	nr	**30.50**
four way intersection	–	0.46	7.45	36.20	nr	**43.65**
Serviseal 240; 240 mm wide	–	0.23	3.73	8.19	m	**11.92**
flat angle	–	0.28	4.54	16.89	nr	**21.43**
vertical angle	–	0.28	4.54	25.73	nr	**30.27**
flat three way intersection	–	0.37	6.00	27.95	nr	**33.95**
four way intersection	–	0.46	7.45	40.79	nr	**48.24**
Serviseal AT240; 240 mm wide	–	0.23	3.73	18.30	m	**22.03**
flat angle	–	0.28	4.54	27.68	nr	**32.22**
vertical angle	–	0.28	4.54	26.51	nr	**31.05**
flat three way intersection	–	0.37	6.00	42.36	nr	**48.36**
four way intersection	–	0.46	7.45	61.83	nr	**69.28**
Serviseal K320; 320 mm wide	–	0.28	4.54	10.77	m	**15.31**
flat angle	–	0.32	5.19	35.91	nr	**41.10**
vertical angle	–	0.32	5.19	19.39	nr	**24.58**
flat three way intersection	–	0.42	6.81	52.79	nr	**59.60**
four way intersection	–	0.51	8.26	66.42	nr	**74.68**
ACCESSORIES CAST IN						
Foundation bolt boxes						
Temporary plywood; for group of 4 nr bolts						
75 × 75 × 150 mm	–	0.42	6.81	0.79	nr	**7.60**
75 × 75 × 250 mm	–	0.42	6.81	0.95	nr	**7.76**
Expanded metal; Expamet Building Products Ltd or other equal and approved						
75 mm dia. × 150 mm long	–	0.28	4.54	0.99	nr	**5.53**
75 mm dia. × 300 mm long	–	0.28	4.54	1.47	nr	**6.01**
100 mm dia. × 450 mm long	–	0.28	4.54	2.10	nr	**6.64**
Foundation bolts and nuts						
Black hexagon						
10 mm dia. × 100 mm long	–	0.23	3.73	0.40	nr	**4.13**
12 mm dia. × 120 mm long	–	0.23	3.73	0.62	nr	**4.35**
16 mm dia. × 160 mm long	–	0.28	4.54	1.70	nr	**6.24**
20 mm dia. × 180 mm long	–	0.28	4.54	1.99	nr	**6.53**

11 IN SITU CONCRETE WORKS

Item	PC £	Labour hours	Labour £	Material £	Unit	Total rate £
Masonry slots						
Stainless steel; dovetail slots; 1.20 mm thick; 18G						
1000 mm long	–	0.25	4.05	6.28	m	**10.33**
100 mm long	–	0.07	1.14	0.28	nr	**1.42**
Stainless steel; metal insert slots; Halfen Ltd Ribslot or other equal and approved; 2.50 mm thick; end caps and foam filling						
41 × 41 mm; ref P3270	–	0.37	6.00	8.11	m	**14.11**
41 × 41 × 100 mm; ref P3250	–	0.09	1.46	1.19	nr	**2.65**
41 × 41 × 150 mm; ref P3251	–	0.09	1.46	0.84	nr	**2.30**
Cramps						
Stainless steel; once bent; one end shot fired into concrete; other end fanged and built into brickwork joint						
200 mm girth	–	0.14	2.49	1.08	nr	**3.57**
Column guards						
White nylon coated steel; Rigifix or other equal and approved; Huntley and Sparks Ltd; plugging; screwing to concrete; 1.50 mm thick						
75 × 75 × 1000 mm	–	0.74	11.99	14.62	nr	**26.61**
Galvanized steel; Rigifix or other equal and approved; Huntley and Sparks Ltd; 3 mm thick						
75 × 75 × 1000 mm	–	0.56	9.07	10.33	nr	**19.40**
Galvanized steel; Rigifix or other equal and approved; Huntley and Sparks Ltd; 4.50 mm thick						
75 × 75 × 1000 mm	–	0.56	9.07	13.90	nr	**22.97**
Stainless steel; HKW or other equal and approved; Halfen Ltd; 5 mm thick						
50 × 50 × 1200 mm	–	0.93	15.07	73.65	nr	**88.72**
50 × 50 × 2000 mm	–	1.11	17.98	121.74	nr	**139.72**
Channels						
Stainless steel; Halfen Ltd or other equal and approved						
ref 38/17/HTA	–	0.32	5.19	42.65	m	**47.84**
ref 41/22/HZA; 80 mm long; including T headed bolts and plate washers	–	0.09	1.46	27.56	nr	**29.02**
Channel ties						
Stainless steel; Halfen Ltd or other equal and approved						
ref HTS – B12; 150 mm projection; including insulation retainer	–	0.03	0.52	0.54	nr	**1.06**
ref HTS – B12; 200 mm projection; including insulation retainer	–	0.03	0.52	0.64	nr	**1.16**

12 PRECAST/COMPOSITE CONCRETE

Item	PC £	Labour hours	Labour £	Material £	Unit	Total rate £
PRECAST/COMPOSITE CONCRETE WORK						
Prestressed precast concrete structural suspended floors; Bison Hollowcore or other equal; supplied and fixed on hard level bearings, to areas of 500 m² per site visit; top surface screeding and ceiling finishes by others						
Floors to dwellings, offices, car parks, shop retail floors, hospitals, school teaching rooms, staff rooms and the like; superimposed load of 5.00 kN/m²						
floor spans up to 3.00 m; 1200 mm × 150 mm	–	–	–	–	m²	48.92
floor spans 3.00 m–6.00 m; 1200 mm × 150 mm	–	–	–	–	m²	50.11
floor spans 6.00 m–7.50 m; 1200 mm × 200 mm	–	–	–	–	m²	53.56
floor spans 7.50 m–9.50 m; 1200 mm × 250 mm	–	–	–	–	m²	57.00
floor spans 9.50 m–12.00 m; 1200 mm × 300 mm	–	–	–	–	m²	58.90
floor spans 12.00 m–12.50 m; 1200 mm × 350 mm	–	–	–	–	m²	60.80
floor spans 12.50 m–14.00 m; 1200 mm × 400 mm	–	–	–	–	m²	67.45
floor spans 14.00 m–15.00 m; 1200 mm × 450 mm	–	–	–	–	m²	68.40
Floors to shop stockrooms, light warehousing, schools, churches or similar places of assembly, light factory accommodation, laboratories and the like; superimposed load of 8.50 kN/m²						
floor spans up to 3.00 m; 1200 mm × 150 mm	–	–	–	–	m²	47.50
floor spans 3.00 m–6.00 m; 1200 mm × 200 mm	–	–	–	–	m²	49.40
floor spans 6.00 m–7.50 m; 1200 mm × 250 mm	–	–	–	–	m²	57.00
Floors to heavy warehousing, factories, stores and the like; superimposed load of 12.50 kN/m²						
floor spans up to 3.00 m; 1200 mm × 150 mm	–	–	–	–	m²	50.35
floor spans 3.00 m–6.00 m; 1200 mm × 250 mm	–	–	–	–	m²	57.00
Prestressed precast concrete staircase, supplied and fixed in conjunction with Bison Hollowcore flooring system or similar; comprising 2 nr 1100 mm wide flights with 7 nr 275 mm treads, 8 nr 185 mm risers and 150 mm waist; 1 nr						
2200 mm × 1400 mm × 150 mm half landing and 1 nr top landing						
3.00 m storey height	–	–	–	–	nr	2185.00
Prestressed precast concrete beam and block floor; cement and sand grout brushed between beams and blocks; 440 × 215 × 100 mm concrete block infill						
Beam and block flooring at ground level						
beam and block flooring at ground level 150 mm thick beams; up to 3.30 m span	–	–	–	–	m²	19.50
beam and block flooring at ground level; 225 mm thick beams; up to 4.20 m span	–	–	–	–	m²	23.40

12 PRECAST/COMPOSITE CONCRETE

Item	PC £	Labour hours	Labour £	Material £	Unit	Total rate £
Composite floor comprising reinforced in situ ready-mixed concrete 30.00 N/mm^2; on and including steel deck permanent shutting; complete with A142 anti-crack mesh; NOTE temporary props may be required, but have not been included in the following rates						
0.9 mm thick re-entrant type deck; 60 mm deep						
150 mm thick suspended slab						
1.50 m–3.00 m high to soffit	–	0.91	13.09	32.08	m^2	**45.17**
3.00 m–4.50 m high to soffit	–	0.94	13.49	32.08	m^2	**45.57**
4.50 m–6.00 m high to soffit	–	0.98	14.12	32.08	m^2	**46.20**
200 mm thick suspended slab						
1.50 m–3.00 m high to soffit	–	0.95	13.63	36.36	m^2	**49.99**
3.00 m–4.50 m high to soffit	–	0.97	14.04	36.36	m^2	**50.40**
4.50 m–6.00 m high to soffit	–	1.00	14.39	36.36	m^2	**50.75**
1.2 mm thick re-entrant type deck						
150 mm thick suspended slab						
1.50 m–3.00 m high to soffit	–	0.91	13.09	34.22	m^2	**47.31**
3.00 m–4.50 m high to soffit	–	0.94	13.49	34.22	m^2	**47.71**
4.50 m–6.00 m high to soffit	–	0.98	14.12	34.22	m^2	**48.34**
200 mm thick suspended slab						
1.50 m–3.00 m high to soffit	–	0.95	13.63	38.51	m^2	**52.14**
3.00 m–4.50 m high to soffit	–	0.97	14.04	38.51	m^2	**52.55**
4.50 m–6.00 m high to soffit	–	1.00	14.39	38.51	m^2	**52.90**
0.9 mm thick trapezoidal deck 60 mm deep						
150 mm thick suspended slab						
1.50 m–3.00 m high to soffit	–	0.91	13.09	27.46	m^2	**40.55**
3.00 m–4.50 m high to soffit	–	0.94	13.49	27.46	m^2	**40.95**
4.50 m–6.00 m high to soffit	–	0.98	14.12	27.46	m^2	**41.58**
200 mm thick suspended slab						
1.50 m–3.00 m high to soffit	–	0.95	13.63	31.74	m^2	**45.37**
3.00 m–4.50 m high to soffit	–	0.97	14.04	31.74	m^2	**45.78**
4.50 m–6.00 m high to soffit	–	1.00	14.39	31.74	m^2	**46.13**
0.9 mm thick trapezoidal deck 80 mm deep						
150 mm thick suspended slab						
1.50 m–3.00 m high to soffit	–	0.91	13.09	30.03	m^2	**43.12**
3.00 m–4.50 m high to soffit	–	0.94	13.49	30.03	m^2	**43.52**
4.50 m–6.00 m high to soffit	–	0.98	14.12	30.03	m^2	**44.15**
200 mm thick suspended slab						
1.50 m–3.00 m high to soffit	–	0.95	13.63	34.32	m^2	**47.95**
3.00 m–4.50 m high to soffit	–	0.97	14.04	34.32	m^2	**48.36**
4.50 m–6.00 m high to soffit	–	1.00	14.39	34.32	m^2	**48.71**

12 PRECAST/COMPOSITE CONCRETE

Item	PC £	Labour hours	Labour £	Material £	Unit	Total rate £
PRECAST/COMPOSITE CONCRETE WORK – cont						
1.2 mm thick trapezoidal deck 80 mm deep						
150 mm thick suspended slab						
1.50 m–3.00 m high to soffit	–	0.91	13.09	29.55	m²	**42.64**
3.00 m–4.50 m high to soffit	–	0.94	13.49	29.55	m²	**43.04**
4.50 m–6.00 m high to soffit	–	0.98	14.12	29.55	m²	**43.67**
200 mm thick suspended slab						
1.50 m–3.00 m high to soffit	–	0.95	13.63	33.84	m²	**47.47**
3.00 m–4.50 m high to soffit	–	0.97	14.04	33.84	m²	**47.88**
4.50 m–6.00 m high to soffit	–	1.00	14.39	33.84	m²	**48.23**
Soffits of coffered or troughed slabs; basic finish						
Cordek Correx trough mould or other equal and approved; 300 mm deep; ribs of mould at 600 mm centres and cross ribs at centres of bay; slab thickness 300–400 mm						
horizontal; height to soffit 1.50–3.00 m	–	2.08	33.68	6.88	m²	**40.56**
horizontal; height to soffit 3.00–4.50 m	–	2.17	35.14	7.00	m²	**42.14**
horizontal; height to soffit 4.50–6.00 m	–	2.25	36.45	7.08	m²	**43.53**

13 PRECAST CONCRETE

Item	PC £	Labour hours	Labour £	Material £	Unit	Total rate £
PRECAST CONCRETE GOODS						
Contractor designed precast concrete staircases and landings; including all associated steel supports and fixing in position						
Straight staircases; 280 mm treads; 170 mm undercut risers						
1200 mm wide; 2750 mm rise	–	–	–	–	nr	**1318.13**
1200 mm wide; 3750 mm rise	–	–	–	–	nr	**1757.50**
Dogleg staircases						
1200 mm wide; one full width half landing; 2750 mm rise	–	–	–	–	nr	**2021.13**
1200 mm wide; one full width half landing; 3750 mm rise	–	–	–	–	nr	**2636.25**
extra for 200 mm concrete landing support walls	–	–	–	–	nr	**615.13**
1800 mm wide; one full width half landing; 2750 mm rise	–	–	–	–	nr	**2855.94**
1800 mm wide; one full width half landing; 3750 mm rise	–	–	–	–	nr	**3734.69**
extra for 200 mm concrete landing support walls	–	–	–	–	nr	**949.05**
Precast concrete sill, lintels, copings						
Lintels; plate; prestressed bedded						
100 mm × 70 mm × 600 mm long	5.64	0.37	6.41	5.80	nr	**12.21**
100 mm × 70 mm × 900 mm long	8.42	0.37	6.41	8.65	nr	**15.06**
100 mm × 70 mm × 1100 mm long	10.32	0.37	6.41	10.60	nr	**17.01**
100 mm × 70 mm × 1200 mm long	11.24	0.37	6.41	11.54	nr	**17.95**
100 mm × 70 mm × 1500 mm long	14.06	0.46	7.95	14.44	nr	**22.39**
100 mm × 70 mm × 1800 mm long	16.86	0.46	7.95	17.31	nr	**25.26**
100 mm × 70 mm × 2100 mm long	19.66	0.56	9.69	20.20	nr	**29.89**
140 mm × 70 mm × 1200 mm long	16.53	0.46	7.95	16.97	nr	**24.92**
140 mm × 70 mm × 1500 mm long	20.66	0.56	9.69	21.22	nr	**30.91**
Lintels; rectangular; reinforced with mild steel bars; bedded						
100 mm × 145 mm × 900 mm long	3.71	0.56	9.69	3.82	nr	**13.51**
100 mm × 145 mm × 1050 mm long	4.34	0.56	9.69	4.47	nr	**14.16**
100 mm × 145 mm × 1200 mm long	4.95	0.56	9.69	5.09	nr	**14.78**
225 mm × 145 mm × 1200 mm long	19.22	0.74	12.80	19.74	nr	**32.54**
225 mm × 225 mm × 1800 mm long	28.74	1.39	24.05	29.51	nr	**53.56**
Lintels; boot; reinforced with mild steel bars; bedded						
250 mm × 225 mm × 1200 mm long	21.29	1.11	19.21	21.87	nr	**41.08**
275 mm × 225 mm × 1800 mm long	35.13	1.67	28.89	36.06	nr	**64.95**
Padstones						
100 mm × 200 mm × 440 mm	19.05	0.28	4.85	19.55	nr	**24.40**
150 mm × 215 mm × 440 mm	19.04	0.37	6.41	19.55	nr	**25.96**
150 mm × 225 mm × 225 mm	14.96	0.56	9.69	15.47	nr	**25.16**

13 PRECAST CONCRETE

Item	PC £	Labour hours	Labour £	Material £	Unit	Total rate £
PRECAST CONCRETE GOODS – cont						
Precast concrete sill, lintels, copings – cont						
Copings; once weathered; once throated; bedded and pointed						
152 mm × 76 mm	5.59	0.65	11.24	5.80	m	**17.04**
178 mm × 64 mm	6.17	0.65	11.24	6.40	m	**17.64**
305 mm × 76 mm	10.42	0.74	12.80	10.84	m	**23.64**
extra for fair ends	–	–	–	4.09	nr	**4.09**
extra for angles	–	–	–	4.65	nr	**4.65**
Copings; twice weathered; twice throated; bedded and pointed						
152 mm × 76 mm	5.59	0.65	11.24	5.80	m	**17.04**
178 mm × 64 mm	6.13	0.65	11.24	6.36	m	**17.60**
305 mm × 76 mm	10.42	0.74	12.80	10.84	m	**23.64**
extra for fair ends	–	–	–	4.09	nr	**4.09**
extra for angles	–	–	–	4.65	nr	**4.65**
Sills; splayed top edge, stooled ends; bedded and pointed						
200 mm × 90 mm	25.13	0.75	12.98	25.82	m	**38.80**
200 mm × 90 mm; slip sill	28.18	0.75	12.98	28.95	m	**41.93**

14 MASONRY

Item	PC £	Labour hours	Labour £	Material £	Unit	Total rate £
BRICK WALLING						
Basic mortar prices						
Mortar materials only						
cement	–	–	–	108.91	tonne	**108.91**
sand	–	–	–	17.15	tonne	**17.15**
lime	–	–	–	147.60	tonne	**147.60**
white cement	–	–	–	184.50	tonne	**184.50**
Cemplas Super mortar plasticizer	–	–	–	6.23	litre	**6.23**
Coloured mortar materials; (excluding cement)						
light	–	–	–	48.89	tonne	**48.89**
medium	–	–	–	50.74	tonne	**50.74**
dark	–	–	–	60.88	tonne	**60.88**
extra dark	–	–	–	60.88	tonne	**60.88**
SUPPLY ONLY BRICK PRICES						
Alternative facing brick prices; prime cost £ per 1000; material rates only						
Ibstock Brick Ltd						
Aldridge Brown Blend	–	–	–	336.00	1000	**336.00**
Aldridge Leicester Anglican Red Rustic	–	–	–	275.80	1000	**275.80**
Ashdown Cottage Mixture	–	–	–	278.60	1000	**278.60**
Ashdown Crowborough Multi	–	–	–	359.80	1000	**359.80**
Ashdown Pevensey Multi	–	–	–	336.70	1000	**336.70**
Cattybrook Bristol Gold	–	–	–	281.40	1000	**281.40**
Chailey Stock	–	–	–	336.70	1000	**336.70**
Dorking Multi	–	–	–	272.30	1000	**272.30**
Funton Second Hard Stock	–	–	–	371.00	1000	**371.00**
Holbook Smooth Red	–	–	–	294.00	1000	**294.00**
Leicester Red Stock	–	–	–	386.40	1000	**386.40**
Roughdales Red Multi Rustic	–	–	–	273.70	1000	**273.70**
Roughdales Trafford Multi Rustic	–	–	–	301.70	1000	**301.70**
Stourbridge Himley Mixed Russet	–	–	–	410.90	1000	**410.90**
Stourbridge Kenilworth Multi	–	–	–	262.50	1000	**262.50**
Stourbridge Pennine Pastone	–	–	–	319.90	1000	**319.90**
Strattford Red Rustic	–	–	–	257.60	1000	**257.60**
Swanage Handmade Restoration	–	–	–	592.20	1000	**592.20**
Tonbridge Handmade Multi	–	–	–	576.80	1000	**576.80**
Hanson Brick Limited, London brick						
Brecken Grey	–	–	–	277.50	1000	**277.50**
Brindle	–	–	–	331.50	1000	**331.50**
Chiltern	–	–	–	330.00	1000	**330.00**
Claydon Red Multi	–	–	–	288.80	1000	**288.80**
Cotswold	–	–	–	323.20	1000	**323.20**
Dapple Light	–	–	–	360.00	1000	**360.00**
Georgian	–	–	–	285.00	1000	**285.00**
Golden Buff	–	–	–	375.00	1000	**375.00**
Heather	–	–	–	330.00	1000	**330.00**
Hereward Light	–	–	–	296.30	1000	**296.30**
Honey Buff	–	–	–	273.70	1000	**273.70**

14 MASONRY

Item	PC £	Labour hours	Labour £	Material £	Unit	Total rate £
BRICK WALLING – cont						
Alternative facing brick prices – cont						
Hanson Brick Limited, London brick – cont						
Ironstone	–	–	–	285.00	1000	285.00
Milton Buff	–	–	–	300.00	1000	300.00
Regency	–	–	–	311.30	1000	311.30
Rustic	–	–	–	372.00	1000	372.00
Sandfaced	–	–	–	317.20	1000	317.20
Saxon Gold	–	–	–	316.50	1000	316.50
Sunset Red	–	–	–	292.50	1000	292.50
Tudor	–	–	–	325.50	1000	325.50
Windsor	–	–	–	286.50	1000	286.50
Selected Regrades	–	–	–	180.00	1000	180.00
Sherbourne Red Pavers	–	–	–	16.50	m²	16.50
Coxmoor Rose Multi Pavers	–	–	–	16.50	m²	16.50
SUPPLY AND LAY PRICES						
Common bricks						
In gauged mortar (1:1:6); prime cost for bricks	240.00	–	–	240.00	1000	240.00
Walls						
half brick thick	14.40	0.84	14.48	17.35	m²	31.83
half brick thick; building against other work; concrete	14.40	0.92	15.88	18.59	m²	34.47
half brick thick; building overhand	14.40	1.04	18.06	17.35	m²	35.41
half brick thick; curved; 6.00 m radii	14.40	1.08	18.69	17.35	m²	36.04
half brick thick; curved; 1.50 m radii	16.80	1.41	24.45	19.94	m²	44.39
one brick thick	28.80	1.41	24.45	34.71	m²	59.16
one brick thick; curved; 6.00 m radii	31.20	1.84	31.76	37.29	m²	69.05
one brick thick; curved; 1.50 m radii	31.20	2.29	39.55	37.91	m²	77.46
one and a half brick thick	43.20	1.92	33.17	52.06	m²	85.23
one and a half brick thick; battering	43.20	2.21	38.15	52.06	m²	90.21
two brick thick	57.60	2.33	40.33	69.41	m²	109.74
two brick thick; battering	57.60	2.75	47.49	69.41	m²	116.90
337 average thick; tapering, one side	43.20	2.41	41.73	52.06	m²	93.79
450 average thick; tapering, one side	57.60	3.12	54.03	69.41	m²	123.44
337 average thick; tapering, both sides	43.20	2.79	48.27	52.06	m²	100.33
450 average thick; tapering, both sides	57.60	3.50	60.57	70.04	m²	130.61
facework one side, half brick thick	14.40	0.92	15.88	17.35	m²	33.23
facework one side, one brick thick	28.80	1.50	26.00	34.71	m²	60.71
facework one side, one and a half brick thick	43.20	2.00	34.56	52.06	m²	86.62
facework one side, two brick thick	57.60	2.41	41.73	69.41	m²	111.14
facework both sides, half brick thick	14.40	1.00	17.28	17.35	m²	34.63
facework both sides, one brick thick	28.80	1.58	27.41	34.71	m²	62.12
facework both sides, one and a half brick thick	43.20	2.08	35.97	52.06	m²	88.03
facework both sides, two brick thick	57.60	2.50	43.29	69.41	m²	112.70
Isolated piers						
one brick thick	28.80	2.36	40.83	34.71	m²	75.54
two brick thick	57.60	3.70	64.01	70.04	m²	134.05
three brick thick	86.40	4.67	80.80	105.36	m²	186.16

14 MASONRY

Item	PC £	Labour hours	Labour £	Material £	Unit	Total rate £
Isolated casings						
half brick thick	14.40	1.20	20.76	17.35	m²	**38.11**
one brick thick	28.80	2.04	35.29	34.71	m²	**70.00**
Chimney stacks						
one brick thick	28.80	2.36	40.83	34.71	m²	**75.54**
two brick thick	57.60	3.70	64.01	70.04	m²	**134.05**
three brick thick	86.40	4.67	80.80	105.36	m²	**186.16**
Projections						
225 mm width; 112 mm depth; vertical	3.20	0.28	4.85	3.66	m	**8.51**
225 mm width; 225 mm depth; vertical	6.40	0.56	9.69	7.32	m	**17.01**
337 mm width; 225 mm depth; vertical	9.60	0.83	14.36	10.98	m	**25.34**
440 mm width; 225 mm depth; vertical	12.80	0.93	16.09	14.65	m	**30.74**
Closing cavities						
width of cavity 50 mm, closing with common brickwork half brick thick; vertical	–	0.28	4.85	0.88	m	**5.73**
width of cavity 50 mm, closing with common brickwork half brick thick; horizontal	–	0.28	4.85	2.72	m	**7.57**
width of cavity 50 mm, closing with common brickwork half brick thick; including damp-proof course; vertical	–	0.37	6.41	1.62	m	**8.03**
width of cavity 50 mm, closing with common brickwork half brick thick; including damp-proof course; horizontal	–	0.32	5.54	3.45	m	**8.99**
width of cavity 75 mm, closing with common brickwork half brick thick; vertical	–	0.28	4.85	1.29	m	**6.14**
width of cavity 75 mm, closing with common brickwork half brick thick; horizontal	–	0.28	4.85	4.01	m	**8.86**
width of cavity 75 mm, closing with common brickwork half brick thick; including damp-proof course; vertical	–	0.37	6.41	2.03	m	**8.44**
width of cavity 75 mm, closing with common brickwork half brick thick; including damp-proof course; horizontal	–	0.32	5.54	4.75	m	**10.29**
Bonding to existing						
half brick thick	–	0.28	4.85	0.97	m	**5.82**
one brick thick	–	0.42	7.27	1.95	m	**9.22**
one and a half brick thick	–	0.65	11.24	2.92	m	**14.16**
two brick thick	–	0.88	15.22	3.90	m	**19.12**
Arches						
height on face 102 mm, width of exposed soffit 102 mm, shape of arch – segmental, one ring	1.60	1.57	24.56	4.91	m	**29.47**
height on face 102 mm, width of exposed soffit 215 mm, shape of arch – segmental, one ring	3.20	2.04	32.70	6.71	m	**39.41**
height on face 102 mm, width of exposed soffit 102 mm, shape of arch – semi-circular, one ring	1.60	1.99	31.83	4.91	m	**36.74**
height on face 102 mm, width of exposed soffit 215 mm, shape of arch – semi-circular, one ring	3.20	2.50	40.65	6.71	m	**47.36**
height on face 215 mm, width of exposed soffit 102 mm, shape of arch – segmental, two ring	3.20	1.99	31.83	6.80	m	**38.63**
height on face 215 mm, width of exposed soffit 215 mm, shape of arch – segmental, two ring	6.40	2.45	39.79	10.49	m	**50.28**

14 MASONRY

Item	PC £	Labour hours	Labour £	Material £	Unit	Total rate £
BRICK WALLING – cont						
Common bricks – cont						
Arches – cont						
height on face 215 mm, width of exposed soffit 102 mm, shape of arch – semi-circular, two ring	3.20	2.68	43.77	6.80	m	**50.57**
height on face 215 mm, width of exposed soffit 215 mm, shape of arch – semi-circular, two ring	6.40	3.05	50.16	10.49	m	**60.65**
ADD or DEDUCT to walls for variation of £10.00/ 1000 in prime cost of common bricks						
half brick thick	–	–	–	0.62	m²	**0.62**
one brick thick	–	–	–	1.23	m²	**1.23**
one and a half brick thick	–	–	–	1.85	m²	**1.85**
two brick thick	–	–	–	2.46	m²	**2.46**
Class B engineering bricks						
In cement: mortar (1:3); prime cost £ per 1000	375.00	–	–	375.00	1000	**375.00**
Walls						
half brick thick	16.74	0.92	15.88	20.09	m²	**35.97**
one brick thick	33.48	1.50	26.00	40.18	m²	**66.18**
one brick thick; building against other work	33.48	1.79	30.99	42.25	m²	**73.24**
one brick thick; curved; 6.00 m radii	33.48	2.00	34.56	40.18	m²	**74.74**
one and a half brick thick	50.22	2.00	34.56	60.27	m²	**94.83**
one and a half brick thick; building against other work	50.22	2.41	41.73	60.27	m²	**102.00**
two brick thick	66.96	2.50	43.29	80.36	m²	**123.65**
337 mm thick; tapering, one side	50.22	2.58	44.69	60.27	m²	**104.96**
450 mm thick; tapering, one side	66.96	3.33	57.62	80.36	m²	**137.98**
337 mm thick; tapering, both sides	50.22	3.00	51.85	60.27	m²	**112.12**
450 mm thick; tapering, both sides	66.96	3.79	65.55	81.05	m²	**146.60**
facework one side, half brick thick	16.74	1.00	17.28	20.09	m²	**37.37**
facework one side, one brick thick	33.48	1.58	27.41	40.18	m²	**67.59**
facework one side, one and a half brick thick	50.22	2.08	35.97	60.27	m²	**96.24**
facework one side, two brick thick	66.96	2.58	44.69	80.36	m²	**125.05**
facework both sides, half brick thick	16.74	1.08	18.69	20.09	m²	**38.78**
facework both sides, one brick thick	33.48	1.67	28.80	40.18	m²	**68.98**
facework both sides, one and a half brick thick	50.22	2.17	37.53	60.27	m²	**97.80**
facework both sides, two brick thick	66.96	2.66	46.09	80.36	m²	**126.45**
Isolated piers						
one brick thick	33.48	2.59	44.81	40.18	m²	**84.99**
two brick thick	66.96	4.07	70.42	81.05	m²	**151.47**
three brick thick	100.44	5.00	86.51	121.91	m²	**208.42**
Isolated casings						
half brick thick	16.74	1.30	22.49	20.09	m²	**42.58**
one brick thick	33.48	2.22	38.41	40.18	m²	**78.59**
Projections						
225 mm width; 112 mm depth; vertical	3.72	0.32	5.54	4.24	m	**9.78**
225 mm width; 225 mm depth; vertical	7.44	0.60	10.38	8.49	m	**18.87**
337 mm width; 225 mm depth; vertical	11.16	0.88	15.22	12.74	m	**27.96**
440 mm width; 225 mm depth; vertical	14.88	1.02	17.65	16.98	m	**34.63**

14 MASONRY

Item	PC £	Labour hours	Labour £	Material £	Unit	Total rate £
Bonding to existing						
half brick thick	–	0.32	5.54	1.11	m	**6.65**
one brick thick	–	0.46	7.95	2.21	m	**10.16**
one and a half brick thick	–	0.65	11.24	3.31	m	**14.55**
two brick thick	–	0.97	16.78	4.42	m	**21.20**
ADD or DEDUCT to walls for variation of £10.00/ 1000 in prime cost of bricks						
half brick thick	–	–	–	0.62	m²	**0.62**
one brick thick	–	–	–	1.23	m²	**1.23**
one and a half brick thick	–	–	–	1.85	m²	**1.85**
two brick thick	–	–	–	2.46	m²	**2.46**
Facing bricks; machine-made facings; in gauged mortar (1:1:6)						
In gauged mortar (1:1:6); prime cost £ per 1000	350.00	–	–	350.00	1000	**350.00**
Walls						
facework one side, half brick thick; stretcher bond	21.00	1.08	18.69	24.46	m²	**43.15**
facework one side, half brick thick, Flemish bond with snapped headers	21.00	1.25	21.65	24.46	m²	**46.11**
facework one side, half brick thick, stretcher bond; building against other work; concrete	21.00	1.17	20.24	25.70	m²	**45.94**
facework one side, half brick thick; Flemish bond with snapped headers; building against other work; concrete	21.00	1.33	23.04	25.70	m²	**48.74**
facework one side, half brick thick, stretcher bond; building overhand	21.00	1.33	23.04	24.46	m²	**47.50**
facework one side, half brick thick; Flemish bond with snapped headers; building overhand	21.00	1.50	26.00	24.46	m²	**50.46**
facework one side, half brick thick; stretcher bond; curved; 6.00 m radii	21.00	1.58	27.41	24.46	m²	**51.87**
facework one side, half brick thick; Flemish bond with snapped headers; curved; 6.00 m radii	21.00	1.79	30.99	24.46	m²	**55.45**
facework one side, half brick thick; stretcher bond; curved; 1.50 m radii	24.50	2.00	34.56	28.23	m²	**62.79**
facework one side, half brick thick; Flemish bond with snapped headers; curved; 1.50 m radii	24.50	2.33	40.33	28.23	m²	**68.56**
facework both sides, one brick thick; two stretcher skins tied together	42.00	1.87	32.39	51.07	m²	**83.46**
facework both sides, one brick thick; Flemish bond	42.00	1.92	33.17	48.91	m²	**82.08**
facework both sides, one brick thick; two stretcher skins tied together; curved; 6.00 m radii	45.50	2.58	44.69	54.83	m²	**99.52**
facework both sides, one brick thick; Flemish bond; curved; 6.00 m radii	45.50	2.66	46.09	52.68	m²	**98.77**
facework both sides, one brick thick; two stretcher skins tied together; curved; 1.50 m radii	49.00	3.20	55.43	59.21	m²	**114.64**
facework both sides, one brick thick; Flemish bond; curved; 1.50 m radii	49.00	3.33	57.62	57.07	m²	**114.69**
Isolated piers						
facework both sides, one brick thick; two stretcher skins tied together	42.00	2.45	42.38	52.29	m²	**94.67**
facework both sides, one brick thick; Flemish bond	42.00	2.50	43.26	52.29	m²	**95.55**

14 MASONRY

Item	PC £	Labour hours	Labour £	Material £	Unit	Total rate £
BRICK WALLING – cont						
Facing bricks – cont						
Isolated casings						
facework one side, half brick thick; stretcher bond	21.00	1.85	32.01	24.46	m²	**56.47**
facework one side, half brick thick; Flemish bond with snapped headers	21.00	2.04	35.29	24.46	m²	**59.75**
Projections						
225 mm width; 112 mm depth; stretcher bond; vertical	4.67	0.28	4.85	5.24	m	**10.09**
225 mm width; 112 mm depth; Flemish bond with snapped headers; vertical	4.67	0.37	6.41	5.24	m	**11.65**
225 mm width; 225 mm depth; Flemish bond; vertical	9.33	0.60	10.38	15.26	m	**25.64**
328 mm width; 112 mm depth; stretcher bond; vertical	7.00	0.56	9.69	7.86	m	**17.55**
328 mm width; 112 mm depth; Flemish bond with snapped headers; vertical	7.00	0.65	11.24	7.86	m	**19.10**
328 mm width; 225 mm depth; Flemish bond; vertical	14.00	1.11	19.21	15.68	m	**34.89**
440 mm width; 112 mm depth; stretcher bond; vertical	9.33	0.83	14.36	10.48	m	**24.84**
440 mm width; 112 mm depth; Flemish bond with snapped headers; vertical	9.33	0.88	15.22	10.48	m	**25.70**
440 mm width; 225 mm depth; Flemish bond; vertical	18.67	1.62	28.02	20.96	m	**48.98**
Arches						
height on face 215 mm, width of exposed soffit 102 mm, shape of arch – flat	4.67	0.93	14.93	6.29	m	**21.22**
height on face 215 mm, width of exposed soffit 215 mm, shape of arch – flat	9.33	1.39	22.89	11.59	m	**34.48**
height on face 215 mm, width of exposed soffit 102 mm, shape of arch – segmental, one ring	4.67	1.76	27.56	8.29	m	**35.85**
height on face 215 mm, width of exposed soffit 215 mm, shape of arch segmental, one ring	9.33	2.13	33.96	13.23	m	**47.19**
height on face 215 mm, width of exposed soffit 102 mm, shape of arch – semi-circular, one ring	4.67	2.68	43.48	8.29	m	**51.77**
height on face 215 mm, width of exposed soffit 215 mm, shape of arch – semi-circular, one ring	9.33	3.61	59.56	13.23	m	**72.79**
height on face 215 mm, width of exposed soffit 102 mm, shape of arch – segmental, two ring	4.67	2.17	34.66	8.29	m	**42.95**
height on face 215 mm, width of exposed soffit 215 mm, shape of arch – segmental; two ring	9.33	2.82	45.90	13.23	m	**59.13**
height on face 215 mm, width of exposed soffit 102 mm, shape of arch – semi-circular, two ring	4.67	3.61	59.56	8.29	m	**67.85**
height on face 215 mm, width of exposed soffit 215 mm, shape of arch – semi-circular, two ring	9.33	5.00	83.62	13.23	m	**96.85**
Arches; cut voussoirs; prime cost £ per 1000	3450.00	–	–	3450.00	1000	**3450.00**
height on face 215 mm, width of exposed soffit 102 mm, shape of arch – segmental, one ring	46.00	1.80	28.25	52.78	m	**81.03**
height on face 215 mm, width of exposed soffit 215 mm, shape of arch – segmental, one ring	92.00	2.27	36.39	102.20	m	**138.59**

14 MASONRY

Item	PC £	Labour hours	Labour £	Material £	Unit	Total rate £
height on face 215 mm, width of exposed soffit 102 mm, shape of arch – semi-circular, one ring	46.00	2.04	32.40	52.78	m	85.18
height on face 215 mm, width of exposed soffit 215 mm, shape of arch – semi-circular, one ring	92.00	2.59	41.92	102.20	m	144.12
height on face 320 mm, width of exposed soffit 102 mm, shape of arch – segmental, one and a half ring	92.00	2.41	38.81	102.41	m	141.22
height on face 320 mm, width of exposed soffit 215 mm, shape of arch – segmental, one and a half ring	184.00	3.15	51.61	208.40	m	260.01
Arches; bullnosed specials; prime cost £ per 1000	2000.00	–	–	2000.00	1000	2000.00
height on face 215 mm, width of exposed soffit 102 mm, shape of arch – flat	26.67	0.97	15.63	29.97	m	45.60
height on face 215 mm, width of exposed soffit 215 mm, shape of arch – flat	54.00	1.43	23.59	59.67	m	83.26
Bullseye windows; 600 mm dia.; facing bricks						
height on face 215 mm, width of exposed soffit 102 mm, two rings	9.80	4.63	77.21	12.30	nr	89.51
height on face 215 mm, width of exposed soffit 215 mm, two rings	19.60	6.48	109.22	23.44	nr	132.66
Bullseye windows; 600 mm; cut voussoirs; prime cost £ per 1000	3450.00	–	–	3450.00	1000	3450.00
height on face 215 mm, width of exposed soffit 102 mm, one ring	120.75	3.89	64.41	131.71	nr	196.12
height on face 215 mm, width of exposed soffit 215 mm, one ring	120.75	5.37	90.02	262.26	nr	352.28
Bullseye windows; 1200 mm dia.; facing bricks						
height on face 215 mm, width of exposed soffit 102 mm, two rings	–	7.22	122.03	25.92	nr	147.95
height on face 215 mm, width of exposed soffit 215 mm, two rings	–	10.36	176.35	48.79	nr	225.14
Bullseye windows; 1200 mm dia.; cut voussoirs	3450.00	–	–	3450.00	1000	3450.00
height on face 215 mm, width of exposed soffit 102 mm, one ring	207.00	6.11	102.82	228.24	nr	331.06
height on face 215 mm, width of exposed soffit 215 mm, one ring	414.00	8.70	147.63	452.17	nr	599.80
ADD or DEDUCT for variation of £10.00 per 1000 in PC of facing bricks in 102 mm high arches with 215 mm soffit	–	–	–	0.28	m	0.28
Facework sills						
150 mm × 102 mm; headers on edge; pointing top and one side; set weathering; horizontal	4.67	0.51	8.83	5.11	m	13.94
150 mm × 102 mm; cant headers on edge; pointing top and one side; set weathering; horizontal; prime cost £ per 1000	26.67	0.56	9.69	28.23	m	37.92
150 mm × 102 mm; bullnosed specials; headers on flat; pointing top and one side; horizontal; prime cost £ per 1000	26.67	0.46	7.95	28.23	m	36.18

14 MASONRY

Item	PC £	Labour hours	Labour £	Material £	Unit	Total rate £
BRICK WALLING – cont						
Facing bricks – cont						
Facework copings						
215 mm × 102 mm; headers on edge; pointing top and both sides; horizontal	–	0.42	7.27	5.33	m	**12.60**
260 mm × 102 mm; headers on edge; pointing top and both sides; horizontal	–	0.65	11.24	7.92	m	**19.16**
215 mm × 102 mm; double bullnose specials; headers on edge; pointing top and both sides; horizontal; prime cost £ per 1000	2000.00	0.46	7.95	27.64	m	**35.59**
260 mm × 102 mm; single bullnose specials; headers on edge; pointing top and both sides; horizontal; prime cost £ per 1000	2000.00	0.65	11.24	55.05	m	**66.29**
ADD or DEDUCT for variation of £10.00 per 1000 in prime cost of facing bricks in copings 215 mm wide, 102 mm high	–	–	–	0.14	m	**0.14**
Extra over facing bricks for; facework ornamental bands and the like, plain bands						
flush; horizontal; 225 mm width; entirely of stretchers; prime cost £/1000	500.00	0.19	3.29	0.57	m	**3.86**
Extra over facing brick for; facework quoins						
flush; mean girth 320 mm; prime cost £ per 1000	500.00	0.28	4.85	0.57	m	**5.42**
Bonding to existing						
facework one side, half brick thick; stretcher bond	–	0.46	7.95	1.35	m	**9.30**
facework one side, half brick thick; Flemish bond with snapped headers	–	0.46	7.95	1.35	m	**9.30**
facework both sides, one brick thick; two stretcher skins tied together	–	0.65	11.24	2.70	m	**13.94**
facework both sides, one brick thick; Flemish bond	–	0.65	11.24	2.70	m	**13.94**
ADD or DEDUCT for variation of £10.00 per 1000 in prime cost of facing bricks; in walls built entirely of facings; in stretcher or Flemish bond						
half brick thick	–	–	–	0.62	m²	**0.62**
one brick thick	–	–	–	1.23	m²	**1.23**
Facing bricks; handmade						
In gauged mortar (1:1:6); prime cost for bricks	500.00	–	–	500.00	1000	**500.00**
Walls						
facework one side, half brick thick; stretcher bond	–	1.08	18.69	34.14	m²	**52.83**
facework one side, half brick thick; Flemish bond with snapped headers	–	1.25	21.65	34.14	m²	**55.79**
facework one side; half brick thick; stretcher bond; building against other work; concrete	–	1.17	20.24	35.38	m²	**55.62**
facework one side, half brick thick; Flemish bond with snapped headers; building against other work; concrete	–	1.33	23.04	35.38	m²	**58.42**
facework one side, half brick thick; stretcher bond; building overhand	–	1.33	23.04	34.14	m²	**57.18**
facework one side, half brick thick; Flemish bond with snapped headers; building overhand	–	1.50	26.00	34.14	m²	**60.14**

14 MASONRY

Item	PC £	Labour hours	Labour £	Material £	Unit	Total rate £
facework one side, half brick thick; stretcher bond; curved; 6.00 m radii	–	1.58	27.41	34.14	m²	**61.55**
facework one side, half brick thick; Flemish bond with snapped headers; curved; 6.00 m radii	–	1.79	30.99	38.18	m²	**69.17**
facework one side, half brick thick; stretcher bond; curved 1.50 m radii	–	2.00	34.56	34.14	m²	**68.70**
facework one side, half brick thick; Flemish bond with snapped headers; curved; 1.50 m radii	–	2.33	40.33	40.87	m²	**81.20**
facework both sides, one brick thick; two stretcher skins tied together	–	1.87	32.39	70.44	m²	**102.83**
facework both sides, one brick thick; Flemish bond	–	1.92	33.17	68.29	m²	**101.46**
facework both sides; one brick thick; two stretcher skins tied together; curved; 6.00 m radii	–	2.58	44.69	75.82	m²	**120.51**
facework both sides, one brick thick; Flemish bond; curved; 6.00 m radii	–	2.66	46.09	73.67	m²	**119.76**
facework both sides, one brick thick; two stretcher skins tied together; curved; 1.50 m radii	–	3.20	55.43	81.82	m²	**137.25**
facework both sides, one brick thick; Flemish bond; curved; 1.50 m radii	–	3.33	57.62	79.67	m²	**137.29**
Isolated piers						
facework both sides, one brick thick; two stretcher skins tied together	–	2.45	42.38	71.66	m²	**114.04**
facework both sides, one brick thick; Flemish bond	–	2.50	43.26	71.66	m²	**114.92**
Isolated casings						
facework one side, half brick thick; stretcher bond	–	1.85	32.01	34.14	m²	**66.15**
facework one side, half brick thick; Flemish bond with snapped headers	–	2.04	35.29	34.14	m²	**69.43**
Projections						
225 mm width; 112 mm depth; stretcher bond; vertical		0.28	4.85	7.39	m	**12.24**
225 mm width; 112 mm depth; Flemish bond with snapped headers; vertical	–	0.37	6.41	7.39	m	**13.80**
225 mm width; 225 mm depth; Flemish bond; vertical	–	0.60	10.38	14.78	m	**25.16**
328 mm width; 112 mm depth; stretcher bond; vertical		0.56	9.69	11.10	m	**20.79**
328 mm width; 112 mm depth; Flemish bond with snapped headers; vertical	–	0.65	11.24	11.10	m	**22.34**
328 mm width; 225 mm depth; Flemish bond; vertical	–	1.11	19.21	22.14	m	**41.35**
440 mm width; 112 mm depth; stretcher bond; vertical	–	0.83	14.36	14.78	m	**29.14**
440 mm width; 112 mm depth; Flemish bond with snapped headers; vertical	–	0.88	15.22	14.78	m	**30.00**
440 mm width; 225 mm depth; Flemish bond; vertical	–	1.62	28.02	29.57	m	**57.59**

14 MASONRY

Item	PC £	Labour hours	Labour £	Material £	Unit	Total rate £
BRICK WALLING – cont						
Facing bricks – cont						
Arches						
height on face 215 mm, width of exposed soffit 102 mm, shape of arch – flat	–	0.93	14.93	8.45	m	**23.38**
height on face 215 mm, width of exposed soffit 215 mm, shape of arch – flat	–	1.39	22.89	16.08	m	**38.97**
height on face 215 mm, width of exposed soffit 102 mm, shape of arch – segmental, one ring	–	1.76	27.56	10.44	m	**38.00**
height on face 215 mm, width of exposed soffit 215 mm, shape of arch – segmental, one ring	–	2.13	33.96	17.54	m	**51.50**
height on face 215 mm, width of exposed soffit 102 mm, shape of arch – semi-circular, one ring	–	2.68	43.48	10.44	m	**53.92**
height on face 215 mm, width of exposed soffit 215 mm, shape of arch – semi-circular, one ring	–	3.61	59.56	17.54	m	**77.10**
height on face 215 mm, width of exposed soffit 102 mm, shape of arch – segmental, two ring	–	2.17	34.66	10.44	m	**45.10**
height on face 215 mm, width of exposed soffit 215 mm, shape of arch – segmental, two ring	–	2.82	45.90	17.54	m	**63.44**
height on face 215 mm, width of exposed soffit 102 mm, shape of arch – semi-circular, two ring	–	3.61	59.56	10.44	m	**70.00**
height on face 215 mm, width of exposed soffit 215 mm, shape of arch – semi-circular, two ring	–	5.00	83.62	17.54	m	**101.16**
Arches; cut voussoirs (PC £ per 1000)	3450.00	–	–	3450.00	1000	**3450.00**
height on face 215 mm, width of exposed soffit 102 mm, shape of arch – segmental, one ring	–	1.80	28.25	52.78	m	**81.03**
height on face 215 mm, width of exposed soffit 215 mm, shape of arch – segmental, one ring	–	2.27	36.39	102.20	m	**138.59**
height on face 215 mm, width of exposed soffit 102 mm, shape of arch – semi-circular, one ring	–	2.04	32.40	52.78	m	**85.18**
height one face 215 mm, width of exposed soffit 215 mm, shape of – arch semi-circular, one ring	–	2.59	41.92	102.20	m	**144.12**
height on face 320 mm, width of exposed soffit 102 mm, shape of arch – segmental, one and a half ring	–	2.41	38.81	102.41	m	**141.22**
height on face 320 mm, width of exposed soffit 215 mm, shape of arch – segmental, one and a half ring	–	3.15	51.61	208.40	m	**260.01**
Arches; bullnosed specials; prime cost £ per 1000	2000.00	–	–	2000.00	1000	**2000.00**
height on face 215 mm, width of exposed soffit 102 mm, shape of arch – flat	–	0.97	15.63	29.97	m	**45.60**
height on face 215 mm, width of exposed soffit 215 mm, shape of arch – flat	–	1.43	23.59	59.67	m	**83.26**
Bullseye windows; 600 mm dia.						
height on face 215 mm, width of exposed soffit 102 mm, two ring	–	4.63	77.21	16.82	nr	**94.03**
height on face 215 mm, width of exposed soffit 215 mm, two ring	–	6.48	109.22	43.62	nr	**152.84**

14 MASONRY

Item	PC £	Labour hours	Labour £	Material £	Unit	Total rate £
Bullseye windows; 600 mm dia.; cut voussoirs; prime cost £ per 1000	3450.00	–	–	3536.25	1000	3536.25
height on face 215 mm, width of exposed soffit 102 mm, one ring	–	3.89	64.41	131.71	nr	196.12
height on face 215 mm, width of exposed soffit 215 mm, one ring	–	5.37	90.02	262.26	nr	352.28
Bullseye windows; 1200 mm dia.						
height on face 215 mm, width of exposed soffit 102 mm, two ring	–	7.22	122.03	34.96	nr	156.99
height on face 215 mm, width of exposed soffit 215 mm, two ring	–	10.36	176.35	66.87	nr	243.22
Bullseye windows; 1200 mm dia.; cut voussoirs; prime cost £ per 1000	3450.00	–	–	3450.00	1000	3450.00
height on face 215 mm, width of exposed soffit 102 mm, one ring	–	6.11	102.82	228.24	nr	331.06
height on face 215 mm, width of exposed soffit 215 mm, one ring	–	8.70	147.63	452.17	nr	599.80
ADD or DEDUCT for variation of £10.00 per 1000 in prime cost of facing bricks in 102 mm high arches with 215 mm soffit	–	–	–	0.28	m	0.28
Facework sills						
150 mm × 102 mm; headers on edge; pointing top and one side; set weathering; horizontal	–	0.51	8.83	7.39	m	16.22
150 mm × 102 mm; cant headers on edge; pointing top and one side; set weathering; horizontal; prime cost £ per 1000	2000.00	0.56	9.69	28.92	m	38.61
150 mm × 102 mm; bullnosed specials; headers on edge; pointing top and one side; horizontal; prime cost £ per 1000	2000.00	0.46	7.95	28.92	m	36.87
Facework copings						
215 mm × 102 mm; headers on edge; pointing top and both sides; horizontal	–	0.42	7.27	7.48	m	14.75
260 mm × 102 mm; headers on edge; pointing top and both sides; horizontal	–	0.65	11.24	11.15	m	22.39
215 mm × 102 mm; double bullnose specials; headers on edge; pointing top and both sides; horizontal; prime cost £ per 1000	2000.00	0.46	7.95	29.01	m	36.96
260 mm × 102 mm; single bullnose specials; headers on edge; pointing top and both sides; horizontal; prime cost £ per 1000	2000.00	0.65	11.24	57.79	m	69.03
ADD or DEDUCT for variation of £10.00 per 1000 in prime cost of facing bricks in copings 215 mm wide, 102 mm high	–	–	–	0.14	m	0.14
Extra over facing bricks for; facework ornamental bands and the like, plain bands						
flush; horizontal; 225 mm width; entirely of stretchers; prime cost £ per 1000	550.00	0.19	3.29	0.72	m	4.01
Extra over facing bricks for; facework quoins						
flush mean girth 320 mm; prime cost £ per 1000	550.00	0.28	4.85	0.69	m	5.54

14 MASONRY

Item	PC £	Labour hours	Labour £	Material £	Unit	Total rate £
BRICK WALLING – cont						
Facing bricks – cont						
Bonding ends to existing						
facework one side, half brick thick; stretcher bond	–	0.46	7.95	1.89	m	9.84
facework one side, half brick thick; Flemish bond with snapped headers	–	0.46	7.95	1.89	m	9.84
facework both sides, one brick thick; two stretcher skins tied together	–	0.65	11.24	3.77	m	15.01
facework both sides, one brick thick; Flemish bond	–	0.65	11.24	3.77	m	15.01
ADD or DEDUCT for variation of £10.00/1000 in prime cost of facing bricks; in walls built entirely of facings; in stretcher or Flemish bond						
half brick thick	–	–	–	0.62	m²	0.62
one brick thick	–	–	–	1.23	m²	1.23
Facing bricks; slips						
50 mm thick; in gauged mortar (1:1:6) built up against concrete including flushing up at back (ties not included)						
walls; prime cost £ per 1000	1200.00	1.85	31.23	77.41	m²	108.64
edges of suspended slabs; 200 mm wide	–	0.56	9.69	15.87	m	25.56
columns; 400 mm wide	–	1.11	19.21	31.73	m	50.94
Class A engineering bricks; and bullnosed specials; in cement: mortar (1:3)						
Facework steps						
215 mm × 102 mm; all headers-on-edge; edges set with bullnosed specials; pointing top and one side; set weathering; horizontal; specials prime cost £ per 1000	2000.00	0.51	8.83	28.95	m	37.78
returned ends pointed	–	0.14	2.42	5.09	nr	7.51
430 mm × 102 mm; all headers-on-edge; edges set with bullnosed specials; pointing top and one side; set weathering; horizontal; engineering bricks prime cost £ per 1000	331.50	0.74	12.80	33.94	m	46.74
returned ends pointed	–	0.19	3.29	6.52	nr	9.81
Facing tile bricks; Ibstock Tilebrick or other equal and approved; in gauged mortar (1:1:6)						
Walls						
standard tile brick; facework one side; half brick thick; stretcher bond	–	0.87	15.05	63.64	m²	78.69
three quarter tile brick; facework one side; half brick thick; stretcher bond	–	0.87	15.05	74.41	m²	89.46
Extra over facing tile bricks for						
fair ends; 79 mm long	–	0.28	4.85	27.66	m	32.51
fair ends; 163 mm long	–	0.28	4.85	27.66	m	32.51
90° × ½ external return	–	0.28	4.85	58.29	m	63.14
90° internal return	–	0.28	4.85	68.77	m	73.62
30° or 45° or 60° external return	–	0.28	4.85	58.29	m	63.14
30° or 45° or 60° internal return	–	0.28	4.85	58.29	m	63.14
angled verge	–	0.28	4.85	32.46	m	37.31

14 MASONRY

Item	PC £	Labour hours	Labour £	Material £	Unit	Total rate £
BLOCK WALLING						
SUPPLY ONLY BLOCK PRICES						
Alternative block prices; material rates only						
Durox Supablocs; Aerated concrete;						
620 mm × 215 mm (7 nr per m^2)						
100 mm	–	–	–	6.76	m^2	**6.76**
130 mm	–	–	–	8.04	m^2	**8.04**
140 mm	–	–	–	11.90	m^2	**11.90**
215 mm	–	–	–	14.55	m^2	**14.55**
Hanson Conbloc blocks: 450 × 215 mm						
Cream fair faced						
100 mm hollow	–	–	–	5.64	m^2	**5.64**
100 mm solid	–	–	–	6.11	m^2	**6.11**
140 mm hollow	–	–	–	8.36	m^2	**8.36**
140 mm solid	–	–	–	9.96	m^2	**9.96**
190 mm hollow	–	–	–	11.92	m^2	**11.92**
190 mm solid	–	–	–	13.14	m^2	**13.14**
215 mm hollow	–	–	–	11.89	m^2	**11.89**
Fenlite						
100 mm solid; 3.50 N/mm^2	–	–	–	5.61	m^2	**5.61**
100 mm solid; 7.00 N/mm^2	–	–	–	5.79	m^2	**5.79**
Standard Dense						
100 mm solid	–	–	–	5.35	m^2	**5.35**
140 mm hollow	–	–	–	7.49	m^2	**7.49**
215 mm hollow	–	–	–	11.50	m^2	**11.50**
Celcon blocks; 450 mm × 215 mm						
100 mm Standard	–	–	–	8.25	m^2	**8.25**
140 mm Standard	–	–	–	11.03	m^2	**11.03**
100 mm coursing brick; 215 mm × 65 mm	–	–	–	1.32	m	**1.32**
275 mm wide × 140 mm thick standard footing	–	–	–	3.13	m	**3.13**
215 mm hollow	–	–	–	16.73	m^2	**16.73**
100 mm Solar	–	–	–	10.38	m^2	**10.38**
215 mm Solar	–	–	–	8.85	m^2	**8.85**
265 mm Solar	–	–	–	28.50	m^2	**28.50**
Forticrete painting quality blocks; 450 mm × 215 mm						
100 mm hollow	–	–	–	8.71	m^2	**8.71**
100 mm solid	–	–	–	9.54	m^2	**9.54**
140 mm hollow	–	–	–	12.04	m^2	**12.04**
140 mm solid	–	–	–	14.04	m^2	**14.04**
190 mm hollow	–	–	–	15.97	m^2	**15.97**
190 mm solid	–	–	–	18.21	m^2	**18.21**
215 mm hollow	–	–	–	16.76	m^2	**16.76**
215 mm solid	–	–	–	20.15	m^2	**20.15**
Lignacite Lignacrete standard blocks;						
450 mm × 215 mm; 7.3 N/mm^2						
100 mm	–	–	–	7.54	m^2	**7.54**
140 mm	–	–	–	10.77	m^2	**10.77**
150 mm	–	–	–	12.77	m^2	**12.77**
190 mm	–	–	–	15.57	m^2	**15.57**
215 mm	–	–	–	16.93	m^2	**16.93**

14 MASONRY

Item	PC £	Labour hours	Labour £	Material £	Unit	Total rate £
BLOCK WALLING – cont						
Tarmac Hemelite; 450 mm × 215 mm						
100 mm solid; 3.50 N/mm²	–	–	–	6.23	m²	**6.23**
100 mm solid; 7.00 N/mm²	–	–	–	6.45	m²	**6.45**
140 mm solid; 7.00 N/mm²	–	–	–	9.17	m²	**9.17**
190 mm solid; 7.00 N/mm²	–	–	–	13.25	m²	**13.25**
215 mm solid; 7.00 N/mm²	–	–	–	16.33	m²	**16.33**
Tarmac Toplite standard blocks; 450 mm × 215 mm						
100 mm	–	–	–	6.66	m²	**6.66**
140 mm	–	–	–	9.32	m²	**9.32**
150 mm	–	–	–	9.99	m²	**9.99**
215 mm	–	–	–	11.01	m²	**11.01**
Tarmac Toplite GTI (thermal) blocks; 450 mm × 215 mm						
115 mm	–	–	–	5.89	m²	**5.89**
125 mm	–	–	–	6.41	m²	**6.41**
130 mm	–	–	–	6.66	m²	**6.66**
140 mm	–	–	–	7.18	m²	**7.18**
150 mm	–	–	–	7.69	m²	**7.69**
215 mm	–	–	–	10.68	m²	**10.68**
Plasmor concrete blocks; 450 mm × 215 mm						
Aglite, standard blocks 10.4 N; 100 mm thick	–	–	–	8.95	m²	**8.95**
Aglite, standard blocks 10.4 N; 140 mm thick	–	–	–	12.55	m²	**12.55**
Aglite, standard blocks 10.4 N; 190 mm thick	–	–	–	20.50	m²	**20.50**
Aglite, standard blocks 10.4 N; 215 mm thick	–	–	–	23.50	m²	**23.50**
Stranlite, standard blocks 10.4 N; 100 mm thick	–	–	–	9.20	m²	**9.20**
Stranlite, standard blocks 10.4 N; 140 mm thick	–	–	–	12.95	m²	**12.95**
Stranlite, standard blocks 10.4 N; 190 mm thick	–	–	–	21.55	m²	**21.55**
Stranlite, standard blocks 10.4 N; 215 mm thick	–	–	–	24.80	m²	**24.80**
Stranlite, paint grade blocks 7.3 N; 100 mm thick	–	–	–	9.30	m²	**9.30**
Stranlite, paint grade blocks 7.3 N; 140 mm thick	–	–	–	13.05	m²	**13.05**
Stranlite, paint grade blocks 7.3 N; 190 mm thick	–	–	–	21.40	m²	**21.40**
Stranlite, paint grade blocks 7.3 N; 215 mm thick	–	–	–	24.70	m²	**24.70**
Stranlite, paint grade blocks 10.4 N; 100 mm thick	–	–	–	9.75	m²	**9.75**
Stranlite, paint grade blocks 10.4 N; 140 mm thick	–	–	–	13.65	m²	**13.65**
Stranlite, paint grade blocks 10.4 N; 190 mm thick	–	–	–	22.25	m²	**22.25**
Stranlite, paint grade blocks 10.4 N; 215 mm thick	–	–	–	26.50	m²	**26.50**
SUPPLY AND LAY PRICES						
Lightweight aerated concrete blocks; Thermalite Turbo blocks or other equal; in gauged mortar (1:2:9)						
Walls						
100 mm thick	6.64	0.41	7.16	7.72	m²	**14.88**
115 mm thick	7.64	0.41	7.16	8.88	m²	**16.04**
125 mm thick	8.30	0.41	7.16	9.65	m²	**16.81**
130 mm thick	8.63	0.41	7.16	10.02	m²	**17.18**
140 mm thick	9.30	0.46	7.94	10.80	m²	**18.74**
150 mm thick	9.96	0.46	7.94	11.57	m²	**19.51**
190 mm thick	12.62	0.50	8.72	14.66	m²	**23.38**

14 MASONRY

Item	PC £	Labour hours	Labour £	Material £	Unit	Total rate £
200 mm thick	13.28	0.50	8.72	15.43	m²	**24.15**
215 mm thick	14.28	0.50	8.72	16.59	m²	**25.31**
Isolated piers or chimney stacks						
190 mm thick	–	0.83	14.36	14.66	m²	**29.02**
215 mm thick	–	0.83	14.36	16.59	m²	**30.95**
Isolated casings						
100 mm thick	–	0.51	8.83	7.72	m²	**16.55**
115 mm thick	–	0.51	8.83	8.88	m²	**17.71**
125 mm thick	–	0.51	8.83	9.65	m²	**18.48**
140 mm thick	–	0.56	9.69	10.80	m²	**20.49**
Extra over for fair face; flush pointing						
walls; one side	–	0.04	0.70	–	m²	**0.70**
walls; both sides	–	0.09	1.56	–	m²	**1.56**
Closing cavities						
width of cavity 50 mm, closing with lightweight blockwork 100 mm thick; vertical	–	0.23	3.98	0.44	m	**4.42**
width of cavity 50 mm, closing with lightweight blockwork 100 mm thick; including damp-proof course; vertical	–	0.28	4.85	1.18	m	**6.03**
width of cavity 75 mm, closing with lightweight blockwork 100 mm thick; vertical	–	0.23	3.98	0.63	m	**4.61**
width of cavity 75 mm, closing with lightweight blockwork 100 mm thick; including damp-proof course; vertical	–	0.28	4.85	1.36	m	**6.21**
Bonding ends to common brickwork						
100 mm thick	–	0.14	2.42	0.89	m	**3.31**
115 mm thick	–	0.14	2.42	1.03	m	**3.45**
125 mm thick	–	0.23	3.98	1.12	m	**5.10**
130 mm thick	–	0.23	3.98	1.17	m	**5.15**
140 mm thick	–	0.23	3.98	1.26	m	**5.24**
150 mm thick	–	0.23	3.98	1.34	m	**5.32**
190 mm thick	–	0.28	4.85	1.69	m	**6.54**
200 mm thick	–	0.28	4.85	1.78	m	**6.63**
215 mm thick	–	0.32	5.54	1.93	m	**7.47**
Lightweight aerated concrete blocks; Thermalite Shield blocks or other equal and approved; in thin joint mortar						
Walls						
50 mm thick	4.84	0.40	6.92	6.25	m²	**13.17**
75 mm thick	5.48	0.40	6.92	6.96	m²	**13.88**
90 mm thick	6.46	0.40	6.92	8.22	m²	**15.14**
100 mm thick	6.46	0.46	7.95	8.35	m²	**16.30**
140 mm thick	9.04	0.51	8.83	11.70	m²	**20.53**
150 mm thick	9.69	0.51	8.83	12.55	m²	**21.38**
190 mm thick	12.27	0.56	9.69	15.87	m²	**25.56**
200 mm thick	12.91	0.60	10.38	16.71	m²	**27.09**
Isolated piers or chimney stacks						
190 mm thick	–	0.60	10.38	15.87	m²	**26.25**

14 MASONRY

Item	PC £	Labour hours	Labour £	Material £	Unit	Total rate £
BLOCK WALLING – cont						
Lightweight aerated concrete blocks – cont						
Isolated casings						
75 mm thick	–	0.35	6.06	7.28	m²	**13.34**
90 mm thick	–	0.35	6.06	8.35	m²	**14.41**
100 mm thick	–	0.35	6.06	8.35	m²	**14.41**
140 mm thick	–	0.38	6.57	11.70	m²	**18.27**
Lightweight aerated concrete blocks; Thermalite Shield blocks or other equal and approved; in gauged mortar (1:2:9)						
Walls						
75 mm thick	5.89	0.38	6.54	6.33	m²	**12.87**
90 mm thick	6.94	0.38	6.54	7.47	m²	**14.01**
100 mm thick	6.94	0.41	7.16	7.51	m²	**14.67**
140 mm thick	9.72	0.46	7.94	10.52	m²	**18.46**
150 mm thick	10.41	0.46	7.94	11.28	m²	**19.22**
190 mm thick	13.19	0.50	8.72	14.28	m²	**23.00**
200 mm thick	13.88	0.50	8.72	15.03	m²	**23.75**
Isolated piers or chimney stacks						
190 mm thick	–	0.83	14.36	14.28	m²	**28.64**
Isolated casings						
75 mm thick	–	0.51	8.83	6.33	m²	**15.16**
90 mm thick	–	0.51	8.83	7.47	m²	**16.30**
100 mm thick	–	0.51	8.83	7.51	m²	**16.34**
140 mm thick	–	0.56	9.69	10.52	m²	**20.21**
Extra over for fair face; flush pointing						
walls; one side	–	0.04	0.70	–	m²	**0.70**
walls; both sides	–	0.09	1.56	–	m²	**1.56**
Closing cavities						
width of cavity 50 mm, closing with lightweight blockwork 100 mm thick; vertical	–	0.23	3.98	0.43	m	**4.41**
width of cavity 50 mm, closing with lightweight blockwork 100 mm thick; including damp-proof course; vertical	–	0.28	4.85	1.17	m	**6.02**
width of cavity 75 mm, closing with lightweight blockwork 100 mm thick; vertical	–	0.23	3.98	0.62	m	**4.60**
width of cavity 75 mm, closing with lightweight blockwork 100 mm thick; including damp-proof course; vertical	–	0.28	4.85	1.35	m	**6.20**
Bonding ends to common brickwork						
75 mm thick	–	0.09	1.56	0.73	m	**2.29**
90 mm thick	–	0.09	1.56	0.86	m	**2.42**
100 mm thick	–	0.14	2.42	0.87	m	**3.29**
140 mm thick	–	0.23	3.98	1.23	m	**5.21**
150 mm thick	–	0.23	3.98	1.30	m	**5.28**
190 mm thick	–	0.28	4.85	1.65	m	**6.50**
200 mm thick	–	0.28	4.85	1.74	m	**6.59**

14 MASONRY

Item	PC £	Labour hours	Labour £	Material £	Unit	Total rate £
Lightweight smooth face aerated concrete blocks; Thermalite Smooth Face blocks or other equal and approved; in gauged mortar (1:2:9); flush pointing one side						
Walls						
100 mm thick	8.77	0.50	8.72	10.07	m²	**18.79**
140 mm thick	12.28	0.58	10.12	14.08	m²	**24.20**
150 mm thick	13.15	0.58	10.12	15.09	m²	**25.21**
190 mm thick	16.66	0.67	11.52	19.11	m²	**30.63**
200 mm thick	17.54	0.67	11.52	20.12	m²	**31.64**
215 mm thick	18.85	0.67	11.52	21.63	m²	**33.15**
100 mm thick; Paintgrade Smooth	12.18	0.55	9.53	13.83	m²	**23.36**
215 mm thick; Paintgrade Smooth	18.86	0.73	12.56	21.64	m²	**34.20**
Isolated piers or chimney stacks						
190 mm thick	–	0.93	16.09	19.11	m²	**35.20**
200 mm thick	–	0.93	16.09	20.12	m²	**36.21**
215 mm thick	–	0.93	16.09	21.63	m²	**37.72**
Isolated casings						
100 mm thick	–	0.69	11.94	10.07	m²	**22.01**
140 mm thick	–	0.74	12.80	14.08	m²	**26.88**
Extra over for fair face flush pointing						
walls; one side	–	0.04	0.70	–	m²	**0.70**
walls; both sides	–	0.09	1.56	–	m²	**1.56**
Bonding ends to common brickwork						
100 mm thick	–	0.23	3.98	1.18	m	**5.16**
140 mm thick	–	0.23	3.98	1.66	m	**5.64**
150 mm thick	–	0.28	4.85	1.76	m	**6.61**
190 mm thick	–	0.32	5.54	2.23	m	**7.77**
200 mm thick	–	0.32	5.54	2.36	m	**7.90**
215 mm thick	–	0.32	5.54	2.54	m	**8.08**
Lightweight smooth face aerated concrete blocks; Thermalite Party Wall blocks or other equal and approved; in gauged mortar (1:2:9); flush pointing one side						
Walls						
100 mm thick	6.46	0.56	9.69	7.35	m²	**17.04**
215 mm thick	13.88	0.74	12.80	15.80	m²	**28.60**
Isolated piers or chimney stacks						
215 mm thick	–	0.93	16.09	15.80	m²	**31.89**
Isolated casings						
100 mm thick	–	0.69	11.94	7.35	m²	**19.29**
Extra over for fair face flush pointing						
walls; both sides	–	0.04	0.70	–	m²	**0.70**
Bonding ends to common brickwork						
100 mm thick	–	0.23	3.98	0.81	m	**4.79**
215 mm thick	–	0.32	5.54	1.76	m	**7.30**

14 MASONRY

Item	PC £	Labour hours	Labour £	Material £	Unit	Total rate £
BLOCK WALLING – cont						
Lightweight aerated high strength concrete blocks (7.00 N/mm^2); Thermalite High Strength 7 blocks or other equal and approved; in cement: mortar (1:3)						
Walls						
100 mm thick	8.30	0.41	7.16	9.61	m^2	**16.77**
140 mm thick	11.62	0.46	7.94	13.44	m^2	**21.38**
150 mm thick	12.46	0.46	7.94	14.41	m^2	**22.35**
190 mm thick	15.77	0.50	8.72	18.25	m^2	**26.97**
200 mm thick	16.60	0.50	8.72	19.21	m^2	**27.93**
215 mm thick	17.85	0.50	8.72	20.65	m^2	**29.37**
Isolated piers or chimney stacks						
190 mm thick	–	0.83	14.36	18.25	m^2	**32.61**
200 mm thick	–	0.83	14.36	19.21	m^2	**33.57**
215 mm thick	–	0.83	14.36	20.65	m^2	**35.01**
Isolated casings						
100 mm thick	–	0.51	8.83	9.61	m^2	**18.44**
140 mm thick	–	0.56	9.69	13.44	m^2	**23.13**
150 mm thick	–	0.56	9.69	14.41	m^2	**24.10**
190 mm thick	–	0.69	11.94	18.25	m^2	**30.19**
200 mm thick	–	0.69	11.94	19.21	m^2	**31.15**
215 mm thick	–	0.69	11.94	20.65	m^2	**32.59**
Extra over for flush pointing						
walls; one side	–	0.04	0.70	–	m^2	**0.70**
walls; both sides	–	0.09	1.56	–	m^2	**1.56**
Bonding ends to common brickwork						
100 mm thick	–	0.23	3.98	1.13	m	**5.11**
140 mm thick	–	0.23	3.98	1.59	m	**5.57**
150 mm thick	–	0.28	4.85	1.70	m	**6.55**
190 mm thick	–	0.32	5.54	2.15	m	**7.69**
200 mm thick	–	0.32	5.54	2.27	m	**7.81**
215 mm thick	–	0.32	5.54	2.44	m	**7.98**
Lightweight aerated high strength concrete blocks (10.00 N/mm^2); Thermalite High Strength 10 blocks; in cement: mortar (1:3)						
Walls						
100 mm thick (NB other thicknesses as a special order item)	–	0.50	8.72	12.32	m^2	**21.04**
Lightweight concrete blocks; Thermalite Trenchblock 3.6 N/mm^2; with tongued and grooved joints; in cement mortar (1:4)						
Walls						
255 mm thick	–	0.60	10.38	18.11	m^2	**28.49**
275 mm thick	–	0.65	11.24	20.49	m^2	**31.73**
305 mm thick	–	0.70	12.12	22.33	m^2	**34.45**
355 mm thick	–	0.75	12.98	25.03	m^2	**38.01**

14 MASONRY

Item	PC £	Labour hours	Labour £	Material £	Unit	Total rate £
Concrete blocks; Thermalite Trenchblock 7.00 N/mm²; with tongued and grooved joints; in cement mortar (1:4)						
Walls						
255 mm thick	21.17	0.70	12.12	24.20	m²	**36.32**
275 mm thick	22.84	0.75	12.98	26.11	m²	**39.09**
305 mm thick	24.91	0.80	13.84	28.45	m²	**42.29**
355 mm thick	29.47	0.85	14.71	33.54	m²	**48.25**
Medium dense smooth faced concrete blocks; Lignacite standard and paint grade 3.60 N/mm² blocks; in gauged mortar (1:2:9); flush pointing one side						
Walls						
100 mm thick	8.42	0.56	9.66	9.68	m²	**19.34**
140 mm thick	12.32	0.65	11.21	14.13	m²	**25.34**
150 mm thick	13.13	0.67	11.52	15.07	m²	**26.59**
190 mm thick	16.75	0.77	13.39	19.21	m²	**32.60**
215 mm thick	17.95	0.85	14.79	20.64	m²	**35.43**
Isolated piers or chimney stacks						
190 mm thick	–	1.14	19.72	19.21	m²	**38.93**
215 mm thick	–	1.26	21.80	20.64	m²	**42.44**
Isolated casings						
100 mm thick	–	0.78	13.50	9.68	m²	**23.18**
140 mm thick	–	0.90	15.57	14.13	m²	**29.70**
Extra over for fair face flush pointing						
walls; both sides	–	0.04	0.70	–	m²	**0.70**
Bonding ends to common brickwork						
100 mm thick	–	0.23	3.98	1.13	m	**5.11**
140 mm thick	–	0.23	3.98	1.66	m	**5.64**
150 mm thick	–	0.28	4.85	1.76	m	**6.61**
190 mm thick	–	0.32	5.54	2.24	m	**7.78**
215 mm thick	–	0.32	5.54	2.43	m	**7.97**
Architectural blockwork						
Lignacite Ltd; 440 × 215 mm face size; solid blocks 17.5 N; laid stretcher bond with class 3 Snowstorm mortar with white Portland cement and recessed joints						
Snowstorm Weathered						
100 mm thick weathered one face	–	1.00	17.30	41.23	m²	**58.53**
100 mm thick; weathered one face and one end	–	0.10	1.73	20.77	m	**22.50**
217 mm × 100 mm thick; cut half block; weathered one face	–	1.05	18.16	58.60	m²	**76.76**
quoins; weathered two external faces; 440 mm × 100 mm × 215 mm with 215 mm external return	–	0.11	1.91	87.85	m	**89.76**

14 MASONRY

Item	PC £	Labour hours	Labour £	Material £	Unit	Total rate £
BLOCK WALLING – cont						
Architectural blockwork – cont						
Snowstorm Split						
100 mm thick; split one face	–	1.00	17.30	42.17	m²	**59.47**
100 mm thick; split one face and weathered one end	–	0.10	1.73	21.19	m	**22.92**
217 mm × 100 mm thick; cut half block; split one face	–	1.05	18.16	69.94	m²	**88.10**
quoins; split two external faces; 440 mm × 215 mm × 100 mm thick with 215 mm external return	–	0.11	1.91	102.03	m	**103.94**
Snowstorm Polished						
Note: polished dimensions reduced by 3 mm						
100 mm thick; polished one face	–	1.10	19.03	74.26	m²	**93.29**
100 mm thick; polished one face one end	–	0.11	1.91	61.45	m	**63.36**
217 mm × 100 mm thick; cut half block; polished one face	–	1.18	20.33	91.76	m²	**112.09**
quoins; polished two external faces; 440 mm × 215 mm × 100 mm thick with 215 mm external return	–	0.12	2.08	151.53	m	**153.61**
Snowstorm Planished						
Note: planished dimensions reduced by 3 mm						
100 mm thick; planished one face	–	1.10	19.03	77.97	m²	**97.00**
100 mm thick; planished one face one end	–	0.11	1.91	63.34	m	**65.25**
100 mm thick; cut half block; planished one face	–	1.18	20.33	95.96	m²	**116.29**
quoins; planished two external faces; 440 mm × 215 mm × 100 mm thick with 215 mm external return	–	0.12	2.08	154.39	m	**156.47**
Snowstorm and Crushed recycled Glass (Polished or Planished). Standard colours						
Note: planished dimensions reduced by 3 mm						
100 mm thick; planished one face	–	1.10	19.03	93.84	m²	**112.87**
100 mm thick; planished one face one end	–	0.11	1.91	70.04	m	**71.95**
100 mm thick; cut half block; planished one face	–	1.18	20.33	111.27	m²	**131.60**
quoins; planished two external faces; 440 mm × 215 mm × 100 mm thick with 215 mm external return	–	0.12	2.08	164.41	m	**166.49**
Midnight Polished (Note 7.3 N strength) 10 mm Black Granite facing bonded to 90 mm thick backing block						
100 mm thick; polished one face	–	1.15	19.90	135.62	m²	**155.52**
100 mm thick; polished one face one end	–	0.12	2.08	215.98	m	**218.06**
100 mm thick; cut half block; polished one face	–	1.19	20.50	189.55	m²	**210.05**
quoins; polished two external faces; 440 mm × 215 mm × 110 mm thick with 215 mm external return	–	0.14	2.42	243.13	m	**245.55**

14 MASONRY

Item	PC £	Labour hours	Labour £	Material £	Unit	Total rate £
Medium dense smooth faced concrete blocks; Lignacite standard and paint grade 7 N blocks; in gauged mortar (1:2:9); flush pointing one side						
Walls						
100 mm thick	9.20	0.56	9.66	9.83	m²	**19.49**
140 mm thick	13.37	0.65	11.21	14.26	m²	**25.47**
150 mm thick	14.83	0.67	11.52	15.80	m²	**27.32**
190 mm thick	18.58	0.77	13.39	19.79	m²	**33.18**
215 mm thick	20.58	0.85	14.79	21.95	m²	**36.74**
Isolated piers or chimney stacks						
190 mm thick	–	1.14	19.72	19.79	m²	**39.51**
215 mm thick	–	1.26	21.80	21.95	m²	**43.75**
Isolated casings						
100 mm thick	–	0.78	13.50	9.83	m²	**23.33**
140 mm thick	–	0.90	15.57	14.26	m²	**29.83**
Dense aggregate concrete blocks; ARC Conbloc or other equal and approved; in cement mortar (1:2:9)						
Walls or partitions or skins of hollow walls						
75 mm thick; solid	5.07	0.50	8.72	5.88	m²	**14.60**
100 mm thick; solid	6.30	0.62	10.74	7.34	m²	**18.08**
140 mm thick; solid	9.38	0.75	12.93	10.89	m²	**23.82**
140 mm thick; hollow	10.59	0.67	11.52	12.23	m²	**23.75**
190 mm thick; hollow	12.51	0.84	14.48	14.53	m²	**29.01**
215 mm thick; hollow	13.04	0.92	15.88	15.23	m²	**31.11**
Isolated piers or chimney stacks						
140 mm thick; hollow	–	1.02	17.65	12.23	m²	**29.88**
190 mm thick; hollow	–	1.34	23.19	14.53	m²	**37.72**
215 mm thick; hollow	–	1.53	26.47	15.23	m²	**41.70**
Isolated casings						
75 mm thick; solid	–	0.69	11.94	5.88	m²	**17.82**
100 mm thick; solid	–	0.74	12.80	7.34	m²	**20.14**
140 mm thick; solid	–	0.93	16.09	10.89	m²	**26.98**
Extra over for fair face; flush pointing						
walls; one side	–	0.09	1.56	–	m²	**1.56**
walls; both sides	–	0.14	2.42	–	m²	**2.42**
Bonding ends to common brickwork						
75 mm thick solid	–	0.14	2.42	0.70	m	**3.12**
100 mm thick solid	–	0.23	3.98	0.87	m	**4.85**
140 mm thick solid	–	0.28	4.85	1.29	m	**6.14**
140 mm thick hollow	–	0.28	4.85	1.45	m	**6.30**
190 mm thick hollow	–	0.32	5.54	1.71	m	**7.25**
215 mm thick hollow	–	0.37	6.41	1.81	m	**8.22**

14 MASONRY

Item	PC £	Labour hours	Labour £	Material £	Unit	Total rate £
BLOCK WALLING – cont						
Dense aggregate concrete blocks; (7.00 N/mm²)						
Forticrete blocks; in cement: mortar (1:3)						
Walls						
75 mm thick; solid	4.36	0.50	8.72	5.15	m²	**13.87**
100 mm thick; hollow	3.56	0.62	10.74	4.38	m²	**15.12**
100 mm thick; solid	3.31	0.62	10.74	4.12	m²	**14.86**
140 mm thick; hollow	5.25	0.67	11.52	6.43	m²	**17.95**
140 mm thick; solid	5.25	0.75	12.93	6.43	m²	**19.36**
190 mm thick; hollow	7.09	0.84	14.48	8.69	m²	**23.17**
190 mm thick; solid	7.09	0.92	15.88	8.68	m²	**24.56**
215 mm thick; hollow	6.43	0.92	15.88	8.08	m²	**23.96**
215 mm thick; solid	7.76	0.94	16.26	9.54	m²	**25.80**
Dwarf support wall						
140 mm thick; solid	–	1.16	20.07	6.43	m²	**26.50**
190 mm thick; solid	–	1.34	23.19	8.68	m²	**31.87**
215 mm thick; solid	–	1.53	26.47	9.54	m²	**36.01**
Isolated piers or chimney stacks						
140 mm thick; hollow	–	1.02	17.65	6.43	m²	**24.08**
190 mm thick; hollow	–	1.34	23.19	8.69	m²	**31.88**
215 mm thick; hollow	–	1.53	26.47	8.08	m²	**34.55**
Isolated casings						
75 mm thick; solid	–	0.69	11.94	5.15	m²	**17.09**
100 mm thick; solid	–	0.74	12.80	4.12	m²	**16.92**
140 mm thick; solid	–	0.93	16.09	6.43	m²	**22.52**
Extra over for fair face; flush pointing						
walls; one side	–	0.09	1.56	–	m²	**1.56**
walls; both sides	–	0.14	2.42	–	m²	**2.42**
Bonding ends to common brickwork						
75 mm thick solid	–	0.14	2.42	0.60	m	**3.02**
100 mm thick solid	–	0.23	3.98	0.49	m	**4.47**
140 mm thick solid	–	0.28	4.85	0.78	m	**5.63**
190 mm thick solid	–	0.32	5.54	1.04	m	**6.58**
215 mm thick solid	–	0.37	6.41	1.15	m	**7.56**
Dense aggregate coloured concrete blocks;						
Forticrete Bathstone or other equal and						
approved; in coloured gauged mortar (1:1:6);						
flush pointing one side						
Walls						
100 mm thick hollow	13.42	0.74	12.80	14.95	m²	**27.75**
100 mm thick solid	13.42	0.74	12.80	14.95	m²	**27.75**
140 mm thick hollow	19.44	0.83	14.36	21.64	m²	**36.00**
140 mm thick solid	19.44	0.93	16.09	21.64	m²	**37.73**
215 mm thick hollow	22.22	1.16	20.07	25.01	m²	**45.08**
Isolated piers or chimney stacks						
140 mm thick solid	–	1.25	21.63	21.64	m²	**43.27**
215 mm thick hollow	–	1.57	27.16	25.01	m²	**52.17**

14 MASONRY

Item	PC £	Labour hours	Labour £	Material £	Unit	Total rate £
Extra over blocks for						
100 mm thick half lintel blocks; ref D14	–	0.23	3.98	11.21	m	**15.19**
140 mm thick half lintel blocks; ref H14	–	0.28	4.85	20.07	m	**24.92**
140 mm thick quoin blocks; ref H16	–	0.32	5.54	16.86	m	**22.40**
140 mm thick cavity closer blocks; ref H17	–	0.32	5.54	18.06	m	**23.60**
140 mm thick sill blocks; ref H21	–	0.28	4.85	13.28	m	**18.13**
Glazed finish blocks; Forticrete Astra-Glaze or other equal and approved; in gauged mortar (1:1:6); joints raked out; gun applied latex grout to joints						
Walls or partitions or skins of hollow walls						
100 mm thick; glazed one side	–	0.93	16.09	72.93	m²	**89.02**
extra; glazed square end return	–	0.37	6.41	21.02	m	**27.43**
100 mm thick; glazed both sides	–	1.11	19.21	91.38	m²	**110.59**
100 mm thick lintel 200 mm high; glazed one side	–	0.83	12.92	20.08	m	**33.00**
Fireborn terracotta blocks or other equal and approved; Ibstock Brick Ltd; in coloured gauged mortar (1:1:6); flush pointing one side						
Walls or partitions or skins of hollow walls						
102.50 mm thick; stretcher bond	–	0.33	5.71	48.88	m²	**54.59**
102.50 mm thick; stack bond	–	0.35	6.06	48.84	m²	**54.90**
GLASS BLOCK WALLING						
NOTE: The following specialist prices for glass block walling; supplied by Roger Wilde Ltd; assume standard blocks in panels of 50 m²; work in straight walls at ground level; and all necessary ancillary fixing; strengthening; easy access; pointing and expansion materials etc.						
Hollow glass block walling; Pittsburgh Corning sealed Thinline or other equal and approved; in cement mortar joints; reinforced with 6 mm dia. stainless steel rods; pointed both sides with mastic or other equal and approved						
Walls; facework both sides						
115 mm × 115 mm × 80 mm Flemish blocks	–	–	–	–	m²	**504.00**
190 mm × 190 mm × 80 mm Flemish; cross reeded or clear blocks	–	–	–	–	m²	**207.00**
240 mm × 240 mm × 80 mm Flemish; cross reeded or clear blocks	–	–	–	–	m²	**324.00**
240 mm × 115 mm × 80 mm Flemish, or clear blocks	–	–	–	–	m²	**234.00**
Fire-rated walls						
190 mm × 190 mm × 100 mm glass blocks; 30 minute fire-rated	–	–	–	–	m²	**499.50**
190 mm × 190 mm × 160 mm glass blocks; 60 minute fire-rated	–	–	–	–	m²	**900.00**

14 MASONRY

Item	PC £	Labour hours	Labour £	Material £	Unit	Total rate £
NATURAL STONE RUBBLE WALLING						
Cotswold Guiting limestone or other equal and approved; laid dry						
Uncoursed random rubble walling						
275 mm thick	–	2.07	38.95	83.51	m²	**122.46**
350 mm thick	–	2.46	46.05	106.25	m²	**152.30**
425 mm thick	–	2.81	52.34	129.04	m²	**181.38**
500 mm thick	–	3.15	58.43	151.81	m²	**210.24**
Cotswold Guiting limestone or other equal and approved; bedded; jointed and pointed in cement: lime mortar (1:2:9)						
Uncoursed random rubble walling; faced and pointed; both sides						
275 mm thick	–	1.98	37.14	87.14	m²	**124.28**
350 mm thick	–	2.18	40.40	110.86	m²	**151.26**
425 mm thick	–	2.39	43.86	134.63	m²	**178.49**
500 mm thick	–	2.59	47.12	158.40	m²	**205.52**
Coursed random rubble walling; rough dressed; faced and pointed one side						
114 mm thick	–	1.48	25.17	53.65	m²	**78.82**
150 mm thick	–	1.76	34.51	54.12	m²	**88.63**
Fair returns on walling						
114 mm wide	–	0.02	0.35	–	m	**0.35**
150 mm wide	–	0.03	0.52	–	m	**0.52**
275 mm wide	–	0.06	1.04	–	m	**1.04**
350 mm wide	–	0.08	1.38	–	m	**1.38**
425 mm wide	–	0.10	1.73	–	m	**1.73**
500 mm wide	–	0.12	2.08	–	m	**2.08**
Fair raking cutting or circular cutting						
114 mm wide	–	0.20	3.52	7.82	m	**11.34**
150 mm wide	–	0.25	4.41	7.82	m	**12.23**
Level uncoursed rubble walling for damp-proof courses and the like						
275 mm wide	–	0.19	3.83	9.34	m	**13.17**
350 mm wide	–	0.20	4.04	11.80	m	**15.84**
425 mm wide	–	0.21	4.24	14.32	m	**18.56**
500 mm wide	–	0.22	4.45	16.81	m	**21.26**
Copings formed of rough stones; faced and pointed all round						
275 mm × 200 mm (average) high	–	0.56	10.73	34.12	m	**44.85**
350 mm × 250 mm (average) high	–	0.75	14.24	47.94	m	**62.18**
425 mm × 300 mm (average) high	–	0.97	18.30	67.59	m	**85.89**
500 mm × 300 mm (average) high	–	1.23	23.02	91.93	m	**114.95**

14 MASONRY

Item	PC £	Labour hours	Labour £	Material £	Unit	Total rate £
STONE WALLING/DRESSINGS						
Reconstructed walling; Bradstone masonry blocks; standard colours; or other equal and approved; laid to pattern or course recommended; bedded; jointed and pointed in approved coloured cement: lime mortar (1:2:9)						
Walls; facing and pointing one side						
Fyfestone Enviromasonry Rustic	–	1.00	17.30	22.98	m²	40.28
Fyfestone Enviromasonry Fairfaced	–	1.00	17.30	17.47	m²	34.77
Fyfestone Enviromasonry Split	–	1.10	19.03	27.97	m²	47.00
Fyfestone Enviromasonry Textured	–	1.10	19.03	30.07	m²	49.10
Fyfestone Enviromasonry Polished	–	2.00	34.60	74.20	m²	108.80
masonry blocks; random uncoursed	–	1.04	17.99	40.83	m²	58.82
extra for:						
returned ends	–	0.37	6.41	32.24	m	38.65
plain L shaped quoins	–	0.12	2.08	40.09	m	42.17
traditional walling; coursed squared	–	1.30	22.49	40.83	m²	63.32
squared coursed rubble	–	1.25	21.63	41.91	m²	63.54
squared random rubble	–	1.30	22.49	41.76	m²	64.25
squared and pitched rock faced walling; coursed	–	1.34	23.19	41.76	m²	64.95
ashlar; 440 × 215 × 100 mm thick	–	1.10	19.03	42.64	m²	61.67
rough hewn rockfaced walling; random	–	1.39	24.05	41.60	m²	65.65
extra for:						
returned ends	–	0.15	2.59	–	m	2.59
Isolated piers or chimney stacks; facing and pointing one side						
Enviromasonry Rustic	–	1.40	24.22	22.98	m²	47.20
masonry blocks; random uncoursed	–	1.43	24.74	40.83	m²	65.57
traditional walling; coursed squared	–	1.80	31.14	40.83	m²	71.97
squared coursed rubble	–	1.76	30.45	41.91	m²	72.36
squared random rubble	–	1.80	31.14	41.76	m²	72.90
squared and pitched rock faced walling; coursed	–	1.90	32.87	41.76	m²	74.63
ashlar; 440 × 215 × 100 mm thick	–	1.54	26.64	42.64	m²	69.28
rough hewn rockfaced walling; random	–	1.94	33.57	41.60	m²	75.17
Isolated casings; facing and pointing one side						
Enviromasonry Rustic	–	1.20	20.76	22.98	m²	43.74
masonry blocks; random uncoursed	–	1.25	21.63	40.83	m²	62.46
traditional walling; coursed squared	–	1.57	27.16	40.83	m²	67.99
squared coursed rubble	–	1.53	26.47	41.91	m²	68.38
squared random rubble	–	1.57	27.16	41.76	m²	68.92
squared and pitched rock faced walling; coursed	–	1.62	28.02	41.76	m²	69.78
ashlar; 440 × 215 × 100 mm thick	–	1.32	22.84	42.64	m²	65.48
rough hewn rockfaced walling; random	–	1.67	28.89	41.60	m²	70.49
Fair returns 100 mm wide						
Enviromasonry Rustic	–	0.10	1.73	–	m²	1.73
masonry blocks; random uncoursed	–	0.11	1.91	–	m²	1.91
traditional walling; coursed squared	–	0.14	2.42	–	m²	2.42
squared coursed rubble	–	0.13	2.24	–	m²	2.24
squared random rubble	–	0.14	2.42	–	m²	2.42
squared and pitched rock faced walling; coursed	–	0.14	2.42	–	m²	2.42
ashlar; 440 × 215 × 100 mm thick	–	0.14	2.42	–	m²	2.42
rough hewn rockfaced walling; random	–	0.15	2.59	–	m²	2.59

14 MASONRY

Item	PC £	Labour hours	Labour £	Material £	Unit	Total rate £
STONE WALLING/DRESSINGS – cont						
Reconstructed walling – cont						
Fair raking cutting or circular cutting						
100 mm wide	–	0.17	2.94	–	m	**2.94**
Quoin						
ashlar; 440 × 215 × 215 × 100 mm thick	–	0.75	12.98	81.98	m	**94.96**
Reconstructed limestone dressings; Bradstone Architectural dressings in weathered Cotswold or North Cerney shades or other equal and approved; bedded, jointed and pointed in approved coloured cement: lime mortar (1:2:9)						
Copings; twice weathered and throated						
305 mm × 76 mm; type A	–	0.37	6.41	24.54	m	**30.95**
extra for						
fair end	–	–	–	12.19	nr	**12.19**
returned mitred fair end	–	–	–	12.19	nr	**12.19**
Copings; once weathered and throated						
305 mm × 76 mm	–	0.37	6.41	24.18	m	**30.59**
356 mm × 76 mm	–	0.37	6.41	22.41	m	**28.82**
extra for						
fair end	–	–	–	12.19	nr	**12.19**
returned mitred fair end	–	–	–	12.19	nr	**12.19**
Pier caps; four times weathered and throated						
305 mm × 305 mm	–	0.23	3.98	14.47	nr	**18.45**
381 mm × 381 mm	–	0.23	3.98	21.47	nr	**25.45**
457 mm × 457 mm	–	0.28	4.85	29.36	nr	**34.21**
533 mm × 533 mm	–	0.28	4.85	40.76	nr	**45.61**
Splayed corbels						
479 mm × 100 mm × 215 mm	–	0.14	2.42	24.01	nr	**26.43**
665 mm × 100 mm × 215 mm	–	0.19	3.29	33.19	nr	**36.48**
100 mm × 140 mm lintels; rectangular; reinforced with mild steel bars						
all lengths to 2.07 m	–	0.26	4.50	38.35	m	**42.85**
100 mm × 215 mm lintels; rectangular; reinforced with mild steel bars						
all lengths to 2.85 m	–	0.30	5.19	40.95	m	**46.14**
Sills to suit standard windows; stooled 100 mm at ends						
150 mm ×140 mm; not exceeding 1.97 m long	–	0.28	4.85	51.75	m	**56.60**
197 mm ×140 mm; not exceeding 1.97 m long	–	0.28	4.85	58.12	m	**62.97**
Window surround; traditional with label moulding; for single light; sill 146 mm × 133 mm; jambs 146 mm × 146 mm; head 146 mm × 105 mm; including all dowels and anchors						
overall size 508 mm × 1479 mm	–	0.83	14.36	178.47	nr	**192.83**
Window surround; traditional with label moulding; three light; for windows 508 mm × 1219 mm; sill 146 mm × 133 mm; jambs 146 mm × 146 mm; head 146 mm × 103 mm; mullions 146 mm × 108 mm; including all dowels and anchors						
overall size 1975 mm × 1479 mm	–	2.17	37.55	420.01	nr	**457.56**

14 MASONRY

Item	PC £	Labour hours	Labour £	Material £	Unit	Total rate £
Door surround; moulded continuous jambs and head with label moulding; including all dowels and anchors						
door 839 mm × 1981 mm in 102 mm × 64 mm frame	–	1.53	26.47	380.48	nr	406.95
Portland Whitbed limestone bedded and jointed in cement–lime–mortar (1:2:9); slurrying with weak lime and stone dust mortar; flush pointing and cleaning on completion (cramps etc. not included)						
Facework; one face plain and rubbed; bedded against backing						
50 mm thick stones	–	–	–	–	m²	300.90
63 mm thick stones	–	–	–	–	m²	342.72
75 mm thick stones	–	–	–	–	m²	387.60
100 mm thick stones	–	–	–	–	m²	413.10
Fair returns on facework						
50 mm wide	–	–	–	–	m	4.08
63 mm wide	–	–	–	–	m	5.10
75 mm wide	–	–	–	–	m	7.14
100 mm wide	–	–	–	–	m	9.18
Fair raking cutting on facework						
50 mm thick	–	–	–	–	m	18.36
63 mm thick	–	–	–	–	m	20.40
75 mm thick	–	–	–	–	m	24.48
100 mm thick	–	–	–	–	m	26.52
Copings; once weathered; and throated; rubbed; set horizontal or raking						
250 mm × 50 mm	–	–	–	–	m	142.80
extra for						
external angle	–	–	–	–	nr	25.50
internal angle	–	–	–	–	nr	25.50
300 mm × 50 mm	–	–	–	–	m	150.96
extra for						
external angle	–	–	–	–	nr	25.50
internal angle	–	–	–	–	nr	30.60
350 mm × 75 mm	–	–	–	–	m	168.30
extra for						
external angle	–	–	–	–	nr	25.50
internal angle	–	–	–	–	nr	32.64
400 mm × 100 mm	–	–	–	–	m	201.96
extra for						
external angle	–	–	–	–	nr	30.60
internal angle	–	–	–	–	nr	42.84
450 mm × 100 mm	–	–	–	–	m	244.80
extra for						
external angle	–	–	–	–	nr	38.76
internal angle	–	–	–	–	nr	53.04

14 MASONRY

Item	PC £	Labour hours	Labour £	Material £	Unit	Total rate £
STONE WALLING/DRESSINGS – cont						
Portland Whitbed limestone bedded and jointed in cement–lime–mortar (1:2:9) – cont						
Copings – cont						
500 mm × 125 mm	–	–	–	–	m	372.30
extra for						
external angle	–	–	–	–	nr	53.04
internal angle	–	–	–	–	nr	66.30
Band courses; plain; rubbed; horizontal						
225 mm × 112 mm	–	–	–	–	m	112.20
300 mm × 112 mm	–	–	–	–	m	149.94
extra for						
stopped ends	–	–	–	–	nr	6.12
external angles	–	–	–	–	nr	6.12
Band courses; moulded 100 mm girth on face; rubbed; horizontal						
125 mm × 75 mm	–	–	–	–	m	127.50
extra for						
stopped ends	–	–	–	–	nr	20.40
external angles	–	–	–	–	nr	25.50
internal angles	–	–	–	–	nr	51.00
150 mm × 75 mm	–	–	–	–	m	147.90
extra for						
stopped ends	–	–	–	–	nr	20.40
external angles	–	–	–	–	nr	35.70
internal angles	–	–	–	–	nr	61.20
200 mm × 100 mm	–	–	–	–	m	168.30
extra for						
stopped ends	–	–	–	–	nr	20.40
external angles	–	–	–	–	nr	51.00
internal angles	–	–	–	–	nr	91.80
250 mm × 150 mm	–	–	–	–	m	244.80
extra for						
stopped ends	–	–	–	–	nr	20.40
external angles	–	–	–	–	nr	51.00
internal angles	–	–	–	–	nr	102.00
300 mm × 250 mm	–	–	–	–	m	397.80
extra for						
stopped ends	–	–	–	–	nr	20.40
external angles	–	–	–	–	nr	71.40
internal angles	–	–	–	–	nr	122.40
Coping apex block; two sunk faces; rubbed						
650 mm × 450 mm × 225 mm	–	–	–	–	nr	499.80
Coping kneeler block; three sunk faces; rubbed						
350 mm × 350 mm × 375 mm	–	–	–	–	nr	397.80
450 mm × 450 mm × 375 mm	–	–	–	–	nr	459.00
Corbel; turned and moulded; rubbed						
225 mm × 225 mm × 375 mm	–	–	–	–	nr	326.40

14 MASONRY

Item	PC £	Labour hours	Labour £	Material £	Unit	Total rate £
Slab surrounds to openings; one face splayed; rubbed						
75 mm × 100 mm	–	–	–	–	m	71.40
75 mm × 200 mm	–	–	–	–	m	96.90
100 mm × 100 mm	–	–	–	–	m	86.70
125 mm × 100 mm	–	–	–	–	m	96.90
125 mm × 150 mm	–	–	–	–	m	117.30
175 mm × 175 mm	–	–	–	–	m	142.80
225 mm × 175 mm	–	–	–	–	m	168.30
300 mm × 175 mm	–	–	–	–	m	204.00
300 mm × 225 mm	–	–	–	–	m	265.20
Slab surrounds to openings; one face sunk splayed; rubbed						
75 mm × 100 mm	–	–	–	–	m	91.80
75 mm × 200 mm	–	–	–	–	m	117.30
100 mm × 100 mm	–	–	–	–	m	107.10
125 mm × 100 mm	–	–	–	–	m	117.30
125 mm × 150 mm	–	–	–	–	m	137.70
175 mm × 175 mm	–	–	–	–	m	163.20
225 mm × 175 mm	–	–	–	–	m	188.70
300 mm × 175 mm	–	–	–	–	m	224.40
300 mm × 225 mm	–	–	–	–	m	285.60
extra for:						
throating	–	–	–	–	m	10.20
rebates and grooves	–	–	–	–	m	22.44
stooling	–	–	–	–	m	38.76
Eurobrick insulated brick cladding systems or other equal and approved; extruded polystyrene foam insulation; brick slips bonded to insulation panels with Eurobrick gun applied adhesive or other equal and approved; pointing with formulated mortar grout						
25 mm insulation to walls						
over 300 mm wide; fixing with proprietary screws and plates to timber	–	1.39	24.31	40.04	m²	64.35
50 mm insulation to walls						
over 300 mm wide; fixing with proprietary screws and plates; to timber	–	1.39	24.31	43.74	m²	68.05
SUNDRY ITEMS AND ACCESSORIES						
Forming cavities						
In hollow walls						
width of cavity 50 mm; polypropylene ties; three wall ties per m²	–	0.05	0.86	0.26	m²	1.12
width of cavity 50 mm; galvanized steel twisted wall ties; three wall ties per m²	–	0.05	0.86	0.75	m²	1.61
width of cavity 50 mm; stainless steel butterfly wall ties; three wall ties per m²	–	0.05	0.86	0.54	m²	1.40

14 MASONRY

Item	PC £	Labour hours	Labour £	Material £	Unit	Total rate £
SUNDRY ITEMS AND ACCESSORIES – cont						
Forming cavities – cont						
In hollow walls – cont						
width of cavity 50 mm; stainless steel twisted wall ties; three wall ties per m²	–	0.05	0.86	0.69	m²	**1.55**
width of cavity 75 mm; polypropylene ties; three wall ties per m²	–	0.05	0.86	0.26	m²	**1.12**
width of cavity 75 mm; galvanized steel twisted wall ties; three wall ties per m²	–	0.05	0.86	0.79	m²	**1.65**
width of cavity 75 mm; stainless steel butterfly wall ties; three wall ties per m²	–	0.05	0.86	0.77	m²	**1.63**
width of cavity 75 mm; stainless steel twisted wall ties; three wall ties per m²	–	0.05	0.86	0.76	m²	**1.62**
Damp-proof courses						
Polythene damp-proof course or other equal and approved; 200 mm laps; in gauged mortar (1:1:6)						
width exceeding 225 mm; horizontal	–	0.23	3.98	0.35	m²	**4.33**
width exceeding 225 mm; forming cavity gutters in hollow walls; horizontal	–	0.37	6.41	0.35	m²	**6.76**
width not exceeding 225 mm; horizontal	–	0.46	7.95	0.35	m²	**8.30**
width not exceeding 225 mm; vertical	–	0.69	11.94	0.35	m²	**12.29**
Icopal Polymeric damp-proof course; Xtra-Load Pro-Build or other equal and approved; 200 mm laps; in gauged morter (1:1:6)						
width exceeding 225 mm; horizontal	–	0.23	3.98	13.45	m²	**17.43**
width exceeding 225 mm; forming cavity gutters in hollow walls; horizontal	–	0.37	6.41	13.45	m²	**19.86**
width not exceeding 225 mm; horizontal	–	0.46	7.95	13.45	m²	**21.40**
width not exceeding 225 mm; vertical	–	0.69	11.94	13.45	m²	**25.39**
Zedex CPT (Co-Polymer Thermoplastic) damp-proof course or other equal and approved; 200 mm laps; in gauged mortar (1:1:6)						
width exceeding 225 mm; horizontal	–	0.23	3.98	5.74	m²	**9.72**
width exceeding 225 mm wide; forming cavity gutters in hollow walls; horizontal	–	0.37	6.41	5.74	m²	**12.15**
width not exceeding 225 mm; horizontal	–	0.46	7.95	5.74	m²	**13.69**
width not exceeding 225 mm; vertical	–	0.69	11.94	5.74	m²	**17.68**
Hyload (pitch polymer) damp-proof course or other equal and approved; 150 mm laps; in gauged mortar (1:1:6)						
width exceeding 225 mm; horizontal	–	0.23	3.98	4.05	m²	**8.03**
width exceeding 225 mm; forming cavity gutters in hollow walls; horizontal	–	0.37	6.41	4.05	m²	**10.46**
width not exceeding 225 mm; horizontal	–	0.46	7.95	4.05	m²	**12.00**
width not exceeding 225 mm; vertical	–	0.69	11.94	4.05	m²	**15.99**

14 MASONRY

Item	PC £	Labour hours	Labour £	Material £	Unit	Total rate £
Nubit bitumen and polyester-based damp-proof course or other equal and approved; 200 mm laps; in gauged mortar (1:1:6)						
width exceeding 225 mm; horizontal	–	0.23	3.98	8.09	m²	**12.07**
width exceeding 225 mm wide; forming cavity gutters in hollow walls; horizontal	–	0.37	6.41	8.09	m²	**14.50**
width not exceeding 225 mm; horizontal	–	0.46	7.95	8.09	m²	**16.04**
width not exceeding 225 mm; vertical	–	0.69	11.94	8.09	m²	**20.03**
Permabit bitumen polymer damp-proof course or other equal and approved; 150 mm laps; in gauged mortar (1:1:6)						
width exceeding 225 mm; horizontal	–	0.23	3.98	8.72	m²	**12.70**
width exceeding 225 mm; forming cavity gutters in hollow walls; horizontal	–	0.37	6.41	8.72	m²	**15.13**
width not exceeding 225 mm; horizontal	–	0.46	7.95	8.72	m²	**16.67**
width not exceeding 225 mm; vertical	–	0.69	11.94	8.72	m²	**20.66**
Alumite aluminium cored bitumen gas retardant damp-proof course or other equal and approved; 200 mm laps; in gauged mortar (1:1;6)						
width exceeding 225 mm; horizontal	–	0.31	5.36	6.06	m²	**11.42**
width exceeding 225 mm; forming cavity gutters in hollow walls; horizontal	–	0.49	8.48	6.06	m²	**14.54**
width not exceeding 225 mm; horizontal	–	0.60	10.38	6.06	m²	**16.44**
width not exceeding 225 mm; vertical	–	0.83	14.36	6.06	m²	**20.42**
Milled lead damp-proof course; BS 1178; 1.80 mm thick (code 4), 175 mm laps; in cement: lime mortar (1:2:9)						
width exceeding 225 mm; horizontal (PC £/kg)	–	1.85	32.01	33.53	m²	**65.54**
width not exceeding 225 mm; horizontal	–	2.78	48.09	33.53	m²	**81.62**
Two courses slates in cement: mortar (1:3)						
width exceeding 225 mm; horizontal	–	1.39	24.05	13.52	m²	**37.57**
width exceeding 225 mm; vertical	–	2.08	35.99	13.52	m²	**49.51**
Synthaprufe damp-proof membrane or other equal and approved; three coats brushed on						
width not exceeding 150 mm; vertical	–	0.31	3.59	6.14	m²	**9.73**
width 150 mm–225 mm; vertical	–	0.30	3.47	6.14	m²	**9.61**
width 225 mm–300 mm; vertical	–	0.28	3.24	6.14	m²	**9.38**
width exceeding 300 mm wide; vertical	–	0.26	3.01	6.14	m²	**9.15**
Joint reinforcement						
Brickforce galvanized steel joint reinforcement or other equal and approved						
width 60 mm; ref GBF40W60B25	–	0.05	0.86	0.87	m	**1.73**
width 100 mm; ref GBF40W100B25	–	0.07	1.21	1.03	m	**2.24**
width 175 mm; ref GBF40W175B25	–	0.10	1.73	2.05	m	**3.78**
Brickforce stainless steel joint reinforcement or other equal and approved						
width 60 mm; ref SBF35W60BSC	–	0.05	0.86	2.61	m	**3.47**
width 100 mm; ref SBF35W100BSC	–	0.07	1.21	3.66	m	**4.87**
width 175 mm; ref SBF35W175BSC	–	0.10	1.73	2.96	m	**4.69**

14 MASONRY

Item	PC £	Labour hours	Labour £	Material £	Unit	Total rate £
SUNDRY ITEMS AND ACCESSORIES – cont						
Joint reinforcement – cont						
Wallforce stainless steel joint reinforcement or other equal and approved						
width 240 mm; ref SWF35W240	–	0.12	2.08	4.29	m	**6.37**
width 260 mm; ref SWF35W260	–	0.13	2.24	5.34	m	**7.58**
width 275 mm; ref SWF35W275	–	0.14	2.42	6.39	m	**8.81**
Weather fillets						
Weather fillets in cement: mortar (1:3)						
50 mm face width	–	0.11	1.91	0.05	m	**1.96**
100 mm face width	–	0.19	3.29	0.18	m	**3.47**
Angle fillets						
Angle fillets in cement: mortar (1:3)						
50 mm face width	–	0.11	1.91	0.05	m	**1.96**
100 mm face width	–	0.19	3.29	0.18	m	**3.47**
Pointing in						
Pointing with mastic						
wood frames or sills	–	0.09	1.30	0.59	m	**1.89**
Pointing with polysulphide sealant						
wood frames or sills	–	0.09	1.30	2.24	m	**3.54**
Wedging and pinning						
To underside of existing construction with slates in cement: mortar (1:3)						
width of wall – one brick thick	–	0.74	12.80	3.00	m	**15.80**
width of wall – one and a half brick thick	–	0.93	16.09	6.01	m	**22.10**
width of wall – two brick thick	–	1.11	19.21	9.00	m	**28.21**
Joints						
Hacking joints and faces of brickwork or blockwork						
to form key for plaster	–	0.24	2.78	–	m²	**2.78**
Raking out joint in brickwork or blockwork for turned-in edge of flashing						
horizontal	–	0.14	2.42	–	m	**2.42**
stepped	–	0.19	3.29	–	m	**3.29**
Raking out and enlarging joint in brickwork or blockwork for nib of asphalt						
horizontal	–	0.19	3.29	–	m	**3.29**
Cutting grooves in brickwork or blockwork						
for water bars and the like	–	0.23	2.67	0.69	m	**3.36**
for nib of asphalt; horizontal	–	0.23	2.67	0.69	m	**3.36**
Preparing to receive new walls						
top existing 215 mm wall	–	0.19	3.29	–	m	**3.29**

14 MASONRY

Item	PC £	Labour hours	Labour £	Material £	Unit	Total rate £
Cleaning and priming both faces; filling with preformed closed cell joint filler and pointing one side with polysulphide sealant; 12 mm deep						
expansion joints; 12 mm wide	–	0.23	3.69	4.31	m	**8.00**
expansion joints; 20 mm wide	–	0.28	4.41	6.34	m	**10.75**
expansion joints; 25 mm wide	–	0.32	4.96	7.70	m	**12.66**
Fire-resisting horizontal expansion joints; filling with joint filler; fixed with high temperature slip adhesive; between top of wall and soffit						
wall not exceeding 215 mm wide; 10 mm wide joint with 30 mm deep filler (one hour fire seal)	–	0.23	3.98	4.98	m	**8.96**
wall not exceeding 215 mm wide; 10 mm wide joint with 30 mm deep filler (two hour fire seal)	–	0.23	3.98	4.98	m	**8.96**
wall not exceeding 215 mm wide; 20 mm wide joint with 45 mm deep filler (two hour fire seal)	–	0.28	4.85	7.69	m	**12.54**
wall not exceeding 215 mm wide; 30 mm wide joint with 75 mm deep filler (three hour fire seal)	–	0.32	5.54	19.79	m	**25.33**
Fire-resisting vertical expansiojn joints; filling with joint filler; fixed with high temperature slip adhesive; with polysulphide sealant one side; between end of wall and concrete						
wall not exceeding 215 mm wide; 20 mm wide joint with 45 mm deep filler (two hour fire seal)	–	0.37	6.11	12.18	m	**18.29**
Slate and tile sills						
Sills; two courses of machine-made plain roofing tiles						
set weathering; bedded and pointed	–	0.56	9.69	5.36	m	**15.05**
Sundries						
Weep holes						
perpend units; plastic	–	0.02	0.35	0.12	nr	**0.47**
Chimney pots; red terracotta; plain or cannon-head; setting and flaunching in cement: mortar (1:3)						
185 mm dia. × 300 mm long	24.55	1.67	28.89	26.55	nr	**55.44**
185 mm dia. × 600 mm long	37.59	1.85	32.01	39.91	nr	**71.92**
185 mm dia. × 900 mm long	71.11	1.85	32.01	1.38	nr	**33.39**
Air bricks						
Air bricks; red terracotta; building into prepared openings						
215 mm × 65 mm	–	0.07	1.21	2.05	nr	**3.26**
215 mm × 140 mm	–	0.07	1.21	2.84	nr	**4.05**
215 mm × 215 mm	–	0.07	1.21	7.62	nr	**8.83**

14 MASONRY

Item	PC £	Labour hours	Labour £	Material £	Unit	Total rate £
SUNDRY ITEMS AND ACCESSORIES – cont						
Halfen channels and brick support (NB Supply only rates)						
cavity brick tie;' Halfen HTS – C 12–225 mm stainless steel A2	–	–	–	0.40	nr	**0.40**
cavity brick tie; Halfen HTS – C 12–225 mm, pre-galvanized	–	–	–	0.20	nr	**0.20**
frame cramp, with slot; Halfen HTS – FS 12–150 mm stainless steel A2	–	–	–	0.31	nr	**0.31**
frame cramp, with slot; Halfen HTS – FS 12–150 mm pre-galvanized	–	–	–	0.18	nr	**0.18**
head restraint; Halfen CHR-V telescopic concealed tube and HSC–9 tie, in A2/ 304 stainless steel	–	–	–	2.23	nr	**2.23**
brick tie channel; Halfen FRS 25/14 2700, supplied in 2700 mm lengths with fixing holes at 110 mm centres. grade 304 stainless steel	–	–	–	6.85	nr	**6.85**
channel tie; Halfen FRS–03 150 mm long safety end to suit 25/14 FRS channel, stainless steel	–	–	–	0.29	nr	**0.29**
ribslot brick tie channel; Halfen – 3000 mm long, stainless steel A2, complete with polystyrene filler (vf) and nail holes, to suit HTS ties	–	–	–	11.66	nr	**11.66**
channel tie; Halfen HTS – B 12–200 mm long, grade 304 stainless steel safety end to suit 28/15, 25/17 HH and ribslot channels	–	–	–	0.40	nr	**0.40**
ribslot brick tie channel; Halfen – 3000 mm long, pre-galvanized. mild steel, complete with polystyrene filler (vf) and nail holes, to suit HTS ties	–	–	–	6.63	nr	**6.63**
channel tie; Halfen HTS – B 12–150 mm long, safety end pre-galvanized to suit 28/15, 25/17 HH and ribslot channels	–	–	–	0.17	nr	**0.17**
brick support angle swystem; Halfen HK4-U–100 cavity/6 kN inc fixings stainless steel, A4	–	–	–	58.87	m	**58.87**
stone support anchor, grout-in; Halfen UMA – 10– 1 – 150 mm stainless steel	–	–	–	1.99	nr	**1.99**
stone restraint anchor, grout-in; Halfen UHA – 5– 1 – 180 mm stainless steel A4	–	–	–	1.01	nr	**1.01**
windpost, brickwork; Halfen BW1–1406 × 2500 mm complete with fixings & ties, stainless steel, A4	–	–	–	206.53	nr	**206.53**
windpost, cavity; Halfen CW2–3544 × 2500 mm stainless steel complete with fixings & ties	–	–	–	121.00	nr	**121.00**
cast-in CE marked channel, cold rolled; Halfen HTA-K – 38/17–200 mm anchor centres, stainless steel A2–70 , complete with combi filler(kf) and nail holes supplied in 3000 mm lengths	–	–	–	17.30	m	**17.30**
'T' bolt and nut to suit; Halfen HS – 38/17 channel – M12 × 50 mm, stainless steel A2–70	–	–	–	2.29	nr	**2.29**

14 MASONRY

Item	PC £	Labour hours	Labour £	Material £	Unit	Total rate £
cast-in CE marked channel, cold rolled; Halfen HTA-K – 38/17–200 mm anchor centres , hot-dip galvanized (fv), complete with combi filler (kf) and nail holes supplies in 3000 mm lengths	–	–	–	11.46	m	**11.46**
T bolt and nut to suit Halfen HS – 38/17 channel – M12 × 50 mm, hot-dip galvanized (fv) 4.6 mild steel	–	–	–	0.87	nr	**0.87**
cast-in CE marked channel, hot rolled; Halfen HTA-K – 40/22 – hot-dip galvanized (fv), complete with combi filler (kf) and nail holes supplied in 3000 mm lengths	–	–	–	19.52	m	**19.52**
T bolt and nut to suit Halfen HS – 40/22 channel – M16 × 50 mm, hot-dip galvanized (fv) 4.6 mild steel	–	–	–	1.62	nr	**1.62**
cast-in channel, hot rolled; Halfen HTA – 52/34 – hot-dip galvanized (fv), complete with polystyrene filler (vf) and nail holes supplied in 6070 mm lengths	–	–	–	53.39	m	**53.39**
T bolt and nut, to suit Halfen HS – 52/34 channel – M16 × 60 mm hot-dip galvanized (fv) 4.6 mild steel	–	–	–	2.76	nr	**2.76**
cast-in CE marked channel, hot rolled; Halfen HTA – 52/34 – stainless steel A4, complete with polystyrene filler (vf) and nail holes supplied in 6070 mm lengths	–	–	–	133.96	m	**133.96**
T bolt and nut, to suit Halfen HS – 52/34 channel – M16 × 60 mm stainless steel A4 (fv) 4.6 mild steel	–	–	–	8.47	nr	**8.47**
reinforcement, female coupler; Halfen HBM-F 16/ M20 810 mm long	–	–	–	4.34	nr	**4.34**
reinforcement, male coupler; Halfen HBM-M 16/ M20 770 mm long	–	–	–	3.97	nr	**3.97**
reinforcement continuity system; Halfen Kwikastrip – M17S b12 C150 1200 mm	–	–	–	22.72	nr	**22.72**
shear dowel set; Halfen CRET – 124 stainless steel A4	–	–	–	67.80	nr	**67.80**
shear dowel set; Halfen CRET – 10–20 × 300 stainless steel A4	–	–	–	14.16	nr	**14.16**
shear sleeve; Halfen CRET – J – 20–160 stainless steel A4	–	–	–	6.79	nr	**6.79**
concrete Balcony Insulated Connection Unit Halfen HIT HP – 80 mm unit for a 200 mm slab for a 1500 mm cantilever balcony (Passivhaus Certified)	–	–	–	78.41	nr	**78.41**
concrete Balcony Insulated Connection Unit Halfen HIT HP – 80 mm Unit for a 200 mm slab for a 2000 mm cantilever balcony (Passivhaus Certified)	–	–	–	115.31	nr	**115.31**
steel Balcony Connection SBC-HBM-M20x4 BZP Cast-in Socket Anchor Assembly, 190 × 110 mm	–	–	–	40.83	nr	**40.83**

14 MASONRY

Item	PC £	Labour hours	Labour £	Material £	Unit	Total rate £
SUNDRY ITEMS AND ACCESSORIES – cont						
Halfen channels and brick support (NB Supply only rates) – cont						
steel balcony connection; Halfen; SBC-TSS–10 /330 × 200 mm thermal separator pad, 4 no. 22 mm holes at 190 × 110 mm	–	–	–	52.31	nr	**52.31**
precast lifting spread anchor CE marked; Halfen TPA – FS – 4.0 Tonne × 180 mm carbon steel (wb)	–	–	–	2.06	nr	**2.06**
precast ring clutch CE marked; Halfen TPA – R1– 5.0T (T)	–	–	–	101.48	nr	**101.48**
External door and window cavity closers; Thermabate or equivalent; inclusive of flange clips; jointing strips; wall fixing ties and adhesive tape						
closing cavities; width of cavity 50 mm–60 mm	–	0.14	2.42	1.89	m	**4.31**
closing cavities; width of cavity 75 mm–84 mm	–	0.14	2.42	2.05	m	**4.47**
closing cavities; width of cavity 90 mm–99 mm	–	0.14	2.42	2.24	m	**4.66**
closing cavities; width of cavity 100 mm–110 mm	–	0.14	2.42	2.24	m	**4.66**
External door and window cavity closers; Kooltherm or equivalent; inclusive of flange clips; jointing strips; wall fixing ties and adhesive tape; 1 hr fire-rating						
closing cavities; width of cavity 50 mm	–	0.14	2.42	2.40	m	**4.82**
closing cavities; width of cavity 75 mm	–	0.14	2.42	2.49	m	**4.91**
closing cavities; width of cavity 100 mm	–	0.14	2.42	2.67	m	**5.09**
closing cavities; width of cavity 125 mm	–	0.14	2.42	3.17	m	**5.59**
closing cavities; width of cavity 150 mm	–	0.14	2.42	3.31	m	**5.73**
Type H cavicloser or other equal and approved; uPVC universal cavity closer, insulator and damp-proof course by Cavity Trays Ltd; built into cavity wall as work proceeds, complete with face closer and ties						
closing cavities; width of cavity 50 mm–100 mm	–	0.07	1.21	4.85	m	**6.06**
Type L durropolyethelene lintel stop ends or other equal and approved; Cavity Trays Ltd; fixing with butyl anchoring strip; building in as the work proceeds						
adjusted to lintel as required	–	0.04	0.70	0.48	nr	**1.18**
Type W polypropylene weeps/vents or other equal and approved; Cavity Trays Ltd; built into cavity wall as work proceeds						
100/115 mm × 65 mm × 10 mm including lock fit wedges	–	0.04	0.70	0.38	nr	**1.08**
extra; extension duct 200/ 225 mm × 65 mm × 10 mm	–	0.07	1.21	0.64	nr	**1.85**

14 MASONRY

Item	PC £	Labour hours	Labour £	Material £	Unit	Total rate £
Type X polypropylene abutment cavity tray or other equal and approved; Cavity Trays Ltd; built into facing brickwork as the work proceeds; complete with Code 4 flashing; intermediate/catchment tray with short leads (requiring soakers); to suit roof of						
17–20° pitch	–	0.05	0.86	4.52	nr	5.38
21–25° pitch	–	0.05	0.86	4.20	nr	5.06
26–45° pitch	–	0.05	0.86	4.02	nr	4.88
Type X polypropylene abutment cavity tray or other equal and approved; Cavity Trays Ltd; built into facing brickwork as the work proceeds; complete with Code 4 flashing; intermediate/catchment tray with long leads (suitable only for corrugated roof tiles); to suit roof of						
17–20° pitch	–	0.05	0.86	6.10	nr	6.96
21–25° pitch	–	0.05	0.86	5.62	nr	6.48
26–45° pitch	–	0.05	0.86	5.18	nr	6.04
Type X polypropylene abutment cavity tray or other equal and approved; Cavity Trays Ltd; built into facing brickwork as the work proceeds; complete with Code 4 flashing; ridge tray with short/long leads; to suit roof of						
17–20° pitch	–	0.05	0.86	10.27	nr	11.13
21–25° pitch	–	0.05	0.86	9.53	nr	10.39
26–45° pitch	–	0.05	0.86	8.48	nr	9.34
Servicised Bituthene MR aluminium faced gas-resistant cavity flashing or other equal and approved; sealed at joints with Servitape 30 mm; in gauged mortar (1:1:6)						
width exceeding 225 mm wide	–	0.79	13.66	21.88	m²	35.54
Expamet stainless steel wall starters or other equal and approved; plugged and screwed						
to suit walls 60 mm–75 mm thick	–	0.23	2.67	1.87	m	4.54
to suit walls 100 mm–115 mm thick	–	0.23	2.67	18.02	m	20.69
to suit walls 125 mm–180 mm thick	–	0.37	4.28	2.56	m	6.84
to suit walls 190 mm–260 mm thick	–	0.46	5.33	3.18	m	8.51
Galvanized steel lintels; Catnic or other equal and approved; built into brickwork or blockwork						
70/125 Range CG open back lintel for cavity wall						
750 mm long	–	0.23	3.98	29.84	nr	33.82
900 mm long	–	0.28	4.85	35.63	nr	40.48
1200 mm long	–	0.32	5.54	46.81	nr	52.35
1500 mm long	–	0.37	6.41	58.90	nr	65.31
1800 mm long	–	0.42	7.27	80.82	nr	88.09
2100 mm long	–	0.46	7.95	95.16	nr	103.11
2400 mm long	–	0.56	9.69	131.38	nr	141.07
70/125 range CUB open back lintel for cavity wall						
2700 mm long	–	0.65	11.24	154.25	nr	165.49
3000 mm long	–	0.74	12.80	214.77	nr	227.57

14 MASONRY

Item	PC £	Labour hours	Labour £	Material £	Unit	Total rate £
SUNDRY ITEMS AND ACCESSORIES – cont						
Galvanized steel lintels – cont						
70/125 range CU open back lintel for cavity wall						
3300 mm long	–	0.83	14.36	264.67	nr	**279.03**
3600 mm long	–	0.93	16.09	297.24	nr	**313.33**
3900 mm long	–	1.02	17.65	318.85	nr	**336.50**
4200 mm long	–	0.46	7.95	349.70	nr	**357.65**
90/125 range CG open back lintel for cavity wall						
750 mm long	–	0.23	3.98	33.18	nr	**37.16**
900 mm long	–	0.28	4.85	39.81	nr	**44.66**
1200 mm long	–	0.32	5.54	52.26	nr	**57.80**
1500 mm long	–	0.37	6.41	65.17	nr	**71.58**
1800 mm long	–	0.42	7.27	82.45	nr	**89.72**
2100 mm long	–	0.46	7.95	97.70	nr	**105.65**
2400 mm long	–	0.56	9.69	138.00	nr	**147.69**
90/125 range CUB open back lintel for cavity wall						
2700 mm long	–	0.65	11.24	159.67	nr	**170.91**
3000 mm long	–	0.74	12.80	229.65	nr	**242.45**
90/125 range CU open back lintel for cavity wall						
3300 mm long	–	0.83	14.36	285.81	nr	**300.17**
3600 mm long	–	0.93	16.09	318.33	nr	**334.42**
3900 mm long	–	1.02	17.65	339.61	nr	**357.26**
4200 mm long	–	0.46	7.95	364.04	nr	**371.99**
CN92 single lintel; for 75 mm internal walls						
1050 mm long	–	0.28	4.85	5.06	nr	**9.91**
1200 mm long	–	0.32	5.54	5.72	nr	**11.26**
CN102 single lintel; for 100 mm internal walls						
1050 mm long	–	0.28	4.85	6.40	nr	**11.25**
1200 mm long	–	0.32	5.54	7.05	nr	**12.59**
CN100 single lintel; for 75 mm internal walls						
1050 mm long	–	0.28	4.85	15.59	nr	**20.44**
1200 mm long	–	0.32	5.54	19.36	nr	**24.90**
CN5XA single lintel; for 100 mm internal walls						
1050 mm long	–	0.28	4.85	19.08	nr	**23.93**
1200 mm long	–	0.32	5.54	19.95	nr	**25.49**
Sundries – stone walling						
Coating backs of stones with brush applied cold bitumen solution; two coats						
limestone facework	–	0.19	2.60	2.05	m²	**4.65**
Cutting grooves in limestone masonry for						
water bars or the like	–	–	–	–	m	**10.20**
Mortices in limestone masonry for						
metal dowel	–	–	–	–	nr	**2.04**
metal cramp	–	–	–	–	nr	**4.08**
Stainless steel cramps and dowels; Halfen-Deha or other equal and approved; one end built into brickwork or set in slot in concrete						

14 MASONRY

Item	PC £	Labour hours	Labour £	Material £	Unit	Total rate £
Dowel						
8 mm dia. × 75 mm long	0.19	0.04	0.70	0.20	nr	**0.90**
10 mm dia. × 150 mm long	0.56	0.04	0.70	0.57	nr	**1.27**
Pattern J tie						
25 mm × 3 mm × 100 mm	0.42	0.06	1.04	0.43	nr	**1.47**
Pattern S cramp; with two 20 mm turndowns						
(190 mm girth)						
25 mm × 3 mm × 150 mm	0.59	0.06	1.04	0.60	nr	**1.64**
Pattern B anchor; with 8 mm × 75 mm loose dowel						
25 mm × 3 mm × 150 mm	0.72	0.09	1.56	0.74	nr	**2.30**
Pattern Q tie						
25 mm × 3 mm × 200 mm	0.73	0.06	1.04	0.75	nr	**1.79**
38 mm × 3 mm × 250 mm	1.48	0.06	1.04	1.52	nr	**2.56**
Pattern P half twist tie						
25 mm × 3 mm × 200 mm	0.79	0.06	1.04	0.81	nr	**1.85**
38 mm × 3 mm × 250 mm	1.24	0.06	1.04	1.27	nr	**2.31**
FLUES AND FLUE LININGS						
Flues						
Scheidel Rite-Vent ICS Plus flue system; suitable for domestic multifuel appliances; stainless steel; twin wall; insulated; for use internally or externally						
80 mm pipes; including one locking band (fixing brackets measured separately)	–	0.90	11.59	100.04	m	**111.63**
extra for						
Appliance Connecter	–	0.80	10.31	15.48	nr	**25.79**
30° Bend	–	1.80	23.20	75.31	nr	**98.51**
45° Bend	–	1.80	23.20	71.56	nr	**94.76**
135° Tee; fully welded	–	2.70	34.79	148.11	nr	**182.90**
Inspection Length	–	0.90	11.59	10.70	nr	**22.29**
Drain Plug and Support	–	1.00	12.88	69.37	nr	**82.25**
Damper	–	0.90	11.59	56.41	nr	**68.00**
Angled Flashing including Storm Collar	–	1.25	16.10	74.73	nr	**90.83**
Stub Terminal	–	1.00	12.88	24.74	nr	**37.62**
Tapered Terminal	–	1.00	12.88	51.55	nr	**64.43**
Floor Support (2 piece)	–	1.50	19.33	40.71	nr	**60.04**
Firestop Floor Support (2 piece)	–	1.50	19.33	22.82	nr	**42.15**
Wall Support (Stainless Steel)	–	1.00	12.88	86.45	nr	**99.33**
Wall Sleeve	–	1.20	15.47	34.31	nr	**49.78**
100 mm pipes; including one locking band (fixing brackets measured separately)	–	1.00	12.88	106.55	m	**119.43**

14 MASONRY

Item	PC £	Labour hours	Labour £	Material £	Unit	Total rate £
FLUES AND FLUE LININGS – cont						
Flues – cont						
Scheidel Rite-Vent ICS Plus flue system – cont						
extra for						
Appliance Connecter	–	0.90	11.59	17.09	nr	**28.68**
30° Bend	–	2.00	25.77	78.78	nr	**104.55**
45° Bend	–	2.00	25.77	74.79	nr	**100.56**
135° Tee; fully welded	–	3.00	38.66	143.80	nr	**182.46**
Inspection Length	–	1.00	12.88	245.23	nr	**258.11**
Drain Plug and Support	–	1.10	14.18	71.83	nr	**86.01**
Damper	–	1.00	12.88	59.39	nr	**72.27**
Angled Flashing including Storm Collar	–	1.40	18.04	83.27	nr	**101.31**
Stub Terminal	–	1.10	14.18	25.14	nr	**39.32**
Tapered Terminal	–	1.10	14.18	55.01	nr	**69.19**
Floor Support (2 piece)	–	1.65	21.26	46.30	nr	**67.56**
Firestop Floor Support (2 piece)	–	1.65	21.26	25.15	nr	**46.41**
Wall Support (Stainless Steel)	–	1.10	14.18	91.36	nr	**105.54**
Wall Sleeve	–	1.35	17.39	38.95	nr	**56.34**
150 mm pipes; including one locking band (fixing brackets measured separately)	–	1.10	14.18	124.39	m	**138.57**
extra for						
Appliance Connecter	–	1.00	12.88	22.03	nr	**34.91**
30° Bend	–	2.20	28.35	94.55	nr	**122.90**
45° Bend	–	2.20	28.35	89.89	nr	**118.24**
135° Tee; fully welded	–	3.30	42.53	166.06	nr	**208.59**
Inspection Length	–	1.10	14.18	257.18	nr	**271.36**
Drain Plug and Support	–	1.20	15.47	91.73	nr	**107.20**
Damper	–	1.10	14.18	77.62	nr	**91.80**
Angled Flashing including Storm Collar	–	1.55	19.98	84.49	nr	**104.47**
Stub Terminal	–	1.20	15.47	27.05	nr	**42.52**
Tapered Terminal	–	1.20	15.47	63.23	nr	**78.70**
Floor Support (2 piece)	–	1.80	23.20	46.30	nr	**69.50**
Firestop Floor Support (2 piece)	–	1.80	23.20	25.15	nr	**48.35**
Wall Support (Stainless Steel)	–	1.20	15.47	101.18	nr	**116.65**
Wall Sleeve	–	1.50	19.33	38.95	nr	**58.28**

15 STRUCTURAL METALWORK

Item	PC £	Labour hours	Labour £	Material £	Unit	Total rate £
FRAMED MEMBERS, FRAMING, FABRICATION						
BASIC STEEL PRICES – MATERIAL SUPPLY ONLY						
TATA Steel have recently revised their pricing structure and now publish prices for extras only. The extras cost needs to be added to their steel Base Price. The Base Price given here is a guide price only and subject to normal trading conditions with regards to discounts or plussages. The Base Price assumes a minimum order of 5 tonnes, S355JR steel and standard lengths.						
Universal beams and columns						
Base price for steel	–	–	–	700.00	tonne	**700.00**
Extra to be added to Base Price for: (kg/m)						
1016 × 305 mm (222, 249, 272, 314, 349, 393, 438, 487)	–	–	–	350.00	tonne	**350.00**
914 × 419 mm (343, 388)	–	–	–	200.00	tonne	**200.00**
914 × 305 mm (201, 224, 253, 289)	–	–	–	130.00	tonne	**130.00**
838 × 292 mm (226)	–	–	–	130.00	tonne	**130.00**
838 × 292 mm (176, 194)	–	–	–	80.00	tonne	**80.00**
762 × 267 mm (197)	–	–	–	110.00	tonne	**110.00**
762 × 267 mm (134, 147, 173)	–	–	–	80.00	tonne	**80.00**
686 × 254 mm (152, 170)	–	–	–	110.00	tonne	**110.00**
686 × 254 mm (125, 140)	–	–	–	80.00	tonne	**80.00**
610 × 305 mm (238)	–	–	–	130.00	tonne	**130.00**
610 × 305 mm (149, 179)	–	–	–	80.00	tonne	**80.00**
610 × 229 mm (125, 140)	–	–	–	80.00	tonne	**80.00**
610 × 229 mm (101, 113)	–	–	–	30.00	tonne	**30.00**
610 × 178 mm (82, 92, 100)	–	–	–	60.00	tonne	**60.00**
533 × 312 mm (219, 272)	–	–	–	160.00	tonne	**160.00**
533 × 312 mm (150, 182)	–	–	–	110.00	tonne	**110.00**
533 × 210 mm (138)	–	–	–	110.00	tonne	**110.00**
533 × 210 mm (101, 109, 122)	–	–	–	30.00	tonne	**30.00**
533 × 210 mm (82, 92)	–	–	–	20.00	tonne	**20.00**
533 × 165 mm (74, 85)	–	–	–	60.00	tonne	**60.00**
533 × 165 mm (66)	–	–	–	40.00	tonne	**40.00**
457 × 191 mm (133, 161)	–	–	–	110.00	tonne	**110.00**
457 × 191 mm (106)	–	–	–	60.00	tonne	**60.00**
457 × 191 mm (74, 82, 89, 98)	–	–	–	30.00	tonne	**30.00**
457 × 191 mm (67)	–	–	–	10.00	tonne	**10.00**
457 × 152 mm (74, 82)	–	–	–	60.00	tonne	**60.00**
457 × 152 mm (67)	–	–	–	40.00	tonne	**40.00**
457 × 152 mm (52, 60)	–	–	–	10.00	tonne	**10.00**
406 × 178 mm (85)	–	–	–	60.00	tonne	**60.00**
406 × 178 mm (74)	–	–	–	30.00	tonne	**30.00**
406 × 178 mm (54, 60, 67)	–	–	–	10.00	tonne	**10.00**
406 × 140 mm (53)	–	–	–	40.00	tonne	**40.00**
406 × 140 mm (46)	–	–	–	10.00	tonne	**10.00**
406 × 140 mm (39)	–	–	–	–	tonne	**-**
356 × 171 mm (45, 51, 57, 67)	–	–	–	10.00	tonne	**10.00**

Prices for Measured Works

15 STRUCTURAL METALWORK

Item	PC £	Labour hours	Labour £	Material £	Unit	Total rate £
FRAMED MEMBERS, FRAMING, FABRICATION – cont						
Universal beams and columns – cont						
Extra to be added to Base Price for: (kg/m) – cont						
305 × 165 mm (40, 46, 54)	–	–	–	10.00	tonne	**10.00**
305 × 127 mm (42, 48)	–	–	–	40.00	tonne	**40.00**
305 × 127 mm (37)	–	–	–	–	tonne	**-**
305 × 102 mm (28, 33)	–	–	–	30.00	tonne	**30.00**
305 × 102 mm (25)	–	–	–	–	tonne	**-**
254 × 102 mm (28)	–	–	–	30.00	tonne	**30.00**
254 × 102 mm (22, 25)	–	–	–	–	tonne	**-**
254 × 146 (43)	–	–	–	10.00	tonne	**10.00**
254 × 146 (31, 37)	–	–	–	–	tonne	**-**
203 × 133 mm (25, 30)	–	–	–	–	tonne	**-**
203 × 102 mm (23)	–	–	–	–	tonne	**-**
178 × 102 mm (19)	–	–	–	10.00	tonne	**10.00**
152 × 89 mm (16)	–	–	–	10.00	tonne	**10.00**
127 × 76 mm (13)	–	–	–	10.00	tonne	**10.00**
Asymmetric Slimflor Beams (ASB) Price on Application						
Universal columns (kg/m)						
356 × 406 mm (634)	–	–	–	300.00	tonne	**300.00**
356 × 406 mm (340, 393, 467, 551)	–	–	–	200.00	tonne	**200.00**
356 × 406 mm (235, 287)	–	–	–	130.00	tonne	**130.00**
356 × 368 mm (202)	–	–	–	130.00	tonne	**130.00**
356 × 368 mm (129, 153, 177)	–	–	–	80.00	tonne	**80.00**
305 × 305 mm (240, 283)	–	–	–	130.00	tonne	**130.00**
305 × 305 mm (137, 158, 198)	–	–	–	80.00	tonne	**80.00**
305 × 305 mm (97, 118)	–	–	–	30.00	tonne	**30.00**
254 × 254 mm (132, 167)	–	–	–	80.00	tonne	**80.00**
254 × 254 mm (73, 89, 107)	–	–	–	30.00	tonne	**30.00**
203 × 203 mm (127)	–	–	–	110.00	tonne	**110.00**
203 × 203 mm (100, 113)	–	–	–	60.00	tonne	**60.00**
203 × 203 mm (71, 86)	–	–	–	30.00	tonne	**30.00**
203 × 203 mm (46, 52, 60)	–	–	–	10.00	tonne	**10.00**
152 × 152 mm (44, 51)	–	–	–	40.00	tonne	**40.00**
152 × 152 mm (23, 30, 37)	–	–	–	–	tonne	**-**
Channels (kg/m)						
430 × 100 mm (64.4)	–	–	–	300.00	tonne	**300.00**
380 × 100 mm (54.0)	–	–	–	225.00	tonne	**225.00**
300 × 100 mm (45.5)	–	–	–	125.00	tonne	**125.00**
300 × 90 mm (41.4)	–	–	–	125.00	tonne	**125.00**
260 × 90 mm (34.8)	–	–	–	100.00	tonne	**100.00**
260 × 75 mm (27.6)	–	–	–	100.00	tonne	**100.00**
230 × 90 mm (32.2)	–	–	–	100.00	tonne	**100.00**
230 × 75 mm (25.7)	–	–	–	100.00	tonne	**100.00**
200 × 90 mm (29.7)	–	–	–	100.00	tonne	**100.00**
200 × 75 mm (23.4)	–	–	–	50.00	tonne	**50.00**
180 × 90 mm (26.1)	–	–	–	100.00	tonne	**100.00**

15 STRUCTURAL METALWORK

Item	PC £	Labour hours	Labour £	Material £	Unit	Total rate £
180 × 75 mm (20.3)	–	–	–	50.00	tonne	**50.00**
150 × 90 mm (23.9)	–	–	–	100.00	tonne	**100.00**
150 × 75 mm (17.9)	–	–	–	50.00	tonne	**50.00**
125 × 65 mm (14.8)	–	–	–	50.00	tonne	**50.00**
Equal angles (mm)						
200 × 200 mm (16,18, 20, 24)	–	–	–	45.00	tonne	**45.00**
150 × 150 mm (10,12,15,18)	–	–	–	15.00	tonne	**15.00**
120 × 120 mm (8, 10, 12, 15)	–	–	–	–	tonne	**-**
Unequal angles (mm)						
200 × 150 mm (12,15, 18)	–	–	–	285.00	tonne	**285.00**
200 × 100 mm (10,12,15)	–	–	–	130.00	tonne	**130.00**
Specification extras						
for Advance275JR steel	–	–	–	30.00	tonne	**30.00**
for Advance275JO steel	–	–	–	40.00	tonne	**40.00**
for Advance275J2 steel	–	–	–	80.00	tonne	**80.00**
for Advance355JO steel	–	–	–	10.00	tonne	**10.00**
for Advance355J2 steel	–	–	–	40.00	tonne	**40.00**
for Advance355 K2 steel	–	–	–	80.00	tonne	**80.00**
Hollow sections						
NOTE: The following prices are for basic quantities of 10 tonnes and over in one size, thickness, length, steelgrade and surface finish and include delivery to outer London.						
circular hollow sections	–	–	–	866.00	tonne	**866.00**
rectangular hollow section	–	–	–	874.00	tonne	**874.00**
square hollow sections	–	–	–	844.00	tonne	**844.00**
SUPPLY AND FIX PRICES						
Framing, fabrication; weldable steel; BS EN 10025: 2004 Grade S275; hot rolled structural steel sections; welded fabrication						
Columns						
weight not exceeding 40 kg/m	–	–	–	–	tonne	**1396.80**
weight not exceeding 40 kg/m; cellular (Fabsec)	–	–	–	–	tonne	**1830.60**
weight not exceeding 40 kg/m; curved	–	–	–	–	tonne	**1828.80**
weight not exceeding 40 kg/m; square hollow section	–	–	–	–	tonne	**1728.90**
weight not exceeding 40 kg/m; circular hollow section	–	–	–	–	tonne	**1823.40**
weight 40–100 kg/m	–	–	–	–	tonne	**1221.30**
weight 40–100 kg/m; cellular (Fabsec)	–	–	–	–	tonne	**1559.70**
weight 40–100 kg/m; curved	–	–	–	–	tonne	**1557.00**
weight 40–100 kg/m; square hollow section	–	–	–	–	tonne	**1485.00**
weight 40–100 kg/m; circular hollow section	–	–	–	–	tonne	**1563.30**
weight exceeding 100 kg/m	–	–	–	–	tonne	**1106.10**
weight exceeding 100 kg/m; cellular (Fabsec)	–	–	–	–	tonne	**1361.70**
weight exceeding 100 kg/m; curved	–	–	–	–	tonne	**1393.20**
weight exceeding 100 kg/m; square hollow section	–	–	–	–	tonne	**1431.90**
weight exceeding 100 kg/m; circular hollow section	–	–	–	–	tonne	**1526.40**

15 STRUCTURAL METALWORK

Item	PC £	Labour hours	Labour £	Material £	Unit	Total rate £
FRAMED MEMBERS, FRAMING, FABRICATION – cont						
Framing, fabrication – cont						
Beams						
weight not exceeding 40 kg/m	–	–	–	–	tonne	1439.10
weight not exceeding 40 kg/m; cellular (Fabsec)	–	–	–	–	tonne	1878.30
weight not exceeding 40 kg/m; curved	–	–	–	–	tonne	1825.20
weight not exceeding 40 kg/m; square hollow section	–	–	–	–	tonne	1878.30
weight not exceeding 40 kg/m; circular hollow section	–	–	–	–	tonne	2213.10
weight 40–100 kg/m	–	–	–	–	tonne	1193.40
weight 40–100 kg/m; cellular (Fabsec)	–	–	–	–	tonne	1549.80
weight 40–100 kg/m; curved	–	–	–	–	tonne	1496.70
weight 40–100 kg/m; square hollow section	–	–	–	–	tonne	1806.30
weight 40–100 kg/m; circular hollow section	–	–	–	–	tonne	1999.80
weight exceeding 100 kg/m	–	–	–	–	tonne	1075.50
weight exceeding 100 kg/m; cellular (Fabsec)	–	–	–	–	tonne	1366.20
weight exceeding 100 kg/m; curved	–	–	–	–	tonne	1396.80
weight exceeding 100 kg/m; square hollow section	–	–	–	–	tonne	1714.50
weight exceeding 100 kg/m; circular hollow section	–	–	–	–	tonne	1947.60
Bracings						
weight not exceeding 40 kg/m	–	–	–	–	tonne	1706.40
weight not exceeding 40 kg/m; square hollow section	–	–	–	–	tonne	2074.50
weight not exceeding 40 kg/m; circular hollow section	–	–	–	–	tonne	2074.50
weight 40–100 kg/m	–	–	–	–	tonne	1589.40
weight 40–100 kg/m; square hollow section	–	–	–	–	tonne	1951.20
weight 40–100 kg/m; circular hollow section	–	–	–	–	tonne	1951.20
weight exceeding 100 kg/m	–	–	–	–	tonne	1506.60
weight exceeding 100 kg/m; square hollow section	–	–	–	–	tonne	1862.10
weight exceeding 100 kg/m; circular hollow section	–	–	–	–	tonne	1862.10
Purlins and cladding rails						
weight not exceeding 40 kg/m	–	–	–	–	tonne	1321.20
weight not exceeding 40 kg/m; square hollow section	–	–	–	–	tonne	2101.50
weight not exceeding 40 kg/m; circular hollow section	–	–	–	–	tonne	2101.50
weight 40–100 kg/m	–	–	–	–	tonne	1180.80
weight 40–100 kg/m; square hollow section	–	–	–	–	tonne	1893.60
weight 40–100 kg/m; circular hollow section	–	–	–	–	tonne	1893.60
weight exceeding 100 kg/m	–	–	–	–	tonne	1092.60
weight exceeding 100 kg/m; square hollow section	–	–	–	–	tonne	1833.30
weight exceeding 100 kg/m; circular hollow section	–	–	–	–	tonne	1833.30

15 STRUCTURAL METALWORK

Item	PC £	Labour hours	Labour £	Material £	Unit	Total rate £
Grillages						
weight not exceeding 40 kg/m	–	–	–	–	tonne	1485.00
weight 40–100 kg/m	–	–	–	–	tonne	1201.50
weight exceeding 100 kg/m	–	–	–	–	tonne	1147.50
Trestles, towers and built up columns						
straight	–	–	–	–	tonne	1991.70
Trusses and built up girders						
straight	–	–	–	–	tonne	1991.70
curved	–	–	–	–	tonne	2442.60
Fittings						
general steel fittings; nuts, bolts, plates etc.	–	–	–	–	tonne	1948.50
Add to the aforementioned prices for						
grade 355 steelwork	–	–	–	–	%	3.00
Framing, erection						
Trial erection	–	–	–	–	tonne	200.00
Permanent erection on site	–	–	–	–	tonne	210.00
Surface preparation						
At works						
blast cleaning	–	–	–	–	m²	2.65
Surface treatment						
At works						
galvanizing	–	–	–	–	tonne	175.00
galvanizing	–	–	–	–	m²	11.85
shotblasting and priming to SA 2.5	–	–	–	–	m²	6.95
touch up primer and one coat of two pack epoxy zinc phosphate primer	–	–	–	–	m²	4.80
intumescent paint fire protection (30 minutes); spray applied	–	–	–	–	m²	11.41
intumescent paint fire protection (60 minutes); spray applied	–	–	–	–	m²	18.90
intumescent paint fire protection (90 minutes); spray applied	–	–	–	–	m²	31.50
intumescent paint fire protection (120 minutes); spray applied	–	–	–	–	m²	47.25
extra over for; separate decorative sealer top coat	–	–	–	–	m²	3.42
On site						
intumescent paint fire protection (30 minutes); spray applied	–	–	–	–	m²	8.57
intumescent paint fire protection (30 minutes) to circular columns etc.; spray applied	–	–	–	–	m²	14.35
intumescent paint fire protection (60 minutes) to UBs etc.; spray applied	–	–	–	–	m²	10.85
intumescent paint fire protection (60 minutes) to circular columns etc.; spray applied	–	–	–	–	m²	18.21
extra for						
separate decorative sealer top coat	–	–	–	–	m²	2.85

15 STRUCTURAL METALWORK

Item	PC £	Labour hours	Labour £	Material £	Unit	Total rate £
FRAMED MEMBERS, FRAMING, FABRICATION – cont						
Metsec Lightweight Steel Framing System (SFS); or other equal and approved; as inner leaf to external wall; studs typically at 600 mm centres; including provision for all openings, abutments, junctions and head details etc.						
Inner leaf; with supports and perimeter sections; 12 mm plasterboard internally; 10 mm cement fibre substrate externally; (insulation and external cladding measured separately)						
100 mm thick steel walling	–	–	–	–	m²	**57.05**
150 mm thick steel walling	–	–	–	–	m²	**61.55**
200 mm thick steel walling	–	–	–	–	m²	**66.06**
Inner leaf; with 16 mm Pyroc sheething board						
100 mm thick steel walling	–	–	–	–	m²	**70.56**
150 mm thick steel walling	–	–	–	–	m²	**75.07**
200 mm thick steel walling	–	–	–	–	m²	**79.57**
16 mm Pyroc sheething board fixed to slab perimeter						
not exceeding 300 mm	–	–	–	–	m	**7.00**
Inner leaf; with 16 mm Pyroc sheething board and Thermawall TW50 insulation supported by Halfen channels type 28/15 fixed to studs at 450 mm centres						
100 mm thick steel walling with 50 mm insulation	–	–	–	–	m²	**81.73**
150 mm thick steel walling with 75 mm insulation	–	–	–	–	m²	**91.11**
200 mm thick steel walling with 100 mm insulatiom	–	–	–	–	m²	**99.58**
16 mm Pyroc sheething board and 40 mm Thermawall TW55 insulation fixed to slab perimeter						
not exceeding 300 mm	–	–	–	–	m	**8.00**
Cold formed galvanized steel; Kingspan Multibeam or other equal and approved						
Cold rolled purlins and cladding rails						
175 × 65 × 1.40 mm gauge purlins or rails; fixed to steelwork	–	0.04	0.65	20.57	m	**21.22**
175 × 65 × 1.60 mm gauge purlins or rails; fixed to steelwork	–	0.04	0.65	21.91	m	**22.56**
175 × 65 × 2.00 mm gauge purlins or rails; fixed to steelwork	–	0.04	0.65	26.52	m	**27.17**
205 × 65 × 1.40 mm gauge purlins or rails; fixed to steelwork	–	0.04	0.65	22.77	m	**23.42**
205 × 65 × 1.60 mm gauge purlins or rails; fixed to steelwork	–	0.04	0.65	24.78	m	**25.43**
205 × 65 × 2.00 mm gauge purlins or rails; fixed to steelwork	–	0.04	0.65	28.22	m	**28.87**
Heavy duty Zed section spacers						
vertically; across cladding rails; fixed to steelwork	–	0.05	0.81	12.15	m	**12.96**

15 STRUCTURAL METALWORK

Item	PC £	Labour hours	Labour £	Material £	Unit	Total rate £
Cleats						
weld-on for 175 mm purlin or rail	–	0.10	1.62	5.96	nr	**7.58**
bolt-on for 175 mm purlin or rail; including fixing bolts	–	0.02	0.33	11.08	m	**11.41**
weld-on for 205 mm purlin or rail	–	0.10	1.62	6.80	nr	**8.42**
bolt-on for 205 mm purlin or rail; including fixing bolts	–	0.02	0.33	12.06	m	**12.39**
Tubular ties						
1500 mm long; bolted diagonally across purlins or cladding rails	–	0.02	0.33	12.03	m	**12.36**
Storage costs						
Costs for storing fabricated steelwork						
storage off site	–	–	–	–	t/week	**18.45**
storage on extending trailers	–	–	–	–	t/week	**30.75**
ISOLATED STRUCTURAL METAL MEMBERS						
Isolated structural member; weldable steel; BS EN 10025: 2004 Grade S275; hot rolled structural steel sections						
Plain member; beams						
weight not exceeding 40 kg/m	–	–	–	–	tonne	**1078.20**
weight 40–100 kg/m	–	–	–	–	tonne	**1049.40**
weight exceeding 100 kg/m	–	–	–	–	tonne	**1023.30**
Metsec open web steel lattice beams or other equal and approved; in single members; raised 3.50 m above ground; ends built in						
Beams; one coat zinc phosphate primer at works						
220 mm deep; to span 6.00 m (11.50 kg/m); ref B22	–	0.19	3.83	24.08	m	**27.91**
270 mm deep; to span 7.00 m (11.50 kg/m); ref B27	–	0.19	3.83	24.08	m	**27.91**
300 mm deep; to span 8.00 m (12.50 kg/m); ref B30	–	0.23	4.64	26.15	m	**30.79**
350 mm deep; to span 9.00 m (14.00 kg/m); ref B35	–	0.23	4.64	29.24	m	**33.88**
350 mm deep; to span 10.00 m (20.00 kg/m); ref D35	–	0.28	5.66	41.64	m	**47.30**
450 mm deep; to span 11.00 m (21.00 kg/m); ref D45	–	0.32	6.47	43.69	m	**50.16**
450 mm deep; to span 12.00 m (32.50 kg/m); ref G45	–	0.46	9.30	67.43	m	**76.73**
Beams; galvanized						
220 mm deep; to span 6.00 m (11.50 kg/m); ref B22	–	0.19	3.83	27.33	m	**31.16**
270 mm deep; to span 7.00 m (11.50 kg/m); ref B27	–	0.19	3.83	27.33	m	**31.16**
300 mm deep; to span 8.00 m (12.50 kg/m); ref B30	–	0.23	4.64	29.66	m	**34.30**
350 mm deep; to span 9.00 m (14.00 kg/m); ref B35	–	0.23	4.64	33.19	m	**37.83**
350 mm deep; to span 10.00 m (20.00 kg/m); ref D35	–	0.28	5.66	47.25	m	**52.91**
450 mm deep; to span 11.00 m (21.00 kg/m); ref D45	–	0.32	6.47	49.61	m	**56.08**
450 mm deep; to span 12.00 m (32.50 kg/m); ref G45	–	0.46	9.30	76.57	m	**85.87**

15 STRUCTURAL METALWORK

Item	PC £	Labour hours	Labour £	Material £	Unit	Total rate £
PROFILED METAL DECKING						
Soffits of slabs; galvanized steel permanent re-entrant type shuttering; including safety net						
Slab thickness not exceeding 200 mm						
0.9 mm decking; height to soffit 1.50–3.00 m	14.33	0.27	5.29	17.19	m²	**22.48**
0.9 mm decking; height to soffit 3.00–4.50 m	14.33	0.30	5.88	17.19	m²	**23.07**
1.2 mm decking; height to soffit 3.00–4.50 m	16.38	0.27	5.29	19.29	m²	**24.58**
Edge trim and restraints to decking						
edge trim 1.2 mm × 300 mm girth	–	0.17	3.24	5.28	m	**8.52**
edge trim 1.2 mm × 350 mm girth	–	0.17	3.24	5.82	m	**9.06**
edge trim 1.2 mm × 400 mm girth	–	0.17	3.24	6.38	m	**9.62**
Bearings to decking; connection to steel work with thru-deck welded shear studs						
1995 × 95 mm high studs at 100 mm centres	–	–	–	10.60	m	**10.60**
1995 × 95 mm high studs at 200 mm centres	–	–	–	5.30	m	**5.30**
1995 × 95 mm high studs at 300 mm centres	–	–	–	3.53	m	**3.53**
19120 × 120 mm high studs at 100 mm centres	–	–	–	10.60	m	**10.60**
19120 × 120 mm high studs at 200 mm centres	–	–	–	5.30	m	**5.30**
19120 × 120 mm high studs at 300 mm centres	–	–	–	3.53	m	**3.53**
Soffits of slabs; galvanized steel permanent trapezoidal type shuttering; including safety net						
Slab thickness not exceeding 200 mm						
0.9 mm decking; height to soffit 1.50–3.00 m	9.93	0.27	5.29	12.68	m²	**17.97**
0.9 mm decking; height to soffit 3.00–4.50 m	9.93	0.30	5.88	12.68	m²	**18.56**
1.2 mm decking; height to soffit 3.00–4.50 m	11.93	0.27	5.29	14.72	m²	**20.01**

16 CARPENTRY

Item	PC £	Labour hours	Labour £	Material £	Unit	Total rate £
TIMBER FRAMING						
SUPPLY ONLY TIMBER PRICES						
Hardwood; Joinery quality; 25 mm thicknes (£/m³)						
Black Walnut	–	–	–	315.19	m³	**315.19**
American White Ash	–	–	–	1365.81	m³	**1365.81**
American White Oak	–	–	–	1891.13	m³	**1891.13**
Pretreatment of timber by vacuum/pressure impregnation, excluding transport costs and any subsequent seasoning						
interior work; minimum salt retention 4.00 kg/m³	–	–	–	59.80	m³	**59.80**
exterior work; minimum salt retention 5.30 kg/m³	–	–	–	68.85	m³	**68.85**
Pretreatment of timber including flame proofing						
all purposes; minimum salt retention 36.00 kg/m³	–	–	–	136.58	m³	**136.58**
Aquaseal timber treatments – (£/25 litres)						
Timbershield	–	–	–	102.63	25litr	**102.63**
Longlife Wood Protector	–	–	–	51.72	25litr	**51.72**
SUPPLY AND FIX PRICES						
Sawn softwood; untreated						
Floor members						
38 mm × 100 mm	–	0.11	1.96	1.65	m	**3.61**
38 mm × 150 mm	–	0.13	2.32	2.28	m	**4.60**
47 mm × 75 mm	–	0.11	1.96	1.05	m	**3.01**
47 mm × 100 mm	–	0.13	2.32	1.34	m	**3.66**
47 mm × 125 mm	–	0.13	2.32	1.73	m	**4.05**
47 mm × 150 mm	–	0.14	2.49	1.98	m	**4.47**
47 mm × 175 mm	–	0.14	2.49	2.35	m	**4.84**
47 mm × 200 mm	–	0.15	2.68	2.54	m	**5.22**
47 mm × 225 mm	–	0.15	2.68	2.92	m	**5.60**
47 mm × 250 mm	–	0.16	2.85	3.30	m	**6.15**
75 mm × 125 mm	–	0.15	2.68	3.97	m	**6.65**
75 mm × 150 mm	–	0.15	2.68	4.31	m	**6.99**
75 mm × 175 mm	–	0.15	2.68	5.18	m	**7.86**
75 mm × 200 mm	–	0.16	2.85	5.80	m	**8.65**
75 mm × 225 mm	–	0.16	2.85	6.32	m	**9.17**
75 mm × 250 mm	–	0.17	3.03	9.59	m	**12.62**
100 mm × 150 mm	–	0.20	3.57	5.54	m	**9.11**
100 mm × 200 mm	–	0.21	3.74	7.38	m	**11.12**
100 mm × 250 mm	–	0.23	4.10	9.25	m	**13.35**
100 mm × 300 mm	–	0.25	4.46	11.94	m	**16.40**
Wall or partition members						
25 mm × 25 mm	–	0.06	1.07	0.65	m	**1.72**
25 mm × 38 mm	–	0.06	1.07	0.73	m	**1.80**
25 mm × 75 mm	–	0.08	1.42	0.91	m	**2.33**
38 mm × 38 mm	–	0.08	1.42	0.85	m	**2.27**
38 mm × 50 mm	–	0.08	1.42	1.08	m	**2.50**
38 mm × 75 mm	–	0.11	1.96	1.32	m	**3.28**
38 mm × 100 mm	–	0.14	2.49	1.65	m	**4.14**
47 mm × 50 mm	–	0.11	1.96	0.83	m	**2.79**
47 mm × 75 mm	–	0.14	2.49	1.09	m	**3.58**

16 CARPENTRY

Item	PC £	Labour hours	Labour £	Material £	Unit	Total rate £
TIMBER FRAMING – cont						
Sawn softwood – cont						
Wall or partition members – cont						
47 mm × 100 mm	–	0.17	3.03	1.37	m	**4.40**
47 mm × 125 mm	–	0.18	3.21	1.76	m	**4.97**
75 mm × 75 mm	–	0.17	3.03	2.44	m	**5.47**
75 mm × 100 mm	–	0.19	3.38	3.31	m	**6.69**
100 mm × 100 mm	–	0.19	3.38	4.03	m	**7.41**
Joist strutting; herringbone						
47 mm × 50 mm; depth of joist 150 mm	–	0.46	8.20	2.01	m	**10.21**
47 mm × 50 mm; depth of joist 175 mm	–	0.46	8.20	2.04	m	**10.24**
47 mm × 50 mm; depth of joist 200 mm	–	0.46	8.20	2.08	m	**10.28**
47 mm × 50 mm; depth of joist 225 mm	–	0.46	8.20	2.11	m	**10.31**
47 mm × 50 mm; depth of joist 250 mm	–	0.46	8.20	2.15	m	**10.35**
Joist strutting; block						
47 mm × 150 mm; depth of joist 150 mm	–	0.28	4.99	2.44	m	**7.43**
47 mm × 175 mm; depth of joist 175 mm	–	0.28	4.99	2.80	m	**7.79**
47 mm × 200 mm; depth of joist 200 mm	–	0.28	4.99	3.00	m	**7.99**
47 mm × 225 mm; depth of joist 225 mm	–	0.28	4.99	3.38	m	**8.37**
47 mm × 250 mm; depth of joist 250 mm	–	0.28	4.99	3.75	m	**8.74**
Cleats						
225 mm × 100 mm × 75 mm	–	0.19	3.38	0.65	nr	**4.03**
Extra for stress grading to above timbers						
general structural (GS) grade	–	–	–	25.38	m³	**25.38**
special structural (SS) grade	–	–	–	50.75	m³	**50.75**
Extra for protecting and flameproofing timber with						
Celgard CF protection or other equal and approved						
small sections	–	–	–	109.24	m³	**109.24**
large sections	–	–	–	105.47	m³	**105.47**
Wrot surfaces						
plain; 50 mm wide	–	0.02	0.36	–	m	**0.36**
plain; 100 mm wide	–	0.03	0.53	–	m	**0.53**
plain; 150 mm wide	–	0.04	0.72	–	m	**0.72**
Sawn softwood; tanalized						
Floor members						
38 mm × 75 mm	–	0.11	1.96	1.45	m	**3.41**
38 mm × 100 mm	–	0.11	1.96	1.81	m	**3.77**
38 mm × 150 mm	–	0.13	2.32	2.52	m	**4.84**
47 mm × 75 mm	–	0.11	1.96	1.22	m	**3.18**
47 mm × 100 mm	–	0.13	2.32	1.56	m	**3.88**
47 mm × 125 mm	–	0.13	2.32	2.01	m	**4.33**
47 mm × 150 mm	–	0.14	2.49	2.32	m	**4.81**
47 mm × 175 mm	–	0.14	2.49	2.74	m	**5.23**
47 mm × 200 mm	–	0.15	2.68	2.99	m	**5.67**
47 mm × 225 mm	–	0.15	2.68	3.42	m	**6.10**
47 mm × 250 mm	–	0.16	2.85	3.85	m	**6.70**
75 mm × 125 mm	–	0.15	2.68	4.39	m	**7.07**
75 mm × 150 mm	–	0.15	2.68	4.80	m	**7.48**
75 mm × 175 mm	–	0.15	2.68	5.76	m	**8.44**
75 mm × 200 mm	–	0.16	2.85	6.47	m	**9.32**

16 CARPENTRY

Item	PC £	Labour hours	Labour £	Material £	Unit	Total rate £
75 mm × 225 mm	–	0.16	2.85	7.07	m	**9.92**
75 mm × 250 mm	–	0.17	3.03	10.43	m	**13.46**
100 mm × 150 mm	–	0.20	3.57	6.20	m	**9.77**
100 mm × 200 mm	–	0.21	3.74	8.26	m	**12.00**
100 mm × 250 mm	–	0.23	4.10	10.35	m	**14.45**
100 mm × 300 mm	–	0.25	4.46	13.27	m	**17.73**
Wall or partition members						
25 mm × 25 mm	–	0.06	1.07	0.67	m	**1.74**
25 mm × 38 mm	–	0.06	1.07	0.78	m	**1.85**
25 mm × 75 mm	–	0.08	1.42	0.99	m	**2.41**
38 mm × 38 mm	–	0.08	1.42	0.91	m	**2.33**
38 mm × 50 mm	–	0.08	1.42	1.16	m	**2.58**
38 mm × 75 mm	–	0.11	1.96	1.45	m	**3.41**
38 mm × 100 mm	–	0.14	2.49	1.81	m	**4.30**
47 mm × 50 mm	–	0.11	1.96	0.94	m	**2.90**
47 mm × 75 mm	–	0.14	2.49	1.25	m	**3.74**
47 mm × 100 mm	–	0.17	3.03	1.60	m	**4.63**
47 mm × 125 mm	–	0.18	3.21	2.04	m	**5.25**
75 mm × 75 mm	–	0.17	3.03	2.69	m	**5.72**
75 mm × 100 mm	–	0.19	3.38	3.65	m	**7.03**
100 mm × 100 mm	–	0.19	3.38	4.47	m	**7.85**
Roof members; flat						
38 mm × 75 mm	–	0.13	2.32	1.45	m	**3.77**
38 mm × 100 mm	–	0.13	2.32	1.81	m	**4.13**
38 mm × 125 mm	–	0.13	2.32	2.17	m	**4.49**
38 mm × 150 mm	–	0.13	2.32	2.52	m	**4.84**
47 mm × 100 mm	–	0.13	2.32	1.56	m	**3.88**
47 mm × 125 mm	–	0.13	2.32	2.01	m	**4.33**
47 mm × 150 mm	–	0.14	2.49	2.32	m	**4.81**
47 mm × 175 mm	–	0.14	2.49	2.74	m	**5.23**
47 mm × 200 mm	–	0.15	2.68	2.99	m	**5.67**
47 mm × 225 mm	–	0.15	2.68	3.42	m	**6.10**
47 mm × 250 mm	–	0.16	2.85	3.85	m	**6.70**
75 mm × 150 mm	–	0.15	2.68	4.80	m	**7.48**
75 mm × 175 mm	–	0.15	2.68	5.76	m	**8.44**
75 mm × 200 mm	–	0.16	2.85	6.47	m	**9.32**
75 mm × 225 mm	–	0.16	2.85	7.07	m	**9.92**
75 mm × 250 mm	–	0.17	3.03	10.43	m	**13.46**
Roof members; pitched						
25 mm × 100 mm	–	0.11	1.96	1.44	m	**3.40**
25 mm × 125 mm	–	0.11	1.96	1.95	m	**3.91**
25 mm × 150 mm	–	0.14	2.49	2.34	m	**4.83**
25 mm × 175 mm	–	0.16	2.85	2.74	m	**5.59**
25 mm × 200 mm	–	0.17	3.03	3.14	m	**6.17**
38 mm × 100 mm	–	0.14	2.49	1.81	m	**4.30**
38 mm × 125 mm	–	0.14	2.49	2.17	m	**4.66**
38 mm × 150 mm	–	0.14	2.49	2.52	m	**5.01**
38 mm × 175 mm	–	0.16	2.85	2.99	m	**5.84**
38 mm × 200 mm	–	0.17	3.03	3.44	m	**6.47**
47 mm × 50 mm	–	0.11	1.96	0.90	m	**2.86**
47 mm × 75 mm	–	0.14	2.49	1.22	m	**3.71**
47 mm × 100 mm	–	0.17	3.03	1.56	m	**4.59**

16 CARPENTRY

Item	PC £	Labour hours	Labour £	Material £	Unit	Total rate £
TIMBER FRAMING – cont						
Sawn softwood – cont						
Roof members – cont						
47 mm × 125 mm	–	0.17	3.03	2.01	m	**5.04**
47 mm × 150 mm	–	0.19	3.38	2.32	m	**5.70**
47 mm × 175 mm	–	0.19	3.38	2.74	m	**6.12**
47 mm × 200 mm	–	0.19	3.38	2.99	m	**6.37**
47 mm × 225 mm	–	0.19	3.38	3.42	m	**6.80**
75 mm × 100 mm	–	0.23	4.10	3.58	m	**7.68**
75 mm × 125 mm	–	0.23	4.10	4.39	m	**8.49**
75 mm × 150 mm	–	0.23	4.10	4.80	m	**8.90**
100 mm × 150 mm	–	0.28	4.99	6.23	m	**11.22**
100 mm × 175 mm	–	0.28	4.99	7.27	m	**12.26**
100 mm × 200 mm	–	0.28	4.99	8.26	m	**13.25**
100 mm × 225 mm	–	0.31	5.52	9.29	m	**14.81**
100 mm × 250 mm	–	0.31	5.52	10.35	m	**15.87**
Plates						
38 mm × 75 mm	–	0.11	1.96	1.50	m	**3.46**
38 mm × 100 mm	–	0.14	2.49	1.81	m	**4.30**
47 mm × 75 mm	–	0.14	2.49	1.22	m	**3.71**
47 mm × 100 mm	–	0.17	3.03	1.56	m	**4.59**
75 mm × 100 mm	–	0.19	3.38	3.58	m	**6.96**
75 mm × 125 mm	–	0.22	3.93	4.35	m	**8.28**
75 mm × 150 mm	–	0.25	4.46	4.77	m	**9.23**
Plates; fixing by bolting						
38 mm × 75 mm	–	0.20	3.57	1.45	m	**5.02**
38 mm × 100 mm	–	0.23	4.10	1.81	m	**5.91**
47 mm × 75 mm	–	0.23	4.10	1.22	m	**5.32**
47 mm × 100 mm	–	0.26	4.63	1.56	m	**6.19**
75 mm × 100 mm	–	0.29	5.17	3.58	m	**8.75**
75 mm × 125 mm	–	0.31	5.52	4.35	m	**9.87**
75 mm × 150 mm	–	0.34	6.06	4.77	m	**10.83**
Joist strutting; herringbone						
47 mm × 50 mm; depth of joist 150 mm	–	0.46	8.20	2.23	m	**10.43**
47 mm × 50 mm; depth of joist 175 mm	–	0.46	8.20	2.28	m	**10.48**
47 mm × 50 mm; depth of joist 200 mm	–	0.46	8.20	2.32	m	**10.52**
47 mm × 50 mm; depth of joist 225 mm	–	0.46	8.20	2.36	m	**10.56**
47 mm × 50 mm; depth of joist 250 mm	–	0.46	8.20	2.40	m	**10.60**
Joist strutting; block						
47 mm × 150 mm; depth of joist 150 mm	–	0.28	4.99	2.77	m	**7.76**
47 mm × 175 mm; depth of joist 175 mm	–	0.28	4.99	3.19	m	**8.18**
47 mm × 200 mm; depth of joist 200 mm	–	0.28	4.99	3.44	m	**8.43**
47 mm × 225 mm; depth of joist 225 mm	–	0.28	4.99	3.88	m	**8.87**
47 mm × 250 mm; depth of joist 250 mm	–	0.28	4.99	4.31	m	**9.30**
Cleats						
225 mm × 100 mm × 75 mm	–	0.19	3.38	0.72	nr	**4.10**
Extra for stress grading to above timbers						
general structural (GS) grade	–	–	–	25.38	m³	**25.38**
special structural (SS) grade	–	–	–	50.75	m³	**50.75**

16 CARPENTRY

Item	PC £	Labour hours	Labour £	Material £	Unit	Total rate £
Extra for protecting and flameproofing timber with Celgard CF protection or other equal and approved						
small sections	–	–	–	109.24	m³	**109.24**
large sections	–	–	–	105.47	m³	**105.47**
Wrot surfaces						
plain; 50 mm wide	–	0.02	0.36	–	m	**0.36**
plain; 100 mm wide	–	0.03	0.53	–	m	**0.53**
plain; 150 mm wide	–	0.04	0.72	–	m	**0.72**
Trussed rafters, stress graded sawn softwood pressure impregnated; raised through two storeys and fixed in position						
W type truss (Fink); 22.5° pitch; 450 mm eaves overhang						
5.00 m span	–	1.48	26.38	42.14	nr	**68.52**
7.60 m span	–	1.62	28.88	59.75	nr	**88.63**
10.00 m span	–	1.85	32.98	93.95	nr	**126.93**
W type truss (Fink); 30° pitch; 450 mm eaves overhang						
5.00 m span	–	1.48	26.38	43.97	nr	**70.35**
7.60 m span	–	1.62	28.88	63.94	nr	**92.82**
10.00 m span	–	1.85	32.98	104.32	nr	**137.30**
W type truss (Fink); 45° pitch; 450 mm eaves overhang						
4.60 m span	–	1.48	26.38	101.59	nr	**127.97**
7.00 m span	–	1.62	28.88	189.31	nr	**218.19**
Mono type truss; 17.5° pitch; 450 mm eaves overhang						
3.30 m span	–	1.30	23.18	27.58	nr	**50.76**
5.60 m span	–	1.48	26.38	54.21	nr	**80.59**
7.00 m span	–	1.71	30.48	65.90	nr	**96.38**
Attic type truss; 45° pitch; 450 mm eaves overhang						
5.00 m span	–	2.91	51.88	47.48	nr	**99.36**
7.60 m span	–	3.05	54.38	72.65	nr	**127.03**
9.00 m span	–	3.24	57.76	223.19	nr	**280.95**
Moelven Toreboda glulam timber beams or other equal and approved; Moelven Laminated Timber Structures; LB grade whitewood; pressure impregnated; phenbol resorcinal adhesive; clean planed finish; fixed						
Laminated roof beams						
approximate rate for glulam beams	–	–	–	–	m³	**3500.00**
56 mm × 225 mm	–	–	–	–	m	**43.98**
66 mm × 315 mm	–	–	–	–	m	**72.58**
90 mm × 315 mm	–	–	–	–	m	**98.98**
90 mm × 405 mm	–	–	–	–	m	**127.26**
115 mm × 405 mm	–	–	–	–	m	**162.59**
115 mm × 495 mm	–	–	–	–	m	**198.50**
115 mm × 630 mm	–	–	–	–	m	**252.37**

16 CARPENTRY

Item	PC £	Labour hours	Labour £	Material £	Unit	Total rate £
TIMBER FRAMING – cont						
Masterboard or other equal and approved; 6 mm thick						
Eaves, verge soffit boards, fascia boards and the like						
over 300 mm wide	7.97	0.65	11.58	9.48	m²	**21.06**
75 mm wide	–	0.19	3.38	0.73	m	**4.11**
150 mm wide	–	0.22	3.93	1.42	m	**5.35**
225 mm wide	–	0.26	4.63	2.11	m	**6.74**
300 mm wide	–	0.28	4.99	2.81	m	**7.80**
Plywood; external quality; 12 mm thick						
Eaves, verge soffit boards, fascia boards and the like						
over 300 mm wide	7.71	0.76	13.55	9.19	m²	**22.74**
75 mm wide	–	0.23	4.10	0.71	m	**4.81**
150 mm wide	–	0.27	4.82	1.38	m	**6.20**
225 mm wide	–	0.31	5.52	2.05	m	**7.57**
300 mm wide	–	0.34	6.06	2.73	m	**8.79**
Plywood; external quality; 15 mm thick						
Eaves, verge soffit boards, fascia boards and the like						
over 300 mm wide	9.75	0.76	13.55	11.45	m²	**25.00**
75 mm wide	–	0.23	4.10	0.87	m	**4.97**
150 mm wide	–	0.27	4.82	1.72	m	**6.54**
225 mm wide	–	0.31	5.52	2.56	m	**8.08**
300 mm wide	–	0.34	6.06	3.40	m	**9.46**
Plywood; external quality; 18 mm thick						
Eaves, verge soffit boards, fascia boards and the like						
over 300 mm wide	11.43	0.76	13.55	13.29	m²	**26.84**
75 mm wide	–	0.23	4.10	1.01	m	**5.11**
150 mm wide	–	0.27	4.82	2.00	m	**6.82**
225 mm wide	–	0.31	5.52	2.97	m	**8.49**
300 mm wide	–	0.34	6.06	3.96	m	**10.02**
Plywood; marine quality; 18 mm thick						
Gutter boards; butt joints						
over 300 mm wide	11.25	0.86	15.33	13.10	m²	**28.43**
150 mm wide	–	0.31	5.52	1.97	m	**7.49**
225 mm wide	–	0.34	6.06	2.96	m	**9.02**
300 mm wide	–	0.38	6.78	3.93	m	**10.71**
Eaves, verge soffit boards, fascias boards and the like						
over 300 mm wide	11.25	0.76	13.55	13.10	m²	**26.65**
75 mm wide	–	0.23	4.10	1.00	m	**5.10**
150 mm wide	–	0.27	4.82	1.97	m	**6.79**
225 mm wide	–	0.31	5.52	2.93	m	**8.45**
300 mm wide	–	0.34	6.06	3.90	m	**9.96**

16 CARPENTRY

Item	PC £	Labour hours	Labour £	Material £	Unit	Total rate £
Plywood; marine quality; 25 mm thick						
Gutter boards; butt joints						
over 300 mm wide	15.64	0.93	16.57	17.93	m²	**34.50**
150 mm wide	–	0.32	5.71	2.69	m	**8.40**
225 mm wide	–	0.37	6.60	4.05	m	**10.65**
300 mm wide	–	0.42	7.48	5.38	m	**12.86**
Eaves, verge soffit boards, fascia baords and the like						
over 300 mm wide	15.64	0.81	14.44	17.93	m²	**32.37**
75 mm wide	–	0.24	4.27	1.36	m	**5.63**
150 mm wide	–	0.29	5.17	2.69	m	**7.86**
225 mm wide	–	0.29	5.17	4.02	m	**9.19**
300 mm wide	–	0.37	6.60	5.34	m	**11.94**
Sawn softwood; untreated						
Gutter boards; butt joints						
19 mm thick; 150 mm wide; sloping	–	1.16	20.68	9.57	m²	**30.25**
19 mm thick; 75 mm wide	–	0.32	5.71	0.74	m	**6.45**
19 mm thick; 150 mm wide	–	0.37	6.60	1.39	m	**7.99**
19 mm thick; 225 mm wide	–	0.42	7.48	2.52	m	**10.00**
25 mm thick; sloping	–	1.16	20.68	15.21	m²	**35.89**
25 mm thick; 75 mm wide	–	0.32	5.71	0.94	m	**6.65**
25 mm thick; 150 mm wide	–	0.37	6.60	2.23	m	**8.83**
25 mm thick; 225 mm wide	–	0.42	7.48	3.54	m	**11.02**
Cesspools with 25 mm thick sides and bottom						
225 mm × 225 mm × 150 mm	–	1.11	19.79	2.96	nr	**22.75**
300 mm × 300 mm × 150 mm	–	1.30	23.18	3.88	nr	**27.06**
Individual supports; firrings						
50 mm wide × 36 mm average depth	–	0.14	2.49	1.97	m	**4.46**
50 mm wide × 50 mm average depth	–	0.14	2.49	3.02	m	**5.51**
50 mm wide × 75 mm average depth	–	0.14	2.49	3.92	m	**6.41**
Individual supports; bearers						
25 mm × 50 mm	–	0.09	1.61	0.89	m	**2.50**
38 mm × 50 mm	–	0.09	1.61	1.15	m	**2.76**
50 mm × 50 mm	–	0.09	1.61	0.86	m	**2.47**
50 mm × 75 mm	–	0.09	1.61	1.12	m	**2.73**
Individual supports; angle fillets						
38 mm × 38 mm	–	0.09	1.61	0.79	m	**2.40**
50 mm × 50 mm	–	0.09	1.61	1.01	m	**2.62**
75 mm × 75 mm	–	0.11	1.96	2.08	m	**4.04**
Individual supports; tilting fillets						
19 mm × 38 mm	–	0.09	1.61	0.47	m	**2.08**
25 mm × 50 mm	–	0.09	1.61	0.78	m	**2.39**
38 mm × 75 mm	–	0.09	1.61	1.19	m	**2.80**
50 mm × 75 mm	–	0.09	1.61	1.53	m	**3.14**
75 mm × 100 mm	–	0.14	2.49	2.84	m	**5.33**
Individual supports; grounds or battens						
13 mm × 19 mm	–	0.04	0.72	0.38	m	**1.10**
13 mm × 32 mm	–	0.04	0.72	0.38	m	**1.10**
25 mm × 50 mm	–	0.04	0.72	0.82	m	**1.54**

16 CARPENTRY

Item	PC £	Labour hours	Labour £	Material £	Unit	Total rate £
TIMBER FRAMING – cont						
Sawn softwood – cont						
Individual supports; grounds or battens; plugged and screwed						
13 mm × 19 mm	–	0.14	2.49	0.37	m	**2.86**
13 mm × 32 mm	–	0.14	2.49	0.37	m	**2.86**
25 mm × 50 mm	–	0.14	2.49	0.81	m	**3.30**
Framed supports; open-spaced grounds or battens; at 300 mm centres one way						
25 mm × 50 mm	–	0.14	2.49	2.71	m²	**5.20**
25 mm × 50 mm; plugged and screwed	–	0.42	7.48	2.69	m²	**10.17**
Framed supports; at 300 mm centres one way and 600 mm centres the other way						
25 mm × 50 mm	–	0.69	12.30	4.06	m²	**16.36**
38 mm × 50 mm	–	0.69	12.30	5.34	m²	**17.64**
50 mm × 50 mm	–	0.69	12.30	3.93	m²	**16.23**
50 mm × 75 mm	–	0.69	12.30	5.21	m²	**17.51**
75 mm × 75 mm	–	0.69	12.30	11.99	m²	**24.29**
Framed supports; at 300 mm centres one way and 600 mm centres the other way; plugged and screwed						
25 mm × 50 mm	–	1.16	20.68	4.16	m²	**24.84**
38 mm × 50 mm	–	1.16	20.68	5.44	m²	**26.12**
50 mm × 50 mm	–	1.16	20.68	4.03	m²	**24.71**
50 mm × 75 mm	–	1.16	20.68	5.31	m²	**25.99**
75 mm × 75 mm	–	1.16	20.68	12.10	m²	**32.78**
Framed supports; at 500 mm centres both ways						
25 mm × 50 mm; to bath panels	–	0.83	14.80	5.29	m²	**20.09**
Framed supports; as bracketing and cradling around steelwork						
25 mm × 50 mm	–	1.30	23.18	5.73	m²	**28.91**
50 mm × 50 mm	–	1.39	24.78	5.55	m²	**30.33**
50 mm × 75 mm	–	1.48	26.38	7.34	m²	**33.72**
Sawn softwood; tanalized						
Gutter boards; butt joints						
19 mm thick; 150 mm; sloping	–	1.16	20.68	10.41	m²	**31.09**
19 mm thick; 75 mm wide	–	0.32	5.71	0.80	m	**6.51**
19 mm thick; 150 mm wide	–	0.37	6.60	1.52	m	**8.12**
19 mm thick; 225 mm wide	–	0.42	7.48	2.71	m	**10.19**
25 mm thick; sloping	–	1.16	20.68	16.32	m²	**37.00**
25 mm thick; 75 mm wide	–	0.32	5.71	1.03	m	**6.74**
25 mm thick; 150 mm wide	–	0.37	6.60	2.41	m	**9.01**
25 mm thick; 225 mm wide	–	0.42	7.48	3.78	m	**11.26**
Cesspools with 25 mm thick sides and bottom						
225 mm × 225 mm × 150 mm	–	1.11	19.79	3.18	nr	**22.97**
300 mm × 300 mm × 150 mm	–	1.30	23.18	4.18	nr	**27.36**
Individual supports; firrings						
50 mm wide × 36 mm average depth	–	0.14	2.49	2.05	m	**4.54**
50 mm wide × 50 mm average depth	–	0.14	2.49	3.14	m	**5.63**
50 mm wide × 75 mm average depth	–	0.14	2.49	4.09	m	**6.58**

16 CARPENTRY

Item	PC £	Labour hours	Labour £	Material £	Unit	Total rate £
Individual supports; bearers						
25 mm × 50 mm	–	0.09	1.61	0.94	m	**2.55**
38 mm × 50 mm	–	0.09	1.61	1.23	m	**2.84**
50 mm × 50 mm	–	0.09	1.61	0.97	m	**2.58**
50 mm × 75 mm	–	0.09	1.61	1.29	m	**2.90**
Individual supports; angle fillets						
38 mm × 38 mm	–	0.09	1.61	0.83	m	**2.44**
50 mm × 50 mm	–	0.09	1.61	1.07	m	**2.68**
75 mm × 75 mm	–	0.11	1.96	2.20	m	**4.16**
Individual supports; tilting fillets						
19 mm × 38 mm	–	0.09	1.61	0.49	m	**2.10**
25 mm × 50 mm	–	0.09	1.61	0.80	m	**2.41**
38 mm × 75 mm	–	0.09	1.61	1.25	m	**2.86**
50 mm × 75 mm	–	0.09	1.61	1.61	m	**3.22**
75 mm × 100 mm	–	0.14	2.49	3.00	m	**5.49**
Individual supports; grounds or battens						
13 mm × 19 mm	–	0.04	0.72	0.39	m	**1.11**
13 mm × 32 mm	–	0.04	0.72	0.40	m	**1.12**
25 mm × 50 mm	–	0.04	0.72	0.88	m	**1.60**
Individual supports; grounds or battens; plugged and screwed						
13 mm × 19 mm	–	0.14	2.49	0.38	m	**2.87**
13 mm × 32 mm	–	0.14	2.49	0.38	m	**2.87**
25 mm × 50 mm	–	0.14	2.49	0.86	m	**3.35**
Framed supports; open-spaced grounds or battens; at 300 mm centres one way						
25 mm × 50 mm	–	0.14	2.49	2.89	m²	**5.38**
25 mm × 50 mm; plugged and screwed	–	0.42	7.48	2.87	m²	**10.35**
Framed supports; at 300 mm centres one way and 600 mm centres the other way						
25 mm × 50 mm	–	0.69	12.30	4.35	m²	**16.65**
38 mm × 50 mm	–	0.69	12.30	5.76	m²	**18.06**
50 mm × 50 mm	–	0.69	12.30	4.48	m²	**16.78**
50 mm × 75 mm	–	0.69	12.30	6.04	m²	**18.34**
75 mm × 75 mm	–	0.69	12.30	13.24	m²	**25.54**
Framed supports; at 300 mm centres one way and 600 mm centres the other way; plugged and screwed						
25 mm × 50 mm	–	1.16	20.68	4.44	m²	**25.12**
38 mm × 50 mm	–	1.16	20.68	5.86	m²	**26.54**
50 mm × 50 mm	–	1.16	20.68	4.58	m²	**25.26**
50 mm × 75 mm	–	1.16	20.68	6.14	m²	**26.82**
75 mm × 75 mm	–	1.16	20.68	13.35	m²	**34.03**
Framed supports; at 500 mm centres both ways						
25 mm × 50 mm; to bath panels	–	0.83	14.80	5.65	m²	**20.45**
Framed supports; as bracketing and cradling around steelwork						
25 mm × 50 mm	–	1.30	23.18	6.13	m²	**29.31**
50 mm × 50 mm	–	1.39	24.78	6.32	m²	**31.10**
50 mm × 75 mm	–	1.48	26.38	8.51	m²	**34.89**

16 CARPENTRY

Item	PC £	Labour hours	Labour £	Material £	Unit	Total rate £
TIMBER FRAMING – cont						
Wrought softwood						
Gutter boards; tongued and grooved joints						
19 mm thick; 150 mm; sloping	–	1.39	24.78	13.16	m²	**37.94**
19 mm thick; 75 mm wide	–	0.37	6.60	0.95	m	**7.55**
19 mm thick; 150 mm wide	–	0.42	7.48	1.93	m	**9.41**
19 mm thick; 225 mm wide	–	0.46	8.20	2.82	m	**11.02**
25 mm thick; sloping	–	1.39	24.78	13.61	m²	**38.39**
25 mm thick; 75 mm wide	–	0.37	6.60	1.07	m	**7.67**
25 mm thick; 150 mm wide	–	0.42	7.48	1.91	m	**9.39**
25 mm thick; 225 mm wide	–	0.46	8.20	2.87	m	**11.07**
Eaves, verge soffit boards, fascia boards and the like						
19 mm thick; over 300 mm wide	–	1.15	20.50	13.20	m²	**33.70**
19 mm thick; 150 mm wide; once grooved	–	0.19	3.38	2.22	m	**5.60**
25 mm thick; 150 mm wide; once grooved	–	0.19	3.38	2.73	m	**6.11**
25 mm thick; 175 mm wide; once grooved	–	0.19	3.38	2.69	m	**6.07**
32 mm thick; 225 mm wide; once grooved	–	0.23	4.10	4.38	m	**8.48**
Wrought softwood; tanalized						
Gutter boards; tongued and grooved joints						
19 mm thick; 150 mm; sloping	–	1.39	24.78	14.00	m²	**38.78**
19 mm thick; 75 mm wide	–	0.37	6.60	1.01	m	**7.61**
19 mm thick; 150 mm wide	–	0.42	7.48	2.06	m	**9.54**
19 mm thick; 225 mm wide	–	0.46	8.20	3.00	m	**11.20**
25 mm thick; sloping	–	1.39	24.78	14.72	m²	**39.50**
25 mm thick; 75 mm wide	–	0.37	6.60	1.15	m	**7.75**
25 mm thick; 150 mm wide	–	0.42	7.48	2.08	m	**9.56**
25 mm thick; 225 mm wide	–	0.46	8.20	3.12	m	**11.32**
Eaves, verge soffit boards, fascia boards and the like						
19 mm thick; over 300 mm wide	–	1.15	20.50	14.05	m²	**34.55**
19 mm thick; 150 mm wide; once grooved	–	0.19	3.38	2.35	m	**5.73**
25 mm thick; 150 mm wide; once grooved	–	0.19	3.38	2.89	m	**6.27**
25 mm thick; 175 mm wide; once grooved	–	0.20	3.57	2.88	m	**6.45**
32 mm thick; 225 mm wide; once grooved	–	0.23	4.10	4.69	m	**8.79**
Straps; mild steel; galvanized						
Standard twisted vertical restraint; fixing to softwood and brick or blockwork						
27.5 mm × 2.5 mm × 400 mm girth	–	0.23	4.10	1.33	nr	**5.43**
27.5 mm × 2.5 mm × 600 mm girth	–	0.24	4.27	1.85	nr	**6.12**
27.5 mm × 2.5 mm × 800 mm girth	–	0.25	4.46	2.67	nr	**7.13**
27.5 mm × 2.5 mm × 1000 mm girth	–	0.28	4.99	3.46	nr	**8.45**
27.5 mm × 2.5 mm × 1200 mm girth	–	0.29	5.17	4.19	nr	**9.36**

16 CARPENTRY

Item	PC £	Labour hours	Labour £	Material £	Unit	Total rate £
Hangers; mild steel; galvanized						
Joist hangers 0.90 mm thick; The Expanded Metal Company Ltd Speedy or other equal and approved; for fixing to softwood; joist sizes						
50 mm wide; all sizes to 225 mm deep	1.78	0.11	1.96	1.99	nr	**3.95**
75 mm wide; all sizes to 225 mm deep	1.86	0.14	2.49	2.15	nr	**4.64**
100 mm wide; all sizes to 225 mm deep	2.00	0.17	3.03	2.38	nr	**5.41**
Joist hangers 2.50 mm thick; for building in; joist sizes						
50 mm × 100 mm	3.28	0.07	1.24	3.64	nr	**4.88**
50 mm × 125 mm	3.29	0.07	1.24	3.66	nr	**4.90**
50 mm × 150 mm	3.09	0.09	1.60	3.50	nr	**5.10**
50 mm × 175 mm	3.23	0.09	1.60	3.65	nr	**5.25**
50 mm × 200 mm	3.76	0.11	1.95	4.08	nr	**6.03**
50 mm × 225 mm	3.81	0.11	1.95	4.32	nr	**6.27**
75 mm × 150 mm	4.76	0.09	1.60	5.29	nr	**6.89**
75 mm × 175 mm	4.47	0.09	1.60	4.98	nr	**6.58**
75 mm × 200 mm	4.76	0.11	1.95	5.35	nr	**7.30**
75 mm × 225 mm	5.11	0.11	1.95	5.72	nr	**7.67**
75 mm × 250 mm	5.41	0.13	2.31	6.10	nr	**8.41**
100 mm × 200 mm	5.92	0.11	1.95	6.60	nr	**8.55**
Metal connectors; mild steel; galvanized						
Round toothed plate; for 10 mm or 12 mm dia. bolts						
38 mm dia.; single sided	–	0.01	0.18	0.49	nr	**0.67**
38 mm dia.; double sided	–	0.01	0.18	0.54	nr	**0.72**
50 mm dia.; single sided	–	0.01	0.18	0.52	nr	**0.70**
50 mm dia.; double sided	–	0.01	0.18	0.58	nr	**0.76**
63 mm dia.; single sided	–	0.01	0.18	0.77	nr	**0.95**
63 mm dia.; double sided	–	0.01	0.18	0.84	nr	**1.02**
75 mm dia.; single sided	–	0.01	0.18	1.13	nr	**1.31**
75 mm dia.; double sided	–	0.01	0.18	1.17	nr	**1.35**
framing anchor	–	0.14	2.49	0.92	nr	**3.41**
Bolts; mild steel; galvanized						
Fixing only bolts; 50 mm–200 mm long						
6 mm dia.	–	0.03	0.53	–	nr	**0.53**
8 mm dia.	–	0.03	0.53	–	nr	**0.53**
10 mm dia.	–	0.04	0.72	–	nr	**0.72**
12 mm dia.	–	0.04	0.72	–	nr	**0.72**
16 mm dia.	–	0.05	0.89	–	nr	**0.89**
20 mm dia.	–	0.05	0.89	–	nr	**0.89**
Bolts						
Expanding bolts; Rawlbolt projecting type or other equal and approved; Rawl Fixings; plated; one nut; one washer						
6 mm dia.; ref M6 10P	–	0.09	1.61	0.59	nr	**2.20**
6 mm dia.; ref M6 25P	–	0.09	1.61	0.70	nr	**2.31**
6 mm dia.; ref M6 60P	–	0.09	1.61	0.70	nr	**2.31**

16 CARPENTRY

Item	PC £	Labour hours	Labour £	Material £	Unit	Total rate £
TIMBER FRAMING – cont						
Bolts – cont						
Expanding bolts – cont						
8 mm dia.; ref M8 25P	–	0.09	1.61	0.69	nr	**2.30**
8 mm dia.; ref M8 60P	–	0.09	1.61	0.70	nr	**2.31**
10 mm dia.; ref M10 15P	–	0.09	1.61	0.92	nr	**2.53**
10 mm dia.; ref M10 30P	–	0.09	1.61	0.96	nr	**2.57**
10 mm dia.; ref M10 60P	–	0.09	1.61	0.94	nr	**2.55**
12 mm dia.; ref M12 15P	–	0.09	1.61	1.70	nr	**3.31**
12 mm dia.; ref M12 30P	–	0.10	1.78	0.15	nr	**1.93**
12 mm dia.; ref M12 75P	–	0.09	1.61	0.20	nr	**1.81**
16 mm dia.; ref M16 35P	–	0.09	1.61	3.62	nr	**5.23**
16 mm dia.; ref M16 75P	–	0.09	1.61	3.74	nr	**5.35**
Expanding bolts; Rawlbolt loose bolt type or other equal; Rawl Fixings; plated; one bolt; one washer						
6 mm dia.; ref M6 10L	–	0.09	1.61	0.64	nr	**2.25**
6 mm dia.; ref M6 25L	–	0.09	1.61	0.64	nr	**2.25**
6 mm dia.; ref M6 40L	–	0.09	1.61	0.75	nr	**2.36**
8 mm dia.; ref M8 25L	–	0.09	1.61	0.83	nr	**2.44**
8 mm dia.; ref M8 40L	–	0.09	1.61	0.87	nr	**2.48**
10 mm dia.; ref M10 10L	–	0.09	1.61	0.88	nr	**2.49**
10 mm dia.; ref M10 25L	–	0.09	1.61	0.97	nr	**2.58**
10 mm dia.; ref M10 50L	–	0.09	1.61	0.97	nr	**2.58**
10 mm dia.; ref M10 75L	–	0.09	1.61	1.08	nr	**2.69**
12 mm dia.; ref M12 10L	–	0.09	1.61	1.69	nr	**3.30**
12 mm dia.; ref M12 25L	–	0.09	1.61	1.75	nr	**3.36**
12 mm dia.; ref M12 40L	–	0.09	1.61	1.74	nr	**3.35**
12 mm dia.; ref M12 60L	–	0.09	1.61	2.20	nr	**3.81**
16 mm dia.; ref M16 30L	–	0.09	1.61	3.70	nr	**5.31**
16 mm dia.; ref M16 60L	–	0.09	1.61	4.33	nr	**5.94**
Truss clips						
Truss clips; fixing to softwood; joist size						
38 mm wide	0.70	0.14	2.49	1.05	nr	**3.54**
50 mm wide	0.66	0.14	2.49	1.01	nr	**3.50**
Sole plate angles; mild steel galvanized						
Sole plate angle; fixing to softwood and concrete						
112 mm × 40 mm × 76 mm	0.80	0.19	3.38	1.92	nr	**5.30**
Chemical anchors						
R-CAS Spin-in epoxy acrylate capsules and standard studs or other equal and approved; Rawl Fixings; with nuts and washers; drilling masonry						
capsule ref 60–408; stud ref 60–448	–	0.25	4.46	0.93	nr	**5.39**
capsule ref 60–410; stud ref 60–454	–	0.28	4.99	1.08	nr	**6.07**
capsule ref 60–412; stud ref 60–460	–	0.31	5.52	1.47	nr	**6.99**
capsule ref 60–416; stud ref 60–472	–	0.34	6.06	1.81	nr	**7.87**
capsule ref 60–420; stud ref 60–478	–	0.36	6.42	2.36	nr	**8.78**
capsule ref 60–424; stud ref 60–484	–	0.40	7.13	7.52	nr	**14.65**

16 CARPENTRY

Item	PC £	Labour hours	Labour £	Material £	Unit	Total rate £
R-CAS Spin-in epoxy acrylate capsules and stainless steel studs or other equal and approved; Rawl Fixings; with nuts and washers; drilling masonry						
capsule ref 60–408; stud ref 60–905	–	0.25	4.46	1.88	nr	**6.34**
capsule ref 60–410; stud ref 60–910	–	0.28	4.99	3.30	nr	**8.29**
capsule ref 60–412; stud ref 60–915	–	0.31	5.52	4.20	nr	**9.72**
capsule ref 60–416; stud ref 60–920	–	0.34	6.06	8.97	nr	**15.03**
capsule ref 60–420; stud ref 60–925	–	0.36	6.42	17.44	nr	**23.86**
capsule ref 60–424; stud ref 60–930	–	0.40	7.13	31.04	nr	**38.17**
R-CAS Spin-in epoxy acrylate capsules and standard internal threaded sockets or other equal and approved; Rawl Fixings; drilling masonry						
capsule ref 60–408; socket ref 60–650	–	0.25	4.46	1.28	nr	**5.74**
capsule ref 60–410; socket ref 60–656	–	0.28	4.99	1.32	nr	**6.31**
capsule ref 60–412; socket ref 60–662	–	0.31	5.52	1.41	nr	**6.93**
capsule ref 60–416; socket ref 60–668	–	0.34	6.06	2.01	nr	**8.07**
capsule ref 60–420; socket ref 60–674	–	0.36	6.42	1.47	nr	**7.89**
capsule ref 60–424; socket ref 60–676	–	0.40	7.13	4.62	nr	**11.75**
R-CAS Spin-in epoxy acrylate capsules and stainless steel internal threaded sockets or other equal and approved; Rawl Fixings; drilling masonry						
capsule ref 60–408; socket ref 60–943	–	0.25	4.46	2.63	nr	**7.09**
capsule ref 60–410; socket ref 60–945	–	0.28	4.99	2.69	nr	**7.68**
capsule ref 60–412; socket ref 60–947	–	0.31	5.52	3.34	nr	**8.86**
capsule ref 60–416; socket ref 60–949	–	0.34	6.06	4.87	nr	**10.93**
capsule ref 60–420; socket ref 60–951	–	0.36	6.42	5.17	nr	**11.59**
capsule ref 60–424; socket ref 60–955	–	0.40	7.13	12.41	nr	**19.54**
R-CAS Spin-in epoxy acrylate capsules, perforated sleeves and standard studs or other equal and approved; Rawl Fixings; in low density material; with nuts and washers; drilling masonry						
capsule ref 60–408; sleeve ref 60–538; stud ref 60–448	–	0.25	4.46	2.32	nr	**6.78**
capsule ref 60–410; sleeve ref 60–544; stud ref 60–454	–	0.28	4.99	2.63	nr	**7.62**
capsule ref 60–412; sleeve ref 60–550; stud ref 60–460	–	0.31	5.52	3.08	nr	**8.60**
capsule ref 60–416; sleeve ref 60–562; stud ref 60–472	–	0.34	6.06	3.43	nr	**9.49**
R-CAS Spin-in epoxy acrylate capsules, perforated sleeves and stainless steel studs or other equal and approved; Rawl Fixings; in low density material; with nuts and washers; drilling masonry						
capsule ref 60–408; sleeve ref 60–538; stud ref 60–905	–	0.25	4.46	3.26	nr	**7.72**
capsule ref 60–410; sleeve ref 60–544; stud ref 60–910	–	0.28	4.99	4.86	nr	**9.85**
capsule ref 60–412; sleeve ref 60–550; stud ref 60–915	–	0.31	5.52	5.81	nr	**11.33**
capsule ref 60–416; sleeve ref 60–562; stud ref 60–920	–	0.34	6.06	10.59	nr	**16.65**

16 CARPENTRY

Item	PC £	Labour hours	Labour £	Material £	Unit	Total rate £
TIMBER FRAMING – cont						
Chemical anchors – cont						
R-CAS Spin-in epoxy acrylate capsules, perforated sleeves and standard internal threaded sockets or other equal and approved; The Rawlplug Company; in low density material; with nuts and washers; drilling masonry						
capsule ref 60–408; sleeve ref 60–538; socket ref 60–650	–	0.25	4.46	2.67	nr	**7.13**
capsule ref 60–410; sleeve ref 60–544; socket ref 60–656	–	0.28	4.99	2.88	nr	**7.87**
capsule ref 60–412; sleeve ref 60–550; socket ref 60–662	–	0.31	5.52	3.02	nr	**8.54**
R-CAS Spin-in epoxy acrylate capsules, perforated sleeves and stainless steel internal threaded sockets or other equal and approved; The Rawlplug Company; in low density material; drilling masonry						
capsule ref 60–416; sleeve ref 60–562; socket ref 60–668	–	0.34	6.06	3.64	nr	**9.70**
capsule ref 60–408; sleeve ref 60–538; socket ref 60–943	–	0.25	4.46	4.03	nr	**8.49**
capsule ref 60–410; sleeve ref 60–544; socket ref 60–945	–	0.28	4.99	4.23	nr	**9.22**
capsule ref 60–412; sleeve ref 60–550; socket ref 60–947	–	0.31	5.52	4.94	nr	**10.46**
capsule ref 60–416; sleeve ref 60–562; socket ref 60–949	–	0.34	6.06	6.50	nr	**12.56**
BOARDING TO FLOORS						
Rigid sheet boarding						
Chipboard boarding and flooring						
Boarding to floors; butt joints						
18 mm thick	4.11	0.28	4.99	4.88	m²	**9.87**
Boarding to floors; tongued and grooved joints						
18 mm thick	5.18	0.30	5.35	6.06	m²	**11.41**
22 mm thick	6.02	0.32	5.71	6.98	m²	**12.69**
Acoustic Chipboard flooring						
Boarding to floors; tongued and grooved joints						
chipboard on blue bat bearers	–	–	–	–	m²	**19.70**
chipboard on New Era levelling system	–	–	–	–	m²	**26.61**

16 CARPENTRY

Item	PC £	Labour hours	Labour £	Material £	Unit	Total rate £
Laminated engineered board flooring; 180 or 240 mm face widths; with 6 mm wear surface down to tongue; pre-finished laquered, oiled or untreated						
Boarding to floors; micro bevel or square edge						
Country laquered; on 10 mm Pro Foam	–	–	–	–	m²	**57.91**
Rustic laquered; on 10 mm Pro Foam	–	–	–	–	m²	**55.02**
Plywood flooring						
Boarding to floors; tongued and grooved joints						
18 mm thick	8.53	0.41	7.31	9.75	m²	**17.06**
22 mm thick	10.35	0.45	8.03	11.76	m²	**19.79**
Strip boarding						
Wrought softwood						
Boarding to floors; butt joints						
19 mm × 75 mm boards	–	0.56	9.98	12.46	m²	**22.44**
19 mm × 125 mm boards	–	0.51	9.09	9.55	m²	**18.64**
22 mm × 150 mm boards	–	0.46	8.20	10.61	m²	**18.81**
25 mm × 100 mm boards	–	0.51	9.09	11.59	m²	**20.68**
25 mm × 150 mm boards	–	0.46	8.20	11.80	m²	**20.00**
Boarding to floors; tongued and grooved joints						
19 mm × 75 mm boards	–	0.65	11.58	13.42	m²	**25.00**
19 mm × 125 mm boards	–	0.60	10.70	10.74	m²	**21.44**
22 mm × 150 mm boards	–	0.56	9.98	10.92	m²	**20.90**
25 mm × 100 mm boards	–	0.60	10.70	13.78	m²	**24.48**
25 mm × 150 mm boards	–	0.56	9.98	12.90	m²	**22.88**
BOARDING TO CEILINGS						
Rigid sheet boarding						
Plywood (Eastern European); internal quality						
Lining to ceilings 4 mm thick						
over 300 mm wide	2.19	0.46	8.20	2.70	m²	**10.90**
not exceeding 300 mm wide	–	0.30	5.35	0.83	m	**6.18**
holes for pipes and the like	–	0.02	0.36	–	nr	**0.36**
Lining to ceilings 6 mm thick						
over 300 mm wide	3.17	0.49	8.73	3.77	m²	**12.50**
not exceeding 300 mm wide	–	0.32	5.71	1.15	m	**6.86**
holes for pipes and the like	–	0.02	0.36	–	nr	**0.36**
Lining to ceilings 12 mm thick						
over 300 mm wide	5.88	0.56	9.98	6.76	m²	**16.74**
not exceeding 300 mm wide	–	0.37	6.60	2.05	m	**8.65**
holes for pipes and the like	–	0.03	0.53	–	nr	**0.53**
Lining to ceilings 18 mm thick						
over 300 mm wide	8.59	0.60	10.70	9.75	m²	**20.45**
not exceeding 300 mm wide	–	0.40	7.13	2.94	m	**10.07**
holes for pipes and the like	–	0.03	0.53	–	nr	**0.53**

16 CARPENTRY

Item	PC £	Labour hours	Labour £	Material £	Unit	Total rate £
BOARDING TO CEILINGS – cont						
Plywood (Eastern European); external quality						
Lining to ceilings 4 mm thick						
over 300 mm wide	5.43	0.46	8.20	6.26	m²	**14.46**
not exceeding 300 mm wide	–	0.30	5.35	1.90	m	**7.25**
holes for pipes and the like	–	0.02	0.36	–	nr	**0.36**
Lining to ceilings 6.5 mm thick						
over 300 mm wide	6.07	0.49	8.73	6.97	m²	**15.70**
not exceeding 300 mm wide	–	0.32	5.71	2.11	m	**7.82**
holes for pipes and the like	–	0.02	0.36	–	nr	**0.36**
Lining to ceilings 9 mm thick						
over 300 mm wide	7.81	0.53	9.45	8.89	m²	**18.34**
not exceeding 300 mm wide	–	0.34	6.06	2.69	m	**8.75**
holes for pipes and the like	–	0.03	0.53	–	nr	**0.53**
Lining to ceilings 12 mm thick						
over 300 mm wide	9.75	0.56	9.98	11.03	m²	**21.01**
not exceeding 300 mm wide	–	0.37	6.60	3.33	m	**9.93**
holes for pipes and the like	–	0.03	0.53	–	nr	**0.53**
Extra over Linings fixed with nails for screwing	–	–	–	–	m²	**1.72**
Non-asbestos board; Masterboard or other equal and approved; sanded finish						
Lining to ceilings 6 mm thick						
over 300 mm wide	7.97	0.41	7.31	8.99	m²	**16.30**
not exceeding 300 mm wide	–	0.25	4.46	2.71	m	**7.17**
holes for pipes and the like	–	0.02	0.36	–	nr	**0.36**
Lining to ceilings 9 mm thick						
over 300 mm wide	18.41	0.42	7.48	20.50	m²	**27.98**
not exceeding 300 mm wide	–	0.27	4.82	6.16	m	**10.98**
holes for pipes and the like	–	0.03	0.53	–	nr	**0.53**
Supalux or other equal and approved; sanded finish						
Lining to ceilings 6 mm thick						
over 300 mm wide	14.42	0.41	7.31	16.10	m²	**23.41**
not exceeding 300 mm wide	–	0.25	4.46	4.84	m	**9.30**
holes for pipes and the like	–	0.03	0.53	–	nr	**0.53**
Lining to ceilings 9 mm thick						
over 300 mm wide	18.48	0.42	7.48	20.57	m²	**28.05**
not exceeding 300 mm wide	–	0.27	4.82	6.18	m	**11.00**
holes for pipes and the like	–	0.03	0.53	–	nr	**0.53**
Lining to ceilings 12 mm thick						
over 300 mm wide	25.09	0.49	8.73	27.86	m²	**36.59**
not exceeding 300 mm wide	–	0.30	5.35	8.36	m	**13.71**
holes for pipes and the like	–	0.04	0.72	–	nr	**0.72**
Extra over Linings fixed with nails for screwing	–	–	–	–	m²	**1.72**

16 CARPENTRY

Item	PC £	Labour hours	Labour £	Material £	Unit	Total rate £
Strip boarding						
Wrought softwood						
Boarding to internal ceilings						
12 mm × 100 mm boards	–	0.93	16.57	12.29	m²	**28.86**
16 mm × 100 mm boards	–	0.93	16.57	13.29	m²	**29.86**
19 mm × 100 mm boards	–	0.93	16.57	15.16	m²	**31.73**
19 mm × 125 mm boards	–	0.88	15.69	13.79	m²	**29.48**
19 mm × 125 mm boards; chevron pattern	–	1.30	23.18	13.79	m²	**36.97**
25 mm × 125 mm boards	–	0.88	15.69	11.90	m²	**27.59**
12 mm × 100 mm boards; knotty pine	–	0.93	16.57	7.56	m²	**24.13**
BOARDING TO ROOFS						
Rigid sheet boarding						
Plywood; external quality; 18 mm thick						
Boarding to roofs; butt joints						
flat to falls	11.43	0.37	6.60	12.94	m²	**19.54**
sloping	11.43	0.40	7.13	12.94	m²	**20.07**
vertical	11.43	0.53	9.45	12.94	m²	**22.39**
Plywood; external quality; 12 mm thick						
Boarding to roofs; butt joints						
flat to falls	7.71	0.37	6.60	8.85	m²	**15.45**
sloping	7.71	0.40	7.13	8.85	m²	**15.98**
vertical	7.71	0.53	9.45	8.85	m²	**18.30**
Strip boarding						
Sawn softwood; untreated						
Boarding to roofs; 150 mm wide boards; butt joints						
19 mm thick; flat; over 600 mm wide	–	0.42	7.48	9.04	m²	**16.52**
19 mm thick; flat; not exceeding 600 mm wide	–	0.55	9.81	5.49	m	**15.30**
19 mm thick; sloping; over 600 mm wide	–	0.46	8.20	9.04	m²	**17.24**
19 mm thick; sloping; not exceeding 600 mm wide	–	0.62	11.05	5.49	m	**16.54**
19 mm thick; sloping; laid diagonally; over 600 mm wide		0.58	10.34	9.04	m²	**19.38**
19 mm thick; sloping; laid diagonally; not exceeding 600 mm wide	–	0.75	13.37	5.49	m	**18.86**
25 mm thick; flat; over 600 mm wide	–	0.42	7.48	14.69	m²	**22.17**
25 mm thick; flat; not exceeding 600 mm wide	–	0.56	9.98	8.89	m	**18.87**
25 mm thick; sloping; over 600 mm wide	–	0.46	8.20	14.69	m²	**22.89**
25 mm thick; sloping; not exceeding 600 mm wide	–	0.62	11.05	8.89	m	**19.94**
25 mm thick; sloping; laid diagonally; over 600 mm wide	–	0.58	10.34	14.69	m²	**25.03**
25 mm thick; sloping; laid diagonally; not exceeding 600 mm wide	–	0.74	13.19	8.89	m	**22.08**

16 CARPENTRY

Item	PC £	Labour hours	Labour £	Material £	Unit	Total rate £
BOARDING TO ROOFS – cont						
Sawn softwood – cont						
Boarding to tops or cheeks of dormers; 150 mm wide boards; butt joints						
19 mm thick; laid diagonally; over 600 mm wide	–	0.74	13.19	9.04	m²	**22.23**
19 mm thick; laid diagonally; not exceeding 600 mm wide	–	0.92	16.40	5.49	m	**21.89**
19 mm thick; laid diagonally; area not exceeding 1.00 m² irrespective of width	–	0.93	16.57	8.59	nr	**25.16**
Sawn softwood; tanalized						
Boarding to roofs; 150mm wide boards; butt joints						
19 mm thick; flat; over 600 mm wide	–	0.42	7.48	9.89	m²	**17.37**
19 mm thick; flat; not exceeding 600 mm wide	–	0.56	9.98	6.01	m	**15.99**
19 mm thick; sloping; over 600 mm wide	–	0.46	8.20	9.89	m²	**18.09**
19 mm thick; sloping; not exceeding 600 mm wide	–	0.62	11.05	6.01	m	**17.06**
19 mm thick; sloping; laid diagonally; over 600 mm wide	–	0.58	10.34	9.89	m²	**20.23**
19 mm thick; sloping; laid diagonally; not exceeding 600 mm wide	–	0.74	13.19	6.01	m	**19.20**
25 mm thick; flat; over 600 mm wide	–	0.42	7.48	15.80	m²	**23.28**
25 mm thick; flat; not exceeding 600 mm wide	–	0.56	9.98	9.39	m	**19.37**
25 mm thick; sloping; over 600 mm wide	–	0.46	8.20	15.80	m²	**24.00**
25 mm thick; sloping; not exceeding 600 mm wide	–	0.62	11.05	9.39	m	**20.44**
25 mm thick; sloping; laid diagonally; over 600 mm wide	–	0.58	10.34	15.80	m²	**26.14**
25 mm thick; sloping; laid diagonally; not exceeding 600 mm wide	–	0.74	13.19	9.39	m	**22.58**
Boarding to tops or cheeks of dormers; 150 mm wide boards; butt joints						
19 mm thick; laid diagonally; over 600 mm wide	–	0.74	13.19	9.89	m²	**23.08**
19 mm thick; laid diagonally; not exceeding 600 mm wide	–	0.92	16.40	6.01	m	**22.41**
19 mm thick; laid diagonally; area not exceeding 1.00 m² irrespective of width	–	0.93	16.57	9.43	nr	**26.00**
Wrought softwood						
Boarding to roofs; tongued and grooved joints						
19 mm thick; flat to falls	–	0.51	9.09	12.64	m²	**21.73**
19 mm thick; sloping	–	0.56	9.98	12.64	m²	**22.62**
19 mm thick; sloping; laid diagonally	–	0.72	12.83	12.64	m²	**25.47**
25 mm thick; flat to falls	–	0.51	9.09	12.73	m²	**21.82**
25 mm thick; sloping	–	0.56	9.98	12.73	m²	**22.71**
Boarding to tops or cheeks of dormers; tongued and grooved joints						
19 mm thick; laid diagonally	–	0.93	16.57	12.64	m²	**29.21**

16 CARPENTRY

Item	PC £	Labour hours	Labour £	Material £	Unit	Total rate £
Wrought softwood; tanalized						
Boarding to roofs; tongued and grooved joints						
19 mm thick; flat to falls	–	0.51	9.09	13.48	m²	**22.57**
19 mm thick; sloping	–	0.56	9.98	13.48	m²	**23.46**
19 mm thick; sloping; laid diagonally	–	0.72	12.83	13.48	m²	**26.31**
25 mm thick; flat to falls	–	0.51	9.09	13.85	m²	**22.94**
25 mm thick; sloping	–	0.56	9.98	13.85	m²	**23.83**
Boarding to tops or cheeks of dormers; tongued and grooved joints						
19 mm thick; laid diagonally	–	0.93	16.57	13.48	m²	**30.05**
CASINGS						
Chipboard (plain)						
Two-sided 15 mm thick pipe casing; to softwood framing (not included)						
300 mm girth	–	0.56	9.98	1.09	m	**11.07**
600 mm girth	–	0.65	11.58	1.96	m	**13.54**
Three-sided 15 mm thick pipe casing; to softwood framing (not included)						
450 mm girth	–	1.16	20.68	1.63	m	**22.31**
900 mm girth	–	1.39	24.78	2.97	m	**27.75**
extra for 400 mm × 400 mm removable access panel; brass cups and screws; additional framing	–	0.93	16.57	1.07	nr	**17.64**
Plywood (Eastern European); internal quality						
Two-sided 6 mm thick pipe casings; to softwood framing (not included)						
300 mm girth	–	0.74	13.19	1.26	m	**14.45**
600 mm girth	–	0.93	16.57	2.31	m	**18.88**
Three-sided 6 mm thick pipe casing; to softwood framing (not included)						
450 mm girth	–	1.06	18.90	1.89	m	**20.79**
900 mm glrth	–	1.25	22.28	3.48	m	**25.76**
Plywood (Eastern European); external quality						
Two-sided 6.5 mm thick pipe casings; to softwood framing (not included)						
300 mm girth	–	0.74	13.19	2.21	m	**15.40**
600 mm girth	–	0.93	16.57	4.22	m	**20.79**
Three-sided 6.5 mm thick pipe casing; to softwood framing (not included)						
450 mm girth	–	1.06	18.90	3.32	m	**22.22**
900 mm girth	–	1.25	22.28	6.37	m	**28.65**
Two-sided 12 mm thick pipe casing; to softwood framing (not included)						
300 mm girth	–	0.69	12.30	3.43	m	**15.73**
600 mm girth	–	0.83	14.80	6.66	m	**21.46**

16 CARPENTRY

Item	PC £	Labour hours	Labour £	Material £	Unit	Total rate £
CASINGS – cont						
Plywood (Eastern European) – cont						
Three-sided 12 mm thick pipe casing; to softwood framing (not included)						
450 mm girth	–	0.93	16.57	5.16	m	**21.73**
900 mm girth	–	1.11	19.79	10.02	m	**29.81**
extra for 400 mm × 400 mm removable access panel; brass cups and screws; additional framing	–	1.00	17.82	1.07	nr	**18.89**
Preformed white melamine faced plywood casings; Pendock Profiles Ltd or other equal and approved; to softwood battens (not included)						
Skirting trunking profile; plain butt joints in the running length						
45 mm × 150 mm; ref TK150	–	0.11	1.96	25.50	m	**27.46**
extra for stop end	–	0.04	0.72	15.88	nr	**16.60**
extra for external corner	–	0.09	1.61	21.93	nr	**23.54**
extra for internal corner	–	0.09	1.61	13.33	nr	**14.94**
Casing profiles						
150 mm × 150 mm; ref MX150/150; 5 mm thick	–	0.11	1.96	21.26	m	**23.22**
extra for stop end	–	0.04	0.72	5.48	nr	**6.20**
extra for external corner	–	0.09	1.61	33.57	nr	**35.18**
extra for internal corner	–	0.09	1.61	13.33	nr	**14.94**

17 SHEET ROOF COVERINGS

Item	PC £	Labour hours	Labour £	Material £	Unit	Total rate £
BUILT UP FELT ROOF COVERINGS						
NOTE: The following items of felt roofing, unless otherwise described, include for conventional lapping, laying and bonding between layers and to base; and laying flat or to falls, crossfalls or to slopes not exceeding 10° – but exclude any insulation etc.						
Reinforced bitumen membranes						
Three layer coverings						
type S1P1 bitumen glass fibre based felt	–	–	–	–	m²	14.35
cover with and bed in hot bitumen 13 mm thick stone chippings	–	–	–	–	m²	4.16
two base layers type S2P3 bitumen polyester-based felt; top layer type S4P4 polyester-based mineral surfaced felt; 10 mm stone chipping covering; bitumen bonded	–	–	–	–	m²	24.44
cover with and bed in hot bitumen 300 mm × 300 mm × 8 mm g.r.p. tiles	–	–	–	–	m²	44.46
Skirtings; three layer; top layer mineral surfaced; dressed over tilting fillet; turned into groove						
not exceeding 200 mm girth	–	–	–	–	m	10.69
200 mm–400 mm girth	–	–	–	–	m	13.21
Coverings to kerbs; three layer						
400 mm–600 mm girth	–	–	–	–	m	17.10
Linings to gutters; three layer						
400 mm–600 mm girth	–	–	–	–	m	20.77
Collars around pipes and the like; three layer mineral surface; 150 mm high						
not exceeding 55 mm nominal size	–	–	–	–	nr	11.34
55 mm–110 mm nominal size	–	–	–	–	nr	11.34
Outlets and dishing to gullies						
300 mm dia.	–	–	–	–	nr	12.30
Polyester-based roofing systems						
Andersons high performance polyester-based roofing system or other equal						
two layer coverings; first layer HT 125 underlay; second layer HT 350; fully bonded to wood; fibre or cork base	–	–	–	–	m²	20.26
top layer mineral surfaced	–	–	–	–	m²	1.72
13 mm thick stone chippings	–	–	–	–	m²	4.16
third layer of type 3B as underlay for concrete or screeded base	–	–	–	–	m²	5.30
working into outlet pipes and the like	–	–	–	–	nr	12.29
Skirtings; two layer; top layer mineral surfaced; dressed over tilting fillet; turned into groove						
not exceeding 200 mm girth	–	–	–	–	m	104.38
200 mm–400 mm girth	–	–	–	–	m	13.51
Coverings to kerbs; two layer						
400 mm–600 mm girth	–	–	–	–	m	17.50
Linings to gutters; three layer						
400 mm–600 mm girth	–	–	–	–	m	18.82

17 SHEET ROOF COVERINGS

Item	PC £	Labour hours	Labour £	Material £	Unit	Total rate £
BUILT UP FELT ROOF COVERINGS – cont						
Polyester-based roofing systems – cont						
Collars around pipes and the like; two layer; 150 mm high						
not exceeding 55 mm nominal size	–	–	–	–	nr	**12.29**
55 mm–110 mm nominal size	–	–	–	–	nr	**12.29**
Ruberoid Challenger SBS high performance roofing or other equal						
two layer coverings; first and second layers Ruberglas 120 GP; fully bonded to wood, fibre or cork base	–	–	–	–	m²	**13.34**
top layer mineral surfaced	–	–	–	–	m²	**4.71**
13 mm thick stone chippings	–	–	–	–	m²	**4.16**
third layer of Rubervent 3G as underlay for concrete or screeded base	–	–	–	–	m²	**5.28**
working into outlet pipes and the like	–	–	–	–	nr	**12.21**
Skirtings; two layer; top layer mineral surfaced; dressed over tilting fillet; turned into groove						
not exceeding 200 mm girth	–	–	–	–	m	**10.17**
200 mm–400 mm girth	–	–	–	–	m	**13.32**
Coverings to kerbs; two layer						
400 mm–600 mm girth	–	–	–	–	m	**17.26**
Linings to gutters; three layer						
400 mm–600 mm girth	–	–	–	–	m	**18.48**
Collars around pipes and the like; two layer, 150 mm high						
not exceeding 55 mm nominal size	–	–	–	–	nr	**12.21**
55 mm–110 mm nominal size	–	–	–	–	nr	**12.21**
Ruberfort HP 350 high performance roofing or other equal						
two layer coverings; first layer Ruberfort HP 180; second layer Ruberfort HP 350; fully bonded; to wood; fibre or cork base	–	–	–	–	m²	**15.71**
top layer mineral surfaced	–	–	–	–	m²	**6.51**
13 mm thick stone chippings	–	–	–	–	m²	**4.16**
third layer of 'Rubervent 3G'; as underlay for concrete or screeded base	–	–	–	–	m²	**5.28**
working into outlet pipes and the like	–	–	–	–	nr	**12.35**
Skirtings; two layer; top layer mineral surface; dressed over tilting fillet; turned into groove						
not exceeding 200 mm girth	–	–	–	–	m	**10.38**
200 mm–400 mm girth	–	–	–	–	m	**13.59**
Coverings to kerbs; two layer						
400 mm–600 mm girth	–	–	–	–	m	**17.61**
Linings to gutters; three layer						
400 mm–600 mm girth	–	–	–	–	m	**22.74**
Collars around pipes and the like; two layer; 150 mm high						
not exceeding 55 mm nominal size	–	–	–	–	nr	**12.35**
55 mm–110 mm nominal size	–	–	–	–	nr	**12.35**

17 SHEET ROOF COVERINGS

Item	PC £	Labour hours	Labour £	Material £	Unit	Total rate £
Ruberoid Superflex Firebloc high performance roofing or other equal (15 year guarantee specification)						
two layer coverings; first layer Superflex 180; second layer Superflex 250; fully bonded to wood; fibre or cork base	–	–	–	–	m²	**19.43**
top layer mineral surfaced	–	–	–	–	m²	**4.63**
13 mm thick stone chippings	–	–	–	–	m²	**4.16**
third layer of Rubervent 3G as underlay for concrete or screeded base	–	–	–	–	m²	**5.28**
working into outlet pipes and the like	–	–	–	–	nr	**14.02**
Skirtings; two layer; top layer mineral surfaced; dressed over tilting fillet; turned into groove						
not exceeding 200 mm girth	–	–	–	–	m	**12.14**
200 mm–400 mm girth	–	–	–	–	m	**16.02**
Coverings to kerbs; two layer						
400 mm–600 mm girth	–	–	–	–	m	**21.38**
Linings to gutters; three layer						
400 mm–600 mm girth	–	–	–	–	m	**23.15**
Collars around pipes and the like; two layer; 150 mm high						
not exceeding 55 mm nominal size	–	–	–	–	nr	**14.02**
55 mm–110 mm nominal size	–	–	–	–	nr	**14.02**
Ruberoid Ultra Prevent high performance roofing or other equal						
two layer coverings; first layer Ultra prevENt underlay; second layer Ultra prevENt mineral surface cap sheet	–	–	–	–	m²	**35.75**
extra over for						
third layer of Rubervent 3G as underlay for concrete or screeded base	–	–	–	–	m²	**5.28**
working into outlet pipes and the like	–	–	–	–	nr	**16.96**
Skirtings; two layer; dressed over tilting fillet; turned into groove						
not exceeding 200 mm girth	–	–	–	–	m	**15.19**
200 mm–400 mm girth	–	–	–	–	m	**20.23**
Coverings to kerbs; two layer						
400 mm–600 mm girth	–	–	–	–	m	**27.93**
Linings to gutters; three layer						
400 mm–600 mm girth	–	–	–	–	m	**29.19**
Collars around pipes and the like; two layer; 150 mm high						
not exceeding 55 mm nominal size	–	–	–	–	nr	**16.94**
55 mm–110 mm nominal size	–	–	–	–	nr	**16.94**
Accessories						
Eaves trim; extruded aluminium alloy; working felt into trim						
Rubertrim; type FL/G; 65 mm face	–	–	–	–	m	**11.67**
extra over for:						
external angle	–	–	–	–	nr	**11.74**

17 SHEET ROOF COVERINGS

Item	PC £	Labour hours	Labour £	Material £	Unit	Total rate £
BUILT UP FELT ROOF COVERINGS – cont						
Accessories – cont						
Roof screed ventilator – aluminium alloy						
Extr-aqua-vent or other equal and approved – set on screed over and including dished sinking and collar	–	–	–	–	nr	36.91
Insulation board underlays						
Vapour barrier						
reinforced; metal lined	–	–	–	–	m²	11.16
Rockwool; Duorock flat insulation board						
140 mm thick (0.25 U-value)	–	–	–	–	m²	30.57
Kingspan Thermaroof TR21 zero OPD urethene insulation board						
50 mm thick	–	–	–	–	m²	18.79
90 mm thick	–	–	–	–	m²	32.37
100 mm thick (0.25 U-value)	–	–	–	–	m²	35.95
Wood fibre boards; impregnated; density 220–350 kg/m³						
12.70 mm thick	–	–	–	–	m²	5.07
Tapered insulation board underlays						
Tapered insulation £/m² prices can vary dramatically depending upon the factors which determine the scheme layout; these primarily being gutter/outlet locations and the length of fall involved. The following guide assumes a U-value of 0.18W/m² K as a benchmark.						
As the required insulation value will vary from project to project the required U-value should be deternined by calculating the buildings energy consumption at the design stage. Due to tapered insulation scheme prices varying by project, the following prices are indicative. Please contact a specialist for a project specific quotation. U-Value must be calculated in accordance with BSENISO 6946:2007 Annex C						
Tapered PIR (Polyisocyanurate) boards; bedded in hot bitumen						
effective thickness achieving 0.18 W/m² K	25.38	–	–	–	m²	47.99
minimum thickness achieving 0.18 W/m² K	30.19	–	–	–	m²	53.28
Tapered PIR (Polyisocyanurate) board; mechanically fastened						
effective thickness achieving 0.18 W/m² K	25.38	–	–	–	m²	50.09
Tapered Aspire; Hybrid EPS/PIR boards; bedded in hot bitumen						
effective thickness achieving 0.18 W/m² K	23.63	–	–	–	m²	47.99
minimum thickness achieving 0.18 W/m² K	28.44	–	–	–	m²	53.28

17 SHEET ROOF COVERINGS

Item	PC £	Labour hours	Labour £	Material £	Unit	Total rate £
Tapered PIR (Polyisocyanurate) board; mechanically fastened						
effective thickness achieving 0.18 W/m^2 K	23.63	–	–	–	m^2	**50.09**
minimum thickness achieving 0.18 W/m^2 K	28.44	–	–	–	m^2	**55.39**
Tapered Rockwool boards; bedded in hot bitumen						
effective thickness achieving 0.18 W/m^2 K	45.33	–	–	–	m^2	**70.61**
minimum thickness achieving 0.18 W/m^2 K	56.00	–	–	–	m^2	**76.50**
Tapered Rockwool boards; mechanically fastened						
effective thickness achieving 0.18 W/m^2 K	45.33	–	–	–	m^2	**72.71**
minimum thickness achieving 0.18 W/m^2 K	56.00	–	–	–	m^2	**78.89**
Tapered EPS (Expanded polystyrene) boards; bedded in hot bitumen						
effective thickness achieving 0.18 W/m^2 K	20.74	–	–	–	m^2	**42.75**
minimum thickness achieving 0.18 W/m^2 K	25.38	–	–	–	m^2	**47.99**
Tapered EPS (Expanded polystyrene) boards; mechanicaly fastened						
effective thickness achieving 0.18 W/m^2 K	20.74	–	–	–	m^2	**44.80**
minimum thickness achieving 0.18 W/m^2 K	25.38	–	–	–	m^2	**50.09**
Insulation board overlays						
Dow Roofmate SL extruded polystyrene foam boards or other equal and approved; Thermal conductivity – 0.028 W/mK						
50 mm thick	–	–	–	–	m^2	**12.44**
140 mm thick	–	–	–	–	m^2	**21.71**
160 mm thick	–	–	–	–	m^2	**23.51**
Dow Roofmate LG extruded polystyrene foam boards or other equal and approved; Thermal conductivity – 0.028 W/mK						
80 mm thick	–	–	–	–	m^2	**44.29**
100 mm thick	–	–	–	–	m^2	**47.47**
120 mm thick	–	–	–	–	m^2	**50.69**
SINGLE LAYER PLASTIC ROOF COVERINGS						
Trocal S PVC roofing or other equal and approved						
Coverings	–	–	–	–	m^2	**15.30**
Skirtings; dressed over metal upstands						
not exceeding 200 mm girth	–	–	–	–	m	**11.88**
200 mm–400 mm girth	–	–	–	–	m	**14.61**
Coverings to kerbs						
400 mm–600 mm girth	–	–	–	–	m	**26.75**
Collars around pipes and the like; 150 mm high						
not exceeding 55 mm nominal size	–	–	–	–	nr	**8.18**
55 mm–110 mm nominal size	–	–	–	–	nr	**8.18**
Trocal metal upstands or other equal and approved						

17 SHEET ROOF COVERINGS

Item	PC £	Labour hours	Labour £	Material £	Unit	Total rate £
SINGLE LAYER PLASTIC ROOF COVERINGS – **cont**						
Sarnafil polymeric waterproofing membrane; **cold roof**						
Roof coverings						
pitch not exceeding 5°; to metal decking or the like	–	–	–	–	m²	**27.19**
pitch not exceeding 5°; to concrete base or the like; prime concrete with spririt priming solution	–	–	–	–	m²	**27.19**
Sarnafil polymeric waterproofing membrane; **1.2 mm thick fleece backed membrane; cold roof**						
Roof coverings						
pitch not exceeding 5°; to metal decking or the like	–	–	–	–	m²	**27.19**
pitch not exceeding 5°; to concrete base or the like; prime concrete with spririt priming solution	–	–	–	–	m²	**27.19**
Sarnafil polymeric waterproofing membrane; **120 mm thick Sarnaform G CFC and HCFC free** **insulation board; vapour control layer; prime** **concrete with spirit priming solution**						
Mechanically fastened system						
Roof coverings						
pitch not exceeding 5°; to metal decking or the like	–	–	–	–	m²	**42.50**
pitch not exceeding 5°; to concrete base or the like	–	–	–	–	m²	**49.64**
Coverings to kerbs; parapet flashing; Sarnatrim 50 mm deep on face 100 mm fixing arm; standard Sarnafil detail 1.1						
not exceeding 200 mm girth	–	–	–	–	m	**23.84**
200 mm–400 mm girth	–	–	–	–	m	**27.15**
400 mm–600 mm girth	–	–	–	–	m	**30.46**
Eaves detail; Sarnatrim drip edge to gutter; standard Sarnafil detail 1.3						
not exceeding 200 mm girth	–	–	–	–	m	**22.92**
Skirtings/Upstands; skirting to brickwork with galvanized steel counter flashing to top edge; standard Sarnafil detail 2.3						
not exceeding 200 mm girth	–	–	–	–	m	**27.87**
200 mm–400 mm girth	–	–	–	–	m	**33.53**
400 mm–600 mm girth	–	–	–	–	m	**39.23**
Skirtings/Upstands; skirting to brickwork with Sarnametal Raglet to chase; standard Sarnafil detail 2.8						
not exceeding 200 mm girth	–	–	–	–	m	**27.87**
200 mm–400 mm girth	–	–	–	–	m	**33.53**
400 mm–600 mm girth	–	–	–	–	m	**39.23**

17 SHEET ROOF COVERINGS

Item	PC £	Labour hours	Labour £	Material £	Unit	Total rate £
Collars around pipe standards, and the like						
50 mm dia. × 150 mm high	–	–	–	–	nr	36.12
100 mm dia. × 150 mm high	–	–	–	–	nr	36.12
Outlets and dishing to gullies						
fix Sarnadrain PVC rainwater outlet; 110 mm dia.; weld membrane to same; fit plastic leafguard	–	–	–	–	nr	85.38
Fully adhered system						
Roof coverings						
pitch not exceeding 5°; to metal decking or the like	–	–	–	–	m²	46.07
pitch not exceeding 5°; to concrete base or the like	–	–	–	–	m²	51.39
Coverings to kerbs; parapet flashing; Sarnatrim 50 mm deep on face 100 mm fixing arm; standard Sarnafil detail 1.1						
not exceeding 200 mm girth	–	–	–	–	m	22.03
200 mm–400 mm girth	–	–	–	–	m	25.36
400 mm–600 mm girth	–	–	–	–	m	28.68
Eaves detail; Sarnametal drip edge to gutter; standard Sarnafil detail 1.3						
not exceeding 200 mm girth	–	–	–	–	m	21.11
Skirtings/Upstands; skirting to brickwork with galvanized steel counter flashing to top edge; standard Sarnafil detail 2.3						
not exceeding 200 mm girth	–	–	–	–	m	26.08
200 mm–400 mm girth	–	–	–	–	m	31.77
400 mm–600 mm girth	–	–	–	–	m	35.57
Skirtings/Upstands; skirting to brickwork with Sarnametal Raglet to chase; standard Sarnafil detail 2.8						
not exceeding 200 mm girth	–	–	–	–	m	27.04
200 mm–400 mm girth	–	–	–	–	m	37.50
400 mm–600 mm girth	–	–	–	–	m	37.50
Collars around pipe standards, and the like						
50 mm dia. × 150 mm high	–	–	–	–	nr	36.29
100 mm dia. × 150 mm high	–	–	–	–	nr	36.29
Outlets and dishing to gullies						
Fix Sarnadrain PVC rainwater outlet; 110 mm dia.; weld membrane to same; fit plastic leafguard	–	–	–	–	nr	85.79
Options						
Extra over for 1.2 mm fleece backed membrane	–	–	–	–	m²	4.67
Landscape roofing						
SarnaVert extensive biodiverse roof; sedum blanket; 100 mm growing medium; aquafrain; 1.5 mm thick membrane; 120 mm thick insulation board; vapour control layer						
Pitch not exceeding 5°; to metal decking or the like	–	–	–	–	m²	112.34
Pitch not exceeding 5°; to concrete base or the like; prime concrete with spirit priming solution	–	–	–	–	m²	116.90

17 SHEET ROOF COVERINGS

Item	PC £	Labour hours	Labour £	Material £	Unit	Total rate £
SINGLE LAYER PLASTIC ROOF COVERINGS – **cont**						
SarnaVert extensive biodiverse roof – cont						
Kerb and eaves; standard Sarnafil details						
Sarnafil kerb; 150 mm above roof level	–	–	–	–	m	**27.82**
Sarnafil eaves detail with gravel stop ne 200 mm girth	–	–	–	–	m	**46.10**
Collars around pipes and the like						
50 mm dia. × 150 mm high	–	–	–	–	nr	**35.76**
100 mm dia. × 150 mm high	–	–	–	–	nr	**35.76**
Sarnafil rainwater outlet	–	–	–	–	nr	**143.87**
SHEET METALS						
Zalutite coated steel flat composite panel cladding; Kingspan or other equal and approved; outer panel 0.7 mm gauge HPS200 colour coated; HCFC free LPCB FM/FW core and 0.4 mm stucco embossed lining panel with bright white polyester paint finish						
Roof cladding; vertical fixing to steel rails (measured elsewhere)						
80 mm wall panel; ref. KS1000RW	–	–	–	–	m^2	**34.65**
Wall cladding; vertical fixing to steel rails (measured elsewhere)						
60 mm wall panel; ref. KS1000RW	–	–	–	–	m^2	**31.50**
70 mm wall panel; ref. KS1000RW	–	–	–	–	m^2	**32.55**
80 mm wall panel; ref. KS1000RW	–	–	–	–	m^2	**34.65**
70 mm wall panel; ref. KS1000MR	–	–	–	–	m^2	**44.10**
80 mm wall panel; ref. KS1000MR	–	–	–	–	m^2	**47.25**
70 mm wall panel; ref. KS900MR	–	–	–	–	m^2	**44.10**
80 mm wall panel; ref. KS900MR	–	–	–	–	m^2	**47.25**
70 mm wall panel ; ref. KS600MR	–	–	–	–	m^2	**68.25**
80 mm wall panel ; ref. KS600MR	–	–	–	–	m^2	**73.50**
Extra over for;						
raking cutting to 60 mm KS1000RW panel including waste	–	–	–	–	m	**21.00**
raking cutting to 70 mm KS1000RW panel including waste	–	–	–	–	m	**21.00**
raking cutting to 80 mm KS1000RW panel including waste	–	–	–	–	m	**21.00**
raking cutting to 70 mm KS1000MR panel including waste	–	–	–	–	m	**21.00**
raking cutting to 80 mm KS1000MR panel including waste	–	–	–	–	m	**21.00**
raking cutting to 70 mm KS900MR panel including waste	–	–	–	–	m	**21.00**
raking cutting to 80 mm KS900MR panel including waste	–	–	–	–	m	**21.00**

17 SHEET ROOF COVERINGS

Item	PC £	Labour hours	Labour £	Material £	Unit	Total rate £
raking cutting to 70 mm KS600MR panel including waste	–	–	–	–	m	**21.00**
raking cutting to 80 mm KS600MR panel including waste	–	–	–	–	m	**21.00**
panel bearers' 1500 mm centres	–	–	–	–	m	**5.25**
vertical tophat joint in HPS200	–	–	–	–	m	**10.50**
vertical tophat joint with cap in HPS200	–	–	–	–	m	**12.60**
cranked KS1000MR panel	–	–	–	–	m	**99.75**
cranked KS900MR panel	–	–	–	–	m	**115.50**
cranked KS600MR panel	–	–	–	–	m	**157.50**
roof penetration; 150 mm dia. opening; with top hat flashing and collar 150 mm high; and silicone joint to roofsheet	–	–	–	–	nr	**49.50**
roof penetration; 250 mm dia. opening; with top hat flashing and collar 150 mm high; and silicone joint to roofsheet	–	–	–	–	nr	**70.44**
GLASS REINFORCED PLASTIC PANEL						
Glass fibre translucent sheeting grade AB class 3 Roof cladding; sloping not exceeding 50°; fixing to timber purlins with drive screws; to suit						
Profile 3 or other equal and approved	10.58	0.18	3.64	14.90	m²	**18.54**
Profile 6 or other equal and approved	10.84	0.23	4.64	15.18	m²	**19.82**
Roof cladding; sloping not exceeding 50°; fixing to timber purlins with hook bolts; to suit						
Profile 3 or other equal and approved	10.58	0.23	4.64	15.56	m²	**20.20**
Profile 6 or other equal and approved	10.84	0.28	5.66	15.84	m²	**21.50**
Longrib 1000 or other equal and approved	12.17	0.28	5.66	17.27	m²	**22.93**
LEAD SHEET COVERINGS/FLASHINGS						
Milled Lead; BS EN 12588; on and including Geotec underlay						
Roof and dormer coverings						
1.80 mm thick (code 4) roof coverings						
flat (in wood roll construction (Prime Cost £ per kg)	3.06	0.90	23.39	40.63	m²	**64.02**
pitched (in wood roll construction)	–	1.00	25.99	40.83	m²	**66.82**
pitched (in welded seam construction)	–	0.90	23.39	40.63	m²	**64.02**
vertical (in welded seam construction)	–	1.00	25.99	38.78	m²	**64.77**
1.80 mm thick (code 4) dormer coverings						
flat (in wood roll construction) (Prime Cost £ per kg)	3.06	0.68	17.55	40.17	m²	**57.72**
pitched (in wood roll construction)	–	0.75	19.50	40.32	m²	**59.82**
pitched (in welded seam construction)	–	0.68	17.55	40.17	m²	**57.72**
vertical (in welded seam construction)	–	1.50	38.99	38.78	m²	**77.77**
2.24 mm thick (code 5) roof coverings						
flat (in wood roll construction) (Prime Cost £ per kg)	3.06	0.94	24.57	48.98	m²	**73.55**
pitched (in wood roll construction)	–	1.05	27.30	49.19	m²	**76.49**
pitched (in welded seam construction)	–	0.94	24.57	48.98	m²	**73.55**
vertical (in welded seam construction)	–	1.05	27.30	47.04	m²	**74.34**

17 SHEET ROOF COVERINGS

Item	PC £	Labour hours	Labour £	Material £	Unit	Total rate £
LEAD SHEET COVERINGS/FLASHINGS – cont						
Roof and dormer coverings – cont						
2.24 mm thick (code 5) dormer coverings						
flat (in wood roll construction) (Prime Cost £ per kg)	3.06	0.71	18.43	48.49	m²	66.92
pitched (in wood roll construction)	–	0.79	20.48	48.65	m²	69.13
pitched (in welded seam construction)	–	0.71	18.43	48.49	m²	66.92
vertical (in welded seam construction)	–	1.57	40.94	47.04	m²	87.98
2.65 mm thick (code 6) roof coverings						
flat (in wood roll construction) (Prime Cost £ per kg)	3.06	0.99	25.74	56.76	m²	82.50
pitched (in wood roll construction)	–	1.10	28.60	56.99	m²	85.59
pitched (in welded seam construction)	–	0.99	25.74	56.76	m²	82.50
vertical (in welded seam construction)	–	1.10	28.60	54.73	m²	83.33
2.65 mm thick (code 6) dormer coverings						
flat (in wood roll construction) (Prime Cost £ per kg)	3.06	0.74	19.31	56.25	m²	75.56
pitched (in wood roll construction)	–	0.82	21.44	56.42	m²	77.86
pitched (in welded seam construction)	–	0.74	19.31	56.25	m²	75.56
vertical (in welded seam construction)	–	1.65	42.89	54.73	m²	97.62
3.15 mm thick (code 7) roof coverings (35.72 kg per m²)						
flat (in wood roll construction) (Prime Cost £ per kg)	3.06	1.06	27.50	66.28	m²	93.78
pitched (in wood roll construction)	–	1.18	30.55	66.52	m²	97.07
pitched (in welded seam construction)	–	1.06	27.50	66.28	m²	93.78
vertical (in welded seam construction)	–	1.18	30.55	64.11	m²	94.66
3.15 mm thick (code 7) dormer coverings						
flat (in wood roll construction) (Prime Cost £ per kg)	3.06	0.79	20.61	65.74	m²	86.35
pitched (in wood roll construction)	–	0.88	22.90	65.92	m²	88.82
pitched (in welded seam construction)	–	0.79	20.61	65.74	m²	86.35
vertical (in welded seam construction)	–	1.76	45.83	64.11	m²	109.94
3.55 mm thick (code 8) roof coverings (40.26 kg per m²)						
flat (in wood roll construction) (Prime Cost £ per kg)	3.06	1.15	29.84	73.98	m²	103.82
pitched (in wood roll construction)	–	1.27	33.14	74.24	m²	107.38
pitched (in welded seam construction)	–	1.15	29.84	73.98	m²	103.82
vertical (in welded seam construction)	–	1.27	33.14	71.62	m²	104.76
3.55 mm thick (code 8) dormer coverings						
flat (in wood roll construction) (Prime Cost £ per kg)	3.06	0.86	22.39	73.39	m²	95.78
pitched (in wood roll construction)	–	0.96	24.85	73.58	m²	98.43
pitched (in welded seam construction)	–	0.86	22.39	73.39	m²	95.78
vertical (in welded seam construction)	–	1.91	49.72	71.62	m²	121.34
Sundries						
patination oil to finished work surfaces	–	0.03	0.65	0.20	m²	0.85
chalk slurry to underside of panels	–	0.33	8.65	1.80	m²	10.45
provision of 45 × 45 mm wood rolls at 600 mm centres (per m)	–	0.10	2.60	0.85	m	3.45
dressing over glazing bars and glass	–	0.25	6.50	0.55	m	7.05
soldered nail head	–	0.01	0.21	0.05	nr	0.26
1.32 mm thick (code 3) lead flashings, etc.						
Soakers						
200 × 200 mm	–	0.02	0.39	0.91	nr	1.30
300 × 300 mm	–	0.02	0.39	2.07	nr	2.46

17 SHEET ROOF COVERINGS

Item	PC £	Labour hours	Labour £	Material £	Unit	Total rate £
1.80 mm thick (code 4) lead flashings, etc.						
Flashings; wedging into grooves						
150 mm girth	–	0.25	6.50	5.26	m	**11.76**
200 mm girth	–	0.25	6.50	7.01	m	**13.51**
240 mm girth	–	0.25	6.50	8.41	m	**14.91**
300 mm girth	–	0.25	6.50	10.51	m	**17.01**
Stepped flashings; wedging into grooves						
180 mm girth	–	0.50	13.00	6.31	m	**19.31**
270 mm girth	–	0.50	13.00	9.46	m	**22.46**
Linings to sloping gutters						
390 mm girth	–	0.40	10.39	13.66	m	**24.05**
450 mm girth	–	0.45	11.70	15.76	m	**27.46**
600 mm girth	–	0.55	14.30	21.02	m	**35.32**
Cappings to hips or ridges						
450 mm girth	–	0.50	13.00	15.76	m	**28.76**
600 mm girth	–	0.60	15.60	21.02	m	**36.62**
Saddle flashings; at intersections of hips and ridges; dressing and bossing						
450 × 450 mm	–	0.50	13.00	9.15	nr	**22.15**
600 × 200 mm	–	0.50	13.00	14.66	nr	**27.66**
Slates; with 150 mm high collar						
450 × 450 mm; to suit 50 mm dia. pipe	–	0.75	19.50	11.00	nr	**30.50**
450 × 450 mm; to suit 100 mm dia. pipe	–	0.75	19.50	11.82	nr	**31.32**
450 × 450 mm; to suit 150 mm dia. pipe	–	0.75	19.50	12.65	nr	**32.15**
2.24 mm thick (code 5) lead flashings, etc.						
Flashings; wedging into grooves						
150 mm girth	–	0.25	6.50	6.46	m	**12.96**
200 mm girth	–	0.25	6.50	8.61	m	**15.11**
240 mm girth	–	0.25	6.50	10.33	m	**16.83**
300 mm girth	–	0.25	6.50	12.91	m	**19.41**
Stepped flashings; wedging into grooves						
180 mm girth	–	0.50	13.00	7.75	m	**20.75**
270 mm girth	–	0.50	13.00	11.62	m	**24.62**
Linings to sloping gutters						
390 mm girth	–	0.40	10.39	16.79	m	**27.18**
450 mm girth	–	0.45	11.70	19.37	m	**31.07**
600 mm girth	–	0.55	14.30	25.82	m	**40.12**
Cappings to hips or ridges						
450 mm girth	–	0.50	13.00	19.37	m	**32.37**
600 mm girth	–	0.60	15.60	25.82	m	**41.42**
Saddle flashings; at intersections of hips and ridges; dressing and bossing						
450 × 450 mm	–	0.50	13.00	9.74	nr	**22.74**
600 × 200 mm	–	0.50	13.00	16.52	nr	**29.52**
Slates; with 150 mm high collar						
450 × 450 mm; to suit 50 mm dia. pipe	–	0.75	19.50	11.27	nr	**30.77**
450 × 450 mm; to suit 100 mm dia. pipe	–	0.75	19.50	12.28	nr	**31.78**
450 × 450 mm; to suit 150 mm dia. pipe	–	0.75	19.50	13.30	nr	**32.80**

17 SHEET ROOF COVERINGS

Item	PC £	Labour hours	Labour £	Material £	Unit	Total rate £
ALUMINIUM SHEET COVERINGS/FLASHINGS						
Aluminium roofing; commercial grade; on and including Geotec underlay						
The following rates are based upon nett 'deck' or 'wall' areas						
Roof, dormer and wall coverings						
0.7 mm thick roof coverings; mill finish						
flat (in wood roll construction) (Prime Cost rate £ per kg)	4.61	1.00	25.99	18.09	m²	**44.08**
eaves detail ED1	–	0.20	5.20	2.24	m	**7.44**
abutment upstands at perimeters	–	0.33	8.58	0.90	m	**9.48**
pitched over 3° (in standing seam construction)	–	0.75	19.50	15.13	m²	**34.63**
vertical (in angled or flat seam construction)	–	0.80	20.80	15.13	m²	**35.93**
0.7 mm thick dormer coverings; mill finish						
flat (in wood roll construction)	–	1.50	38.99	17.83	m²	**56.82**
eaves detail ED1	–	0.20	5.20	2.24	m	**7.44**
pitched over 3° (in standing seam construction)	–	1.25	32.49	15.13	m²	**47.62**
vertical (in angled or flat seam construction)	–	1.35	35.10	15.13	m²	**50.23**
0.7 mm thick roof coverings; Pvf2 finish						
flat (in wood roll construction) (Prime Cost rate £ per kg)	5.73	1.00	25.99	21.22	m²	**47.21**
eaves detail ED1	–	0.20	5.20	2.79	m	**7.99**
abutment upstands at perimeters	–	0.33	8.58	1.12	m	**9.70**
pitched over 3° (in standing seam construction)	–	0.75	19.50	18.42	m²	**37.92**
vertical (in angled or flat seam construction)	–	0.80	20.80	18.42	m²	**39.22**
0.7 mm thick dormer coverings; Pvf2 finish						
flat (in wood roll construction)	–	1.50	38.99	21.22	m²	**60.21**
eaves detail ED1	–	0.20	5.20	2.79	m	**7.99**
pitched over 3° (in standing seam construction)	–	1.25	32.49	18.42	m²	**50.91**
vertical (in angled or flat seam construction)	–	1.35	35.10	18.42	m²	**53.52**
0.7 mm thick aluminium flashings, etc.						
Flashings; wedging into grooves; mill finish						
150 mm girth (PC per kg)	4.61	0.25	6.50	1.35	m	**7.85**
240 mm girth	–	0.25	6.50	2.15	m	**8.65**
300 mm girth	–	0.25	6.50	2.69	m	**9.19**
Stepped flashings; wedging into grooves; mill finish						
180 mm girth	–	0.50	13.00	1.62	m	**14.62**
270 mm girth	–	0.50	13.00	2.42	m	**15.42**
Flashings; wedging into grooves; Pvf2 finish						
150 mm girth (PC per kg)	5.73	0.25	6.50	1.68	m	**8.18**
240 mm girth	–	0.25	6.50	2.68	m	**9.18**
300 mm girth	–	0.25	6.50	3.35	m	**9.85**
Stepped flashings; wedging into grooves; Pvf2 finish						
180 mm girth	–	0.50	13.00	2.01	m	**15.01**
270 mm girth	–	0.50	13.00	3.02	m	**16.02**
Sundries						
provision of square batten roll at 500 mm centres (per m)	–	0.10	2.60	1.08	m	**3.68**

17 SHEET ROOF COVERINGS

Item	PC £	Labour hours	Labour £	Material £	Unit	Total rate £
Standing seam aluminium roofing Kalzip; 65 mm seam, 400 cover width, Ref BS AW 3004 standard natural aluminium, stucco embossed finish, 0.9 mm thick; ST clips fixed with stainless steel fasteners; 37 Plus 180 mm Glassfibre Insulation compressed to 165 mm (0.25 U-value); vapour control layer, clear reinforced polyethelyne 530 MNs/g all laps sealed; Liner Sheets, profiled steel, 1000 mm cover width, bright white polyester paint finish Ref TR35/200S, 0.7 mm thick, fixed with stainless steel fasteners.						
roof coverings (twin skin construction); pitch not less than 1.5°; fixed to cold rolled purlins (not included	–	–	–	–	m²	**56.38**
Eaves details						
40 × 20 mm extruded aluminium drip angle fixed to Kalzip sheet using aluminium blind sealed rivets; black solid rubber eaves filler blocks; ST clips fixed with stainless steel fasteners	–	–	–	–	m	**18.27**
0.90 mm thick stucco embossed natural aluminium external eaves closure; 375 mm girth twice bent	–	–	–	–	m	**7.67**
0.70 mm thick bright white polyester liner sheet closure internal flashing; 200 mm girth once bent with stainless steel fasteners, black solid rubber profiled liner small flute filler sealed top and bottom with sealant tape	–	–	–	–	m	**8.83**
Verge details						
extruded aluminium gable end channel fixed to Kalzip seam using aluminium blind rivets; Extruded aluminium gable end clips fixed to ST clips with stainless steel fasteners; extruded aluminium gable tolerence clip hooked over gable end channel	–	–	–	–	m	**14.90**
0.90 mm thick stucco embossed natural aluminium external verge closure 600 mm girth four times bent, fixed to extruded aluminium gable tolerence clip and vertical cladding with stainless steel fasteners, black profiled filler blocks to vertical clddding	–	–	–	–	m	**14.96**
0.70 mm thick bright white polyester liner sheet closure internal flashing 200 mm girth once bent fixed with stainless steel fasteners, black solid rubber profiled filler blocks sealed top and bottom with sealant tape	–	–	–	–	m	**8.99**

17 SHEET ROOF COVERINGS

Item	PC £	Labour hours	Labour £	Material £	Unit	Total rate £
ALUMINIUM SHEET COVERINGS/FLASHINGS – cont						
Standing seam aluminium roofing – cont						
Duo-Ridge details						
2 nr Extruded aluminium zed sections fixed to Kalzip seams using aluminium blind sealed rivets; 2 nr natural aluminium stucco embossed U Type ridge closures fixed to Kalzip seams using aluminium blind sealed rivets; 2 nr black solid rubber ridge filler blocks, 2 nr ST clips fixed with stainless steel fasteners; fix seam of Kalzip sheet to ST clips using aluminium blind sealed rivets (for fixed point); turn up Kalzip 400 sheets both sides	–	–	–	–	m	28.86
0.90 mm thick stucco embossed natural aluminium external ridge closure; 600 mm girth three times bent, fixed to U Type Ridge closure with stainless steel fasteners	–	–	–	–	m	11.47
0.70 mm thick bright white polyester liner sheet closure flashing 600 mm girth once bent fixed with stainless steel fasteners, black solid rubber profiled filler blocks sealed top and bottom with sealant tape	–	–	–	–	m	9.99
Accessories						
extra over for;						
smooth curving Kalzip sheets	–	–	–	–	m²	6.63
crimp curving liner (below 52.5 m convex radius)	–	–	–	–	sheet	11.19
polyster coating Kalzip sheets	–	–	–	–	m²	3.45
PvDF coating Kalzip sheets	–	–	–	–	m²	3.98
vapour control layer, foil encapsulated polythene 4300 MNs/g	–	–	–	–	m²	0.55
200 mm thick thermal insulation quilt	–	–	–	–	m²	0.17
30 mm thick semi-rigid acoustic insulation slab	–	–	–	–	m²	4.13
1.0 mm Flashings etc.; fixing/wedging into grooves						
flashing; 500 mm girth	–	–	–	–	m	12.86
flashing; 750 mm girth	–	–	–	–	m	16.99
flashing; 1000 mm girth	–	–	–	–	m	23.90
1.2 mm Flashings etc.; fixing/wedging into grooves						
flashing; 500 mm girth	–	–	–	–	m	14.15
flashing; 750 mm girth	–	–	–	–	m	18.69
flashing; 1000 mm girth	–	–	–	–	m	23.62
1.4 mm Flashings etc.; fixing/wedging into grooves						
flashing; 500 mm girth	–	–	–	–	m	16.27
flashing; 750 mm girth	–	–	–	–	m	21.49
flashing; 1000 mm girth	–	–	–	–	m	32.29

17 SHEET ROOF COVERINGS

Item	PC £	Labour hours	Labour £	Material £	Unit	Total rate £
Aluminium Alumasc Skyline coping system; polyester powder coated						
Coping; fixing straps plugged and screwed to brickwork						
362 mm wide; for parapet wall 241–300 mm wide	–	0.50	7.64	24.25	m	**31.89**
extra over for;						
90° angle	–	0.25	3.82	59.20	nr	**63.02**
90° tee junction	–	0.35	5.34	65.18	nr	**70.52**
stop end	–	0.15	2.29	30.19	nr	**32.48**
stop end upstand	–	0.20	3.05	33.13	nr	**36.18**
COPPER STRIP SHEET COVERINGS/FLASHINGS						
Copper roofing; BS EN 504; on and including Geotec underlay						
The following rates are based upon nett deck or wall areas						
Roof and dormer coverings						
0.6 mm thick roof coverings; mill finish						
flat (in wood roll construction) (Prime Cost £ per kg)	5.36	1.10	28.60	48.71	m²	**77.31**
eaves detail ED1	–	0.20	5.20	5.91	m	**11.11**
abutment upstands at perimeters	–	0.33	8.58	2.96	m	**11.54**
pitched over 3° (in standing seam construction)	–	0.85	22.10	39.84	m²	**61.94**
vertical (in angled or flat seam construction)	–	0.90	23.39	39.84	m²	**63.23**
0.6 mm thick dormer coverings; mill finish						
flat (in wood roll construction)	5.36	1.60	41.59	48.71	m²	**90.30**
eaves detail ED1	–	0.20	5.20	5.91	m	**11.11**
pitched over 3° (in standing seam construction)	–	1.25	32.49	39.84	m²	**72.33**
vertical (in angled or flat seam construction)	–	1.35	35.10	39.84	m²	**74.94**
0.6 mm thick roof coverings; oxid finish						
flat (in wood roll construction) (Prime Cost £ per kg)	6.58	1.10	28.60	59.12	m²	**87.72**
eaves detail ED1	–	0.20	5.20	7.26	m	**12.46**
abutment upstands at perimeters	–	0.33	8.58	3.63	m	**12.21**
pitched over 3° (in standing seam construction)	–	0.85	22.10	48.55	m²	**70.65**
vertical (in angled or flat seam construction)	–	0.80	20.80	48.55	m²	**69.35**
0.6 mm thick dormer coverings; oxid finish						
flat (in wood roll construction)	6.58	1.50	38.99	59.12	m²	**98.11**
eaves detail ED1	–	0.20	5.20	7.26	m	**12.46**
pitched over 3° (in standing seam construction)	–	1.25	32.49	48.55	m²	**81.04**
vertical (in angled or flat seam construction)	–	1.35	35.10	48.55	m²	**83.65**
0.6 mm thick roof coverings; KME pre-patinated finish						
flat (in wood roll construction)	55.09	1.10	28.60	87.27	m²	**115.87**
eaves detail ED1	–	0.20	5.20	11.05	m	**16.25**
abutment upstands at perimeters	–	0.33	8.58	5.52	m	**14.10**
pitched over 3° (in standing seam construction)	–	0.85	22.10	70.71	m²	**92.81**
vertical (in angled or flat seam construction)	–	0.90	23.39	70.71	m²	**94.10**

17 SHEET ROOF COVERINGS

Item	PC £	Labour hours	Labour £	Material £	Unit	Total rate £
COPPER STRIP SHEET COVERINGS/ FLASHINGS – cont						
Roof and dormer coverings – cont						
0.6 mm thick dormer coverings; KME pre-patinated finish						
flat (in wood roll construction)	55.09	1.50	38.99	87.27	m²	**126.26**
eaves detail ED1	–	0.20	5.20	11.05	m	**16.25**
pitched over 3° (in standing seam construction)	–	1.25	32.49	70.71	m²	**103.20**
vertical (in angled or flat seam construction)	–	1.35	35.10	70.71	m²	**105.81**
0.7 mm thick roof coverings; mill finish						
flat (in wood roll construction) (Prime Cost £ per kg)	5.36	1.00	25.99	55.16	m²	**81.15**
eaves detail ED1	–	0.20	5.20	6.77	m	**11.97**
abutment upstands at perimeters	–	0.33	8.58	3.39	m	**11.97**
pitched over 3° (in standing seam construction)	–	0.75	19.50	45.00	m²	**64.50**
vertical (in angled or flat seam construction)	–	0.80	20.80	45.00	m²	**65.80**
0.7 mm thick dormer coverings; mill finish						
flat (in wood roll construction)	5.36	1.50	38.99	55.16	m²	**94.15**
eaves detail ED1	–	0.20	5.20	6.77	m	**11.97**
pitched over 3° (in standing seam construction)	–	1.25	32.49	44.97	m²	**77.46**
vertical (in angled or flat seam construction)	–	1.35	35.10	44.95	m²	**80.05**
0.7 mm thick roof coverings; oxid finish						
flat (in wood roll construction) (Prime Cost £ per kg)	6.58	1.00	25.99	74.96	m²	**100.95**
eaves detail ED1	–	0.20	5.20	8.32	m	**13.52**
abutment upstands at perimeters	–	0.33	8.58	4.16	m	**12.74**
pitched over 3° (in standing seam construction)	–	0.75	19.50	61.10	m²	**80.60**
vertical (in angled or flat seam construction)	–	0.80	20.80	54.49	m²	**75.29**
0.7 mm thick dormer coverings; oxid finish						
flat (in wood roll construction)	6.58	1.50	38.99	67.04	m²	**106.03**
eaves detail ED1	–	0.20	5.20	8.32	m	**13.52**
pitched over 3° (in standing seam construction)	–	1.25	32.49	54.49	m²	**86.98**
vertical (in angled or flat seam construction)	–	1.35	35.10	54.49	m²	**89.59**
0.7 mm thick roof coverings; KME pre-patinated finish						
flat (in wood roll construction)	63.38	1.10	28.60	100.15	m²	**128.75**
eaves detail ED1	–	0.20	5.20	12.71	m	**17.91**
abutment upstands at perimeters	–	0.33	8.58	6.35	m	**14.93**
pitched over 3° (in standing seam construction)	–	0.85	22.10	81.09	m²	**103.19**
vertical (in angled or flat seam construction)	–	0.90	23.39	81.09	m²	**104.48**
0.7 mm thick dormer coverings; KME pre-patinated finish						
flat (in wood roll construction)	63.38	1.50	38.99	100.15	m²	**139.14**
eaves detail ED1	–	0.20	5.20	12.71	m	**17.91**
pitched over 3° (in standing seam construction)	–	1.25	32.49	81.09	m²	**113.58**
vertical (in angled or flat seam construction)	–	1.35	35.10	81.09	m²	**116.19**

17 SHEET ROOF COVERINGS

Item	PC £	Labour hours	Labour £	Material £	Unit	Total rate £
0.6 mm thick copper flashings, etc.						
Flashings; wedging into grooves; mill finish						
150 mm girth (Prime Cost £ per kg)	5.63	0.25	6.50	3.55	m	**10.05**
240 mm girth	–	0.25	6.50	7.10	m	**13.60**
300 mm girth	–	0.25	6.50	8.87	m	**15.37**
Stepped flashings; wedging into grooves; mill finish						
180 mm girth	–	0.50	13.00	5.32	m	**18.32**
270 mm girth	–	0.50	13.00	7.98	m	**20.98**
Flashings; wedging into grooves; oxid finish						
150 mm girth (Prime Cost £ per kg)	6.91	0.25	6.50	5.44	m	**11.94**
240 mm girth	–	0.25	6.50	8.71	m	**15.21**
300 mm girth	–	0.25	6.50	10.89	m	**17.39**
Stepped flashings; wedging into grooves; oxide finish						
180 mm girth	–	0.50	13.00	6.53	m	**19.53**
270 mm girth	–	0.50	13.00	9.80	m	**22.80**
Flashings; wedging into grooves; KME pre-patinated finish						
150 mm girth (Prime Cost £ per kg)	57.84	0.25	6.50	8.28	m	**14.78**
240 mm girth	–	0.25	6.50	13.26	m	**19.76**
300 mm girth	–	0.25	6.50	16.57	m	**23.07**
Stepped flashings; wedging into grooves; KME pre-patinated finish						
180 mm girth	–	0.50	13.00	9.94	m	**22.94**
270 mm girth	–	0.50	13.00	14.91	m	**27.91**
0.7 mm thick copper flashings, etc.						
Flashings; wedging into grooves; mill finish						
150 mm girth (Prime Cost £ per kg)	5.36	0.25	6.50	5.08	m	**11.58**
240 mm girth	–	0.25	6.50	8.13	m	**14.63**
300 mm girth	–	0.25	6.50	10.16	m	**16.66**
Stepped flashings; wedging into grooves; mill finish						
180 mm girth	–	0.50	13.00	6.10	m	**19.10**
270 mm girth	–	0.50	13.00	9.15	m	**22.15**
Flashings; wedging into grooves; oxid finish						
150 mm girth (Prime Cost £ per kg)	6.58	0.25	6.50	6.24	m	**12.74**
240 mm girth	–	0.25	6.50	9.98	m	**16.48**
300 mm girth	–	0.25	6.50	12.47	m	**18.97**
Stepped flashings; wedging into grooves; oxid finish						
180 mm girth	–	0.50	13.00	7.48	m	**20.48**
270 mm girth	–	0.50	13.00	11.23	m	**24.23**
Flashings; wedging into grooves; KME pre-patinated finish						
150 mm girth (Prime Cost £ per kg)	63.38	0.25	6.50	9.53	m	**16.03**
240 mm girth	–	0.25	6.50	15.25	m	**21.75**
300 mm girth	–	0.25	6.50	19.06	m	**25.56**
Stepped flashings; wedging into grooves; KME pre-patinated finish						
180 mm girth	–	0.50	13.00	11.44	m	**24.44**
270 mm girth	–	0.50	13.00	17.16	m	**30.16**

17 SHEET ROOF COVERINGS

Item	PC £	Labour hours	Labour £	Material £	Unit	Total rate £
COPPER STRIP SHEET COVERINGS/ FLASHINGS – cont						
0.7 mm thick copper flashings, etc. – cont						
Sundries						
provision of square batten roll at 500 mm centres (per m)	–	0.10	2.60	1.08	m	**3.68**
ZINC STRIP SHEET COVERINGS/FLASHINGS						
Zinc roofing; BS 849; on and including Delta Trella underlay						
The following rates are based upon nett deck or wall areas						
Natural Bright Rheinzink						
Roof, dormer and wall coverings						
0.7 mm thick roof coverings						
flat (in wood roll construction) (Prime Cost £ per kg)	2.83	1.00	25.99	25.35	m²	**51.34**
eaves detail ED1	–	0.20	5.20	4.28	m	**9.48**
abutment upstands at perimeters	–	0.33	8.58	2.14	m	**10.72**
pitched over 3° (in standing seam construction)	–	0.75	19.50	21.06	m²	**40.56**
0.7 mm thick dormer coverings						
flat (in wood roll construction) (Prime Cost £ per kg)	2.83	1.50	38.99	25.35	m²	**64.34**
eaves detail ED1	–	0.20	5.20	4.28	m	**9.48**
pitched over 3° (in standing seam construction)	–	1.25	32.49	21.06	m²	**53.55**
0.8 mm thick wall coverings						
vertical (in angled or flat seam construction)	–	0.80	20.80	23.58	m²	**44.38**
0.8 mm thick dormer coverings						
vertical (in angled or flat seam construction)	–	1.35	35.10	23.66	m²	**58.76**
0.8 mm thick zinc flashings, etc.; Natural Bright Rheinzink						
Flashings; wedging into grooves						
150 mm girth	–	0.25	6.50	2.46	m	**8.96**
240 mm girth	–	0.25	6.50	3.93	m	**10.43**
300 mm girth	–	0.25	6.50	4.91	m	**11.41**
Stepped flashings; wedging into grooves						
180 mm girth	–	0.50	13.00	2.95	m	**15.95**
270 mm girth	–	0.50	13.00	4.42	m	**17.42**
Integral box gutter						
900 mm girth; 2 × bent; 2 × welted	–	1.00	25.99	20.31	m	**46.30**
Valley gutter						
600 mm girth; 2 × bent; 2 × welted	–	0.75	19.50	12.18	m	**31.68**
Hips and ridges						
450 mm girth; 2 × bent; 2 × welted	–	1.00	25.99	7.37	m	**33.36**

17 SHEET ROOF COVERINGS

Item	PC £	Labour hours	Labour £	Material £	Unit	Total rate £
Natural Bright Rheinzink PRO						
Roof, dormer and wall coverings						
0.7 mm thick roof coverings						
flat (in wood roll construction) (Prime Cost £ per kg)	3.85	1.00	25.99	34.54	m²	**60.53**
eaves detail ED1	–	0.20	5.20	5.83	m	**11.03**
abutment upstands at perimeters	–	0.33	8.58	2.91	m	**11.49**
pitched over 3° (in standing seam construction)	–	0.75	19.50	28.69	m²	**48.19**
0.7 mm thick dormer coverings						
flat (in wood roll construction) (Prime Cost £ per kg)	3.85	1.50	38.99	34.54	m²	**73.53**
eaves detail ED1	–	0.20	5.20	5.83	m	**11.03**
pitched over 3° (in standing seam construction)	–	1.25	32.49	28.68	m²	**61.17**
0.8 mm thick wall coverings						
vertical (in angled or flat seam construction)	–	0.80	20.80	32.14	m²	**52.94**
0.8 mm thick dormer coverings						
vertical (in angled or flat seam construction)	–	1.35	35.10	32.14	m²	**67.24**
0.8 mm thick zinc flashings, etc.; Natural Bright Rheinzink PRO						
Flashings; wedging into grooves						
150 mm girth	–	0.25	6.50	3.35	m	**9.85**
240 mm girth	–	0.25	6.50	5.36	m	**11.86**
300 mm girth	–	0.25	6.50	6.69	m	**13.19**
Stepped flashings; wedging into grooves						
180 mm girth	–	0.50	13.00	4.01	m	**17.01**
270 mm girth	–	0.50	13.00	6.03	m	**19.03**
Integral box gutter						
900 mm girth; 2 × bent; 2 × welted	–	1.00	25.99	27.67	m	**53.66**
Valley gutter						
600 mm girth; 2 × bent; 2 × welted	–	0.75	19.50	16.60	m	**36.10**
Hips and ridges						
450 mm girth; 2 × bent; 2 × welted	–	1.00	25.99	10.04	m	**36.03**
Pre-weathered Rheinzink						
Roof, dormer and wall coverings						
0.7 mm thick roof coverings; pre-weathered Rheinzink						
flat (in wood roll construction) (Prime Cost £ per kg)	3.27	1.00	25.99	29.27	m²	**55.26**
eaves detail ED1	–	0.20	5.20	4.94	m	**10.14**
abutment upstands at perimeters	–	0.33	8.58	2.47	m	**11.05**
pitched over 3° (in standing seam construction)	–	0.75	19.50	24.32	m²	**43.82**
0.7 mm thick dormer coverings; pre-weathered Rheinzink						
flat (in wood roll construction) (Prime Cost £ per kg)	3.27	1.50	38.99	29.27	m²	**68.26**
eaves detail ED1	–	0.20	5.20	4.94	m	**10.14**
pitched over 3° (in standing seam construction)	–	1.25	32.49	24.32	m²	**56.81**
0.8 mm thick wall coverings; pre-weathered Rheinzink						
vertical (in angled or flat seam construction)	–	0.80	20.80	27.24	m²	**48.04**
0.8 mm thick dormer coverings; pre-weathered Rheinzink						
vertical (in angled or flat seam construction)	–	1.35	35.10	27.24	m²	**62.34**

17 SHEET ROOF COVERINGS

Item	PC £	Labour hours	Labour £	Material £	Unit	Total rate £
ZINC STRIP SHEET COVERINGS/FLASHINGS – cont						
0.8 mm thick zinc flashings, etc.; pre-weathered Rheinzink						
Flashings; wedging into grooves						
150 mm girth	–	0.25	6.50	2.84	m	**9.34**
240 mm girth	–	0.25	6.50	4.54	m	**11.04**
300 mm girth	–	0.25	6.50	5.67	m	**12.17**
Stepped flashings; wedging into grooves						
180 mm girth	–	0.50	13.00	3.40	m	**16.40**
270 mm girth	–	0.50	13.00	5.11	m	**18.11**
Integral box gutter						
900 mm girth; 2 × bent; 2 × welted	–	1.00	25.99	23.46	m	**49.45**
Valley gutter						
600 mm girth; 2 × bent; 2 × welted	–	0.75	19.50	14.07	m	**33.57**
Hips and ridges						
450 mm girth; 2 × bent; 2 × welted	–	1.00	25.99	8.51	m	**34.50**
Pre-weathered Rheinzink PRO						
Roof, dormer and wall coverings						
0.7 mm thick roof coverings						
flat (in wood roll construction) (Prime Cost £ per kg)	4.29	1.00	25.99	38.47	m^2	**64.46**
eaves detail ED1	–	0.20	5.20	6.49	m	**11.69**
abutment upstands at perimeters	–	0.33	8.58	3.25	m	**11.83**
pitched over 3° (in standing seam construction)	–	0.75	19.50	31.96	m^2	**51.46**
0.7 mm thick dormer coverings						
flat (in wood roll construction) (Prime Cost £ per kg)	4.29	1.50	38.99	38.47	m^2	**77.46**
eaves detail ED1	–	0.20	5.20	6.49	m	**11.69**
pitched over 3° (in standing seam construction)	–	1.25	32.49	31.96	m^2	**64.45**
0.8 mm thick wall coverings						
vertical (in angled or flat seam construction)	–	0.80	20.80	35.79	m^2	**56.59**
0.8 mm thick dormer coverings						
vertical (in angled or flat seam construction)	–	1.35	35.10	35.79	m^2	**70.89**
0.8 mm thick zinc flashings, etc.; Pre-weathered Rheinzink PRO						
Flashings; wedging into grooves						
150 mm girth	–	0.25	6.50	3.73	m	**10.23**
240 mm girth	–	0.25	6.50	5.97	m	**12.47**
300 mm girth	–	0.25	6.50	7.45	m	**13.95**
Stepped flashings; wedging into grooves						
180 mm girth	–	0.50	13.00	4.47	m	**17.47**
270 mm girth	–	0.50	13.00	6.71	m	**19.71**
Integral box gutter						
900 mm girth; 2 × bent; 2 × welted	–	–	–	–	m	**30.82**
Valley gutter						
600 mm girth; 2 × bent; 2 × welted	–	0.75	19.50	18.49	m	**37.99**
Hips and ridges						
450 mm girth; 2 × bent; 2 × welted	–	1.00	25.99	11.18	m	**37.17**

17 SHEET ROOF COVERINGS

Item	PC £	Labour hours	Labour £	Material £	Unit	Total rate £
VM Natural Bright						
Roof, dormer and wall coverings						
0.7 mm thick roof coverings						
flat (in wood roll construction) (Prime Cost £ per kg)	2.83	1.00	25.99	25.35	m^2	**51.34**
eaves detail ED1	–	0.20	5.20	4.28	m	**9.48**
abutment upstands at perimeters	–	0.33	8.58	2.14	m	**10.72**
pitched over 3° (in standing seam construction)	–	0.75	19.50	21.06	m^2	**40.56**
0.7 mm thick dormer coverings						
flat (in wood roll construction) (Prime Cost £ per kg)	2.83	1.50	38.99	25.35	m^2	**64.34**
eaves detail ED1	–	0.20	5.20	4.28	m	**9.48**
pitched over 3° (in standing seam construction)	–	1.25	32.49	21.06	m^2	**53.55**
0.8 mm thick wall coverings						
vertical (in angled or flat seam construction)	–	0.80	20.80	23.58	m^2	**44.38**
0.8 mm thick dormer coverings						
vertical (in angled or flat seam construction)	–	1.35	35.10	23.58	m^2	**58.68**
0.8 mm thick zinc flashings, etc.; VM Natural Bright						
Flashings; wedging into grooves						
150 mm girth	–	0.25	6.50	2.46	m	**8.96**
240 mm girth	–	0.25	6.50	3.93	m	**10.43**
300 mm girth	–	0.25	6.50	4.91	m	**11.41**
Stepped flashings; wedging into grooves						
180 mm girth	–	0.50	13.00	2.95	m	**15.95**
270 mm girth	–	0.50	13.00	4.42	m	**17.42**
Integral box gutter						
900 mm girth; 2 × bent; 2 × welted	–	1.00	25.99	20.31	m	**46.30**
Valley gutter						
600 mm girth; 2 × bent; 2 × welted	–	0.75	19.50	12.18	m	**31.68**
Hips and ridges						
450 mm girth; 2 × bent; 2 × welted	–	1.00	25.99	6.69	m	**32.68**
VM Natural Bright PLUS						
Roof, dormer and wall coverings						
0.7 mm thick roof coverings						
flat (in wood roll construction) (Prime Cost £ per kg)	3.85	1.00	25.99	34.54	m^2	**60.53**
eaves detail ED1	–	0.20	5.20	5.83	m	**11.03**
abutment upstands at perimeters	–	0.33	8.58	2.91	m	**11.49**
pitched over 3° (in standing seam construction)	–	0.75	19.50	28.69	m^2	**48.19**
0.7 mm thick dormer coverings						
flat (in wood roll construction) (Prime Cost £ per kg)	3.85	1.50	38.99	34.54	m^2	**73.53**
eaves detail ED1	–	0.20	5.20	5.83	m	**11.03**
pitched over 3° (in standing seam construction)	–	1.25	32.49	28.69	m^2	**61.18**
0.8 mm thick wall coverings						
vertical (in angled or flat seam construction)	–	0.80	20.80	32.14	m^2	**52.94**
0.8 mm thick dormer coverings						
vertical (in angled or flat seam construction)	–	1.35	35.10	32.14	m^2	**67.24**

17 SHEET ROOF COVERINGS

Item	PC £	Labour hours	Labour £	Material £	Unit	Total rate £
ZINC STRIP SHEET COVERINGS/FLASHINGS – **cont**						
0.8 mm thick zinc flashings, etc.; VM Natural **Bright PLUS**						
Flashings; wedging into grooves						
150 mm girth	–	0.25	6.50	3.35	m	**9.85**
240 mm girth	–	0.25	6.50	5.36	m	**11.86**
300 mm girth	–	0.25	6.50	6.69	m	**13.19**
Stepped flashings; wedging into grooves						
180 mm girth	–	0.50	13.00	4.01	m	**17.01**
270 mm girth	–	0.50	13.00	6.03	m	**19.03**
Integral box gutter						
900 mm girth; 2 × bent; 2 × welted	–	1.00	25.99	27.67	m	**53.66**
Valley gutter						
600 mm girth; 2 × bent; 2 × welted	–	0.75	19.50	16.60	m	**36.10**
Hips and ridges						
450 mm girth; 2 × bent; 2 × welted	–	1.00	25.99	10.04	m	**36.03**
VM Quartz (pre-weathered)						
Roof, dormer and wall coverings						
0.7 mm thick roof coverings						
flat (in wood roll construction) (Prime Cost £ per kg)	3.51	1.00	25.99	31.50	m²	**57.49**
eaves detail ED1	–	0.20	5.20	5.32	m	**10.52**
abutment upstands at perimeters	–	0.33	8.58	2.66	m	**11.24**
pitched over 3° (in standing seam construction)	–	0.75	19.50	26.17	m²	**45.67**
0.7 mm thick dormer coverings						
flat (in wood roll construction) (Prime Cost £ per kg)	3.51	1.50	38.99	31.50	m²	**70.49**
eaves detail ED1	–	0.20	5.20	5.32	m	**10.52**
pitched over 3° (in standing seam construction)	–	1.25	32.49	26.17	m²	**58.66**
0.8 mm thick wall coverings						
vertical (in angled or flat seam construction)	–	0.80	20.80	29.31	m²	**50.11**
0.8 mm thick dormer coverings						
vertical (in angled or flat seam construction)	–	1.35	35.10	29.31	m²	**64.41**
0.8 mm thick zinc flashings, etc.; VM Quartz **(pre-weathered)**						
Flashings; wedging into grooves						
150 mm girth	–	0.25	6.50	3.05	m	**9.55**
240 mm girth	–	0.25	6.50	4.89	m	**11.39**
300 mm girth	–	0.25	6.50	6.11	m	**12.61**
Stepped flashings; wedging into grooves						
180 mm girth	–	0.50	13.00	3.66	m	**16.66**
270 mm girth	–	0.50	13.00	5.50	m	**18.50**
Integral box gutter						
900 mm girth; 2 × bent; 2 × welted	–	1.00	25.99	25.24	m	**51.23**
Valley gutter						
600 mm girth; 2 × bent; 2 × welted	–	0.75	19.50	15.14	m	**34.64**
Hips and ridges						
450 mm girth; 2 × bent; 2 × welted	–	1.00	25.99	9.16	m	**35.15**

17 SHEET ROOF COVERINGS

Item	PC £	Labour hours	Labour £	Material £	Unit	Total rate £
VM Quartz (pre-weathered) PLUS						
Roof, dormer and wall coverings						
0.7 mm thick roof coverings						
flat (in wood roll construction) (Prime Cost £ per kg)	4.53	1.00	25.99	40.70	m²	**66.69**
eaves detail ED1	–	0.20	5.20	6.87	m	**12.07**
abutment upstands at perimeters	–	0.33	8.58	3.43	m	**12.01**
pitched over 3° (in standing seam construction)	–	0.75	19.50	33.81	m²	**53.31**
0.7 mm thick dormer coverings						
flat (in wood roll construction) (Prime Cost £ per kg)	4.53	1.50	38.99	40.70	m²	**79.69**
eaves detail ED1	–	0.20	5.20	6.87	m	**12.07**
pitched over 3° (in standing seam construction)	–	1.25	32.49	33.81	m²	**66.30**
0.8 mm thick wall coverings						
vertical (in angled or flat seam construction)	–	0.80	20.80	37.87	m²	**58.67**
0.8 mm thick dormer coverings						
vertical (in angled or flat seam construction)	–	1.35	35.10	37.87	m²	**72.97**
0.8 mm thick zinc flashings, etc.; VM Quartz (pre-weathered) PLUS						
Flashings; wedging into grooves						
150 mm girth	–	0.25	6.50	3.94	m	**10.44**
240 mm girth	–	0.25	6.50	6.31	m	**12.81**
300 mm girth	–	0.25	6.50	7.89	m	**14.39**
Stepped flashings; wedging into grooves						
180 mm girth	–	0.50	13.00	4.73	m	**17.73**
270 mm girth	–	0.50	13.00	7.10	m	**20.10**
Integral box gutter						
900 mm girth; 2 × bent; 2 × welted	–	1.00	25.99	32.61	m	**58.60**
Valley gutter						
600 mm girth; 2 × bent; 2 × welted	–	0.75	19.50	19.56	m	**39.06**
Hips and ridges						
450 mm girth; 2 × bent; 2 × welted	–	1.00	25.99	11.83	m	**37.82**
Sundries and accessories						
Klober breather membrane/underlay	–	0.10	2.60	3.08	m²	**5.68**
Delta Trela Chestwig underlay	–	0.10	2.60	5.28	m²	**7.88**
Delta Trela Football Studs underlay	–	0.10	2.60	0.98	m²	**3.58**
Trapezoidal batten roll at 500 mm centres (per m)	–	0.10	2.60	0.98	m	**3.58**
Zinflash; 0.6 mm thick lead look flashing (no patination oil required)						
Flashings; wedging into grooves						
150 mm girth	–	0.25	6.50	4.36	m	**10.86**
250 mm girth	–	0.25	6.50	7.26	m	**13.76**
300 mm girth	–	0.25	6.50	8.72	m	**15.22**
380 mm girth	–	0.25	6.50	11.04	m	**17.54**
450 mm girth	–	0.25	6.50	13.08	m	**19.58**
Stepped flashings; wedging into grooves						
150 mm girth	–	0.50	13.00	4.36	m	**17.36**
250 mm girth	–	0.50	13.00	7.26	m	**20.26**
300 mm girth	–	0.50	13.00	8.72	m	**21.72**
380 mm girth	–	0.50	13.00	11.04	m	**24.04**
450 mm girth	–	0.50	13.00	13.08	m	**26.08**

17 SHEET ROOF COVERINGS

Item	PC £	Labour hours	Labour £	Material £	Unit	Total rate £
STAINLESS STEEL SHEET COVERINGS/ FLASHINGS						
Terne-coated stainless steel roofing; Associated Lead Mills Ltd; or other equal and approved: on and including Metmatt underlay						
The following rates are based upon nett deck or wall areas						
Roof, dormer and wall coverings in Uginox grade 316; marine						
0.4 mm thick roof coverings						
flat (in wood roll construction) (Prime Cost £ per kg)	6.58	1.00	25.99	33.58	m²	**59.57**
eaves detail ED1	–	0.20	5.20	3.96	m	**9.16**
abutment upstands at perimeters	–	0.33	8.58	1.98	m	**10.56**
pitched over 3° (in standing seam construction)	–	0.75	19.50	27.63	m²	**47.13**
0.5 mm thick dormer coverings						
flat (in wood roll construction) (Prime Cost £ per kg)	6.14	1.50	38.99	38.94	m²	**77.93**
eaves detail ED1	–	0.20	5.20	3.69	m	**8.89**
pitched over 3° (in standing seam construction)	–	1.25	32.49	31.93	m²	**64.42**
0.5 mm thick wall coverings						
vertical (in angled or flat seam construction)	6.14	0.80	20.80	31.93	m²	**52.73**
vertical (with Coulisseau joint construction)	–	1.25	32.49	32.96	m²	**65.45**
0.5 mm thick Uginox grade 316 flashings, etc.						
Flashings; wedging into grooves						
150 mm girth (Prime Cost £ per kg)	6.14	0.25	6.50	3.51	m	**10.01**
240 mm girth	–	0.25	6.50	5.62	m	**12.12**
300 mm girth	–	0.25	6.50	7.02	m	**13.52**
Stepped flashings; wedging into grooves						
180 mm girth	–	0.50	13.00	4.21	m	**17.21**
270 mm girth	–	0.50	13.00	6.32	m	**19.32**
Fan apron						
250 mm girth	–	0.25	6.50	5.85	m	**12.35**
Integral box gutter						
900 mm girth; 2 × bent; 2 × welted	–	1.00	25.99	24.10	m	**50.09**
Valley gutter						
600 mm girth; 2 × bent; 2 × welted	–	0.75	19.50	17.13	m	**36.63**
Hips and ridges						
450 mm girth; 2 × bent; 2 × welted	–	1.00	25.99	10.53	m	**36.52**
Roof, dormer and wall coverings in Ugitop grade 304						
0.4 mm thick roof coverings						
flat (in wood roll construction) (Prime Cost £ per kg)	4.97	1.00	25.99	28.80	m²	**54.79**
eaves detail ED1	–	0.20	5.20	2.99	m	**8.19**
abutment upstands at perimeters	–	0.33	8.58	1.50	m	**10.08**
pitched over 3° (in standing seam construction)	–	0.75	19.50	21.82	m²	**41.32**

17 SHEET ROOF COVERINGS

Item	PC £	Labour hours	Labour £	Material £	Unit	Total rate £
0.5 mm thick dormer coverings						
flat (in wood roll construction) (Prime Cost £ per kg)	4.73	1.50	38.99	30.87	m²	**69.86**
eaves detail ED1	–	0.20	5.20	2.84	m	**8.04**
pitched over 3° (in standing seam construction)	–	1.25	32.49	25.47	m²	**57.96**
0.5 mm thick wall coverings						
vertical (in angled or flat seam construction)	–	0.80	20.80	25.47	m²	**46.27**
vertical (with Coulisseau joint construction)	–	1.25	32.49	26.26	m²	**58.75**
0.5 mm thick Ugitop grade 304 flashings, etc.						
Flashings; wedging into grooves						
150 mm girth (Prime Cost £ per kg)	4.73	0.25	6.50	2.59	m	**9.09**
240 mm girth	–	0.25	6.50	4.55	m	**11.05**
300 mm girth	–	0.25	6.50	5.68	m	**12.18**
Stepped flashings; wedging into grooves						
180 mm girth	–	0.50	13.00	3.41	m	**16.41**
270 mm girth	–	0.50	13.00	5.11	m	**18.11**
Fan apron						
250 mm girth	–	0.25	6.50	4.74	m	**11.24**
Integral box gutter						
900 mm girth; 2 × bent; 2 × welted	–	1.00	25.99	19.51	m	**45.50**
Valley gutter						
600 mm girth; 2 × bent; 2 × welted	–	0.75	19.50	13.87	m	**33.37**
Hips and ridges						
450 mm girth; 2 × bent; 2 × welted	–	1.00	25.99	8.52	m	**34.51**
Roof, dormer and wall coverings in Ugitop grade 316						
0.4 mm thick roof coverings						
flat (in wood roll construction) (Prime Cost £ per kg)	6.14	1.00	25.99	34.63	m²	**60.62**
eaves detail ED1	–	0.20	5.20	3.69	m	**8.89**
abutment upstands at perimeters	–	0.33	8.58	1.85	m	**10.43**
pitched over 3° (in standing seam construction)	–	0.75	19.50	26.02	m²	**45.52**
0.5 mm thick dormer coverings						
flat (in wood roll construction) (Prime Cost £ per kg)	5.85	1.50	38.99	37.29	m²	**76.28**
eaves detail ED1	–	0.20	5.20	3.52	m	**8.72**
pitched over 3° (in standing seam construction)	–	1.25	32.49	30.61	m²	**63.10**
0.5 mm thick wall coverings						
vertical (in angled or flat seam construction)	–	0.80	20.80	30.61	m²	**51.41**
vertical (with Coulisseau joint construction)	–	1.25	32.49	26.26	m²	**58.75**
0.5 mm thick Ugitop grade 316 flashings, etc.						
Flashings; wedging into grooves						
150 mm girth	–	0.25	6.50	3.51	m	**10.01**
240 mm girth	–	0.25	6.50	5.62	m	**12.12**
300 mm girth	–	0.25	6.50	7.02	m	**13.52**
Stepped flashings; wedging into grooves						
180 mm girth	–	0.50	13.00	4.21	m	**17.21**
270 mm girth	–	0.50	13.00	6.32	m	**19.32**

17 SHEET ROOF COVERINGS

Item	PC £	Labour hours	Labour £	Material £	Unit	Total rate £
STAINLESS STEEL SHEET COVERINGS/ FLASHINGS – cont						
0.5 mm thick Ugitop grade 316 flashings, etc. – cont						
Fan apron						
250 mm girth	–	0.25	6.50	5.85	m	**12.35**
Integral box gutter						
900 mm girth; 2 × bent; 2 × welted	–	1.00	25.99	24.10	m	**50.09**
Valley gutter						
600 mm girth; 2 × bent; 2 × welted	–	0.75	19.50	17.13	m	**36.63**
Hips and ridges						
450 mm girth; 2 × bent; 2 × welted	–	1.00	25.99	10.53	m	**36.52**
Sundries						
provision of square batten roll at 500 mm centres						
(per m)	–	0.10	2.60	1.08	m	**3.68**
FIBRE BITUMEN THERMOPLASTIC SHEET COVERINGS/FLASHINGS						
Glass fibre reinforced bitumen strip slates; Ruberglas 105 or other equal and approved; 1000 mm × 336 mm mineral finish; to external quality plywood boarding (boarding not included)						
Roof coverings	9.79	0.23	4.64	11.16	m²	**15.80**
Wall coverings	9.79	0.37	7.47	11.16	m²	**18.63**
Extra over coverings for;						
double course at eaves; felt soaker	–	0.19	3.83	7.47	m	**11.30**
verges; felt soaker	–	0.14	2.83	6.20	m	**9.03**
valley slate; cut to shape; felt soaker and cutting						
both sides	–	0.42	8.49	9.74	m	**18.23**
ridge slate; cut to shape	–	0.28	5.66	6.20	m	**11.86**
hip slate; cut to shape; felt soaker and cutting						
both sides	–	0.42	8.49	9.68	m	**18.17**
holes for pipes and the like	–	0.48	9.70	–	nr	**9.70**
Bostik Findley Flashband Plus sealing strips and flashings or other equal and approved; special grey finish						
Flashings; wedging at top if required; pressure bonded; to walls						
100 mm girth	–	0.23	3.52	1.37	m	**4.89**
150 mm girth	–	0.31	4.74	1.82	m	**6.56**
225 mm girth	–	0.37	5.65	2.55	m	**8.20**
300 mm girth	–	0.42	6.42	2.95	m	**9.37**
450 mm girth	–	0.45	6.88	4.17	m	**11.05**
600 mm girth	–	0.47	7.26	5.04	m	**12.30**

18 TILE AND SLATE ROOF AND WALL COVERINGS

Item	PC £	Labour hours	Labour £	Material £	Unit	Total rate £
ALTERNATIVE SUPPLY ONLY TILE PRICES						
Clay tiles; plain, interlocking and pantiles						
Dreadnought						
Handformed Classic Staffordshire Blue	–	–	–	600.00	1000	**600.00**
Handformed Classic Purple Brown	–	–	–	560.00	1000	**560.00**
Handformed Classic Bronze	–	–	–	560.00	1000	**560.00**
Handformed Classic Deep Red	–	–	–	540.00	1000	**540.00**
Red smooth/sandfaced	–	–	–	280.00	1000	**280.00**
Country brown smooth/sandfaced	–	–	–	312.00	1000	**312.00**
Brown Antique smooth/sandfaced	–	–	–	324.00	1000	**324.00**
Blue/Dark Heather	–	–	–	340.00	1000	**340.00**
Sandtoft pantiles						
Bridgewater Double Roman	–	–	–	5820.00	1000	**5820.00**
Gaelic	–	–	–	2290.00	1000	**2290.00**
Arcadia	–	–	–	1395.00	1000	**1395.00**
William Blyth pantiles						
Barco Bold Roll	–	–	–	760.00	1000	**760.00**
Celtic (French)	–	–	–	860.80	1000	**860.80**
Concrete tiles; plain and interlocking						
Marley Eternit roof tiles						
Anglia	–	–	–	610.00	1000	**610.00**
Ashmore	–	–	–	735.00	1000	**735.00**
Duo Modern	–	–	–	–	1000	**-**
Pewter Mendip	–	–	–	1050.00	1000	**1050.00**
Malvern	–	–	–	980.00	1000	**980.00**
Plain	–	–	–	350.00	1000	**350.00**
Redland roof tiles						
Redland 49	–	–	–	702.00	1000	**702.00**
50 Double Roman	–	–	–	610.00	1000	**610.00**
Mini Stoneworld	–	–	–	912.00	1000	**912.00**
Grovebury	–	–	–	1024.00	1000	**1024.00**
PLAIN TILING						
SUPPLY AND FIX PRICES						
NOTE: The following items of tile roofing unless otherwise described, include for conventional fixing assuming normal exposure with appropriate nails and/or rivets or clips to pressure impregnated softwood battens fixed with galvanized nails; prices also include for all bedding and pointing at verges, beneath ridge tiles, etc.						

18 TILE AND SLATE ROOF AND WALL COVERINGS

Item	PC £	Labour hours	Labour £	Material £	Unit	Total rate £
PLAIN TILING – cont						
Clay interlocking plain tiles; Sandtoft 20/20 natural red faced or other equal; 75 mm lap; on 25 mm × 38 mm battens and type 1F reinforced underlay						
Tiles 370 mm × 223 mm (Prime Cost £ per 1000)	816.00	–	–	816.00	1000	**816.00**
roof coverings	12.24	0.42	8.49	17.03	m²	**25.52**
extra over coverings for;						
fixing every tile	–	0.02	0.40	0.83	m²	**1.23**
double course at eaves	–	0.28	5.66	12.16	m	**17.82**
verges; extra single undercloak course of plain tiles	–	0.28	5.66	5.02	m	**10.68**
open valleys; cutting both sides	–	0.17	3.43	3.52	m	**6.95**
dry ridge tiles	–	0.56	11.32	14.47	m	**25.79**
dry hips; cutting both sides	–	0.69	13.94	12.17	m	**26.11**
holes for pipes and the like	–	0.19	3.83	–	nr	**3.83**
Clay pantiles; Sandtoft Old English; red sand faced or other equal; 75 mm lap; on 25 mm × 38 mm battens and type 1F reinforced underlay						
Pantiles 342 mm × 241 mm (Prime Cost £ per 1000)	977.60	–	–	977.60	1000	**977.60**
roof coverings	15.64	0.42	8.49	21.09	m²	**29.58**
extra over coverings for;						
fixing every tile	–	0.02	0.40	2.06	m²	**2.46**
other colours	–	–	–	1.14	m²	**1.14**
double course at eaves	–	0.31	6.26	4.90	m	**11.16**
verges; extra single undercloak course of plain tiles	–	0.28	5.66	13.44	m	**19.10**
open valleys; cutting both sides	–	0.17	3.43	4.21	m	**7.64**
ridge tiles; tile slips	–	0.56	11.32	40.71	m	**52.03**
hips; cutting both sides	–	0.69	13.94	44.93	m	**58.87**
holes for pipes and the like	–	0.19	3.83	–	nr	**3.83**
Clay pantiles; William Blyth's Lincoln natural or other equal; 75 mm lap; on 19 mm × 38 mm battens and type 1F reinforced underlay						
Pantiles 343 mm × 280 mm (Prime Cost £ per 1000)	1028.00	–	–	1028.00	1000	**1028.00**
roof coverings	15.93	0.42	8.49	21.41	m²	**29.90**
extra over coverings for;						
fixing every tile	–	0.02	0.40	2.06	m²	**2.46**
other colours	–	–	–	1.40	m²	**1.40**
double course at eaves	–	0.31	6.26	5.11	m	**11.37**
verges; extra single undercloak course of plain tiles	–	0.28	5.66	11.91	m	**17.57**
open valleys; cutting both sides	–	0.17	3.43	4.43	m	**7.86**
ridge tiles; tile slips	–	0.56	11.32	24.05	m	**35.37**
hips; cutting both sides	–	0.69	13.94	28.47	m	**42.41**
holes for pipes and the like	–	0.19	3.83	–	nr	**3.83**

18 TILE AND SLATE ROOF AND WALL COVERINGS

Item	PC £	Labour hours	Labour £	Material £	Unit	Total rate £
Clay plain tiles; Hinton, Perry and Davenhill Dreadnought smooth red machine-made or other equal; on 19 mm × 38 mm battens and type 1F reinforced underlay						
Tiles 265 mm × 165 mm (Prime Cost £ per 1000)	280.00	–	–	280.00	1000	**280.00**
roof coverings; to 64 mm lap	16.80	0.97	19.60	26.96	m²	**46.56**
wall coverings; to 38 mm lap	14.84	1.16	23.43	23.32	m²	**46.75**
extra over coverings for;						
ornamental tiles	–	–	–	18.98	m²	**18.98**
double course at eaves	–	0.23	4.64	3.26	m	**7.90**
verges	–	0.28	5.66	0.93	m	**6.59**
swept valleys; cutting both sides	–	0.60	12.13	4.82	m	**16.95**
bonnet hips; cutting both sides	–	0.74	14.95	48.29	m	**63.24**
external vertical angle tiles; supplementary nail fixings	–	0.37	7.47	61.24	m	**68.71**
half round ridge tiles	–	0.56	11.32	11.11	m	**22.43**
holes for pipes and the like	–	0.19	3.83	–	nr	**3.83**
Concrete plain tiles; BS EN 490 group A; on 25 mm × 38 mm battens and type 1F reinforced underlay						
Tiles 267 mm × 165 mm (Prime Cost £ per 1000)	358.70	–	–	358.70	1000	**358.70**
roof coverings; to 64 mm lap	21.52	0.97	19.60	32.04	m²	**51.64**
wall coverings; to 38 mm lap	19.01	1.16	23.43	27.81	m²	**51.24**
extra over coverings for;						
ornamental tiles	–	–	–	19.76	m²	**19.76**
double course at eaves	–	0.23	4.64	3.76	m	**8.40**
verges	–	0.31	6.26	1.29	m	**7.55**
swept valleys; cutting both sides	–	0.60	12.13	35.25	m	**47.38**
bonnet hips; cutting both sides	–	0.74	14.95	35.32	m	**50.27**
external vertical angle tiles; supplementary nail fixings	–	0.37	7.47	25.33	m	**32.80**
half round ridge tiles	–	0.46	9.30	8.97	m	**18.27**
third round hip tiles; cutting both sides	–	0.46	9.30	11.29	m	**20.59**
holes for pipes and the like	–	0.19	3.83	–	nr	**3.83**

18 TILE AND SLATE ROOF AND WALL COVERINGS

Item	PC £	Labour hours	Labour £	Material £	Unit	Total rate £
INTERLOCKING TILING						
SUPPLY AND FIX PRICES						
NOTE: The following items of tile roofing unless otherwise described, include for conventional fixing assuming normal exposure with appropriate nails and/or rivets or clips to pressure impregnated softwood battens fixed with galvanized nails; prices also include for all bedding and pointing at verges, beneath ridge tiles, etc.						
Clay interlocking plain tiles; Sandtoft 20/20 natural red faced or other equal; 75 mm lap; on 25 mm × 38 mm battens and type 1F reinforced underlay						
Tiles 370 mm × 223 mm (Prime Cost £ per 1000)	816.00	–	–	816.00	1000	**816.00**
roof coverings	12.24	0.42	8.49	17.03	m²	**25.52**
extra over coverings for;						
fixing every tile	–	0.02	0.40	0.83	m²	**1.23**
double course at eaves	–	0.28	5.66	12.16	m	**17.82**
verges; extra single undercloak course of plain tiles	–	0.28	5.66	5.02	m	**10.68**
open valleys; cutting both sides	–	0.17	3.43	3.52	m	**6.95**
dry ridge tiles	–	0.56	11.32	14.47	m	**25.79**
dry hips; cutting both sides	–	0.69	13.94	12.17	m	**26.11**
holes for pipes and the like	–	0.19	3.83	–	nr	**3.83**
Clay pantiles; Sandtoft Old English; red sand faced or other equal; 75 mm lap; on 25 mm × 38 mm battens and type 1F reinforced underlay						
Tiles 342 mm × 241 mm (Prime Cost £ per 1000)	977.60	–	–	977.60	1000	**977.60**
roof coverings	15.64	0.42	8.49	21.09	m²	**29.58**
extra over coverings for;						
fixing every tile	–	0.02	0.40	2.06	m²	**2.46**
other colours	–	–	–	1.14	m²	**1.14**
double course at eaves	–	0.31	6.26	4.90	m	**11.16**
verges; extra single undercloak course of plain tiles	–	0.28	5.66	13.44	m	**19.10**
open valleys; cutting both sides	–	0.17	3.43	4.21	m	**7.64**
ridge tiles; tile slips	–	0.56	11.32	40.71	m	**52.03**
hips; cutting both sides	–	0.69	13.94	44.93	m	**58.87**
holes for pipes and the like	–	0.19	3.83	–	nr	**3.83**
Clay pantiles; William Blyth's Lincoln natural or other equal; 75 mm lap; on 19 mm × 38 mm battens and type 1F reinforced underlay						
Tiles 343 mm × 280 mm (Prime Cost £ per 1000)	1028.00	–	–	1028.00	1000	**1028.00**
roof coverings	16.73	0.42	8.49	21.41	m²	**29.90**

18 TILE AND SLATE ROOF AND WALL COVERINGS

Item	PC £	Labour hours	Labour £	Material £	Unit	Total rate £
extra over coverings for;						
fixing every tile	–	0.02	0.40	2.06	m²	**2.46**
other colours	–	–	–	1.40	m²	**1.40**
double course at eaves	–	0.31	6.26	5.11	m	**11.37**
verges; extra single undercloak course of plain tiles	–	0.28	5.66	11.91	m	**17.57**
open valleys; cutting both sides	–	0.17	3.43	4.43	m	**7.86**
ridge tiles; tile slips	–	0.56	11.32	24.05	m	**35.37**
hips; cutting both sides	–	0.69	13.94	28.47	m	**42.41**
holes for pipes and the like	–	0.19	3.83	–	nr	**3.83**
Concrete interlocking tiles; Marley Eternit Anglia granule finish tiles or other equal; 75 mm lap; on 25 mm × 38 mm battens and type 1F reinforced underlay						
Tiles 387 mm × 230 mm (Prime Cost £ per 1000)	579.50	–	–	579.50	1000	**579.50**
roof coverings	9.10	0.42	8.49	13.55	m²	**22.04**
extra over coverings for;						
fixing every tile	–	0.02	0.40	0.35	m²	**0.75**
eaves; eaves filler	–	0.04	0.81	11.57	m	**12.38**
verges; 150 mm wide asbestos free strip undercloak	–	0.21	4.24	1.98	m	**6.22**
valley trough tiles; cutting both sides	–	0.51	10.30	26.51	m	**36.81**
segmental ridge tiles; tile slips	–	0.51	10.30	13.33	m	**23.63**
segmental hip tiles; tile slips; cutting both sides	–	0.65	13.13	15.19	m	**28.32**
dry ridge tiles; segmental including batten sections; unions and filler pieces	–	0.28	5.66	19.79	m	**25.45**
segmental mono-ridge tiles	–	0.51	10.30	21.17	m	**31.47**
gas ridge terminal	–	0.46	9.30	71.80	nr	**81.10**
holes for pipes and the like	–	0.19	3.83	–	nr	**3.83**
Concrete interlocking tiles; Marley Eternit Ludlow Major granule finish tiles or other equal; 75 mm lap; on 25 mm × 38 mm battens and type 1F reinforced underlay						
Tiles 420 mm × 330 mm (Prime Cost £ per 1000)	840.80	–	–	840.80	1000	**840.80**
roof coverings	8.65	0.32	6.47	11.94	m²	**18.41**
extra over coverings for;						
fixing every tile	–	0.02	0.40	0.35	m²	**0.75**
eaves; eaves filler	–	0.04	0.81	0.33	m	**1.14**
verges; 150 mm wide asbestos free strip undercloak	–	0.21	4.24	1.98	m	**6.22**
dry verge system; extruded white PVC	–	0.14	2.83	12.35	m	**15.18**
segmental ridge cap to dry verge	–	0.02	0.40	4.09	m	**4.49**
valley trough tiles; cutting both sides	–	0.51	10.30	27.07	m	**37.37**
segmental ridge tiles	–	0.46	9.30	8.42	m	**17.72**
segmental hip tiles; cutting both sides	–	0.60	12.13	11.13	m	**23.26**
dry ridge tiles; segmental including batten sections; unions and filler pieces	–	0.28	5.66	19.83	m	**25.49**
segmental mono-ridge tiles	–	0.46	9.30	18.30	m	**27.60**
gas ridge terminal	–	0.46	9.30	71.80	nr	**81.10**
holes for pipes and the like	–	0.19	3.83	–	nr	**3.83**

18 TILE AND SLATE ROOF AND WALL COVERINGS

Item	PC £	Labour hours	Labour £	Material £	Unit	Total rate £
INTERLOCKING TILING – cont						
Concrete interlocking tiles; Marley Eternit						
Mendip granule finish double pantiles or other						
equal; 75 mm lap; on 22 mm × 38 mm battens and						
type 1F reinforced underlay						
Tiles 420 mm × 330 mm (Prime Cost £ per 1000)	855.00	–	–	855.00	1000	**855.00**
roof coverings	8.71	0.32	6.47	12.00	m²	**18.47**
extra over coverings for;						
fixing every tile	–	0.02	0.40	0.35	m²	**0.75**
eaves; eaves filler	–	0.02	0.40	11.37	m	**11.77**
verges; 150 mm wide asbestos free strip						
undercloak	–	0.21	4.24	1.98	m	**6.22**
dry verge system; extruded white PVC	–	0.14	2.83	12.35	m	**15.18**
segmental ridge cap to dry verge	–	0.02	0.40	4.09	m	**4.49**
valley trough tiles; cutting both sides	–	0.51	10.30	27.10	m	**37.40**
segmental ridge tiles	–	0.51	10.30	13.33	m	**23.63**
segmental hip tiles; cutting both sides	–	0.65	13.13	16.08	m	**29.21**
dry ridge tiles; segmental including batten						
sections; unions and filler pieces	–	0.28	5.66	19.83	m	**25.49**
segmental mono-ridge tiles	–	0.46	9.30	20.76	m	**30.06**
gas ridge terminal	–	0.46	9.30	71.80	nr	**81.10**
holes for pipes and the like	–	0.19	3.83	–	nr	**3.83**
Concrete interlocking tiles; Marley Eternit						
Modern smooth finish tiles or other equal; 75 mm						
lap; on 25 mm × 38 mm battens and type 1F						
reinforced underlay						
Tiles 420 mm × 220 mm (Prime Cost £ per 1000)	874.00	–	–	874.00	1000	**874.00**
roof coverings	9.09	0.32	6.47	12.70	m²	**19.17**
extra over coverings for;						
fixing every tile	–	0.02	0.40	0.35	m²	**0.75**
verges; 150 wide asbestos free strip undercloak	–	0.21	4.24	1.98	m	**6.22**
dry verge system; extruded white PVC	–	0.19	3.83	12.35	m	**16.18**
Modern ridge cap to dry verge	–	0.02	0.40	4.09	m	**4.49**
valley trough tiles; cutting both sides	–	0.51	10.30	27.14	m	**37.44**
Modern ridge tiles	–	0.46	9.30	10.69	m	**19.99**
Modern hip tiles; cutting both sides	–	0.60	12.13	13.51	m	**25.64**
dry ridge tiles; Modern; including batten sections;						
unions and filler pieces	–	0.28	5.66	22.10	m	**27.76**
Modern mono-ridge tiles	–	0.46	9.30	18.30	m	**27.60**
gas ridge terminal	–	0.46	9.30	71.80	nr	**81.10**
holes for pipes and the like	–	0.19	3.83	–	nr	**3.83**
Concrete interlocking tiles; Marley Eternit						
Ecologic Ludlow Major granule finish tiles or						
other equal; 75 mm lap; on 25 mm × 38 mm						
battens and type 1F reinforced underlay						
Tiles 420 mm × 330 mm (Prime Cost £ per 1000)	902.50	–	–	902.50	1000	**902.50**
roof coverings	8.84	0.32	6.47	12.60	m²	**19.07**
extra over coverings for:						
fixing every tile	–	0.02	0.40	0.35	m²	**0.75**

18 TILE AND SLATE ROOF AND WALL COVERINGS

Item	PC £	Labour hours	Labour £	Material £	Unit	Total rate £
eaves; eaves filler	–	0.04	0.81	0.33	m	1.14
verges; 150 mm wide asbestos free strip						
undercloak	–	0.21	4.24	1.98	m	6.22
dry verge system; extruded white PVC	–	0.14	2.83	12.35	m	15.18
segmental ridge cap to dry verge	–	0.02	0.40	4.09	m	4.49
valley trough tiles; cutting both sides	–	0.51	10.30	27.20	m	37.50
segmental ridge tiles	–	0.46	9.30	8.42	m	17.72
segmental hip tiles; cutting both sides	–	0.60	12.13	11.33	m	23.46
dry ridge tiles; segmental including batten						
sections; unions and filler pieces	–	0.28	5.66	19.83	m	25.49
segmental mono-ridge tiles	–	0.46	9.30	18.30	m	27.60
gas ridge terminal	–	0.46	9.30	71.80	nr	81.10
holes for pipes and the like	–	0.19	3.83	–	nr	3.83
Concrete interlocking tiles; Marley Eternit Wessex smooth finish tiles or other equal; 75 mm lap; on 25 mm × 38 mm battens and type 1F reinforced underlay						
Tiles 413 mm × 330 mm (Prime Cost £ per 1000)	1330.00	–	–	1330.00	1000	1330.00
roof coverings	13.17	0.32	6.47	17.56	m²	24.03
extra over coverings for;						
fixing every tile	–	0.02	0.40	0.35	m²	0.75
verges; 150 mm wide asbestos free strip						
undercloak	–	0.21	4.24	1.98	m	6.22
dry verge system; extruded white PVC	–	0.19	3.83	12.35	m	16.18
Modern ridge cap to dry verge	–	0.02	0.40	4.09	m	4.49
valley trough tiles; cutting both sides	–	0.51	10.30	28.13	m	38.43
Modern ridge tiles	–	0.46	9.30	10.69	m	19.99
Modern hip tiles; cutting both sides	–	0.60	12.13	14.99	m	27.12
dry ridge tiles; Modern; including batten sections;						
unions and filler pieces	–	0.28	5.66	22.07	m	27.73
Modern mono-ridge tiles	–	0.46	9.30	18.30	m	27.60
gas ridge terminal	–	0.46	9.30	71.80	nr	81.10
holes for pipes and the like	–	0.19	3.83	–	nr	3.83
Concrete interlocking tiles; Redland Norfolk smooth finish pantiles or other equal; 75 mm lap; on 25 mm × 38 mm battens and type 1F reinforced underlay						
Tiles 381 mm × 229 mm (Prime Cost £ per 1000)	619.60	–	–	619.60	1000	619.60
roof coverings	11.58	0.42	8.49	15.67	m²	24.16
extra over coverings for;						
fixing every tile	–	0.04	0.81	0.15	m²	0.96
eaves; eaves filler	–	0.04	0.81	1.28	m	2.09
verges; extra single undercloak course of plain						
tiles	–	0.28	5.66	7.20	m	12.86
valley trough tiles; cutting both sides	–	0.56	11.32	36.69	m	48.01
universal ridge tiles	–	0.46	9.30	13.85	m	23.15
universal hip tiles; cutting both sides	–	0.60	12.13	17.18	m	29.31
universal gas flue ridge tile	–	0.46	9.30	78.63	nr	87.93
universal ridge vent tile with 110 mm dia. adaptor	–	0.50	10.11	93.58	nr	103.69
holes for pipes and the like	–	0.19	3.83	–	nr	3.83

18 TILE AND SLATE ROOF AND WALL COVERINGS

Item	PC £	Labour hours	Labour £	Material £	Unit	Total rate £
INTERLOCKING TILING – cont						
Concrete interlocking tiles; Redland Renown granule finish tiles or other equal and approved; 418 mm × 330 mm; to 75 mm lap; on 25 mm × 38 mm battens and type 1F reinforced underlay						
Tiles 418 mm × 330 mm (Prime Cost £ per 1000)	855.10	–	–	855.10	1000	855.10
roof coverings	8.29	0.32	6.47	12.42	m²	18.89
extra over coverings for;						
fixing every tile	–	0.02	0.40	0.18	m²	0.58
verges; extra single undercloak course of plain tiles	–	0.23	4.64	4.15	m	8.79
cloaked verge system	–	0.14	2.83	8.51	m	11.34
valley trough tiles; cutting both sides	–	0.51	10.30	36.12	m	46.42
universal ridge tiles	–	0.46	9.30	13.85	m	23.15
universal hip tiles; cutting both sides	–	0.60	12.13	16.60	m	28.73
dry ridge system; universal ridge tiles	–	0.23	4.64	45.65	m	50.29
universal half round mono-pitch ridge tiles	–	0.51	10.30	30.01	m	40.31
universal gas flue ridge tile	–	0.46	9.30	78.63	nr	87.93
universal ridge vent tile with 110 mm dia. adaptor	–	0.46	9.30	93.58	nr	102.88
holes for pipes and the like	–	0.19	3.83	–	nr	3.83
Concrete interlocking tiles; Redland Regent granule finish bold roll tiles or other equal; 75 mm lap; on 25 mm × 38 mm battens and type 1F reinforced underlay						
Tiles 418 mm × 332 mm (Prime Cost £ per 1000)	884.90	–	–	884.90	1000	884.90
roof coverings	8.58	0.32	6.47	12.65	m²	19.12
extra over coverings for;						
fixing every tile	–	0.03	0.60	0.59	m²	1.19
eaves; eaves filler	–	0.04	0.81	1.04	m	1.85
verges; extra single undercloak course of plain tiles	–	0.23	4.64	3.45	m	8.09
cloaked verge system	–	0.14	2.83	8.42	m	11.25
valley trough tiles; cutting both sides	–	0.51	10.30	36.21	m	46.51
universal ridge tiles	–	0.46	9.30	13.85	m	23.15
universal hip tiles; cutting both sides	–	0.60	12.13	16.71	m	28.84
dry ridge system; universal ridge tiles	–	0.23	4.64	45.93	m	50.57
universal half round mono-pitch ridge tiles	–	0.51	10.30	30.01	m	40.31
universal gas flue ridge tile	–	0.46	9.30	78.63	nr	87.93
universal ridge vent tile with 110 mm dia. adaptor	–	0.46	9.30	93.58	nr	102.88
holes for pipes and the like	–	0.19	3.83	–	nr	3.83

18 TILE AND SLATE ROOF AND WALL COVERINGS

Item	PC £	Labour hours	Labour £	Material £	Unit	Total rate £
Concrete interlocking slates; Redland Mini Stonewold concrete slates or other equal; 75 mm lap; on 25 mm × 38 mm battens and type 1F reinforced underlay						
Slates 418 mm × 334 mm (Prime Cost £ per 1000)	912.00	–	–	912.00	1000	**912.00**
roof coverings	9.76	0.32	6.47	14.06	m²	**20.53**
extra over coverings for;						
fixing every tile	–	0.02	0.40	1.13	m²	**1.53**
eaves; eaves filler	–	0.02	0.40	6.23	m	**6.63**
verges; extra single undercloak course of plain tiles	–	0.28	5.66	4.27	m	**9.93**
ambi-dry verge system	–	0.19	3.83	12.28	m	**16.11**
ambi-dry verge eave/ridge end piece	–	0.02	0.40	4.40	m	**4.80**
valley trough tiles; cutting both sides	–	0.51	10.30	36.16	m	**46.46**
universal angle ridge tiles	–	0.46	9.30	10.25	m	**19.55**
universal hip tiles; cutting both sides	–	0.60	12.13	17.37	m	**29.50**
dry ridge system; universal angle ridge tiles	–	0.23	4.64	16.16	m	**20.80**
universal mono-pitch angle ridge tiles	–	0.51	10.30	20.32	m	**30.62**
universal gas flue angle ridge tile	–	0.46	9.30	78.63	nr	**87.93**
universal angle ridge vent tile with 110 mm dia. adaptor	–	0.46	9.30	91.02	nr	**100.32**
holes for pipes and the like	–	0.19	3.83	–	nr	**3.83**
Concrete interlocking slates; Redland Richmond 10 Slates laid broken bonded, normally from right to left; smooth finish tiles or other equal; 75 mm lap; on 25 mm × 38 mm battens and type 1F reinforced underlay						
Slates 418 mm × 330 mm (Prime Cost £ per 1000)	1083.20	–	–	1083.20	1000	**1083.20**
roof coverings	11.60	0.32	6.47	16.74	m²	**23.21**
extra over coverings for;						
fixing every tile	–	0.02	0.40	1.13	m²	**1.53**
eaves; eaves filler	–	0.02	0.40	6.10	m	**6.50**
verges; extra single undercloak course of plain tiles	–	0.23	4.64	4.15	m	**8.79**
ambi-dry verge system	–	0.19	3.83	12.28	m	**16.11**
ambi-dry verge eave/ridge end piece	–	0.02	0.40	4.40	m	**4.80**
universal valley trough tiles; cutting both sides	–	0.56	11.32	38.03	m	**49.35**
universal hip tiles; cutting both sides	–	0.60	12.13	13.75	m	**25.88**
universal angle ridge tiles	–	0.46	9.30	10.25	m	**19.55**
dry ridge system; universal angle ridge tiles	–	0.23	4.64	16.16	m	**20.80**
universal mono-pitch angle ridge tiles	–	0.51	10.30	20.32	m	**30.62**
gas ridge terminal	–	0.46	9.30	77.72	nr	**87.02**
ridge vent with 110 mm dia. flexible adaptor	–	0.46	9.30	90.33	nr	**99.63**
holes for pipes and the like	–	0.19	3.83	–	nr	**3.83**

18 TILE AND SLATE ROOF AND WALL COVERINGS

Item	PC £	Labour hours	Labour £	Material £	Unit	Total rate £
INTERLOCKING TILING – cont						
Concrete interlocking slates; Redland Stonewold II smooth finish tiles or other equal; 75 mm lap; on 25 mm × 38 mm battens and type 1F reinforced underlay						
Slates 430 mm × 380 mm (Prime Cost £ per 1000)	2205.80	–	–	2205.80	1000	2205.80
roof coverings	21.40	0.32	6.47	26.74	m²	33.21
extra over coverings for;						
fixing every tile	–	0.02	0.40	1.13	m²	1.53
eaves; eaves filler	–	0.02	0.40	6.10	m	6.50
verges; extra single undercloak course of plain tiles	–	0.28	5.66	4.15	m	9.81
ambi-dry verge system	–	0.19	3.83	12.28	m	16.11
ambi-dry verge eave/ridge end piece	–	0.02	0.40	4.40	m	4.80
valley trough tiles; cutting both sides	–	0.51	10.30	40.48	m	50.78
universal angle ridge tiles	–	0.46	9.30	10.25	m	19.55
universal hip tiles; cutting both sides	–	0.60	12.13	17.37	m	29.50
dry ridge system; universal angle ridge tiles	–	0.23	4.64	16.16	m	20.80
universal mono-pitch angle ridge tiles	–	0.51	10.30	20.32	m	30.62
universal gas flue angle ridge tile	–	0.46	9.30	78.63	nr	87.93
universal angle ridge vent tile with 110 mm dia. adaptor	–	0.46	9.30	91.02	nr	100.32
holes for pipes and the like	–	0.19	3.83	–	nr	3.83
Concrete interlocking slates; Redland Cambrian Slates made from 60% recycled Welsh slate or other equal; 50 mm lap; on 25 mm × 38 mm battens and type 1F reinforced underlay						
Slates 300 mm × 336 mm (Prime Cost £ per 1000)	2344.00	–	–	2344.00	1000	2344.00
roof coverings	37.27	0.32	6.47	46.66	m²	53.13
extra over coverings for;						
fixing every tile	–	0.02	0.40	1.04	m²	1.44
eaves; eaves filler	–	0.02	0.40	6.68	m	7.08
verges; extra single undercloak course of plain tiles	–	0.23	4.64	13.92	m	18.56
ambi-dry verge system	–	0.19	3.83	12.28	m	16.11
ambi-dry verge eave/ridge end piece	–	0.02	0.40	4.40	m	4.80
universal valley trough tiles; cutting both sides	–	0.56	11.32	38.03	m	49.35
universal hip tiles; cutting both sides	–	0.60	12.13	17.46	m	29.59
universal angle ridge tiles	–	0.46	9.30	10.25	m	19.55
dry ridge system; universal angle ridge tiles	–	0.23	4.64	16.16	m	20.80
universal mono-pitch angle ridge tiles	–	0.51	10.30	20.32	m	30.62
gas ridge terminal	–	0.46	9.30	77.72	nr	87.02
ridge vent with 110 mm dia. flexible adaptor	–	0.46	9.30	90.33	nr	99.63
holes for pipes and the like	–	0.19	3.83	–	nr	3.83

18 TILE AND SLATE ROOF AND WALL COVERINGS

Item	PC £	Labour hours	Labour £	Material £	Unit	Total rate £
Sundries and accessories						
Hip irons						
galvanized mild steel; fixing with screws	–	0.09	1.81	2.69	nr	**4.50**
Rytons Clip strip or other equal and approved; continuous soffit ventilator						
51 mm wide; plastic; code CS351	–	0.28	5.66	0.88	m	**6.54**
Rytons over fascia ventilator or other equal and approved; continuous eaves ventilator						
40 mm wide; plastic; code OFV890	–	0.09	1.81	1.23	m	**3.04**
Rytons roof ventilator or other equal and approved; to suit rafters at 600 mm centres						
250 mm deep × 43 mm high; plastic; code TV600	–	0.09	1.81	1.39	m	**3.20**
Rytons push and lock ventilators or other equal and approved; circular						
83 mm dia.; plastic; code PL235	–	0.04	0.72	0.32	nr	**1.04**
Fixing only						
lead soakers (supply cost not included)	–	0.07	1.07	–	nr	**1.07**
Pressure impregnated softwood counter battens; 25 mm × 50 mm						
450 mm centres	–	0.06	1.21	2.05	m²	**3.26**
600 mm centres	–	0.04	0.81	1.55	m²	**2.36**
Underlay; BS EN 13707 type 1B; bitumen felt weighing 14 kg/10 m²; 75 mm laps						
To sloping or vertical surfaces	0.52	0.02	0.40	0.82	m²	**1.22**
Underlay; BS EN 13707 type 1F; reinforced bitumen felt weighing 22.50 kg/10 m²; 75 mm laps						
To sloping or vertical surfaces	0.66	0.02	0.40	0.96	m²	**1.36**
Underlay; Visqueen Tilene 200P or other equal and approved; micro-perforated sheet; 75 mm laps						
To sloping or vertical surfaces	1.46	0.02	0.40	1.78	m²	**2.18**
Underlay; reinforced breather membrane; 75 mm laps						
To sloping or vertical surfaces	1.10	0.02	0.40	1.41	m²	**1.81**
Underlay; Anticon or other equal and approved sarking membrane; polyethylene; 75 mm laps						
To sloping or vertical surfaces	0.69	0.02	0.40	1.03	m²	**1.43**

18 TILE AND SLATE ROOF AND WALL COVERINGS

Item	PC £	Labour hours	Labour £	Material £	Unit	Total rate £
INTERLOCKING TILING – cont						
UV Breather membranes; Web Dynamics Ltd						
Web UV 10 standard grade breather membrane draped between open rafters; 100 mm vertical laps, 150 mm horizontal laps, secured with galvanized nails	0.72	0.03	0.60	0.89	m²	**1.49**
Web UV 15 professional grade breather membrane draped between open rafters or on sarking board; 100 mm vertical laps, 150 mm horizontal laps secured with galvanized nails	0.71	0.03	0.60	0.88	m²	**1.48**
Web UV 25 heavy dutyl grade breather membrane draped between open rafters or on sarking board; 100 mm vertical laps, 150 mm horizontal laps secured with galvanized nails	1.47	0.04	0.71	1.73	m²	**2.44**

18 TILE AND SLATE ROOF AND WALL COVERINGS

Item	PC £	Labour hours	Labour £	Material £	Unit	Total rate £
FIBRE CEMENT SLATING						
Asbestos-free artificial slates; Eternit Garsdale/ E2000T or other equal and approved; to 75 mm lap; on 19 mm × 50 mm battens and type 1F reinforced underlay						
Coverings; 500 mm × 250 mm slates						
roof coverings	–	0.60	12.13	20.39	m²	**32.52**
wall coverings	–	0.74	14.95	20.39	m²	**35.34**
Coverings; 600 mm × 300 mm slates						
roof coverings	–	0.46	9.30	16.65	m²	**25.95**
wall coverings	–	0.60	12.13	16.65	m²	**28.78**
Extra over slate coverings for						
double course at eaves	–	0.23	4.64	4.18	m	**8.82**
verges; extra single undercloak course	–	0.31	6.26	0.86	m	**7.12**
open valleys; cutting both sides	–	0.19	3.83	3.43	m	**7.26**
stop end	–	0.09	1.81	10.56	nr	**12.37**
roll top ridge tiles	–	0.56	11.32	28.11	m	**39.43**
stop end	–	0.09	1.81	15.57	nr	**17.38**
mono-pitch ridge tiles	–	0.46	9.30	32.84	m	**42.14**
stop end	–	0.09	1.81	35.70	nr	**37.51**
duo-pitch ridge tiles	–	0.46	9.30	26.61	m	**35.91**
stop end	–	0.09	1.81	26.18	nr	**27.99**
half round hip tiles; cutting both sides	–	0.19	3.83	69.26	m	**73.09**
holes for pipes and the like	–	0.19	3.83	–	nr	**3.83**
NATURAL SLATING						
NOTE: The following items of slate roofing unless otherwise described, include for conventional fixing assuming normal exposure with appropriate nails and/or rivets or clips to pressure impregnated softwood battens fixed with galvanized nails; prices also include for all bedding and pointing at verges; beneath verge tiles etc.						
Natural slates; BS EN 12326 Part 2; Spanish blue grey; uniform size; to 75 mm lap; on 25 mm × 50 mm battens and type 1F reinforced underlay						
Coverings; 400 mm × 250 mm slates (Prime Cost £ per 1000)	524.00	–	–	524.00	1000	**524.00**
roof coverings	12.84	0.73	14.75	20.71	m²	**35.46**
wall coverings	12.84	1.06	21.41	20.71	m²	**42.12**
Coverings; 500 mm × 250 mm slates (Prime Cost £ per 1000)	856.00	–	–	856.00	1000	**856.00**
roof coverings	16.01	0.60	12.13	22.64	m²	**34.77**
wall coverings	16.01	0.88	17.77	22.64	m²	**40.41**
Coverings; 600 mm × 300 mm slates (Prime Cost £ per 1000)	1360.00	–	–	1360.00	1000	**1360.00**
roof coverings	17.14	0.50	10.11	22.91	m²	**33.02**
wall coverings	17.14	0.69	13.94	22.91	m²	**36.85**

18 TILE AND SLATE ROOF AND WALL COVERINGS

Item	PC £	Labour hours	Labour £	Material £	Unit	Total rate £
NATURAL SLATING – cont						
Natural slates – cont						
Extra over coverings for;						
double course at eaves	–	0.28	5.66	5.83	m	**11.49**
verges; extra single undercloak course	–	0.39	7.88	2.99	m	**10.87**
open valleys; cutting both sides	–	0.20	4.04	11.71	m	**15.75**
blue/black glass reinforced concrete 152 mm half round ridge tiles	–	0.46	9.30	13.27	m	**22.57**
blue/black glass reinforced concrete 125 mm × 125 mm plain angle ridge tiles	–	0.46	9.30	13.27	m	**22.57**
mitred hips; cutting both sides	–	0.20	4.04	11.71	m	**15.75**
blue/black glass reinforced concrete 152 mm half round hip tiles; cutting both sides	–	0.65	13.13	24.98	m	**38.11**
blue/black glass reinforced concrete 125 mm × 125 mm plain angle hip tiles; cutting both sides	–	0.65	13.13	24.97	m	**38.10**
holes for pipes and the like	–	0.19	3.83	–	nr	**3.83**
Natural slates; BS EN 12326 Part 2; Welsh blue grey; 7 mm nominal thickness; uniform size; to 75 mm lap; on 25 mm × 50 mm battens and type 1F reinforced underlay						
Coverings; 400 mm × 250 mm slates (Prime Cost £ per 1000)	1400.00	–	–	1400.00	1000	**1400.00**
roof coverings	34.30	0.70	14.15	43.80	m²	**57.95**
wall coverings	34.30	1.00	20.20	43.80	m²	**64.00**
Coverings; 500 mm × 250 mm slate (Prime Cost £ per 1000)	2373.00	–	–	2373.00	1000	**2373.00**
roof coverings	44.38	0.60	12.13	53.17	m²	**65.30**
wall coverings	44.38	0.80	16.16	53.17	m²	**69.33**
Coverings; 500 mm × 300 mm slates (Prime Cost £ per 1000)	2261.00	–	–	2261.00	1000	**2261.00**
roof coverings	35.27	0.60	12.13	43.37	m²	**55.50**
wall coverings	35.27	0.75	15.15	43.37	m²	**58.52**
Coverings; 600 mm × 300 mm slates (Prime Cost £ per 1000)	2695.00	–	–	2695.00	1000	**2695.00**
roof coverings	33.96	0.50	10.11	41.02	m²	**51.13**
wall coverings	33.96	0.65	13.13	41.02	m²	**54.15**
Extra over coverings for;						
double course at eaves	–	0.25	5.05	10.58	m	**15.63**
verges; extra single undercloak course	–	0.35	7.07	5.87	m	**12.94**
open valleys; cutting both sides	–	0.20	4.04	23.21	m	**27.25**
blue/black glazed ware 152 mm half round ridge tiles	–	0.46	9.30	8.39	m	**17.69**
blue/black glazed ware 125 mm × 125 mm plain angle ridge tiles	–	0.46	9.30	23.98	m	**33.28**
mitred hips; cutting both sides	–	0.20	4.04	23.21	m	**27.25**
blue/black glazed ware 152 mm half round hip tiles; cutting both sides	–	0.65	13.13	31.60	m	**44.73**

18 TILE AND SLATE ROOF AND WALL COVERINGS

Item	PC £	Labour hours	Labour £	Material £	Unit	Total rate £
blue/black glazed ware 125 mm × 125 mm plain angle hip tiles; cutting both sides	–	0.65	13.13	47.19	m	**60.32**
holes for pipes and the like	–	0.19	3.83	–	nr	**3.83**
Natural slates; Westmoreland green; random lengths; 457 mm–229 mm proportionate widths to 75 mm lap; in diminishing courses; on 25 mm × 50 mm battens and type 1F underlay						
Coverings (Prime Cost £ per tonne; approximately 17.5 m²)	2210.00	–	–	2210.00	tonne	**2210.00**
roof coverings	125.97	1.00	20.20	141.61	m²	**161.81**
wall coverings	125.97	1.30	26.26	141.61	m²	**167.87**
Extra over coverings for;						
double course at eaves	–	0.60	12.13	24.89	m	**37.02**
verges; extra single undercloak course slates 152 mm wide	–	0.67	13.54	21.61	m	**35.15**
holes for pipes and the like	–	0.25	5.05	–	nr	**5.05**
NATURAL OR ARTIFICIAL STONE SLATING						
Reconstructed stone slates; Hardrow Slates or other equal and approved; standard colours; or similar; 75 mm lap; on 25 mm × 50 mm battens and type 1F reinforced underlay						
Coverings; 457 mm × 305 mm slates (Prime Cost £ per 1000)	1197.00	–	–	1226.93	1000	**1226.93**
roof coverings	20.59	0.74	14.95	28.34	m²	**43.29**
wall coverings	20.59	0.93	18.79	28.34	m²	**47.13**
Coverings; 457 mm × 457 mm slates (Prime Cost £ per 1000)	1795.50	–	–	1840.39	1000	**1840.39**
roof coverings	20.65	0.60	12.13	28.21	m²	**40.34**
wall coverings	20.65	0.79	15.96	28.21	m²	**44.17**
Extra over 457 mm × 305 mm coverings for;						
double course at eaves	–	0.28	5.66	5.28	m	**10.94**
verges; pointed	–	0.39	7.88	0.08	m	**7.96**
open valleys; cutting both sides	–	0.20	4.04	12.88	m	**16.92**
ridge tiles	–	0.46	9.30	39.39	m	**48.69**
hip tiles; cutting both sides	–	0.65	13.13	30.63	m	**43.76**
holes for pipes and the like	–	0.19	3.83	–	nr	**3.83**
Reconstructed stone slates; Bradstone Cotswold style or other equal and approved; random lengths 550 mm–300 mm; proportional widths; to 80 mm lap; in diminishing courses; on 25 mm × 50 mm battens and type 1F reinforced underlay						
Roof coverings (all-in rate inclusive of eaves and verges) (Prime Cost £ per m²)	28.70	0.97	19.60	37.07	m²	**56.67**

18 TILE AND SLATE ROOF AND WALL COVERINGS

Item	PC £	Labour hours	Labour £	Material £	Unit	Total rate £
NATURAL OR ARTIFICIAL STONE SLATING – cont						
Reconstructed stone slates – cont						
Extra over coverings for;						
open valleys/mitred hips; cutting both sides	–	0.42	8.49	13.90	m²	**22.39**
ridge tiles	–	0.61	12.32	18.43	m	**30.75**
hip tiles; cutting both sides	–	0.97	19.60	31.40	m	**51.00**
holes for pipes and the like	–	0.28	5.66	–	nr	**5.66**
Reconstructed stone slates; Bradstone Moordale style or other equal and approved; random lengths 550 mm–450 mm; proportional widths; to 80 mm lap; in diminishing course; on 25 mm × 50 mm battens and type 1F reinforced underlay						
Roof coverings (all-in rate inclusive of eaves and verges), (Prime Cost £ per m²)	27.00	0.97	19.60	35.25	m²	**54.85**
Extra over coverings for;						
open valleys/mitred hips; cutting both sides	–	0.42	8.49	13.08	m²	**21.57**
ridge tiles	–	0.61	12.32	18.43	m	**30.75**
holes for pipes and the like	–	0.28	5.66	–	nr	**5.66**
TIMBER SHINGLES						
Red Cedar sawn shingles preservative treated; uniform length 400 mm; to 125 mm gauge; on 25 mm × 38 mm battens and type 1F reinforced underlay						
Shingles; uniform length 400 mm (Prime Cost £ per bundle)	44.00	–	–	44.00	bundle	**44.00**
roof coverings; 125 mm gauge (2.28 m² per bundle)	19.32	0.97	19.60	28.87	m²	**48.47**
wall coverings; 190 mm gauge (3.47 m² per bundle)	12.67	0.74	14.95	19.31	m²	**34.26**
extra over for;						
double course at eaves	–	0.19	3.83	2.71	m	**6.54**
open valleys; cutting both sides	–	0.19	3.83	5.08	m	**8.91**
preformed ridge capping	–	0.28	5.66	13.62	m	**19.28**
preformed hip capping; cutting both sides	–	0.46	9.30	18.71	m	**28.01**
double starter course to cappings	–	0.09	1.81	1.40	m	**3.21**
holes for pipes and the like	–	0.14	2.83	–	nr	**2.83**

19 WATERPROOFING

Item	PC £	Labour hours	Labour £	Material £	Unit	Total rate £
MASTIC ASPHALT ROOFING						
Mastic asphalt to BS 6925 Type R 988						
20 mm thick two coat coverings; felt isolating membrane; to concrete (or timber) base; flat or to falls or slopes not exceeding 10° from horizontal						
over 300 mm wide	–	–	–	–	m²	15.84
225 mm–300 mm wide	–	–	–	–	m²	24.47
150 mm–225 mm wide	–	–	–	–	m²	28.58
not exceeding 150 mm wide	–	–	–	–	m²	36.79
Add to the above for covering with						
10 mm thick limestone chippings in hot bitumen	–	–	–	–	m²	2.58
coverings with solar reflective paint	–	–	–	–	m²	2.91
300 mm × 300 mm × 8 mm g.r.p. tiles in hot bitumen	–	–	–	–	m²	43.82
Cutting to line; jointing to old asphalt	–	–	–	–	m	5.02
13 mm thick two coat skirtings to brickwork base						
not exceeding 150 mm girth	–	–	–	–	m	10.80
150 mm–225 mm girth	–	–	–	–	m	12.39
225 mm–300 mm girth	–	–	–	–	m	15.17
13 mm thick three coat skirtings; expanded metal lathing reinforcement nailed to timber base						
not exceeding 150 mm girth	–	–	–	–	m	18.15
150 mm–225 mm girth	–	–	–	–	m	21.63
225 mm–300 mm girth	–	–	–	–	m	25.30
13 mm thick two coat fascias to concrete base						
not exceeding 150 mm girth	–	–	–	–	m	10.80
150 mm–225 mm girth	–	–	–	–	m	12.39
20 mm thick two coat linings to channels to concrete base						
not exceeding 150 mm girth	–	–	–	–	m	23.74
150 mm–225 mm girth	–	–	–	–	m	27.00
225 mm–300 mm girth	–	–	–	–	m	27.78
20 mm thick two coat lining to cesspools						
250 mm × 150 mm × 150 mm deep	–	–	–	–	nr	23.27
Collars around pipes, standards and like members	–	–	–	–	nr	16.65
Accessories						
Eaves trim; extruded aluminium alloy; working asphalt into trim						
Alutrim; type A roof edging or other equal and approved	–	–	–	–	m	9.92
extra; angle	–	–	–	–	nr	5.55
Roof screed ventilator – aluminium alloy						
Extr-aqua-vent or other equal and approved; set on screed over and including dished sinking; working collar around ventilator	–	–	–	–	nr	19.23
Bituminous lightweight insulating roof screeds						
Bit-Ag or similar roof screed or other equal and approved; to falls or crossfalls; bitumen felt vapour barrier; over 300 mm wide						
75 mm (average) thick	–	–	–	–	m²	45.99
100 mm (average) thick	–	–	–	–	m²	58.28

19 WATERPROOFING

Item	PC £	Labour hours	Labour £	Material £	Unit	Total rate £
APPLIED LIQUID APPLIED TANKING						
Tanking and damp-proofing						
Synthaprufe or other equal and approved; blinding with sand; horizontal on slabs						
two coats	–	0.19	2.60	2.88	m²	5.48
three coats	–	0.26	3.56	4.24	m²	7.80
One coat Vandex Super 0.75 kg/m² slurry or other equal and approved; one consolidating coat of Vandex BB75 1 kg/m² slurry or other equal and approved; horizontal on beds						
over 225 mm wide	–	0.32	4.38	7.48	m²	11.86
Intergritank; Methacrylate resin based structural waterproffing membrane; in two separate colour coded coats; minimumm 2 mm overalll dry film finish; on a primed substrate						
over 250 mm wide						
100 m²–499 m²	–	–	–	–	m²	44.89
500 m²–1999 m²	–	–	–	–	m²	37.05
over 2000 m²	–	–	–	–	m²	29.93
SPECIALIST WATERPROOF RENDERING						
Sika waterproof rendering or other equal; steel trowelled						
20 mm work to walls; three coat; to concrete base						
over 300 mm wide	–	–	–	–	m²	34.56
not exceeding 300 mm wide	–	–	–	–	m²	52.36
25 mm work to walls; three coat; to concrete base						
over 300 mm wide	–	–	–	–	m²	40.84
not exceeding 300 mm wide	–	–	–	–	m²	62.83
40 mm work to walls; four coat; to concrete base						
over 300 mm wide	–	–	–	–	m²	60.22
not exceeding 300 mm wide	–	–	–	–	m²	94.26
Sto External render only system; comprising glassfibre mesh reinforcement embedded in 10 mm Sto Levell Cote with Sto Armat Classic Basecoat Render and Stolit K 1.5 Decorative Topcoat Render (white)						
15 mm thick work to walls; two coats; to brickwork or blockwork base						
over 300 mm wide	–	–	–	–	m²	41.72
extra over for;						
bellcast bead	–	–	–	–	m	3.97
external angle with PVC mesh angle bead	–	–	–	–	m	3.66
internal angle with Sto Armor angle	–	–	–	–	m	3.66
render stop bead	–	–	–	–	m	3.66

19 WATERPROOFING

Item	PC £	Labour hours	Labour £	Material £	Unit	Total rate £
K-Rend render or similar through-colour render system						
18 mm thick work to walls; two coats; to brickwork or blockwork base; first coat 8 mm standard base coat; second coat 10 mm K-rend silicone WP/FT						
over 300 mm wide	–	–	–	–	m²	52.91
MASTIC ASPHALT TANKING AND DAMP-PROOF MEMBRANES						
Mastic asphalt to BS 6925 Type T 1097						
13 mm thick one coat coverings to concrete base; flat; subsequently covered						
over 300 mm wide	–	–	–	–	m²	11.91
225 mm–300 mm wide	–	–	–	–	m²	34.20
150 mm–225 mm wide	–	–	–	–	m²	37.49
not exceeding 150 mm wide	–	–	–	–	m²	46.84
20 mm thick two coat coverings to concrete base; flat; subsequently covered						
over 300 mm wide	–	–	–	–	m²	14.98
225 mm–300 mm wide	–	–	–	–	m²	30.87
150 mm–225 mm wide	–	–	–	–	m²	43.18
not exceeding 150 mm wide	–	–	–	–	m²	50.45
30 mm thick three coat coverings to concrete base; flat; subsequently covered						
over 300 mm wide	–	–	–	–	m²	24.04
225 mm–300 mm wide	–	–	–	–	m²	49.55
150 mm–225 mm wide	–	–	–	–	m²	53.76
not exceeding 150 mm wide	–	–	–	–	m²	65.50
13 mm thick two coat coverings to brickwork base; vertical; subsequently covered						
over 300 mm wide	–	–	–	–	m²	33.07
225 mm–300 mm wide	–	–	–	–	m²	47.56
150 mm–225 mm wide	–	–	–	–	m²	51.35
not exceeding 150 mm wide	–	–	–	–	m²	67.10
20 mm thick three coat coverings to brickwork base; vertical; subsequently covered						
over 300 mm wide	–	–	–	–	m²	53.51
225 mm–300 mm wide	–	–	–	–	m²	64.06
150 mm–225 mm wide	–	–	–	–	m²	70.29
not exceeding 150 mm wide	–	–	–	–	m²	91.14
Turning into groove 20 mm deep	–	–	–	–	m	0.63
Internal angle fillets; subsequently covered	–	–	–	–	m	3.72

19 WATERPROOFING

Item	PC £	Labour hours	Labour £	Material £	Unit	Total rate £
FLEXIBLE SHEET TANKING AND DAMP-PROOF MEMBRANES						
Sheet tanking						
Preprufe pre-applied self adhesive waterproofing membranes for use below concrete slabs or behind concrete walls. HPE film with a pressure sensitive adhesive and weather resistant protective coating						
Preprufe 300R heavy duty grade for use below slabs and on rafts						
over 300 mm wide; horizontal	–	0.10	1.37	13.19	m²	**14.56**
not exceeding 300 mm wide; horizontal	–	0.11	1.51	5.52	m	**7.03**
Preprufe 160R thinner grade for use blindside, zero property line applications against soil retention						
over 300 mm wide; horizontal	–	0.10	1.37	11.59	m²	**12.96**
not exceeding 300 mm wide; horizontal	–	0.11	1.51	4.86	m	**6.37**
Preprufe 800PA reinforced cross laminated HDPE film for use in below ground car parks, basements, underground reservoirs and tanks						
over 300 mm wide; horizontal	–	0.15	2.05	7.67	m²	**9.72**
not exceeding 300 mm wide; horizontal	–	0.17	2.26	3.22	m	**5.48**
Visqueen self-adhesive damp-proof membrane						
over 300 mm wide; horizontal	–	–	–	–	m²	**5.54**
not exceeding 300 mm wide; horizontal	–	–	–	–	m	**2.13**
Tanking primer for self-adhesive dpm						
over 300 mm wide; horizontal	–	–	–	–	m²	**3.83**
not exceeding 300 mm wide; horizontal	–	–	–	–	m	**1.75**
Bituthene sheeting or other equal and approved; lapped joints; horizontal on slabs						
8000 grade	–	0.10	1.37	6.18	m²	**7.55**
5000HD heavy duty grade	–	0.12	1.64	5.65	m²	**7.29**
Bituthene sheeting or other equal and approved; lapped joints; dressed up vertical face of concrete						
8000 grade	–	0.17	2.33	6.18	m²	**8.51**
RIW Structureseal tanking and damp-proof membrane; or other equal and approved						
over 300 mm wide; horizontal	–	–	–	–	m²	**6.58**
Structureseal Fillet						
40 mm × 40 mm	–	–	–	–	m	**4.97**
Ruberoid Plasfrufe 2000SA self-adhesive damp-proof membrane						
over 300 mm wide; horizontal	–	–	–	–	m²	**10.54**
not exceeding 300 mm wide; horizontal	–	–	–	–	m	**4.04**
extra for 50 mm thick sand blinding	–	–	–	–	m²	**1.85**
Servi-pak protection board or other equal and approved; butt jointed; taped joints; to horizontal surfaces;						
3 mm thick	–	0.14	1.92	8.17	m²	**10.09**
6 mm thick	–	0.14	1.92	9.34	m²	**11.26**
12 mm thick	–	0.19	2.60	15.99	m²	**18.59**

19 WATERPROOFING

Item	PC £	Labour hours	Labour £	Material £	Unit	Total rate £
Servi-pak protection board or other equal and approved; butt jointed; taped joints; to vertical surfaces						
3 mm thick	–	0.19	2.60	8.17	m²	**10.77**
6 mm thick	–	0.19	2.60	9.34	m²	**11.94**
12 mm thick	–	0.23	3.15	15.99	m²	**19.14**
Bituthene reinforcing strip or other equal and approved; 70 mm wide						
Bitutape 4000	–	0.09	1.23	0.49	m	**1.72**
Expandite Famflex hot bitumen bonded waterproof tanking or other equal and approved; 150 mm laps						
horizontal; over 300 mm wide	–	0.37	5.06	12.98	m²	**18.04**
vertical; over 300 mm wide	–	0.60	8.21	12.98	m²	**21.19**

20 PROPRIETARY LININGS AND PARTITIONS

Item	PC £	Labour hours	Labour £	Material £	Unit	Total rate £
PLASTERBOARD DRY LINING WALLS AND PARTITIONS						
SUPPLY ONLY SHEET LINING – ALTERNATIVE MATERIAL PRICES						
Fibreboard; 19 mm Decorative faced						
Ash	–	–	–	10.83	m²	**10.83**
Beech	–	–	–	10.35	m²	**10.35**
Oak	–	–	–	10.66	m²	**10.66**
Edgings; self adhesive						
22 mm Ash	–	–	–	0.31	m	**0.31**
22 mm Beech	–	–	–	0.31	m	**0.31**
22 mm Oak	–	–	–	0.31	m	**0.31**
Chipboard Standard Grade						
12 mm	–	–	–	2.04	m²	**2.04**
18 mm	–	–	–	2.88	m²	**2.88**
22 mm	–	–	–	3.53	m²	**3.53**
25 mm	–	–	–	4.04	m²	**4.04**
Chipboard; melamine faced						
15 mm	–	–	–	3.00	m²	**3.00**
18 mm	–	–	–	3.27	m²	**3.27**
Medium density fibreboard; external quality						
6 mm	–	–	–	4.25	m²	**4.25**
9 mm	–	–	–	5.64	m²	**5.64**
19 mm	–	–	–	9.15	m²	**9.15**
25 mm	–	–	–	12.77	m²	**12.77**
Wallboard plank						
9.5 mm	–	–	–	1.67	m²	**1.67**
12.5 mm	–	–	–	1.67	m²	**1.67**
15 mm	–	–	–	2.00	m²	**2.00**
Moisture-resistant board						
12.5 mm	–	–	–	2.73	m²	**2.73**
15 mm	–	–	–	3.28	m²	**3.28**
Fireline board						
12.5 mm	–	–	–	2.13	m²	**2.13**
15 mm	–	–	–	2.56	m²	**2.56**
SUPPLY AND FIX PRICES						
Dry wall linings						
Linings; Gyproc GypLyner IWL independent walling system or other equal; comprising 48 mm wide metal I stud frame; 50 mmm wide metal C stud floor and head channels; plugged and screwed to concrete						
62.5 mm partition; outer skin of 12.50 mm thick tapered edge wallboard one side; joints filled with joint filler and joint tape to receive direct decoration						
average height 2.00 m	–	0.53	9.45	5.39	m²	**14.84**
average height 3.00 m	–	0.53	9.45	5.31	m²	**14.76**
average height 4.00 m	–	0.53	9.45	5.29	m²	**14.74**

20 PROPRIETARY LININGS AND PARTITIONS

Item	PC £	Labour hours	Labour £	Material £	Unit	Total rate £
Rate per m (SMM7 measurement rule)						
height 2.10 m–2.40 m	–	1.30	23.18	12.18	m	**35.36**
height 2.40 m–2.70 m	–	1.45	25.85	14.58	m	**40.43**
height 2.70 m–3.00 m	–	1.60	28.53	15.92	m	**44.45**
height 3.00 m–3.30 m	–	1.75	31.20	17.31	m	**48.51**
height 3.30 m–3.60 m	–	1.90	33.88	18.74	m	**52.62**
height 3.60 m–3.90 m	–	2.05	36.54	20.16	m	**56.70**
height 3.90 m–4.20 m	–	2.20	39.22	21.60	m	**60.82**
62.5 mm partition; outer skin of 15.00 mm thick tapered edge wallboard one side; joints filled with joint filler and joint tape to receive direct decoration						
average height 2.00 m	–	0.53	9.45	5.97	m²	**15.42**
average height 3.00 m	–	0.53	9.45	5.78	m²	**15.23**
average height 4.00 m	–	0.53	9.45	5.56	m²	**15.01**
Rate per m (SMM7 measurement rule)						
height 2.10 m–2.40 m	–	1.30	23.18	13.05	m	**36.23**
height 2.40 m–2.70 m	–	1.45	25.85	15.55	m	**41.40**
height 2.70 m–3.00 m	–	1.60	28.53	17.00	m	**45.53**
height 3.00 m–3.30 m	–	1.75	31.20	18.50	m	**49.70**
height 3.30 m–3.60 m	–	1.90	33.88	20.04	m	**53.92**
height 3.60 m–3.90 m	–	2.05	36.54	21.58	m	**58.12**
height 3.90 m–4.20 m	–	2.20	39.22	23.11	m	**62.33**
62.5 mm partition; outer skin of 12.50 mm thick tapered edge Fireline board one side; joints filled with joint filler and joint tape to receive direct decoration						
average height 2.00 m	–	0.53	9.45	5.99	m²	**15.44**
average height 3.00 m	–	0.53	9.45	5.71	m²	**15.16**
average height 4.00 m	–	0.53	9.45	5.60	m²	**15.05**
Rate per m (SMM7 measurement rule)						
height 2.10 m–2.40 m	–	1.30	23.41	13.34	m	**36.75**
height 2.40 m–2.70 m	–	1.45	26.09	15.88	m	**41.97**
height 2.70 m–3.00 m	–	1.60	28.76	17.36	m	**46.12**
height 3.00 m–3.30 m	–	1.75	31.44	18.90	m	**50.34**
height 3.30 m–3.60 m	–	1.90	34.11	20.47	m	**54.58**
height 3.60 m–3.90 m	–	2.05	36.79	22.05	m	**58.84**
height 3.90 m–4.20 m	–	2.20	39.45	23.63	m	**63.08**
62.5 mm partition; outer skin of 15.00 mm thick tapered edge Fireline board one side; joints filled with joint filler and joint tape to receive direct decoration						
average height 2.00 m	–	0.53	9.45	6.45	m²	**15.90**
average height 3.00 m	–	0.53	9.45	6.17	m²	**15.62**
average height 4.00 m	–	0.53	9.45	6.06	m²	**15.51**
Rate per m (SMM7 measurement rule)						
height 2.10 m–2.40 m	–	1.30	23.41	14.43	m	**37.84**
height 2.40 m–2.70 m	–	1.45	26.09	17.12	m	**43.21**
height 2.70 m–3.00 m	–	1.60	28.76	18.75	m	**47.51**
height 3.00 m–3.30 m	–	1.75	31.44	20.42	m	**51.86**
height 3.30 m–3.60 m	–	1.90	34.11	22.13	m	**56.24**
height 3.60 m–3.90 m	–	2.05	36.79	23.84	m	**60.63**
height 3.90 m–4.20 m	–	2.20	39.45	25.55	m	**65.00**

20 PROPRIETARY LININGS AND PARTITIONS

Item	PC £	Labour hours	Labour £	Material £	Unit	Total rate £
PLASTERBOARD DRY LINING WALLS AND PARTITIONS – cont						
Linings – cont						
62.5 mm partition; outer skin of 12.50 mm thick tapered edge wallboard one side; filling cavity with 50 mm Isover insulation; wallboard joints filled with joint filler and joint tape to receive direct decoration						
average height 2.00 m	–	0.65	11.58	6.81	m²	**18.39**
average height 3.00 m	–	0.65	11.58	6.53	m²	**18.11**
average height 4.00 m	–	0.65	11.58	6.47	m²	**18.05**
Rate per m (SMM7 measurement rule)						
height 2.10 m–2.40 m	–	1.42	25.55	15.36	m	**40.91**
height 2.40 m–2.70 m	–	1.60	28.76	18.15	m	**46.91**
height 2.70 m–3.00 m	–	1.65	29.42	19.91	m	**49.33**
height 3.00 m–3.30 m	–	1.92	34.38	21.69	m	**56.07**
height 3.30 m–3.60 m	–	2.08	37.32	23.51	m	**60.83**
height 3.60 m–3.90 m	–	2.25	40.26	25.34	m	**65.60**
height 3.90 m–4.20 m	–	2.41	43.20	27.17	m	**70.37**
62.5 mm partition; outer skin of 15.00 mm thick tapered edge wallboard one side; filling cavity with 50 mm Isover slabs; wallboard joints filled with joint filler and joint tape to receive direct decoration						
average height 2.00 m	–	0.65	11.58	7.17	m²	**18.75**
average height 3.00 m	–	0.65	11.58	6.90	m²	**18.48**
average height 4.00 m	–	0.65	11.58	6.83	m²	**18.41**
Rate per m (SMM7 measurement rule)						
height 2.10 m–2.40 m	–	1.42	25.55	16.24	m	**41.79**
height 2.40 m–2.70 m	–	1.60	28.76	19.14	m	**47.90**
height 2.70 m–3.00 m	–	1.65	29.42	20.98	m	**50.40**
height 3.00 m–3.30 m	–	1.92	34.38	22.88	m	**57.26**
height 3.30 m–3.60 m	–	2.08	37.32	24.82	m	**62.14**
height 3.60 m–3.90 m	–	2.25	40.26	26.75	m	**67.01**
height 3.90 m–4.20 m	–	2.41	43.20	28.69	m	**71.89**
Gypsum plasterboard to BS EN 520; fixing on dabs or with nails; joints filled with joint filler and joint tape to receive direct decoration; to softwood base (measured elsewhere)						
Plain grade tapered edge wallboard						
9.50 mm board to walls						
average height 2.00 m	–	0.16	2.85	2.50	m²	**5.35**
average height 3.00 m	–	0.16	2.77	2.50	m²	**5.27**
average height 4.00 m	–	0.15	2.68	2.50	m²	**5.18**
Rate per m (SMM7 measurement rule)						
wall height 2.40 m–2.70 m	–	0.43	7.90	6.74	m	**14.64**
wall height 2.70 m–3.00 m	–	0.46	8.44	7.50	m	**15.94**
wall height 3.00 m–3.30 m	–	0.50	9.15	8.25	m	**17.40**
wall height 3.30 m–3.60 m	–	0.59	10.75	9.06	m	**19.81**

20 PROPRIETARY LININGS AND PARTITIONS

Item	PC £	Labour hours	Labour £	Material £	Unit	Total rate £
9.50 mm board to reveals and soffits of openings and recesses						
not exceeding 300 mm wide	–	0.14	2.61	1.44	m	**4.05**
300 mm–600 mm wide	–	0.22	4.04	2.01	m	**6.05**
9.50 mm board to faces of columns – 4 nr faces						
not exceeding 300 mm total girth	–	0.20	3.57	2.33	m	**5.90**
300 mm–600 mm total girth	–	0.20	3.57	2.90	m	**6.47**
600 mm–900 mm total girth	–	0.20	3.57	3.48	m	**7.05**
900 mm–1200 mm total girth	–	0.20	3.57	4.07	m	**7.64**
12.50 mm board to walls						
average height 2.00 m	–	0.16	2.85	3.72	m²	**6.57**
average height 3.00 m	–	0.16	2.77	3.72	m²	**6.49**
average height 4.00 m	–	0.15	2.68	3.72	m²	**6.40**
Rate per m (SMM7 measurement rule)						
wall height 2.40 m–2.70 m	–	0.45	8.03	10.18	m	**18.21**
wall height 2.70 m–3.00 m	–	0.50	8.92	11.31	m	**20.23**
wall height 3.00 m–3.30 m	–	0.55	9.81	12.53	m	**22.34**
wall height 3.30 m–3.60 m	–	0.60	10.70	13.62	m	**24.32**
12.50 mm board to reveals and soffits of openings and recesses						
not exceeding 300 mm wide	–	0.19	3.63	1.85	m	**5.48**
300 mm–600 mm wide	–	0.37	7.07	2.78	m	**9.85**
12.50 mm board to faces of columns – 4 nr faces						
not exceeding 300 mm total girth	–	0.30	5.35	2.69	m	**8.04**
300 mm–600 mm total girth	–	0.30	5.35	3.72	m	**9.07**
600 mm–900 mm total girth	–	0.30	5.35	4.70	m	**10.05**
900 mm–1200 mm total girth	–	0.30	5.35	5.68	m	**11.03**
Tapered edge wallboard TEN						
12.50 mm board to walls						
average height 2.00 m	–	0.16	2.85	2.81	m²	**5.66**
average height 3.00 m	–	0.16	2.77	2.81	m²	**5.58**
average height 4.00 m	–	0.15	2.68	2.81	m²	**5.49**
Rate per m (SMM7 measurement rule)						
wall height 2.40 m–2.70 m	–	0.47	8.62	7.71	m	**16.33**
wall height 2.70 m–3.00 m	–	0.56	10.34	8.57	m	**18.91**
wall height 3.00 m–3.30 m	–	0.65	12.06	9.44	m	**21.50**
wall height 3.30 m–3.60 m	–	0.78	14.50	10.34	m	**24.84**
12.50 mm board to reveals and soffits of openings and recesses						
not exceeding 300 mm wide	–	0.19	3.63	1.57	m	**5.20**
300 mm–600 mm wide	–	0.37	7.07	2.23	m	**9.30**
12.50 mm board to faces of columns – 4 nr faces						
not exceeding 300 mm total girth	–	0.25	5.05	3.37	m	**8.42**
300 mm–600 mm total girth	–	0.20	3.57	4.02	m	**7.59**
600 mm–900 mm total girth	–	0.20	3.57	4.77	m	**8.34**
900 mm–1200 mm total girth	–	0.20	3.57	5.61	m	**9.18**

20 PROPRIETARY LININGS AND PARTITIONS

Item	PC £	Labour hours	Labour £	Material £	Unit	Total rate £
PLASTERBOARD DRY LINING WALLS AND PARTITIONS – cont						
Tapered edge plank						
19 mm plank to walls						
average height 2.00 m	–	0.16	2.85	4.47	m²	**7.32**
average height 3.00 m	–	0.16	2.77	4.47	m²	**7.24**
average height 4.00 m	–	0.15	2.68	4.47	m²	**7.15**
Rate per m (SMM7 measurement rule)						
wall height 2.40 m–2.70 m	–	1.02	19.61	12.21	m	**31.82**
wall height 2.70 m–3.00 m	–	1.20	23.05	13.57	m	**36.62**
wall height 3.00 m–3.30 m	–	1.30	25.07	14.93	m	**40.00**
wall height 3.30 m–3.60 m	–	1.53	29.42	16.34	m	**45.76**
19 mm plank to reveals and soffits of openings and recesses						
not exceeding 300 mm wide	–	0.20	3.80	2.07	m	**5.87**
300 mm–600 mm wide	–	0.42	7.96	3.23	m	**11.19**
19 mm plank to faces of columns – 4 nr faces						
not exceeding 300 mm total girth	–	0.80	16.16	3.77	m	**19.93**
300 mm–600 mm total girth	–	0.80	14.26	5.11	m	**19.37**
600 mm–900 mm total girth	–	0.80	14.26	6.46	m	**20.72**
900 mm–1200 mm total girth	–	0.80	14.26	7.70	m	**21.96**
ThermaLine Plus board (0.19 W/mK)						
27 mm board to walls						
average height 2.00 m	–	0.16	2.85	7.06	m²	**9.91**
average height 3.00 m	–	0.16	2.77	7.06	m²	**9.83**
average height 4.00 m	–	0.15	2.68	7.06	m²	**9.74**
Rate per m (SMM7 measurement rule)						
wall height 2.40 m–2.70 m	–	1.06	18.90	19.21	m	**38.11**
wall height 2.70 m–3.00 m	–	1.23	21.92	21.34	m	**43.26**
wall height 3.00 m–3.30 m	–	1.34	23.89	23.48	m	**47.37**
wall height 3.30 m–3.60 m	–	1.62	28.88	25.67	m	**54.55**
27 mm board to reveals and soffits of openings and recesses						
not exceeding 300 mm wide	–	0.21	3.74	2.85	m	**6.59**
300 mm–600 mm wide	–	0.43	7.67	4.79	m	**12.46**
27 mm board to faces of columns – 4 nr faces						
not exceeding 300 mm total girth	–	0.52	9.27	5.73	m	**15.00**
300 mm–600 mm total girth	–	0.65	11.58	5.73	m	**17.31**
600 mm–900 mm total girth	–	0.75	13.37	9.64	m	**23.01**
900 mm–1200 mm total girth	–	1.00	17.82	9.70	m	**27.52**
48 mm board to walls						
average height 2.00 m	–	0.16	2.85	9.51	m²	**12.36**
average height 3.00 m	–	0.16	2.77	9.51	m²	**12.28**
average height 4.00 m	–	0.15	2.68	9.51	m²	**12.19**
Rate per m (SMM7 measurement rule)						
wall height 2.40 m–2.70 m	–	1.06	18.90	25.93	m	**44.83**
wall height 2.70 m–3.00 m	–	1.30	23.18	28.83	m	**52.01**
wall height 3.00 m–3.30 m	–	1.43	25.49	31.72	m	**57.21**
wall height 3.30 m–3.60 m	–	1.71	30.48	34.66	m	**65.14**

20 PROPRIETARY LININGS AND PARTITIONS

Item	PC £	Labour hours	Labour £	Material £	Unit	Total rate £
48 mm board to reveals and soffits of openings and recesses						
not exceeding 300 mm wide	–	0.23	4.10	3.60	m	7.70
300 mm–600 mm wide	–	0.46	8.20	6.29	m	14.49
48 mm board to faces of columns – 4 nr faces						
not exceeding 300 mm total girth	–	0.52	9.27	7.20	m	16.47
300 mm–600 mm total girth	–	0.65	11.58	7.20	m	18.78
600 mm–900 mm total girth	–	0.75	13.37	12.58	m	25.95
900 mm–1200 mm total girth	–	1.00	17.82	12.63	m	30.45
ThermaLine Super boards						
50 mm board to walls						
average height 2.00 m	–	0.40	7.13	12.95	m²	20.08
average height 3.00 m	–	0.43	7.67	12.95	m²	20.62
average height 4.00 m	–	0.47	8.47	12.95	m²	21.42
Rate per m (SMM7 measurement rule)						
wall height 2.40 m–2.70 m	–	1.06	18.90	34.92	m	53.82
wall height 2.70 m–3.00 m	–	1.30	23.18	38.83	m	62.01
wall height 3.00 m–3.30 m	–	1.43	25.49	42.71	m	68.20
wall height 3.30 m–3.60 m	–	1.71	30.48	46.65	m	77.13
50 mm board to reveals and soffits of openings and recesses						
not exceeding 300 mm wide	–	0.52	9.27	4.60	m	13.87
300 mm–600 mm wide	–	0.46	8.20	8.29	m	16.49
50 mm board to faces of columns – 4 nr faces						
not exceeding 300 mm total girth	–	0.52	9.27	5.62	m	14.89
300–600 mm total girth	–	0.65	11.58	9.28	m	20.86
600 mm–900 mm total girth	–	0.75	13.37	13.01	m	26.38
900 mm–1200 mm total girth	–	1.00	17.82	16.75	m	34.57
60 mm board to walls						
average height 2.00 m	–	0.40	7.13	15.24	m²	22.37
average height 3.00 m	–	0.43	7.67	15.24	m²	22.91
average height 4.00 m	–	0.47	8.47	15.24	m²	23.71
Rate per m (SMM7 measurement rule)						
wall height 2.40 m–2.70 m	–	1.06	18.90	41.13	m	60.03
wall height 2.70 m–3.00 m	–	1.30	23.18	45.71	m	68.89
wall height 3.00 m–3.30 m	–	1.43	25.49	50.30	m	75.79
wall height 3.30 m–3.60 m	–	1.71	30.48	54.92	m	85.40
60 mm board to reveals and soffits of openings and recesses						
not exceeding 300 mm wide	–	0.23	4.10	5.29	m	9.39
300 mm–600 mm wide	–	0.46	8.20	9.68	m	17.88
60 mm board to faces of columns – 4 nr faces						
not exceeding 300 mm total girth	–	0.52	9.27	6.30	m	15.57
300–600 mm total girth	–	0.65	11.58	10.65	m	22.23
600 mm–900 mm total girth	–	0.75	13.37	15.08	m	28.45
900 mm–1200 mm total girth	–	1.00	17.82	19.51	m	37.33
80 mm board to walls						
average height 2.00 m	–	0.40	7.13	18.49	m²	25.62
average height 3.00 m	–	0.43	7.67	18.49	m²	26.16
average height 4.00 m	–	0.47	8.47	18.49	m²	26.96

20 PROPRIETARY LININGS AND PARTITIONS

Item	PC £	Labour hours	Labour £	Material £	Unit	Total rate £
PLASTERBOARD DRY LINING WALLS AND PARTITIONS – cont						
ThermaLine Super boards – cont						
Rate per m (SMM7 measurement rule)						
wall height 2.40 m–2.70 m	–	1.06	18.90	49.92	m	**68.82**
wall height 2.70 m–3.00 m	–	1.30	23.18	55.48	m	**78.66**
wall height 3.00 m–3.30 m	–	1.43	25.49	61.03	m	**86.52**
wall height 3.30 m–3.60 m	–	1.71	30.48	66.64	m	**97.12**
80 mm board to reveals and soffits of openings and recesses						
not exceeding 300 mm wide	–	0.23	4.10	6.26	m	**10.36**
300 mm–600 mm wide	–	0.46	8.20	11.62	m	**19.82**
80 mm board to faces of columns – 4 nr faces						
not exceeding 300 mm total girth	–	0.52	9.27	7.29	m	**16.56**
300–600 mm total girth	–	0.65	11.58	12.61	m	**24.19**
600 mm–900 mm total girth	–	0.75	13.37	18.01	m	**31.38**
900 mm–1200 mm total girth	–	1.00	17.82	23.41	m	**41.23**
90 mm board to walls						
average height 2.00 m	–	0.40	7.13	21.16	m²	**28.29**
average height 3.00 m	–	0.43	7.67	21.16	m²	**28.83**
average height 4.00 m	–	0.47	8.47	21.16	m²	**29.63**
Rate per m (SMM7 measurement rule)						
wall height 2.40 m–2.70 m	–	1.06	18.90	57.09	m	**75.99**
wall height 2.70 m–3.00 m	–	1.30	23.18	63.46	m	**86.64**
wall height 3.00 m–3.30 m	–	1.43	25.49	69.81	m	**95.30**
wall height 3.30 m–3.60 m	–	1.71	30.48	76.21	m	**106.69**
90 mm board to reveals and soffits of openings and recesses						
not exceeding 300 mm wide	–	0.23	4.10	7.06	m	**11.16**
300 mm–600 mm wide	–	0.46	8.20	13.22	m	**21.42**
90 mm board to faces of columns – 4 nr faces						
not exceeding 300 mm total girth	–	0.52	9.27	8.08	m	**17.35**
300–600 mm total girth	–	0.65	11.58	14.21	m	**25.79**
600 mm–900 mm total girth	–	0.75	13.37	20.40	m	**33.77**
900 mm–1200 mm total girth	–	1.00	17.82	26.60	m	**44.42**
Kingspan Kooltherm K18 insulated plasterboard; fixing with nails; joints filled with joint filler and joint tape to receive direct decoration; to softwood base (measured elsewhere)						
12.5 mm plasterboard bonded to CFC/HCFC free rigid phenolic insulation; overall thickness (0.21 W/m.K)						
32.5 mm thick panel	–	0.23	4.01	7.84	m²	**11.85**
37.5 mm thick panel	–	0.23	4.01	8.05	m²	**12.06**
42.5 mm thick panel	–	0.23	4.01	8.26	m²	**12.27**
52.5 mm thick panel	–	0.25	4.46	8.69	m²	**13.15**
62.5 mm thick panel	–	0.25	4.46	9.79	m²	**14.25**
72.5 mm thick panel	–	0.25	4.46	10.23	m²	**14.69**
82.5 mm thick panel	–	0.25	4.46	11.94	m²	**16.40**
92.5 mm thick panel	–	0.28	4.90	13.64	m²	**18.54**
112.5 mm thick panel	–	0.28	4.90	17.05	m²	**21.95**

20 PROPRIETARY LININGS AND PARTITIONS

Item	PC £	Labour hours	Labour £	Material £	Unit	Total rate £
White plastic faced gypsum plasterboard to BS EN 520; industrial grade square edge wallboard; fixing on dabs or with screws; butt joints; to softwood base						
12.50 mm board to walls						
average height 2.00 m	–	0.21	3.74	4.87	m²	**8.61**
average height 3.00 m	–	0.25	4.46	4.87	m²	**9.33**
average height 4.00 m	–	0.31	5.49	4.87	m²	**10.36**
Rate per m (SMM7 measurement rule)						
wall height 2.40 m–2.70 m	–	0.69	12.30	13.16	m	**25.46**
wall height 2.70 m–3.00 m	–	0.83	14.80	14.62	m	**29.42**
wall height 3.00 m–3.30 m	–	0.97	17.29	16.07	m	**33.36**
wall height 3.30 m–3.60 m	–	1.11	19.79	17.53	m	**37.32**
12.50 mm board to reveals and soffits of openings and recesses						
not exceeding 300 mm wide	–	0.15	2.68	1.48	m	**4.16**
300 mm–600 mm wide	–	0.30	5.35	2.92	m	**8.27**
12.50 mm board to faces of columns – 4 nr						
not exceeding 300 mm total girth	–	0.30	5.35	1.56	m	**6.91**
300–600 mm total girth	–	0.39	6.95	2.62	m	**9.57**
600 mm–900 mm total girth	–	0.60	10.70	4.46	m	**15.16**
900 mm–1200 mm total girth	–	0.75	13.37	5.90	m	**19.27**
Plasterboard jointing system; filling joint with jointing compounds						
To walls and ceilings						
to suit 9.50 mm or 12.50 mm thick boards	–	0.09	1.61	1.25	m	**2.86**
Angle trim; plasterboard edge support system						
To walls and ceilings						
to suit 9.50 mm or 12.50 mm thick boards	–	0.09	1.61	1.22	m	**2.83**

20 PROPRIETARY LININGS AND PARTITIONS

Item	PC £	Labour hours	Labour £	Material £	Unit	Total rate £
PLASTERBOARD DRY LINING WALLS AND PARTITIONS – cont						
Gyproc SoundBloc tapered edge plasterboard with higher density core; fixing on dabs or with nails; joints filled with joint filler and joint tape to receive direct decoration; to softwood base						
12.50 mm board to walls						
average height 2.00 m	–	0.36	6.42	4.17	m²	**10.59**
average height 3.00 m	–	0.38	6.78	4.17	m²	**10.95**
average height 4.00 m	–	0.40	7.13	4.23	m²	**11.36**
Rate per m (SMM7 measurement rule)						
wall height 2.40 m–2.70 m	–	0.97	17.29	11.25	m	**28.54**
wall height 2.70 m–3.00 m	–	1.11	19.79	12.52	m	**32.31**
wall height 3.00 m–3.30 m	–	1.32	23.53	13.77	m	**37.30**
wall height 3.30 m–3.60 m	–	1.43	25.49	15.08	m	**40.57**
12.50 mm board to ceilings						
over 300 mm wide	–	0.41	7.31	4.17	m²	**11.48**
15.00 mm board to walls						
average height 2.00 m	–	0.36	6.42	4.89	m²	**11.31**
average height 3.00 m	–	0.38	6.78	4.89	m²	**11.67**
average height 4.00 m	–	0.40	7.13	4.95	m²	**12.08**
Rate per m (SMM7 measurement rule)						
wall height 2.40 m–2.70 m	–	1.00	17.82	13.33	m	**31.15**
wall height 2.70 m–3.00 m	–	1.14	20.33	14.81	m	**35.14**
wall height 3.00 m–3.30 m	–	1.27	22.64	16.30	m	**38.94**
wall height 3.30 m–3.60 m	–	1.46	26.02	17.83	m	**43.85**
15.00 mm board to reveals and soffits of openings and recesses						
not exceeding 300 mm wide	–	0.20	3.57	2.19	m	**5.76**
300 mm–600 mm wide	–	0.38	6.78	3.48	m	**10.26**
15.00 mm board to ceilings						
over 300 mm wide	–	0.43	7.67	4.94	m²	**12.61**
Two layers of gypsum plasterboard to BS 1230; plain grade square and tapered edge wallboard; fixing on dabs or with nails; joints filled with joint filler and joint tape; top layer to receive direct decoration; to softwood base						
19 mm two layer board to walls						
average height 2.00 m	–	0.48	8.56	4.44	m²	**13.00**
average height 3.00 m	–	0.51	9.09	4.44	m²	**13.53**
average height 4.00 m	–	0.54	9.62	4.51	m²	**14.13**
Rate per m (SMM7 measurement rule)						
wall height 2.40 m–2.70 m	–	1.30	23.18	11.97	m	**35.15**
wall height 2.70 m–3.00 m	–	1.48	26.38	13.31	m	**39.69**
wall height 3.00 m–3.30 m	–	1.67	29.78	14.66	m	**44.44**
wall height 3.30 m–3.60 m	–	1.94	34.58	16.04	m	**50.62**
19 mm two layer board to reveals and soffits of openings and recesses						
not exceeding 300 mm wide	–	0.28	4.99	2.07	m	**7.06**
300 mm–600 mm wide	–	0.56	9.98	3.20	m	**13.18**

20 PROPRIETARY LININGS AND PARTITIONS

Item	PC £	Labour hours	Labour £	Material £	Unit	Total rate £
19 mm two layer board to faces of columns – 4 nr						
not exceeding 300 mm total girth	–	0.50	8.92	4.61	m	**13.53**
300–600 mm total girth	–	0.69	12.30	4.22	m	**16.52**
600 mm–900 mm total girth	–	0.75	13.37	5.31	m	**18.68**
900 mm–1200 mm total girth	–	1.00	17.82	7.66	m	**25.48**
25 mm two layer board to walls						
average height 2.00 m	–	0.48	8.56	5.66	m²	**14.22**
average height 3.00 m	–	0.51	9.09	5.66	m²	**14.75**
average height 4.00 m	–	0.54	9.62	5.73	m²	**15.35**
Rate per m (SMM7 measurement rule)						
wall height 2.40 m–2.70 m	–	1.39	24.78	15.41	m	**40.19**
wall height 2.70 m–3.00 m	–	1.57	27.99	17.13	m	**45.12**
wall height 3.00 m–3.30 m	–	1.76	31.38	18.85	m	**50.23**
wall height 3.30 m–3.60 m	–	2.04	36.37	20.61	m	**56.98**
25 mm two layer board to reveals and soffits of openings and recesses						
not exceeding 300 mm wide	–	0.28	4.99	2.48	m	**7.47**
300 mm–600 mm wide	–	0.56	9.98	3.98	m	**13.96**
25 mm two layer board to faces of columns – 4 nr						
not exceeding 300 mm total girth	–	0.35	6.24	3.34	m	**9.58**
300–600 mm total girth	–	0.69	12.30	5.03	m	**17.33**
600 mm–900 mm total girth	–	1.00	17.82	6.46	m	**24.28**
900 mm–1200 mm total girth	–	1.25	22.32	7.72	m	**30.04**
Gyproc Dri-Wall dry lining system or other equal or approved; plain grade tapered edge wallboard; fixed to walls with adhesive; joints filled with joint filler and joint tape; to receive direct decoration						
9.50 mm board to walls						
average height 2.00 m	–	0.41	7.33	3.43	m²	**10.76**
average height 3.00 m	–	0.43	7.72	3.43	m²	**11.15**
average height 4.00 m	–	0.51	9.02	3.43	m²	**12.45**
Rate per m (SMM7 measurement rule)						
wall height 2.40 m–2.70 m	–	1.11	19.79	9.27	m	**29.06**
wall height 2.70 m–3.00 m	–	1.28	22.82	10.28	m	**33.10**
wall height 3.00 m–3.30 m	–	1.43	25.49	11.30	m	**36.79**
wall height 3.30 m–3.60 m	–	1.67	29.78	12.36	m	**42.14**
9.50 mm board to reveals and soffits of openings and recesses						
not exceeding 300 mm wide	–	0.23	4.34	1.67	m	**6.01**
300 mm–600 mm wide	–	0.46	8.67	2.52	m	**11.19**
9.50 mm board to faces of columns – 4 nr faces						
300 mm total girth	–	0.45	8.03	1.71	m	**9.74**
300–600 mm total girth	–	0.58	10.34	3.34	m	**13.68**
600 mm–900 mm total girth	–	0.75	13.37	3.93	m	**17.30**
900 mm–1200 mm total girth	–	0.95	16.93	4.48	m	**21.41**
Angle; with joint tape bedded and covered with Jointex or other equal						
internal	–	0.05	1.01	0.41	m	**1.42**
external	–	0.11	2.22	0.41	m	**2.63**

20 PROPRIETARY LININGS AND PARTITIONS

Item	PC £	Labour hours	Labour £	Material £	Unit	Total rate £
PLASTERBOARD DRY LINING WALLS AND PARTITIONS – cont						
Gyproc Dri-Wall M/F dry lining system or other equal; mild steel furrings fixed to walls with adhesive; tapered edge wallboard screwed to furrings; joints filled with joint filler and joint tape						
9.50 mm board to walls						
average height 2.00 m	–	0.55	9.81	5.45	m²	**15.26**
average height 3.00 m	–	0.57	10.25	5.45	m²	**15.70**
average height 4.00 m	–	0.62	11.00	5.47	m²	**16.47**
Rate per m (SMM7 measurement rule)						
wall height 2.40 m–2.70 m	–	1.48	26.38	14.73	m	**41.11**
wall height 2.70 m–3.00 m	–	1.69	31.80	16.36	m	**48.16**
wall height 3.00 m–3.30 m	–	1.90	33.88	18.00	m	**51.88**
wall height 3.30 m–3.60 m	–	2.22	39.58	19.69	m	**59.27**
9.50 mm board to reveals and soffits of openings and recesses						
not exceeding 300 mm wide	–	0.23	4.34	1.45	m	**5.79**
300 mm–600 mm wide	–	0.46	8.67	2.08	m	**10.75**
12.50 mm board to walls						
average height 2.00 m	–	0.55	9.81	6.67	m²	**16.48**
average height 3.00 m	–	0.57	10.25	6.67	m²	**16.92**
average height 4.00 m	–	0.62	11.00	6.69	m²	**17.69**
Rate per m (SMM7 measurement rule)						
wall height 2.40 m–2.70 m	–	1.48	26.38	18.03	m	**44.41**
wall height 2.70 m–3.00 m	–	1.69	31.80	20.02	m	**51.82**
wall height 3.00 m–3.30 m	–	1.90	33.88	22.03	m	**55.91**
wall height 3.30 m–3.60 m	–	2.22	39.58	24.08	m	**63.66**
12.50 mm board to reveals and soffits of openings and recesses						
not exceeding 300 mm wide	–	0.23	4.34	1.81	m	**6.15**
300 mm–600 mm wide	–	0.46	8.67	2.81	m	**11.48**
Lafarge plasterboard to BS 1230; fixing on dabs or with screws; joints filled with joint filler and joint tape to receive direct decoration; to softwood						
Megadeco wallboard						
12.50 mm board to walls						
average height 2.00 m	–	0.36	6.42	3.43	m²	**9.85**
average height 3.00 m	–	0.38	6.75	3.43	m²	**10.18**
average height 4.00 m	–	0.40	7.13	3.43	m²	**10.56**
Rate per m (SMM7 measurement rule)						
wall height 2.40 m–2.70 m	–	0.97	17.29	9.28	m	**26.57**
wall height 2.70 m–3.00 m	–	1.11	21.45	10.31	m	**31.76**
wall height 3.00 m–3.30 m	–	1.25	22.28	11.35	m	**33.63**
wall height 3.30 m–3.60 m	–	1.43	25.49	12.37	m	**37.86**
12.50 mm board to ceilings						
over 300 mm wide	–	0.41	7.78	3.43	m²	**11.21**

20 PROPRIETARY LININGS AND PARTITIONS

Item	PC £	Labour hours	Labour £	Material £	Unit	Total rate £
Gypsum cladding; Glasroc Firecase S board or other equal; fixed with adhesive; joints pointed in adhesive						
25 mm thick column linings, faces = 4; 2 hour fire protection rating						
not exceeding 300 mm girth	–	0.30	5.35	11.85	m	**17.20**
300 mm–600 mm girth	–	0.30	5.35	16.12	m	**21.47**
600 mm–900 mm girth	–	0.45	8.03	20.40	m	**28.43**
900 mm–1200 mm girth	–	0.55	9.81	24.66	m	**34.47**
1200 mm–1500 mm girth	–	0.60	10.70	28.94	m	**39.64**
30 mm thick beam linings, faces = 3; 2 hour fire protection rating						
not exceeding 300 mm girth	–	0.60	10.70	10.74	m	**21.44**
300 mm–600 mm girth	–	0.60	10.70	15.57	m	**26.27**
600 mm–900 mm girth	–	0.75	13.37	20.40	m	**33.77**
900 mm–1200 mm girth	–	0.90	16.04	25.21	m	**41.25**
1200 mm–1500 mm girth	–	1.20	21.39	30.04	m	**51.43**
Vermiculite gypsum cladding; Vermiculux board or other equal approved; fixed with adhesive; joints pointed in adhesive						
25 mm thick column linings, faces = 4; 2 hour fire protection rating						
not exceeding 300 mm girth	–	0.20	3.57	8.44	m	**12.01**
300 mm–600 mm girth	–	0.30	5.35	16.88	m	**22.23**
600 mm–900 mm girth	–	0.45	8.03	25.06	m	**33.09**
900 mm–1200 mm girth	–	0.60	10.70	33.42	m	**44.12**
1200 mm–1500 mm girth	–	0.70	12.47	41.68	m	**54.15**
30 mm thick beam linings, faces = 3; 2 hour fire protection rating						
not exceeding 300 mm girth	–	0.50	8.92	11.10	m	**20.02**
300 mm–600 mm girth	–	0.60	10.70	22.20	m	**32.90**
600 mm–900 mm girth	–	0.90	16.04	35.40	m	**51.44**
900 mm–1200 mm girth	–	1.00	17.82	43.98	m	**61.80**
1200 mm–1500 mm girth	–	1.20	21.39	54.91	m	**76.30**
55 mm thick column linings, faces = 4 ; 4 hour fire protection rating						
not exceeding 300 mm girth	–	0.60	10.70	23.66	m	**34.36**
300 mm–600 mm girth	–	0.75	13.37	47.30	m	**60.67**
600 mm–900 mm girth	–	0.80	14.26	70.69	m	**84.95**
900 mm–1200 mm girth	–	1.00	17.82	94.26	m	**112.08**
1200 mm–1500 mm girth	–	1.02	18.18	117.73	m	**135.91**
60 mm thick beam linings, faces = 3; 4 hour fire protection rating						
not exceeding 300 mm girth	–	0.60	10.70	25.56	m	**36.26**
300 mm–600 mm girth	–	0.75	13.37	51.13	m	**64.50**
600 mm–900 mm girth	–	0.80	14.26	76.42	m	**90.68**
900 mm–1200 mm girth	–	1.00	17.82	101.90	m	**119.72**
1200 mm–1500 mm girth	–	1.02	18.18	127.28	m	**145.46**

20 PROPRIETARY LININGS AND PARTITIONS

Item	PC £	Labour hours	Labour £	Material £	Unit	Total rate £
PLASTERBOARD DRY LINING WALLS AND PARTITIONS – cont						
Vermiculite gypsum cladding – cont						
Add to the above rates for working at height						
for work 3.50 m–5.00 m high	–	–	–	–	%	2.50
for work 5.00 m–6.50 m high	–	–	–	–	%	5.00
for work 6.50 m–8.00 m high	–	–	–	–	%	12.50
for work over 8.00 m high	–	–	–	–	%	17.50
Cutting and fitting around steel joints, angles, trunking, ducting, ventilators, pipes, tubes, etc.						
not exceeding 0.30 m girth	–	0.28	4.99	–	nr	4.99
0.30 m–1 m girth	–	0.37	6.60	–	nr	6.60
1 m–2 m girth	–	0.51	9.09	–	nr	9.09
over 2 m girth	–	0.42	7.48	–	m	7.48
PARTITIONING						
Gyproc metal stud proprietary partitions or other equal; metal Gypframe C studs at 600 mm centres; floor and ceiling channels						
73 mm partition; 48 mm studs and channels; one layer of 12.5 mm Gyproc Wallboard each side; joints filled with joint filler and joint tape to receive direct decoration						
average height 2.00 m	–	0.63	12.63	9.51	m²	22.14
average height 3.00 m	–	0.58	11.79	8.78	m²	20.57
Rate per m (SMM7 measurement rule)						
height 2.10 m–2.40 m	–	1.50	30.31	22.83	m	53.14
height 2.40 m–2.70 m	–	1.69	34.14	24.72	m	58.86
height 2.70 m–3.00 m	–	1.75	35.35	26.34	m	61.69
78 mm partition; 48 mm studs and channels; one layer of 15 mm Gyproc Wallboard each side; joints filled with joint filler and joint tape to receive direct decoration						
average height 2.00 m	–	0.65	13.13	10.27	m²	23.40
average height 3.00 m	–	0.61	12.32	9.54	m²	21.86
Rate per m (SMM7 measurement rule)						
height 2.10 m–2.40 m	–	1.57	31.82	24.64	m	56.46
height 2.40 m–2.70 m	–	1.77	35.76	26.76	m	62.52
height 2.70 m–3.00 m	–	1.99	40.20	31.44	m	71.64
95 mm partition; 70 mm studs and channels; one layer of 12.5 mm Gyproc Wallboard each side; joints filled with joint filler and joint tape to receive direct decoration						
average height 2.00 m	–	0.63	12.63	9.73	m²	22.36
average height 3.00 m	–	0.58	11.79	9.11	m²	20.90
Rate per m (SMM7 measurement rule)						
height 2.10 m–2.40 m	–	1.50	30.31	23.33	m	53.64
height 2.40 m–2.70 m	–	1.69	34.14	25.35	m	59.49
height 2.70 m–3.00 m	–	1.75	35.35	27.34	m	62.69
height 3.00 m–3.30 m	–	2.00	40.41	29.26	m	69.67

20 PROPRIETARY LININGS AND PARTITIONS

Item	PC £	Labour hours	Labour £	Material £	Unit	Total rate £
100 mm partition; 70 mm studs and channels; one layer of 15 mm Gyproc Wallboard each side; joints filled with joint filler and joint tape to receive direct decoration						
average height 2.00 m	–	0.65	13.13	10.47	m²	**23.60**
average height 3.00 m	–	0.61	12.32	10.74	m²	**23.06**
Rate per m (SMM7 measurement rule)						
height 2.10 m–2.40 m	–	1.57	31.82	25.15	m	**56.97**
height 2.40 m–2.70 m	–	1.77	35.76	27.40	m	**63.16**
height 2.70 m–3.00 m	–	1.99	40.20	32.23	m	**72.43**
height 3.00 m–3.30 m	–	2.10	42.42	34.19	m	**76.61**
120 mm partition; 70 mm studs and channels; two layers of 12.5 mm Gyproc Wallboard each side; joints filled with joint filler and joint tape to receive direct decoration						
average height 2.00 m	–	0.73	14.75	13.57	m²	**28.32**
average height 3.00 m	–	0.68	13.81	11.44	m²	**25.25**
Rate per m (SMM7 measurement rule)						
height 2.10 m–2.40 m	–	1.75	35.35	32.56	m	**67.91**
height 2.40 m–2.70 m	–	1.99	40.20	35.73	m	**75.93**
height 2.70 m–3.00 m	–	2.05	41.42	38.87	m	**80.29**
height 3.00 m–3.30 m	–	2.50	50.50	41.95	m	**92.45**
130 mm partition; 70 mm studs and channels; two layers of 15 mm Gyproc Wallboard each side; joints filled with joint filler and joint tape to receive direct decoration						
average height 2.00 m	–	0.80	16.16	15.09	m²	**31.25**
average height 3.00 m	–	0.75	15.15	14.52	m²	**29.67**
Rate per m (SMM7 measurement rule)						
height 2.10 m–2.40 m	–	1.82	36.77	36.20	m	**72.97**
height 2.40 m–2.70 m	–	2.07	41.82	27.63	m	**69.45**
height 2.70 m–3.00 m	–	2.29	46.27	46.03	m	**92.30**
height 3.00 m–3.30 m	–	2.40	48.48	49.38	m	**97.86**
Gypwall metal stud proprietary partitions or other equal; metal GypWall Acoustic C studs at 600 mm centres; deep flange floor and ceiling channels						
95 mm partition; 70 mm studs and channels; one layer of 12.5 mm Gyproc Soundbloc each side; joints filled with joint filler and joint tape to receive direct decoration						
average height 2.00 m	–	0.63	12.63	16.66	m²	**29.29**
average height 3.00 m	–	0.58	11.79	16.08	m²	**27.87**
Rate per m (SMM7 measurement rule)						
height 2.10 m–2.40 m	–	1.50	30.31	39.96	m	**70.27**
height 2.40 m–2.70 m	–	1.69	34.14	43.76	m	**77.90**
height 2.70 m–3.00 m	–	1.75	35.35	48.26	m	**83.61**

20 PROPRIETARY LININGS AND PARTITIONS

Item	PC £	Labour hours	Labour £	Material £	Unit	Total rate £
PLASTERBOARD DRY LINING WALLS AND PARTITIONS – cont						
Gypwall metal stud proprietary partitions or other equal – cont						
100 mm partition; 70 mm studs and channels; one layer of 15 mm Gyproc Soundbloc each side; joints filled with joint filler and joint tape to receive direct decoration						
average height 2.00 m	–	0.63	12.63	18.15	m²	**30.78**
average height 3.00 m	–	0.58	11.79	17.59	m²	**29.38**
Rate per m (SMM7 measurement rule)						
height 2.10 m–2.40 m	–	1.57	31.82	43.56	m	**75.38**
height 2.40 m–2.70 m	–	1.77	35.76	47.81	m	**83.57**
height 2.70 m–3.00 m	–	1.99	40.20	51.75	m	**91.95**
120 mm partition; 70 mm studs and channels; two layers of 12.5 mm Gyproc Soundbloc each side; joints filled with joint filler and joint tape to receive direct decoration						
average height 2.00 m	–	0.80	16.16	24.03	m²	**40.19**
average height 3.00 m	–	0.75	15.15	23.46	m²	**38.61**
Rate per m (SMM7 measurement rule)						
height 2.10 m–2.40 m	–	2.00	40.41	57.96	m	**98.37**
height 2.40 m–2.70 m	–	2.20	44.44	63.90	m	**108.34**
height 2.70 m–3.00 m	–	2.30	46.46	70.64	m	**117.10**
130 mm partition; 70 mm studs and channels; two layers of 15 mm Gyproc Soundbloc each side; joints filled with joint filler and joint tape to receive direct decoration						
average height 2.00 m	–	0.80	16.16	27.02	m²	**43.18**
average height 3.00 m	–	0.75	15.15	26.46	m²	**41.61**
Rate per m (SMM7 measurement rule)						
height 2.10 m–2.40 m	–	2.00	40.41	65.16	m	**105.57**
height 2.40 m–2.70 m	–	2.20	44.44	72.00	m	**116.44**
height 2.70 m–3.00 m	–	2.30	46.46	79.64	m	**126.10**
Gyproc metal stud proprietary partitions or other equal; metal Gypframe C studs at 600 mm centres; floor and ceiling channels						
95 mm partition; 70 mm studs and channels; one layer of 12.5 mm Gyproc Fireline each side; joints filled with joint filler and joint tape to receive direct decoration						
average height 2.00 m	–	0.80	16.16	18.82	m²	**34.98**
average height 3.00 m	–	0.75	15.15	18.25	m²	**33.40**
Rate per m (SMM7 measurement rule)						
height 2.10 m–2.40 m	–	1.50	30.31	33.71	m	**64.02**
height 2.40 m–2.70 m	–	1.70	34.35	36.73	m	**71.08**
height 2.70 m–3.00 m	–	1.75	35.35	39.44	m	**74.79**

20 PROPRIETARY LININGS AND PARTITIONS

Item	PC £	Labour hours	Labour £	Material £	Unit	Total rate £
100 mm partition; 70 mm studs and channels; one layer of 15 mm Gyproc Fireline each side; joints filled with joint filler and joint tape to receive direct decoration						
average height 2.00 m	–	0.80	16.16	20.74	m²	**36.90**
average height 3.00 m	–	0.75	15.15	20.17	m²	**35.32**
Rate per m (SMM7 measurement rule)						
height 2.10 m–2.40 m	–	1.50	30.31	36.03	m	**66.34**
height 2.40 m–2.70 m	–	1.70	34.35	39.32	m	**73.67**
height 2.70 m–3.00 m	–	1.75	35.35	42.32	m	**77.67**
120 mm partition; 70 mm studs and channels; two layers of 12.5 mm Gyproc Fireline each side; joints filled with joint filler and joint tape to receive direct decoration						
average height 2.00 m	–	1.10	22.22	18.90	m²	**41.12**
average height 3.00 m	–	1.15	23.24	17.58	m²	**40.82**
Rate per m (SMM7 measurement rule)						
height 2.10 m–2.40 m	–	2.00	40.41	45.37	m	**85.78**
height 2.40 m–2.70 m	–	2.20	44.44	49.84	m	**94.28**
height 2.70 m–3.00 m	–	2.30	46.46	52.75	m	**99.21**
130 mm partition; 70 mm studs and channels; two layers of 15 mm Gyproc Fireline each side; joints filled with joint filler and joint tape to receive direct decoration						
average height 2.00 m	–	1.10	22.22	20.83	m²	**43.05**
average height 3.00 m	–	1.15	23.24	19.51	m²	**42.75**
Rate per m (SMM7 measurement rule)						
height 2.10 m–2.40 m	–	2.00	40.41	49.99	m	**90.40**
height 2.40 m–2.70 m	–	2.20	44.44	55.03	m	**99.47**
height 2.70 m–3.00 m	–	2.30	46.46	59.49	m	**105.95**
Gypwall Rapid/db Plus metal stud housing partitioning system; or other equal; floor and ceiling channels plugged and screwed to concrete or nailed to timber						
75 mm partition; 43 mm studs at 600 mm centres; one layer of 15 mm SoundBloc Rapid each side; joints filled with joint filler and joint tape to receive direct decoration						
average height 2.00 m	–	0.63	11.14	14.17	m²	**25.31**
average height 3.00 m	–	0.83	14.85	18.70	m²	**33.55**
Rate per m (SMM7 measurement rule)						
height 2.10 m–2.40 m	–	1.50	26.74	33.99	m	**60.73**
height 2.10 m–2.40 m	–	1.75	31.20	39.44	m	**70.64**
height 2.40 m–2.70 m	–	2.00	35.65	43.61	m	**79.26**
Angles, corners and similar						
T-junctions	–	0.17	3.12	5.76	m	**8.88**
splayed corners	–	0.13	2.22	3.86	m	**6.08**
corners	–	0.10	1.78	1.96	m	**3.74**
fair ends	–	0.11	1.96	0.60	m	**2.56**

20 PROPRIETARY LININGS AND PARTITIONS

Item	PC £	Labour hours	Labour £	Material £	Unit	Total rate £
PLASTERBOARD DRY LINING WALLS AND PARTITIONS – cont						
Cavity insulation to stud partitions pinned to board						
25 mm Isover APR						
average height 2.00 m	–	0.06	1.12	0.89	m	**2.01**
average height 3.00 m	–	0.07	1.25	0.89	m	**2.14**
Rate per m (SMM7 measurement rule)						
height 2.10 m–2.40 m	–	0.15	2.68	2.13	m	**4.81**
height 2.40 m–2.70 m	–	0.16	2.85	2.40	m	**5.25**
height 3.00 m–3.30 m	–	–	–	–	m	**-**
50 mm Isover APR						
average height 2.00 m	–	0.06	1.12	1.46	m	**2.58**
average height 3.00 m	–	0.07	1.25	1.46	m	**2.71**
Rate per m (SMM7 measurement rule)						
height 2.10 m–2.40 m	–	0.15	2.68	3.51	m	**6.19**
height 2.40 m–2.70 m	–	0.16	2.85	3.95	m	**6.80**
height 3.00 m–3.30 m	–	0.17	3.03	4.82	m	**7.85**
GLASS REINFORCED GYPSUM LININGS						
Glass reinforced gypsum Glasroc Multi-board or other equal and approved; fixing with nails; joints filled with joint filler and joint tape; finishing with Jointex or other equal and approved to receive decoration; to softwood base						
10 mm board to walls						
average wall height 2.00 m	–	0.37	6.54	17.16	m^2	**23.70**
average wall height 3.00 m	–	0.36	6.48	17.16	m^2	**23.64**
average wall height 4.00 m	–	0.36	6.42	17.16	m^2	**23.58**
Rate per m (SMM7 measurement rule)						
wall height 2.40 m–2.70 m	–	0.93	16.57	46.33	m	**62.90**
wall height 2.70 m–3.00 m	–	1.06	18.90	51.49	m	**70.39**
wall height 3.00 m–3.30 m	–	1.07	19.01	56.63	m	**75.64**
wall height 3.30 m–3.60 m	–	1.39	24.78	61.84	m	**86.62**
12.50 mm board to walls						
average wall height 2.00 m	–	0.39	6.87	22.39	m^2	**29.26**
average wall height 3.00 m	–	0.38	6.81	22.39	m^2	**29.20**
average wall height 4.00 m	–	0.38	6.73	22.39	m^2	**29.12**
Rate per m (SMM7 measurement rule)						
wall height 2.40 m–2.70 m	–	0.97	17.29	60.56	m	**77.85**
wall height 2.70 m–3.00 m	–	1.11	19.79	67.29	m	**87.08**
wall height 3.00 m–3.30 m	–	1.25	22.28	74.03	m	**96.31**
wall height 3.30 m–3.60 m	–	1.14	20.38	80.81	m	**101.19**

20 PROPRIETARY LININGS AND PARTITIONS

Item	PC £	Labour hours	Labour £	Material £	Unit	Total rate £
RIGID BOARD LININGS						
Blockboard (Birch faced)						
Lining to walls 18 mm thick						
over 600 wide	6.56	0.46	8.20	7.00	m²	**15.20**
not exceeding 600 wide	–	0.60	10.70	4.24	m	**14.94**
holes for pipes and the like	–	0.04	0.72	–	nr	**0.72**
Chipboard (plain)						
Lining to walls 12 mm thick						
over 600 mm wide	2.41	0.35	6.24	2.76	m²	**9.00**
not exceeding 600 mm wide	–	0.40	7.13	1.59	m	**8.72**
holes for pipes and the like	–	0.02	0.36	–	nr	**0.36**
Lining to walls 15 mm thick						
over 600 mm wide	2.84	0.37	6.60	3.20	m²	**9.80**
not exceeding 600 mm wide	–	0.44	7.84	1.96	m	**9.80**
holes for pipes and the like	–	0.03	0.53	–	nr	**0.53**
Lining to walls 18 mm thick						
over 600 mm wide	3.41	0.39	6.95	3.84	m²	**10.79**
not exceeding 600 mm wide	–	0.50	8.92	2.31	m	**11.23**
holes for pipes and the like	–	0.04	0.72	–	nr	**0.72**
Fire-retardant chipboard/MDF; Class 1 spread of flame						
Lining to walls 12 mm thick						
over 600 mm wide	–	0.35	6.24	6.16	m²	**12.40**
not exceeding 600 mm wide	–	0.40	7.13	3.74	m	**10.87**
holes for pipes and the like	–	0.02	0.36	–	nr	**0.36**
Lining to walls 18 mm thick						
over 600 mm wide	–	0.39	6.95	9.15	m²	**16.10**
not exceeding 600 mm wide	–	0.50	8.92	5.54	m	**14.46**
holes for pipes and the like	–	0.04	0.72	–	nr	**0.72**
Lining to walls 25 mm thick						
over 600 mm wide	–	0.41	7.31	12.86	m²	**20.17**
not exceeding 600 mm wide	–	0.56	9.98	7.76	m	**17.74**
holes for pipes and the like	–	0.05	0.89	–	nr	**0.89**
Chipboard Melamine faced; white matt finish; laminated masking strips						
Lining to walls 15 mm thick						
over 600 mm wide	3.66	0.97	17.29	4.33	m²	**21.62**
not exceeding 600 mm wide	–	1.26	22.46	2.65	m	**25.11**
holes for pipes and the like	–	0.06	1.07	–	nr	**1.07**
Insulation board to BS EN 622						
Lining to walls 12 mm thick						
over 600 mm wide	1.95	0.22	3.93	2.28	m²	**6.21**
not exceeding 600 mm wide	–	0.26	4.63	1.40	m	**6.03**
holes for pipes and the like	–	0.01	0.18	–	nr	**0.18**

20 PROPRIETARY LININGS AND PARTITIONS

Item	PC £	Labour hours	Labour £	Material £	Unit	Total rate £
RIGID BOARD LININGS – cont						
Non-asbestos board; Masterboard or other equal and approved; sanded finish						
Lining to walls 6 mm thick						
over 600 mm wide	8.56	0.31	5.52	8.99	m²	**14.51**
not exceeding 600 mm wide	–	0.38	6.78	5.40	m	**12.18**
Lining to walls 9 mm thick						
over 600 mm wide	19.79	0.33	5.88	20.50	m²	**26.38**
not exceeding 600 mm wide	–	0.38	6.78	12.31	m	**19.09**
Supalux or other equal and approved; sanded finish						
Lining to walls 6 mm thick						
over 600 mm wide	15.50	0.31	5.52	16.10	m²	**21.62**
not exceeding 600 mm wide	–	0.38	6.78	9.68	m	**16.46**
Lining to walls 9 mm thick						
over 600 mm wide	19.87	0.33	5.88	20.57	m²	**26.45**
not exceeding 600 mm wide	–	0.38	6.78	12.36	m	**19.14**
Lining to walls 12 mm thick						
over 600 mm wide	26.98	0.37	6.60	27.86	m²	**34.46**
not exceeding 600 mm wide	–	0.44	7.84	16.73	m	**24.57**
Non-asbestos board; Monolux 40 or other equal and approved; 6 mm × 50 mm Supalux cover fillets or other equal and approved one side						
Lining to walls 19 mm thick						
over 600 mm wide	57.66	0.65	11.58	61.75	m²	**73.33**
not exceeding 600 mm wide	–	0.92	16.40	38.67	m	**55.07**
Lining to walls 25 mm thick						
over 600 mm wide	69.14	0.69	12.30	73.51	m²	**85.81**
not exceeding 600 mm wide	41.48	0.98	17.47	45.73	m	**63.20**
Plywood (Eastern European); internal quality						
Lining to walls 4 mm thick						
over 600 mm wide	2.35	0.34	6.06	2.70	m²	**8.76**
not exceeding 600 mm wide	–	0.44	7.84	1.66	m	**9.50**
Lining to walls 6 mm thick						
over 600 mm wide	3.40	0.37	6.60	3.77	m²	**10.37**
not exceeding 600 mm wide	–	0.48	8.56	2.31	m	**10.87**
Lining to walls 12 mm thick						
over 600 mm wide	6.33	0.43	7.67	6.76	m²	**14.43**
not exceeding 600 mm wide	–	0.56	9.98	4.10	m	**14.08**
Lining to walls 18 mm thick						
over 600 mm wide	9.23	0.46	8.20	9.75	m²	**17.95**
not exceeding 600 mm wide	–	0.60	10.70	5.89	m	**16.59**

20 PROPRIETARY LININGS AND PARTITIONS

Item	PC £	Labour hours	Labour £	Material £	Unit	Total rate £
Plywood (Eastern European); external quality						
Lining to walls 4 mm thick						
over 600 mm wide	5.83	0.34	6.06	6.26	m²	**12.32**
not exceeding 600 mm wide	–	0.44	7.84	3.79	m	**11.63**
Lining to walls 6.5 mm thick						
over 600 mm wide	6.52	0.37	6.60	6.97	m²	**13.57**
not exceeding 600 mm wide	–	0.48	8.56	4.22	m	**12.78**
Lining to walls 9 mm thick						
over 600 mm wide	8.40	0.40	7.13	8.89	m²	**16.02**
not exceeding 600 mm wide	–	0.52	9.27	5.37	m	**14.64**
Lining to walls 12 mm thick						
over 600 mm wide	10.48	0.43	7.67	11.03	m²	**18.70**
not exceeding 600 mm wide	–	0.56	9.98	6.66	m	**16.64**
holes for pipes and the like	–	0.03	0.53	–	nr	**0.53**
Extra over wall linings fixed with nails for screwing	–	–	–	–	m²	**1.72**
Internal quality American Cherry veneered plywood; 6 mm thick						
Lining to walls						
over 600 mm wide	7.65	0.41	7.31	8.05	m²	**15.36**
not exceeding 600 mm wide	–	0.54	9.62	4.91	m	**14.53**
Supalux or other equal fire-resisting boards						
Lining to walls; 6 mm thick						
over 600 mm wide	–	0.31	5.52	15.73	m²	**21.25**
not exceeding 600 mm wide	–	0.38	6.78	9.52	m	**16.30**
Lining to walls; 9 mm thick						
over 600 mm wide	–	0.33	5.88	20.10	m²	**25.98**
not exceeding 600 mm wide	–	0.40	7.13	12.15	m	**19.28**
Lining to walls; 12 mm thick						
over 600 mm wide	–	0.37	6.60	27.21	m²	**33.81**
not exceeding 600 mm wide	–	0.44	7.84	16.41	m	**24.25**
Glazed hardboard to BS EN 622; on and including 38 mm × 38 mm sawn softwood framing						
3.20 mm thick panel						
to side of bath	–	1.67	29.78	7.15	nr	**36.93**
to end of bath	–	0.65	11.58	2.17	nr	**13.75**
RIGID SHEET ACOUSTIC PANEL LININGS						
Perforated steel acoustic wall panels; Eckel type HD EFP or other equal and approved; polyurethene enamel finish; fibrous glass acoustic insulation						
Walls						
average height 3.00 m; fixed to timber or masonry	–	–	–	–	m²	**147.90**

20 PROPRIETARY LININGS AND PARTITIONS

Item	PC £	Labour hours	Labour £	Material £	Unit	Total rate £
DEMOUNTABLE PARTITIONS						
Fire-resistant concertina partition						
Insulated panel and two-hour fire wall system for warehouses etc., comprising white polyester coated galvanized steel frame and 0.55 mm galvanized steel panels either side of rockwool infill						
100 mm thick wall: 31 Rw dB acoustic rating	–	–	–	–	m²	35.15
150 mm thick wall: 31 Rw dB acoustic rating	–	–	–	–	m²	37.79
intumescent mastic sealant; bedding frames at						
perimeter of metal fire walls	–	–	–	–	m	3.08
Room divider moveable wall						
Laminated both sides top hung movable acoustic panel wall with concealed uPVC vertical edge profiles, 1106 m × 3000 mm panels and two point panel support system						
105 mm thick wall: 47 Rw dB acoustic rating	–	–	–	–	m²	325.14
105 mm thick wall: 50 Rw dB acoustic rating	–	–	–	–	m²	351.50
105 mm thick wall: 53 Rw dB acoustic rating	–	–	–	–	m²	377.86
TERRAZZO FACED PARTITIONS						
Terrazzo faced partitions; polished on two faces						
Precast reinforced terrazzo faced WC partitions						
38 mm thick; over 300 mm wide	–	–	–	–	m²	285.00
50 mm thick; over 300 mm wide	–	–	–	–	m²	294.50
Wall post; once rebated						
64 mm × 102 mm	–	–	–	–	m	131.10
64 mm × 152 mm	–	–	–	–	m	142.50
Centre post; twice rebated						
64 mm × 102 mm	–	–	–	–	m	137.75
64 mm × 152 mm	–	–	–	–	m	152.00
Lintel; once rebated						
64 mm × 102 mm	–	–	–	–	m	142.50
Pair of brass topped plates or sockets cast into posts for fixings (not included)	–	–	–	–	nr	33.25
Brass indicator bolt lugs cast into posts for fixings (not included)	–	–	–	–	nr	14.25
TIMBER FLOORING AND WALL LININGS						
Wrought softwood						
Boarding to internal walls; tongued and grooved and V-jointed						
12 mm × 100 mm boards	–	0.74	13.19	12.29	m²	25.48
16 mm × 100 mm boards	–	0.74	13.19	13.29	m²	26.48
19 mm × 100 mm boards	–	0.74	13.19	15.16	m²	28.35
19 mm × 125 mm boards	–	0.69	12.30	13.79	m²	26.09
19 mm × 125 mm boards; chevron pattern	–	1.11	19.79	13.79	m²	33.58
25 mm × 125 mm boards	–	0.69	12.30	11.90	m²	24.20
12 mm × 100 mm boards; knotty pine	–	0.74	13.19	7.56	m²	20.75

20 PROPRIETARY LININGS AND PARTITIONS

Item	PC £	Labour hours	Labour £	Material £	Unit	Total rate £
Wood strip; 22 mm thick; Junckers All in Beech Sylva Sport Premium pre-treated or other equal and approved; tongued and grooved joints; on bearers etc.; level fixing to cement and sand base						
Strip flooring; over 300 mm wide						
on 45 × 45 mm blue bat bearers	–	–	–	–	m²	**57.08**
on 10 mm Pro Foam	–	–	–	–	m²	**65.20**
on Uno bat 50 mm bearers	–	–	–	–	m²	**59.71**
on New Era levelling system	–	–	–	–	m²	**61.64**
on Uno bat 62 mm bearers	–	–	–	–	m²	**60.67**
on Duo bat 110 mm bearers	–	–	–	–	m²	**101.83**
Wood strip; 22 mm thick; Junckers pre-treated or other equal and approved flooring systems; tongued and grooved joints; on bearers etc.; level fixing to cement and sand base						
Strip flooring; over 300 mm wide						
Sylva Squash Beech untreated on blue bat bearers	–	–	–	–	m²	**59.85**
Classic Beech clip system	–	–	–	–	m²	**79.29**
Harmoni Oak clip system	–	–	–	–	m²	**83.01**
Classic Beech on blue bat bearers	–	–	–	–	m²	**80.12**
Harmoni Oak on blue bat bearers	–	–	–	–	m²	**82.05**
Unfinished wood strip; 22 mm thick; Havwoods or other equal and approved; tongued and grooved joints; secret fixed; laid on semi-sprung bearers; fixing to cement and sand base; sanded and sealed						
Strip flooring; over 300 mm wide						
Prime Iroko	–	–	–	–	m²	**77.22**
Prime Maple	–	–	–	–	m²	**77.22**
American Oak	–	–	–	–	m²	**84.65**
FRAMED PANEL CUBICLE PARTITIONS						
Toilet cubicle partitions and IPS systems; Amwells or other equal and approved; standard colours and ironmongery; assembling and screwing to floor and wall						
Axis MFC standard cubic system, for use in small offices, retail and community halls etc.; standard cubicle set; 800 mm × 1500 mm × 1980 mm high per cubicle, with polished aluminium framing; 19 mm melamine-faced chipboard divisions and doors						
one cubicle set; 2 nr panels; 1 nr door	–	3.50	106.91	336.78	nr	**443.69**
range of 3 cubicle sets; 4 nr panels; 3 nr doors	–	10.75	328.36	809.70	nr	**1138.06**
range of 6 cubicle sets; 7 nr panels; 6 nr doors	–	21.25	649.08	1519.05	nr	**2168.13**
reduction of 1 nr panel for end unit adjoining side wall	–	–	–	–76.96	nr	**–76.96**

20 PROPRIETARY LININGS AND PARTITIONS

Item	PC £	Labour hours	Labour £	Material £	Unit	Total rate £
FRAMED PANEL CUBICLE PARTITIONS – cont						
Toilet cubicle partitions and IPS systems – cont						
Splash SGL cubic system for heavy use environments like schools, swimming pools and prisons; standard cubicle set; 800 mm × 1500 mm × 1980 mm high per cubicle, with polished aluminium framing; 19 mm melamine-faced chipboard divisions and doors						
one cubicle set; 2 nr panels; 1 nr door	–	3.50	106.91	692.56	nr	**799.47**
range of 3 cubicle sets; 4 nr panels; 3 nr doors	–	10.75	328.36	1598.01	nr	**1926.37**
range of 6 cubicle sets; 7 nr panels; 6 nr door	–	21.25	649.08	2956.15	nr	**3605.23**
reduction of 1 nr panel for end unit adjoining side wall	–	–	–	–212.87	nr	**–212.87**
Minima Stylish stainless steel framed cubicle system in MFC, HPL, SGL, Real Wood veneer or Glass; 800 mm × 1500 mm × 2100 mm high per cubicle, with satin polished stainless steel framing; 18 mm high pressure laminated (HPL) chipboard divisions and doors						
one cubicle set; 2 nr panels; 1 nr door	–	3.50	106.91	809.49	nr	**916.40**
range of 3 cubicle sets; 4 nr panels; 3 nr doors	–	10.75	328.36	1804.28	nr	**2132.64**
range of 6 cubicle sets; 7 nr panels; 6 nr doors	–	21.25	649.08	3296.44	nr	**3945.52**
reduction of 1 nr panel for end unit adjoining side wall	–	–	–	–179.89	nr	**–179.89**
Urban Flush fronted, floor to ceiling option cubicle system in either 13 mm or 20 mm panels, material options HPL, SGL or Real Wood Veneer: cubicle set; 800 mm × 1500 mm × 2400 mm high per cubicle, with sating finished stainless steel ironmongery; 20 mm divisions, doors and pilasters						
one cubicle set; 2 nr panels; 1 nr door	–	4.00	122.18	1232.22	nr	**1354.40**
range of 3 cubicle sets; 4 nr panels; 3 nr doors	–	11.75	358.90	3031.11	nr	**3390.01**
range of 6 cubicle sets; 7 nr panels; 6 nr doors	–	23.40	714.75	5729.42	nr	**6444.17**
reduction of 1 nr panel for end unit adjoining side wall	–	–	–	–238.86	nr	**–238.86**
Sylan High end specification flush fronted system floor to ceiling 44 mm doors, no visible fixings. Finishes in Real Wood Veneer, HPL, high Gloss paint and lacquer; cubicle set; 800 mm × 1500 mm × 2400 mm high per cubicle, with satin finished stainless steel ironmongery; 30 mm high pressure laminated (HPL) chipboard divisions and 44 mm solid cored real wood veneered doors and pilasters						
one cubicle set; 2 nr panels; 1 nr door	–	5.00	152.73	2290.57	nr	**2443.30**
range of 3 cubicle sets; 4 nr panels; 3 nr doors	–	15.00	458.18	5395.62	nr	**5853.80**
range of 6 cubicle sets; 7 nr panels; 6 nr doors	–	30.00	916.35	10052.71	nr	**10969.06**
reduction of 1 nr panel for end unit adjoining side wall	–	–	–	–397.75	nr	**–397.75**

20 PROPRIETARY LININGS AND PARTITIONS

Item	PC £	Labour hours	Labour £	Material £	Unit	Total rate £
IPS Panel Systems; 12.50 mm melamine faced chipboard; fixed to existing timber subframe						
IPS back panel system to accomodate urinals; to conceal pipework; approx 2.70 m high × 2.10 m wide	–	–	–	–	nr	**716.34**
IPS back panel system to accomodate wash hand basins; to conceal pipework; approx 2.70 m high × 5.70 m wide	–	–	–	–	nr	**1412.28**
IPS back panel system to accomodate wash hand basins; to conceal pipework; approx 2.70 m high × 2.10 m wide	–	–	–	–	nr	**797.99**

21 CLADDING AND COVERING

Item	PC £	Labour hours	Labour £	Material £	Unit	Total rate £
PATENT GLAZING						
Patent glazing; aluminium alloy bars 2.55 m long at 622 mm centres; fixed to supports						
Roof cladding						
single glazed with 6.4 mm laminated glass	–	–	–	–	m²	130.00
single glazed with 7 mm thick Georgian wired cast glass	–	–	–	–	m²	135.00
thermally broken and double glazed with low-e clear toughened and laminated double glazed units; aluminium finished RAL matt colour	–	–	–	–	m²	355.00
Extra for opening roof vents						
600 mm × 900 mm top hung opening roof vent; manually operated	–	–	–	–	nr	400.00
600 mm × 900 mm top hung opening roof vent; electrically operated	–	–	–	–	nr	510.00
Skylight						
Self-supporting hipped or gable ended lantern/ skylight thermally broken and double glazed with low-e clear toughened and laminated double glazed units; aluminium finished RAL matt colour	–	–	–	–	m²	710.00
Associated code 4 lead flashings						
top flashing; 210 mm girth	–	–	–	–	m	56.00
bottom flashing; 240 mm girth	–	–	–	–	m	64.00
end flashing; 300 mm girth	–	–	–	–	m	70.00
Wall cladding						
single glazed with 6.4 mm laminated glass	–	–	–	–	m²	130.00
single glazed with 7 mm thick Georgian wired cast glass	–	–	–	–	m²	145.00
thermally broken and double glazed with low-e clear toughened and laminated double glazed units; aluminium finished RAL matt colour	–	–	–	–	m²	370.00
Extra for aluminium alloy perimeter members						
38 mm × 38 mm × 3 mm angle jamb	–	–	–	–	m	20.00
pressed sill member	–	–	–	–	m	40.00
pressed channel head and PVC case	–	–	–	–	m	40.00
CURTAIN WALLING						
NOTE: Toughened and heat soak tested glass costs are comparable with heat strengthened glass. We have removed all reference to toughened glass in this issue due to the ever present risk of nickel sulphide inclusions. Consultation with mechanical services engineers is encouraged to ensure Part L compliance and user comfort criteria are understood, i.e. when to use high performance coated glass and when to add in external shading.						

21 CLADDING AND COVERING

Item	PC £	Labour hours	Labour £	Material £	Unit	Total rate £
Stick curtain walling system; proprietary solution from system supplier; e.g. Schuco or equivalent Polyester powder coated solid colour matt finish or natural anodized finish mullions spaced 1.5 m apart and spanning typical storey height of 3.8 m. Floor to ceiling glass sealed units with 8.8 mm low-e coated laminated inner pane, air filled cavity and 8 mm clear monolithic heat strengthened outer pane, retained by external pressure plates and caps. Rates to include glass fronted solid spandrel panels, all brackets, membranes, fire stopping between floors, trade contractor preliminaries, including external access equipment						
flat system	–	–	–	–	m²	375.00
extra over for						
neutral selective high performance coating on surface #2 in lieu of low-e coating on surface #3 (G-values circa 0.25–0.30); for assisting in solar control	–	–	–	–	m²	30.00
inner laminated glass to be heat strengthened laminated to mitigate thermal fracture risk	–	–	–	–	m²	30.00
outer glass to be heat strengthened laminated in lieu of monolithic heat strengthened	–	–	–	–	m²	30.00
ceramic fritting glass on surface #2 for visual and/or performance requirements (minimal G-value improvement)	–	–	–	–	m²	40.00
flush glass finish without external face caps, achieved by concealed toggle fixings locating within perimeter channels within sealed units including silicone sealing between glass panes	–	–	–	–	m²	40.00
typical coping detail, including pressed aluminium profiles, membranes, seals, etc.	–	–	–	–	m	225.00
typical sill detail, including pressed aluminium profiles, membranes, seals, etc.	–	–	–	–	m	180.00
intermediate transoms (per transom)	–	–	–	–	m	40.00

21 CLADDING AND COVERING

Item	PC £	Labour hours	Labour £	Material £	Unit	Total rate £
CURTAIN WALLING – cont						
Unitized curtain walling system; proprietary solution from system supplier e.g. Schuco or equivalent alternatively pre-designed system solution from specialist facade contractor Polyester powder coated solid colour matt finish or natural anodized curtain walling Element widths of 1.5 m spanning typical storey height of 3.8 m. Floor to ceiling glass sealed units with 8.8 mm low-e coated laminated inner pane, air filled cavity and 8 mm monolithic heat strengthened outer pane, retained by external beading system. Rates include 1.1 m solid spandrel panels, all brackets, membranes, fire stopping between floors, trade contractor preliminaries, including external access equipment						
flat system	–	–	–	–	m²	**775.00**
extra over for						
neutral selective high performance coating on surface #2 in lieu of low-e coating on surface #3 (G-values circa 0.25–0.30); for assisting in solar control	–	–	–	–	m²	**30.00**
inner laminated glass to be heat strengthened laminated to mitigate thermal fracture risk	–	–	–	–	m²	**30.00**
flush glass finish without external face caps, achieved by carrier frames with glass sealed units factory silicone bonded; often referred to as SSG (Structural Silicone Glazing)	–	–	–	–	m²	**50.00**
typical coping detail, including pressed aluminium profiles, membranes, seals, etc.	–	–	–	–	m	**225.00**
typical sill detail, including pressed aluminium profiles, membranes, seals, etc.	–	–	–	–	m	**180.00**
Other curtain walling systems/costs Unitized curtain walling system; bespoke project specific solution via specialist facade contractor mostly based in mainland Europe						
generally as described for system supplier solution but profiles catered to specific performance and visual requirements, thus additional design development. Note: These rates are subject to currency fluctuations between £ and €.	–	–	–	–	m²	**775.00**
flush glass finish finish (SEG) assuming direct bonding in the factory but with base chassis design to accommodate carrier frame for glass replacement, avoiding site bending	–	–	–	–	m²	**25.00**
project specific unitized curtain walling generally requires project specific performance testing. This rate is for a single wall type	–	–	–	–	nr	**75000.00**

21 CLADDING AND COVERING

Item	PC £	Labour hours	Labour £	Material £	Unit	Total rate £
visual mock-ups are often required for project specific unitized curtain walling solutions and in cases for proprietary unitized and stick curtain walling projects. This rate is for a single wall type. all curtain walling should be site hose tested. The rate depends upon the quantum of joints to be tested, generally 5%. Assume 5 days @ £1000 extra over for	–	–	–	–	nr	25000.00
	–	–	–	–	nr	5000.00
neutral selective high performance coating on surface #2 in lieu of low-e coating on surface #3 (G-values circa 0.25–0.30); for assisting in solar control	–	–	–	–	m²	30.00
inner laminated glass to be heat strengthened laminated to mitigate thermal fracture risk	–	–	–	–	m²	30.00
outer glass to be heat strengthened laminated in lieu of monolithic heat strengthened	–	–	–	–	m²	30.00
alternative solutions for achieving Part L and comfort criteria, include increasing the amount of solidity	–	–	–	–	m²	75.00
PROFILED SHEET CLADDING						
Lightweight galvanized steel roof tiles; Decra Roof Systems; or other equal; coated finish						
Zinc coated tile panels, dry fixed						
Classic profile; clay or concrete tile appearance	–	0.23	4.64	14.70	m²	19.34
Stratos profile; slate or concrete tiles appearance	–	0.23	4.64	16.23	m²	20.87
Accessories for roof cladding						
pitched D ridge	–	0.09	1.81	7.75	m	9.56
barge cover (handed)	–	0.09	1.81	5.42	m	7.23
in line air vent	–	0.09	1.81	46.89	nr	48.70
in line soil vent	–	0.09	1.81	67.68	nr	69.49
gas flue terminal	–	0.19	3.83	87.88	nr	91.71
Galvanized steel strip troughed sheets; Corus Products or other equal						
Roof cladding or decking; sloping not exceeding 50°; fixing to steel purlins with plastic headed self-tapping screws						
0.7 mm thick; 46 profile	–	–	–	–	m²	11.03
0.7 mm thick; 60 profile	–	–	–	–	m²	12.07
0.7 mm thick; 100 profile	–	–	–	–	m²	13.11

21 CLADDING AND COVERING

Item	PC £	Labour hours	Labour £	Material £	Unit	Total rate £
PROFILED SHEET CLADDING – cont						
Galvanized steel strip troughed sheets; PMF Strip Mill Products or other equal						
Roof cladding; sloping not exceeding 50°; fixing to steel purlins with plastic headed self-tapping screws						
0.7 mm thick type HPS200 13.5/3 corrugated	–	–	–	–	m²	12.43
0.7 mm thick type HPS200 R32/1000	–	–	–	–	m²	11.28
0.7 mm thick type Arcline 40; plasticol finished	–	–	–	–	m²	16.59
Extra over last for aluminium roof cladding or decking	–	–	–	–	m²	6.81
Accessories for roof cladding						
HPS200 Drip flashing; 250 mm girth	–	–	–	–	m	3.86
HPS200 Ridge flashing; 375 mm girth	–	–	–	–	m	5.05
HPS200 Gable flashing; 500 mm girth	–	–	–	–	m	6.42
HPS200 Internal angle; 625 mm girth	–	–	–	–	m	7.41
Asbestos-free corrugated sheets; Eternit 2000 or other equal						
Roof cladding; sloping not exceeding 50°; fixing to steel purlins with hook bolts						
Profile 3; natural grey	–	0.23	4.64	16.68	m²	21.32
Profile 3; coloured	–	0.23	4.64	19.10	m²	23.74
Profile 6; natural grey	–	0.28	5.66	13.14	m²	18.80
Profile 6; coloured	–	0.28	5.66	14.87	m²	20.53
Profile 6; natural grey; insulated 80 glass fibre infill; lining panel	–	0.46	9.30	26.01	m²	35.31
Profile 6; coloured; insulated 80 glass fibre infill; lining panel	–	0.46	9.30	30.08	m²	39.38
Accessories; to Profile 3 cladding; natural grey						
eaves filler	–	0.09	1.81	10.38	m	12.19
external corner piece	–	0.11	2.22	7.67	m	9.89
apron flashing	–	0.11	2.22	10.38	m	12.60
plain wing or close fitting two piece adjustable capping to ridge	–	0.16	3.23	9.74	m	12.97
ventilating two piece adjustable capping to ridge	–	0.16	3.23	14.98	m	18.21
Accessories; to Profile 6 cladding; natural grey						
eaves filler	–	0.09	1.81	6.26	m	8.07
external corner piece	–	0.11	2.22	7.12	m	9.34
apron flashing	–	0.11	2.22	6.96	m	9.18
underglazing flashing	–	0.11	2.22	9.17	m	11.39
plain cranked crown to ridge	–	0.16	3.23	13.77	m	17.00
plain wing or close fitting two piece adjustable capping to ridge	–	0.16	3.23	12.52	m	15.75
ventilating two piece adjustable capping to ridge	–	0.16	3.23	16.00	m	19.23
GRP Transluscent rooflights; factory assembled						
Rooflight; vertical fixing to steel purlins (measured elsewhere)						
double skin; class 3 over 1	–	–	–	–	m²	47.25
triple skin; class 3 over 1	–	–	–	–	m²	52.50

21 CLADDING AND COVERING

Item	PC £	Labour hours	Labour £	Material £	Unit	Total rate £
Wall cladding; Gasell Profiles Ltd or equal and approved; steel GA50–30 profiled sheeting to outer face; steel; GA600 lining to inner face; including profile fillers; sealing						
Coverings; fixing to and including vertical and horizontal secondary supports						
250 mm girth	–	–	–	–	m²	**117.00**
Extended, hard skinned, foamed PVC-UE profiled sections; Swish Celuka or other equal and approved; Class 1 fire-rated to BS 476; Part 7; in white finish						
Wall cladding; vertical; fixing to timber						
100 mm shiplap profiles; Code 001	–	0.35	6.24	50.98	m²	**57.22**
150 mm shiplap profiles; Code 002	–	0.32	5.71	45.17	m²	**50.88**
125 mm feather-edged profiles; Code C208	–	0.34	6.06	49.23	m²	**55.29**
vertical angles	–	0.19	3.38	5.23	m	**8.61**
raking cutting	–	0.14	2.49	–	m	**2.49**
holes for pipes and the like	–	0.03	0.53	–	nr	**0.53**
RIGID SHEET CLADDING						
Resoplan sheet or other equal and approved; Eternit UK Ltd; flexible neoprene gasket joints; fixing with stainless steel screws and coloured caps						
6 mm thick cladding to walls						
over 300 mm wide	–	1.94	34.58	49.18	m²	**83.76**
not exceeding 300 mm wide	–	0.65	11.58	18.00	m	**29.58**
Eternit 2000 Glasal sheet or other equal and approved; Eternit UK Ltd; flexible neoprene gasket joints; fixing with stainless steel screws and coloured caps						
7.50 mm thick cladding to walls						
over 300 mm wide	–	1.94	34.58	43.33	m²	**77.91**
not exceeding 300 mm wide	–	0.65	11.58	16.24	m	**27.82**
external angle trim	–	0.09	1.61	8.37	m	**9.98**
7.50 mm thick cladding to eaves, verge soffit boards, fascia boards or the like						
100 mm wide	–	0.46	8.20	8.30	m	**16.50**
200 mm wide	–	0.56	9.98	12.27	m	**22.25**
300 mm wide	–	0.65	11.58	16.24	m	**27.82**

21 CLADDING AND COVERING

Item	PC £	Labour hours	Labour £	Material £	Unit	Total rate £
WEATHERBOARDING						
Prodema ProdEX high density resin-bonded cellulose fibre weatherboarding panels; including secondary supports and fixing						
Walls						
8 mm Panels face fixed on to timber battens	–	–	–	–	m²	**152.00**
8 mm Panels face fixed on to aluminium rails	–	–	–	–	m²	**175.75**
8 mm Panels adhesive fixed on to timber battens or aluminium rails	–	–	–	–	m²	**180.50**
10 mm Panels secret fixed on to helping hand aluminium system	–	–	–	–	m²	**213.75**
RAINSCREEN						
Timber						
Western Red Cedar tongued and grooved wall cladding on and including treated softwood battens on breather membrane, 10 mm Eternit Blueclad board and 50 mm insulation board; the whole fixed to Metsec frame system; including sealing all joints etc.						
26 mm thick cladding to walls; boards laid horizontally	–	–	–	–	m²	**79.09**
Aluminium						
Reynobond rainscreen cladding; aluminium composite material cassettes with thermoplastic cores, back ventilated, including insulation, vapour control membrane and aluminium support system						
4 mm thick cladding; fixed to walls	–	–	–	–	m²	**140.60**
Clay						
Terracotta clay rainscreen cladding; including insulation, vapour control membrane and aluminium support system						
400 × 200 × 30 mm tile cladding; fixed to walls	–	–	–	–	m²	**254.84**

22 GENERAL JOINERY

Item	PC £	Labour hours	Labour £	Material £	Unit	Total rate £
SOFTWOOD						
Wrought softwood						
Skirtings, picture rails, dado rails and the like; splayed or moulded						
19 mm × 44 mm; splayed	–	0.09	1.61	1.85	m	**3.46**
19 mm × 44 mm; moulded	–	0.09	1.61	1.96	m	**3.57**
19 mm × 69 mm; splayed	–	0.09	1.61	1.99	m	**3.60**
19 mm × 69 mm; moulded	–	0.09	1.61	1.99	m	**3.60**
19 mm × 94 mm; splayed	–	0.09	1.61	2.21	m	**3.82**
19 mm × 94 mm; moulded	–	0.09	1.61	2.21	m	**3.82**
19 mm × 144 mm; moulded	–	0.11	1.96	2.59	m	**4.55**
19 mm × 169 mm; moulded	–	0.11	1.96	2.75	m	**4.71**
25 mm × 50 mm; moulded	–	0.09	1.61	1.93	m	**3.54**
25 mm × 69 mm; splayed	–	0.09	1.61	2.12	m	**3.73**
25 mm × 94 mm; splayed	–	0.09	1.61	2.37	m	**3.98**
25 mm × 144 mm; splayed	–	0.11	1.96	2.85	m	**4.81**
25 mm × 144 mm; moulded	–	0.11	1.96	2.85	m	**4.81**
25 mm × 169 mm; moulded	–	0.11	1.96	3.16	m	**5.12**
25 mm × 219 mm; moulded	–	0.13	2.32	4.09	m	**6.41**
returned ends	–	0.14	2.49	–	nr	**2.49**
mitres	–	0.09	1.61	–	nr	**1.61**
Architraves, cover fillets and the like; half round; splayed or moulded						
13 mm × 25 mm; half round	–	0.11	1.96	1.73	m	**3.69**
13 mm × 50 mm; moulded	–	0.11	1.96	1.85	m	**3.81**
16 mm × 32 mm; half round	–	0.11	1.96	1.93	m	**3.89**
16 mm × 38 mm; moulded	–	0.11	1.96	1.93	m	**3.89**
16 mm × 50 mm; moulded	–	0.11	1.96	1.93	m	**3.89**
19 mm × 50 mm; splayed	–	0.11	1.96	1.93	m	**3.89**
19 mm × 63 mm; splayed	–	0.11	1.96	1.99	m	**3.95**
19 mm × 69 mm; splayed	–	0.11	1.96	2.10	m	**4.06**
25 mm × 44 mm; splayed	–	0.11	1.96	1.90	m	**3.86**
25 mm × 50 mm; moulded	–	0.11	1.96	2.00	m	**3.96**
25 mm × 63 mm; splayed	–	0.11	1.96	2.07	m	**4.03**
25 mm × 69 mm; splayed	–	0.11	1.96	2.35	m	**4.31**
32 mm × 88 mm; moulded	–	0.11	1.96	2.37	m	**4.33**
38 mm × 38 mm; moulded	–	0.11	1.96	2.10	m	**4.06**
50 mm × 50 mm; moulded	–	0.11	1.96	2.44	m	**4.40**
returned ends	–	0.14	2.49	–	nr	**2.49**
mitres	–	0.09	1.61	–	nr	**1.61**
Stops; screwed on						
16 mm × 38 mm	–	0.09	1.61	0.83	m	**2.44**
16 mm × 50 mm	–	0.09	1.61	0.90	m	**2.51**
19 mm × 38 mm	–	0.09	1.61	0.83	m	**2.44**
25 mm × 38 mm	–	0.09	1.61	0.91	m	**2.52**
25 mm × 50 mm	–	0.09	1.61	0.93	m	**2.54**

22 GENERAL JOINERY

Item	PC £	Labour hours	Labour £	Material £	Unit	Total rate £
SOFTWOOD – cont						
Wrought softwood – cont						
Glazing beads and the like						
13 mm × 16 mm	–	0.04	0.72	1.00	m	**1.72**
13 mm × 19 mm	–	0.04	0.72	1.00	m	**1.72**
13 mm × 25 mm	–	0.04	0.72	1.04	m	**1.76**
13 mm × 25 mm; screwed	–	0.04	0.72	1.68	m	**2.40**
13 mm × 25 mm; fixing with brass cups and screws	–	0.04	0.72	2.16	m	**2.88**
16 mm × 25 mm; screwed	–	0.04	0.72	1.68	m	**2.40**
16 mm quadrant	–	0.04	0.72	1.54	m	**2.26**
19 mm quadrant or scotia	–	0.04	0.72	1.54	m	**2.26**
19 mm × 36 mm; screwed	–	0.04	0.72	1.70	m	**2.42**
25 mm × 38 mm; screwed	–	0.04	0.72	1.78	m	**2.50**
25 mm quadrant or scotia	–	0.04	0.72	1.62	m	**2.34**
38 mm scotia	–	0.04	0.72	1.96	m	**2.68**
50 mm scotia	–	0.04	0.72	2.30	m	**3.02**
Isolated shelves, worktops, seats and the like						
19 mm × 150 mm	–	0.15	2.68	2.13	m	**4.81**
19 mm × 200 mm	–	0.20	3.57	2.94	m	**6.51**
25 mm × 150 mm	–	0.15	2.68	2.44	m	**5.12**
25 mm × 200 mm	–	0.20	3.57	3.46	m	**7.03**
32 mm × 150 mm	–	0.15	2.68	2.85	m	**5.53**
32 mm × 200 mm	–	0.20	3.57	3.88	m	**7.45**
Isolated shelves, worktops, seats and the like; cross-tongued joints						
19 mm × 300 mm	–	0.26	4.63	8.78	m	**13.41**
19 mm × 450 mm	–	0.31	5.52	13.23	m	**18.75**
19 mm × 600 mm	–	0.37	6.60	17.13	m	**23.73**
25 mm × 300 mm	–	0.26	4.63	9.42	m	**14.05**
25 mm × 450 mm	–	0.31	5.52	14.27	m	**19.79**
25 mm × 600 mm	–	0.37	6.60	18.56	m	**25.16**
32 mm × 300 mm	–	0.26	4.63	9.97	m	**14.60**
32 mm × 450 mm	–	0.31	5.52	15.17	m	**20.69**
32 mm × 600 mm	–	0.37	6.60	19.79	m	**26.39**
Isolated shelves, worktops, seats and the like; slatted with 50 wide slats at 75 mm centres						
19 mm thick	–	0.60	10.70	21.68	m	**32.38**
25 mm thick	–	0.60	10.70	22.15	m	**32.85**
32 mm thick	–	0.60	10.70	22.58	m	**33.28**
Window boards, nosings, bed moulds and the like; rebated and rounded						
19 mm × 75 mm	–	0.17	3.03	2.84	m	**5.87**
19 mm × 150 mm	–	0.19	3.38	3.48	m	**6.86**
19 mm × 225 mm; in one width	–	0.24	4.27	4.27	m	**8.54**
19 mm × 300 mm; cross-tongued joints	–	0.28	4.99	9.70	m	**14.69**
25 mm × 75 mm	–	0.17	3.03	2.99	m	**6.02**
25 mm × 150 mm	–	0.19	3.38	3.81	m	**7.19**
25 mm × 225 mm; in one width	–	0.24	4.27	4.78	m	**9.05**

22 GENERAL JOINERY

Item	PC £	Labour hours	Labour £	Material £	Unit	Total rate £
25 mm × 300 mm; cross-tongued joints	–	0.28	4.99	10.48	m	**15.47**
32 mm × 75 mm	–	0.17	3.03	3.13	m	**6.16**
32 mm × 150 mm	–	0.19	3.38	4.10	m	**7.48**
32 mm × 225 mm; in one width	–	0.24	4.27	5.21	m	**9.48**
32 mm × 300 mm; cross-tongued joints	–	0.28	4.99	11.13	m	**16.12**
38 mm × 75 mm	–	0.17	3.03	3.41	m	**6.44**
38 mm × 150 mm	–	0.19	3.38	4.70	m	**8.08**
38 mm × 225 mm; in one width	–	0.24	4.27	6.07	m	**10.34**
38 mm × 300 mm; cross-tongued joints	–	0.28	4.99	12.45	m	**17.44**
returned and fitted ends	–	0.14	2.49	–	nr	**2.49**
Handrails; mopstick						
50 mm dia.	–	0.23	4.10	6.12	m	**10.22**
Handrails; rounded						
44 mm × 50 mm	–	0.23	4.10	5.93	m	**10.03**
50 mm × 75 mm	–	0.25	4.46	6.48	m	**10.94**
63 mm × 87 mm	–	0.28	4.99	7.20	m	**12.19**
75 mm × 100 mm	–	0.32	5.71	8.80	m	**14.51**
Handrails; moulded						
44 mm × 50 mm	–	0.23	4.10	5.93	m	**10.03**
50 mm × 75 mm	–	0.25	4.46	6.48	m	**10.94**
63 mm × 87 mm	–	0.28	4.99	7.20	m	**12.19**
75 mm × 100 mm	–	0.32	5.71	8.80	m	**14.51**
Sundries on softwood						
Extra over fixing with nails for						
gluing and pinning	–	0.02	0.31	0.03	m	**0.34**
masonry nails	–	0.02	0.32	0.09	m	**0.41**
steel screws	–	0.02	0.30	0.07	m	**0.37**
self-tapping screws	–	0.02	0.31	0.07	m	**0.38**
steel screws; gluing	–	0.03	0.52	0.07	m	**0.59**
steel screws; sinking; filling heads	–	0.04	0.66	0.07	m	**0.73**
steel screws; sinking; pellating over	–	0.08	1.44	0.07	m	**1.51**
brass cups and screws	–	0.10	1.77	0.18	m	**1.95**
Extra over for						
countersinking	–	0.01	0.27	–	m	**0.27**
pellating	–	0.07	1.25	–	m	**1.25**
Head or nut in softwood						
let in flush	–	0.04	0.66	–	nr	**0.66**
Head or nut in hardwood						
let in flush	–	0.06	0.98	–	nr	**0.98**
let in over; pellated	–	0.13	2.30	–	nr	**2.30**

22 GENERAL JOINERY

Item	PC £	Labour hours	Labour £	Material £	Unit	Total rate £
HARDWOOD						
Selected Sapele						
Skirtings, picture rails, dado rails and the like; splayed or moulded						
19 mm × 44 mm; splayed	2.57	0.13	2.32	2.78	m	**5.10**
19 mm × 44 mm; moulded	2.57	0.13	2.32	2.78	m	**5.10**
19 mm × 69 mm; splayed	3.00	0.13	2.32	3.21	m	**5.53**
19 mm × 69 mm; moulded	3.00	0.13	2.32	3.21	m	**5.53**
19 mm × 94 mm; splayed	3.49	0.13	2.32	3.72	m	**6.04**
19 mm × 94 mm; moulded	3.49	0.13	2.32	3.72	m	**6.04**
19 mm × 144 mm; moulded	4.64	0.15	2.68	4.90	m	**7.58**
19 mm × 169 mm; moulded	5.14	0.15	2.68	5.41	m	**8.09**
25 mm × 44 mm; moulded	2.89	0.13	2.32	3.11	m	**5.43**
25 mm × 69 mm; splayed	3.45	0.13	2.32	3.67	m	**5.99**
25 mm × 94 mm; splayed	4.22	0.13	2.32	4.47	m	**6.79**
25 mm × 144 mm; splayed	5.54	0.15	2.68	5.82	m	**8.50**
25 mm × 144 mm; moulded	5.54	0.15	2.68	5.82	m	**8.50**
25 mm × 169 mm; moulded	6.21	0.15	2.68	6.51	m	**9.19**
25 mm × 219 mm; moulded	7.10	0.17	3.03	7.41	m	**10.44**
returned ends	–	0.20	3.57	–	nr	**3.57**
mitres	–	0.14	2.49	–	nr	**2.49**
Architraves, cover fillets and the like; half round; splayed or moulded						
13 mm × 25 mm; half round	1.56	0.15	2.68	1.74	m	**4.42**
13 mm × 50 mm; moulded	2.50	0.15	2.68	2.70	m	**5.38**
16 mm × 32 mm; half round	1.58	0.15	2.68	1.76	m	**4.44**
16 mm × 38 mm; moulded	2.41	0.15	2.68	2.61	m	**5.29**
16 mm × 50 mm; moulded	2.57	0.15	2.68	2.78	m	**5.46**
19 mm × 50 mm; splayed	2.57	0.15	2.68	2.78	m	**5.46**
19 mm × 63 mm; splayed	2.79	0.15	2.68	3.00	m	**5.68**
19 mm × 69 mm; splayed	3.00	0.15	2.68	3.21	m	**5.89**
25 mm × 44 mm; splayed	2.78	0.15	2.68	2.99	m	**5.67**
25 mm × 50 mm; moulded	2.89	0.15	2.68	3.11	m	**5.79**
25 mm × 63 mm; splayed	3.23	0.15	2.68	3.44	m	**6.12**
25 mm × 69 mm; splayed	3.45	0.15	2.68	3.67	m	**6.35**
32 mm × 88 mm; moulded	4.20	0.15	2.68	4.44	m	**7.12**
38 mm × 38 mm; moulded	3.30	0.15	2.68	3.52	m	**6.20**
50 mm × 50 mm; moulded	4.31	0.15	2.68	4.56	m	**7.24**
returned ends	–	0.20	3.57	–	nr	**3.57**
mitres	–	0.14	2.49	–	nr	**2.49**
Stops; screwed on						
16 mm × 38 mm	1.06	0.14	2.49	1.09	m	**3.58**
16 mm × 50 mm	1.14	0.14	2.49	1.17	m	**3.66**
19 mm × 38 mm	1.06	0.14	2.49	1.09	m	**3.58**
25 mm × 38 mm	1.34	0.14	2.49	1.37	m	**3.86**
25 mm × 50 mm	1.56	0.14	2.49	1.60	m	**4.09**
Glazing beads and the like						
13 mm × 16 mm	1.39	0.06	1.07	1.42	m	**2.49**
13 mm × 19 mm	1.39	0.06	1.07	1.42	m	**2.49**
13 mm × 25 mm	1.49	0.06	1.07	1.53	m	**2.60**
13 mm × 25 mm; screwed	2.02	0.06	1.07	2.07	m	**3.14**

22 GENERAL JOINERY

Item	PC £	Labour hours	Labour £	Material £	Unit	Total rate £
13 mm × 25 mm; fixing with brass cups and screws	2.49	0.06	1.07	2.55	m	**3.62**
16 mm × 25 mm; screwed	2.02	0.06	1.07	2.07	m	**3.14**
16 mm quadrant	1.95	0.06	1.07	2.00	m	**3.07**
19 mm quadrant or scotia	1.95	0.06	1.07	2.00	m	**3.07**
19 mm × 36 mm; screwed	2.49	0.06	1.07	2.55	m	**3.62**
25 mm × 38 mm; screwed	2.72	0.06	1.07	2.79	m	**3.86**
25 mm quadrant or scotia	2.23	0.06	1.07	2.29	m	**3.36**
38 mm scotia	3.30	0.06	1.07	3.38	m	**4.45**
50 mm scotia	4.31	0.06	1.07	4.42	m	**5.49**
Isolated shelves; worktops, seats and the like						
19 mm × 150 mm	4.73	0.20	3.57	4.85	m	**8.42**
19 mm × 200 mm	5.59	0.28	4.99	5.73	m	**10.72**
25 mm × 150 mm	5.54	0.20	3.57	5.68	m	**9.25**
25 mm × 200 mm	6.65	0.28	4.99	6.82	m	**11.81**
32 mm × 150 mm	6.25	0.20	3.57	6.41	m	**9.98**
32 mm × 200 mm	7.56	0.28	4.99	7.75	m	**12.74**
Isolated shelves, worktops, seats and the like; cross-tongued joints						
19 mm × 300 mm	12.79	0.35	6.24	13.11	m	**19.35**
19 mm × 450 mm	20.02	0.42	7.48	20.52	m	**28.00**
19 mm × 600 mm	26.60	0.51	9.09	27.27	m	**36.36**
25 mm × 300 mm	14.36	0.35	6.24	14.72	m	**20.96**
25 mm × 450 mm	22.56	0.42	7.48	23.12	m	**30.60**
25 mm × 600 mm	29.99	0.51	9.09	30.74	m	**39.83**
32 mm × 300 mm	15.70	0.35	6.24	16.09	m	**22.33**
32 mm × 450 mm	24.73	0.42	7.48	25.35	m	**32.83**
32 mm × 600 mm	32.88	0.51	9.09	33.70	m	**42.79**
Isolated shelves, worktops, seats and the like; slatted with 50 wide slats at 75 mm centres						
19 mm thick	36.85	0.80	14.26	38.29	m²	**52.55**
25 mm thick	39.45	0.80	14.26	40.96	m²	**55.22**
32 mm thick	41.68	0.80	14.26	43.24	m²	**57.50**
Window boards, nosings, bed moulds and the like; rebated and rounded						
19 mm × 75 mm	3.64	0.22	3.93	4.05	m	**7.98**
19 mm × 150 mm	5.29	0.25	4.46	5.74	m	**10.20**
19 mm × 225 mm; in one width	6.50	0.33	5.88	6.97	m	**12.85**
19 mm × 300 mm; cross-tongued joints	13.23	0.37	6.60	13.87	m	**20.47**
25 mm × 75 mm	4.01	0.22	3.93	4.43	m	**8.36**
25 mm × 150 mm	5.89	0.25	4.46	6.36	m	**10.82**
25 mm × 225 mm; in one width	7.73	0.33	5.88	8.24	m	**14.12**
25 mm × 300 mm; cross-tongued joints	15.37	0.37	6.60	16.07	m	**22.67**
32 mm × 75 mm	4.36	0.22	3.93	4.79	m	**8.72**
32 mm × 150 mm	6.55	0.25	4.46	7.03	m	**11.49**
32 mm × 225 mm; in one width	8.74	0.33	5.88	9.27	m	**15.15**
32 mm × 300 mm; cross-tongued joints	16.91	0.37	6.60	17.64	m	**24.24**
returned and fitted ends	–	0.21	3.74	–	nr	**3.74**

22 GENERAL JOINERY

Item	PC £	Labour hours	Labour £	Material £	Unit	Total rate £
HARDWOOD – cont						
Selected Sapele – cont						
Handrails; rounded						
44 mm × 50 mm	8.03	0.31	5.52	8.23	m	**13.75**
50 mm × 75 mm	9.68	0.33	5.88	9.92	m	**15.80**
63 mm × 87 mm	11.36	0.37	6.60	11.64	m	**18.24**
75 mm × 100 mm	14.11	0.42	7.48	14.46	m	**21.94**
Handrails; moulded						
44 mm × 50 mm	8.93	0.31	5.52	9.15	m	**14.67**
50 mm × 75 mm	10.58	0.33	5.88	10.84	m	**16.72**
63 mm × 87 mm	12.26	0.37	6.60	12.57	m	**19.17**
75 mm × 100 mm	15.02	0.42	7.48	15.40	m	**22.88**
Sundries on hardwood						
Extra over fixing with nails for						
gluing and pinning	–	0.02	0.31	0.03	m	**0.34**
masonry nails	–	0.02	0.32	0.09	m	**0.41**
steel screws	–	0.02	0.30	0.07	m	**0.37**
self-tapping screws	–	0.02	0.31	0.07	m	**0.38**
steel screws; gluing	–	0.03	0.52	0.07	m	**0.59**
steel screws; sinking; filling heads	–	0.04	0.66	0.07	m	**0.73**
steel screws; sinking; pellating over	–	0.08	1.44	0.07	m	**1.51**
brass cups and screws	–	0.10	1.77	0.18	m	**1.95**
Extra over for						
countersinking	–	0.01	0.27	–	m	**0.27**
pellating	–	0.07	1.25	–	m	**1.25**
Head or nut in softwood						
let in flush	–	0.04	0.66	–	nr	**0.66**
Head or nut in hardwood						
let in flush	–	0.06	0.98	–	nr	**0.98**
let in over; pellated	–	0.13	2.30	–	nr	**2.30**
MEDIUM DENSITY FIBREBOARD						
Medium density fibreboard; Sapele veneered one side; 18 mm thick						
Window boards and the like; rebated; hardwood lipped on one edge						
18 mm × 200 mm	–	0.25	4.46	9.50	m	**13.96**
18 mm × 250 mm	–	0.28	4.99	9.99	m	**14.98**
18 mm × 300 mm	–	0.31	5.52	10.23	m	**15.75**
18 mm × 350 mm	–	0.33	5.88	10.97	m	**16.85**
returned and fitted ends	–	0.20	3.57	1.82	nr	**5.39**

22 GENERAL JOINERY

Item	PC £	Labour hours	Labour £	Material £	Unit	Total rate £
Medium density fibreboard; American White Ash veneered one side; 18 mm thick						
Window boards and the like; rebated; hardwood lipped on one edge						
18 mm × 200 mm	–	0.25	4.46	9.87	m	**14.33**
18 mm × 250 mm	–	0.28	4.99	10.48	m	**15.47**
18 mm × 300 mm	–	0.31	5.52	10.78	m	**16.30**
18 mm × 350 mm	–	0.33	5.88	11.70	m	**17.58**
returned and fitted ends	–	0.20	3.57	1.82	nr	**5.39**
Medium density fibreboard						
Skirtings, picture rails, dado rails and the like; splayed or moulded						
18 mm × 50 mm; splayed	–	0.09	1.61	1.81	m	**3.42**
18 mm × 50 mm; moulded	–	0.09	1.61	1.81	m	**3.42**
18 mm × 75 mm; splayed	–	0.09	1.61	1.89	m	**3.50**
18 mm × 75 mm; moulded	–	0.09	1.61	1.89	m	**3.50**
18 mm × 100 mm; splayed	–	0.09	1.61	1.96	m	**3.57**
18 mm × 100 mm; moulded	–	0.09	1.61	1.96	m	**3.57**
18 mm × 150 mm; moulded	–	0.11	1.96	2.12	m	**4.08**
18 mm × 175 mm; moulded	–	0.11	1.96	2.20	m	**4.16**
22 mm × 100 mm; splayed	–	0.09	1.61	3.11	m	**4.72**
25 mm × 50 mm; moulded	–	0.09	1.61	1.90	m	**3.51**
25 mm × 75 mm; splayed	–	0.09	1.61	2.00	m	**3.61**
25 mm × 100 mm; splayed	–	0.09	1.61	2.12	m	**3.73**
25 mm × 150 mm; splayed	–	0.11	1.96	2.38	m	**4.34**
25 mm × 150 mm; moulded	–	0.11	1.96	2.38	m	**4.34**
25 mm × 175 mm; moulded	–	0.11	1.96	2.50	m	**4.46**
25 mm × 225 mm; moulded	–	0.13	2.32	2.68	m	**5.00**
returned ends	–	0.14	2.49	–	nr	**2.49**
mitres	–	0.09	1.61	–	nr	**1.61**
Architraves, cover fillets and the like; half round; splayed or moulded						
12 mm × 25 mm; half round	–	0.11	1.96	1.73	m	**3.69**
12 mm × 50 mm; moulded	–	0.11	1.96	1.78	m	**3.74**
15 mm × 32 mm; half round	–	0.11	1.96	1.74	m	**3.70**
15 mm × 38 mm; moulded	–	0.11	1.96	1.75	m	**3.71**
15 mm × 50 mm; moulded	–	0.11	1.96	1.78	m	**3.74**
18 mm × 50 mm; splayed	–	0.11	1.96	1.78	m	**3.74**
18 mm × 63 mm; splayed	–	0.11	1.96	1.87	m	**3.83**
18 mm × 75 mm; splayed	–	0.11	1.96	1.90	m	**3.86**
25 mm × 44 mm; splayed	–	0.11	1.96	1.90	m	**3.86**
25 mm × 50 mm; moulded	–	0.11	1.96	1.90	m	**3.86**
25 mm × 63 mm; splayed	–	0.11	1.96	1.96	m	**3.92**
25 mm × 75 mm; splayed	–	0.11	1.96	2.02	m	**3.98**
30 mm × 88 mm; moulded	–	0.11	1.96	2.59	m	**4.55**
38 mm × 38 mm; moulded	–	0.11	1.96	2.23	m	**4.19**
50 mm × 50 mm; moulded	–	0.11	1.96	2.35	m	**4.31**
returned ends	–	0.14	2.49	–	nr	**2.49**
mitres	–	0.09	1.61	–	nr	**1.61**

22 GENERAL JOINERY

Item	PC £	Labour hours	Labour £	Material £	Unit	Total rate £
MEDIUM DENSITY FIBREBOARD – cont						
Medium density fibreboard – cont						
Stops; screwed on						
15 mm × 38 mm	–	0.09	1.61	0.95	m	**2.56**
15 mm × 50 mm	–	0.09	1.61	0.98	m	**2.59**
18 mm × 38 mm	–	0.09	1.61	0.97	m	**2.58**
25 mm × 38 mm	–	0.09	1.61	1.03	m	**2.64**
25 mm × 50 mm	–	0.09	1.61	1.08	m	**2.69**
Glazing beads and the like						
12 mm × 16 mm	–	0.04	0.72	1.09	m	**1.81**
12 mm × 19 mm	–	0.04	0.72	1.10	m	**1.82**
12 mm × 25 mm	–	0.04	0.72	1.11	m	**1.83**
12 mm × 25 mm; screwed	–	0.04	0.72	1.58	m	**2.30**
12 mm × 25 mm; fixing with brass cups and screws	–	0.04	0.72	1.81	m	**2.53**
15 mm × 25 mm; screwed	–	0.04	0.72	1.65	m	**2.37**
15 mm quadrant	–	0.04	0.72	1.57	m	**2.29**
18 mm quadrant or scotia	–	0.04	0.72	1.58	m	**2.30**
18 mm × 36 mm; screwed	–	0.04	0.72	1.68	m	**2.40**
25 mm × 38 mm; screwed	–	0.04	0.72	1.75	m	**2.47**
25 mm quadrant or scotia	–	0.04	0.72	1.64	m	**2.36**
38 mm scotia	–	0.04	0.72	1.56	m	**2.28**
50 mm scotia	–	0.04	0.72	1.81	m	**2.53**
Isolated shelves, worktops, seats and the like						
18 mm × 150 mm	–	0.15	2.68	2.06	m	**4.74**
18 mm × 200 mm	–	0.20	3.57	2.17	m	**5.74**
25 mm × 150 mm	–	0.15	2.68	2.37	m	**5.05**
25 mm × 200 mm	–	0.20	3.57	2.52	m	**6.09**
30 mm × 150 mm	–	0.15	2.68	3.31	m	**5.99**
30 mm × 200 mm	–	0.20	3.57	3.67	m	**7.24**
Isolated shelves, worktops, seats and the like; cross-tongued joints						
18 mm × 300 mm	–	0.26	4.63	6.71	m	**11.34**
18 mm × 450 mm	–	0.31	5.52	7.66	m	**13.18**
18 mm × 600 mm	–	0.37	6.60	12.81	m	**19.41**
25 mm × 300 mm	–	0.26	4.63	7.07	m	**11.70**
25 mm × 450 mm	–	0.31	5.52	8.65	m	**14.17**
25 mm × 600 mm	–	0.37	6.60	12.48	m	**19.08**
30 mm × 300 mm	–	0.26	4.63	8.16	m	**12.79**
30 mm × 450 mm	–	0.31	5.52	9.71	m	**15.23**
30 mm × 600 mm	–	0.37	6.60	14.13	m	**20.73**
Isolated shelves, worktops, seats and the like; slatted with 50 wide slats at 75 mm centres						
18 mm thick	–	0.60	10.70	21.29	m	**31.99**
25 mm thick	–	0.60	10.70	22.57	m	**33.27**
30 mm thick	–	0.60	10.70	23.74	m	**34.44**
Window boards, nosings, bed moulds and the like; rebated and rounded						
18 mm × 75 mm	–	0.17	3.03	2.15	m	**5.18**
18 mm × 150 mm	–	0.19	3.38	2.41	m	**5.79**
18 mm × 225 mm	–	0.24	4.27	2.60	m	**6.87**

22 GENERAL JOINERY

Item	PC £	Labour hours	Labour £	Material £	Unit	Total rate £
18 mm × 300 mm	–	0.28	4.99	2.86	m	7.85
25 mm × 75 mm	–	0.17	3.03	2.24	m	5.27
25 mm × 150 mm	–	0.19	3.38	2.60	m	5.98
25 mm × 225 mm	–	0.24	4.27	2.90	m	7.17
25 mm × 300 mm	–	0.28	4.99	3.25	m	8.24
30 mm × 75 mm	–	0.17	3.03	3.00	m	6.03
30 mm × 150 mm	–	0.19	3.38	3.67	m	7.05
30 mm × 225 mm	–	0.24	4.27	4.19	m	8.46
30 mm × 300 mm	–	0.28	4.99	4.80	m	9.79
38 mm × 75 mm	–	0.17	3.03	3.38	m	6.41
38 mm × 150 mm	–	0.19	3.38	4.18	m	7.56
38 mm × 225 mm	–	0.24	4.27	4.80	m	9.07
38 mm × 300 mm	–	0.28	4.99	5.54	m	10.53
returned and fitted ends	–	–	–	0.65	nr	0.65
Pin-boards; medium board						
Sundeala A pin-board or other equal and approved; fixed with adhesive to backing (not included); over 300 mm wide						
6 mm thick	–	0.56	9.98	6.10	m²	16.08
Colourboard; 9 mm thick	–	0.56	9.98	10.44	m²	20.42
ASSOCIATED METALWORK						
Metalwork; mild steel						
Angle section bearers; for building in						
90 mm × 90 mm × 6 mm	–	0.31	6.26	–	m	6.26
120 mm × 120 mm × 8 mm	–	0.32	6.47	–	m	6.47
200 mm × 150 mm × 12 mm	–	0.37	7.47	9.91	m	17.38
Metalwork; mild steel; galvanized						
Waterbars; groove in timber						
6 mm × 30 mm	–	0.46	8.20	4.49	m	12.69
6 mm × 40 mm	–	0.46	8.20	5.64	m	13.84
6 mm × 50 mm	–	0.46	8.20	4.16	m	12.36
Angle section bearers; for building in						
90 mm × 90 mm × 6 mm	–	0.31	6.26	13.10	m	19.36
120 mm × 120 mm × 8 mm	–	0.32	6.47	23.11	m	29.58
200 mm × 150 mm × 12 mm	–	0.37	7.47	51.52	m	58.99
Dowels; mortice in timber						
8 mm dia. × 100 mm long	–	0.04	0.72	0.66	nr	1.38
10 mm dia. × 50 mm long	–	0.04	0.72	1.04	nr	1.76
Cramps						
25 mm × 3 mm × 230 mm girth; one end bent, holed and screwed to softwood; other end fishtailed for building in	–	0.06	1.07	1.56	nr	2.63
Metalwork; stainless steel						
Angle section bearers; for building in						
90 mm × 90 mm × 6 mm	–	0.31	6.26	30.49	m	36.75
120 mm × 120 mm × 8 mm	–	0.32	6.47	53.82	m	60.29
200 mm × 150 mm × 12 mm	–	0.37	7.47	118.04	m	125.51

23 WINDOWS, SCREENS AND LIGHTS

Item	PC £	Labour hours	Labour £	Material £	Unit	Total rate £
WINDOWS						
SUPPLY ONLY PRICES						
NOTE: The following supply only prices are for purpose-made components, to which fixings, sealants etc. labour and overheads and profit need to be added, before they may be used to arrive at a guide price for a complete window. The reader is then referred to the following SUPPLY AND FIX pages for fixing costs based on the overall window size.						
Purpose-made window casements; treated wrought softwood						
Casements; rebated; moulded						
44 mm thick	–	–	–	42.09	m²	**42.09**
57 mm thick	–	–	–	44.19	m²	**44.19**
Casements; rebated; moulded; in medium panes						
44 mm thick	–	–	–	67.28	m²	**67.28**
57 mm thick	–	–	–	70.10	m²	**70.10**
Casements; rebated; moulded; with semi-circular head						
44 mm thick	–	–	–	87.97	m²	**87.97**
57 mm thick	–	–	–	90.72	m²	**90.72**
Casements; rebated; moulded; to bullseye window						
44 mm thick; 600 mm dia.	–	–	–	139.44	nr	**139.44**
44 mm thick; 900 mm dia.	–	–	–	166.11	nr	**166.11**
57 mm thick; 600 mm dia.	–	–	–	146.01	nr	**146.01**
57 mm thick; 900 mm dia.	–	–	–	20.09	nr	**20.09**
Fitting and hanging casements (in factory)						
square or rectangular	–	–	–	9.84	nr	**9.84**
semi-circular	–	–	–	15.99	nr	**15.99**
bullseye	–	–	–	20.09	nr	**20.09**
Purpose-made window casements; selected Sapele						
Casements; rebated; moulded						
44 mm thick	–	–	–	47.84	m²	**47.84**
57 mm thick	–	–	–	52.41	m²	**52.41**
Casements; rebated; moulded; in medium panes						
44 mm thick	–	–	–	77.47	m²	**77.47**
57 mm thick	–	–	–	83.58	m²	**83.58**
Casements; rebated; moulded with semi-circular head						
44 mm thick	–	–	–	97.58	m²	**97.58**
57 mm thick	–	–	–	103.55	m²	**103.55**
Casements; rebated; moulded; to bullseye window						
44 mm thick; 600 mm dia.	–	–	–	172.47	nr	**172.47**
44 mm thick; 900 mm dia.	–	–	–	207.54	nr	**207.54**
57 mm thick; 600 mm dia.	–	–	–	186.69	nr	**186.69**
57 mm thick; 900 mm dia.	–	–	–	225.66	nr	**225.66**

23 WINDOWS, SCREENS AND LIGHTS

Item	PC £	Labour hours	Labour £	Material £	Unit	Total rate £
Fitting and hanging casements (in factory)						
square or rectangular	–	–	–	10.66	nr	**10.66**
semi-circular	–	–	–	17.62	nr	**17.62**
bullseye	–	–	–	22.54	nr	**22.54**
Purpose-made window frames; treated wrought softwood						
Frames; rounded; rebated check grooved						
44 mm × 69 mm	–	–	–	11.44	m	**11.44**
44 mm × 94 mm	–	–	–	11.99	m	**11.99**
44 mm × 119 mm	–	–	–	12.54	m	**12.54**
57 mm × 94 mm	–	–	–	12.57	m	**12.57**
69 mm × 144 mm	–	–	–	16.56	m	**16.56**
90 mm × 140 mm	–	–	–	23.50	m	**23.50**
Mullions and transoms; twice rounded, rebated and check grooved						
57 mm × 69 mm	–	–	–	13.33	m	**13.33**
57 mm × 94 mm	–	–	–	14.00	m	**14.00**
69 mm × 94 mm	–	–	–	15.84	m	**15.84**
69 mm × 144 mm	–	–	–	23.26	m	**23.26**
Sill; sunk weathered, rebated and grooved						
69 mm × 94 mm	–	–	–	27.80	m	**27.80**
69 mm × 144 mm	–	–	–	29.42	m	**29.42**
Purpose-made window frames; selected Sapele						
Frames; rounded; rebated check grooved						
44 mm × 69 mm	–	–	–	14.61	m	**14.61**
44 mm × 94 mm	–	–	–	15.71	m	**15.71**
44 mm × 119 mm	–	–	–	16.82	m	**16.82**
57 mm × 94 mm	–	–	–	18.39	m	**18.39**
69 mm × 144 mm	–	–	–	26.34	m	**26.34**
90 mm × 140 mm	–	–	–	37.42	m	**37.42**
Mullions and transoms; twice rounded, rebated and check grooved						
57 mm × 69 mm	–	–	–	16.60	m	**16.60**
57 mm × 94 mm	–	–	–	19.31	m	**19.31**
69 mm × 94 mm	–	–	–	23.17	m	**23.17**
69 mm × 144 mm	–	–	–	35.20	m	**35.20**
Sill; sunk weathered, rebated and grooved						
69 mm × 94 mm	–	–	–	32.80	m	**32.80**
69 mm × 144 mm	–	–	–	36.36	m	**36.36**

23 WINDOWS, SCREENS AND LIGHTS

Item	PC £	Labour hours	Labour £	Material £	Unit	Total rate £
WINDOWS – cont						
SUPPLY AND FIX PRICES						
Standard windows; treated wrought softwood; Jeld-Wen or other equal and approved Side hung casement windows; factory glazed with low-e 24 mm double glazing (U-value = 1.6 W/m² K); with 140 mm wide softwood sills; opening casements and ventilators hung on rustproof hinges; fitted with aluminized lacquered finish casement stays and fasteners						
488 mm × 750 mm; ref LEWN07V	148.84	0.65	11.58	152.65	nr	**164.23**
488 mm × 900 mm; ref LEWN09V	151.42	0.74	13.19	155.31	nr	**168.50**
630 mm × 750 mm; ref LEW107C	134.56	0.74	13.19	138.02	nr	**151.21**
630 mm × 750 mm; ref LEW107V	165.03	0.74	13.19	169.26	nr	**182.45**
630 mm × 900 mm; ref LEW109V	170.99	0.83	14.80	175.37	nr	**190.17**
630 mm × 900 mm; ref LEW109CH	145.11	0.74	13.19	148.84	nr	**162.03**
630 mm × 1050 mm; ref LEW110C	150.44	0.93	16.57	154.33	nr	**170.90**
630 mm × 1050 mm; ref LEW110V	178.91	0.74	13.19	183.52	nr	**196.71**
915 mm × 900 mm; ref LEW2NO9W	205.71	1.02	18.18	210.96	nr	**229.14**
915 mm × 1050 mm; ref LEW2N1OW	216.11	1.06	18.90	221.65	nr	**240.55**
915 mm × 1200 mm; ref LEW2N12W	233.85	1.11	19.79	239.83	nr	**259.62**
915 mm × 1350 mm; ref LEW2N13W	245.16	1.25	22.28	251.42	nr	**273.70**
915 mm × 1500 mm; ref LEW2N15W	278.17	1.30	23.18	285.30	nr	**308.48**
1200 mm × 750 mm; ref LEW2O7C	211.43	1.06	18.90	216.85	nr	**235.75**
1200 mm × 750 mm; ref LEW2O7CV	267.42	1.06	18.90	274.24	nr	**293.14**
1200 mm × 900 mm; ref LEW2O9C	224.92	1.11	19.79	230.68	nr	**250.47**
1200 mm × 900 mm; ref LEW2O9W	236.68	1.11	19.79	242.73	nr	**262.52**
1200 mm × 900 mm; ref LEW2O9CV	279.44	1.11	19.79	286.56	nr	**306.35**
1200 mm × 1050 mm; ref LEW210C	238.11	1.25	22.28	244.23	nr	**266.51**
1200 mm × 1050 mm; ref LEW210W	250.87	1.25	22.28	257.32	nr	**279.60**
1200 mm × 1050 mm; ref LEW210T	296.01	1.25	22.28	303.57	nr	**325.85**
1200 mm × 1050 mm; ref LEW210CV	290.31	1.25	22.28	297.73	nr	**320.01**
1200 mm × 1200 mm; ref LEW212C	256.01	1.34	23.89	262.60	nr	**286.49**
1200 mm × 1200 mm; ref LEW212W	265.36	1.34	23.89	272.20	nr	**296.09**
1200 mm × 1200 mm; ref LEW212TX	344.84	1.34	23.89	353.67	nr	**377.56**
1200 mm × 1200 mm; ref LEW212CV	305.82	1.34	23.89	313.67	nr	**337.56**
1200 mm × 1350 mm; ref LEW213W	284.87	1.43	25.49	292.19	nr	**317.68**
1200 mm × 1350 mm; ref LEW213CV	338.22	1.43	25.49	346.88	nr	**372.37**
1200 mm × 1500 mm; ref LEW215W	330.36	1.57	27.99	338.84	nr	**366.83**
1770 mm × 750 mm; ref LEW307CC	310.50	1.30	23.18	318.46	nr	**341.64**
1770 mm × 900 mm; ref LEW309CC	330.37	1.57	27.99	338.82	nr	**366.81**
1770 mm × 1050 mm; ref LEW310C	313.12	1.62	28.88	321.14	nr	**350.02**
1770 mm × 1050 mm; ref LEW310T	369.64	1.57	27.99	379.09	nr	**407.08**
1770 mm × 1050 mm; ref LEW310CC	346.70	1.30	23.18	355.57	nr	**378.75**
1770 mm × 1050 mm; ref LEW310CW	365.08	1.30	23.18	374.40	nr	**397.58**
1770 mm × 1200 mm; ref LEW312C	334.40	1.67	29.78	343.00	nr	**372.78**
1770 mm × 1200 mm; ref LEW312T	391.57	1.67	29.78	401.58	nr	**431.36**
1770 mm × 1200 mm; ref LEW312CC	373.65	1.67	29.78	383.22	nr	**413.00**
1770 mm × 1200 mm; ref LEW312CW	387.01	1.67	29.78	396.92	nr	**426.70**

23 WINDOWS, SCREENS AND LIGHTS

Item	PC £	Labour hours	Labour £	Material £	Unit	Total rate £
1770 mm × 1200 mm; ref LEW312CVC	435.96	1.67	29.78	447.09	nr	**476.87**
1770 mm × 1350 mm; ref LEW313CC	423.10	1.76	31.38	433.91	nr	**465.29**
1770 mm × 1350 mm; ref LEW313CW	421.41	1.76	31.38	432.18	nr	**463.56**
1770 mm × 1350 mm; ref LEW313CVC	474.68	1.76	31.38	486.78	nr	**518.16**
1770 mm × 1500 mm; ref LEW315T	503.44	1.85	32.98	516.26	nr	**549.24**
2340 mm × 1050 mm; ref LEW410CWC	481.58	1.80	32.09	493.86	nr	**525.95**
2340 mm × 1200 mm; ref LEW412CWC	511.59	1.90	33.88	524.65	nr	**558.53**
2340 mm × 1350 mm; ref LEW413CWC	562.54	2.04	36.37	576.91	nr	**613.28**
Top hung casement windows; factory glazed with low-e 24 mm double glazing (U-value = 1.6 W/m^2 K); with 140 mm wide softwood sills; opening casements and ventilators hung on rustproof hinges; fitted with aluminized lacquered finish casement stays						
630 mm × 750 mm; ref LEW107 A	145.06	0.74	13.19	148.79	nr	**161.98**
630 mm × 900 mm; ref LEW109 A	153.12	0.83	14.80	157.05	nr	**171.85**
630 mm × 1050 mm; ref LEW110 A	162.21	0.93	16.57	166.40	nr	**182.97**
915 mm × 750 mm; ref LEW2N07 A	176.18	0.97	17.29	180.68	nr	**197.97**
915 mm × 900 mm; ref LEW2N09 A	197.66	1.02	18.18	202.70	nr	**220.88**
915 mm × 1050 mm; ref LEW2N10 A	210.30	1.06	18.90	215.69	nr	**234.59**
915 mm × 1350 mm; ref LEW2N13 AS	265.31	1.25	22.28	272.11	nr	**294.39**
1200 mm × 750 mm; ref LEW207 A	207.29	1.06	18.90	212.61	nr	**231.51**
1200 mm × 900 mm; ref LEW209 A	226.23	1.11	19.79	232.02	nr	**251.81**
1200 mm × 1050 mm; ref LEW210 A	241.87	1.25	22.28	248.08	nr	**270.36**
1200 mm × 1200 mm; ref LEW212 A	262.14	1.34	23.89	268.87	nr	**292.76**
1200 mm × 1350 mm; ref LEW213 AS	306.08	1.43	25.49	313.93	nr	**339.42**
1200 mm × 1500 mm; ref LEW215 AS	334.02	1.57	27.99	342.58	nr	**370.57**
1770 mm × 1050 mm; ref LEW310 AE	334.87	1.57	27.99	343.45	nr	**371.44**
1770 mm × 1200 mm; ref LEW312 AE	356.88	1.67	29.78	366.01	nr	**395.79**
High performance Hi-Profile top-hung reversible windows; factory glazed with low-e 24 mm double glazing (U-value = 1.4 W/m^2 K); weather stripping; opening panes hung on rustproof hinges; fitted with aluminized lacquered espagnolette bolts						
600 mm × 900 mm; ref LEXC0609 AR	232.38	0.83	14.80	238.29	nr	**253.09**
600 mm × 1050 mm; ref LEXC0610 AR	244.08	0.93	16.57	250.32	nr	**266.89**
600 mm × 1200 mm; ref LEXC0612 AR	256.75	1.03	18.36	263.33	nr	**281.69**
600 mm × 1350 mm; ref LEXC0613 AR	269.02	1.11	19.79	275.92	nr	**295.71**
1200 mm × 900 mm; ref LEXC1209 AGR	363.65	1.11	19.79	372.87	nr	**392.66**
1200 mm × 1050 mm; ref LEXC1210 AGR	384.14	1.25	22.28	393.92	nr	**416.20**
1200 mm × 1200 mm; ref LEXC1212 AGR	405.00	1.34	23.89	415.32	nr	**439.21**
1200 mm × 1350 mm; ref LEXC1213 AGR	425.76	1.43	25.49	436.61	nr	**462.10**
1800 mm × 900 mm; ref LEXC1809 AGAR	557.88	1.57	27.99	572.03	nr	**600.02**
1800 mm × 1050 mm; ref LEXC1810 AGAR	589.95	1.62	28.88	604.90	nr	**633.78**
1800 mm × 1200 mm; ref LEXC1812 AGAR	622.75	1.67	29.78	638.54	nr	**668.32**
1800 mm × 1350 mm; ref LEXC1813 AGAR	655.37	1.76	31.38	671.99	nr	**703.37**

23 WINDOWS, SCREENS AND LIGHTS

Item	PC £	Labour hours	Labour £	Material £	Unit	Total rate £
WINDOWS – cont						
Standard windows – cont						
High performance double hung sash windows with glazing bars; factory glazed with low-e 24 mm double glazing (U-value = 1.6 W/m² K); solid frames; 63 mm × 175 mm softwood sills; standard flush external linings; spiral spring balances and sash catch						
635 mm × 1050 mm; ref LESV0610B	383.69	1.85	32.98	393.42	nr	**426.40**
635 mm × 1350 mm; ref LESV0613B	422.88	2.04	36.37	433.63	nr	**470.00**
635 mm × 1650 mm; ref LESV0616B	469.63	2.27	40.47	481.57	nr	**522.04**
860 mm × 1050 mm; ref LESV0810B	430.82	2.13	37.98	441.72	nr	**479.70**
860 mm × 1350 mm; ref LESV0813B	474.58	2.41	42.97	486.61	nr	**529.58**
860 mm × 1650 mm; ref LESV0816B	539.25	2.78	49.56	552.93	nr	**602.49**
1085 mm × 1050 mm; ref LESV1010B	502.87	2.41	42.97	515.57	nr	**558.54**
1085 mm × 1350 mm; ref LESV1013B	534.03	2.78	49.56	547.56	nr	**597.12**
1085 mm × 1650 mm; ref LESV1016B	626.55	3.42	60.97	642.42	nr	**703.39**
1725 mm × 1050 mm; ref LESV1710B	862.49	3.42	60.97	884.22	nr	**945.19**
1725 mm × 1350 mm; ref LESV1713B	962.74	4.26	75.94	987.01	nr	**1062.95**
1725 mm × 1650 mm; ref LESV1716B	1104.82	4.35	77.55	1132.68	nr	**1210.23**
Standard windows; Jeld-Wen Hardwood or other equal and approved; factory applied preservative stain base coat						
Side hung casement windows; factory glazed with low-e 24 mm double glazing (U-value = 1.6 W/m² K); 45 mm × 140 mm hardwood sills; weather stripping; opening sashes on canopy hinges; fitted with fasteners; brown finish ironmongery						
630 mm × 750 mm; ref LEW107CH	199.38	0.88	15.69	204.46	nr	**220.15**
630 mm × 900 mm; ref LEW109CH	210.02	1.11	19.79	215.36	nr	**235.15**
630 mm × 900 mm; ref LEW109VH	230.37	0.88	15.69	236.23	nr	**251.92**
630 mm × 1050 mm; ref LEW2110VH	235.64	1.20	21.39	241.63	nr	**263.02**
915 mm × 900 mm; ref LEWN09WH	346.19	1.39	24.78	354.98	nr	**379.76**
915 mm × 1050 mm; ref LEWN10WH	359.41	1.48	26.38	368.56	nr	**394.94**
915 mm × 1200 mm; ref LEWN12WH	373.36	1.57	27.99	382.86	nr	**410.85**
1200 mm × 900 mm; ref LEW209CH	302.58	1.57	27.99	310.25	nr	**338.24**
1200 mm × 900 mm; ref LEW209WH	383.10	1.57	27.99	392.78	nr	**420.77**
1200 mm × 1050 mm; ref LEW210CH	315.79	1.67	29.78	323.79	nr	**353.57**
1200 mm × 1050 mm; ref LEW210WH	386.02	1.67	29.78	395.77	nr	**425.55**
1200 mm × 1200 mm; ref LEW212CH	329.74	1.80	32.09	338.12	nr	**370.21**
1200 mm × 1200 mm; ref LEW212WH	323.86	1.80	32.09	332.09	nr	**364.18**
1200 mm × 1350 mm; ref LEW213WH	329.14	1.94	34.58	337.50	nr	**372.08**
1200 mm × 1550 mm; ref LEW215WH	334.40	2.04	36.37	342.89	nr	**379.26**
1770 mm × 1050 mm; ref LEW310CCH	562.32	2.08	37.08	576.54	nr	**613.62**
1770 mm × 1200 mm ; ref LEW312CCH	584.94	2.22	39.58	599.74	nr	**639.32**
2339 mm × 1200 mm; ref LEW412CMCH	1235.57	2.41	42.97	1266.65	nr	**1309.62**

23 WINDOWS, SCREENS AND LIGHTS

Item	PC £	Labour hours	Labour £	Material £	Unit	Total rate £
Top hung casement windows; factory glazed with low-e 24 mm double glazing (U-value = 1.6W/m² K); 45 mm × 140 mm hardwood sills; weather stripping; opening sashes on canopy hinges; fitted with fasteners; brown finish ironmongery						
630 mm × 900 mm; ref LEW109 AH	204.40	0.88	15.69	209.61	nr	**225.30**
630 mm × 1050 mm; ref LEW110 AH	215.54	1.20	21.39	221.06	nr	**242.45**
915 mm × 900 mm; ref LEW2N09 AH	251.44	1.39	24.78	257.83	nr	**282.61**
915 mm × 1050 mm; ref LEW2N10 AH	262.59	1.48	26.38	269.29	nr	**295.67**
915 mm × 1350 mm; ref LEW2N13 ASH	322.66	1.67	29.78	330.90	nr	**360.68**
1200 mm × 1050 mm; ref LEW210 AH	296.56	1.57	27.99	304.11	nr	**332.10**
1200 mm × 1350 mm; ref LEW213 ASH	362.51	1.67	29.78	371.71	nr	**401.49**
1770 mm × 1050 mm; ref LEW310 AEH	379.06	1.80	32.09	388.70	nr	**420.79**
Purpose-made double hung sash windows; treated wrought softwood						
Cased frames of 100 mm × 25 mm grooved inner linings; 114 mm × 25 mm grooved outer linings; 125 mm × 38 mm twice rebated head linings; 125 mm × 32 mm twice rebated grooved pulley stiles; 150 mm × 13 mm linings; 50 mm × 19 mm parting slips; 25 mm × 19 mm inside beads; 150 mm × 75 mm Oak twice sunk weathered throated sill; 50 mm thick rebated and moulded sashes; moulded horns						
over 1.25 m² each; both sashes in medium panes; including spiral spring balances	309.10	2.08	37.08	374.30	m²	**411.38**
up to 1.25 m² each; both sashes in medium panes; including spiral spring balances with cased mullions	351.34	2.31	41.18	417.60	m²	**458.78**
Purpose-made double hung sash windows; selected Sapele						
Cased frames of 100 mm × 25 mm grooved inner linings; 114 mm × 25 mm grooved outer linings; 125 mm × 38 mm twice rebated head linings; 125 mm × 32 mm twice rebated grooved pulley stiles; 150 mm × 13 mm linings; 50 mm × 19 mm parting slips; 25 mm × 19 mm inside beads; 150 mm × 75 mm Oak twice sunk weathered throated sill; 50 mm thick rebated and moulded sashes; moulded horns						
over 1.25 m² each; both sashes in medium panes; including spiral sash balances	344.00	2.78	49.56	410.07	m²	**459.63**
over 1.25 m² each; both sashes in medium panes; including spiral sash balances with cased mullions	369.12	3.08	54.91	435.82	m²	**490.73**

23 WINDOWS, SCREENS AND LIGHTS

Item	PC £	Labour hours	Labour £	Material £	Unit	Total rate £
WINDOWS – cont						
Clements EB24 range of factory finished steel fixed light; casement and fanlight windows and doors; with a U-value of 2.0 W/m² K (part L compliant); to EN ISO 9001 2000 ; polyester powder coated; factory glazed with low-e double glazing; fixed in position; including lugs plugged and screwed to brickwork or blockwork						
Basic fixed light including easy-glaze ali snap-on beads						
508 mm × 292 mm	128.70	2.00	61.09	131.99	nr	193.08
508 mm × 457 mm	140.40	2.00	61.09	143.98	nr	205.07
508 mm × 628 mm	152.10	2.00	61.09	156.00	nr	217.09
508 mm × 923 mm	175.50	2.00	61.09	180.02	nr	241.11
508 mm × 1218 mm	198.90	2.50	76.36	204.04	nr	280.40
Basic Tilt and Turn window; including easy-glaze ali snap-on beads						
508 mm × 292 mm	304.20	2.00	61.09	311.88	nr	372.97
508 mm × 457 mm	315.90	2.00	61.09	323.87	nr	384.96
508 mm × 628 mm	327.60	2.00	61.09	335.89	nr	396.98
508 mm × 923 mm; including fixed light	386.10	2.00	61.09	395.89	nr	456.98
508 mm × 1218 mm; including fixed light	409.50	2.50	76.36	419.90	nr	496.26
Basic casement; including easy-glaze snap-on beads						
508 mm × 628 mm	362.70	2.00	61.09	371.87	nr	432.96
508 mm × 923 mm	386.10	2.00	61.09	395.89	nr	456.98
508 mm × 1218 mm	409.50	2.50	76.36	419.90	nr	496.26
Double door						
1143 mm × 2057 mm	2293.20	3.50	106.91	2350.77	nr	2457.68
Extra for						
pressed steel sills; to suit above windows	35.10	0.50	8.92	36.01	m	44.93
G + bar	70.20	–	–	71.96	m	71.96
simulated leaded light	70.20	–	–	71.96	m	71.96
Thermally broken composite double glazed aluminium/timber windows; Velfac 200 or other approved; with a maximum glazing U-value of 1.5 W/m² K; argon filled cavity; low-e glazing with laminated glass unless otherwise specified; including multipoint espagnolette locking mechanisms and other ironmongery						
Standard fixed casement windows						
900 mm × 900 mm single fixed pane; low-e glass 6/14/4	183.26	2.70	48.13	190.79	nr	238.92
900 mm × 2000 mm single fixed pane; low-e glass 6/14/4	391.00	5.94	105.89	405.21	nr	511.10
1200 mm × 1200 mm single fixed pane; low-e glass 6/14/4	250.61	4.75	84.68	260.65	nr	345.33

23 WINDOWS, SCREENS AND LIGHTS

Item	PC £	Labour hours	Labour £	Material £	Unit	Total rate £
1200 mm × 2200 mm three fixed panes; low-e glass 6/14/4	501.23	7.90	140.84	519.10	nr	**659.94**
2200 mm × 2200 mm single fixed pane; low-e glass 6/12/6	636.17	12.00	213.93	659.73	nr	**873.66**
Outward opening standard sash casement windows						
900 mm × 900 mm top hung sash; low-e glass 6/14/4	435.17	2.67	47.60	449.00	nr	**496.60**
900 mm × 2200 mm with small top hung sash; fixed lower pane; low-e glass 6/14/4	435.17	5.90	105.18	448.80	nr	**553.98**
900 mm × 3000 mm with small top hung sash; fixed lower pane; low-e glass 6/14/4	551.73	8.00	142.62	571.48	nr	**714.10**
1600 mm × 1600 mm with two sidehung sashes; low-e glass 4/16/4	510.87	2.60	46.35	528.37	nr	**574.72**
1600 mm × 1600 mm with two sidehung projecting sashes; low-e glass 6/14/4	568.70	2.60	46.35	585.34	nr	**631.69**
1800 mm × 900 mm with two sidehung projecting sashes; low-e glass 6/14/4	376.77	5.50	98.05	387.94	nr	**485.99**
1800 mm × 3000 mm with two sidehung projecting sashes; two top hung sashes; low-e glass 6/14/4	1170.00	15.00	267.40	1206.94	nr	**1474.34**
2000 mm × 1600 mm with one sidehung sash next to a tophung projecting sash over a fixed sash; low-e glass 6/16/4	626.53	9.50	169.36	646.19	nr	**815.55**
1200 mm × 2200 mm with fixed lower sash and tophung projecting upper sash; lower low-e upper low-e glass 4 toughened/16/6.4; upper low-e glass 6/14/4	496.41	7.90	140.84	514.59	nr	**655.43**
1200 mm × 2200 mm with fixed lower sash and fully reversible upper sash; lower low-e upper low-e glass 6 toughened/14/4; upper low-e glass 4/16/4	539.78	7.90	140.84	559.05	nr	**699.89**
1800 mm × 900 mm with two sidehung projecting sashes; low-e glass 6/14/4; 60 minute fire integrity	568.70	5.50	98.05	586.09	nr	**684.14**
Outward opening standard doors						
1800 mm × 2200 mm French casement patio door; low-e toughened glass 4/16/4; deadlock; handles; cylinder	2475.00	13.00	231.75	2544.49	nr	**2776.24**
Alternative cavity fill						
Extra for Krypton cavity fill (over Argon)	–	–	–	19.76	m²	**19.76**
uPVC windows; Profile 22 or other equal and approved; reinforced where appropriate with aluminium alloy; including standard ironmongery; sills and factory glazed with low-e 24 mm double glazing; fixed in position; including lugs plugged and screwed to brickwork or blockwork						
Casement/fixed light; including e.p.d.m. glazing gaskets and weather seals						
630 mm × 900 mm; ref P109C	64.39	1.25	38.18	66.13	nr	**104.31**
630 mm × 1200 mm; ref P112V	73.80	1.50	45.82	75.81	nr	**121.63**
1200 mm × 1200 mm; ref P212C	114.08	1.75	53.45	117.14	nr	**170.59**
1770 mm × 1200 mm; ref P312CC	210.19	2.00	61.09	215.65	nr	**276.74**

23 WINDOWS, SCREENS AND LIGHTS

Item	PC £	Labour hours	Labour £	Material £	Unit	Total rate £
WINDOWS – cont						
uPVC windows – cont						
Casement/fixed light; including vents; e.p.d.m. glazing gaskets and weather seals						
630 mm × 900 mm; ref P109V	36.59	1.25	38.18	37.64	nr	**75.82**
630 mm × 1200 mm; ref P112C	46.81	1.50	45.82	48.14	nr	**93.96**
1200 mm × 1200 mm; ref P212W	69.23	1.75	53.45	71.17	nr	**124.62**
1200 mm × 1200 mm; ref P212CV	116.10	1.75	53.45	119.21	nr	**172.66**
1770 mm × 1200 mm; ref P312WW	153.28	2.00	61.09	157.32	nr	**218.41**
Secured by Design accreditation						
Casement/fixed light; including e.p.d.m. glazing gaskets and weather seals						
630 mm × 900 mm; ref P109C	67.34	1.25	38.18	69.16	nr	**107.34**
630 mm × 1200 mm; ref P112V	77.05	1.50	45.82	79.15	nr	**124.97**
1200 mm × 1200 mm; ref P212C	121.00	1.75	53.45	124.22	nr	**177.67**
1770 mm × 1200 mm; ref P312CC	217.89	2.00	61.09	223.54	nr	**284.63**
Casement/fixed light; including vents; e.p.d.m. glazing gaskets and weather seals						
630 mm × 900 mm; ref P109V	38.07	1.25	38.18	39.16	nr	**77.34**
630 mm × 1200 mm; ref P112C	45.67	1.50	45.82	46.98	nr	**92.80**
1200 mm × 1200 mm; ref P212W	74.68	1.75	53.45	76.74	nr	**130.19**
1200 mm × 1200 mm; ref P212CV	121.47	1.75	53.45	124.70	nr	**178.15**
1770 mm × 1200 mm; ref P312WW	155.48	2.00	61.09	159.57	nr	**220.66**
WER A rating						
Casement/fixed light; including e.p.d.m. glazing gaskets and weather seals						
630 mm × 900 mm; ref P109C	74.04	1.25	38.18	76.02	nr	**114.20**
630 mm × 1200 mm; ref P112V	84.35	1.50	45.82	86.62	nr	**132.44**
1200 mm × 1200 mm; ref P212C	127.80	1.75	53.45	131.20	nr	**184.65**
1770 mm × 1200 mm; ref P312CC	240.80	2.00	61.09	247.01	nr	**308.10**
Casement/fixed light; including vents; e.p.d.m. glazing gaskets and weather seals						
630 mm × 900 mm; ref P109V	41.77	1.25	38.18	42.95	nr	**81.13**
630 mm × 1200 mm; ref P112C	51.86	1.50	45.82	53.32	nr	**99.14**
1200 mm × 1200 mm; ref P212W	78.14	1.75	53.45	80.29	nr	**133.74**
1200 mm × 1200 mm; ref P212CV	132.05	1.75	53.45	135.56	nr	**189.01**
1770 mm × 1200 mm; ref P312WW	168.17	2.00	61.09	172.57	nr	**233.66**
WER C rating						
Casement/fixed light; including e.p.d.m. glazing gaskets and weather seals						
630 mm × 900 mm; ref P109C	64.82	1.25	38.18	66.57	nr	**104.75**
630 mm × 1200 mm; ref P112V	74.21	1.50	45.82	76.23	nr	**122.05**
1200 mm × 1200 mm; ref P212C	114.93	1.75	53.45	118.01	nr	**171.46**
1770 mm × 1200 mm; ref P312CC	211.67	2.00	61.09	217.17	nr	**278.26**

23 WINDOWS, SCREENS AND LIGHTS

Item	PC £	Labour hours	Labour £	Material £	Unit	Total rate £
Casement/fixed light; including vents; e.p.d.m. glazing gaskets and weather seals						
630 mm × 900 mm; ref P109V	36.93	1.25	38.18	37.99	nr	**76.17**
630 mm × 1200 mm; ref P112C	46.80	1.50	45.82	48.13	nr	**93.95**
1200 mm × 1200 mm; ref P212W	70.77	1.75	53.45	72.73	nr	**126.18**
1200 mm × 1200 mm; ref P212CV	116.88	1.75	53.45	120.01	nr	**173.46**
1770 mm × 1200 mm; ref P312WW	151.69	2.00	61.09	155.69	nr	**216.78**
Colour finish uPVC windows						
Casement/fixed light; including e.p.d.m. glazing gaskets and weather seals						
630 mm × 900 mm; ref P109C	73.53	1.25	38.18	75.50	nr	**113.68**
630 mm × 1200 mm; ref P112V	87.27	1.50	45.82	89.63	nr	**135.45**
1200 mm × 1200 mm; ref P212C	135.66	1.75	53.45	139.26	nr	**192.71**
1770 mm × 1200 mm; ref P312CC	266.18	2.00	61.09	273.04	nr	**334.13**
Casement/fixed light; including vents; e.p.d.m. glazing gaskets and weather seals						
630 mm × 900 mm; ref P109V	46.39	1.25	38.18	47.68	nr	**85.86**
630 mm × 1200 mm; ref P112C	52.79	1.50	45.82	54.27	nr	**100.09**
1200 mm × 1200 mm; ref P212W	85.94	1.75	53.45	88.28	nr	**141.73**
1200 mm × 1200 mm; ref P212CV	146.13	1.75	53.45	149.99	nr	**203.44**
1770 mm × 1200 mm; ref P312WW	188.24	2.00	61.09	193.15	nr	**254.24**
uPVC windows; Profile 22 or other equal and approved; reinforced where appropriate with aluminium alloy; in refurbishment work, including standard ironmongery; sills and factory glazed with low-e 24 mm double glazing; removing existing windows and fixing new in position; including lugs plugged and screwed to brickwork or blockwork						
Casement/fixed light; including e.p.d.m. glazing gaskets and weather seals						
630 mm × 900 mm; ref P109C	64.39	2.50	76.36	66.13	nr	**142.49**
630 mm × 1200 mm; ref P112V	46.81	2.50	76.36	48.14	nr	**124.50**
1200 mm × 1200 mm; ref P212C	114.08	3.00	91.64	117.14	nr	**208.78**
1770 mm × 1200 mm; ref P312CC	210.19	3.25	99.27	215.65	nr	**314.92**
Casement/fixed light; including vents; e.p.d.m. glazing gaskets and weather seals						
630 mm × 900 mm; ref P109V	36.59	2.50	76.36	37.64	nr	**114.00**
630 mm × 1200 mm; ref P112C	73.80	2.75	84.00	75.81	nr	**159.81**
1200 mm × 1200 mm; ref P212W	69.23	3.00	91.64	71.17	nr	**162.81**
1200 mm × 1200 mm; ref P212CV	116.10	3.00	91.64	119.21	nr	**210.85**
1770 mm × 1200 mm; ref P312WW	153.28	3.25	99.27	157.32	nr	**256.59**
1770 mm × 1200 mm; ref P312CV	141.27	3.25	99.27	145.00	nr	**244.27**

23 WINDOWS, SCREENS AND LIGHTS

Item	PC £	Labour hours	Labour £	Material £	Unit	Total rate £
WINDOWS – cont						
Aluminium windows; Schuco AWS 50 (or similar) proprietary system or equal and approved						
Polyester powder coated solid colour matt finish or natural anodized window system of glass sealed units with 6.4 mm low-e coated laminated inner pane, air filled cavity and 6 mm clear annealed outer pane. Rates to include all brackets, membranes, sills, silicone seals, trade contractor preliminaries, including external access equipment						
Ribbon construction windows 1.5 m high	–	–	–	–	m²	**399.00**
Punched hole windows fixing into prepared apertures by others	–	–	–	–	m²	**427.50**
Extra for						
1.25 m wide × 1.5 m high opening vents, assuming tilt and turn operation	–	–	–	–	m²	**128.25**
neutral selective high performance coating in lieu of low-e, for assisting in solar control	–	–	–	–	m²	**33.25**
outer glass pane to be toughened and heat soak tested or heat strengthened in lieu of annealed	–	–	–	–	m²	**23.75**
inner laminated glass to be toughened and heat soak tested laminated, or heat strengthened laminated	–	–	–	–	m²	**47.50**
ROOFLIGHTS						
Rooflights, skylights, roof windows and frames; pre-glazed; treated Nordic Red Pine and aluminium trimmed Velux windows or other equal and approved; type U flashings and soakers (for tiles and pantiles up to 45 mm deep), and sealed double glazing unit (trimming opening not included)						
Roof windows; U-value = 1.4 w/m²k						
550 mm × 780 mm; ref GGL–3073-C02	223.20	1.85	32.98	228.87	nr	**261.85**
550 mm × 980 mm; ref GGL–3073-C04	232.80	2.08	37.08	238.71	nr	**275.79**
660 mm × 1180 mm; ref GGL–3073-F06	269.60	2.31	41.18	276.45	nr	**317.63**
780 mm × 980 mm; ref GGL–3073-M04	256.00	2.31	41.18	262.51	nr	**303.69**
780 mm × 1180 mm; ref GGL–3073-M06	294.40	2.78	49.56	301.93	nr	**351.49**
780 mm × 1400 mm; ref GGL–3073-M08	307.20	2.31	41.18	315.03	nr	**356.21**
940 mm × 1600 mm; ref GGL–3073-P10	379.20	2.78	49.56	388.85	nr	**438.41**
1140 mm × 1180 mm; ref GGL–3073-S06	363.20	2.78	49.56	372.45	nr	**422.01**
1340 mm × 980 mm; ref GGL–3073-U04	363.20	2.78	49.56	372.45	nr	**422.01**
extra for electric powered windows plugged in to adjacent power supply	–	–	–	225.50	nr	**225.50**

23 WINDOWS, SCREENS AND LIGHTS

Item	PC £	Labour hours	Labour £	Material £	Unit	Total rate £
Rooflights, skylights, roof windows and frames; uPVC; plugged and screwed to concrete; or screwed to timber						
Rooflight; Cox Suntube range or other equal and approved; double skin polycarbonate dome						
230 mm dia.; for flat roof using felt or membrane	234.59	2.50	44.57	240.86	nr	**285.43**
230 mm dia.; for up to 30° pitch roof with standard tiles	259.37	3.00	53.48	266.19	nr	**319.67**
230 mm dia.; for up to 30° pitch roof with bold roll tiles	240.78	3.00	53.48	247.18	nr	**300.66**
300 mm dia.; for flat roof using felt or membrane	380.65	2.50	44.57	390.58	nr	**435.15**
300 mm dia.; for up to 30° pitch roof with standard tiles	380.65	3.00	53.48	390.50	nr	**443.98**
300 mm dia.; for up to 30° pitch roof with bold roll tiles	408.09	3.00	53.48	418.66	nr	**472.14**
Rooflight; Cox Galaxy range or other equal and approved; double skin polycarbonate dome only; fitting to existing kerb						
600 mm × 600 mm	108.88	1.50	26.74	111.83	nr	**138.57**
900 mm × 900 mm	200.95	1.75	31.20	206.27	nr	**237.47**
1200 mm × 1800 mm	589.56	2.00	35.65	604.68	nr	**640.33**
Rooflight; Cox Galaxy range or other equal and approved; triple skin polycarbonate dome only; fitting to existing kerb						
600 mm × 600 mm rooflight	168.19	1.50	26.74	172.62	nr	**199.36**
900 mm × 900 mm rooflight	301.86	1.75	31.20	309.70	nr	**340.90**
1200 mm × 1800 mm rooflight	912.66	2.00	35.65	935.86	nr	**971.51**
Rooflight; Cox Trade range or other equal and approved; double skin polycarbonate fixed light dome on 150 mm PVC upstand						
600 mm × 600 mm	186.78	2.00	35.65	191.72	nr	**227.37**
900 mm × 900 mm	301.86	2.25	40.11	309.74	nr	**349.85**
1200 mm × 1800 mm	597.52	2.50	44.57	612.88	nr	**657.45**
Rooflight; Cox Trade range or other equal and approved; double skin polycarbonate manual opening light dome on 150 mm PVC upstand						
600 mm × 600 mm	292.12	2.50	44.57	299.99	nr	**344.56**
900 mm × 900 mm	423.13	2.75	49.03	434.35	nr	**483.38**
1200 mm × 1800 mm	795.81	3.10	55.27	816.42	nr	**871.69**
Rooflight; Cox Trade range or other equal and approved; double skin polycarbonate electric opening light dome on 150 mm PVC upstand (not including adjacent power supply)						
600 mm × 600 mm	496.61	2.75	49.03	509.59	nr	**558.62**
900 mm × 900 mm	661.26	3.00	53.48	678.43	nr	**731.91**
1200 mm × 1800 mm	1109.18	3.50	62.39	1137.62	nr	**1200.01**
Rooflight; Cox Trade range or other equal and approved; triple skin polycarbonate fixed light dome on 150 mm PVC upstand						
600 mm × 600 mm	219.53	2.00	35.65	225.28	nr	**260.93**
900 mm × 900 mm	361.17	2.00	35.65	370.46	nr	**406.11**
1200 m × 1800 mm	732.08	2.50	44.57	750.79	nr	**795.36**

23 WINDOWS, SCREENS AND LIGHTS

Item	PC £	Labour hours	Labour £	Material £	Unit	Total rate £
ROOFLIGHTS – cont						
Rooflights, skylights, roof windows and frames – cont						
Rooflight; Cox Trade range or other equal and approved; triple skin polycarbonate manual opening light dome on 150 mm PVC upstand						
600 mm × 600 mm	324.88	2.50	44.57	333.57	nr	**378.14**
900 mm × 900 mm	482.44	2.50	44.57	495.06	nr	**539.63**
1200 m × 1800 mm	930.37	3.10	55.27	954.34	nr	**1009.61**
Rooflight; Cox Trade range or other equal and approved; triple skin polycarbonate electric opening light dome on 150 mm PVC upstand						
600 mm × 600 mm light	529.36	2.75	49.03	543.16	nr	**592.19**
900 mm × 900 mm light	720.57	2.75	49.03	739.15	nr	**788.18**
1200 mm × 1800 mm light	1243.74	3.50	62.39	1275.54	nr	**1337.93**
Rooflight; Cox 2000 range or other equal and approved double skin polycarbonate fixed light dome on 235 mm solid core PVC upstand						
600 mm × 600 mm	597.52	2.00	35.65	612.72	nr	**648.37**
900 mm × 900 mm	810.38	2.50	44.57	830.91	nr	**875.48**
1200 mm × 1800 mm	2161.50	3.00	53.48	2215.83	nr	**2269.31**
Rooflight; Cox 2000 range or other equal and approved double skin polycarbonate manual opening light dome on 235 mm solid core PVC upstand						
600 mm × 600 mm	820.60	2.50	44.57	841.68	nr	**886.25**
900 mm × 900 mm	1033.46	3.00	53.48	1059.86	nr	**1113.34**
1200 mm × 1800 mm	2589.95	3.60	64.18	2655.29	nr	**2719.47**
Rooflight; Cox 2000 range or other equal and approved double skin polycarbonate electric opening light dome on 235 mm solid core PVC upstand (not including adjacent power supply)						
600 mm × 600 mm	1079.97	2.75	49.03	1107.53	nr	**1156.56**
900 mm × 900 mm	1292.82	3.25	57.94	1325.70	nr	**1383.64**
1200 mm × 1800 mm	2915.72	4.00	71.31	2989.21	nr	**3060.52**
Rooflight; Cox 2000 range or other equal and approved triple skin polycarbonate fixed light dome on 235 mm solid core PVC upstand						
600 mm × 600 mm	804.66	2.00	35.65	825.04	nr	**860.69**
900 mm × 900 mm	1175.63	2.50	44.57	1205.28	nr	**1249.85**
1200 mm × 1800 mm	2887.50	3.00	53.48	2959.98	nr	**3013.46**
Rooflight; Cox 2000 range or other equal and approved triple skin polycarbonate manual opening light dome on 235 mm solid core PVC upstand						
600 mm × 600 mm	1027.74	2.50	44.57	1054.00	nr	**1098.57**
900 mm × 900 mm	1398.70	3.00	53.48	1434.23	nr	**1487.71**
1200 mm × 1800 mm	3315.95	3.60	64.18	3399.44	nr	**3463.62**

23 WINDOWS, SCREENS AND LIGHTS

Item	PC £	Labour hours	Labour £	Material £	Unit	Total rate £
Rooflight; Cox 2000 range or other equal and approved triple skin polycarbonate electric opening light dome on 235 mm solid core PVC upstand (not including adjacent power supply)						
600 mm × 600 mm	1287.11	2.75	49.03	1319.85	nr	**1368.88**
900 mm × 900 mm	1658.07	3.25	57.94	1700.09	nr	**1758.03**
1200 mm × 1800 mm	3641.72	4.00	71.31	3733.36	nr	**3804.67**
LOUVRES						
Louvres, Brise Soleils and frames; polyester powder coated aluminium; fixing in position including brackets Brise soleil, to mitigate the effects of solar gain. This rate assumes a single natural anodized extruded aluminium fin, with brackets and orientated either horizontally or vertically. The quantity of fins per storey height should be calculated to achieve desired shading						
300 mm deep	–	–	–	–	m	**125.00**

24 DOORS, SHUTTERS AND HATCHES

Item	PC £	Labour hours	Labour £	Material £	Unit	Total rate £
EXTERNAL DOORS AND FRAMES						
EXTERNAL DOORS						
NOTE: door ironmogery is not included unless stated.						
Doors; standard matchboarded; wrought softwood						
Matchboarded, framed, ledged and braced doors; 44 mm thick overall; 19 mm thick tongued, grooved and V-jointed boarding; one side vertical boarding						
762 mm × 1981 mm	67.03	1.67	29.78	68.71	nr	**98.49**
838 mm × 1981 mm	73.24	1.67	29.78	75.07	nr	**104.85**
Flush door; external quality; skeleton or cellular core; plywood faced both sides; lipped all round						
762 mm × 1981 mm × 54 mm	85.69	1.62	28.88	87.83	nr	**116.71**
838 mm × 1981 mm × 54 mm	87.54	1.62	28.88	89.73	nr	**118.61**
Fire doors						
Flush door; half-hour fire-resisting; external quality; skeleton or cellular core; plywood faced both sides; lipped on all four edges						
762 mm × 1981 mm × 44 mm	180.57	1.71	30.48	185.08	nr	**215.56**
838 mm × 1981 mm × 44 mm	182.41	1.71	30.48	186.97	nr	**217.45**
Flush door; half-hour fire-resisting; external quality with 6 mm Georgian wired polished plate glass opening; skeleton or cellular core; plywood faced both sides; lipped on all four edges; including glazing beads						
762 mm × 1981 mm × 44 mm	180.57	1.71	30.48	185.08	nr	**215.56**
838 mm × 1981 mm × 44 mm	182.41	1.71	30.48	186.97	nr	**217.45**
726 mm × 2040 mm × 44 mm	180.57	1.71	30.48	185.08	nr	**215.56**
826 mm × 2040 mm × 44 mm	182.49	1.71	30.48	187.05	nr	**217.53**
926 mm × 2040 mm × 44 mm	192.56	1.71	30.48	197.37	nr	**227.85**
External softwood door frame composite standard joinery sets						
External door frame composite set; 56 mm × 78 mm wide (finished); for external doors						
762 mm × 1981 mm × 44 mm	49.50	0.75	13.37	50.87	nr	**64.24**
813 mm × 1981 mm × 44 mm	50.82	0.75	13.37	52.22	nr	**65.59**
838 mm × 1981 mm × 44 mm	50.48	0.75	13.37	51.88	nr	**65.25**

24 DOORS, SHUTTERS AND HATCHES

Item	PC £	Labour hours	Labour £	Material £	Unit	Total rate £
Doorsets; Anti-Vandal Security door and frame units; Bastion Security Ltd or other equal and approved; to BS 5051; factory primed; fixing with frame anchors to masonry; cutting mortices; external						
46 mm thick insulated door with Birch grade plywood; sheet steel bonded into door core; 2 mm thick polyester coated laminate finish; hardwood lippings all edges; 95 mm × 65 mm hardwood frame; polyester coated standard ironmongery; weather stripping all round; low projecting aluminium threshold; plugging; screwing						
for 980 mm × 2100 mm structural opening; single doorsets; panic bolt	–	–	–	–	nr	**1350.00**
for 1830 mm × 2100 mm structural opening; double doorsets; panic bolt	–	–	–	–	nr	**2250.00**
Doorsets; galvanized steel door and frame units; treated softwood frame, primed hardwood sill; fixing in position; plugged and screwed to brickwork or blockwork						
Door and frame						
838 mm × 1981 mm	–	2.78	49.56	387.58	nr	**437.14**
Doorsets; steel security door and frame; Hormann or other equal and approved; including ironmongery, weather seals and all necessary fixing accessories						
Horman ref E55–1 doorset						
to suit structural opening 1100 × 2105 mm; fire-rating 30 minutes; acoustic rating 38 dB; including stainless steel ironmongery	–	–	–	–	nr	**1537.81**
to suit structural opening 2000 × 2105 mm; fire-rating 30 minutes; acoustic rating 38 dB; including stainless steel ironmongery	–	–	–	–	nr	**2284.75**
Doorsets; steel bullet-resistant door and frame units; Wormald Doors or other equal and approved; Medite laquered panels; ironmongery						
Door and frame						
1000 mm × 2060 mm overall; fixed to masonry	–	–	–	–	nr	**3207.44**

24 DOORS, SHUTTERS AND HATCHES

Item	PC £	Labour hours	Labour £	Material £	Unit	Total rate £
EXTERNAL DOORS AND FRAMES – cont						
uPVC doors; Profile 22 or other equal and approved; reinforced where appropriate with aluminium alloy; including standard ironmongery; thresholds and factory glazed with low-e 24 mm double glazing; fixed in position; including lugs plugged and screwed to brickwork or blockwork						
Fixed light; including e.p.d.m. glazing gaskets and weather seals						
2100 × 900 uPVC door with midrail; half glazed	199.33	1.50	45.82	204.46	nr	**250.28**
2100 × 900 uPVC door with midrail; glazed	250.68	1.50	45.82	257.12	nr	**302.94**
Secured By Design accrediation						
Fixed light; including e.p.d.m. glazing gaskets and weather seals						
2100 × 900 uPVC door with midrail; half glazed	248.63	2.00	61.09	254.99	nr	**316.08**
2100 × 900 uPVC door with midrail; glazed	250.68	1.50	45.82	257.12	nr	**302.94**
WER A rating						
Fixed light; including e.p.d.m. glazing gaskets and weather seals						
2100 × 900 uPVC door with midrail; half glazed	205.18	2.00	61.09	210.45	nr	**271.54**
2100 × 900 uPVC door with midrail; glazed	238.76	1.50	45.82	244.89	nr	**290.71**
WER C rating						
Fixed light; including e.p.d.m. glazing gaskets and weather seals						
2100 × 900 uPVC door with midrail; half glazed	190.68	2.00	61.09	195.59	nr	**256.68**
2100 × 900 uPVC door with midrail; glazed	229.59	1.50	45.82	235.50	nr	**281.32**
Colour finish						
Fixed light; including e.p.d.m. glazing gaskets and weather seals						
2100 × 900 uPVC door with midrail; half glazed	296.51	2.00	61.09	304.07	nr	**365.16**
2100 × 900 uPVC door with midrail; glazed	339.42	1.50	45.82	348.07	nr	**393.89**
Sliding/folding; aluminium double glazed sliding patio doors; Crittal Luminaire or equal and approved; white acrylic finish; with and including 18 thick annealed double glazing; fixed in position; including lugs plugged and screwed to brickwork or blockwork						
Patio doors						
1800 mm × 2100 mm; ref PF1821	1487.50	2.31	41.18	1525.09	nr	**1566.27**
2400 mm × 2100 mm; ref PF2421	2422.50	2.78	49.56	2483.46	nr	**2533.02**
2700 mm × 2100 mm; ref PF2721	2163.25	3.24	57.76	2217.73	nr	**2275.49**

24 DOORS, SHUTTERS AND HATCHES

Item	PC £	Labour hours	Labour £	Material £	Unit	Total rate £
External softwood door frame composite standard joinery sets						
External door frame composite set; 56 mm × 78 mm wide (finished); for external doors						
762 mm × 1981 mm × 44 mm	49.50	0.75	13.37	50.87	nr	**64.24**
813 mm × 1981 mm × 44 mm	50.82	0.75	13.37	52.22	nr	**65.59**
838 mm × 1981 mm × 44 mm	50.48	0.75	13.37	51.88	nr	**65.25**
External door frame composite set; 56 mm × 78 mm wide (finished); with 45 mm × 140 mm (finished) hardwood sill; for external doors						
686 mm × 1981 mm × 44 mm	61.34	1.00	17.82	63.01	nr	**80.83**
762 mm × 1981 mm × 44 mm	63.50	1.00	17.82	65.22	nr	**83.04**
838 mm × 1981 mm × 44 mm	74.87	1.00	17.82	76.88	nr	**94.70**
826 mm × 2040 mm × 44 mm	65.66	1.00	17.82	67.43	nr	**85.25**
INTERNAL DOORS AND FRAMES						
INTERNAL DOORS						
NOTE: door ironmogery is not included.						
Moulded panel doors; white based coated facings suitable for paint finish only; two, four or six panel options						
526 mm × 2040 mm × 40 mm	31.44	0.75	13.37	32.23	nr	**45.60**
626 mm × 2040 mm × 40 mm	31.44	0.75	13.37	32.23	nr	**45.60**
726 mm × 2040 mm × 40 mm	31.44	0.75	13.37	32.23	nr	**45.60**
826 mm × 2040 mm × 40 mm	34.51	0.75	13.37	35.37	nr	**48.74**
926 mm × 2040 mm × 40 mm	39.11	0.75	13.37	40.09	nr	**53.46**
Doors; standard flush; softwood composition						
Flush door; internal quality; skeleton or cellular core; hardboard faced both sides; lipped on two long edges; Jeld-Wen or other equal and approved						
457 mm × 1981 mm × 35 mm	30.36	1.16	20.68	31.12	nr	**51.80**
533 mm × 1981 mm × 35 mm	30.36	1.16	20.68	31.12	nr	**51.80**
610 mm × 1981 mm × 35 mm	30.36	1.16	20.68	31.12	nr	**51.80**
686 mm × 1981 mm × 35 mm	30.36	1.16	20.68	31.12	nr	**51.80**
762 mm × 1981 mm × 35 mm	30.36	1.16	20.68	31.12	nr	**51.80**
838 mm × 1981 mm × 35 mm	34.41	1.16	20.68	35.27	nr	**55.95**
626 mm × 2040 mm × 40 mm	32.38	1.16	20.68	33.19	nr	**53.87**
726 mm × 2040 mm × 40 mm	32.38	1.16	20.68	33.19	nr	**53.87**
826 mm × 2040 mm × 40 mm	32.38	1.16	20.68	33.19	nr	**53.87**
926 mm × 2040 mm × 40 mm	34.41	1.16	20.68	35.27	nr	**55.95**
Flush door; internal quality; skeleton or cellular core; faced both sides; lipped on two long edges; Jeld-Wen paint grade veneer or other equal and approved						
457 mm × 1981 mm × 35 mm	34.41	1.16	20.68	35.27	nr	**55.95**
533 mm × 1981 mm × 35 mm	34.41	1.16	20.68	35.27	nr	**55.95**
610 mm × 1981 mm × 35 mm	34.41	1.16	20.68	35.27	nr	**55.95**
686 mm × 1981 mm × 35 mm	34.41	1.16	20.68	35.27	nr	**55.95**
762 mm × 1981 mm × 35 mm	34.41	1.16	20.68	35.27	nr	**55.95**
838 mm × 1981 mm × 35 mm	36.43	1.16	20.68	37.34	nr	**58.02**

24 DOORS, SHUTTERS AND HATCHES

Item	PC £	Labour hours	Labour £	Material £	Unit	Total rate £
INTERNAL DOORS AND FRAMES – cont						
Doors – cont						
Flush door – cont						
526 mm × 2040 mm × 40 mm	35.42	1.16	20.68	36.31	nr	**56.99**
626 mm × 2040 mm × 40 mm	35.42	1.16	20.68	36.31	nr	**56.99**
726 mm × 2040 mm × 40 mm	36.43	1.16	20.68	37.34	nr	**58.02**
826 mm × 2040 mm × 40 mm	42.50	1.16	20.68	43.56	nr	**64.24**
Flush door; internal quality; skeleton or cellular core; chipboard veneered; faced both sides; lipped on two long edges; Jeld-Wen Sapele veneered or other equal and approved						
457 mm × 1981 mm × 35 mm	56.67	1.25	22.28	58.09	nr	**80.37**
533 mm × 1981 mm × 35 mm	56.67	1.25	22.28	58.09	nr	**80.37**
610 mm × 1981 mm × 35 mm	56.67	1.25	22.28	58.09	nr	**80.37**
686 mm × 1981 mm × 35 mm	56.67	1.25	22.28	58.09	nr	**80.37**
762 mm × 1981 mm × 35 mm	56.67	1.25	22.28	58.09	nr	**80.37**
838 mm × 1981 mm × 35 mm	62.74	1.25	22.28	64.31	nr	**86.59**
526 mm × 2040 mm × 40 mm	62.74	1.25	22.28	64.31	nr	**86.59**
626 mm × 2040 mm × 40 mm	62.74	1.25	22.28	64.31	nr	**86.59**
726 mm × 2040 mm × 40 mm	62.74	1.25	22.28	64.31	nr	**86.59**
826 mm × 2040 mm × 40 mm	66.79	1.25	22.28	68.46	nr	**90.74**
926 mm × 2040 mm × 40 mm	68.82	1.25	22.28	70.54	nr	**92.82**
Doors; purpose-made panelled; wrought softwood						
Panelled doors; one open panel for glass; including glazing beads						
686 mm × 1981 mm × 44 mm	74.88	1.62	28.88	76.75	nr	**105.63**
762 mm × 1981 mm × 44 mm	75.50	1.62	28.88	77.39	nr	**106.27**
838 mm × 1981 mm × 44 mm	76.13	1.62	28.88	78.03	nr	**106.91**
Panelled doors; two open panels for glass; including glazing beads						
686 mm × 1981 mm × 44 mm	104.70	1.62	28.88	107.32	nr	**136.20**
762 mm × 1981 mm × 44 mm	105.60	1.62	28.88	108.24	nr	**137.12**
838 mm × 1981 mm × 44 mm	106.50	1.62	28.88	109.16	nr	**138.04**
Panelled doors; four 19 mm thick plywood panels; mouldings worked on solid both sides						
686 mm × 1981 mm × 44 mm	158.98	1.62	28.88	162.95	nr	**191.83**
762 mm × 1981 mm × 44 mm	161.04	1.62	28.88	165.07	nr	**193.95**
838 mm × 1981 mm × 44 mm	163.10	1.62	28.88	167.18	nr	**196.06**
Panelled doors; six 25 mm thick panels raised and fielded; mouldings worked on solid both sides						
686 mm × 1981 mm × 44 mm	292.54	1.94	34.58	299.85	nr	**334.43**
762 mm × 1981 mm × 44 mm	295.46	1.94	34.58	302.85	nr	**337.43**
838 mm × 1981 mm × 44 mm	298.38	1.94	34.58	305.84	nr	**340.42**
rebated edges beaded	–	–	–	1.80	m	**1.80**
rounded edges or heels	–	–	–	0.41	m	**0.41**
weatherboard fixed to bottom rail	–	0.23	4.10	6.41	m	**10.51**
stopped groove for weatherboard	–	–	–	2.05	m	**2.05**

24 DOORS, SHUTTERS AND HATCHES

Item	PC £	Labour hours	Labour £	Material £	Unit	Total rate £
Doors; purpose-made panelled; selected Sapele						
Panelled doors; one open panel for glass; including glazing beads						
686 mm × 1981 mm × 44 mm	101.38	2.31	41.18	103.91	nr	**145.09**
762 mm × 1981 mm × 44 mm	102.74	2.31	41.18	105.31	nr	**146.49**
838 mm × 1981 mm × 44 mm	104.13	2.31	41.18	106.73	nr	**147.91**
686 mm × 1981 mm × 57 mm	108.36	2.54	45.28	111.07	nr	**156.35**
762 mm × 1981 mm × 57 mm	109.99	2.54	45.28	112.74	nr	**158.02**
838 mm × 1981 mm × 57 mm	111.62	2.54	45.28	114.41	nr	**159.69**
Panelled doors; 250 mm wide cross-tongued intermediate rail; two open panels for glass; mouldings worked on the solid one side; 19 mm × 13 mm beads one side; fixing with brass cups and screws						
686 mm × 1981 mm × 44 mm	154.98	2.31	41.18	158.85	nr	**200.03**
762 mm × 1981 mm × 44 mm	157.60	2.31	41.18	161.54	nr	**202.72**
838 mm × 1981 mm × 44 mm	165.26	2.31	41.18	169.39	nr	**210.57**
686 mm × 1981 mm × 57 mm	165.26	2.54	45.28	169.39	nr	**214.67**
762 mm × 1981 mm × 57 mm	168.35	2.54	45.28	172.56	nr	**217.84**
838 mm × 1981 mm × 57 mm	171.53	2.54	45.28	175.82	nr	**221.10**
Panelled doors; four panels; (19 mm thick for 44 mm doors, 25 mm thick for 57 mm doors); mouldings worked on solid both sides						
686 mm × 1981 mm × 44 mm	217.78	2.31	41.18	223.22	nr	**264.40**
762 mm × 1981 mm × 44 mm	234.76	2.31	41.18	240.63	nr	**281.81**
838 mm × 1981 mm × 44 mm	246.37	2.31	41.18	252.53	nr	**293.71**
686 mm × 1981 mm × 57 mm	222.18	2.54	45.28	227.73	nr	**273.01**
762 mm × 1981 mm × 57 mm	240.57	2.54	45.28	246.58	nr	**291.86**
838 mm × 1981 mm × 57 mm	226.59	2.54	45.28	232.25	nr	**277.53**
Panelled doors; 150 mm wide stiles in one width; 430 mm wide cross-tongued bottom rail; six panels raised and fielded one side; (19 mm thick for 44 mm doors, 25 mm thick for 57 mm doors); mouldings worked on solid both sides						
686 mm × 1981 mm × 44 mm	371.77	2.31	41.18	381.06	nr	**422.24**
762 mm × 1981 mm × 44 mm	410.26	2.31	41.18	420.52	nr	**461.70**
838 mm × 1981 mm × 44 mm	418.06	2.31	41.18	428.51	nr	**469.69**
686 mm × 1981 mm × 57 mm	395.88	2.54	45.28	405.78	nr	**451.06**
762 mm × 1981 mm × 57 mm	436.52	2.54	45.28	447.43	nr	**492.71**
838 mm × 1981 mm × 57 mm	446.46	2.54	45.28	457.62	nr	**502.90**
rebated edges beaded	–	–	–	2.30	m	**2.30**
rounded edges or heels	–	–	–	0.60	m	**0.60**
weatherboard fixed to bottom rail	–	0.31	5.52	8.66	m	**14.18**
stopped groove for weatherboard	–	–	–	2.13	m	**2.13**

24 DOORS, SHUTTERS AND HATCHES

Item	PC £	Labour hours	Labour £	Material £	Unit	Total rate £
INTERNAL DOORS AND FRAMES – cont						
Internal white foiled moisture-resistant MDF door lining composite standard joinery set						
22 mm × 77 mm wide (finished) set; with loose stops; for internal doors						
610 mm × 1981 mm × 35 mm	7.92	0.70	12.47	8.25	nr	**20.72**
686 mm × 1981 mm × 35 mm	7.92	0.70	12.47	8.25	nr	**20.72**
762 mm × 1981 mm × 35 mm	7.92	0.70	12.47	8.25	nr	**20.72**
838 mm × 1981 mm × 35 mm	8.72	0.70	12.47	9.07	nr	**21.54**
864 mm × 1981 mm × 35 mm	7.92	0.70	12.47	8.25	nr	**20.72**
22 mm × 150 mm wide (finished) set; with loose stops; for internal doors						
610 mm × 1981 mm × 35 mm	8.72	0.70	12.47	9.07	nr	**21.54**
686 mm × 1981 mm × 35 mm	8.72	0.70	12.47	9.07	nr	**21.54**
762 mm × 1981 mm × 35 mm	8.72	0.70	12.47	9.07	nr	**21.54**
838 mm × 1981 mm × 35 mm	8.72	0.70	12.47	9.07	nr	**21.54**
864 mm × 1981 mm × 35 mm	8.72	0.70	12.47	9.07	nr	**21.54**
Internal softwood door lining set; with loose stops						
32 mm × 115 mm wide (finished) set; with loose stops; for internal doors						
686/762/838 wide doors	11.90	0.70	12.47	12.33	nr	**24.80**
32 mm × 138 mm wide (finished) set; with loose stops; for internal doors						
686/762/838 wide doors	13.52	0.70	12.47	13.99	nr	**26.46**
Internal softwood fire door door lining set; with loose stops						
38 mm × 115 mm wide (finished) set; with loose stops; for internal doors						
686/762/838 wide doors	17.93	0.70	12.47	18.51	nr	**30.98**
38 mm × 138 mm wide (finished) set; with loose stops; for internal doors						
686/762/838 wide doors	21.28	0.70	12.47	21.95	nr	**34.42**
Door frames and door linings, sets; purpose-made; wrought softwood						
Jambs and heads; as linings						
32 mm × 63 mm	–	0.16	2.85	4.41	m	**7.26**
32 mm × 100 mm	–	0.16	2.85	4.97	m	**7.82**
32 mm × 140 mm	–	0.16	2.85	5.29	m	**8.14**
Jambs and heads; as frames; rebated, rounded and grooved						
44 mm × 75 mm	–	0.16	2.85	7.10	m	**9.95**
44 mm × 100 mm	–	0.16	2.85	7.65	m	**10.50**
44 mm × 115 mm	–	0.16	2.85	7.68	m	**10.53**
44 mm × 140 mm	–	0.19	3.38	8.05	m	**11.43**
57 mm × 100 mm	–	0.19	3.38	8.16	m	**11.54**
57 mm × 125 mm	–	0.19	3.38	8.62	m	**12.00**
69 mm × 88 mm	–	0.19	3.38	8.31	m	**11.69**
69 mm × 100 mm	–	0.19	3.38	8.90	m	**12.28**
69 mm × 125 mm	–	0.20	3.57	9.45	m	**13.02**
69 mm × 150 mm	–	0.20	3.57	10.00	m	**13.57**

24 DOORS, SHUTTERS AND HATCHES

Item	PC £	Labour hours	Labour £	Material £	Unit	Total rate £
94 mm × 100 mm	–	0.23	4.10	13.50	m	**17.60**
94 mm × 150 mm	–	0.23	4.10	15.73	m	**19.83**
Mullions and transoms; in linings						
32 mm × 63 mm	–	0.11	1.96	5.88	m	**7.84**
32 mm × 100 mm	–	0.11	1.96	6.45	m	**8.41**
32 mm × 140 mm	–	0.11	1.96	6.72	m	**8.68**
Mullions and transoms; in frames; twice rebated, rounded and grooved						
44 mm × 75 mm	–	0.11	1.96	8.92	m	**10.88**
44 mm × 100 mm	–	0.11	1.96	9.29	m	**11.25**
44 mm × 115 mm	–	0.11	1.96	9.29	m	**11.25**
44 mm × 140 mm	–	0.13	2.32	9.66	m	**11.98**
57 mm × 100 mm	–	0.13	2.32	9.78	m	**12.10**
57 mm × 125 mm	–	0.13	2.32	10.23	m	**12.55**
69 mm × 88 mm	–	0.13	2.32	9.66	m	**11.98**
69 mm × 100 mm	–	0.13	2.32	10.22	m	**12.54**
Add 5% to the above material prices for selected softwood for staining						
Door frames and door linings, sets; purpose-made; medium density fireboard						
Jambs and heads; as linings						
18 mm × 126 mm	–	0.16	2.85	5.31	m	**8.16**
22 mm × 126 mm	–	0.16	2.85	5.50	m	**8.35**
25 mm × 126 mm	–	0.16	2.85	5.61	m	**8.46**
Door frames and door linings, sets; purpose-made; selected Sapele						
Jambs and heads; as linings						
32 mm × 63 mm	6.66	0.21	3.74	6.86	m	**10.60**
32 mm × 100 mm	8.22	0.21	3.74	8.47	m	**12.21**
32 mm × 140 mm	9.00	0.21	3.74	9.30	m	**13.04**
Jambs and heads; as frames; rebated, rounded and grooved						
44 mm × 75 mm	10.90	0.21	3.74	11.21	m	**14.95**
44 mm × 100 mm	12.42	0.21	3.74	12.77	m	**16.51**
44 mm × 115 mm	12.84	0.21	3.74	13.23	m	**16.97**
44 mm × 140 mm	13.46	0.25	4.46	13.86	m	**18.32**
57 mm × 100 mm	13.76	0.25	4.46	14.18	m	**18.64**
57 mm × 125 mm	15.06	0.25	4.46	15.50	m	**19.96**
69 mm × 88 mm	13.74	0.25	4.46	14.12	m	**18.58**
69 mm × 100 mm	15.29	0.25	4.46	15.73	m	**20.19**
69 mm × 125 mm	16.83	0.28	4.99	17.32	m	**22.31**
69 mm × 150 mm	18.38	0.28	4.99	18.90	m	**23.89**
94 mm × 100 mm	21.46	0.28	4.99	22.07	m	**27.06**
94 mm × 150 mm	26.33	0.28	4.99	27.05	m	**32.04**
Mullions and transoms; in linings						
32 mm × 63 mm	8.35	0.15	2.68	8.56	m	**11.24**
32 mm × 100 mm	9.91	0.15	2.68	10.16	m	**12.84**
32 mm × 140 mm	10.70	0.15	2.68	10.97	m	**13.65**

24 DOORS, SHUTTERS AND HATCHES

Item	PC £	Labour hours	Labour £	Material £	Unit	Total rate £
INTERNAL DOORS AND FRAMES – cont						
Door frames and door linings, sets – cont						
Mullions and transoms; in frames; twice rebated, rounded and grooved						
44 mm × 75 mm	13.42	0.15	2.68	13.76	m	**16.44**
44 mm × 100 mm	14.45	0.15	2.68	14.81	m	**17.49**
44 mm × 115 mm	14.86	0.15	2.68	15.23	m	**17.91**
44 mm × 140 mm	15.48	0.17	3.03	15.87	m	**18.90**
57 mm × 100 mm	15.79	0.17	3.03	16.18	m	**19.21**
57 mm × 125 mm	17.09	0.17	3.03	17.52	m	**20.55**
69 mm × 88 mm	15.48	0.17	3.03	15.87	m	**18.90**
69 mm × 100 mm	17.42	0.17	3.03	17.86	m	**20.89**
Sills; once sunk weathered; once rebated, three times grooved						
63 mm × 175 mm	38.56	0.31	5.52	39.52	m	**45.04**
75 mm × 125 mm	37.14	0.31	5.52	38.07	m	**43.59**
75 mm × 150 mm	38.90	0.31	5.52	39.87	m	**45.39**
Door frames and door linings, sets; European Oak						
Sills; once sunk weathered; once rebated, three times grooved						
63 mm × 175 mm	63.39	0.31	5.52	64.97	m	**70.49**
75 mm × 125 mm	62.58	0.31	5.52	64.14	m	**69.66**
75 mm × 150 mm	68.42	0.31	5.52	70.13	m	**75.65**
Fire doors						
Flush door; half-hour fire-resisting (FD30); hardboard faced both sides; Jeld-Wen or other equal and approved						
762 mm × 1981 mm × 44 mm	54.65	1.62	28.88	56.02	nr	**84.90**
838 mm × 1981 mm × 44 mm	58.70	1.62	28.88	60.17	nr	**89.05**
726 mm × 2040 mm × 44 mm	56.67	1.62	28.88	58.09	nr	**86.97**
826 mm × 2040 mm × 44 mm	56.67	1.62	28.88	58.09	nr	**86.97**
926 mm × 2040 mm × 44 mm	56.67	1.62	28.88	58.09	nr	**86.97**
Flush door; half-hour fire-resisting (FD30); chipboard veneered; faced both sides; lipped on two long edges; Jeld-Wen paint grade veneer or other equal and approved						
610 mm × 1981 mm × 44 mm	47.57	1.62	28.88	48.76	nr	**77.64**
686 mm × 1981 mm × 44 mm	47.57	1.62	28.88	48.76	nr	**77.64**
762 mm × 1981 mm × 44 mm	47.57	1.62	28.88	48.76	nr	**77.64**
838 mm × 1981 mm × 44 mm	47.57	1.62	28.88	48.76	nr	**77.64**
526 mm × 2040 mm × 44 mm	48.58	1.62	28.88	49.79	nr	**78.67**
626 mm × 2040 mm × 44 mm	48.58	1.62	28.88	49.79	nr	**78.67**
726 mm × 2040 mm × 44 mm	49.58	1.62	28.88	50.82	nr	**79.70**
826 mm × 2040 mm × 44 mm	55.66	1.62	28.88	57.05	nr	**85.93**

24 DOORS, SHUTTERS AND HATCHES

Item	PC £	Labour hours	Labour £	Material £	Unit	Total rate £
Moulded panel doors; half-hour fire-resisting (FD30); white based coated facings suitable for paint finish only; two, four or six panel options; Premdor Fireshield or other equal and approved						
526 mm × 2040 mm × 44 mm	80.51	1.10	19.61	82.52	nr	**102.13**
626 mm × 2040 mm × 44 mm	80.51	1.10	19.61	82.52	nr	**102.13**
726 mm × 2040 mm × 44 mm	80.51	1.10	19.61	82.52	nr	**102.13**
826 mm × 2040 mm × 44 mm	85.11	1.10	19.61	87.24	nr	**106.85**
926 mm × 2040 mm × 44 mm	88.95	1.10	19.61	91.17	nr	**110.78**
Flush door; half-hour fire-resisting (FD30); faced both sides; lipped on two long edges; Jeld-Wen Sapele veneered or other equal and approved						
610 mm × 1981 mm × 44 mm	87.03	1.71	30.48	89.21	nr	**119.69**
686 mm × 1981 mm × 44 mm	87.03	1.71	30.48	89.21	nr	**119.69**
762 mm × 1981 mm × 44 mm	87.03	1.71	30.48	89.21	nr	**119.69**
838 mm × 1981 mm × 44 mm	93.10	1.71	30.48	95.43	nr	**125.91**
726 mm × 2040 mm × 44 mm	93.10	1.71	30.48	95.43	nr	**125.91**
826 mm × 2040 mm × 44 mm	97.15	1.71	30.48	99.58	nr	**130.06**
926 mm × 2040 mm × 44 mm	99.18	1.71	30.48	101.66	nr	**132.14**
Flush door; half-hour fire-resisting (FD30); chipboard for painting; hardwood lipping two long edges; Premdor Fireshield or other equal and approved;						
526 mm × 2040 mm × 44 mm	42.17	1.62	28.88	43.22	nr	**72.10**
626 mm × 2040 mm × 44 mm	42.17	1.62	28.88	43.22	nr	**72.10**
726 mm × 2040 mm × 44 mm	42.17	1.62	28.88	43.22	nr	**72.10**
826 mm × 2040 mm × 44 mm	42.95	1.62	28.88	44.02	nr	**72.90**
926 mm × 2040 mm × 44 mm	54.45	1.62	28.88	55.81	nr	**84.69**
826 mm × 2040 mm × 44 mm; single side vision panel 150 mm × 700 mm; factory fitted clear fire-rated glass	169.46	1.71	30.48	173.70	nr	**204.18**
826 mm × 2040 mm × 44 mm; two side vision panels 150 mm × 700 mm; factory fitted clear fire-rated glass	219.31	1.71	30.48	224.79	nr	**255.27**
926 mm × 2040 mm × 44 mm; single side vision panel 150 mm × 700 mm; factory fitted clear fire-rated glass	174.07	1.71	30.48	178.42	nr	**208.90**
926 mm × 2040 mm × 44 mm; two side vision panels 150 mm × 700 mm; factory fitted clear fire-rated glass	223.91	1.71	30.48	229.51	nr	**259.99**
Flush door; half-hour fire-resisting (FD30); White Oak veneer; hardwood lipping all edges; Premdor Fireshield or other equal and approved;						
526 mm × 2040 mm × 44 mm	78.98	1.62	28.88	80.95	nr	**109.83**
626 mm × 2040 mm × 44 mm	78.98	1.62	28.88	80.95	nr	**109.83**
726 mm × 2040 mm × 44 mm	78.98	1.62	28.88	80.95	nr	**109.83**
826 mm × 2040 mm × 44 mm	82.05	1.62	28.88	84.10	nr	**112.98**
926 mm × 2040 mm × 44 mm	99.69	1.62	28.88	102.18	nr	**131.06**
826 mm × 2040 mm × 44 mm; single side vision panel 508 mm × 1649 mm; factory fitted clear fire-rated glass	239.25	1.71	30.48	245.23	nr	**275.71**

24 DOORS, SHUTTERS AND HATCHES

Item	PC £	Labour hours	Labour £	Material £	Unit	Total rate £
INTERNAL DOORS AND FRAMES – cont						
Fire doors – cont						
Flush door – cont						
826 mm × 2040 mm × 44 mm; two side vision panels 150 mm × 775 mm and 150 mm × 700 mm; factory fitted clear fire-rated glass	269.15	1.71	30.48	275.88	nr	**306.36**
926 mm × 2040 mm × 44 mm; single side vision panel 508 mm × 1649 mm; factory fitted clear fire-rated glass	246.91	1.71	30.48	253.08	nr	**283.56**
926 mm × 2040 mm × 44 mm; two side vision panels 150 mm × 775 mm and 150 mm × 700 mm; factory fitted clear fire-rated glass	275.29	1.71	30.48	282.17	nr	**312.65**
Flush door; one-hour fire-resisting (FD60); chipboard for painting; hardwood lipping two long edges; Premdor Firemaster or other equal and approved						
626 mm × 2040 mm × 54 mm	149.53	1.71	30.48	153.27	nr	**183.75**
726 mm × 2040 mm × 54 mm	149.53	1.71	30.48	153.27	nr	**183.75**
826 mm × 2040 mm × 54 mm	149.53	1.71	30.48	153.27	nr	**183.75**
926 mm × 2040 mm × 54 mm	164.86	1.71	30.48	168.98	nr	**199.46**
Moulded panel doors; half-hour fire-resisting (FD60); white based coated facings suitable for paint finish only; two, four or six panel options; Premdor Firemaster or other equal and approved						
626 mm × 2040 mm × 54 mm	233.88	1.71	30.48	239.73	nr	**270.21**
726 mm × 2040 mm × 54 mm	233.88	1.71	30.48	239.73	nr	**270.21**
826 mm × 2040 mm × 54 mm	235.41	1.71	30.48	241.30	nr	**271.78**
926 mm × 2040 mm × 54 mm	240.02	1.71	30.48	246.02	nr	**276.50**
Flush door; one-hour fire-resisting (FD60); White Oak veneer; hardwood lipping all edges; Premdor Firemaster or other equal and approved						
626 mm × 2040 mm × 54 mm	181.73	1.94	34.58	186.27	nr	**220.85**
726 mm × 2040 mm × 54 mm	181.73	1.94	34.58	186.27	nr	**220.85**
826 mm × 2040 mm × 54 mm	191.70	1.94	34.58	196.49	nr	**231.07**
926 mm × 2040 mm × 54 mm	207.04	1.94	34.58	212.22	nr	**246.80**
Flush door; one-hour fire-resisting (FD60); Steamed Beech veneer; hardwood lipping all edges; Premdor Firemaster or other equal and approved						
626 mm × 2040 mm × 54 mm	207.04	1.94	34.58	212.22	nr	**246.80**
726 mmx 2040 mm × 54 mm	207.04	1.94	34.58	212.22	nr	**246.80**
826 mm × 2040 mm × 54 mm	220.07	1.94	34.58	225.57	nr	**260.15**
926 mm × 2040 mm × 54 mm	232.34	1.94	34.58	238.15	nr	**272.73**

24 DOORS, SHUTTERS AND HATCHES

Item	PC £	Labour hours	Labour £	Material £	Unit	Total rate £
Fire-resisting door frame; internal and external; fitted with 15 mm × 4 mm intumescent strips; 12 mm deep rebates; screwed to masonry/concrete						
Softwood frames; no sill – open in or out						
Door size						
762 mm × 1981 mm × 44 mm; FD30; intumescent strip only	60.89	0.75	13.37	62.55	nr	**75.92**
838 mm × 1981 mm × 44 mm; FD30; intumescent strip only	62.56	0.75	13.37	64.26	nr	**77.63**
826 mm × 2040 mm × 44 mm; FD30; intumescent strip only	71.78	0.75	13.37	73.71	nr	**87.08**
762 mm × 1981 mm × 44 mm; FD30; intumescent strip/smoke seal	64.79	0.75	13.37	66.54	nr	**79.91**
838 mm × 1981 mm × 44 mm; FD30; intumescent strip/smoke seal	66.21	0.75	13.37	68.00	nr	**81.37**
826 mm × 2040 mm × 44 mm; FD30; intumescent strip/smoke seal	66.21	0.75	13.37	68.00	nr	**81.37**
Hardwood frames; no sill – open in or out						
762 mm × 1981 mm × 44 mm; FD60; intumescent strip/smoke seal	122.82	0.75	13.37	126.02	nr	**139.39**
838 mm × 1981 mm × 44 mm; FD60; intumescent strip/smoke seal	125.58	0.75	13.37	128.85	nr	**142.22**
826 mm × 2040 mm × 44 mm; FD60; intumescent strip/smoke seal	127.48	0.75	13.37	130.80	nr	**144.17**
Intumescent strips						
Factory installed intumescent smoke seals to fire doors; 3 edges, jambs and head; site fixed intumescent strips;	–	–	–	20.75	door	**20.75**
fire and smoke intumescent strips 15 mm × 4 mm – FD30 doors	–	0.15	2.68	0.57	m	**3.25**
fire and smoke intumescent strips 20 mm × 4 mm – FD60 doors	–	0.15	2.68	0.81	m	**3.49**
Bedding and pointing frames						
Pointing wood frames or sills with mastic						
one side	–	0.09	1.30	0.59	m	**1.89**
both sides	–	0.19	2.74	1.18	m	**3.92**
Pointing wood frames or sills with polysulphide sealant						
one side	–	0.09	1.30	2.24	m	**3.54**
both sides	–	0.19	2.74	4.49	m	**7.23**
Bedding wood frames in cement: mortar (1:3) and point						
one side	–	0.07	1.21	0.08	m	**1.29**
both sides	–	0.09	1.56	0.10	m	**1.66**
one side in mortar; other side in mastic	–	0.19	3.02	0.66	m	**3.68**

24 DOORS, SHUTTERS AND HATCHES

Item	PC £	Labour hours	Labour £	Material £	Unit	Total rate £
SHUTTER DOORS AND PANEL DOORS						
Insulated rolling shutters; Bolton Gate Company Ltd or other equal and approved; electrically operated, self-coiling; galvanized finish; fixing by bolting						
Shutter doors						
2400 mm × 2400 mm clear opening	–	–	–	–	nr	2075.75
3000 mm × 3000 mm clear opening	–	–	–	–	nr	2213.50
4300 mm × 4200 mm clear opening	–	–	–	–	nr	3313.60
Rolling shutters and collapsible gates; steel counter shutters; push-up, self-coiling; polyester power coated; fixing by bolting						
Shutter doors						
3000 mm × 1000 mm	–	–	–	–	nr	1049.85
4000 mm × 1000 mm; in two panels	–	–	–	–	nr	1825.82
Rolling shutters with collapsible gates; galvanized steel; one-hour fire-resisting; self-coiling; activated by fusible link; fixing with bolts						
Shutter doors						
1000 mm × 2750 mm	–	–	–	–	nr	1255.25
1500 mm × 2750 mm	–	–	–	–	nr	1318.02
2400 mm × 2750 mm	–	–	–	–	nr	1557.66
Translucent GRP stacking door (78% translucency); U-value = 2.6W/m^2 K; electrically operated; Envirodoor Ltd; fuly enclosed aluminium track system with SAA finish; manual override, lock interlock, stop and return safety cage, deadmans down button, anti-flip device, photoelectric cell and beam deflectors; fixing by bolting; standard panel finishes						
Stacking doors						
HT40 stacking door 2500 mm × 3000 mm; 40 mm thick × 500 mm high twin walled GRP translucent panels	–	–	–	–	nr	5000.00
HT40 stacking door 4500 mm × 6000 mm; 40 mm thick × 500 mm high twin walled translucent panels	–	–	–	–	nr	9200.00
HT60-N stacking door 6000 mm × 6000 mm: 60 mm thick × 500 mm high twin walled translucent panels	–	–	–	–	nr	13000.00
HT60-H stacking door 7000 mm × 8000 mm: 60 mm thick × 1000 mm high twin walled translucent panels	–	–	–	–	nr	20000.00
HT80–400 stacking door 10000 mm × 10000 mm; 80 mm thick × 1000 mm high twin walled translucent panels	–	–	–	–	nr	47500.00

24 DOORS, SHUTTERS AND HATCHES

Item	PC £	Labour hours	Labour £	Material £	Unit	Total rate £
Doors; galvanized steel up and over type garage doors; Catnic Horizon 90 or other equal and approved; spring counter balanced; fixed to timber frame (not included)						
Garage door						
2135 mm × 1980 mm	307.67	3.70	65.96	315.47	nr	**381.43**
2135 mm × 2135 mm	352.98	3.70	65.96	361.92	nr	**427.88**
2400 mm × 2135 mm	446.79	3.70	65.96	458.09	nr	**524.05**
3965 mm × 2135 mm	1063.71	5.55	98.94	1090.57	nr	**1189.51**
Grilles; Galaxy nylon rolling counter grille or other equal and approved; Bolton Brady Ltd; colour, off-white; self-coiling; fixing by bolting						
Grilles						
3000 mm × 1000 mm	–	–	–	–	nr	**874.00**
4000 mm × 1000 mm	–	–	–	–	nr	**1377.50**
Sliding/folding partitions; Alco Beldan Ltd or equal and approved						
Sliding/folding partitions						
ref. NW100 Moveable Wall; 5000 mm (wide) × 2495 mm (high) comprising 4 nr. 954 mm (wide) standard panels and 1 nr. 954 mm (wide) telescopic panel; sealing; fixing	–	–	–	–	nr	**8500.00**
ASSOCIATED IRONMONGERY						
NOTE: Ironmongery is largely a matter of selection and specification and prices vary considerably; indicative prices for reasonable quantities of good quality ironmongery are given below.						
Allgood Ltd or other equal; to softwood						
Bolts						
75 × 35 mm Modric anodized aluminium straight barrel bolt	8.38	0.30	5.35	8.59	nr	**13.94**
150 × 35 mm Modric anodized aluminium straight barrel bolt	9.56	0.30	5.35	9.80	nr	**15.15**
75 × 35 mm Modric anodized aluminium necked barrel bolt	9.38	0.30	5.35	9.61	nr	**14.96**
150 × 35 mm Modric anodized aluminium necked barrel bolt	11.99	0.30	5.35	12.29	nr	**17.64**
11 mm Easiclean socket for wood or stone	4.08	0.10	1.78	4.18	nr	**5.96**
Security hinge bolt chubb WS12	8.48	0.50	8.92	8.69	nr	**17.61**
203 × 19 × 11 mm lever action flush bolt set, with coil spring and intumescent pack for FD30 and FD60 fire doors	28.79	0.60	10.70	29.51	nr	**40.21**

24 DOORS, SHUTTERS AND HATCHES

Item	PC £	Labour hours	Labour £	Material £	Unit	Total rate £
ASSOCIATED IRONMONGERY – cont						
Allgood Ltd or other equal – cont						
Bolts – cont						
609 × 19 mm lever action flush bolt set, with coil spring and intumescent pack for FD30 and FD60 fire doors	83.87	0.60	10.70	85.97	nr	**96.67**
Stainless steel indicating bolt complete with outside indicator and emergency release	60.13	0.60	10.70	61.63	nr	**72.33**
Catches						
Magnetic catch	0.40	0.20	3.57	0.41	nr	**3.98**
Door closers and furniture						
13 mm stainless steel rebate component for 7104/08/78/79/86	24.94	0.60	10.70	25.56	nr	**36.26**
70 × 70 mm Modric anodized aluminium electrically powered hold open wall magnet (excluding power supply and connection)	117.34	0.40	7.13	120.27	nr	**127.40**
Modric anodized aluminium bathroom configuration with quadaxial assembly, turn, release and optional indicator	45.72	0.80	14.26	46.86	nr	**61.12**
Concealed jamb door closer check action	136.88	1.00	17.82	140.30	nr	**158.12**
75 × 57 × 170 mm Modric anodized aluminium Door coordinator for pairs of rebated leaves, CE Marked to BS EN1158 3–5-3/5–1-1–0	30.29	0.80	14.26	31.05	nr	**45.31**
290 × 48 × 50 mm Modric anodized aluminium rectangular overhead door closer with adjustable power and adjustable backcheck intumescent protected arm heavy duty U.L. & certifire listed & CE marked to BS EN1154 4–8-2/4–1-1–3 and kite marked	87.93	1.00	17.82	90.13	nr	**107.95**
Stainless steel overhead door closer. Projecting armset, Power EN 2–5, CE marked, c/w backcheck, latch action and speed control. Max door width 1100 mm, max door weight 100 kg	87.69	1.00	17.82	89.88	nr	**107.70**
288 × 45 × 32 mm fully concealed overhead door closer complete with track and arm for single action doors, adjustable power, latch action and backcheck. Certifire approved	146.91	0.80	14.26	150.58	nr	**164.84**
75 × 45 mm heavy duty floor pivot set with thrust roller bearing 200 kg load capacity. Complete with forged steel intumescent protected double action strap with 10 mm height adjustment and matching cover plate	223.88	2.30	41.00	229.48	nr	**270.48**
Cavalier floor spring unit with cover plate and loose box for concrete, adjustable power 25 mm offset strap and top centre, intumescent pack. Certfire listed	237.41	2.30	41.00	243.35	nr	**284.35**
Double action pivot set for door maximum width 1100 mm and maximum weight 80 kg	74.95	2.30	41.00	76.82	nr	**117.82**

24 DOORS, SHUTTERS AND HATCHES

Item	PC £	Labour hours	Labour £	Material £	Unit	Total rate £
Surface vertical rod push bar panic bolt, reversible, to suit doors 2500 x 1100 mm maximum, silver finish, CE marked to EN1125 class 3–7-5-1–1-3-2–2-A	125.25	1.50	26.74	128.38	nr	**155.12**
Rim push bar panic latch, reversible, to suit doors 1100 mm wide maximum, silver finish, CE marked to EN1125 class 3–7-5-1–1-3-2–2-A	89.30	1.30	23.18	91.53	nr	**114.71**
76 × 51 × 13 mm adjustable heavy roller catch, stainless steel forend and strike complete with satin nickel plate roller bolt	7.06	0.60	10.70	7.24	nr	**17.94**
External access device for use with XX10280/2 panic hardware to suit door thickness 45–55 mm, complete with SS3006N lever, SS755 rose, SS796 profile escutcheon and spindle. For use with MA7420 A51 or MA7420 A55 profile cylinders	29.67	1.30	23.18	30.41	nr	**53.59**
142 × 22 mm Ø Concealed jamb door closer light duty	15.88	0.80	14.26	16.28	nr	**30.54**
80 × 40 × 45 mm Emergency release door stop with holdback facility	74.95	1.00	17.82	76.82	nr	**94.64**
Modric anodized aluminium quadaxial lever MA3503 assembly tested to BS EN1906 4/7/-/1/1/4/0/U	27.79	0.80	14.26	28.48	pair	**42.74**
Modric anodized aluminium quadaxial lever MA3502 assembly tested to BS EN1906 4/7/-/1/1/4/0/U	29.62	0.80	14.26	30.36	pair	**44.62**
Modric anodized aluminium quadaxial lever assembly tested to BS EN1906 4/7/-/1/1/4/0/U with Biocote® anti-bacterial protection	53.41	0.80	14.26	54.75	pair	**69.01**
Modric stainless steel quadaxial lever assembly tested to BS EN1906 4/7/-/1/1/4/0/U	50.85	0.80	14.26	52.12	pair	**66.38**
152 × 38 × 13 mm Modric anodized aluminium security door chain leather covered	39.62	0.40	7.13	40.61	nr	**47.74**
50 Ø × 3 mm Modric anodized aluminium circular covered rose for profile cylinder	3.90	0.10	1.78	4.00	nr	**5.78**
50 Ø × 3 mm Modric anodized aluminium circular covered rose with indicator and emergency release	8.04	0.15	2.68	8.24	nr	**10.92**
50 Ø × 3 mm Modric anodized aluminium circular covered rose with heavy turn, 5–8 mm spindle	14.09	0.15	2.68	14.44	nr	**17.12**
Budget lock escutcheon – satin stainless steel 316	7.68	0.10	1.78	7.87	nr	**9.65**
50 Ø × 3 mm Stainless steel circular covered rose for profile cylinder	6.24	0.10	1.78	6.40	nr	**8.18**
50 Ø × 3 mm Stainless steel circular covered rose with indicator and emergency release	8.42	0.15	2.68	8.63	nr	**11.31**
50 Ø × 3 mm Stainless steel circular covered rose with heavy turn, 5–8 mm spindle	17.34	0.15	2.68	17.77	nr	**20.45**
330 × 76 × 1.6 mm Modric anodized aluminium push plate	3.07	0.15	2.68	3.15	nr	**5.83**
330 × 76 × 1.6 mm Stainless steel push plate	7.11	0.15	2.68	7.29	nr	**9.97**

Prices for Measured Works

24 DOORS, SHUTTERS AND HATCHES

Item	PC £	Labour hours	Labour £	Material £	Unit	Total rate £
ASSOCIATED IRONMONGERY – cont						
Allgood Ltd or other equal – cont						
Door closers and furniture – cont						
800 × 150 × 1.5 mm Modric anodized aluminium kicking plate, drilled and countersunk with screws	7.56	0.25	4.46	7.75	nr	**12.21**
900 × 150 × 1.5 mm Modric anodized aluminium kicking plate, drilled and countersunk with screws	8.50	0.25	4.46	8.71	nr	**13.17**
1000 × 150 × 1.5 mm Modric anodized aluminium kicking plate, drilled & countersunk with screws	9.43	0.25	4.46	9.67	nr	**14.13**
800 × 150 × 1.5 mm Stainless steel kicking plate, drilled and countersunk with screws	13.38	0.25	4.46	13.71	nr	**18.17**
900 × 150 × 1.5 mm Stainless steel kicking plate, drilled and countersunk with screws	15.05	0.25	4.46	15.43	nr	**19.89**
1000 × 150 × 1.5 mm Stainless steel kicking plate, drilled & countersunk with screws	16.72	0.25	4.46	17.14	nr	**21.60**
610 × 70 × 19 mm Ø Modric anodized aluminium grab handle bolt through fixing for doors 10 to 55 mm thick	27.33	0.40	7.13	28.01	nr	**35.14**
400 × 19 mm Ø Stainless steel D line straight pull handle with M8 threaded holes, fixing centres 300 mm	42.30	0.33	5.88	47.98	nr	**53.86**
Hinges						
100 × 75 × 3 mm Stainless steel triple knuckle concealed twin bearings, button tipped butt hinges, jig drilled for metal doors/frames, complete with M6x12MT 'undercut' machine screws, stainless steel 316 CE marked to EN1935	16.80	0.25	4.46	17.22	pair	**21.68**
100 × 100 × 3 mm Stainless steel triple knuckle concealed twin Newton bearings, button tipped hinges, jig drilled, stainless steel grade 316 CE marked to EN1935	29.81	0.25	4.46	30.56	pair	**35.02**
Latches						
Modric anodized aluminium round cylinder for rim night latch, 2 keyed satin nickel plated	23.60	0.40	7.13	24.19	nr	**31.32**
93 × 75 mm Cylinder rim non-deadlocking night latch case only 60 mm backset	20.23	0.40	7.13	20.74	nr	**27.87**
71 series mortice latch, case only, low friction latchbolt, griptight follower, heavy spring for levers. Radius forend and sq strike. CE marked to BS EN12209 3/X/8/1/0G/-/B/02/0	15.29	0.80	14.26	15.67	nr	**29.93**
Modric anodized aluminium latch configuration with quadaxial assembly	27.61	0.80	14.26	28.30	nr	**42.56**
Modric anodized aluminium Nightlatch configuration with quadaxial assembly and single cylinder	51.20	0.80	14.26	52.48	nr	**66.74**

24 DOORS, SHUTTERS AND HATCHES

Item	PC £	Labour hours	Labour £	Material £	Unit	Total rate £
Locks						
44 mm case Bright zinc plated steel mortice budget lock with slotted strike plate 33 mm backset	25.95	0.80	14.26	26.60	nr	**40.86**
76 × 58 mm b/s Stainless steel cubicle mortice deadlock with 8 mm follower	12.29	0.80	14.26	12.60	nr	**26.86**
'A' length European profile double cylinder lock, 2 keyed satin nickel plated	24.20	0.80	14.26	24.80	nr	**39.06**
'A' length European profile cylinder and large turn, 2 keyed satin nickel plated	27.51	0.80	14.26	28.20	nr	**42.46**
'A' length European profile cylinder and large turn, 2 keyed under master key, satin nickel plated	25.30	0.80	14.26	25.93	nr	**40.19**
'A' length European profile single cylinder, 2 keyed satin nickel plated	18.71	0.80	14.26	19.18	nr	**33.44**
'A' length European profile single cylinder, 2 keyed under master key, satin nickel plated	18.71	0.80	14.26	19.18	nr	**33.44**
93 × 60 mm b/s 71 series profile cylinder mortice deadlock, case only. Single throw 22 mm deadbolt. Radius forend and square strike. CE marked to BS EN12209 3/X/8/1/0/G/4/B/A/0/0	15.29	0.80	14.26	15.67	nr	**29.93**
92 × 60 mm b/s 71 series bathroom lock, case only, low friction latchbolt, griptight follower, heavy spring for levers, twin 8 mm followers at 78 mm centres. Radius forend and square strike. CE marked to BS EN12209 3/X/8/0/0/G-/B/0/2/0	18.12	0.80	14.26	18.57	nr	**32.83**
93 × 60 mm b/s 71 series profile cylinder mortice lock, case only, low friction latchbolt, griptight follower. Heavy spring for levers, 22 mm throw deadbolt, cylinder withdraws bolt bolts. Radius forend and square strike. CE marked to BS EN12209 3/X/8/1/0G/4/B/A2/0	18.12	0.80	14.26	18.57	nr	**32.83**
92 × 60 mm b/s 71 series profile cylinder emergency lock, case only. Low friction latchbolt, griptight follower, heavy spring for lever, single throw 22 mm deadbolt, lever can withdraw both bolts. Radius forend and strike	59.59	0.80	14.26	61.08	nr	**75.34**
Modric anodized aluminium lock configuration with quadaxial assembly and cylinder with turn	78.89	0.80	14.26	80.86	nr	**95.12**
Sundries						
76 mm Ø Modric anodized aluminium circular sex symbol male	4.29	0.08	1.42	4.40	nr	**5.82**
76 mm Ø Modric anodized aluminium circular symbol fire door keep locked	4.29	0.08	1.42	4.40	nr	**5.82**
76 mm Ø Modric anodized aluminium circular symbol fire door keep shut	4.29	0.10	1.78	4.40	nr	**6.18**
38 × 47 mm Ø Modric anodized aluminium heavy circular floor door stop with cover	8.65	0.10	1.78	8.87	nr	**10.65**
38 × 47 mm Ø Stainless steel heavy circular floor door stop with cover	15.47	0.10	1.78	15.86	nr	**17.64**
63 × 19 mm Ø Modric anodised aluminium circular heavy duty skirting buffer with thief-resistant insert	5.66	0.10	1.78	5.80	nr	**7.58**

24 DOORS, SHUTTERS AND HATCHES

Item	PC £	Labour hours	Labour £	Material £	Unit	Total rate £
ASSOCIATED IRONMONGERY – cont						
Allgood Ltd or other equal – cont						
Sundries – cont						
102 × 25 mm Ø Stainless steel circular heavy duty skirting buffer with thief-resistant insert	9.81	0.10	1.78	10.06	nr	**11.84**
152 mm Cabin hook satin chrome on brass	20.51	0.15	2.68	21.02	nr	**23.70**
14 mm Ø × 145 × 94 mm Toilet roll holder, length 145 mm, colour white, satin stainless steel 316	57.11	0.15	2.68	60.08	nr	**62.76**
Towel rail with bushes, fixing centres 450 mm, satin stainless steel 316	77.61	0.25	4.46	84.17	nr	**88.63**
Toilet brush holder with toilet brush, with bushes, satin stainless steel 316	120.51	0.20	3.57	126.60	nr	**130.17**
Bathline 850 mm lift up support rail	182.54	0.75	13.37	187.10	set	**200.47**
Bathline 600 × 95 × 35 mm support rail with concealed fixing roses	93.57	0.50	8.92	95.91	set	**104.83**
Bathline 400 × 250 × 35 mm backrest rail with concealed fixing roses	93.57	0.50	8.92	95.91	set	**104.83**
Allgood Ltd or other equal; to hardwood						
Bolts						
75 × 35 mm Modric anodized aluminium straight barrel bolt	8.38	0.40	7.13	8.59	nr	**15.72**
150 × 35 mm Modric anodized aluminium straight barrel bolt	9.56	0.40	7.13	9.80	nr	**16.93**
75 × 35 mm Modric anodized aluminium necked barrel bolt	9.38	0.40	7.13	9.61	nr	**16.74**
150 × 35 mm Modric anodized aluminium necked barrel bolt	11.99	0.40	7.13	12.29	nr	**19.42**
11 mm Easiclean socket for wood or stone	4.08	0.15	2.68	4.18	nr	**6.86**
Security hinge bolt chubb WS12	8.48	0.65	11.58	8.69	nr	**20.27**
203 × 19 × 11 mm lever action flush bolt set with coil spring and intumescent pack for FD30 and FD60 fire doors	28.79	0.80	14.26	29.51	nr	**43.77**
609 × 19 mm lever action flush bolt set with coil spring and intumescent pack for FD30 and FD60 fire doors	83.87	0.80	14.26	85.97	nr	**100.23**
Stainless steel indicating bolt complete with outside indicator and emergency release	60.13	0.80	14.26	61.63	nr	**75.89**
Catches						
Magnetic catch	0.40	0.25	4.46	0.41	nr	**4.87**
Door closers and furniture						
13 mm stainless steel rebate component for 7104/08/78/79/86	24.94	0.80	14.26	25.56	nr	**39.82**
70 × 70 mm Modric anodized aluminium electrically powered hold open wall magnet. CE marked to BS EN1155:1997 & A1:2002 3–5-6/ 3–1-1–3	117.34	0.55	9.81	120.27	nr	**130.08**
Modric anodized aluminium bathroom configuration with quadaxial assembly, turn, release and optional indicator	45.72	1.05	18.72	46.86	nr	**65.58**

24 DOORS, SHUTTERS AND HATCHES

Item	PC £	Labour hours	Labour £	Material £	Unit	Total rate £
495 × 17.5 × 16 mm concealed overhead door restraining stay with aluminium channel. Stainless steel plated arm and bracket with adjustable friction slide. Block and spring buffer to cussion door opening. Not for use with door closing devices	23.52	1.35	24.07	24.11	nr	**48.18**
Concealed jamb door closer check action	136.88	1.35	24.07	140.30	nr	**164.37**
75 × 57 × 170 mm Modric anodized aluminium door coordinator for pairs of rebated leaves, CE marked to BS EN1158 3–5-3/5–1-1–0	30.29	1.05	18.72	31.05	nr	**49.77**
290 × 48 × 50 mm Modric anodized aluminium rectangular overhead door closer with adjustable power and adjustable backcheck intumescent protected arm heavy duty U.L. & certifire listed & CE marked to BS EN1154 4–8-2/4–1-1–3 and Kitemarked.	87.93	1.35	24.07	90.13	nr	**114.20**
Stainless steel overhead door closer. Projecting armset, Power EN 2–5, CE marked, c/w Backcheck, Latch action and Speed control. Max door width 1100 mm, max door weight 100 kg	87.69	1.35	24.07	89.88	nr	**113.95**
288 × 45 × 32 mm fully concealed overhead door closer complete with track and arm for single action doors, adjustable power, latch action and backcheck. Certifire approved	146.91	1.05	18.72	150.58	nr	**169.30**
75 × 45 mm heavy duty floor pivot set with thrust roller bearing 200 kg load capacity. Complete with forged steel intumescent protected double action strap with 10 mm height adjustment and matching cover plate	223.88	3.05	54.38	229.48	nr	**283.86**
Cavalier floor spring unit with cover plate and loose box for concrete, adjustable power 25 mm offset strap and top centre with intumescent pack. Certfire listed	237.40	2.30	41.00	243.34	nr	**284.34**
Double action pivot set for door maximum width 1100 mm and maximum weight 80 kg	74.95	3.05	54.38	76.82	nr	**131.20**
Surface vertical rod push bar panic bolt, reversible, to suit doors 2500 × 1100 mm maximum, silver finish, CE marked to EN1125 class 3-7-5-1–1-3-2–2-A	125.25	2.00	35.65	128.38	nr	**164.03**
Rim push bar panic latch, reversible, to suit doors 1100 mm wide maximum, silver finish, CE marked to EN1125 class 3–7-5-1–1-3-2–2-A	89.30	1.75	31.20	91.53	nr	**122.73**
76 × 51 × 13 mm adjustable heavy roller catch, stainless steel forend and strike complete with satin nickel roller bolt satin chrome	7.06	0.80	14.26	7.24	nr	**21.50**
External access device for use with XX10280/2 panic hardware to suit door thickness 45–55 mm, complete with SS3006N lever, SS755 rose, SS796 profile escutcheon and spindle. For use with MA7420 A51 or MA7420 A55 profile cylinders	29.67	1.75	31.20	30.41	nr	**61.61**
142 × 22 mm Ø Concealed jamb door closer light duty	15.88	1.05	18.72	16.28	nr	**35.00**

24 DOORS, SHUTTERS AND HATCHES

Item	PC £	Labour hours	Labour £	Material £	Unit	Total rate £
ASSOCIATED IRONMONGERY – cont						
Allgood Ltd or other equal – cont						
Door closers and furniture – cont						
80 × 40 × 45 mm Emergency release door stop with holdback facility	74.95	1.35	24.07	76.82	nr	**100.89**
Modric anodized aluminium quadaxial lever MA3503 assembly tested to BS EN1906 4/7/-/1/1/4/0/U	27.79	1.05	18.72	28.48	pair	**47.20**
Modric anodized aluminium quadaxial lever MA3502 assembly tested to BS EN1906 4/7/-/1/1/4/0/U	29.62	1.05	18.72	30.36	pair	**49.08**
Modric anodized aluminium quadaxial lever assembly tested to BS EN1906 4/7/-/1/1/4/0/U with Biocote® anti-bacterial protection	53.41	1.05	18.72	54.75	pair	**73.47**
Modric stainless steel quadaxial lever assembly Tested to BS EN1906 4/7/-/1/1/4/0/U	50.85	1.05	18.72	52.12	pair	**70.84**
152 × 38 × 13 mm Modric anodized aluminium security door chain leather covered	39.62	0.55	9.81	40.61	nr	**50.42**
50 Ø × 3 mm Modric anodized aluminium circular covered rose for profile cylinder	3.90	0.15	2.68	4.00	nr	**6.68**
50 Ø × 3 mm Modric anodized aluminium circular covered rose with indicator and emergency release	8.04	0.20	3.57	8.24	nr	**11.81**
50 Ø × 3 mm Modric anodized aluminium circular covered rose with heavy turn, 5–8 mm spindle	14.09	0.20	3.57	14.44	nr	**18.01**
Budget lock escutcheon – satin stainless steel 316	7.68	0.15	2.68	7.87	nr	**10.55**
50 Ø × 3 mm Stainless steel circular covered rose for profile cylinder	6.24	0.15	2.68	6.40	nr	**9.08**
50 Ø × 3 mm Stainless steel circular covered rose with indicator and emergency release	8.42	0.20	3.57	8.63	nr	**12.20**
50 Ø × 3 mm Stainless steel circular covered rose with heavy turn, 5–8 mm spindle	17.34	0.20	3.57	17.77	nr	**21.34**
330 × 76 × 1.6 mm Modric anodized aluminium push plate	3.07	0.20	3.57	3.15	nr	**6.72**
330 × 76 × 1.6 mm Stainless steel push plate	7.11	0.20	3.57	7.29	nr	**10.86**
800 × 150 × 1.5 mm Modric anodized aluminium kicking plate, drilled and countersunk with screws.	7.56	0.35	6.24	7.75	nr	**13.99**
900 × 150 × 1.5 mm Modric anodized aluminium kicking plate, drilled and countersunk with screws.	8.50	0.35	6.24	8.71	nr	**14.95**
1000 × 150 × 1.5 mm Modric anodized aluminium kicking plate, drilled & countersunk with screws.	9.43	0.35	6.24	9.67	nr	**15.91**
800 × 150 × 1.5 mm Stainless steel kicking plate, drilled and countersunk with screws.	13.38	0.35	6.24	13.71	nr	**19.95**
900 × 150 × 1.5 mm Stainless steel kicking plate, drilled and countersunk with screws.	15.05	0.35	6.24	15.43	nr	**21.67**
1000 × 150 × 1.5 mm Stainless steel kicking plate, drilled & countersunk with screws.	16.72	0.35	6.24	17.14	nr	**23.38**
610 × 70 × 19 mm Ø Modric anodized aluminium grab handle bolt through fixing for doors 10 to 55 mm thick	27.33	0.55	9.81	28.01	nr	**37.82**
400 × 19 mm Ø Stainless steel D line straight pull handle with M8 threaded holes, fixing centres 300 mm	42.30	0.45	8.03	47.98	nr	**56.01**

24 DOORS, SHUTTERS AND HATCHES

Item	PC £	Labour hours	Labour £	Material £	Unit	Total rate £
Hinges						
100 × 75 × 3 mm Stainless steel triple knuckle concealed twin Newton bearings, button tipped butt hinges, jig drilled for metal doors/frames, complete with M6x12MT 'undercut' machine screws, stainless steel 316 CE marked to EN1935	16.80	0.35	6.24	17.22	pair	**23.46**
100 × 100 × 3 mm Stainless steel triple knuckle concealed twin Newton bearings, button tipped hinges, jig drilled, stainless steel grade 316 CE marked to EN1935	29.81	0.35	6.24	30.56	pair	**36.80**
Latches						
Modric anodized aluminium round cylinder for rim night latch, 2 keyed satin nickel plated	23.60	0.55	9.81	24.19	nr	**34.00**
93 × 75 mm Cylinder rim non-deadlocking night latch case only 60 mm backset	20.23	0.55	9.81	20.74	nr	**30.55**
71 series mortice latch, case only, low friction latchbolt, griptight follower, heavy spring for levers. Radius forend and sq strike. CE marked to BS EN12209 3/X/8/1/0G/-/B/02/0	15.29	1.05	18.72	15.67	nr	**34.39**
Modric anodized aluminium latch configuration with quadaxial assembly	27.61	1.05	18.72	28.30	nr	**47.02**
Modric anodized aluminium Nightlatch configuration with quadaxial assembly and single cylinder	51.20	1.05	18.72	52.48	nr	**71.20**
Locks						
44 mm case Bright zinc plated steel mortice budget lock with slotted strike plate 33 mm backset	25.95	1.05	18.72	26.60	nr	**45.32**
76 × 58 mm b/s Stainless steel cubicle mortice deadlock with 8 mm follower	12.29	1.05	18.72	12.60	nr	**31.32**
'A' length European profile double cylinder lock, 2 keyed satin nickel plated	24.20	1.05	18.72	24.80	nr	**43.52**
'A' length European profile cylinder and large turn, 2 keyed satin nickel plated	27.51	1.05	18.72	28.20	nr	**46.92**
'A' length European profile cylinder and large turn, 2 keyed under master key, satin nickel plated	25.30	1.05	18.72	25.93	nr	**44.65**
'A' length European profile single cylinder, 2 keyed satin nickel plated	18.71	1.05	18.72	19.18	nr	**37.90**
'A' length European profile single cylinder, 2 keyed under master key, satin nickel plated	18.71	1.05	18.72	19.18	nr	**37.90**
93 × 60 mm b/s 71 series profile cylinder mortice deadlock, case only. Single throw 22 mm deadbolt. Radius forend and square strike. CE marked to BS EN12209 3/X/8/1/0/G/4/B/A/0/0	15.29	1.05	18.72	15.67	nr	**34.39**
92 × 60 mm b/s 71 series bathroom lock, case only, low friction latchbolt, griptight follower, heavy spring for levers, twin 8 mm followers at 78 mm centres. Radius forend and square strike. CE marked to BS EN12209 3/X/8/0/0/G-/B/0/2/0	18.12	1.05	18.72	18.57	nr	**37.29**

24 DOORS, SHUTTERS AND HATCHES

Item	PC £	Labour hours	Labour £	Material £	Unit	Total rate £
ASSOCIATED IRONMONGERY – cont						
Allgood Ltd or other equal – cont						
Locks – cont						
93 × 60 mm b/s 71 series profile cylinder mortice lock, case only, low friction latchbolt, griptight follower. Heavy spring for levers, 22 mm throw deadbolt, cylinder withdraws bolt bolts. Radius forend and square strike. CE marked to BS EN12209 3/X/8/1/0G/4/B/A2/0	18.12	1.05	18.72	18.57	nr	**37.29**
92 × 60 mm b/s 71 series profile cylinder emergency lock, case only. Low friction latchbolt, griptight follower, heavy spring for lever, single throw 22 mm deadbolt, lever can withdraw both bolts. Radius forend and strike	59.59	1.05	18.72	61.08	nr	**79.80**
Modric anodized aluminium lock configuration with quadaxial assembly and cylinder with turn	78.89	1.05	18.72	80.86	nr	**99.58**
Sundries						
76 mm Ø Modric anodized aluminium circular sex symbol male	4.29	0.10	1.78	4.40	nr	**6.18**
76 mm Ø Modric anodized aluminium circular symbol fire door keep locked	4.29	0.10	1.78	4.40	nr	**6.18**
76 mm Ø Modric anodized aluminium circular symbol fire door keep shut	4.29	0.15	2.68	4.40	nr	**7.08**
38 × 47 mm Ø Modric anodized aluminium heavy circular floor door stop with cover	8.65	0.15	2.68	8.87	nr	**11.55**
38 × 47 mm Ø Stainless steel heavy circular floor door stop with cover	15.47	0.15	2.68	15.86	nr	**18.54**
63 × 19 mm Ø Modric ancdised aluminium Circular heavy duty skirting buffer with thief-resistant insert	5.66	0.15	2.68	5.80	nr	**8.48**
102 × 25 mm Ø Stainless steel circular heavy duty skirting buffer with thief-resistant insert	9.81	0.15	2.68	10.06	nr	**12.74**
152 mm Cabin hook satin chrome on brass	20.51	0.20	3.57	21.02	nr	**24.59**
14 mm Ø × 145 × 94 mm Toilet roll holder, length 145 mm, colour white, satin stainless steel 316	57.11	0.20	3.57	60.08	nr	**63.65**
Towel rail with bushes, fixing centres 450 mm, satin stainless steel 316	77.61	0.35	6.24	84.17	nr	**90.41**
Toilet brush holder with toilet brush, with bushes, satin stainless steel 316	120.51	0.25	4.46	126.60	nr	**131.06**
Bathline 850 mm lift up support rail	182.54	0.75	13.37	187.10	set	**200.47**
Bathline 600 × 95 × 35 mm support rail with concealed fixing roses	93.57	0.50	8.92	95.91	set	**104.83**
Bathline 400 × 250 × 35 mm backrest rail with concealed fixing roses	93.57	0.50	8.92	95.91	set	**104.83**

24 DOORS, SHUTTERS AND HATCHES

Item	PC £	Labour hours	Labour £	Material £	Unit	Total rate £
ASSA ABBLOY Ltd typical door ironmongery sets						
These are standard doorsets compilations which ASSA ABBLOY most commonly supply						
Classroom doorset (fire-rated)	–	2.00	35.65	258.81	nr	**294.46**
Ironmongery schedule and cost						
UNION PowerLOAD 603 bushed bearing butt hinges (1.5 pair) with a fixed in and radius corners, 100 mm × 88 mm, satin stainless steel, grade 13 BS EN 1935	5.88	–	–	–	nr	–
ASSA ABBLOY DC700 A overhead cam motion backcheck closure with slide arm size 2–6. Full cover standard silver	100.44	–	–	–	nr	–
UNION keyULTRA euro profile key and turn cylinder, turn/key, 35 mm/35 mm, satin chrome, BS EN 1303 Security Grade 2 when fitted with security escutcheon (sold separately)	23.97	–	–	–	nr	–
Union 2S21 euro-profile sashlock, 72 mm centres, 55 mm backset, satin stainless steel, Grade 3 category of use	7.29	–	–	–	nr	–
Union 1000 01 style, 19 mm dia. lever on 8 mm round rose, sprung, satin stainless steel grade 304, bolt through connections	5.03	–	–	–	pair	–
7012F ASSA finger guard 2.05 m push side, silver	59.60	–	–	–	nr	–
UNION 1000EE Euro Escutcheon, 54 mm dia. × 8 mm projection, satin stainless steel	0.40	–	–	–	nr	–
Kick Plate 900 mm × 150 mm, satin stainless steel softened and finished edges with radius corners	11.12	–	–	–	nr	–
'Fire Door Keep Shut' 75 mm dia. sign, satin stainless steel	3.09	–	–	–	nr	–
UNION DS100 Floor Mounted Door Stop, half moon design, 45 mm dia. × 24 mm high, satin stainless steel	1.51	–	–	–	nr	–
Non-locking office doorset (fire-rated)	–	2.00	35.65	164.38	nr	**200.03**
Ironmongery schedule and cost						
UNION PowerLOAD 603 bushed bearing butt hinges (1.5 pair) with a fixed pin and radius corners, 100 mm × 88 mm, satin stainless steel, grade 13 BS EN 1935	17.65	–	–	–	nr	–
ASSA ABBLOY DC700 A overhead cam motion backcheck closure with slide arm size 2–6. Full cover standard silver	100.44	–	–	–	nr	–
UNION keyULTRA euro profile key and turn cylinder, turn/key, 35 mm/35 mm, satin chrome, BS EN 1303 Security Grade 2 when fitted with security escutcheon (sold separately)	23.97	–	–	–	nr	–
UNION 2C23 DIN mortice latch, 55 mm backset, satin stainless steel, radius forend, Grade 3 category of use, BS EN 12209, CE marked	7.32	–	–	–	nr	–

24 DOORS, SHUTTERS AND HATCHES

Item	PC £	Labour hours	Labour £	Material £	Unit	Total rate £
ASSOCIATED IRONMONGERY – cont						
ASSA ABBLOY Ltd typical door ironmongery sets – cont						
Ironmongery schedule and cost – cont						
Union 1000 01 style, 19 mm dia. lever on 8 mm round rose, sprung, satin stainless steel grade 304, bolt through connections	5.03	–	–	–	pair	–
Kick Plate 900 mm × 150 mm, satin stainless steel softened & finished edges with radius corners	11.12	–	–	–	nr	–
'Fire Door Keep Shut' 75 mm dia. sign, satin stainless steel	3.09	–	–	–	nr	–
UNION DS100 Floor Mounted Door Stop, half moon design, 45 mm dia. × 24 mm high, satin stainless steel	1.51	–	–	–	nr	–
Office/Store locking doorset (fire-rated)	–	2.00	35.65	189.73	nr	**225.38**
Ironmongery schedule and cost						
UNION PowerLOAD 603 bushed bearing butt hinges (1.5 pair) with a fixed pin and radius corners, 100 mm × 88 mm, satin stainless steel, grade 13 BS EN 1935	17.65	–	–	–	nr	–
ASSA ABBLOY DC700 A overhead cam motion backcheck closure with slide arm size 2–6. Full cover standard silver	100.44	–	–	–	nr	–
UNION keyULTRA euro profile key and turn cylinder, turn/key, 35 mm/35 mm, satin chrome, BS EN 1303 Security Grade 2 when fitted with security escutcheon (sold separately)	23.97	–	–	–	nr	–
Union 2S21 euro-profile sashlock, 72 mm centres, 55 mm backset, satin stainless steel, Grade 3 category of use	7.29	–	–	–	nr	–
Union 1000 01 style, 19 mm dia. lever on 8 mm round rose, sprung, satin stainless steel grade 304, bolt through connections	5.03	–	–	–	pair	–
Kick Plate 200 mm × 1.5 mm, satin stainless steel softened & finished edges with radius corners	11.12	–	–	–	nr	–
'Fire Door Keep Shut' 75 mm dia. sign, satin stainless steel	3.09	–	–	–	nr	–
UNION DS100 Floor Mounted Door Stop, half moon design, 45 mm dia. × 24 mm high, satin stainless steel	1.51	–	–	–	nr	–
Common room locking doorset (fire-rated)	–	2.50	44.57	264.50	nr	**309.07**
Ironmongery schedule and cost						
UNION PowerLOAD 603 bushed bearing butt hinges (1.5 pair) with a fixed pin and radius corners, 100 mm × 88 mm, satin stainless steel, Grade 13 BS EN 1935	17.65	–	–	–	nr	–
ASSA ABBLOY DC700 A overhead cam motion backcheck closure with slide arm size 2–6. Full cover standard silver	100.44	–	–	–	nr	–

24 DOORS, SHUTTERS AND HATCHES

Item	PC £	Labour hours	Labour £	Material £	Unit	Total rate £
UNION keyULTRA euro profile key and turn cylinder, turn/key (Large), 35 mm/35 mm, 6pin, satin chrome, BS EN 1303 Security Grade 2 when fitted with security escutcheon (sold separately)	31.77	–	–	–	nr	–
UNION 2C22 DIN euro-profile deadlock, 55 mm backset, satin stainless steel, radius forend, Grade 3 category of use, BS EN 12209, CE marked	10.65	–	–	–	nr	–
UNION 01 style pull handle 19 mm dia. × 425 mm centres, satin stainless steel, Grade 304, bolt through fixing	3.20	–	–	–	nr	–
Kick Plate 200 mm × 1.5 mm, satin stainless steel softened & finished edges with radius corners	11.12	–	–	–	nr	–
Push Plate 450 mm × 100 mm × 1.5 mm satin stainless steel, softened & finished edges with radius corners	1.25	–	–	–	nr	–
7012F ASSA finger guard 2.05 m push side, silver	59.60	–	–	–	nr	–
UNION 1000EE Euro Escutcheon, 54 mm dia. × 8 mm projection, satin stainless steel	0.40	–	–	–	nr	–
UNION DS100 Floor Mounted Door Stop, Half Moon Design, 45 mm dia. × 24 mm high, Satin Stainless Steel	1.51	–	–	–	nr	–
'Fire Door Keep Shut' 75 mm dia. sign, satin stainless steel	3.09	–	–	–	nr	–
PULL or PUSH signage, 75 mm dia., satin stainless steel	1.38	–	–	–	nr	–
Maintenance/Plant room doorset (fire-rated)	–	1.75	31.20	181.44	nr	**212.64**
Ironmongery schedule and cost						
1½ pairs Union Powerload 603 brushed bearing butt hinge with a fixed pin; 100 mm × 88 mm; satin stainless steel grade 13 BS EN 1935	17.65	–	–	–	nr	–
ASSA ABBLOY DC700 A overhead cam motion backcheck closure with slide arm size 2–6. Full cover standard silver	100.44	–	–	–	nr	–
UNION 2C22 DIN euro-profile deadlock, 55 mm backset, satin stainless steel, radius forend, Grade 3 category of use, BS EN 12209, CE marked	10.65	–	–	–	nr	–
UNION keyULTRA euro profile single cylinder, 35 mm/35 mm, satin chrome, BS EN 1303 Security Grade 2 when fitted with security escutcheon	31.77	–	–	–	nr	–
Kick Plate 900 mm × 150 mm, satin stainless steel softened & finished edges With Radius Corner	11.12	–	–	–	nr	–
'Fire Door Keep Locked' 75 mm dia. sign, satin stainless steel	3.09	–	–	–	nr	–
Euro Rim Cylinder Pull, stainless steel	0.38	–	–	–	nr	–

24 DOORS, SHUTTERS AND HATCHES

Item	PC £	Labour hours	Labour £	Material £	Unit	Total rate £
ASSOCIATED IRONMONGERY – cont						
ASSA ABBLOY Ltd typical door ironmongery sets – cont						
Ironmongery schedule and cost – cont						
UNION 1000EE Euro Escutcheon, 54 mm dia. × 8 mm projection, satin stainless steel	0.40	–	–	–	nr	–
UNION DS100 Floor Mounted Door Stop, half moon design, 45 mm dia. × 24 mm high, satin stainless steel	1.51	–	–	–	nr	–
Standard bathroom doorset (Unisex)	–	1.78	31.83	144.46	nr	**176.29**
Ironmongery schedule and cost						
1½ pairs Union Powerload 603 brushed bearing butt hinge with a fixed pin; 100 mm × 88 mm; satin stainless steel Grade 13 BS EN 1935	17.65	–	–	–	nr	–
ASSA ABBLOY DC700 A overhead cam motion backcheck closure with slide arm size 2–6. Full cover standard silver	100.44	–	–	–	nr	–
UNION 2C27 DIN bathroom lock, 78 mm centres, 55 mm backset, satin stainless steel, radius forend, Grade 3 category of use, BS EN 12209, CE marked	7.57	–	–	–	nr	–
Union 1000 01 style, 19 mm dia. lever on 8 mm round rose, sprung, satin stainless steel grade 304, bolt through connections	5.03	–	–	–	pair	–
UNION 1000ER emergency release, 54 mm dia. × 8 mm projection, satin stainless steel	2.41	–	–	–	nr	–
UNION 1000T standard turn, 54 mm dia. × 8 mm projection, satin stainless steel	3.59	–	–	–	nr	–
UNION DS100 Floor Mounted Door Stop, half moon design, 45 mm dia. × 24 mm high, satin stainless steel	1.51	–	–	–	nr	–
UNION US200 Unisex Symbol, 75 mm dia., satin stainless steel, Grade 304	1.38	–	–	–	nr	–
UNION HC100 hat & coat hook with a buffer, satin stainless steel	1.36	–	–	–	nr	–
Accessible toilet doorset	–	2.55	45.46	62.59	nr	**108.05**
Ironmongery schedule and cost						
1½ pairs Union Powerload 603 brushed bearing butt hinge with a fixed pin; 100 mm × 88 mm; satin stainless steel Grade 13 BS EN 1935	17.65	–	–	–	nr	–
UNION 2C27 DIN bathroom lock, 78 mm centres, 55 mm backset, satin stainless steel, radius forend, Grade 3 category of use, BS EN 12209, CE marked	7.57	–	–	–	nr	–
Union 1000 01 style, 19 mm dia. lever on 8 mm round rose, sprung, satin stainless steel grade 304, bolt through connections	5.03	–	–	–	pair	–

24 DOORS, SHUTTERS AND HATCHES

Item	PC £	Labour hours	Labour £	Material £	Unit	Total rate £
Kick plates 900 mm × 150 mm, stainless steel	11.12	–	–	–	nr	–
UNION WS200 Disabled Symbol, 75 mm dia., satin stainless steel, Grade 304	1.38	–	–	–	nr	–
UNION 1000ER emergency release, 54 mm dia. × 8 mm projection, satin stainless steel	2.41	–	–	–	nr	–
UNION 1000LT Large Turn, 54 mm dia. × 8 mm projection, satin stainless steel	1.92	–	–	–	nr	–
UNION DS100 Floor Mounted Door Stop, half moon design, 45 mm dia. × 24 mm high, satin stainless steel	1.51	–	–	–	nr	–
UNION HC100 hat & coat hook with a buffer, satin stainless steel	1.36	–	–	–	nr	–
Single Lobby/Corridor doorset (fire-rated)	–	2.25	40.11	220.20	nr	**260.31**
Ironmongery schedule and cost						
UNION PowerLOAD 603 bushed bearing butt hinges (1.5 pair) with a fixed pin and radius corners, 100 mm × 88 mm, satin stainless steel, Grade 13 BS EN 1935	17.65	–	–	–	nr	–
UNION 01 style pull handle 19 mm dia. × 425 mm centres, satin stainless steel, Grade 304, bolt through fixing	3.20	–	–	–	nr	–
ASSA ABBLOY DC700 A overhead cam motion backcheck closure with slide arm size 2–6. Full cover standard silver	100.44	–	–	–	nr	–
7012F ASSA finger guard 2.05 m push side, silver	59.60	–	–	–	nr	–
Kick Plate 900 mm × 150 mm, satin stainless steel softened & finished edges with radius corners	11.12	–	–	–	nr	–
'Fire Door Keep Shut' 75 mm dia. sign, satin stainless steel	3.09	–	–	–	nr	–
PULL or PUSH signage, 75 mm dia., satin stainless steel	1.38	–	–	–	nr	–
UNION DS100 Floor Mounted Door Stop, half moon design, 45 mm dia. × 24 mm high, satin stainless steel	1.51	–	–	–	nr	–
Double Lobby/Corridor doorset (fire-rated)	–	4.00	71.31	454.82	nr	**526.13**
Ironmongery schedule and cost						
UNION PowerLOAD 603 bushed bearing butt hinges (3 pair) with a fixed pin and radius corners, 100 mm × 88 mm, satin stainless steel, Grade 13 BS EN 1935	35.30	–	–	–	nr	–
UNION 8899 Electro Mag closer hold open/swing free, fixed strength size 4, painted silver, app 1, 61 and 66, BS EN 1155, CE marked, Certifire approved. Supplied with a curve cover c/w matching arms	107.48	–	–	–	nr	–
UNION 01 style pull handle 19 mm dia. × 425 mm centres, satin stainless steel, Grade 304, bolt through fixing	3.20	–	–	–	nr	–

24 DOORS, SHUTTERS AND HATCHES

Item	PC £	Labour hours	Labour £	Material £	Unit	Total rate £
ASSOCIATED IRONMONGERY – cont						
ASSA ABBLOY Ltd typical door ironmongery sets – cont						
Ironmongery schedule and cost – cont						
Push Plate 450 mm × 100 mm × 1.5 mm satin stainless steel, softened & finished edges with radius corners	1.25	–	–	–	nr	–
7012F ASSA finger guard 2.05 m push side, silver	59.60	–	–	–	nr	–
Kick Plate 900 mm × 150 mm, satin stainless steel softened & finished edges with radius corners	11.12	–	–	–	nr	–
'Fire Door Keep Shut' 75 mm dia. sign, satin stainless steel	3.09	–	–	–	nr	–
PULL or PUSH signage, 75 mm dia., satin stainless steel	1.38	–	–	–	nr	–
UNION DS100 Floor Mounted Door Stop, half moon design, 45 mm dia. × 24 mm high, satin stainless steel	1.51	–	–	–	nr	–
External Rebated Double Fire Escape doorset	–	4.00	71.31	996.76	nr	**1068.07**
Ironmongery schedule and cost						
UNION PowerLOAD 603 bushed bearing butt hinges (1.5 pair) with a fixed pin and radius corners, 100 mm × 88 mm, satin stainless steel, Grade 13 BS EN 1935	17.65	–	–	–	nr	–
UNION satin stainless steel, Ball Bearing Dog Bolt Hinge	89.36	–	–	–	nr	–
ASSA ABBLOY DC700 A overhead cam motion backcheck closure with slide arm size 2–6. Full cover standard silver	100.44	–	–	–	nr	–
UNION 883T panic exit double rebated doorset, 3 point locking with pullman latches, touch bar design, 900 mm maximum door width, silver, EN1125, CE marked	126.22	–	–	–	nr	–
UNION 885 outside access device (OAD), lever handle to suit 880 series only, silver, uses a UNION 2X50 5 pin euro profile cylinder to be ordered separately	30.78	–	–	–	nr	–
UNION keyULTRA euro profile single cylinder, 35 mm/35 mm, satin chrome, BS EN 1303 Security Grade 2 when fitted with security escutcheon	31.77	–	–	–	nr	–
Generic door door limiting stays	27.16	–	–	–	nr	–
Kick Plate 900 mm × 150 mm, satin stainless steel softened & finished edges with radius corners	11.12	–	–	–	nr	–
'Fire Door Keep Shut' 75 mm dia. sign, satin stainless steel	3.09	–	–	–	nr	–
SP(102)FA1201 – Aluminium	18.25	–	–	–	nr	–

24 DOORS, SHUTTERS AND HATCHES

Item	PC £	Labour hours	Labour £	Material £	Unit	Total rate £
Double Aluminium Entrance doorset	–	5.00	89.13	457.86	-	**546.99**
Ironmongery schedule and cost						
Adams Rite SENTINAL Radius Weatherstrip Faceplate, satin anodized aluminium	3.66	–	–	–	nr	–
Adams Rite 4700 Series Faceplate Radius w/s, satin anodized aluminium	3.06	–	–	–	nr	–
Adams Rite SENTINEL 6 Deadlock, 30 mm Backset BS EN 12209	9.96	–	–	–	nr	–
ADAMS RITE SENTINEL Armoured Strike Plate	3.49	–	–	–	nr	–
ADAMS RITE SENTINEL Radius Armoured Trim Strike, satin anodized aluminium	6.73	–	–	–	nr	–
ADAMS RITE 4596 Euro Paddle Handle, Pull to Left 45 mm Door Satin Anodized Aluminium	51.21	–	–	–	nr	–
ADAMS RITE 4750 Deadlatch EuroProfile 28 mm backset body only – left hand	26.81	–	–	–	nr	–
ADAMS RITE 2 Point Latch Push to Left, satin anodized aluminium	99.48	–	–	–	nr	–
ADAMS RITE Transom ARC–51N EN3 Spring Closer HO side load 70 mm pivot	49.31	–	–	–	nr	–
ADAMS RITE 7108 12 v DC Fail Safe Escape Electric Release, Satin Anodized Aluminium. Specifer to confirm depth of section as could clash with mechanism of 4781	116.73	–	–	–	nr	–
ADAMS RITE Sentinal Euro Profile Security Escutcheon	26.96	–	–	–	nr	–
External Toilet Lobby doorset (fire-rated)	–	2.50	44.57	265.51	nr	**310.08**
Ironmongery schedule and cost						
1½ pairs Union Powerload 603 brushed bearing butt hinge with a fixed pin; 100 mm × 88 mm; satin stainless steel Grade 13 BS EN 1935	17.65	–	–	–	nr	–
ASSA ABBLOY DC700 A overhead cam motion backcheck closure with slide arm size 2–6. Full cover standard silver	100.44	–	–	–	nr	–
UNION 2C22 DIN euro-profile deadlock, 55 mm backset, satin stainless steel, radius forend, Grade 3 category of use, BS EN 12209, CE marked	10.65	–	–	–	nr	–
UNION 01 style pull handle 19 mm dia. × 425 mm centres, satin stainless steel, Grade 304, bolt through fixing	3.20	–	–	–	nr	–
UNION keyULTRA euro profile single cylinder, 35 mm/35 mm, satin chrome, BS EN 1303 Security Grade 2 when fitted with security escutcheon	31.77	–	–	–	nr	–
Push Plate 450 mm × 100 mm × 1.5 mm satin stainless steel, softened & finished edges with radius corners	1.25	–	–	–	nr	–
Kick Plate 900 mm × 150 mm, satin stainless steel softened & finished edges with radius corner	11.12	–	–	–	nr	–

Prices for Measured Works

24 DOORS, SHUTTERS AND HATCHES

Item	PC £	Labour hours	Labour £	Material £	Unit	Total rate £
ASSOCIATED IRONMONGERY – cont						
ASSA ABBLOY Ltd typical door ironmongery sets – cont						
Ironmongery schedule and cost – cont						
7012F ASSA finger guard 2.05 m push side, silver	59.60	–	–	–	nr	–
'Fire Door Keep Locked' 75 mm dia. sign, satin stainless steel	3.09	–	–	–	nr	–
UNION 1000EE Euro Escutcheon, 54 mm dia. × 8 mm projection, satin stainless steel	0.40	–	–	–	nr	–
UNION DS100 Floor Mounted Door Stop, half moon design, 45 mm dia. × 24 mm high, satin stainless steel	1.51	–	–	–	nr	–
PULL or PUSH signage, 75 mm dia., satin stainless steel	1.38	–	–	–	nr	–
UNION US200 Unisex Symbol, 75 mm dia., satin stainless steel, Grade 304	1.38	–	–	–	nr	–
Internal Toilet Lobby doorset (non-fire-rated)	–	2.00	35.65	152.78	nr	**188.43**
Ironmongery schedule and cost						
1½ pairs Union Powerload 603 brushed bearing butt hinge with a fixed pin; 100 mm × 88 mm; satin stainless steel Grade 13 BS EN 1935	17.65	–	–	–	nr	–
ASSA ABBLOY DC700 A overhead cam motion backcheck closure with slide arm size 2–6. Full cover standard silver	100.44	–	–	–	nr	–
UNION 01 style pull handle 19 mm dia. × 425 mm centres, satin stainless steel, Grade 304, bolt through fixing	3.20	–	–	–	nr	–
Push Plate 450 mm × 100 mm × 1.5 mm satin stainless steel, softened & finished edges with radius corners	1.25	–	–	–	nr	–
Kick Plate 900 mm × 150 mm, satin stainless steel softened & finished edges with radius corners	11.12	–	–	–	nr	–
UNION DS100 Floor Mounted Door Stop, half moon design, 45 mm dia. × 24 mm high, satin stainless steel	1.51	–	–	–	nr	–
PULL or PUSH signage, 75 mm dia., satin stainless steel	1.38	–	–	–	nr	–
Double Classroom Store doorset (non-fire-rated)	–	1.50	26.74	222.51	nr	**249.25**
Ironmongery schedule and cost						
UNION PowerLOAD 603 bushed bearing butt hinges (3 pair) with a fixed pin and radius corners, 100 mm × 88 mm, satin stainless steel, Grade 13 BS EN 1935	17.65	–	–	–	nr	–
Union 2S21 euro-profile sashlock, 72 mm centres, 55 mm backset, satin stainless steel, Grade 3 category of use	7.29	–	–	–	nr	–

24 DOORS, SHUTTERS AND HATCHES

Item	PC £	Labour hours	Labour £	Material £	Unit	Total rate £
UNION keyULTRA euro profile key and turn cylinder, turn/key (Large), 35 mm/35 mm , 6pin, satin chrome, BS EN 1303 Security Grade 2 when fitted with security escutcheon (sold separately)	31.77	–	–	–	nr	–
Union 1000 01 style, 19 mm dia. lever on 8 mm round rose, sprung, satin stainless steel grade 304, bolt through connections	5.03	–	–	–	pair	–
UNION 1000EE Euro Escutcheon, 54 mm dia. × 8 mm projection, Satin Stainless Steel	0.40	–	–	–	nr	–
Kick Plate 900 mm × 150 mm, satin stainless steel softened & finished edges with radius corners	11.12	–	–	–	nr	–
UNION 8060 aluminium lever action flush bolt 200 mm satin stainless steel satin chrome	4.78	–	–	–	pair	–
Sliding door gear; Hillaldam Coburn Ltd or other equal and approved; Commercial/Light industrial; for top hung timber/metal doors, weight not exceeding 365 kg						
Sliding door gear						
bottom guide; fixed to concrete in groove	19.34	0.46	8.20	19.82	m	**28.02**
top track	26.28	0.23	4.10	26.94	m	**31.04**
detachable locking bar	32.88	0.31	5.52	33.70	nr	**39.22**
hangers; timber doors	53.05	0.46	8.20	54.38	nr	**62.58**
hangers; metal doors	33.90	0.46	8.20	34.75	nr	**42.95**
head brackets; open, soffit fixing; screwing to timber	6.76	0.32	5.71	6.95	nr	**12.66**
head brackets; open, side fixing; bolting to masonry	7.09	0.46	8.20	8.03	nr	**16.23**
door guide to timber door	5.94	0.23	4.10	6.09	nr	**10.19**
door stop; rubber buffers; to masonry	26.21	0.69	12.30	26.87	nr	**39.17**
drop bolt; screwing to timber	22.92	0.46	8.20	23.49	nr	**31.69**
bow handle; to timber	9.09	0.23	4.10	9.32	nr	**13.42**
Sundries						
rubber door stop; plugged and screwed to concrete	4.95	0.09	1.61	5.07	nr	**6.68**

25 STAIRS, WALKWAYS AND BALUSTRADES

Item	PC £	Labour hours	Labour £	Material £	Unit	Total rate £
STAIRS						
Standard staircases; wrought softwood (Parana pine)						
Stairs; 25 mm thick treads with rounded nosings; 9 mm thick plywood risers; 32 mm thick strings; bullnose bottom tread; 50 mm × 75 mm hardwood handrail; 32 mm square plain balusters; 100 mm square plain newel posts						
straight flight; 838 mm wide; 2676 mm going; 2600 mm rise; with two newel posts	–	6.48	115.52	339.38	nr	**454.90**
straight flight with turn; 838 mm wide; 2676 mm going; 2600 mm rise; with two newel posts; three top treads winding	–	6.48	115.52	442.68	nr	**558.20**
dogleg staircase; 838 mm wide; 2676 mm going; 2600 mm rise; with two newel posts; quarter space landing third riser from top	–	6.48	115.52	414.96	nr	**530.48**
dogleg staircase; 838 mm wide; 2676 mm going; 2600 mm rise; with two newel posts; half space landing third riser from top	–	7.40	131.92	510.70	nr	**642.62**
Hardwood staircases; purpose-made; assembled at works						
Fixing only complete staircase including landings, balustrades, etc.						
plugging and screwing to brickwork or blockwork	–	13.88	247.45	2.08	nr	**249.53**
The following are supply only prices for purpose-made staircase components in selected Sapele supplied as part of an assembled staircase and may be used to arrive at a guide price for a complete hardwood staircase						
Board landings; cross-tongued joints; 100 mm × 50 mm sawn softwood bearers						
25 mm thick	–	–	–	76.93	m²	**76.93**
32 mm thick	–	–	–	86.47	m²	**86.47**
Treads; cross-tongued joints and risers; rounded nosings; tongued, grooved, glued and blocked together; one 175 mm × 50 mm sawn softwood carriage						
25 mm treads; 19 mm risers	–	–	–	154.60	m²	**154.60**
ends; quadrant	–	–	–	47.09	nr	**47.09**
ends; housed to hardwood	–	–	–	0.87	nr	**0.87**
32 mm treads; 25 mm risers	–	–	–	160.17	m²	**160.17**
ends; quadrant	–	–	–	60.54	nr	**60.54**
ends; housed to hardwood	–	–	–	0.87	nr	**0.87**

25 STAIRS, WALKWAYS AND BALUSTRADES

Item	PC £	Labour hours	Labour £	Material £	Unit	Total rate £
Winders; cross-tongued joints and risers in one width; rounded nosings; tongued, grooved glued and blocked together; one 175 mm × 50 mm sawn softwood carriage						
25 mm treads; 19 mm risers	–	–	–	214.91	m²	**214.91**
32 mm treads; 25 mm risers	–	–	–	219.97	m²	**219.97**
wide ends; housed to hardwood	–	–	–	1.73	nr	**1.73**
narrow ends; housed to hardwood	–	–	–	1.30	nr	**1.30**
Closed strings; in one width; 230 mm wide; rounded twice						
32 mm thick	–	–	–	28.31	m	**28.31**
38 mm thick	–	–	–	30.82	m	**30.82**
50 mm thick	–	–	–	34.34	m	**34.34**
Closed strings; cross-tongued joints; 280 mm wide; once rounded						
32 mm thick	–	–	–	36.75	m	**36.75**
extra for short ramp	–	–	–	18.88	nr	**18.88**
38 mm thick	–	–	–	40.08	m	**40.08**
extra for short ramp	–	–	–	21.46	nr	**21.46**
50 mm thick	–	–	–	44.73	m	**44.73**
extra for short ramp	–	–	–	26.57	nr	**26.57**
The following labours are irrespective of timber width						
ends; fitted	–	–	–	1.13	nr	**1.13**
ends; framed	–	–	–	6.58	nr	**6.58**
extra for tongued heading joint	–	–	–	3.25	nr	**3.25**
Closed strings; ramped; crossed-tongued joints 280 mm wide; once rounded						
32 mm thick	–	–	–	36.75	m	**36.75**
44 mm thick	–	–	–	40.08	m	**40.08**
57 mm thick	–	–	–	44.73	m	**44.73**
Apron linings; in one width 230 mm wide						
19 mm thick	–	–	–	9.77	m	**9.77**
25 mm thick	–	–	–	11.51	m	**11.51**
The following are supply only prices for purpose-made staircase components in selected American Oak; supplied as part of an assembled staircase						
Board landings; cross-tongued joints; 100 mm × 50 mm sawn softwood bearers						
25 mm thick	–	–	–	113.49	m²	**113.49**
32 mm thick	–	–	–	137.04	m²	**137.04**
Treads; cross-tongued joints and risers; rounded nosings; tongued, grooved, glued and blocked together; one 175 mm × 50 mm sawn softwood carriage						
25 mm treads; 19 mm risers	–	–	–	188.40	m²	**188.40**
ends; quadrant	–	–	–	94.12	nr	**94.12**
ends; housed to hardwood	–	–	–	1.15	nr	**1.15**
32 mm treads; 25 mm risers	–	–	–	215.88	m²	**215.88**
ends; quadrant	–	–	–	115.85	nr	**115.85**
ends; housed to hardwood	–	–	–	1.15	nr	**1.15**

25 STAIRS, WALKWAYS AND BALUSTRADES

Item	PC £	Labour hours	Labour £	Material £	Unit	Total rate £
STAIRS – cont						
The following are supply only prices for purpose-made staircase components in selected American Oak – cont						
Winders; cross-tongued joints and risers in one width; rounded nosings; tongued, grooved glued and blocked together; one 175 mm × 50 mm sawn softwood carriage						
25 mm treads; 19 mm risers	–	–	–	238.81	m²	**238.81**
32 mm treads; 25 mm risers	–	–	–	260.58	m²	**260.58**
wide ends; housed to hardwood	–	–	–	2.32	nr	**2.32**
narrow ends; housed to hardwood	–	–	–	1.74	nr	**1.74**
Closed strings; in one width; 230 mm wide; rounded twice						
32 mm thick	–	–	–	44.51	m	**44.51**
44 mm thick	–	–	–	51.37	m	**51.37**
57 mm thick	–	–	–	70.53	m	**70.53**
Closed strings; cross-tongued joints; 280 mm wide; once rounded						
32 mm thick	–	–	–	56.55	m	**56.55**
extra for short ramp	–	–	–	32.33	nr	**32.33**
38 mm thick	–	–	–	65.54	m	**65.54**
extra for short ramp	–	–	–	36.84	nr	**36.84**
50 mm thick	–	–	–	89.85	m	**89.85**
extra for short ramp	–	–	–	48.99	nr	**48.99**
Closed strings; ramped; crossed-tongued joints 280 mm wide; once rounded						
32 mm thick	–	–	–	65.04	m	**65.04**
44 mm thick	–	–	–	75.38	m	**75.38**
57 mm thick	–	–	–	103.34	m	**103.34**
Apron linings; in one width 230 mm wide						
19 mm thick	–	–	–	15.15	m	**15.15**
25 mm thick	–	–	–	18.57	m	**18.57**
Handrails; rounded						
40 mm × 50 mm	–	–	–	12.28	m	**12.28**
50 mm × 75 mm	–	–	–	15.74	m	**15.74**
57 mm × 87 mm	–	–	–	23.64	m	**23.64**
69 mm × 100 mm	–	–	–	31.82	m	**31.82**
Handrails; moulded						
40 mm × 50 mm	–	–	–	13.47	m	**13.47**
50 mm × 75 mm	–	–	–	16.93	m	**16.93**
57 mm × 87 mm	–	–	–	24.83	m	**24.83**
69 mm × 100 mm	–	–	–	32.99	m	**32.99**
Add to above for						
grooved once	–	–	–	0.60	m	**0.60**
ends; framed	–	–	–	6.09	nr	**6.09**
ends; framed on rake	–	–	–	7.71	nr	**7.71**

25 STAIRS, WALKWAYS AND BALUSTRADES

Item	PC £	Labour hours	Labour £	Material £	Unit	Total rate £
Heading joints to handrail; mitred or raked						
overall size not exceeding 50 mm × 75 mm	–	–	–	32.47	nr	**32.47**
overall size not exceeding 69 mm × 100 mm	–	–	–	38.56	nr	**38.56**
Knee piece to handrail; mitred or raked						
overall size not exceeding 69 mm × 100 mm	–	–	–	69.00	nr	**69.00**
Balusters; stiffeners						
25 mm × 25 mm	–	–	–	2.81	m	**2.81**
32 mm × 32 mm	–	–	–	3.54	m	**3.54**
44 mm × 44 mm	–	–	–	5.58	m	**5.58**
ends; housed	–	–	–	1.41	nr	**1.41**
Sub-rails						
32 mm × 63 mm	–	–	–	7.56	m	**7.56**
ends; framed joint to newel	–	–	–	6.09	nr	**6.09**
Knee rails						
32 mm × 140 mm	–	–	–	13.14	m	**13.14**
ends; framed joint to newel	–	–	–	6.09	nr	**6.09**
Newel posts						
44 mm × 94 mm; half newel	–	–	–	9.86	m	**9.86**
69 mm × 69 mm	–	–	–	16.79	m	**16.79**
94 mm × 94 mm	–	–	–	41.90	m	**41.90**
Newel caps; splayed on four sides						
62.50 mm × 125 mm × 50 mm	–	–	–	8.46	nr	**8.46**
100 mm × 100 mm × 50 mm	–	–	–	8.93	nr	**8.93**
125 mm × 125 mm × 50 mm	–	–	–	9.82	nr	**9.82**
Spiral staircases, balustrades and handrails; mild steel; galvanized and polyester powder coated						
Staircase						
2080 mm dia. × 3695 mm high; 18 nr treads; 16 mm dia. intermediate balusters; 1040 mm × 1350 mm landing unit with matching balustrade both sides; fixing with 16 mm dia. resin anchors to masonry at landing and with 12 mm dia. expanding bolts to concrete at base	–	–	–	–	nr	**6500.00**

25 STAIRS, WALKWAYS AND BALUSTRADES

Item	PC £	Labour hours	Labour £	Material £	Unit	Total rate £
BALUSTRADES						
Standard balustrades; wrought softwood						
Landing balustrade; 50 mm × 75 mm hardwood handrail; 32 mm square plain balusters; one end of handrail jointed to newel post; other end built into wall; balusters housed in at bottom (newel post and mortices both not included)						
3.00 m long	–	3.70	65.96	85.67	nr	**151.63**
Steel						
Balustrades; galvanized mild steel CHS posts and top rail, with one infill rail						
1100 mm high	–	–	–	–	m	**178.13**
Balustrades; painted mild steel flat bar posts and CHS top rail, with 3 nr. stainless steel infills						
1100 mm high	–	–	–	–	m	**249.38**
Stainless steel						
Balustrades; stainless steel flat bar posts and circular handrail, with 3 nr. stainless steel infills						
1100 mm high	–	–	–	–	m	**299.25**
Balustrades; stainless steel 50 mm Ø posts and circular handrail, with 10 mm thick toughened glass infill panels						
1100 mm high	–	–	–	–	m	**498.75**
Glass						
Balustrades; laminated glass; with stainless steel cap channel to top and including all necessary support fixings						
1100 mm high	–	–	–	–	m	**498.75**
Softwood handrails						
Handrails; rounded						
40 mm × 50 mm	–	–	–	10.61	m	**10.61**
50 mm × 75 mm	–	–	–	12.79	m	**12.79**
57 mm × 87 mm	–	–	–	15.02	m	**15.02**
69 mm × 100 mm	–	–	–	18.65	m	**18.65**
Handrails; moulded						
40 mm × 50 mm	–	–	–	11.80	m	**11.80**
50 mm × 75 mm	–	–	–	13.98	m	**13.98**
57 mm × 87 mm	–	–	–	16.22	m	**16.22**
69 mm × 100 mm	–	–	–	19.84	m	**19.84**
Add to above for						
grooved once	–	–	–	0.52	m	**0.52**
ends; framed	–	–	–	4.96	nr	**4.96**
ends; framed on rake	–	–	–	6.09	nr	**6.09**
Heading joints to handrail; mitred or raked						
overall size not exceeding 50 mm × 75 mm	–	–	–	24.35	nr	**24.35**
overall size not exceeding 69 mm × 100 mm	–	–	–	30.44	nr	**30.44**

25 STAIRS, WALKWAYS AND BALUSTRADES

Item	PC £	Labour hours	Labour £	Material £	Unit	Total rate £
Knee piece to handrail; mitred or raked						
overall size not exceeding 69 mm × 100 mm	–	–	–	64.94	nr	**64.94**
Balusters; stiffeners						
25 mm × 25 mm	–	–	–	2.72	m	**2.72**
32 mm × 32 mm	–	–	–	3.10	m	**3.10**
44 mm × 44 mm	–	–	–	4.06	m	**4.06**
ends; housed	–	–	–	1.22	nr	**1.22**
Sub-rails						
32 mm × 63 mm	–	–	–	6.25	m	**6.25**
ends; framed joint to newel	–	–	–	5.27	nr	**5.27**
Knee rails						
32 mm × 140 mm	–	–	–	10.38	m	**10.38**
ends; framed joint to newel	–	–	–	5.27	nr	**5.27**
Newel posts						
44 mm × 94 mm; half newel	–	–	–	7.21	m	**7.21**
69 mm × 69 mm	–	–	–	7.81	m	**7.81**
94 mm × 94 mm	–	–	–	15.98	m	**15.98**
Newel caps; splayed on four sides						
62.50 mm × 125 mm × 50 mm	–	–	–	7.63	nr	**7.63**
100 mm × 100 mm × 50 mm	–	–	–	7.78	nr	**7.78**
125 mm × 125 mm × 50 mm	–	–	–	8.18	nr	**8.18**
Metal handrails						
Wallrails; painted mild steel CHS wall rail; with wall rose bracket						
42 mm dia.	–	–	–	–	m	**71.25**
Wallrails; stainless steel circular wall rail; with wall rose bracket						
42 mm dia.	–	–	–	–	m	**92.63**
WALKWAYS						
Flooring; metalwork						
Chequer plate flooring; galvanized mild steel; over 300 mm wide; bolted to steel supports						
6 mm thick	–	–	–	–	m²	**213.75**
8 mm thick	–	–	–	–	m²	**228.00**
Open mesh flooring; galvanized; over 300 mm wide; bolted to steel supports						
8 mm thick	–	–	–	–	m²	**213.75**
Surface treatment						
At works						
galvanizing	–	–	–	–	tonne	**285.00**
shotblasting	–	–	–	–	m²	**3.56**
touch up primer and one coat of two pack epoxy zinc phosphate or chromate primer	–	–	–	–	m²	**7.13**

25 STAIRS, WALKWAYS AND BALUSTRADES

Item	PC £	Labour hours	Labour £	Material £	Unit	Total rate £
LADDERS						
Loft ladders; fixing with screws to timber lining (not included)						
Loft ladders						
Youngman Easiway 3 section aluminium ladder; 2.3 m to 3.0 m ceiling height	–	0.93	16.57	100.48	nr	**117.05**
Youngman Eco folding 3 section timber ladder; 2.8 m ceiling height	–	2.00	35.65	126.91	nr	**162.56**
Youngman Spacemaker aluminium sliding ladder; 2.6 m ceiling height	–	1.25	22.28	55.02	nr	**77.30**
Youngman Deluxe 2 section aluminium ladder; 3.25 m ceiling height; spring assisted	–	2.50	44.57	296.03	nr	**340.60**
Access ladders; mild steel						
Ladders						
400 mm wide; 3850 mm long (overall); 12 mm dia. rungs; 65 mm × 15 mm strings; 50 mm × 5 mm safety hoops; fixing with expanded bolts; to masonry; mortices; welded fabrication	–	–	–	–	nr	**1125.00**

27 GLAZING

Item	PC £	Labour hours	Labour £	Material £	Unit	Total rate £
SUPPLY ONLY RATES FOR GLASS						
BASIC GLASS PRICES SUPPLY ONLY						
Ordinary transluscent/patterned glass						
3 mm	–	–	–	24.41	m²	**24.41**
4 mm	–	–	–	25.90	m²	**25.90**
5 mm	–	–	–	31.49	m²	**31.49**
6 mm	–	–	–	34.57	m²	**34.57**
Obscured ground sheet glass – patterned						
4 mm white	–	–	–	36.61	m²	**36.61**
6 mm white	–	–	–	40.29	m²	**40.29**
Rough cast						
6 mm	–	–	–	34.33	m²	**34.33**
Ordinary Georgian wired						
7 mm cast	–	–	–	34.99	m²	**34.99**
6 mm polish	–	–	–	53.82	m²	**53.82**
Cetuff toughened; float						
4 mm	–	–	–	28.68	m²	**28.68**
5 mm	–	–	–	38.00	m²	**38.00**
6 mm	–	–	–	41.83	m²	**41.83**
10 mm	–	–	–	68.68	m²	**68.68**
Clear laminated; safety						
4.40 mm	–	–	–	34.86	m²	**34.86**
6.40 mm	–	–	–	40.79	m²	**40.79**
GENERAL GLAZING						
SUPPLY AND FIX PRICES						
NOTE: The following measured rates are provided by a glazing contractor and assume in excess of 500 m², within 20 miles of the suppliers branch.						
Standard plain glass; BS EN 14449; clear float; panes area 0.15 m²–4.00 m²						
3 mm thick; glazed with						
screwed beads	–	–	–	–	m²	**39.98**
4 mm thick; glazed with						
screwed beads	–	–	–	–	m²	**42.41**
5 mm thick; glazed with						
screwed beads	–	–	–	–	m²	**51.64**
6 mm thick; glazed with						
screwed beads	–	–	–	–	m²	**56.58**
Standard plain glass; BS EN 14449; obscure patterned; panes area 0.15 m²–4.00 m²						
4 mm thick; glazed with						
screwed beads	–	–	–	–	m²	**59.96**
6 mm thick; glazed with						
screwed beads	–	–	–	–	m²	**65.98**

27 GLAZING

Item	PC £	Labour hours	Labour £	Material £	Unit	Total rate £
GENERAL GLAZING – cont						
Standard plain glass; BS EN 14449; rough cast; panes area 0.15 m²–4.00 m²						
6 mm thick; glazed with						
screwed beads	–	–	–	–	m²	54.26
Standard plain glass; BS EN 14449; Georgian wired cast; panes area 0.15 m²–4.00 m²						
7 mm thick; glazed with						
screwed beads	–	–	–	–	m²	55.30
extra for lining up wired glass	–	–	–	–	m²	3.52
Standard plain glass; BS EN 14449; Georgian wired polished; panes area 0.15 m²–4.00 m²						
6 mm thick; glazed with						
screwed beads	–	–	–	–	m²	67.56
extra for lining up wired glass	–	–	–	–	m²	3.52
Factory-made double hermetically sealed units; to wood or metal with screwed or clipped beads						
Two panes; BS EN 14449; clear float glass; 4 mm thick; 6 mm air space						
0.35 m²–2.00 m²	–	–	–	–	m²	107.21
Two panes; BS 952; clear float glass; 6 mm thick; 6 mm air space						
0.35 m²–2.0 m²	–	–	–	–	m²	124.85
2.00 m²–4.00 m²	–	–	–	–	m²	187.74
Factory-made double hermetically sealed units; with inner pane of Pilkington's K low emissivity coated glass; to wood or metal with screwed or clipped beads						
Two panes; BS EN 14449; clear float glass; 4 mm thick; 6 mm air space						
0.35 m²–2.00 m²	–	–	–	–	m²	130.41
Two panes; BS EN 14449; clear float glass; 6 mm thick; 6 mm air space						
0.35 m²–2.0 m²	–	–	–	–	m²	151.84
2.00 m²–4.00 m²	–	–	–	–	m²	228.36
Factory-made triple hermetically sealed units; with inner pane of Pilkington's K low emissivity coated glass; to wood or metal with screwed or clipped beads						
Three panes; BS EN 14449; clear float glass; 4 mm thick; 6 mm air spaces						
0.35 m²–2.00 m²	–	–	–	–	m²	209.97

27 GLAZING

Item	PC £	Labour hours	Labour £	Material £	Unit	Total rate £
Three panes; BS EN 14449; clear float glass; 6 mm thick; 6 mm air spaces						
0.35 m²–2.0 m²	–	–	–	–	m²	244.50
2.00 m²–4.00 m²	–	–	–	–	m²	367.68
FIRE-RESISTING GLAZING						
Special glass; BS EN 14449; Pyran half-hour fire-resisting glass or other equal or approved 6.50 mm thick rectangular panes; glazed with screwed hardwood beads and Sealmaster Fireglaze intumescent compound or other equal and approved to rebated frame						
300 mm × 400 mm pane	–	0.37	11.31	48.47	nr	59.78
400 mm × 800 mm pane	–	0.46	14.05	125.32	nr	139.37
500 mm × 1400 mm pane	–	0.74	22.60	270.02	nr	292.62
600 mm × 1800 mm pane	–	0.93	28.40	432.09	nr	460.49
Special glass; BS EN 14449; Pyrostop one-hour fire-resisting glass or other equal and approved 15 mm thick regular panes; glazed with screwed hardwood beads and Sealmaster Fireglaze intumescent liner and compound or other equal and approved both sides						
300 mm × 400 mm pane	–	1.11	33.91	90.16	nr	124.07
400 mm × 800 mm pane	–	1.39	42.46	183.87	nr	226.33
500 mm × 1400 mm pane	–	1.85	56.51	383.67	nr	440.18
600 mm × 1800 mm pane	–	2.31	70.56	576.29	nr	646.85
TOUGHENED AND LAMINATED GLAZING						
Special glass; BS EN 14449; toughened clear float; panes area 0.15 m²–4.00 m² 4 mm thick; glazed with						
screwed beads	–	–	–	–	m²	42.52
5 mm thick; glazed with						
screwed beads	–	–	–	–	m²	56.47
6 mm thick; glazed with						
screwed beads	–	–	–	–	m²	62.12
10 mm thick; glazed with						
screwed beads	–	–	–	–	m²	103.08
Special glass; BS EN 14449; clear laminated safety glass; panes area 0.15 m²–4.00 m² 4.40 mm thick; glazed with						
screwed beads	–	–	–	–	m²	57.03
6.40 mm thick; glazed with						
screwed beads	–	–	–	–	m²	68.22

27 GLAZING

Item	PC £	Labour hours	Labour £	Material £	Unit	Total rate £
TOUGHENED AND LAMINATED GLAZING – cont						
Special glass; BS EN 14449; clear laminated security glass						
7.50 mm thick regular panes; glazed with screwed hardwood beads and Intergens intumescent strip						
300 mm × 400 mm pane	–	0.37	11.31	28.80	nr	**40.11**
400 mm × 800 mm pane	–	0.46	14.05	73.69	nr	**87.74**
500 mm × 1400 mm pane	–	0.74	22.60	157.33	nr	**179.93**
600 mm × 1800 mm pane	–	0.93	28.40	251.19	nr	**279.59**
MIRRORS AND LOUVRES						
Mirror panels; BS EN 14449; silvered; insulation backing						
4 mm thick float; fixing with adhesive						
1000 mm × 1000 mm	–	–	–	–	nr	**44.89**
1000 mm × 2000 mm	–	–	–	–	nr	**89.85**
1000 mm × 4000 mm	–	–	–	–	nr	**327.47**
Glass louvres; BS EN 14449; with long edges ground or smooth						
6 mm thick float						
150 mm wide	–	–	–	–	m	**22.37**
7 mm thick Georgian wired cast						
150 mm wide	–	–	–	–	m	**35.22**
6 mm thick Georgian wire polished						
150 mm wide	–	–	–	–	m	**49.48**
ADDITIONAL LABOURS						
Drill holes in glass						
Drill holes 6 mm to 15 mm dia. to glass thickness:						
not exceeding 6 mm thick	–	–	–	3.40	nr	**3.40**
not exceeding 10 mm thick	–	–	–	4.38	nr	**4.38**
not exceeding 12 mm thick	–	–	–	5.42	nr	**5.42**
not exceeding 19 mm thick	–	–	–	6.84	nr	**6.84**
not exceeding 25 mm thick	–	–	–	8.48	nr	**8.48**
Drill holes 16 mm to 38 mm dia. to glass thickness:						
not exceeding 6 mm thick	–	–	–	4.89	nr	**4.89**
not exceeding 10 mm thick	–	–	–	6.46	nr	**6.46**
not exceeding 12 mm thick	–	–	–	7.71	nr	**7.71**
not exceeding 19 mm thick	–	–	–	9.72	nr	**9.72**
not exceeding 25 mm thick	–	–	–	12.02	nr	**12.02**
Drill holes over 38 mm dia. to glass thickness:						
not exceeding 6 mm thick	–	–	–	9.72	nr	**9.72**
not exceeding 10 mm thick	–	–	–	11.73	nr	**11.73**
not exceeding 12 mm thick	–	–	–	13.88	nr	**13.88**
not exceeding 19 mm thick	–	–	–	17.08	nr	**17.08**
not exceeding 25 mm thick	–	–	–	21.31	nr	**21.31**

27 GLAZING

Item	PC £	Labour hours	Labour £	Material £	Unit	Total rate £
Other works to glass						
Curved cutting to glass						
to 4 mm thick panes	–	–	–	4.67	m	**4.67**
to 6 mm thick panes	–	–	–	4.67	m	**4.67**
to 6 mm thick wired panes	–	–	–	7.12	m	**7.12**
Intumescant paste to glazed panels for die doors;						
per side treated	–	–	–	9.87	m	**9.87**
Imitation washleather/black velvet bedding to						
edge of glass	–	–	–	1.70	m	**1.70**

28 FLOOR, WALL, CEILING AND ROOF FINISHINGS

Item	PC £	Labour hours	Labour £	Material £	Unit	Total rate £
FLOORS; SCREEDS (CEMENT: SAND; CONCRETE; GRANOLITHIC)						
Cement and sand (1:3) screeds; steel trowelled						
Work to floors; one coat level; to concrete base; screeded; over 600 mm wide						
25 mm thick	–	–	–	–	m²	8.45
50 mm thick	–	–	–	–	m²	10.01
75 mm thick	–	–	–	–	m²	13.25
100 mm thick	–	–	–	–	m²	16.48
Add to the above for work to falls and crossfalls and to slopes						
not exceeding 15° from horizontal	–	0.02	0.40	–	m²	0.40
over 15° from horizontal	–	0.09	1.81	–	m²	1.81
water repellent additive incorporated in the mix	–	0.02	0.40	3.43	m²	3.83
oil repellent additive incorporated in the mix	–	0.07	1.41	3.64	m²	5.05
Fine concrete (1: 4–5) levelling screeds; steel trowelled						
Work to floors; one coat; level; to concrete base; over 600 mm wide						
50 mm thick	–	–	–	–	m²	10.01
75 mm thick	–	–	–	–	m²	13.25
extra over last for isolation joint to perimeter	–	–	–	–	m	1.09
Early drying floor screed; RMC Mortars Readyscreed; or other equal and approved; steel trowelled						
Work to floors; one coat; level; to concrete base; over 600 mm wide						
100 mm thick	–	–	–	–	m²	17.57
extra over last for galvanized chicken wire anticrack reinforcement	–	–	–	–	m²	0.83
Granolithic paving; cement and granite chippings 5 to dust (1:1:2); steel trowelled						
Work to floors; one coat; level; laid on concrete while green; bonded; over 600 mm wide						
25 mm thick	–	–	–	–	m²	22.79
38 mm thick	–	–	–	–	m²	25.36
Work to floors; two coats; laid on hacked concrete with slurry; over 600 mm wide						
50 mm thick	–	–	–	–	m²	28.05
75 mm thick	–	–	–	–	m²	34.52
Work to landings; one coat; level; laid on concrete while green; bonded; over 600 mm wide						
25 mm thick	–	–	–	–	m²	34.08
38 mm thick	–	–	–	–	m²	38.04

28 FLOOR, WALL, CEILING AND ROOF FINISHINGS

Item	PC £	Labour hours	Labour £	Material £	Unit	Total rate £
Work to landings; two coats; laid on hacked concrete with slurry; over 600 mm wide						
50 mm thick	–	–	–	–	m²	**42.07**
75 mm thick	–	–	–	–	m²	**51.79**
Add to the above over 600 mm wide for						
liquid hardening additive incorporated in the mix	–	0.04	0.81	0.39	m²	**1.20**
oil-repellent additive incorporated in the mix	–	0.07	1.41	3.64	m²	**5.05**
25 mm work to treads; one coat; to concrete base						
225 mm wide	–	0.83	16.77	8.58	m	**25.35**
275 mm wide	–	0.83	16.77	9.60	m	**26.37**
returned end	–	0.17	3.43	–	nr	**3.43**
13 mm skirtings; rounded top edge and coved bottom junction; to brickwork or blockwork base						
75 mm wide on face	–	0.51	10.30	0.41	m	**10.71**
150 mm wide on face	–	0.69	13.94	7.55	m	**21.49**
ends; fair	–	0.04	0.81	–	nr	**0.81**
angles	–	0.06	1.21	–	nr	**1.21**
13 mm outer margin to stairs; to follow profile of and with rounded nosing to treads and risers; fair edge and arris at bottom, to concrete base						
75 mm wide	–	0.83	16.77	4.12	m	**20.89**
angles	–	0.06	1.21	–	nr	**1.21**
13 mm wall string to stairs; fair edge and arris on top; coved bottom junction with treads and risers; to brickwork or blockwork base						
275 mm (extreme) wide	–	0.74	14.95	7.21	m	**22.16**
ends	–	0.04	0.81	–	nr	**0.81**
angles	–	0.06	1.21	–	nr	**1.21**
ramps	–	0.07	1.41	–	nr	**1.41**
ramped and wreathed corners	–	0.09	1.81	–	nr	**1.81**
13 mm outer string to stairs; rounded nosing on top at junction with treads and risers; fair edge and arris at bottom; to concrete base						
300 mm (extreme) wide	–	0.74	14.95	8.92	m	**23.87**
ends	–	0.04	0.81	–	nr	**0.81**
angles	–	0.06	1.21	–	nr	**1.21**
ramps	–	0.07	1.41	–	nr	**1.41**
ramps and wreathed corners	–	0.09	1.81	–	nr	**1.81**
19 mm thick skirtings; rounded top edge and coved bottom junction; to brickwork or blockwork base						
75 mm wide on face	–	0.51	10.30	7.55	m	**17.85**
150 mm wide on face	–	0.69	13.94	11.66	m	**25.60**
ends; fair	–	0.04	0.81	–	nr	**0.81**
angles	–	0.06	1.21	–	nr	**1.21**
19 mm riser; one rounded nosing; to concrete base						
150 mm high; plain	–	0.83	16.77	6.52	m	**23.29**
150 mm high; undercut	–	0.83	16.77	6.52	m	**23.29**
180 mm high; plain	–	0.83	16.77	8.92	m	**25.69**
180 mm high; undercut	–	0.83	16.77	8.92	m	**25.69**

28 FLOOR, WALL, CEILING AND ROOF FINISHINGS

Item	PC £	Labour hours	Labour £	Material £	Unit	Total rate £
FLOORS; SCREEDS (RESIN)						
Latex self levelling floor screeds; steel trowelled						
Work to floors; level; to concrete base; over 600 mm wide						
3 mm thick; one coat	–	–	–	–	m²	3.75
5 mm thick; two coats	–	–	–	–	m²	5.22
Resin flooring; Altro or other equal and approved; steel trowelled						
Work to floors; level; to concrete base; over 600 mm wide						
AltroTect clear; up to 3 mm thick; two coat	–	–	–	–	m²	6.75
AltroTect coloured; up to 3 mm thick; two coats	–	–	–	–	m²	7.20
AltroFlow EP 1000; up to 3 mm thick; two coats	–	–	–	–	m²	15.80
AltroFlow EP 2000; up to 3 mm thick; two coats	–	–	–	–	m²	25.00
AltroFlow EP 3000; up to 3 mm thick; two coats	–	–	–	–	m²	32.00
AltroScreed Quartz EP; 4–6 mm thick	–	–	–	–	m²	40.00
Altro SoloSafe; 5–6 mm thick	–	–	–	–	m²	58.00
Altro MultiScreed; 4 mm thick	–	–	–	–	m²	35.00
Altrocrete PU (Polyurethane); 2–9 mm thick	–	–	–	–	m²	34.00
Isocrete K screeds or other equal; steel trowelled						
Work to floors; level; to concrete base; over 600 mm wide						
35 mm thick; plus polymer bonder coat	–	–	–	–	m²	12.34
40 mm thick	–	–	–	–	m²	11.39
45 mm thick	–	–	–	–	m²	12.04
50 mm thick	–	–	–	–	m²	12.69
Work to floors; to falls or cross-falls; to concrete base; over 600 mm wide						
55 mm (average) thick	–	–	–	–	m²	13.33
60 mm (average) thick	–	–	–	–	m²	13.99
65 mm (average) thick	–	–	–	–	m²	14.63
75 mm (average) thick	–	–	–	–	m²	15.93
90 mm (average) thick	–	–	–	–	m²	17.87
Isocrete K screeds; quick drying; or other equal and approved; steel trowelled						
Work to floors; level or to floors n.e. 15° from the horizontal; to concrete base; over 600 mm wide						
55 mm thick	–	–	–	–	m²	17.26
75 mm thick	–	–	–	–	m²	21.57
Isocrete pumpable Self Level Plus screeds; or other equal and approved; protected with Corex type polythene; knifed off prior to laying floor finish; flat smooth finish						
Work to floors; level or to floors n.e. 15° from the horizontal; to concrete base; over 600 mm wide						
20 mm thick	–	–	–	–	m²	21.03
50 mm thick	–	–	–	–	m²	28.05

28 FLOOR, WALL, CEILING AND ROOF FINISHINGS

Item	PC £	Labour hours	Labour £	Material £	Unit	Total rate £
FLOORS; MASTIC ASPHALT						
Mastic asphalt flooring to BS 6925 Type F 1076; black						
20 mm thick; one coat coverings; felt isolating membrane; to concrete base; flat						
over 300 mm wide	–	–	–	–	m²	17.50
225 mm–300 mm wide	–	–	–	–	m²	32.51
150 mm–225 mm wide	–	–	–	–	m²	35.70
not exceeding 150 mm wide	–	–	–	–	m²	43.70
25 mm thick; one coat coverings; felt isolating membrane; to concrete base; flat						
over 300 mm wide	–	–	–	–	m²	20.31
225 mm–300 mm wide	–	–	–	–	m²	34.67
150 mm–225 mm wide	–	–	–	–	m²	37.79
not exceeding 150 mm wide	–	–	–	–	m²	45.81
20 mm three coat skirtings to brickwork base						
not exceeding 150 mm girth	–	–	–	–	m	17.87
150 mm–225 mm girth	–	–	–	–	m	21.85
225 mm–300 mm girth	–	–	–	–	m	25.83
Mastic asphalt flooring; acid-resisting; black						
20 mm thick; one coat coverings; felt isolating membrane; to concrete base flat						
over 300 mm wide	–	–	–	–	m²	20.49
225 mm–300 mm wide	–	–	–	–	m²	37.48
150 mm–225 mm wide	–	–	–	–	m²	38.71
not exceeding 150 mm wide	–	–	–	–	m²	46.70
25 mm thick; one coat coverings; felt isolating membrane; to concrete base; flat						
over 300 mm wide	–	–	–	–	m²	24.21
225 mm–300 mm wide	–	–	–	–	m²	38.53
150 mm–225 mm wide	–	–	–	–	m²	41.71
not exceeding 150 mm wide	–	–	–	–	m²	49.72
20 mm thick; three coat skirtings to brickwork base						
not exceeding 150 mm girth	–	–	–	–	m	18.06
150 mm–225 mm girth	–	–	–	–	m	21.04
225 mm–300 mm girth	–	–	–	–	m	23.88
Mastic asphalt flooring to BS 6925 Type F 1451; red						
20 mm thick; one coat coverings; felt isolating membrane; to concrete base; flat						
over 300 mm wide	–	–	–	–	m²	28.67
225 mm–300 mm wide	–	–	–	–	m²	47.37
150 mm–225 mm wide	–	–	–	–	m²	51.16
not exceeding 150 mm wide	–	–	–	–	m²	61.21
20 mm thick; three coat skirtings to brickwork base						
not exceeding 150 mm girth	–	–	–	–	m	22.51
150 mm–225 mm girth	–	–	–	–	m	28.67

28 FLOOR, WALL, CEILING AND ROOF FINISHINGS

Item	PC £	Labour hours	Labour £	Material £	Unit	Total rate £
FLOORS; EDGE FIXED CARPETING; CARPET TILES						
Fitted carpeting; Wilton wool/nylon or other equal and approved; 80/20 velvet pile; heavy domestic plain						
Work to floors						
over 600 mm wide	24.00	0.37	5.06	27.06	m²	32.12
Work to treads and risers						
not exceeding 600 mm wide	14.40	0.74	10.13	16.24	m	26.37
Fitted carpet; Forbo Flooring; Flocked flooring						
Work to floors						
Flotex classic textile; 2.0 m wide roll	21.25	0.45	8.03	23.57	m²	31.60
Flotex HD textile; 2.0 m wide roll	23.80	0.45	8.03	26.38	m²	34.41
Carpet tiles; Forbo Flooring; 500 mm × 500 mm carpet tiles; adhesive applied to subfloor						
Work to floors						
Teviot; low level loop pile; 2.50 mm thick	10.20	0.45	8.03	11.40	m²	19.43
Barcode; low level loop pile; 2.50 mm thick	15.30	0.45	8.03	17.02	m²	25.05
Circulate; random lay, batchless	19.55	0.45	8.03	21.70	m²	29.73
Arran; textured loop; 4.00 mm thick	23.80	0.45	8.03	26.38	m²	34.41
Alignment; textured cut and loop; 4.00 mm thick	23.80	0.45	8.03	26.38	m²	34.41
Fitted carpeting; Gradus broadloom polypropylene or other equal and approved						
Work to floors over 600 mm wide						
Bodega/Pacific/Predator ranges, loop pile	9.25	0.37	5.06	10.42	m²	15.48
Genus/Volnay ranges, cut pile	15.62	0.33	4.56	17.61	m²	22.17
Work to treads and risers						
not exceeding 600 mm wide	9.37	0.74	10.13	10.57	m	20.70
Underlay to carpeting						
Work to floors						
over 600 mm wide	1.77	0.07	0.95	1.91	m²	2.86
raking cutting	–	0.07	0.81	–	m	0.81
Carpet tiles to cement and sand base						
Work to floors over 600 mm wide						
Heuga 580 heavy duty loop pile	14.15	0.28	5.66	15.23	m²	20.89
Gradus Adventure range; bitumen backing; tufted pattern loop pile	18.00	0.28	5.66	19.37	m²	25.03
Latour/Predator ranges; bitumen backing; loop pile	9.46	0.28	5.66	10.18	m²	15.84
Sundries						
Carpet gripper fixed to floor; standard edging						
22 mm wide	–	0.04	0.46	0.22	m	0.68

28 FLOOR, WALL, CEILING AND ROOF FINISHINGS

Item	PC £	Labour hours	Labour £	Material £	Unit	Total rate £
Stair nosings; aluminium; Gradus or equivalent						
Medium duty hard aluminium alloy stair tread nosings; plugged and screwed in concrete						
56 mm × 32 mm; ref AS11	8.03	0.23	3.31	8.49	m	**11.80**
84 mm × 32 mm; ref AS12	11.13	0.28	4.04	11.75	m	**15.79**
Heavy duty aluminium alloy stair tread nosings; plugged and screwed to concrete						
48 mm × 38 mm; ref HE1	9.38	0.28	4.04	9.91	m	**13.95**
82 mm × 38 mm; ref HE2	13.02	0.32	4.61	13.72	m	**18.33**
Door entrance mats						
Entrance mat systems; aluminium wiper bar; laying in position; 18 mm thick						
Gradus Topguard; 900 mm × 550 mm	126.85	0.46	5.33	130.02	nr	**135.35**
Gradus Topguard; 1200 mm × 750 mm	228.32	0.46	5.33	234.03	nr	**239.36**
Gradus Topguard; 2400 mm × 1200 mm	730.63	0.93	10.77	748.90	nr	**759.67**
Nuway Tuftiguard 500 × 1300	–	–	–	–	nr	**221.81**
Nuway Tuftiguard 1000 × 1600	–	–	–	–	nr	**465.46**
Nuway Tuftiguard 1800 × 3000	–	–	–	–	nr	**1842.75**
extra for brass wipers	–	–	–	181.22	m²	**181.22**
Nuway Tuftiguard; single sided aluminium scraper bar	–	–	–	–	m²	**341.25**
Nuway Tuftiguard; double sided aluminium rigid scraper bars	–	–	–	–	m²	**487.50**
Coral Classic textile secondary and circulation matting system; 2.0 m wide roll	–	–	–	–	m²	**58.50**
Coral Brush Activ secondary and circulation matting system; 2.0 m wide roll	–	–	–	–	m²	**58.50**
Coral Duo secondary matting system; 2.0 m wide roll	–	–	–	–	m²	**58.50**
Matwells						
Polished stainless steel matwell; angle rim with lugs; bedding in screed						
Stainless steel frame bed in screed						
500 × 1300 mm	–	–	–	–	nr	**44.36**
Stainless steel frame bed in screed						
1000 × 1600 mm	–	–	–	–	nr	**88.72**
Stainless steel frame bed in screed						
1100 × 1400 mm	–	–	–	–	nr	**85.31**
Stainless steel frame bed in screed						
1800 × 3000 mm	–	–	–	–	nr	**163.80**
Polished aluminium matwell; angle rim with lugs brazed on; bedding in screed						
900 mm × 550 mm; constructed with 25 × 25 × 3 mm angle	25.91	0.93	10.77	26.56	nr	**37.33**
1200 mm × 750 mm; constructed with 34 × 26 × 6 mm angle	43.36	1.00	11.58	44.44	nr	**56.02**
2400 mm × 1200 mm; constructed with 50 × 50 × 6 mm angle	166.10	1.50	17.37	170.25	nr	**187.62**

28 FLOOR, WALL, CEILING AND ROOF FINISHINGS

Item	PC £	Labour hours	Labour £	Material £	Unit	Total rate £
FLOORS; EDGE FIXED CARPETING; CARPET TILES – cont						
Matwells – cont						
Polished brass matwell; comprising angle rim with lugs brazed on; bedding in screed						
900 mm × 550 mm; constructed with						
25 × 25 × 5 mm angle	57.93	0.93	10.77	59.38	nr	**70.15**
1200 mm × 750 mm; constructed with						
38 × 38 × 6 mm angle	177.82	1.00	11.58	182.27	nr	**193.85**
2400 mm x1200 mm; constructed with						
38 × 38 × 6 mm angle	328.28	1.50	17.37	336.49	nr	**353.86**
FLOORS; GRANITE; MARBLE; SLATE; TERRAZZO						
ALTERNATIVE TILE MATERIALS						
Prime cost rates for Dennis Ruabon clay floor quarries or equivalent (£/1000)						
194 mm × 194 mm × 12.5 mm; square; red	–	–	–	560.00	1000	**560.00**
194 mm × 194 mm × 12.5 mm; polygon; red	–	–	–	41.68	m²	**41.68**
150 mm × 150 mm × 12.5 mm; square; heatherbrown	–	–	–	640.00	1000	**640.00**
150 mm × 150 mm × 12.5 mm; studded square; heatherbrown or red	–	–	–	928.00	1000	**928.00**
150 mm × 150 mm × 12.50 mm; polygon; red	–	–	–	59.55	m²	**59.55**
SUPPLY AND FIX PRICES						
Clay floor quarries; BS EN 10545; class 1; Dennis Ruabon tiles or other equal; level bedding 10 mm thick and jointing in cement and sand (1:3); butt joints; straight both ways; flush pointing with grout; to cement and sand base						
Work to floors; over 600 mm wide						
150 mm × 150 mm × 12.50 mm thick; heatherbrown	26.88	0.74	14.95	35.83	m²	**50.78**
150 mm × 150 mm × 12.50 mm thick; red	23.52	0.74	14.95	32.23	m²	**47.18**
194 mm × 194 mm × 12.50 mm thick; heatherbrown	20.00	0.60	12.13	28.43	m²	**40.56**
Works to floors; in staircase areas or plant rooms						
150 mm × 150 mm × 12.50 mm thick; heatherbrown	26.88	0.83	16.77	35.83	m²	**52.60**
150 mm × 150 mm × 12.50 mm thick; red	24.70	0.83	16.77	32.23	m²	**49.00**
194 mm × 194 mm × 12.50 mm thick; heatherbrown	20.00	0.69	13.94	28.43	m²	**42.37**
Work to floors; not exceeding 600 mm wide						
150 mm × 150 mm × 12.50 mm thick; heatherbrown	–	0.37	7.47	9.57	m	**17.04**
150 mm × 150 mm × 12.50 mm thick; red	–	0.37	7.47	8.47	m	**15.94**
194 mm × 194 mm × 12.50 mm thick; heatherbrown	–	0.31	6.26	7.14	m	**13.40**
fair square cutting against flush edges of existing finishes	–	0.11	1.50	2.37	m	**3.87**
raking cutting	–	0.19	2.63	2.67	m	**5.30**
cutting around pipes; not exceeding 0.30 m girth	–	0.14	2.02	–	nr	**2.02**
extra for cutting and fitting into recessed manhole cover 600 mm × 600 mm	–	0.93	13.41	–	nr	**13.41**

28 FLOOR, WALL, CEILING AND ROOF FINISHINGS

Item	PC £	Labour hours	Labour £	Material £	Unit	Total rate £
Work to sills; 150 mm wide; rounded edge tiles						
200 mm × 150 mm × 22 mm thick; interior;						
heatherbrown or red	–	0.31	6.26	8.99	m	**15.25**
150 mm × 173 mm × 58 mm thick; exterior;						
heatherbrown or red	–	0.32	6.47	37.42	m	**43.89**
fitted end	–	0.14	2.02	–	nr	**2.02**
Coved skirtings; 150 mm high; rounded top edge						
150 mm × 150 mm × 12.50 mm thick; ref. CBTR;						
heatherbrown or red	–	0.23	4.64	10.28	m	**14.92**
ends	–	0.04	0.57	–	nr	**0.57**
angles	–	0.14	2.02	2.75	nr	**4.77**
50 mm × 50 mm × 5.50 mm thick slip-resistant mosaic floor tiles, Series 2 or other equal and approved; Langley London Ltd; fixing with adhesive; butt joints; straight both ways; flush pointing with white grout; to cement and sand base						
Work to floors						
over 600 mm wide	25.54	1.76	35.56	27.71	m²	**63.27**
not exceeding 600 mm wide	–	1.38	27.88	17.08	m	**44.96**
Dakota mahogany granite cladding; polished finish; jointed and pointed in coloured mortar (1:2:8)						
20 mm work to floors; level; to cement and sand base						
over 600 mm wide	–	–	–	–	m²	**318.35**
20 mm × 300 mm treads; plain nosings	–	–	–	–	m	**181.16**
raking, cutting	–	–	–	–	m	**32.49**
polished edges	–	–	–	–	m	**40.84**
birdsmouth	–	–	–	–	m	**42.15**
Riven Welsh slate floor tiles; level; bedding 10 mm thick and jointing in cement and sand (1:3); butt joints; straight both ways; flush pointing with coloured mortar; to cement and sand base						
Work to floors; over 600 mm wide						
250 mm × 250 mm × 12 mm–15 mm thick	28.85	0.56	11.32	41.74	m²	**53.06**
Work to floors; not exceeding 600 mm wide						
250 mm × 250 mm × 12 mm–15 mm thick	–	0.56	11.32	25.19	m	**36.51**
Roman Travertine marble cladding; polished finish; jointed and pointed in coloured mortar (1:2:8)						
20 mm thick work to floors; level; to cement and sand base						
over 600 mm wide	–	–	–	–	m²	**207.90**
20 mm × 300 mm treads; plain nosings	–	–	–	–	m	**125.24**
raking cutting	–	–	–	–	m	**24.28**
polished edges	–	–	–	–	m	**22.32**
birdsmouth	–	–	–	–	m	**44.74**

28 FLOOR, WALL, CEILING AND ROOF FINISHINGS

Item	PC £	Labour hours	Labour £	Material £	Unit	Total rate £
FLOORS; GRANITE; MARBLE; SLATE; TERRAZZO – cont						
Terrazzo tiles; BS EN 13748; aggregate size random ground grouted and polished to 80's grit finish; standard colour range; 3 mm joints symmetrical layout; bedding in 42 mm cement semi-dry mix (1:4); grouting with neat matching cement						
300 mm × 300 mm × 28 mm (nominal) Terrazzo tile units; hydraulically pressed, mechanically vibrated, steam cured; to floors on concrete base (not included); sealed with penetrating case hardener or other equal and approved; 2 coats applied immediately after final polishing						
plain; laid level	–	–	–	–	m²	39.33
plain; to slopes exceeding 15° from horizontal	–	–	–	–	m²	47.94
to small areas/toilets	–	–	–	–	m²	90.02
Accessories						
plastic division strips; 6 mm × 38 mm; set into floor tiling above crack inducing joints, to the nearest full tile module	–	–	–	–	m	2.72
Specially made terrazzo precast units; BS EN 13748–1; aggregate size random; standard colour range; 3 mm joints; grouting with neat matching cement						
Standard tread and riser square combined terrazzo units (with riser cast down) or other equal and approved; 280 mm wide; 150 mm high; 40 mm thick; machine-made; vibrated and fully machine polished; incorporating 1 nr. Ferodo anti-slip insert ref. OT40D or other equal and approved cast-in during manufacture; one end polished only						
fixed with cement: sand (1:4) mortar on prepared backgrounds (not included); grouted in neat tinted cement; wiped clean on completion of fixing	–	–	–	–	m	205.96
Standard tread square terrazzo units or other equal and approved; 40 mm thick; 280 mm wide; factory polished; incorporating 1 nr. Ferodo anti-slip insert ref. OT40D or other equal and approved						
fixed with cement: sand (1:4) mortar on prepared backgrounds (not included); grouted in neat tinted cement; wiped clean on completion of fixing	–	–	–	–	m	123.92
extra over for 55 × 55 mm contrasting colour to step nosing	–	–	–	–	m	43.31
Standard riser square terrazzo units or other equal and approved; 40 mm thick; 150 mm high; factory polished						
fixed with cement: sand (1:4) mortar on prepared backgrounds (not included); grouted in neat tinted cement; wiped clean on completion of fixing	–	–	–	–	m	76.89

28 FLOOR, WALL, CEILING AND ROOF FINISHINGS

Item	PC £	Labour hours	Labour £	Material £	Unit	Total rate £
Standard coved terrazzo skirting units or other equal and approved; 904 mm long; 150 mm high; nominal finish; 23 mm thick; with square top edge						
fixed with cement: sand (1:4) mortar on prepared backgrounds (by others); grouted in neat tinted cement; wiped clean on completion of fixing	–	–	–	–	m	**72.24**
extra over for special internal/external angle pieces to match	–	–	–	–	m	**20.51**
extra over for special polished ends	–	–	–	–	nr	**6.76**
FLOORS; RUBBER; VINYL						
Linoleum sheet; Forbo Flooring; laid level; fixing with adhesive; welded joints; to cement and sand base						
Work to floors; over 600 mm wide						
Marmoleum Real marbled sheet, 2.0 m wide × 2.50 mm thick	15.30	0.37	7.47	16.63	m²	**24.10**
Marmoleum Dual marbled tile 333 mm × 333 mm × 3.20 mm thick	18.70	0.45	9.09	20.28	m²	**29.37**
Decibel 17dB marbled acoustic sheet, 2.0 m wide; 3.50 mm thick	18.70	0.45	9.09	20.28	m²	**29.37**
High performance vinyl sheet; Forbo Flooring Eternal; laid level with welded seams; fixing with adhesive; to cement and sand base						
Work to floors; over 600 mm wide						
2.0 m wide roll × 2.00 mm thick	12.75	0.35	7.07	13.88	m²	**20.95**
Slip-resistant vinyl sheet; Forbo Flooring Surestep; laid level with welded seams; fixing with adhesive; to cement and sand base						
Work to floors; over 600 mm wide						
2.0 m wide roll × 2.00 mm thick	12.75	0.37	7.47	13.88	m²	**21.35**
Anti-static vinyl tiles; Forbo Flooring Colorex SD; laid level; fixing with adhesive; to cement and sand base						
Work to floors; over 600 mm wide						
615 mm × 615 mm tiles × 2.00 mm thick	18.70	0.50	10.11	20.28	m²	**30.39**
Acoustic vinyl sheet; Forbo Flooring Sarlon Traffic 19db; level with welded seams; fixing with adhesive; to cement and sand base						
Work to floors; over 600 mm wide						
2.0 m wide roll × 3.40 mm thick	15.30	0.37	7.47	16.63	m²	**24.10**

Prices for Measured Works

28 FLOOR, WALL, CEILING AND ROOF FINISHINGS

Item	PC £	Labour hours	Labour £	Material £	Unit	Total rate £
FLOORS; RUBBER; VINYL – cont						
Vinyl sheet; Altro Safety range; with welded seams; level; fixing with adhesive where appropriate; to cement and sand base						
Work to floors; over 600 mm wide						
2.00 mm thick; Impressionist II	11.87	0.74	14.95	15.67	m²	**30.62**
2.00 mm thick; Marine	11.89	0.56	11.32	13.63	m²	**24.95**
2.00 mm thick; Suprema	12.71	0.74	14.95	16.57	m²	**31.52**
2.00 mm thick; Timbersafe II	12.65	0.74	14.95	16.51	m²	**31.46**
2.00 mm thick; Walkway	9.67	0.56	11.32	11.23	m²	**22.55**
2.20 mm thick; Xpresslay (NB No adhesive required)	8.17	0.81	16.44	8.79	m²	**25.23**
2.50 mm thick; Classic 25	13.83	0.65	13.13	15.17	m²	**28.30**
2.50 mm thick; Maxis	15.16	0.74	14.95	19.21	m²	**34.16**
2.50 mm thick; Designer	14.32	0.74	14.95	18.31	m²	**33.26**
3.50 mm thick; Stronghold 30	16.65	0.74	14.95	20.81	m²	**35.76**
Vinyl sheet; Altro Sports surfaces range; with welded seams; level; fixing with adhesive; to cement and sand base						
Work to floors; over 600 mm wide						
4.00 mm thick; Mondoflex	15.03	0.70	14.15	17.00	m²	**31.15**
4.50 mm thick; Mondo Sportflex	18.38	0.70	14.15	20.61	m²	**34.76**
Homogeneous Vinyl sheet; Marleyflor Plus or other equal; level; with welded seams; fixing with adhesive; level; to cement and sand base						
Work to floors; over 600 mm wide						
2.00 mm thick	5.75	0.42	8.49	6.54	m²	**15.03**
100 mm high skirtings	–	0.11	2.22	1.60	m	**3.82**
Safety sheet; Marleyflor Granite Multisafe or other equal; level; with welded seams; fixing with adhesive; level; to cement and sand base						
Work to floors; over 600 mm wide						
2.00 mm thick	11.41	0.42	8.49	12.64	m²	**21.13**
Vinyl sheet; Marleyflor Omnisports or other equal; level; with welded seams; fixing with adhesive; level; to cement and sand base						
Work to floors; over 600 mm wide						
7.65 mm thick; Pro	22.74	0.90	18.18	24.83	m²	**43.01**
8.75 mm thick; Competition	26.65	1.00	20.20	29.04	m²	**49.24**
Vinyl semi-flexible tiles; Marleyflor Homogeneous tiles range; level; fixing with adhesive; butt joints; straight both ways; to cement and sand base						
Work to floors; over 600 mm wide						
300 mm × 300 mm × 2.00 mm thick; Vylon Plus	4.86	0.23	4.64	5.59	m²	**10.23**
500 mm × 500 mm × 2.00 mm thick; Marleyflor Plus	5.97	0.20	4.04	6.79	m²	**10.83**

28 FLOOR, WALL, CEILING AND ROOF FINISHINGS

Item	PC £	Labour hours	Labour £	Material £	Unit	Total rate £
Vinyl tiles; Polyflex Plus; level; fixing with adhesive; butt joints; straight both ways; to cement and sand base						
Work to floors; over 600 mm wide						
300 mm × 300 mm × 2.00 mm thick	5.28	0.23	4.64	6.04	m²	**10.68**
Vinyl tiles; Polyflex Camaro; level; fixing with adhesive; butt joints; straight both ways; to cement and sand base						
Work to floors; over 600 mm wide						
300 mm × 300 mm × 2.00 mm thick	11.89	0.30	6.06	13.15	m²	**19.21**
Vinyl tiles; Polyflor XL; level; fixing with adhesive; butt joints; straight both ways; to cement and sand base						
Work to floors; over 600 mm wide						
300 mm × 300 mm × 2.00 mm thick	6.59	0.32	6.47	7.44	m²	**13.91**
Vinyl sheet; Polyflor XL; level; fixing with adhesive; butt joints; straight both ways; to cement and sand base						
Work to floors; over 600 mm wide						
2.00 mm thick	4.52	0.28	5.56	5.22	m²	**10.78**
Vinyl sheet; Polysafe Standard; level; fixing with adhesive; welded seams; to cement and sand base						
Work to floors; over 600 mm wide						
2.00 mm thick	8.30	0.28	5.56	9.29	m²	**14.85**
2.50 mm thick	12.24	0.28	5.56	13.53	m²	**19.09**
Vinyl sheet; Polysafe hydro; level; fixing with adhesive; welded seams; to cement and sand base						
Work to floors; over 600 mm wide						
2.00 mm thick	10.73	0.28	5.56	11.90	m²	**17.46**
Luxury mineral vinyl tiles; Marley I D Naturelle; level; fixing with adhesive; butt joints; straight both ways; to cement and sand base						
Work to floors; over 600 mm wide						
330 mm × 330 mm × 2.00 mm thick	7.97	0.23	4.64	8.94	m²	**13.58**
Acoustic vinyl tiles; Marley Tapiflex 243; level; fixing with adhesive; butt joints; straight both ways; to cement and sand base						
Work to floors; over 600 mm wide						
500 mm × 500 mm × 2.00 mm thick	10.77	0.20	4.04	11.94	m²	**15.98**

28 FLOOR, WALL, CEILING AND ROOF FINISHINGS

Item	PC £	Labour hours	Labour £	Material £	Unit	Total rate £
FLOORS; RUBBER; VINYL – cont						
Linoleum tiles; Marley Veneto XF; level; fixing with adhesive; butt joints; straight both ways; to cement and sand base						
Work to floors; over 600 mm wide						
500 mm × 500 mm × 2.50 mm thick	12.76	0.20	4.04	14.08	m²	**18.12**
Cork tiles Wicanders Cork-Master; level; fixing with adhesive; butt joints; straight both ways; to cement and sand base						
Work to floors; over 600 mm wide						
300 mm × 300 mm × 8.00 mm thick; contract quality	16.80	0.40	8.08	20.30	m²	**28.38**
300 mm × 300 mm × 6.00 mm thick; high density	17.20	0.40	8.08	20.73	m²	**28.81**
300 mm × 300 mm × 4.00 mm thick; domestic	12.36	0.37	7.47	15.52	m²	**22.99**
Rubber studded tiles; Altro Mondopave; level; fixing with adhesive; butt joints; straight to cement and sand base						
Work to floors; over 600 mm wide						
500 mm × 500 mm × 2.50 mm thick; type MRB; black	16.08	0.56	11.32	22.00	m²	**33.32**
500 mm × 500 mm × 4.00 mm thick; type MRB; black	18.87	0.56	11.32	25.00	m²	**36.32**
Work to landings; over 600 mm wide						
500 mm × 500 mm × 4.00 mm thick; type MRB; black	18.87	0.74	14.95	25.00	m²	**39.95**
Work to stairs						
tread; 275 mm wide	–	0.46	9.30	7.43	m	**16.73**
riser; 180 mm wide	–	0.56	11.32	5.27	m	**16.59**
Sundry floor sheeting underlays						
For floor finishings; over 600 mm wide						
building paper to BS 1521; class A; 75 mm lap (laying only)	–	0.05	0.57	–	m²	**0.57**
3.20 mm thick hardboard	–	0.19	3.83	1.44	m²	**5.27**
6.00 mm thick plywood	–	0.28	5.66	5.65	m²	**11.31**
Skirtings; plastic; Gradus or equivalent						
Set-in skirtings						
100 mm high; ref. SI100	–	0.11	2.22	2.05	m	**4.27**
150 mm high; ref. SI150	–	0.22	4.45	3.36	m	**7.81**
Set-on skirtings						
100 mm high; ref. SO100	–	0.22	4.45	1.67	m	**6.12**
150 mm high; ref. SO150	–	0.40	8.08	3.15	m	**11.23**

28 FLOOR, WALL, CEILING AND ROOF FINISHINGS

Item	PC £	Labour hours	Labour £	Material £	Unit	Total rate £
Stair nosings; aluminium; Gradus or equivalent						
Medium duty hard aluminium alloy stair tread nosings; plugged and screwed in concrete						
56 mm × 32 mm; ref AS11	8.23	0.23	3.31	8.49	m	**11.80**
84 mm × 32 mm; ref AS12	11.41	0.28	4.04	11.75	m	**15.79**
Heavy duty aluminium alloy stair tread nosings; plugged and screwed to concrete						
48 mm × 38 mm; ref HE1	9.62	0.28	4.04	9.91	m	**13.95**
82 mm × 38 mm; ref HE2	13.34	0.32	4.61	13.72	m	**18.33**
FLOORS; WOOD BLOCK						
Wood blocks; Havwoods or other equal and approved; 25 mm thick; level; laid to herringbone pattern with 2 block border; fixing with adhesive; to cement: sand base; sanded and sealed						
Work to floors; over 300 mm wide						
Merbau	–	–	–	–	m²	**100.39**
Iroko	–	–	–	–	m²	**100.39**
American Oak	–	–	–	–	m²	**105.21**
European Oak	–	–	–	–	m²	**110.04**
Add to wood block flooring over 300 mm wide for						
buff; one coat seal	–	–	–	–	m²	**3.86**
buff; two coats seal	–	–	–	–	m²	**6.27**
sand; three coats for seal or oil	–	–	–	–	m²	**17.37**
FLOORS; RAISED ACCESS FLOORS						
Raised flooring system; laid on or fixed to concrete floor						
Full access system; 150 mm high overall; pedestal supports						
PSA light grade; steel finish	–	–	–	–	m²	**27.20**
PSA medium grade; steel finish	–	–	–	–	m²	**27.20**
PSA heavy grade; steel finish	–	–	–	–	m²	**39.11**
Extra for						
factory applied needlepunch carpet	–	–	–	–	m²	**12.75**
factory applied anti-static vinyl	–	–	–	–	m²	**21.25**
factory applied black PVC edge strips	–	–	–	–	m	**3.86**
ramps; 3.00 m × 1.40 m (no finish)	–	–	–	–	nr	**595.00**
steps (no finish)	–	–	–	–	m	**34.00**
forming cut-out for electrical boxes	–	–	–	–	nr	**3.40**
supply and lay protection to raised floor; 2440 × 1220 polypropylene sheets with taped joints	–	–	–	–	m²	**1.49**

28 FLOOR, WALL, CEILING AND ROOF FINISHINGS

Item	PC £	Labour hours	Labour £	Material £	Unit	Total rate £
WALLS; PLASTERED; RENDERED; ROUGHCAST						
Cement and sand (1:3) beds and backings						
10 mm thick work to walls; one coat; to brickwork or blockwork base						
over 600 mm wide	–	–	–	–	m²	13.51
not exceeding 600 mm wide	–	–	–	–	m	6.76
13 mm thick; work to walls; two coats; to brickwork or blockwork base						
over 600 mm wide	–	–	–	–	m²	16.25
not exceeding 600 mm wide	–	–	–	–	m	8.14
15 mm thick work to walls; two coats; to brickwork or blockwork base						
over 600 mm wide	–	–	–	–	m²	17.51
not exceeding 600 mm wide	–	–	–	–	m	8.77
Cement and sand (1:3); steel trowelled						
13 mm thick work to walls; two coats; to brickwork or blockwork base						
over 600 mm wide	–	–	–	–	m²	14.14
not exceeding 600 mm wide	–	–	–	–	m	7.07
16 mm thick work to walls; two coats; to brickwork or blockwork base						
over 600 mm wide	–	–	–	–	m²	15.84
not exceeding 600 mm wide	–	–	–	–	m	7.93
19 mm thick work to walls; two coats; to brickwork or blockwork base						
over 600 mm wide	–	–	–	–	m²	18.32
not exceeding 600 mm wide	–	–	–	–	m	9.16
ADD to above						
over 600 mm wide in water-repellent cement	–	–	–	–	m²	3.75
finishing coat in colour cement	–	–	–	–	m²	7.99
Cement-lime-sand (1:2:9); steel trowelled						
19 mm thick work to walls; two coats; to brickwork or blockwork base						
over 600 mm wide	–	–	–	–	m²	17.77
not exceeding 600 mm wide	–	–	–	–	m	8.89
Cement-lime-sand (1:1:6); steel trowelled						
13 mm thick work to walls; two coats; to brickwork or blockwork base						
over 600 mm wide	–	–	–	–	m²	14.54
not exceeding 600 mm wide	–	–	–	–	m	7.16
Add to the above over 600 mm wide for						
waterproof additive	–	–	–	–	m²	2.45

28 FLOOR, WALL, CEILING AND ROOF FINISHINGS

Item	PC £	Labour hours	Labour £	Material £	Unit	Total rate £
Plaster; first 11 mm coat of Thistle Hardwall plaster; second 2 mm finishing coat of Thistle Multi Finish plaster; steel trowelled						
13 mm thick work to walls; two coats; to brickwork or blockwork base						
over 600 mm wide	–	–	–	–	m²	11.21
over 600 mm wide; in staircase areas or plant rooms	–	–	–	–	m²	13.47
not exceeding 600 mm wide	–	–	–	–	m	5.96
13 mm thick work to isolated brickwork or blockwork columns; two coats						
over 600 mm wide	–	–	–	–	m²	21.12
not exceeding 600 mm wide	–	–	–	–	m	10.56
Plaster; first 11 mm coat of Thistle Browning plaster; second finishing coat of 2 mm Thistle Multi Finish plaster; steel trowelled finish						
13 mm thick; work to walls; two coats; to brickwork or blockwork base						
over 600 mm wide	–	–	–	–	m²	11.21
over 600 mm wide; in staircase areas or plant rooms	–	–	–	–	m²	13.46
not exceeding 600 mm wide	–	–	–	–	m	5.96
13 mm thick work to isolated brickwork or blockwork columns; two coats						
over 600 mm wide	–	–	–	–	m²	21.12
not exceeding 600 mm wide	–	–	–	–	m	9.36
Plaster; first 8 mm or 11 mm coat of Thistle Bonding plaster; second 2 mm finishing coat of Thistle Multi Finish plaster; steel trowelled finish						
13 mm thick work to walls; two coats; to concrete base						
over 600 mm wide	–	–	–	–	m²	12.55
over 600 mm wide; in staircase areas or plant rooms	–	–	–	–	m²	14.87
not exceeding 600 mm wide	–	–	–	–	m	5.75
13 mm thick work to isolated piers or columns; two coats; to concrete base						
over 600 mm wide	–	–	–	–	m²	22.43
not exceeding 600 mm wide	–	–	–	–	m	10.50
Plaster; one coat Snowplast plaster or other equal and approved; steel trowelled						
13 mm thick work to walls; one coat; to brickwork or blockwork base						
over 600 mm wide	–	–	–	–	m²	12.40
over 600 mm wide; in staircase areas or plant rooms	–	–	–	–	m²	14.73
not exceeding 600 mm wide	–	–	–	–	m	6.21
13 thick work to isolated columns; one coat						
over 600 mm wide	–	–	–	–	m²	15.00
not exceeding 600 mm wide	–	–	–	–	m	7.53

28 FLOOR, WALL, CEILING AND ROOF FINISHINGS

Item	PC £	Labour hours	Labour £	Material £	Unit	Total rate £
WALLS; PLASTERED; RENDERED; ROUGHCAST – cont						
Plaster; first coat of Limelite renovating plaster; finishing coat of Limelite finishing plaster; or other equal and approved; steel trowelled						
13 mm thick work to walls; two coats; to brickwork or blockwork base						
over 600 mm wide	–	–	–	–	m²	17.19
over 600 mm wide; in staircase areas or plant rooms	–	–	–	–	m²	18.87
not exceeding 600 mm wide	–	–	–	–	m	8.60
Dubbing out existing walls with undercoat plaster; average 6 mm thick						
over 600 mm wide	–	–	–	–	m²	5.16
not exceeding 600 mm wide	–	–	–	–	m	2.61
Dubbing out existing walls with undercoat plaster; average 12 mm thick						
over 600 mm wide	–	–	–	–	m²	10.31
not exceeding 600 mm wide	–	–	–	–	m	5.16
Plaster; first coat of Thistle X-ray plaster or other equal and approved; finishing coat of Thistle X-ray finishing plaster or other equal and approved; steel trowelled						
17 mm thick work to walls; two coats; to brickwork or blockwork base						
over 600 mm wide	–	–	–	–	m²	55.79
over 600 mm wide; in staircase areas or plant rooms	–	–	–	–	m²	59.94
not exceeding 600 mm wide	–	–	–	–	m	22.31
17 mm thick work to isolated columns; two coats						
over 600 mm wide	–	–	–	–	m²	90.50
not exceeding 600 mm wide	–	–	–	–	m	36.17
Plaster; one coat Thistle projection plaster or other equal and approved; steel trowelled						
13 mm thick work to walls; one coat; to brickwork or blockwork base						
over 600 mm wide	–	–	–	–	m²	11.96
over 600 mm wide; in staircase areas or plant rooms	–	–	–	–	m²	13.68
not exceeding 600 mm wide	–	–	–	–	m	5.97
10 mm thick work to isolated columns; one coat						
over 600 mm wide	–	–	–	–	m²	14.56
not exceeding 600 mm wide	–	–	–	–	m	7.26

28 FLOOR, WALL, CEILING AND ROOF FINISHINGS

Item	PC £	Labour hours	Labour £	Material £	Unit	Total rate £
Plaster; first 11 mm coat of Thistle Bonding plaster; second 2 mm finishing coat of Thistle Multi Finish plaster; steel trowelled						
13 mm thick work to swept soffit of metal lathing arch former						
not exceeding 600 mm wide	–	–	–	–	m	**9.38**
300 mm–400 mm wide	–	–	–	–	m	**12.54**
13 mm thick work to vertical face of metal lathing arch former						
not exceeding 0.50 m^2 per side	–	–	–	–	nr	**13.32**
0.50 m^2–1 m^2 per side	–	–	–	–	nr	**19.98**
Cemrend self-coloured render or other equal and approved; one coat; to brickwork or blockwork base						
20 mm thick work to walls; to brickwork or blockwork base						
over 600 mm wide	–	–	–	–	m^2	**29.93**
not exceeding 600 mm wide	–	–	–	–	m	**17.57**
Tyrolean decorative rendering or similar; 13 mm thick first coat of cement-lime-sand (1:1:6); finishing three coats of Cullamix or other equal and approved; applied with approved hand operated machine external						
To walls; four coats; to brickwork or blockwork base						
over 600 mm wide	–	–	–	–	m^2	**28.59**
not exceeding 600 mm wide	–	–	–	–	m	**14.27**
Drydash (pebbledash) finish of Derbyshire Spar chippings or other equal and approved on and including cement-lime-sand (1:2:9) backing						
18 mm thick work to walls; two coats; to brickwork or blockwork base						
over 600 mm wide	–	–	–	–	m^2	**24.78**
not exceeding 600 mm wide	–	–	–	–	m	**12.40**
Plaster; one coat Thistle board finish or other equal and approved; steel trowelled (prices included within plasterboard rates)						
3 mm thick work to walls or ceilings; one coat; to plasterboard base						
over 600 mm wide	–	–	–	–	m^2	**5.06**
over 600 mm wide; in staircase areas or plant rooms	–	–	–	–	m^2	**6.07**
not exceeding 600 mm wide	–	–	–	–	m	**2.02**

28 FLOOR, WALL, CEILING AND ROOF FINISHINGS

Item	PC £	Labour hours	Labour £	Material £	Unit	Total rate £
WALLS; PLASTERED; RENDERED; ROUGHCAST – cont						
Plaster; one coat Thistle board finish or other and approved; steel trowelled 3 mm work to walls or ceilings; one coat on and including gypsum plasterboard; BS 1230; fixing with nails; 3 mm joints filled with plaster and jute scrim cloth; to softwood base; plain grade baseboard or lath with rounded edges						
9.50 mm thick boards to walls						
over 600 mm wide	–	0.97	13.27	3.19	m²	**16.46**
not exceeding 600 mm wide	–	0.37	5.33	0.91	m	**6.24**
9.50 mm thick boards to walls; in staircase areas or plant rooms						
over 600 mm wide	–	1.06	14.57	3.19	m²	**17.76**
not exceeding 600 mm wide	–	0.46	6.63	0.91	m	**7.54**
9.50 mm thick boards to isolated columns						
over 600 mm wide	–	1.06	14.57	3.19	m²	**17.76**
not exceeding 600 mm wide	–	0.56	8.07	0.91	m	**8.98**
12.50 mm thick boards to walls; in staircase areas or plant rooms						
over 600 mm wide	–	1.12	15.44	3.19	m²	**18.63**
not exceeding 600 mm wide	–	0.50	7.21	0.91	m	**8.12**
12.50 mm thick boards to isolated columns						
over 600 mm wide	–	1.12	15.44	3.19	m²	**18.63**
not exceeding 600 mm wide	–	0.59	8.51	0.91	m	**9.42**
Accessories						
Expamet render beads or other equal and approved; white PVC nosings; to brickwork or blockwork base						
external stop bead; ref 573	–	0.07	1.00	5.29	m	**6.29**
Expamet render beads or other equal and approved; stainless steel; to brickwork or blockwork base						
stop bead; ref 546	–	0.07	1.00	4.34	m	**5.34**
stop bead; ref 547	–	0.07	1.00	4.33	m	**5.33**
Expamet plaster beads or other equal and approved; to brickwork or blockwork base						
angle bead; ref 550	–	0.08	1.15	1.12	m	**2.27**
architrave bead; ref 579	–	0.10	1.45	3.05	m	**4.50**
stop bead; ref 562	–	0.07	1.00	1.37	m	**2.37**
stop bead; ref 563	–	0.07	1.00	1.91	m	**2.91**
movement bead; ref 588	–	0.09	1.30	10.75	m	**12.05**
Expamet plaster beads or other equal and approved; stainless steel; to brickwork or blockwork base						
angle bead; ref 545	–	0.08	1.15	4.87	m	**6.02**
stop bead; ref 534	–	0.07	1.00	4.34	m	**5.34**
stop bead; ref 533	–	0.07	1.00	4.34	m	**5.34**
Expamet thin coat plaster beads or other equal and approved; galvanized steel; to timber base						
angle bead; ref 553	–	0.07	1.00	1.05	m	**2.05**
stop bead; ref 560	–	0.06	0.86	1.59	m	**2.45**
stop bead; ref 561	–	0.06	0.86	1.59	m	**2.45**

28 FLOOR, WALL, CEILING AND ROOF FINISHINGS

Item	PC £	Labour hours	Labour £	Material £	Unit	Total rate £
WALLS; INSULATED RENDER						
Sto Therm Classic M-system insulation render						
70 mm EPS insulation fixed with adhesive to SFS structure (measured separately) with horizontal PVC intermediate track and vertical T-spines; with glassfibre mesh reinforcement embedded in Sto Armat Classic Basecoat Render and Stolit K 1.5 Decorative Topcoat Render (white)						
over 300 mm wide	–	–	–	–	m²	**49.52**
70 mm EPS insulation mechanically fixed to SFS structure (measured separately) with horizontal PVC intermediate track and vertical T-spines; with glassfibre mesh reinforcement embedded in Sto Armat Classic Basecoat Render and Stolit K 1.5 Decorative Topcoat Render (white)						
over 300 mm wide	–	–	–	–	m²	**54.48**
rendered heads and reveals not exceeding 100 mm wide; including angle beads	–	–	–	–	m	**13.68**
Extra for						
aluminium starter track at base of insulated render system	–	–	–	–	m	**8.32**
external angle with PVC mesh angle bead	–	–	–	–	m	**3.66**
internal angle with Sto Armor angle	–	–	–	–	m	**3.66**
render stop bead	–	–	–	–	m	**3.66**
Sto seal tape to all vertical abutments	–	–	–	–	m	**3.41**
Sto Armor mat HD mesh reinforcement to areas prone to physical damage (e.g. 1800 mm high adjoining floor level)						
over 300 mm wide	–	–	–	–	m²	**10.96**
WALLS; CERAMIC TILING; MARBLE						
Glazed ceramic wall tiles; BS EN 10545; fixing with adhesive; butt joints; straight both ways; flush pointing with white grout; to plaster base						
Work to walls; over 600 mm wide						
152 mm × 152 mm × 5.50 mm thick; white	11.76	0.56	11.32	13.29	m²	**24.61**
152 mm × 152 mm × 5.50 mm thick; light colours	12.60	0.56	11.32	14.16	m²	**25.48**
152 mm × 152 mm × 5.50 mm thick; dark colours	14.70	0.56	11.32	16.31	m²	**27.63**
extra for RE or REX tile	–	–	–	5.60	m²	**5.60**
200 mm × 100 mm × 6.50 mm thick; white and light colours	10.50	0.56	11.32	12.00	m²	**23.32**
250 mm × 200 mm × 7.00 mm thick; white and light colours	12.60	0.56	11.32	14.16	m²	**25.48**
Work to walls; in staircase areas or plant rooms						
152 mm × 152 mm × 5.50 mm thick; white	–	0.62	12.53	13.29	m²	**25.82**

28 FLOOR, WALL, CEILING AND ROOF FINISHINGS

Item	PC £	Labour hours	Labour £	Material £	Unit	Total rate £
WALLS; CERAMIC TILING; MARBLE – cont						
Glazed ceramic wall tiles – cont						
Work to walls; not exceeding 600 mm wide						
152 mm × 152 mm × 5.50 mm thick; white	–	0.56	11.32	7.94	m	**19.26**
152 mm × 152 mm × 5.50 mm thick; light colours	–	0.56	11.32	8.86	m	**20.18**
152 mm × 152 mm × 5.50 mm thick; dark colours	–	0.56	11.32	10.15	m	**21.47**
200 mm × 100 mm × 6.50 mm thick; white and light colours	–	0.56	11.32	7.16	m	**18.48**
250 mm × 200 mm × 7.00 mm thick; white and light colours	–	0.46	9.30	8.46	m	**17.76**
cutting around pipes; not exceeding 0.30 m girth	–	0.09	1.30	–	nr	**1.30**
Work to sills; 150 mm wide; rounded edge tiles						
152 mm × 152 mm × 5.50 mm thick; white	–	0.23	4.64	1.99	m	**6.63**
fitted end	–	0.09	1.30	–	nr	**1.30**
198 mm × 64.50 mm × 6 mm thick wall tiles; fixing with adhesive; butt joints; straight both ways; flush pointing with white grout; to plaster base						
Work to walls						
over 600 mm wide	26.46	1.67	33.74	28.36	m²	**62.10**
not exceeding 600 mm wide	–	1.30	26.26	16.98	m	**43.24**
20 mm × 20 mm × 5.50 mm thick glazed mosaic wall tiles; fixing with adhesive; butt joints; straight both ways; flush pointing with white grout; to plaster base						
Work to walls						
over 600 mm wide	30.87	1.76	35.56	32.95	m²	**68.51**
not exceeding 600 mm wide	–	1.38	27.88	20.36	m	**48.24**
Dakota mahogany granite cladding; polished finish; jointed and pointed in coloured mortar (1:2:8)						
20 mm thick work to walls; to cement and sand base						
over 600 mm wide	–	–	–	–	m²	**324.77**
not exceeding 600 mm wide	–	–	–	–	m	**146.13**
40 mm thick work to walls; to cement and sand base						
over 600 mm wide	–	–	–	–	m²	**539.24**
not exceeding 600 mm wide	–	–	–	–	m	**242.64**
Roman Travertine marble cladding; polished finish; jointed and pointed in coloured mortar (1:2:8)						
20 mm thick work to walls; to cement and sand base						
over 600 mm wide	–	–	–	–	m²	**247.21**
not exceeding 600 mm wide	–	–	–	–	m	**111.83**
40 mm thick work to walls; to cement and sand base						
over 600 mm wide	–	–	–	–	m²	**340.31**
not exceeding 600 mm wide	–	–	–	–	m	**153.13**

28 FLOOR, WALL, CEILING AND ROOF FINISHINGS

Item	PC £	Labour hours	Labour £	Material £	Unit	Total rate £
CEILINGS; PLASTERED; RENDERED						
Plaster; first 8 mm or 11 mm coat of Thistle Bonding plaster; second 2 mm finishing coat of Thistle Multi Finish plaster; steel trowelled finish						
10 mm thick work to ceilings; two coats; to concrete base						
over 600 mm wide	–	–	–	–	m²	10.74
over 600 wide; 3.50 m–5.00 m high	–	–	–	–	m²	12.89
over 600 mm wide; in staircase areas or plant rooms	–	–	–	–	m²	14.26
not exceeding 600 mm wide	–	–	–	–	m	6.00
10 mm thick work to isolated beams; two coats; to concrete base						
over 600 mm wide	–	–	–	–	m²	21.48
over 600 mm wide; 3.50 m–5.00 m high	–	–	–	–	m²	22.91
not exceeding 600 mm wide	–	–	–	–	m	10.80
Plaster; first 11 mm coat of Thistle Bonding plaster; second coat 2 mm finishing coat of Thistle Multi Finish plaster; steel trowelled						
13 mm thick work to ceilings; three coats to metal lathing base						
over 600 mm wide	–	–	–	–	m²	13.03
over 600 mm wide; in staircase areas or plant rooms	–	–	–	–	m²	15.63
not exceeding 600 mm wide	–	–	–	–	m	7.03
Plaster; one coat Thistle board finish or other and approved; steel trowelled 3 mm work to ceilings; one coat on and including gypsum plasterboard; BS 1230; fixing with nails; 3 mm joints filled with plaster and jute scrim cloth; to softwood base; plain grade baseboard or lath with rounded edges						
9.50 mm thick boards to ceilings						
over 600 mm wide	–	0.89	12.12	3.19	m²	15.31
over 600 mm wide; 3.50 m–5.00 m high	–	1.03	14.13	3.19	m²	17.32
not exceeding 600 mm wide	–	0.43	6.20	0.91	m	7.11
9.50 mm thick boards to ceilings; in staircase areas or plant rooms						
over 600 mm wide	–	0.98	13.42	3.19	m²	16.61
not exceeding 600 mm wide	–	0.47	6.78	0.91	m	7.69
9.50 mm thick boards to isolated beams						
over 600 mm wide	–	1.05	14.42	3.19	m²	17.61
not exceeding 600 mm wide	–	0.50	7.21	0.91	m	8.12
12.50 mm thick boards to ceilings						
over 600 mm wide	–	0.95	12.99	3.19	m²	16.18
over 600 mm wide; 3.50 m–5.00 m high	–	1.06	14.57	3.19	m²	17.76
not exceeding 600 mm wide	–	0.45	6.49	0.91	m	7.40

28 FLOOR, WALL, CEILING AND ROOF FINISHINGS

Item	PC £	Labour hours	Labour £	Material £	Unit	Total rate £
CEILINGS; PLASTERED; RENDERED – cont						
Plaster – cont						
12.50 mm thick boards to ceilings; in staircase areas or plant rooms						
over 600 mm wide	–	1.06	14.57	3.19	m²	17.76
not exceeding 600 mm wide	–	0.51	7.35	0.91	m	8.26
12.50 mm thick boards to isolated beams						
over 600 mm wide	–	1.15	15.87	3.19	m²	19.06
not exceeding 600 mm wide	–	0.56	8.07	0.91	m	8.98
Plaster; one coat Thistle board finish or other equal and approved; steel trowelled (prices included within plasterboard rates)						
3 mm thick work to walls or ceilings; one coat; to plasterboard base						
over 600 mm wide	–	–	–	–	m²	5.06
over 600 mm wide; in staircase areas or plant rooms	–	–	–	–	m²	6.07
not exceeding 600 mm wide	–	–	–	–	m	2.02
Cement-lime-sand (1:1:6); steel trowelled						
19 mm thick work to ceilings; three coats; to metal lathing base						
over 600 mm wide	–	–	–	–	m²	17.09
not exceeding 600 mm wide	–	–	–	–	m	9.99
CEILINGS; FIBROUS PLASTER						
Fibrous plaster; fixing with screws; plugging; countersinking; stopping; filling and pointing joints with plaster						
16 mm thick plain slab coverings to ceilings						
over 300 mm wide	–	–	–	–	m²	111.29
not exceeding 300 mm wide	–	–	–	–	m	37.43
Coves; not exceeding 150 mm girth						
per 25 mm girth	–	–	–	–	m	5.37
Coves; 150 mm–300 mm girth						
per 25 mm girth	–	–	–	–	m	6.57
Cornices						
per 25 mm girth	–	–	–	–	m	6.67
Cornice enrichments						
per 25 mm girth; depending on degree of enrichments	–	–	–	–	m	7.89

28 FLOOR, WALL, CEILING AND ROOF FINISHINGS

Item	PC £	Labour hours	Labour £	Material £	Unit	Total rate £
Fibrous plaster; fixing with plaster wadding filling and pointing joints with plaster; to steel base						
16 mm thick plain slab coverings to ceilings						
over 300 mm wide	–	–	–	–	m²	111.29
not exceeding 300 mm wide	–	–	–	–	m	37.43
16 mm thick plain casings to stanchions						
per 25 mm girth	–	–	–	–	m	3.34
16 mm thick plain casings to beams						
per 25 mm girth	–	–	–	–	m	3.34
Gyproc cove or other equal and approved; fixing with adhesive; filling and pointing joints with plaster						
Cove						
125 mm girth	–	0.19	2.74	1.31	m	4.05
angles	–	0.03	0.43	0.81	nr	1.24
METAL MESH LATHING FOR PLASTERED COATINGS						
Accessories						
Preformed galvanized expanded steel semi-circular arch-frames; Expamet or other equal and approved; to suit walls up to 230 mm thick						
for 760 mm opening; ref ESC 30	72.56	0.46	5.98	74.37	nr	80.35
for 840 mm opening; ref ESC 32	73.75	0.46	5.98	75.59	nr	81.57
for 920 mm opening; ref ESC 36	89.13	0.46	5.98	91.36	nr	97.34
for 1220 mm opening; ref ESC 48	111.08	0.46	5.98	113.86	nr	119.84
Lathing; Expamet BB expanded metal lathing or other equal and approved; BS EN 13658; 50 mm laps						
6 mm thick mesh linings to ceilings; fixing with staples; to softwood base; over 300 mm wide						
ref BB263; 0.500 mm thick	6.88	0.56	7.28	7.05	m²	14.33
ref BB264; 0.675 mm thick	9.63	0.56	7.28	9.87	m²	17.15
6 mm thick mesh linings to ceilings; fixing with wire; to steelwork; over 300 mm wide						
ref BB263; 0.500 mm thick	–	0.59	7.69	7.05	m²	14.74
ref BB264; 0.675 mm thick	–	0.59	7.69	9.87	m²	17.56
6 mm thick mesh linings to ceilings; fixing with wire; to steelwork; not exceeding 300 mm wide						
ref BB263; 0.500 mm thick	–	0.37	4.80	7.05	m²	11.85
ref BB264; 0.675 mm thick	–	0.37	4.80	9.87	m²	14.67
raking cutting	–	0.19	2.74	–	m	2.74
cutting and fitting around pipes; not exceeding 0.30 m girth	–	0.28	4.04	–	nr	4.04

28 FLOOR, WALL, CEILING AND ROOF FINISHINGS

Item	PC £	Labour hours	Labour £	Material £	Unit	Total rate £
METAL MESH LATHING FOR PLASTERED COATINGS – cont						
Lathing; Expamet Riblath or Spraylath or other equal and approved stiffened expanded metal lathing or similar; 50 mm laps						
10 mm thick mesh lining to walls; fixing with nails; to softwood base; over 300 mm wide						
Riblath ref 269; 0.30 mm thick	14.45	0.46	5.98	14.90	m²	**20.88**
Riblath ref 271; 0.50 mm thick	11.38	0.46	5.98	11.76	m²	**17.74**
10 mm thick mesh lining to walls; fixing with nails; to softwood base; not exceeding 300 mm wide						
Riblath ref 269; 0.30 mm thick	–	0.28	3.64	4.51	m²	**8.15**
Riblath ref 271; 0.50 mm thick	–	0.28	3.64	3.56	m²	**7.20**
10 mm thick mesh lining to walls; fixing to brick or blockwork; over 300 mm wide						
Red-rib ref 274; 0.50 mm thick	14.45	0.37	4.80	15.83	m²	**20.63**
Stainless steel Riblath ref 267; 0.30 mm thick	28.08	0.37	4.80	29.81	m²	**34.61**
10 mm thick mesh lining to ceilings; fixing with wire; to steelwork; over 300 mm wide						
Riblath ref 269; 0.30 mm thick	–	0.59	7.69	15.21	m²	**22.90**
Riblath ref 271; 0.50 mm thick	–	0.59	7.69	12.06	m²	**19.75**
SPRAYED MINERAL FIBRE COATINGS						
Prepare and apply by spray Mandolite CP2 fire protection or other equal and approved on structural steel/metalwork						
16 mm thick (one hour) fire protection						
to walls and columns	–	–	–	–	m²	**8.48**
to ceilings and beams	–	–	–	–	m²	**9.35**
to isolated metalwork	–	–	–	–	m²	**18.65**
22 mm thick (one and a half hour) fire protection						
to walls and columns	–	–	–	–	m²	**9.86**
to ceilings and beams	–	–	–	–	m²	**10.94**
to isolated metalwork	–	–	–	–	m²	**21.87**
28 mm thick (two hour) fire protection						
to walls and columns	–	–	–	–	m²	**11.56**
to ceilings and beams	–	–	–	–	m²	**12.63**
to isolated metalwork	–	–	–	–	m²	**25.25**
52 mm thick (four hour) fire protection						
to walls and columns	–	–	–	–	m²	**17.49**
to ceilings and beams	–	–	–	–	m²	**19.47**
to isolated metalwork	–	–	–	–	m²	**38.74**
Prepare and apply by spray; cementitious Pyrok WF26 render or other equal and approved; on expanded metal lathing (not included)						
15 mm thick						
to ceilings and beams	–	–	–	–	m²	**26.83**

29 DECORATION

Item	PC £	Labour hours	Labour £	Material £	Unit	Total rate £
BASIC PAINT PRICES – MATERIAL ONLY						
Paints						
matt emulsion	–	–	–	8.20	5 litre	**8.20**
gloss	–	–	–	22.95	5 litre	**22.95**
eggshell gloss	–	–	–	20.96	5 litre	**20.96**
oil-based undercoat	–	–	–	20.65	5 litre	**20.65**
Weathershield gloss	–	–	–	38.17	5 litre	**38.17**
Weathershield undercoat	–	–	–	41.25	5 litre	**41.25**
Sandtex masonry paint						
brilliant white	–	–	–	11.05	5 litre	**11.05**
coloured	–	–	–	19.64	5 litre	**19.64**
Primer/undercoats						
acrylic	–	–	–	20.60	5 litre	**20.60**
red oxide	–	–	–	22.71	5 litre	**22.71**
water-based	–	–	–	18.58	5 litre	**18.58**
zinc phosphate	–	–	–	35.24	5 litre	**35.24**
masonry sealer	–	–	–	15.67	5 litre	**15.67**
MDF primer	–	–	–	41.64	5 litre	**41.64**
knotting solution	–	–	–	54.53	5 litre	**54.53**
Special paints						
solar reflective aluminium	–	–	–	36.67	5 litre	**36.67**
anti-graffiti	–	–	–	122.09	5 litre	**122.09**
bituminous emulsion	–	–	–	12.87	5 litre	**12.87**
Hammerite	–	–	–	42.48	5 litre	**42.48**
fire retardant						
undercoat	–	–	–	49.62	5 litre	**49.62**
top coat	–	–	–	66.09	5 litre	**66.09**
Stains and Preservatives						
Cuprinol						
Clear	–	–	–	22.95	5 litre	**22.95**
Boiled linseed oil	–	–	–	28.67	5 litre	**28.67**
Sadolin						
Extra	–	–	–	43.93	5 litre	**43.93**
New Base	–	–	–	19.38	5 litre	**19.38**
Sikkens						
Cetol HLS	–	–	–	46.25	5 litre	**46.25**
Cetol TS	–	–	–	66.52	5 litre	**66.52**
Cetol Filter 7	–	–	–	70.36	5 litre	**70.36**
Protim Solignum						
Architectural	–	–	–	60.11	5 litre	**60.11**
Green	–	–	–	23.39	5 litre	**23.39**
Cedar	–	–	–	18.25	5 litre	**18.25**
Varnishes						
polyurethane	–	–	–	33.57	5 litre	**33.57**

29 DECORATION

Item	PC £	Labour hours	Labour £	Material £	Unit	Total rate £
PAINTING AND CLEAR FINISHES – EXTERNAL						
SUPPLY AND FIX PRICES						
NOTE: The following prices include for preparing surfaces. Painting woodwork also includes for knotting prior to applying the priming coat and for all stopping of nail holes etc.						
Two coats of cement paint, Sandtex Matt or other equal and approved						
Brick or block walls						
over 300 mm girth	–	0.26	3.75	1.41	m²	**5.16**
Cement render or concrete walls						
over 300 mm girth	–	0.23	3.31	0.93	m²	**4.24**
Roughcast walls						
over 300 mm girth	–	0.40	5.76	0.93	m²	**6.69**
One coat sealer and two coats of external grade emulsion paint, Dulux Weathershield or other equal and approved						
Brick or block walls						
over 300 mm girth	–	0.43	6.20	6.66	m²	**12.86**
Cement render or concrete walls						
over 300 mm girth	–	0.35	5.04	4.44	m²	**9.48**
Concrete soffits						
over 300 mm girth	–	0.40	5.76	4.44	m²	**10.20**
One coat sealer (applied by brush) and two coats of external grade emulsion paint, Dulux Weathershield or other equal and approved (spray applied)						
Roughcast						
over 300 mm girth	–	0.29	4.18	9.03	m²	**13.21**
One coat sealer and two coats of anti-graffiti paint (spray applied)						
Brick or block walls						
over 300 mm girth	–	0.01	0.08	3.93	m²	**4.01**
Cement render or concrete walls						
over 300 mm girth	–	0.01	0.08	4.66	m²	**4.74**
2.5 mm of Vandalene or similar anti-climb paint (spray applied)						
General surfaces						
over 300 mm girth	–	0.01	0.08	3.87	m²	**3.95**
Two coats solar reflective aluminium paint; on bituminous roofing						
General surfaces						
over 300 mm girth	–	0.44	6.34	12.40	m²	**18.74**

29 DECORATION

Item	PC £	Labour hours	Labour £	Material £	Unit	Total rate £
Touch up primer; two undercoats and one finishing coat of gloss oil paint; on wood surfaces						
General surfaces						
over 300 mm girth	–	0.35	5.04	1.88	m²	**6.92**
isolated surfaces not exceeding 300 mm girth	–	0.15	2.16	0.50	m	**2.66**
isolated areas not exceeding 1.00 m²; irrespective of girth	–	0.27	3.90	1.01	nr	**4.91**
Glazed windows and screens						
panes; area not exceeding 0.10 m²	–	0.59	8.51	1.66	m²	**10.17**
panes; area 0.10 m²–0.50 m²	–	0.59	8.51	1.39	m²	**9.90**
panes; area 0.50 m²–1.00 m²	–	0.47	6.78	1.23	m²	**8.01**
panes; area over 1.00 m²	–	0.35	5.04	1.01	m²	**6.05**
Glazed windows and screens; multi-coloured work						
panes; area not exceeding 0.10 m²	–	0.68	9.80	1.66	m²	**11.46**
panes; area 0.10 m²–0.50 m²	–	0.55	7.92	1.45	m²	**9.37**
panes; area 0.50 m²–1.00 m²	–	0.47	6.78	1.23	m²	**8.01**
panes; area over 1.00 m²	–	0.41	5.90	1.01	m²	**6.91**
Knot; one coat primer; two undercoats and one finishing coat of gloss oil paint; on wood surfaces						
General surfaces						
over 300 mm girth	–	0.46	6.63	2.12	m²	**8.75**
isolated surfaces not exceeding 300 mm girth	–	0.19	2.74	3.62	m	**6.36**
isolated areas not exceeding 1.00 m²; irrespective of girth	–	0.35	5.04	1.38	nr	**6.42**
Glazed windows and screens						
panes; area not exceeding 0.10 m²	–	0.78	11.24	2.37	m²	**13.61**
panes; area 0.10 m²–0.50 m²	–	0.62	8.94	2.11	m²	**11.05**
panes; area 0.50 m²–1.00 m²	–	0.55	7.92	1.62	m²	**9.54**
panes; area over 1.00 m²	–	0.46	6.63	1.13	m²	**7.76**
Glazed windows and screens; multi-coloured work						
panes; area not exceeding 0.10 m²	–	0.89	12.82	2.37	m²	**15.19**
panes; area 0.10 m²–0.50 m²	–	0.72	10.37	2.12	m²	**12.49**
panes; area 0.50 m²–1.00 m²	–	0.64	9.23	1.62	m²	**10.85**
panes; area over 1.00 m²	–	0.54	7.78	1.13	m²	**8.91**
Touch up primer; two undercoats and one finishing coat of gloss oil paint; on iron or steel surfaces						
General surfaces						
over 300 mm girth	–	0.35	5.04	1.63	m²	**6.67**
isolated surfaces not exceeding 300 mm girth	–	0.14	2.02	0.44	m	**2.46**
isolated areas not exceeding 1.00 m²; irrespective of girth	–	0.26	3.75	0.91	nr	**4.66**
Glazed windows and screens						
panes; area not exceeding 0.10 m²	–	0.59	8.51	1.65	m²	**10.16**
panes; area 0.10 m²–0.50 m²	–	0.47	6.78	1.42	m²	**8.20**
panes; area 0.50 m²–1.00 m²	–	0.41	5.90	1.21	m²	**7.11**
panes; area over 1.00 m²	–	0.35	5.04	0.98	m²	**6.02**

29 DECORATION

Item	PC £	Labour hours	Labour £	Material £	Unit	Total rate £
PAINTING AND CLEAR FINISHES – EXTERNAL – cont						
Touch up primer – cont						
Structural steelwork						
over 300 mm girth	–	0.40	5.76	1.70	m²	**7.46**
Members of roof trusses						
over 300 mm girth	–	0.54	7.78	1.93	m²	**9.71**
Ornamental railings and the like; each side measured overall						
over 300 mm girth	–	0.60	8.65	1.98	m²	**10.63**
Eaves gutters						
over 300 mm girth	–	0.64	9.23	2.17	m²	**11.40**
not exceeding 300 mm girth	–	0.25	3.61	0.91	m	**4.52**
Pipes or conduits						
over 300 mm girth	–	0.54	7.78	2.17	m²	**9.95**
not exceeding 300 mm girth	–	0.21	3.02	0.73	m	**3.75**
One coat primer; two undercoats and one finishing coat of gloss oil paint; on iron or steel surfaces						
General surfaces						
over 300 mm girth	–	0.43	6.20	1.86	m²	**8.06**
isolated surfaces not exceeding 300 mm girth	–	0.18	2.59	0.48	m	**3.07**
isolated areas not exceeding 1.00 m²; irrespective of girth	–	0.32	4.61	0.96	nr	**5.57**
Glazed windows and screens						
panes; area not exceeding 0.10 m²	–	0.71	10.23	1.71	m²	**11.94**
panes; area 0.10 m²–0.50 m²	–	0.56	8.07	1.49	m²	**9.56**
panes; area 0.50 m²–1.00 m²	–	0.50	7.21	1.26	m²	**8.47**
panes; area over 1.00 m²	–	0.43	6.20	0.96	m²	**7.16**
Structural steelwork						
over 300 mm girth	–	0.48	6.92	1.94	m²	**8.86**
Members of roof trusses						
over 300 mm girth	–	0.64	9.23	2.15	m²	**11.38**
Ornamental railings and the like; each side measured overall						
over 300 mm girth	–	0.72	10.37	2.15	m²	**12.52**
Eaves gutters						
over 300 mm girth	–	0.76	10.96	2.44	m²	**13.40**
not exceeding 300 mm girth	–	0.31	4.47	0.84	m	**5.31**
Pipes or conduits						
over 300 mm girth	–	0.64	9.23	2.44	m²	**11.67**
not exceeding 300 mm girth	–	0.25	3.61	3.67	m	**7.28**
One coat of Andrews Hammerite paint or other equal and approved; on iron or steel surfaces						
General surfaces						
over 300 mm girth	–	0.15	2.16	1.34	m²	**3.50**
isolated surfaces not exceeding 300 mm girth	–	0.08	1.15	0.42	m	**1.57**
isolated areas not exceeding 1.00 m²; irrespective of girth	–	0.11	1.59	0.77	nr	**2.36**

29 DECORATION

Item	PC £	Labour hours	Labour £	Material £	Unit	Total rate £
Glazed windows and screens						
panes; area not exceeding 0.10 m^2	–	0.25	3.61	0.99	m^2	**4.60**
panes; area 0.10 m^2–0.50 m^2	–	0.19	2.74	1.13	m^2	**3.87**
panes; area 0.50 m^2–1.00 m^2	–	0.18	2.59	1.03	m^2	**3.62**
panes; area over 1.00 m^2	–	0.15	2.16	1.03	m^2	**3.19**
Structural steelwork						
over 300 mm girth	–	0.17	2.45	1.24	m^2	**3.69**
Members of roof trusses						
over 300 mm girth	–	0.23	3.31	1.34	m^2	**4.65**
Ornamental railings and the like; each side measured overall						
over 300 mm girth	–	0.26	3.75	1.34	m^2	**5.09**
Eaves gutters						
over 300 mm girth	–	0.27	3.90	1.46	m^2	**5.36**
not exceeding 300 mm girth	–	0.08	1.15	0.72	m	**1.87**
Pipes or conduits						
over 300 mm girth	–	0.26	3.75	1.24	m^2	**4.99**
not exceeding 300 mm girth	–	0.08	1.15	0.58	m	**1.73**
Two coats of creosote; on wood surfaces						
General surfaces						
over 300 mm girth	–	0.16	2.31	0.37	m^2	**2.68**
isolated surfaces not exceeding 300 mm girth	–	0.01	0.08	0.23	m	**0.31**
Two coats of Solignum wood preservative or other equal and approved; on wood surfaces						
General surfaces						
over 300 mm girth	–	0.14	2.02	0.99	m^2	**3.01**
isolated surfaces not exceeding 300 mm girth	–	0.05	0.72	0.33	m	**1.05**
Three coats of polyurethane; on wood surfaces						
General surfaces						
over 300 mm girth	–	0.29	4.18	2.10	m^2	**6.28**
isolated surfaces not exceeding 300 mm girth	–	0.11	1.59	1.05	m	**2.64**
isolated areas not exceeding 1.00 m^2; irrespective of girth	–	0.21	3.02	1.21	nr	**4.23**
Two coats of New Base primer or other equal and approved; and two coats of Extra or other equal and approved; Sadolin Ltd; pigmented; on wood surfaces						
General surfaces						
over 300 mm girth	–	0.43	6.20	3.12	m^2	**9.32**
isolated surfaces not exceeding 300 mm girth	–	0.13	1.88	0.54	m	**2.42**
Glazed windows and screens						
panes; area not exceeding 0.10 m^2	–	0.71	10.23	2.22	m^2	**12.45**
panes; area 0.10 m^2–0.50 m^2	–	0.57	8.21	2.10	m^2	**10.31**
panes; area 0.50 m^2–1.00 m^2	–	0.50	7.21	1.98	m^2	**9.19**
panes; area over 1.00 m^2	–	0.43	6.20	1.60	m^2	**7.80**

29 DECORATION

Item	PC £	Labour hours	Labour £	Material £	Unit	Total rate £
PAINTING AND CLEAR FINISHES – EXTERNAL – cont						
Two coats Sikkens Cetol Filter 7 exterior stain or other equal and approved; on wood surfaces						
General surfaces						
over 300 mm girth	–	0.20	2.88	3.54	m²	**6.42**
isolated surfaces not exceeding 300 mm girth	–	0.09	1.30	1.23	m	**2.53**
isolated areas not exceeding 1.00 m²; irrespective of girth	–	0.14	2.02	1.80	nr	**3.82**
PAINTING AND CLEAR FINISHES – INTERNAL						
SUPPLY AND FIX PRICES						
NOTE: The following prices include for preparing surfaces. Painting woodwork also includes for knotting prior to applying the priming coat and for all stopping of nail holes etc.						
Touch up primer; two undercoats and one finishing coat of gloss oil paint; on wood surfaces						
General surfaces						
over 300 mm girth	–	0.35	5.04	1.88	m²	**6.92**
isolated surfaces not exceeding 300 mm girth	–	0.15	2.16	0.50	m	**2.66**
isolated areas not exceeding 1.00 m²; irrespective of girth	–	0.27	3.90	1.01	nr	**4.91**
One coat primer; on wood surfaces before fixing						
General surfaces						
over 300 mm girth	–	0.08	1.15	0.80	m²	**1.95**
isolated surfaces not exceeding 300 mm girth	–	0.02	0.29	0.29	m	**0.58**
isolated areas not exceeding 1.00 m²; irrespective of girth	–	0.06	0.86	0.30	nr	**1.16**
One coat polyurethane sealer; on wood surfaces before fixing						
General surfaces						
over 300 mm girth	–	0.10	1.45	0.72	m²	**2.17**
isolated surfaces not exceeding 300 mm girth	–	0.03	0.43	0.26	m	**0.69**
isolated areas not exceeding 1.00 m²; irrespective of girth	–	0.08	1.15	0.34	nr	**1.49**
One coat of Sikkens Cetol HLS stain or other equal and approved; on wood surfaces before fixing						
General surfaces						
over 300 mm girth	–	0.11	1.59	0.86	m²	**2.45**
isolated surfaces not exceeding 300 mm girth	–	0.03	0.43	0.63	m	**1.06**
isolated areas not exceeding 1.00 m²; irrespective of girth	–	0.08	1.15	0.41	nr	**1.56**

29 DECORATION

Item	PC £	Labour hours	Labour £	Material £	Unit	Total rate £
One coat of Sikkens Cetol TS interior stain or other equal and approved; on wood surfaces before fixing						
General surfaces						
over 300 mm girth	–	0.11	1.59	1.20	m²	**2.79**
isolated surfaces not exceeding 300 mm girth	–	0.03	0.43	0.54	m	**0.97**
isolated areas not exceeding 1.00 m²; irrespective of girth	–	0.08	1.15	0.57	nr	**1.72**
One coat Cuprinol clear wood preservative or other equal and approved; on wood surfaces before fixing						
General surfaces						
over 300 mm girth	–	0.08	1.15	0.58	m²	**1.73**
isolated surfaces not exceeding 300 mm girth	–	0.02	0.29	0.22	m	**0.51**
isolated areas not exceeding 1.00 m²; irrespective of girth	–	0.05	0.72	0.27	nr	**0.99**
One coat HCC Protective Coatings Ltd Permacor urethane alkyd gloss finishing coat or other equal and approved; on previously primed steelwork						
Members of roof trusses						
over 300 mm girth	–	0.01	0.15	0.75	m²	**0.90**
Two coats emulsion paint						
Brick or block walls						
over 300 mm girth	–	0.21	3.02	0.50	m²	**3.52**
Cement render or concrete						
over 300 mm girth	–	0.20	2.88	0.46	m²	**3.34**
isolated surfaces not exceeding 300 mm girth	–	0.10	1.45	0.17	m	**1.62**
Plaster walls or plaster/plasterboard ceilings						
over 300 mm girth	–	0.18	2.59	0.45	m²	**3.04**
over 300 mm girth; in multi colours	–	0.24	3.45	0.52	m²	**3.97**
over 300 mm girth; in staircase areas	–	0.21	3.02	0.50	m²	**3.52**
cutting in edges on flush surfaces	–	0.08	1.15	–	m	**1.15**
Plaster/plasterboard ceilings						
over 300 mm girth; 3.50 m–5.00 m high	–	0.21	3.02	0.45	m²	**3.47**
One mist and two coats emulsion paint						
Brick or block walls						
over 300 mm girth	–	0.19	2.74	0.66	m²	**3.40**
Cement render or concrete						
over 300 mm girth	–	0.19	2.74	0.62	m²	**3.36**
Plaster walls or plaster/plasterboard ceilings						
over 300 mm girth	–	0.18	2.59	0.62	m²	**3.21**
over 300 mm girth; in multi colours	–	0.25	3.61	0.63	m²	**4.24**
over 300 mm girth; in staircase areas	–	0.21	3.02	0.62	m²	**3.64**
cutting in edges on flush surfaces	–	0.09	1.30	–	m	**1.30**
Plaster/plasterboard ceilings						
over 300 mm girth; 3.50 m–5.00 m high	–	0.21	3.02	0.62	m²	**3.64**

29 DECORATION

Item	PC £	Labour hours	Labour £	Material £	Unit	Total rate £
PAINTING AND CLEAR FINISHES – INTERNAL – cont						
One mist Supermatt; one full Supermatt and one full coat of quick drying Acrylic Eggshell						
Brick or block walls						
over 300 mm girth	–	0.19	2.74	1.42	m²	**4.16**
Cement render or concrete						
over 300 mm girth	–	0.19	2.74	1.32	m²	**4.06**
Plaster walls or plaster/plasterboard ceilings						
over 300 mm girth	–	0.18	2.59	1.32	m²	**3.91**
over 300 mm girth; in multi colours	–	0.25	3.61	1.32	m²	**4.93**
over 300 mm girth; in staircase areas	–	0.21	3.02	1.32	m²	**4.34**
cutting in edges on flush surfaces	–	0.09	1.30	–	m	**1.30**
Plaster/plasterboard ceilings						
over 300 mm girth; 3.50 m–5.00 m high		0.21	3.02	1.32	m²	**4.34**
One coat primer and two coats of Keim Ecosil paint						
Brick or block walls						
over 300 mm girth	–	0.25	3.61	2.71	m²	**6.32**
Cement render or concrete						
over 300 mm girth	–	0.25	3.61	2.37	m²	**5.98**
Plaster walls or plaster/plasterboard ceilings						
over 300 mm girth	–	0.20	2.88	2.37	m²	**5.25**
over 300 mm girth; in staircase areas	–	0.21	3.02	2.37	m²	**5.39**
cutting in edges on flush surfaces	–	0.09	1.30	–	m	**1.30**
Plaster/plasterboard ceilings						
over 300 mm girth; 3.50 m–5.00 m high	–	0.25	3.61	2.51	m²	**6.12**
One coat Tretol No 10 Sealer or other equal and approved; two coats Tretol sprayed Supercover Spraytone emulsion paint or other equal and approved						
Plaster walls or plaster/plasterboard ceilings						
over 300 mm girth	–	–	–	–	m²	**5.12**
Textured plastic; Artex or other equal and approved finish						
Plasterboard ceilings						
over 300 mm girth	–	0.19	2.74	2.19	m²	**4.93**
Concrete walls or ceilings						
over 300 mm girth	–	0.23	3.31	2.01	m²	**5.32**
Touch up primer; one undercoat and one finishing coat of gloss oil paint; on wood surfaces						
General surfaces						
over 300 mm girth	–	0.20	2.88	1.88	m²	**4.76**
isolated surfaces not exceeding 300 mm girth	–	0.08	1.15	0.66	m	**1.81**
isolated areas not exceeding 1.00 m²; irrespective of girth	–	0.18	2.59	1.01	nr	**3.60**

29 DECORATION

Item	PC £	Labour hours	Labour £	Material £	Unit	Total rate £
Glazed windows and screens						
panes; area not exceeding 0.10 m²	–	0.38	5.47	1.49	m²	**6.96**
panes; area 0.10 m²–0.50 m²	–	0.31	4.47	1.14	m²	**5.61**
panes; area 0.50 m²–1.00 m²	–	0.26	3.75	0.91	m²	**4.66**
panes; area over 1.00 m²	–	0.23	3.31	0.78	m²	**4.09**
Knot; one coat primer; stop; one undercoat and one finishing coat of gloss oil paint; on wood surfaces						
General surfaces						
over 300 mm girth	–	0.33	4.76	1.87	m²	**6.63**
isolated surfaces not exceeding 300 mm girth	–	0.13	1.88	0.63	m	**2.51**
isolated areas not exceeding 0.50 m²; irrespective of girth	–	0.25	3.61	1.22	nr	**4.83**
Glazed windows and screens						
panes; area not exceeding 0.10 m²	–	0.56	8.07	1.87	m²	**9.94**
panes; area 0.10 m²–0.50 m²	–	0.45	6.49	1.56	m²	**8.05**
panes; area 0.50 m²–1.00 m²	–	0.40	5.76	1.56	m²	**7.32**
panes; area over 1.00 m²	–	0.33	4.76	1.15	m²	**5.91**
One coat primer; one undercoat and one finishing coat of gloss oil paint						
Plaster surfaces						
over 300 mm girth	–	0.30	4.33	2.38	m²	**6.71**
One coat primer; two undercoats and one finishing coat of gloss oil paint						
Plaster surfaces						
over 300 mm girth	–	0.40	5.76	3.13	m²	**8.89**
One coat primer; two undercoats and one finishing coat of eggshell paint						
Plaster surfaces						
over 300 mm girth	–	0.40	5.76	3.29	m²	**9.05**
Touch up primer; one undercoat and one finishing coat of gloss paint; on iron or steel surfaces						
General surfaces						
over 300 mm girth	–	0.23	3.31	1.41	m²	**4.72**
isolated surfaces not exceeding 300 mm girth	–	0.09	1.30	0.48	m	**1.78**
isolated areas not exceeding 0.50 m²; irrespective of girth	–	0.18	2.59	0.78	nr	**3.37**
Glazed windows and screens						
panes; area not exceeding 0.10 m²	–	0.38	5.47	1.47	m²	**6.94**
panes; area 0.10 m²–0.50 m²	–	0.31	4.47	1.15	m²	**5.62**
panes; area 0.50 m²–1.00 m²	–	0.26	3.75	0.89	m²	**4.64**
panes; area over 1.00 m²	–	0.23	3.31	0.75	m²	**4.06**

29 DECORATION

Item	PC £	Labour hours	Labour £	Material £	Unit	Total rate £
PAINTING AND CLEAR FINISHES – INTERNAL – cont						
Touch up primer – cont						
Structural steelwork						
over 300 mm girth	–	0.25	3.61	1.49	m²	**5.10**
Members of roof trusses						
over 300 mm girth	–	0.34	4.90	1.69	m²	**6.59**
Ornamental railings and the like; each side measured overall						
over 300 mm girth	–	0.40	5.76	1.87	m²	**7.63**
Iron or steel radiators						
over 300 mm girth	–	0.23	3.31	1.55	m²	**4.86**
Pipes or conduits						
over 300 mm girth	–	0.34	4.90	1.62	m²	**6.52**
not exceeding 300 mm girth	–	0.13	1.88	0.53	m	**2.41**
One coat primer; one undercoat and one finishing coat of gloss oil paint; on iron or steel surfaces						
General surfaces						
over 300 mm girth	–	0.30	4.33	1.42	m²	**5.75**
isolated surfaces not exceeding 300 mm girth	–	0.12	1.73	0.81	m	**2.54**
isolated areas not exceeding 0.50 m²; irrespective of girth	–	0.23	3.31	1.36	nr	**4.67**
Glazed windows and screens						
panes; area not exceeding 0.10 m²	–	0.50	7.21	2.22	m²	**9.43**
panes; area 0.10 m²–0.50 m²	–	0.40	5.76	1.77	m²	**7.53**
panes; area 0.50 m²–1.00 m²	–	0.34	4.90	1.53	m²	**6.43**
panes; area over 1.00 m²	–	0.30	4.33	1.36	m²	**5.69**
Structural steelwork						
over 300 mm girth	–	0.33	4.76	2.17	m²	**6.93**
Members of roof trusses						
over 300 mm girth	–	0.45	6.49	2.31	m²	**8.80**
Ornamental railings and the like; each side measured overall						
over 300 mm girth	–	0.51	7.35	2.74	m²	**10.09**
Iron or steel radiators						
over 300 mm girth	–	0.30	4.33	2.31	m²	**6.64**
Pipes or conduits						
over 300 mm girth	–	0.45	6.49	2.31	m²	**8.80**
not exceeding 300 mm girth	–	0.18	2.59	0.77	m	**3.36**
Two coats of bituminous paint; on iron or steel surfaces						
General surfaces						
over 300 mm girth	–	0.23	3.31	0.68	m²	**3.99**
Inside of galvanized steel cistern						
over 300 mm girth	–	0.34	4.90	0.81	m²	**5.71**

29 DECORATION

Item	PC £	Labour hours	Labour £	Material £	Unit	Total rate £
Two coats bituminous paint; first coat blinded with clean sand prior to second coat; on concrete surfaces						
General surfaces						
over 300 mm girth	–	0.79	11.39	2.03	m²	**13.42**
Mordant solution; one coat HCC Protective Coatings Ltd Permacor Alkyd MIO or other equal and approved; one coat Permatex Epoxy Gloss finishing coat or other equal and approved on galvanized steelwork						
Structural steelwork						
over 300 mm girth	–	0.44	6.34	2.60	m²	**8.94**
One coat HCC Protective Coatings Ltd Epoxy Zinc Primer or other equal and approved; two coats Permacor Alkyd MIO or other equal and approved; one coat Permacor Epoxy Gloss finishing coat or other equal and approved on steelwork						
Structural steelwork						
over 300 mm girth	–	0.63	9.08	4.90	m²	**13.98**
Steel protection; HCC Protective Coatings Ltd Unitherm or other equal and approved; two coats to steelwork						
Structural steelwork						
over 300 mm girth	–	0.99	14.27	1.74	m²	**16.01**
Two coats of epoxy anti-slip floor paint; on screeded concrete surfaces						
General surfaces						
over 300 mm girth	–	0.25	3.61	11.26	m²	**14.87**
Nitoflor Lithurin floor hardener and dust proofer or other equal and approved; Fosroc Expandite Ltd; two coats; on concrete surfaces						
General surfaces						
over 300 mm girth	–	0.24	2.78	0.48	m²	**3.26**
Two coats of boiled linseed oil; on hardwood surfaces						
General surfaces						
over 300 mm girth	–	0.18	2.59	2.29	m²	**4.88**
isolated surfaces not exceeding 300 mm girth	–	0.07	1.00	0.74	m	**1.74**
isolated areas not exceeding 0.50 m²; irrespective of girth	–	0.13	1.88	1.32	nr	**3.20**

29 DECORATION

Item	PC £	Labour hours	Labour £	Material £	Unit	Total rate £
PAINTING AND CLEAR FINISHES – INTERNAL – cont						
Two coats polyurethane varnish; on wood surfaces						
General surfaces						
over 300 mm girth	–	0.18	2.59	1.28	m²	**3.87**
isolated surfaces not exceeding 300 mm girth	–	0.07	1.00	0.48	m	**1.48**
isolated areas not exceeding 0.50 m²; irrespective of girth	–	0.13	1.88	0.18	nr	**2.06**
Three coats polyurethane varnish; on wood surfaces						
General surfaces						
over 300 mm girth	–	0.26	3.75	1.93	m²	**5.68**
isolated surfaces not exceeding 300 mm girth	–	0.10	1.45	0.62	m	**2.07**
isolated areas not exceeding 0.50 m²; irrespective of girth	–	0.19	2.74	1.09	nr	**3.83**
One undercoat; and one finishing coat; of Albi clear flame retardant surface coating or other equal and approved; on wood surfaces						
General surfaces						
over 300 mm girth	–	0.34	4.90	4.79	m²	**9.69**
isolated surfaces not exceeding 300 mm girth	–	0.14	2.02	1.66	m	**3.68**
isolated areas not exceeding 0.50 m²; irrespective of girth	–	0.19	2.74	3.63	nr	**6.37**
Two undercoats; and one finishing coat; of Albi clear flame retardant surface coating or other equal and approved; on wood surfaces						
General surfaces						
over 300 mm girth	–	0.40	5.76	6.05	m²	**11.81**
isolated surfaces not exceeding 300 mm girth	–	0.20	2.88	2.43	m	**5.31**
isolated areas not exceeding 0.50 m²; irrespective of girth	–	0.33	4.76	3.23	nr	**7.99**
Seal and wax polish; dull gloss finish on wood surfaces						
General surfaces						
over 300 mm girth	PC	–	–	–	m²	**9.80**
isolated surfaces not exceeding 300 mm girth	–	–	–	–	m	**4.42**
isolated areas not exceeding 0.50 m²; irrespective of girth	–	–	–	–	nr	**6.86**
One coat of Sadolin Extra or other equal and approved; clear or pigmented; one further coat of Holdex clear interior silk matt lacquer or similar						
General surfaces						
over 300 mm girth	–	0.25	3.61	4.12	m²	**7.73**
isolated surfaces not exceeding 300 mm girth	–	0.10	1.45	1.93	m	**3.38**
isolated areas not exceeding 0.50 m²; irrespective of girth	–	0.20	2.88	2.00	nr	**4.88**

29 DECORATION

Item	PC £	Labour hours	Labour £	Material £	Unit	Total rate £
Glazed windows and screens						
panes; area not exceeding 0.10 m²	–	0.42	6.06	2.36	m²	**8.42**
panes; area 0.10 m²–0.50 m²	–	0.33	4.76	2.19	m²	**6.95**
panes; area 0.50 m²–1.00 m²	–	0.29	4.18	2.03	m²	**6.21**
panes; area over 1.00 m²	–	0.25	3.61	1.93	m²	**5.54**
Two coats of Sadolin Extra or other equal and approved; clear or pigmented; two further coats of PV67 clear interior silk matt lacquer or similar						
General surfaces						
over 300 mm girth	–	0.40	5.76	7.56	m²	**13.32**
isolated surfaces not exceeding 300 mm girth	–	0.16	2.31	3.78	m	**6.09**
isolated areas not exceeding 0.50 m²; irrespective of girth	–	0.30	4.33	4.32	nr	**8.65**
Glazed windows and screens						
panes; area not exceeding 0.10 m²	–	0.66	9.51	4.63	m²	**14.14**
panes; area 0.10 m²–0.50 m²	–	0.52	7.49	4.32	m²	**11.81**
panes; area 0.50 m²–1.00 m²	–	0.45	6.49	4.00	m²	**10.49**
panes; area over 1.00 m²	–	0.40	5.76	3.78	m²	**9.54**
Two coats of Sikkens Cetol TS interior stain or other equal and approved; on wood surfaces						
General surfaces						
over 300 mm girth	–	0.19	2.74	2.12	m²	**4.86**
isolated surfaces not exceeding 300 mm girth	–	0.08	1.15	0.76	m	**1.91**
isolated areas not exceeding 0.50 m²; irrespective of girth	–	0.13	1.88	1.17	nr	**3.05**
Body in and wax polish; dull gloss finish; on hardwood surfaces						
General surfaces						
over 300 mm girth	–	–	–	–	m²	**11.03**
isolated surfaces not exceeding 300 mm girth	–	–	–	–	m	**4.97**
isolated areas not exceeding 0.50 m²; irrespective of girth	–	–	–	–	nr	**7.73**
Stain; body in and wax polish; dull gloss finish; on hardwood surfaces						
General surfaces						
over 300 mm girth	–	–	–	–	m²	**14.77**
isolated surfaces not exceeding 300 mm girth	–	–	–	–	m	**6.65**
isolated areas not exceeding 0.50 m²; irrespective of girth	–	–	–	–	nr	**10.35**
Seal; two coats of synthetic resin lacquer; decorative flatted finish; wire down, wax and burnish; on wood surfaces						
General surfaces						
over 300 mm girth	–	–	–	–	m²	**18.60**
isolated surfaces not exceeding 300 mm girth	–	–	–	–	m	**8.70**
isolated areas not exceeding 0.50 m²; irrespective of girth	–	–	–	–	nr	**13.09**

29 DECORATION

Item	PC £	Labour hours	Labour £	Material £	Unit	Total rate £
PAINTING AND CLEAR FINISHES – INTERNAL – cont						
Stain; body in and fully French polish; full gloss finish; on hardwood surfaces						
General surfaces						
over 300 mm girth	–	–	–	–	m²	**21.53**
isolated surfaces not exceeding 300 mm girth	–	–	–	–	m	**9.68**
isolated areas not exceeding 0.50 m²; irrespective of girth	–	–	–	–	nr	**15.07**
Stain; fill grain and fully French polish; full gloss finish; on hardwood surfaces						
General surfaces						
over 300 mm girth	–	–	–	–	m²	**32.01**
isolated surfaces not exceeding 300 mm girth	–	–	–	–	m	**14.40**
isolated areas not exceeding 0.50 m²; irrespective of girth	–	–	–	–	nr	**22.40**
Stain black; body in and fully French polish; ebonized finish; on hardwood surfaces						
General surfaces						
over 300 mm girth	–	–	–	–	m²	**36.51**
isolated surfaces not exceeding 300 mm girth	–	–	–	–	m	**16.42**
isolated areas not exceeding 0.50 m²; irrespective of girth	–	–	–	–	nr	**25.56**
DECORATIVE PAPERS OR FABRICS						
Lining paper; and hanging						
Plaster walls or columns						
over 1.00 m² (PC £ per roll)	2.00	0.19	2.74	0.28	m²	**3.02**
Plaster ceilings or beams						
over 1.00 m² girth (PC £ per roll)	2.00	0.23	3.31	0.28	m²	**3.59**
Decorative paper-backed vinyl wallpaper; and hanging						
Plaster walls or columns						
over 1.00 m² girth (PC £ per roll)	10.00	0.23	3.31	1.76	m²	**5.07**
PVC Wall lining; Altro Whiterock; or other equal and approved; fixed directly to plastered brick or blockwork						
Work to walls over 300 mm wide						
Standard white	17.65	1.00	14.41	34.51	m²	**48.92**
Satins	24.23	1.00	14.41	41.59	m²	**56.00**
Stone	41.74	1.30	18.74	60.44	m²	**79.18**
Illusions	41.74	1.30	18.74	60.44	m²	**79.18**
Chameleon	44.12	1.30	18.74	63.01	m²	**81.75**

30 SUSPENDED CEILINGS

Item	PC £	Labour hours	Labour £	Material £	Unit	Total rate £
SUSPENDED CEILINGS						
Gyproc M/F suspended ceiling system or other equal and approved; hangers plugged and screwed to concrete soffit, 900 mm × 1800 mm × 12.50 mm tapered edge wallboard infill; joints filled with joint filler and taped to receive direct decoration						
Lining to ceilings; hangers 150 mm–500 mm long						
over 600 mm wide	–	–	–	–	m²	27.98
not exceeding 600 mm wide in isolated strips	–	–	–	–	m	28.35
Edge treatments						
20 × 20 mm SAS perimeter shadow gap; screwed to plasterboard	–	–	–	–	m	5.09
20 × 20 mm SAS shadow gap around 450 mm dia. column; including 15 × 44 mm batten plugged and screwed to concrete	–	–	–	–	nr	60.47
Vertical bulkhead; including additional hangers						
over 600 mm wide	–	–	–	–	m²	35.32
not exceeding 600 mm wide in isolated strips	–	–	–	–	m	34.83
Rockfon, or other equal and approved; Z demountable suspended concealed ceiling system; 400 mm long hangers plugged and screwed to concrete soffit						
Lining to ceilings; 600 mm × 600 mm × 20 mm Sonar suspended ceiling tiles						
over 600 mm wide	–	–	–	–	m²	35.81
not exceeding 600 mm wide in isolated strips	–	–	–	–	m	21.11
edge trim; shadow-line trim	–	–	–	–	m	4.18
Vertical bulkhead, as upstand to rooflight well; including additional hangers; perimeter trim						
300 mm × 600 mm wide	–	–	–	–	m	39.20
Ecophon, or other equal and approved; Z demountable suspended concealed ceiling system; 400 mm long hangers plugged and screwed to concrete soffit						
Lining to ceilings; 600 mm × 600 mm × 20 mm Gedina ET15 suspended ceiling tiles						
over 600 mm wide	–	–	–	–	m²	31.07
not exceeding 600 mm wide in isolated strips	–	–	–	–	m	19.58
edge trim; shadow-line trim	–	–	–	–	m	3.87
Vertical bulkhead, as upstand to rooflight well; including additional hangers; perimeter trim						
300 mm × 600 mm wide	–	–	–	–	m	36.07

30 SUSPENDED CEILINGS

Item	PC £	Labour hours	Labour £	Material £	Unit	Total rate £
SUSPENDED CEILINGS – cont						
Ecophon, or other equal and approved – cont						
Lining to ceilings; 600 mm × 600 mm × 20 mm Hygiene Performance washable suspended ceiling tiles						
over 600 mm wide	–	–	–	–	m²	**41.49**
not exceeding 600 mm wide in isolated strips	–	–	–	–	m	**34.01**
edge trim; shadow-line trim	–	–	–	–	m	**5.47**
Vertical bulkhead, as upstand to rooflight well; including additional hangers; perimeter trim						
not exceeding 600 mm wide in isolated strips	–	–	–	–	m	**37.45**
Lining to ceilings; 1200 mm × 1200 mm × 20 mm Focus DG suspended ceiling tiles						
over 600 mm wide	–	–	–	–	m²	**37.64**
not exceeding 600 mm wide in isolated strips	–	–	–	–	m	**22.31**
edge trim; shadow-line trim	–	–	–	–	m	**3.87**
vertical bulkhead, as upstand to rooflight well; including additional hangers; perimeter trim						
300 mm × 600 mm wide	–	–	–	–	m	**38.01**
Z demountable suspended ceiling system or other equal and approved; hangers plugged and screwed to concrete soffit, 600 mm × 600 mm × 19 mm Echostop glass reinforced fibrous plaster lightweight plain bevelled edge tiles						
Lining to ceilings; hangers 150 mm–500 mm long						
over 600 mm wide	–	–	–	–	m²	**76.91**
not exceeding 600 mm wide in isolated strips	–	–	–	–	m	**54.84**
Concealed galvanized steel suspension system; hangers plugged and screwed to concrete soffit, Burgess white stove enamelled perforated mild steel tiles 600 mm × 600 mm						
Lining to ceilings; hangers average 150 mm–500 mm long						
over 600 mm wide	–	–	–	–	m²	**36.98**
not exceeding 600 mm wide in isolated strips	–	–	–	–	m	**32.43**
Concealed galvanized steel Trulok suspension system or other equal and approved; hangers plugged and screwed to concrete; Armstrong Ultima Microlok BE Plain 300 mm × 300 mm × 18 mm mineral ceiling tiles						
Linings to ceilings; hangers 500 mm–1000 mm long						
over 600 mm wide	–	–	–	–	m²	**23.21**
over 600 mm wide; 3.50 m–5.00 m high	–	–	–	–	m²	**24.08**
over 600 mm wide; in staircase areas or plant rooms	–	–	–	–	m²	**30.33**

30 SUSPENDED CEILINGS

Item	PC £	Labour hours	Labour £	Material £	Unit	Total rate £
not exceeding 600 mm wide in isolated strips	–	–	–	–	m	**21.38**
cutting and fitting around modular downlighter including yoke	–	–	–	–	nr	**13.98**
24 mm × 19 mm white finished angle edge trim	–	–	–	–	m	**3.46**
Vertical bulkhead; including additional hangers						
over 600 mm wide	–	–	–	–	m²	**42.89**
not exceeding 600 mm wide in isolated strips	–	–	–	–	m	**38.68**
Metal; SAS system 330; EMAC suspension system; 100 mm Omega C profiles at 1500 mm centres filled in with 1400 mm × 250 mm perforated metal tiles with 18 mm thick × 80 kg/m³ density foil wrapped tissue-faced acoustic pad adhered above; ceiling to achieve 40d Dnwc with 0.7 absorption coefficient						
Linings to ceilings; hangers 500 mm–700 mm long						
over 600 mm wide	–	–	–	–	m²	**38.70**
not exceeding 600 mm wide in isolated strips	–	–	–	–	m	**19.75**
cutting and reinforcing to receive a recessed light maximum 1300 mm × 500 mm.	–	–	–	–	nr	**11.06**
edge trim; to perimeter	–	–	–	–	m	**9.87**
edge trim around 450 mm dia. column	–	–	–	–	nr	**37.91**
Galvanized steel suspension system; hangers plugged and screwed to concrete soffit, Luxalon stove enamelled aluminium linear panel ceiling, type 80B or other equal and approved, complete with mineral insulation						
Linings to ceilings; hangers 500 mm–700 mm long						
over 600 mm wide	–	–	–	–	m²	**69.80**
not exceeding 600 mm wide in isolated strips	–	–	–	–	m	**33.84**
SOLID SUSPENDED CEILINGS						
Gypsum plasterboard; BS EN 520; plain grade tapered edge wallboard; fixing on dabs or with nails; joints left open to receive Artex finish or other equal and approved; to softwood base						
9.50 mm board to ceilings						
over 600 mm wide	–	0.23	4.10	1.89	m²	**5.99**
9.50 mm board to beams						
girth not exceeding 600 mm	–	0.28	4.99	1.16	m²	**6.15**
girth 600 mm–1200 mm	–	0.37	6.60	2.29	m²	**8.89**
12.50 mm board to ceilings						
over 300 mm wide	–	0.31	5.52	1.97	m²	**7.49**
12.50 mm board to beams						
girth not exceeding 600 mm	–	0.28	4.99	1.21	m²	**6.20**
girth 600 mm–1200 mm	–	0.37	6.60	2.37	m²	**8.97**

30 SUSPENDED CEILINGS

Item	PC £	Labour hours	Labour £	Material £	Unit	Total rate £
SOLID SUSPENDED CEILINGS – cont						
Gypsum plasterboard to BS EN 520; fixing on dabs or with nails; joints filled with joint filler and joint tape to receive direct decoration; to softwood base (measured elsewhere)						
Plain grade tapered edge wallboard						
9.50 mm board to ceilings						
over 600 mm wide	–	0.29	5.41	2.50	m²	**7.91**
9.50 mm board to faces of beams – 3 nr						
not exceeding 600 mm total girth	–	0.46	8.67	2.87	m	**11.54**
600 mm–1200 mm total girth	–	0.67	12.54	4.02	m	**16.56**
1200 mm–1800 mm total girth	–	0.75	14.08	5.17	m	**19.25**
12.50 mm board to ceilings						
over 600 mm wide	–	0.31	5.76	3.77	m²	**9.53**
12.50 mm board to faces of beams – 3 nr						
not exceeding 600 mm total girth	–	0.46	8.67	3.66	m	**12.33**
600 mm–1200 mm total girth	–	0.67	12.54	5.57	m	**18.11**
1200 mm–1800 mm total girth	–	0.75	14.08	7.46	m	**21.54**
external angle; with joint tape bedded and covered with Jointex or other equal and approved	–	0.11	2.22	0.41	m	**2.63**
Tapered edge wallboard TEN						
12.50 mm board to ceilings						
over 600 mm wide	–	0.41	7.78	2.86	m²	**10.64**
12.50 mm board to faces of beams – 3 nr						
not exceeding 600 mm total girth	–	0.56	10.70	3.12	m	**13.82**
600 mm–1200 mm total girth	–	1.02	19.61	4.47	m	**24.08**
1200 mm–1800 mm total girth	–	1.30	25.19	5.82	m	**31.01**
external angle; with joint tape bedded and covered with Jointex or other equal and approved	–	0.11	2.22	0.41	m	**2.63**
Tapered edge plank						
19 mm plank to ceilings						
over 600 mm wide	–	0.43	8.14	4.52	m²	**12.66**
19 mm plank to faces of beams – 3 nr						
not exceeding 600 mm total girth	–	0.60	11.41	4.11	m	**15.52**
600 mm–1200 mm total girth	–	1.06	20.33	6.47	m	**26.80**
1200 mm–1800 mm total girth	–	1.34	25.91	8.82	m	**34.73**
ThermaLine Plus board						
27 mm board to ceilings						
over 600 mm wide	–	0.46	8.20	7.11	m²	**15.31**
27 mm board to faces of beams – 3 nr						
not exceeding 600 mm total girth	–	0.56	9.98	5.67	m	**15.65**
600 mm–1200 mm total girth	–	1.06	18.90	9.57	m	**28.47**
1200 mm–1800 mm total girth	–	1.43	25.49	13.48	m	**38.97**

30 SUSPENDED CEILINGS

Item	PC £	Labour hours	Labour £	Material £	Unit	Total rate £
48 mm board to ceilings						
over 600 mm wide	–	0.49	8.73	9.61	m²	**18.34**
48 mm board to faces of beams – 3 nr						
not exceeding 600 mm total girth	–	0.58	10.34	7.32	m	**17.66**
600 mm–1200 mm total girth	–	1.17	20.86	12.75	m	**33.61**
1200 mm–1800 mm total girth	–	1.57	27.99	18.18	m	**46.17**
ThermaLine Super boards						
50 mm board to ceilings						
over 600 mm wide	–	0.49	8.73	12.95	m²	**21.68**
50 mm board to faces of beams – 3 nr						
not exceeding 600 mm total girth	–	0.58	10.34	9.32	m	**19.66**
600 mm–1200 mm total girth	–	1.17	20.86	16.75	m	**37.61**
1200 mm–1800 mm total girth	–	1.57	27.99	24.18	m	**52.17**
60 mm board to ceilings						
over 600 mm wide	–	0.49	8.73	15.24	m²	**23.97**
60 mm board to faces of beams – 3 nr						
not exceeding 600 mm total girth	–	0.58	10.34	10.69	m	**21.03**
600 mm–1200 mm total girth	–	1.17	20.86	19.51	m	**40.37**
1200 mm–1800 mm total girth	–	1.57	27.99	28.31	m	**56.30**

31 INSULATION, FIRE STOPPING AND FIRE PROTECTION

Item	PC £	Labour hours	Labour £	Material £	Unit	Total rate £
INSULATION						
Alternative insulation supply only prices, delivered in full loads						
Crown FrameTherm Roll 40 (Thermal conductivity 0.040 W/mK)						
90 mm thick	–	–	–	0.99	m²	**0.99**
140 mm thick	–	–	–	1.43	m²	**1.43**
Crown FrameTherm Roll 35 (Thermal conductivity 0.035 W/mK)						
90 mm thick	–	–	–	2.35	m²	**2.35**
140 mm thick	–	–	–	3.34	m²	**3.34**
Crown Factoryclad 40 (Thermal conductivity 0.040 W/mK)						
80 mm thick	–	–	–	0.67	m²	**0.67**
100 mm thick	–	–	–	0.81	m²	**0.81**
Crown Factoryclad 37 (Thermal conductivity 0.037 W/mK)						
100 mm thick	–	–	–	1.54	m²	**1.54**
120 mm thick	–	–	–	1.81	m²	**1.81**
Crown Factoryclad 32 (Thermal conductivity 0.032 W/mK)						
100 mm thick	–	–	–	3.58	m²	**3.58**
Thermafleece EcoRoll (0.039W/mK); 75% sheep's wool, 15% recycled polyester and 10% polyester binder with a high recycled content						
50 mm thick	–	–	–	2.86	m²	**2.86**
75 mm thick	–	–	–	4.27	m²	**4.27**
100 mm thick	–	–	–	5.69	m²	**5.69**
140 mm thick	–	–	–	7.97	m²	**7.97**
Thermafleece TF35 high density wool insulating batts (0.035 W/mK); 60% British wool, 30% recycled polyester and 10% polyester binder with a high recycled content						
50 mm thick	–	–	–	6.18	m²	**6.18**
70 mm thick	–	–	–	8.65	m²	**8.65**
SUPPLY AND FIX PRICES						
Sisalkraft building papers/vapour barriers or other equal and approved						
Building paper; 150 mm laps; fixed to softwood						
Moistop grade 728 (class A1F)	–	0.08	1.10	1.06	m²	**2.16**
Vapour barrier/reflective insulation 150 mm laps; fixed to softwood						
Insulex grade 714; single sided	–	0.08	1.10	1.24	m²	**2.34**

31 INSULATION, FIRE STOPPING AND FIRE PROTECTION

Item	PC £	Labour hours	Labour £	Material £	Unit	Total rate £
Mat or quilt insulation						
Glass fibre roll; Crown Loft Roll 44 (Thermal conductivity 0.044 W/mK) or other equal; laid loose						
100 mm thick	0.89	0.09	1.23	0.91	m²	**2.14**
150 mm thick	1.34	0.10	1.37	1.37	m²	**2.74**
200 mm thick	1.78	0.11	1.51	1.82	m²	**3.33**
Glass fibre quilt; Isover Modular roll (Thermal conductivity 0.043 W/mK) or other equal and approved; laid loose						
100 mm thick	1.48	0.09	1.23	1.52	m²	**2.75**
150 mm thick	2.18	0.10	1.37	2.23	m²	**3.60**
170 mm thick	2.70	0.12	1.64	2.77	m²	**4.41**
200 mm thick	3.24	0.15	2.05	3.32	m²	**5.37**
Mineral fibre quilt; Isover Acoustic Partition Roll (APR 1200) or other equal; pinned vertically to softwood						
25 mm thick	0.85	0.08	1.10	1.12	m²	**2.22**
50 mm thick	1.39	0.09	1.23	1.67	m²	**2.90**
75 mm thick	2.55	0.10	1.37	2.86	m²	**4.23**
100 mm thick	3.74	0.15	2.05	4.08	m²	**6.13**
Crown Dritherm Cavity Slab 37 (Thermal conductivity 0.037 W/mK) glass fibre batt or other equal; as full or partial cavity fill; including cutting and fitting around wall ties and retaining discs						
50 mm thick	2.54	0.12	1.64	2.85	m²	**4.49**
75 mm thick	2.82	0.13	1.78	3.14	m²	**4.92**
100 mm thick	2.69	0.14	1.92	3.00	m²	**4.92**
Crown Dritherm Cavity Slab 34 (Thermal conductivity 0.034 W/mK) glass fibre batt or other equal; as full or partial cavity fill; including cutting and fitting around wall ties and retaining discs						
65 mm thick	1.84	0.12	1.64	2.13	m²	**3.77**
75 mm thick	2.13	0.13	1.78	2.43	m²	**4.21**
85 mm thick	3.23	0.13	1.78	3.56	m²	**5.34**
100 mm thick	3.23	0.14	1.92	3.56	m²	**5.48**
Crown Dritherm Cavity Slab 32 (Thermal conductivity 0.032 W/mK) glass fibre batt or other equal; as full or partial cavity fill; including cutting and fitting around wall ties and retaining discs						
65 mm thick	2.39	0.12	1.64	2.71	m²	**4.35**
75 mm thick	2.78	0.13	1.78	3.11	m²	**4.89**
85 mm thick	3.16	0.13	1.78	3.48	m²	**5.26**
100 mm thick	3.64	0.14	1.92	3.98	m²	**5.90**
Crown Frametherm Roll 40 (Thermal conductivity 0.040 W/mK) glass fibre semi-rigid or rigid batt or other equal; pinned vertically in timber frame construction						
90 mm thick	2.97	0.14	1.92	3.04	m²	**4.96**
140 mm thick	4.32	0.16	2.19	4.43	m²	**6.62**

31 INSULATION, FIRE STOPPING AND FIRE PROTECTION

Item	PC £	Labour hours	Labour £	Material £	Unit	Total rate £
INSULATION – cont						
Mat or quilt insulation – cont						
Crown Rafter Roll 32 (Thermal conductivity 0.032 W/mK) glass fibre flanged building roll; pinned vertically or to slope between timber framing						
50 mm thick	2.78	0.13	1.78	2.85	m²	**4.63**
75 mm thick	3.97	0.14	1.92	4.07	m²	**5.99**
100 mm thick	5.12	0.15	2.05	5.25	m²	**7.30**
Sheepswool Mix						
Thermafleece EcoRoll (0.039W/mK); 75% sheep's wool, 15% recycled polyester and 10% polyester binder with a high recycled content						
50 mm thick	–	0.15	2.05	2.93	m²	**4.98**
75 mm thick	–	0.14	1.92	4.38	m²	**6.30**
100 mm thick	–	0.15	2.05	5.83	m²	**7.88**
140 mm thick	–	0.17	2.40	8.17	m²	**10.57**
Board or slab insulation						
Kay-Cel (Thermal conductivity 0.033 W/mK) expanded polystyrene board standard grade SD/N or other equal; fixed with adhesive						
20 mm thick	–	0.14	2.49	1.15	m²	**3.64**
25 mm thick	–	0.14	2.49	1.25	m²	**3.74**
30 mm thick	–	0.14	2.49	1.35	m²	**3.84**
40 mm thick	–	0.15	2.68	1.56	m²	**4.24**
50 mm thick	–	0.16	2.85	1.76	m²	**4.61**
60 mm thick	–	0.17	3.03	1.97	m²	**5.00**
75 mm thick	–	0.18	3.21	2.28	m²	**5.49**
100 mm thick	–	0.19	3.38	2.80	m²	**6.18**
KIngspan Kooltherm K8 (Thermal conductivity 0.022 W/mK) Premium performance rigid thermoset modified resin insulant manufactured with a blowing agent that has zero Ozone Depletion Potential (ODP) and low Global Warming Potential (GWP) or other equal; as partial cavity fill; including cutting and fitting around wall ties and retaining discs						
40 mm thick	–	0.15	2.68	5.11	m²	**7.79**
50 mm thick	–	0.15	2.68	6.37	m²	**9.05**
60 mm thick	–	0.15	2.68	7.55	m²	**10.23**
75 mm thick	–	0.15	2.68	9.35	m²	**12.03**
Kingspan Thermawall TW50 (Thermal conductivity 0.022 W/mK) High performance rigid thermoset polyisocyanurate (PIR) insulant or other equal and approved; as partial cavity fill; including cutting and fitting around wall ties and retaining discs						
25 mm thick	–	0.17	3.03	4.78	m²	**7.81**
50 mm thick	–	0.17	3.03	8.05	m²	**11.08**
75 mm thick	–	0.18	3.21	12.07	m²	**15.28**
100 mm thick	–	0.19	3.38	15.62	m²	**19.00**

31 INSULATION, FIRE STOPPING AND FIRE PROTECTION

Item	PC £	Labour hours	Labour £	Material £	Unit	Total rate £
Kingspan Thermafloor TF70 (Thermal conductivity 0.022 W/mK) high performance rigid urethane floor insulation for solid concrete and suspended ground floors						
50 mm thick	–	0.17	3.12	8.65	m²	**11.77**
75 mm thick	–	0.17	3.12	12.96	m²	**16.08**
100 mm thick	–	0.17	3.12	16.84	m²	**19.96**
115 mm thick	–	0.20	3.57	7.52	m²	**11.09**
125 mm thick	–	0.17	3.12	20.36	m²	**23.48**
150 mm thick	–	0.20	3.57	24.95	m²	**28.52**
Kingspan Thermafloor TF73 (Thermal conductivity 0.029 W/mK) high performance rigid extruded polystyrene insulation for floating and suspended ground floors; upper face 18 mm moisture-resistant chipboard						
40 mm thick	–	0.17	3.12	11.38	m²	**14.50**
50 mm thick	–	0.17	3.12	12.67	m²	**15.79**
60 mm thick	–	0.20	3.57	15.38	m²	**18.95**
80 mm thick	–	0.17	3.12	18.26	m²	**21.38**
Styrofoam Floormate 500 (Thermal conductivity 0.033 W/mK) extruded polystyrene foam or other equal and approved						
50 mm thick	–	0.46	8.20	4.49	m²	**12.69**
80 mm thick	–	0.46	8.20	7.25	m²	**15.45**
120 mm thick	–	0.46	8.20	10.62	m²	**18.82**
Dow Chemicals Styrofoam SP or other equal and approved; cold bridging insulation fixed with adhesive to brick, block or concrete base						
Insulation to walls						
50 mm thick	–	0.33	5.88	4.73	m²	**10.61**
75 mm thick	–	0.35	6.24	7.78	m²	**14.02**
Insulation to isolated columns						
50 mm thick	–	0.41	7.31	4.73	m²	**12.04**
75 mm thick	–	0.43	7.67	7.78	m²	**15.45**
Insulation to ceilings						
50 mm thick	–	0.36	6.42	4.73	m²	**11.15**
75 mm thick	–	0.39	6.95	7.78	m²	**14.73**
Insulation to isolated beams						
50 mm thick	–	0.43	7.67	4.73	m²	**12.40**
75 mm thick	–	0.46	8.20	7.78	m²	**15.98**
Sheepswool Mix insulation to walls						
Thermafleece TF35 high density wool insulating batts (0.035 W/mK); 60% British wool, 30% recycled polyester and 10% polyester binder with a high recycled content						
50 mm thick	–	0.35	6.09	6.18	m²	**12.27**
70 mm thick	–	0.38	6.52	8.65	m²	**15.17**

31 INSULATION, FIRE STOPPING AND FIRE PROTECTION

Item	PC £	Labour hours	Labour £	Material £	Unit	Total rate £
FIRE STOPPING						
Fire stopping						
Cape Firecheck channel; intumescent coatings on cut mitres; fixing with brass cups and screws						
19 mm × 44 mm or 19 mm × 50 mm	–	0.56	9.98	13.19	m	**23.17**
Sealmaster intumescent fire and smoke seals or other equal and approved; pinned into groove in timber						
type N30; for single leaf half hour door	–	0.28	4.99	2.57	m	**7.56**
type N60; for single leaf one hour door	–	0.31	5.52	3.36	m	**8.88**
type IMN or IMP; for meeting or pivot stiles of pair of one hour doors; per stile	–	0.31	5.52	3.36	m	**8.88**
intumescent plugs in timber; including boring	–	0.09	1.61	0.38	nr	**1.99**
Rockwool fire stops or other equal and approved; between top of brick/block wall and concrete soffit						
30 mm deep × 100 mm wide	–	0.07	1.25	3.42	m	**4.67**
30 mm deep × 150 mm wide	–	0.09	1.61	5.19	m	**6.80**
30 mm deep × 200 mm wide	–	0.11	1.96	6.93	m	**8.89**
60 mm deep × 100 mm wide	–	0.08	1.42	4.50	m	**5.92**
60 mm deep × 150 mm wide	–	0.10	1.78	6.73	m	**8.51**
60 mm deep × 200 mm wide	–	0.12	2.14	9.06	m	**11.20**
90 mm deep × 100 mm wide	–	0.10	1.78	7.20	m	**8.98**
90 mm deep × 150 mm wide	–	0.12	2.14	10.75	m	**12.89**
90 mm deep × 200 mm wide	–	0.14	2.49	14.38	m	**16.87**
Fire protection compound						
Quelfire QF4, fire protection compound or other equal and approved; filling around pipes, ducts and the like; including all necessary formwork						
300 mm × 300 mm × 250 mm; pipes – 2	–	0.93	14.44	12.43	nr	**26.87**
500 mm × 500 mm × 250 mm; pipes – 2	–	1.16	17.54	37.29	nr	**54.83**
Fire barriers						
Rockwool fire barrier or other equal and approved; between top of suspended ceiling and concrete soffit						
one 50 mm layer × 900 mm wide; half hour	–	0.56	9.98	19.70	m²	**29.68**
two 50 mm layers × 900 mm wide; one hour	–	0.83	14.80	39.21	m²	**54.01**
three 50 mm layers × 900 mm wide; two hour	–	1.10	19.61	56.21	m²	**75.82**
Corofil C144 fire barrier to edge of slab; fixed with non-flammable contact adhesive						
to suit void 30 mm wide × 100 mm deep; one hour	–	–	–	–	m	**10.77**
Lamatherm fire barrier or other equal and approved; to void below raised access floors						
75 mm thick × 300 mm high; half hour	–	0.17	3.03	6.57	m	**9.60**
75 mm thick × 600 mm high; half hour	–	0.17	3.03	14.40	m	**17.43**
90 mm thick × 300 mm high; half hour	–	0.17	3.03	9.23	m	**12.26**
90 mm thick × 600 mm high; half hour	–	0.17	3.03	19.22	m	**22.25**

32 FURNITURE, FITTINGS AND EQUIPMENT

Item	PC £	Labour hours	Labour £	Material £	Unit	Total rate £
GENERAL FIXTURES, FURNISHINGS AND EQUIPMENT						
SUPPLY ONLY PRICES						
NOTE: The fixing of general fixtures will vary considerably dependent upon the size of the fixture and the method of fixing employed. Prices for fixing like sized kitchen fittings may be suitable for certain fixtures, although adjustment to those rates will almost invariably be necessary.						
The following supply only prices are for purpose-made fittings components in various materials supplied as part of an assembled fitting and therefore may be used to arrive at a guide price for a complete fitting.						
Fitting components; medium density fibreboard						
Backs, fronts, sides or divisions; over 300 mm wide						
12 mm thick	–	–	–	18.99	m^2	**18.99**
18 mm thick	–	–	–	20.12	m^2	**20.12**
25 mm thick	–	–	–	22.41	m^2	**22.41**
Shelves or worktops; over 300 mm wide						
18 mm thick	–	–	–	20.12	m^2	**20.12**
25 mm thick	–	–	–	22.41	m^2	**22.41**
Flush doors; lipped on four edges						
450 mm × 750 mm × 18 mm	–	–	–	29.15	nr	**29.15**
450 mm × 750 mm × 25 mm	–	–	–	29.70	nr	**29.70**
600 mm × 900 mm × 18 mm	–	–	–	34.41	nr	**34.41**
600 mm × 900 mm × 25 mm	–	–	–	35.30	nr	**35.30**
Fitting components; moisture-resistant medium density fibreboard						
Backs, fronts, sides or divisions; over 300 mm wide						
12 mm thick	–	–	–	21.26	m^2	**21.26**
18 mm thick	–	–	–	23.54	m^2	**23.54**
25 mm thick	–	–	–	25.81	m^2	**25.81**
Shelves or worktops; over 300 mm wide						
18 mm thick	–	–	–	23.54	m^2	**23.54**
25 mm thick	–	–	–	25.81	m^2	**25.81**
Flush doors; lipped on four edges						
450 mm × 750 mm × 18 mm	–	–	–	29.70	nr	**29.70**
450 mm × 750 mm × 25 mm	–	–	–	30.55	nr	**30.55**
600 mm × 900 mm × 18 mm	–	–	–	35.30	nr	**35.30**
600 mm × 900 mm × 25 mm	–	–	–	36.67	nr	**36.67**

32 FURNITURE, FITTINGS AND EQUIPMENT

Item	PC £	Labour hours	Labour £	Material £	Unit	Total rate £
GENERAL FIXTURES, FURNISHINGS AND EQUIPMENT – cont						
Fitting components; medium density fibreboard; melamine faced both sides						
Backs, fronts, sides or divisions; over 300 mm wide						
12 mm thick	–	–	–	25.06	m²	**25.06**
18 mm thick	–	–	–	27.90	m²	**27.90**
Shelves or worktops; over 300 mm wide						
18 mm thick	–	–	–	27.90	m²	**27.90**
Flush doors; lipped on four edges						
450 mm × 750 mm × 18 mm	–	–	–	20.85	nr	**20.85**
600 mm × 900 mm × 25 mm	–	–	–	26.45	nr	**26.45**
Fitting components; medium density fibreboard; formica faced both sides						
Backs, fronts, sides or divisions; over 300 mm wide						
12 mm thick	–	–	–	76.30	m²	**76.30**
18 mm thick	–	–	–	79.35	m²	**79.35**
Shelves or worktops; over 300 mm wide						
18 mm thick	–	–	–	79.35	m²	**79.35**
Flush doors; lipped on four edges						
450 mm × 750 mm × 18 mm	–	–	–	42.01	nr	**42.01**
600 mm × 900 mm × 25 mm	–	–	–	42.93	nr	**42.93**
Fitting components; wrought softwood						
Backs, fronts, sides or divisions; cross-tongued joints; over 300 mm wide						
25 mm thick	–	–	–	36.47	m²	**36.47**
Shelves or worktops; cross-tongued joints; over 300 mm wide						
25 mm thick	–	–	–	36.47	m²	**36.47**
Bearers						
19 mm × 38 mm	–	–	–	1.91	m	**1.91**
25 mm × 50 mm	–	–	–	2.11	m	**2.11**
44 mm × 44 mm	–	–	–	2.26	m	**2.26**
44 mm × 75 mm	–	–	–	2.60	m	**2.60**
Bearers; framed; to backs, fronts or sides						
19 mm × 38 mm	–	–	–	4.37	m	**4.37**
25 mm × 50 mm	–	–	–	4.73	m	**4.73**
50 mm × 50 mm	–	–	–	6.10	m	**6.10**
50 mm × 75 mm	–	–	–	7.00	m	**7.00**
Add 5% to the above material prices for selected softwood staining						
Fitting components; selected Sapele						
Backs, fronts, sides or divisions; cross-tongued joints; over 300 mm wide						
25 mm thick	–	–	–	57.40	m²	**57.40**

32 FURNITURE, FITTINGS AND EQUIPMENT

Item	PC £	Labour hours	Labour £	Material £	Unit	Total rate £
Shelves or worktops; cross-tongued joints; over 300 mm wide						
25 mm thick	–	–	–	57.40	m²	**57.40**
Bearers						
19 mm × 38 mm	–	–	–	2.88	m	**2.88**
25 mm × 50 mm	–	–	–	3.59	m	**3.59**
50 mm × 50 mm	–	–	–	4.03	m	**4.03**
50 mm × 75 mm	–	–	–	5.24	m	**5.24**
Bearers; framed; to backs, fronts or sides						
19 mm × 38 mm	–	–	–	5.56	m	**5.56**
25 mm × 50 mm	–	–	–	6.17	m	**6.17**
50 mm × 50 mm	–	–	–	8.26	m	**8.26**
50 mm × 75 mm	–	–	–	10.36	m	**10.36**
Fitting components; Iroko						
Backs, fronts, sides or divisions; cross-tongued joints; over 300 mm wide						
25 mm thick	–	–	–	63.27	m²	**63.27**
Shelves or worktops; cross-tongued joints; over 300 mm wide						
25 mm thick	–	–	–	63.27	m²	**63.27**
Draining boards; cross-tongued joints; over 300 mm wide						
25 mm thick	–	–	–	79.27	m²	**79.27**
stopped flutes	–	–	–	4.10	m	**4.10**
grooves; cross-grain	–	–	–	0.60	m	**0.60**
Bearers						
19 mm × 38 mm	–	–	–	3.14	m	**3.14**
25 mm × 50 mm	–	–	–	4.02	m	**4.02**
50 mm × 50 mm	–	–	–	4.57	m	**4.57**
50 mm × 75 mm	–	–	–	6.02	m	**6.02**
Bearers; framed; to backs, fronts or sides						
19 mm × 38 mm	–	–	–	5.75	m	**5.75**
25 mm × 50 mm	–	–	–	6.41	m	**6.41**
50 mm × 50 mm	–	–	–	8.64	m	**8.64**
50 mm × 75 mm	–	–	–	11.22	m	**11.22**
Lockers and cupboards; Welconstruct Distribution or other equal and approved						
Standard clothes lockers; steel body and door within reinforced 19G frame, powder coated finish, cam locks						
1 compartment; placing in position						
300 mm × 300 mm × 1800 mm	–	0.23	2.67	50.40	nr	**53.07**
380 mm × 380 mm × 1800 mm	–	0.23	2.67	72.13	nr	**74.80**
450 mm × 450 mm × 1800 mm	–	0.28	3.24	73.64	nr	**76.88**

32 FURNITURE, FITTINGS AND EQUIPMENT

Item	PC £	Labour hours	Labour £	Material £	Unit	Total rate £
GENERAL FIXTURES, FURNISHINGS AND EQUIPMENT – cont						
Lockers and cupboards – cont						
Compartment lockers; steel body and door within reinforced 19G frame, powder coated finish, cam locks						
2 compartments; placing in position						
300 mm × 300 mm × 1800 mm	–	0.23	2.67	58.28	nr	**60.95**
380 mm × 380 mm × 1800 mm	–	0.23	2.67	70.27	nr	**72.94**
450 mm × 450 mm × 1800 mm	–	0.28	3.24	76.22	nr	**79.46**
4 compartments; placing in position						
300 mm × 300 mm × 1800 mm	–	0.23	2.67	68.53	nr	**71.20**
380 mm × 380 mm × 1800 mm	–	0.23	2.67	82.28	nr	**84.95**
450 mm × 450 mm × 1800 mm	–	0.28	3.24	82.28	nr	**85.52**
Timber clothes lockers; veneered MDF finish, routed door, cam locks						
1 compartment; placing in position						
380 mm × 380 mm × 1830 mm	–	0.28	3.24	179.51	nr	**182.75**
4 compartments; placing in position						
380 mm × 380 mm × 1830 mm	–	0.28	3.24	267.56	nr	**270.80**
Vandal-resistant lockers						
1030 high mm × 370 mm × 560 mm; one compartment	–	0.23	2.67	190.30	nr	**192.97**
1930 mm × 370 mm × 560 mm; two compartments	–	0.23	2.67	295.19	nr	**297.86**
850 mm × 740 mm × 560 mm; 2 high × 2 wide	–	0.23	2.67	387.51	nr	**390.18**
1930 mm × 740 mm × 560 mm; 5 high × 2 wide	–	0.23	2.67	877.33	nr	**880.00**
Shelving support systems; The Welconstruct Company or other equal and approved standard duty; maximum bayload of 2000 kg						
Shelving support systems; steel body; stove enamelled finish; assembling						
open initial bay; 5 shelves; placing in position						
1000 mm × 300 mm × 1850 mm	–	0.69	9.45	139.68	nr	**149.13**
1000 mm × 600 mm × 1850 mm	–	0.69	9.45	177.09	nr	**186.54**
open extension bay; 5 shelves; placing in position						
1000 mm × 300 mm × 1850 mm	–	0.83	11.37	90.70	nr	**102.07**
1000 mm × 600 mm × 1850 mm	–	0.83	11.37	124.56	nr	**135.93**
closed initial bay; 5 shelves; placing in position						
1000 mm × 300 mm × 1850 mm	–	0.69	9.45	195.82	nr	**205.27**
1000 mm × 600 mm × 1850 mm	–	0.69	9.45	253.59	nr	**263.04**
closed extension bay; 5 shelves; placing in position						
1000 mm × 300 mm × 1850 mm	–	0.83	11.37	145.49	nr	**156.86**
1000 mm × 600 mm × 1850 mm	–	0.83	11.37	189.56	nr	**200.93**
extra for pair of doors; fixing in position						
1000 mm × 1850 mm	–	0.75	10.27	275.20	nr	**285.47**
Cloakroom racks; The Welconstruct Company or other equal and approved						

32 FURNITURE, FITTINGS AND EQUIPMENT

Item	PC £	Labour hours	Labour £	Material £	Unit	Total rate £
Cloakroom racks; 40 mm × 40 mm square tube framing, polyester powder coated finish; beech slatted seats and rails to one side only; placing in position						
1675 mm × 325 mm × 1500 mm; 5 nr coat hooks	–	0.30	4.11	313.45	nr	**317.56**
1825 mm × 325 mm × 1500 mm; 15 nr coat hangers	–	0.30	4.11	359.17	nr	**363.28**
Extra for						
shoe baskets	–	–	–	70.58	nr	**70.58**
mesh bottom shelf	–	–	–	49.29	nr	**49.29**
Cloakroom racks; 40 mm × 40 mm square tube framing, polyester powder coated finish; beech slatted seats and rails to both sides; placing in position						
1675 mm × 600 mm × 1500 mm; 10 nr coat hooks	–	0.40	5.47	429.84	nr	**435.31**
1825 mm × 600 mm × 1500 mm; 30 nr coat hangers	–	0.40	5.47	448.13	nr	**453.60**
Extra for						
shoe baskets	–	–	–	88.21	nr	**88.21**
mesh bottom shelf	–	–	–	59.94	nr	**59.94**
6 mm thick rectangular glass mirrors; silver backed; fixed with chromium plated domed headed screws; to background requiring plugging						
Mirror with polished edges						
365 mm × 254 mm	8.40	0.74	10.66	8.83	nr	**19.49**
400 mm × 300 mm	10.95	0.74	10.66	11.44	nr	**22.10**
560 mm × 380 mm	18.97	0.83	11.96	19.66	nr	**31.62**
640 mm × 460 mm	24.79	0.93	13.41	25.63	nr	**39.04**
Mirror with bevelled edges						
365 mm × 254 mm	14.96	0.74	10.66	15.55	nr	**26.21**
400 mm × 300 mm	17.51	0.74	10.66	18.17	nr	**28.83**
560 mm × 380 mm	29.18	0.83	11.96	30.12	nr	**42.08**
640 mm × 460 mm	36.47	0.93	13.41	37.60	nr	**51.01**
Internal blinds						
Roller blinds; Luxaflex EOS type 10 roller; Compact Fabric; plain type material; 1219 mm drop; fixing with screws						
1016 mm wide	33.41	0.93	10.77	34.25	nr	**45.02**
2031 mm wide	49.31	1.45	16.80	50.54	nr	**67.34**
2843 mm wide	61.24	1.97	22.82	62.77	nr	**85.59**
Roller blinds; Luxaflex EOS type 10 roller; Compact Fabric; fire-resisting material; 1219 mm drop; fixing with screws						
1016 mm wide	43.75	0.93	10.77	44.84	nr	**55.61**
2031 mm wide	65.22	1.45	16.80	66.85	nr	**83.65**
2843 mm wide	82.72	1.97	22.82	84.79	nr	**107.61**

32 FURNITURE, FITTINGS AND EQUIPMENT

Item	PC £	Labour hours	Labour £	Material £	Unit	Total rate £
GENERAL FIXTURES, FURNISHINGS AND EQUIPMENT – cont						
Internal blinds – cont						
Roller blinds; Luxaflex EOS type 10 roller; Light-resistant; blackout material; 1219 mm drop; fixing with screws						
1016 mm wide	56.48	0.93	10.77	57.89	nr	**68.66**
2031 mm wide	94.65	1.45	16.80	97.02	nr	**113.82**
2843 mm wide	128.06	1.97	22.82	131.26	nr	**154.08**
Roller blinds; Luxaflex Lite-master Crank Op; 100% blackout; 1219 mm drop; fixing with screws						
1016 mm wide	157.49	1.96	22.70	161.43	nr	**184.13**
2031 mm wide	210.77	2.75	31.85	216.04	nr	**247.89**
2843 mm wide	272.02	3.53	40.89	278.82	nr	**319.71**
Vertical louvre blinds; 89 mm wide louvres; Luxaflex EOS type; Florida Fabric; 1219 mm drop; fixing with screws						
1016 mm wide	45.34	0.82	9.50	46.47	nr	**55.97**
2031 mm wide	69.19	1.30	15.06	70.92	nr	**85.98**
3046 mm wide	94.65	1.77	20.50	97.02	nr	**117.52**
Vertical louvre blinds; 127 mm wide louvres; Luxaflex EOS type; Florida Fabric; 1219 mm drop; fixing with screws						
1016 mm wide	38.17	0.88	10.19	39.12	nr	**49.31**
2031 mm wide	58.06	1.35	15.64	59.51	nr	**75.15**
3046 mm wide	77.95	1.81	20.96	79.90	nr	**100.86**
Venetian blinds; 80 mm wide louvres; Levolux 480 type; solid slat; stock colour; manual hand crank; 2700 mm drop; standard brackets fixed to suitable grounds						
875 mm wide	–	–	–	–	nr	**144.00**
1150 mm wide	–	–	–	–	nr	**166.50**
1500 mm wide	–	–	–	–	nr	**198.00**
2500 mm wide	–	–	–	–	nr	**261.00**
3000 mm wide	–	–	–	–	nr	**283.50**
3500 mm wide	–	–	–	–	nr	**328.50**
4000 mm wide	–	–	–	–	nr	**369.00**
Door entrance mats						
Entrance mats; aluminium wiper bar; laying in position; 18 mm thick						
Gradus Topguard; 900 mm × 550 mm	–	0.46	5.33	130.02	nr	**135.35**
Gradus Topguard; 1200 mm × 750 mm	–	0.46	5.33	234.03	nr	**239.36**
Gradus Topguard; 2400 mm × 1200 mm	–	0.93	10.77	748.90	nr	**759.67**
Nuway Tuftiguard; 500 mm × 1300 mm	–	–	–	–	nr	**221.81**
Nuway Tuftiguard; 1000 mm × 1600 mm	–	–	–	–	nr	**465.46**
Nuway Tuftiguard; 1800 mm × 3000 mm	–	–	–	–	nr	**1842.75**
extra for brass wipers	–	–	–	181.22	m²	**181.22**

32 FURNITURE, FITTINGS AND EQUIPMENT

Item	PC £	Labour hours	Labour £	Material £	Unit	Total rate £
Nuway Tuftiguard; single sided aluminium scraper bar	–	–	–	–	m²	**341.25**
Nuway Tuftiguard; double sided aluminium rigid scraper bars	–	–	–	–	m²	**487.50**
Coral Classic textile secondary and circulation matting system for moisture removal; 2.0 m wide roll	–	–	–	–	m²	**58.50**
Coral Brush Activ secondary and circulation matting system for soil and moisture removal; 2.0 m wide roll	–	–	–	–	m²	**58.50**
Coral Duo secondary matting system ribbed; 2.0 m wide roll	–	–	–	–	m²	**58.50**
Matwells						
Polished stainless steel matwell; angle rim with lugs; bedding in screed						
Stainless steel frame bed in screed 500 mm × 1300 mm	–	–	–	–	nr	**44.36**
Stainless steel frame bed in screed 1000 mm × 1600 mm	–	–	–	–	nr	**88.72**
Stainless steel frame bed in screed 1100 mm × 1400 mm	–	–	–	–	nr	**85.31**
Stainless steel frame bed in screed 1800 mm × 3000 mm	–	–	–	–	nr	**163.80**
Polished aluminium matwell; angle rim with lugs brazed on; bedding in screed						
900 mm × 550 mm; constructed with 25 × 25 × 3 mm angle	25.91	0.93	10.77	26.56	nr	**37.33**
1200 mm × 750 mm; constructed with 34 × 26 × 6 mm angle	43.36	1.00	11.58	44.44	nr	**56.02**
2400 mm × 1200 mm; constructed with 50 × 50 × 6 mm angle	166.10	1.50	17.37	170.25	nr	**187.62**
Polished brass matwell; comprising angle rim with lugs brazed on; bedding in screed						
900 mm × 550 mm; constructed with 25 × 25 × 5 mm angle	57.93	0.93	10.77	59.38	nr	**70.15**
1200 mm × 750 mm; constructed with 38 × 38 × 6 mm angle	177.82	1.00	11.58	182.27	nr	**193.85**
2400 mm x1200 mm; constructed with 38 × 38 × 6 mm angle	328.28	1.50	17.37	336.49	nr	**353.86**

32 FURNITURE, FITTINGS AND EQUIPMENT

Item	PC £	Labour hours	Labour £	Material £	Unit	Total rate £
KITCHEN FITTINGS						
SUPPLY AND FIX PRICES						
NOTE: Kitchen fittings vary considerably. PC supply prices for reasonable quantities for a moderately priced range of kitchen fittings have been shown.						
Kitchen units						
Supplying and fixing to backgrounds requiring plugging; including any pre-assembly						
Wall units						
300 mm × 300 mm × 720 mm	60.00	1.11	16.00	61.60	nr	**77.60**
500 mm × 300 mm × 720 mm	72.00	1.16	16.72	73.90	nr	**90.62**
600 mm × 300 mm × 720 mm	78.00	1.30	18.74	80.05	nr	**98.79**
800 mm × 300 mm × 720 mm	120.00	1.48	21.33	123.10	nr	**144.43**
Floor units with drawers						
500 mm × 600 mm × 870 mm	105.00	1.16	16.72	107.73	nr	**124.45**
600 mm × 600 mm × 870 mm	120.00	1.30	18.74	123.10	nr	**141.84**
1000 mm × 600 mm × 870 mm	180.00	1.57	22.62	184.60	nr	**207.22**
Sink units (excluding sink top)						
1000 mm × 600 mm × 870 mm	165.00	1.48	21.33	169.23	nr	**190.56**
Kitchen worktop						
Laminated plastics worktops; single rolled edge; prices include for fixing						
38 mm thick; 600 mm wide	30.45	0.37	5.33	31.25	m	**36.58**
extra for forming hole for inset sink	–	0.69	9.94	–	nr	**9.94**
extra for jointing strip at corner intersection of worktops	–	0.14	2.02	6.15	nr	**8.17**
extra for butt and scribe joint at corner intersection of worktops	–	4.16	59.95	–	nr	**59.95**

32 FURNITURE, FITTINGS AND EQUIPMENT

Item	PC £	Labour hours	Labour £	Material £	Unit	Total rate £
SANITARY APPLIANCES AND FITTINGS						
Sinks; Armitage Shanks or equal						
Sinks; white glazed fireclay; BS 6465; pointing all round with silicone sealant						
Belfast sink; 46 cm × 38 cm × 21 cm; pair of Nuastyle 21 basin taps with dual indices, chrome handle; wall mounts 38 mm slotted waste, chain and plug, screw stay; pair of 40.5 cm aluminium alloy build-in brackets with 35.5 cm studs	160.19	2.78	42.46	209.33	nr	**251.79**
Belfast sink; 61 cm × 38 cm × 21 cm; pair of Nuastyle 21 basin taps with dual indices, chrome handle; wall mounts; 38 mm slotted waste, chain and plug, screw stay; pair of 40.5 cm aluminium alloy build-in brackets with 35.5 cm studs	190.65	2.78	42.46	240.71	nr	**283.17**
Belfast sink; 76 cm × 38 cm × 21 cm; pair of Nuastyle 21 basin taps with dual indices, chrome handle; wall mounts; 38 mm slotted waste, chain and plug, screw stay; pair of 40.5 cm aluminium alloy build-in brackets with 35.5 cm studs	267.77	2.78	42.46	319.90	nr	**362.36**
Lavatory basins; Armitage Shanks or equal						
Basins; white vitreous china; BS 6465 Part 3; pointing all round with silcone sealant						
Portman 21 40 cm basin; with overflow, chain hole and two tapholes ; pair of Nuastyle 21 basin taps with dual indices; slotted basin waste with plastic plug, chain waste and plug; 32 × 75 mm seal plastic standard bottle trap; pair of Portman concealed brackets with waste support; Isovalve 15 mm plastic servicing valve with outlet for copper	75.22	2.13	32.53	106.36	nr	**138.89**
Portman 21 50 cm basin; with overflow, chain hole and two tapholes ; pair of Nuastyle 21 basin taps with dual indices; slotted basin waste with plastic plug, chain waste and plug; 32 × 75 mm seal plastic standard bottle trap; pair of Portman concealed brackets with waste support; Isovalve 15 mm plastic servicing valve with outlet for copper	91.82	2.13	32.53	123.48	nr	**156.01**
Portman 21 60 cm basin; with overflow, chain hole and two tapholes ; pair of Nuastyle 21 basin taps with dual indices; slotted basin waste with plastic plug, chain waste and plug; 32 × 75 mm seal plastic standard bottle trap; pair of Portman concealed brackets with waste support; Isovalve 15 mm plastic servicing valve with outlet for copper	120.79	2.13	32.53	153.27	nr	**185.80**

32 FURNITURE, FITTINGS AND EQUIPMENT

Item	PC £	Labour hours	Labour £	Material £	Unit	Total rate £
SANITARY APPLIANCES AND FITTINGS – cont						
Lavatory basins – cont						
Basins – cont						
Tiffany 50 cm pedestal basin; with two tapholes; Millenia STD dual control one taphole standard basin mixer with pop-up waste; pair of Millenia STD handles; Full pedestal; Isovalve 15 mm plastic servicing valve with outlet for copper	138.34	2.31	35.28	155.78	nr	**191.06**
Tiffany or similar 55 cm pedestal basin; with two tapholes; Millenia STD dual control one taphole standard basin mixer with pop-up waste; pair of Millenia STD handles; Full pedestal; Isovalve 15 mm plastic servicing valve with outlet for copper	135.44	2.31	35.28	152.81	nr	**188.09**
Tiffany 60 cm pedestal basin; with two tapholes; Millenia STD dual control one taphole standard basin mixer with pop-up waste; pair of Millenia STD handles; Full pedestal; Isovalve 15 mm plastic servicing valve with outlet for copper	141.26	2.31	35.28	160.55	nr	**195.83**
Montana 51 cm pedestal basin; with one taphole; Millenia STD dual control one taphole standard basin mixer with pop-up waste; pair of Millenia STD handles; Full pedestal; Isovalve 15 mm plastic servicing valve with outlet for copper	130.90	2.31	35.28	148.11	nr	**183.39**
Montana 58 cm pedestal basin; with one taphole; Millenia STD dual control one taphole standard basin mixer with pop-up waste; pair of Millenia STD handles; Full pedestal; Isovalve 15 mm plastic servicing valve with outlet for copper	133.95	2.31	35.28	151.31	nr	**186.59**
Drinking fountains; Armitage Shanks or equal						
White vitreous china fountains; pointing all round with silicone sealant						
Aqualon wall mounted drinking fountain; Aqualon self-closing valve with fittings and plastic waste; 32 × 75 mm seal plastic standard bottle trap	225.26	2.31	35.28	237.64	nr	**272.92**
Polished stainless steel fountains; pointing all round with silicone selant						
Purita wall mounted drinking fountain with self-closing valve and fittings; 32 mm unslotted basin strainer waste	178.37	2.31	35.28	183.17	nr	**218.45**
Purita pedestal mounted drinking fountain 90 cm high with self-closing valve and fittings; 32 mm unslotted basin strainer waste	459.37	2.78	42.46	471.96	nr	**514.42**

32 FURNITURE, FITTINGS AND EQUIPMENT

Item	PC £	Labour hours	Labour £	Material £	Unit	Total rate £
Baths; Armitage Shanks or equal						
Pointing all round with silicone sealant						
Sandringham acrylic rectangular bath with chrome plated grips and two tapholes; standard pair of standard bath taps with chrome handles; bath chain waste with plastic plug and overflow; cast brass P trap with plain outlet and overflow connection; 170 cm long × 70 cm wide; white or coloured	111.55	3.50	53.45	114.34	nr	**167.79**
Nisa lowline heavy gauge steel rectangular bath with chrome plated grips and two tapholes; standard pair of standard bath taps with chrome handles; bath chain waste with plastic plug and overflow; cast brass P trap with plain outlet and overflow connection; 170 cm long × 70 cm wide; white or coloured	283.21	3.50	53.45	290.29	nr	**343.74**
Water closets; Armitage Shanks or equal						
White vitreous china pans and cisterns; pointing all round base with silicone sealant						
Wentworth close coupled washdown closet pan with horizontal outlet; Orion 3 plastic toilet seat and cover; Panketa pan connector 14° finned; Universal close coupled bottom inlet cistern with syphon	128.44	3.05	46.59	137.20	nr	**183.79**
Tiffany back to wall washdown closet pan with horizontal outlet; Saturn plastic toilet seat and cover; Panketa pan connector 14° finned; Conceala 2 6 litre low level side inlet cistern with syphon and lever	169.00	3.05	46.59	178.78	nr	**225.37**
Extra for; Panketa pan connector 90° finned	–	–	–	1.30	nr	**1.30**
Tiffany close coupled washdown closet pan with horizontal outlet; Saturn plastic toilet seat and cover; Panketa pan connector 14° finned; Tiffany 7½ litre close coupled cistern with dual flush valve	174.60	3.05	46.59	184.52	nr	**231.11**
Extra for; Panketa pan connector 90° finned	–	–	–	1.30	nr	**1.30**
Cameo close coupled washdown closet pan with horizontal outlet; Accolade/Cameo plastic toilet seat and cover; Panketa pan connector 14° finned; Cameo 6 litre close coupled cistern with dual flush valve	217.74	3.05	46.59	228.74	nr	**275.33**
Extra for; Panketa pan connector 90° finned	–	–	–	1.30	nr	**1.30**

32 FURNITURE, FITTINGS AND EQUIPMENT

Item	PC £	Labour hours	Labour £	Material £	Unit	Total rate £
SANITARY APPLIANCES AND FITTINGS – cont						
Wall urinals; Armitage Shanks or equal						
White vitreous china bowls and cisterns; pointing all round with silicone sealant						
Single Sanura 40 cm urinal bowl; Sanura top inlet spreader; pair of wall hangers for urinal bowl; 38 mm plastic domed waste; 38 × 75 mm seal plastic standard bottle trap; Conceala 4½ litres capacity auto cistern and cover; Sanura concealed flushpipe for single urinal bowl; screwing	151.15	3.70	56.51	171.97	nr	**228.48**
Single Sanura 40 cm urinal bowl; Sanura top inlet spreader; pair of wall hangers for urinal bowl; 38 mm plastic domed waste; 38 × 75 mm seal plastic standard bottle trap; Mura 4½ litres capacity auto cistern and cover; Sanura/Mura exposed flushpipe for single urinal bowl	172.19	3.70	56.51	193.55	nr	**250.06**
Single Sanura 50 cm urinal bowl; Sanura top inlet spreader; pair of wall hangers for urinal bowl; 38 mm plastic domed waste; 38 × 75 mm seal plastic standard bottle trap; Conceala 4½ litres capacity auto cistern and cover; Sanura concealed flushpipe for single urinal bowl	206.40	3.70	56.51	228.69	nr	**285.20**
Single Sanura 50 cm urinal bowl; Sanura top inlet spreader; pair of wall hangers for urinal bowl; 38 mm plastic domed waste; 38 × 75 mm seal plastic standard bottle trap; Mura 4½ litres capacity auto cistern and cover; Sanura/Mura exposed flushpipe for single urinal bowl	227.44	3.70	56.51	250.25	nr	**306.76**
Range of 2 nr Sanura 40 cm urinal bowls; Sanura top inlet spreader; pairs of wall hangers for urinal bowls; 38 mm plastic domed wastes; 38 × 75 mm seal plastic standard bottle traps; Conceala 9 litres capacity auto cistern and cover; Sanura concealed flushpipe for range of 2 nr urinal bowls	254.47	6.95	106.15	294.93	nr	**401.08**
Range of 2 nr Sanura 50 cm urinal bowls; Sanura top inlet spreader; pairs of wall hangers for urinal bowls; 38 mm plastic domed wastes; 38 × 75 mm seal plastic standard bottle traps; Conceala 9 litres capacity auto cistern and cover; Sanura concealed flushpipe for range of 2 nr urinal bowls	364.97	6.95	106.15	408.35	nr	**514.50**
Range of 3 nr Sanura 40 cm urinal bowls; Sanura top inlet spreader; pairs of wall hangers for urinal bowls; 38 mm plastic domed wastes; 38 × 75 mm seal plastic standard bottle traps; Conceala 9 litres capacity auto cistern and cover; Sanura concealed flushpipe for range of 3 nr urinal bowls	354.27	10.15	155.02	414.27	nr	**569.29**
Range of 3 nr Sanura 50 cm urinal bowls; Sanura top inlet spreader; pairs of wall hangers for urinal bowls; 38 mm plastic domed wastes; 38 × 75 mm seal plastic standard bottle traps; Conceala 9 litres capacity auto cistern and cover; Sanura concealed flushpipe for range of 3 nr urinal bowls	520.03	10.15	155.02	584.18	nr	**739.20**

32 FURNITURE, FITTINGS AND EQUIPMENT

Item	PC £	Labour hours	Labour £	Material £	Unit	Total rate £
Range of 4 nr Sanura 40 cm urinal bowls; Sanura top inlet spreader; pairs of wall hangers for urinal bowls; 38 mm plastic domed wastes; 38 × 75 mm seal plastic standard bottle traps; Conceala 9 litres capacity auto cistern and cover; Sanura concealed flushpipe for range of 4 nr urinal bowls; screwing	457.73	13.40	204.65	537.37	nr	**742.02**
Range of 4 nr Sanura 50 cm urinal bowls; Sanura top inlet spreader; pairs of wall hangers for urinal bowls; 38 mm plastic domed wastes; 38 × 75 mm seal plastic standard bottle traps; Conceala 9 litres capacity auto cistern and cover; Sanura concealed flushpipe for range of 4 nr urinal bowls	678.74	13.40	204.65	764.20	nr	**968.85**
Range of 5 nr Sanura 40 cm urinal bowls; Sanura top inlet spreader; pairs of wall hangers for urinal bowls; 38 mm plastic domed wastes; 38 × 75 mm seal plastic standard bottle traps; Conceala 9 litres capacity auto cistern and cover; Sanura concealed flushpipe for range of 5 nr urinal bowls	557.57	16.65	254.29	656.75	nr	**911.04**
Range of 5 nr Sanura 50 cm urinal bowls; Sanura top inlet spreader; pairs of wall hangers for urinal bowls; 38 mm plastic domed wastes; 38 × 75 mm seal plastic standard bottle traps; Conceala 9 litres capacity auto cistern and cover; Sanura concealed flushpipe for range of 5 nr urinal bowls	833.83	16.65	254.29	940.29	nr	**1194.58**
White vitreous china division panels; pointing all round with silicone sealant						
Urinal division with screw and hanger	47.37	0.70	10.69	49.37	nr	**60.06**
Bidets; Armitage Shanks or equal						
Tiffany back to wall bidet with one taphole; vitreous china; chromium plated pop-up waste and mixer tap with hand wheels; 58 cm × 39 cm; white or coloured	251.44	3.50	53.45	257.73	nr	**311.18**
Shower tray and fittings						
Simplicity shower tray; acrylic; with outlet and grated waste; chain and plug; bedding and pointing in waterproof cement mortar						
760 mm × 760 mm; white or coloured	39.90	3.00	45.82	40.90	nr	**86.72**
Shower fitting; riser pipe with mixing valve and shower rose; chromium plated; plugging and screwing mixing valve and pipe bracket						
15 mm dia. riser pipe; 127 mm dia. shower rose	214.70	5.00	76.36	220.07	nr	**296.43**
Corner fitting shower enclosure						
Bliss flat top hinged door with front panel and clear glass side panel	422.72	3.00	34.75	433.29	nr	**468.04**

32 FURNITURE, FITTINGS AND EQUIPMENT

Item	PC £	Labour hours	Labour £	Material £	Unit	Total rate £
SANITARY APPLIANCES AND FITTINGS – cont						
Miscellaneous fittings; Magrini Ltd or equal						
Vertical nappy changing unit						
ref KBCS; screwing	–	0.60	8.65	173.48	nr	**182.13**
Horizontal nappy changing unit						
ref KBHS; screwing	–	0.60	8.65	173.48	nr	**182.13**
Stay Safe baby seat						
ref KBPS; screwing	–	0.55	7.92	59.77	nr	**67.69**
Miscellaneous fittings; Pressalit Ltd or equal and approved						
Grab rails						
300 mm long ref RT100000; screw fix to wall	–	0.50	7.21	44.17	nr	**51.38**
450 mm long ref RT101000; screw fix to wall	–	0.50	7.21	51.09	nr	**58.30**
600 mm long ref RT102000; screw fix to wall	–	0.50	7.21	58.60	nr	**65.81**
800 mm long ref RT103000; screw fix to wall	–	0.50	7.21	65.89	nr	**73.10**
1000 mm long ref RT104000; screw fix to wall	–	0.50	7.21	75.96	nr	**83.17**
Angled grab rails						
900 mm long, angled 135° ref RT110000; screw fix to wall	–	0.50	7.21	95.47	nr	**102.68**
1300 mm long, angled 90° ref RT119000; screw fix to wall	–	0.75	10.80	149.54	nr	**160.34**
Hinged grab rails						
600 mm long ref R3016000; screw fix to wall	–	0.35	5.04	155.06	nr	**160.10**
600 mm long with spring counter balance ref RF016000; screw fix to wall	–	0.35	5.04	216.39	nr	**221.43**
850 mm long ref R3010000; screw fix to wall	–	0.35	5.04	188.29	nr	**193.33**
850 mm long with spring counter balance ref RF010000; screw fix to wall	–	0.35	5.04	232.25	nr	**237.29**
Shower seat; wall mounted; with padded seat and back	–	1.50	21.62	176.84	nr	**198.46**
NOTICES AND SIGNS						
Plain script; in gloss oil paint; on painted or varnished surfaces						
Capital letters; lower case letters or numerals						
per coat; per 25 mm high	–	0.09	1.30	–	nr	**1.30**

33 DRAINAGE ABOVE GROUND

Item	PC £	Labour hours	Labour £	Material £	Unit	Total rate £
RAINWATER INSTALLATIONS						
Aluminium						
Pipes and fittings; BS EN 612; ears cast on;						
polyester powder coated finish						
63 mm dia. pipes; plugged and screwed	11.70	0.34	4.38	12.82	m	17.20
extra for						
fittings with one end	–	0.20	2.57	6.63	nr	9.20
fittings with two ends	–	0.39	5.02	6.71	nr	11.73
fittings with three ends	–	0.56	7.22	10.43	nr	17.65
shoe	6.99	0.20	2.57	7.59	nr	10.16
bend	7.45	0.39	5.02	7.68	nr	12.70
single branch	9.72	0.56	7.22	10.43	nr	17.65
offset 228 projection	17.20	0.39	6.95	18.05	nr	25.00
offset 304 projection	19.17	0.39	5.02	20.08	nr	25.10
access pipe	21.34	0.39	5.02	21.89	nr	26.91
connection to clay pipes; cement and sand						
(1:2) joint	–	0.14	1.80	0.13	nr	1.93
76.50 mm dia. pipes; plugged and screwed	13.63	0.37	4.77	14.80	m	19.57
extra for						
shoe	9.60	0.23	2.96	10.26	nr	13.22
bend	9.42	0.42	5.41	9.71	nr	15.12
single branch	11.70	0.60	7.73	12.47	nr	20.20
offset 228 projection	19.02	0.42	5.41	19.92	nr	25.33
offset 304 projection	21.03	0.42	5.41	21.99	nr	27.40
access pipe	29.22	0.42	5.41	29.97	nr	35.38
connection to clay pipes; cement and sand						
(1:2) joint	–	0.16	2.06	0.13	nr	2.19
100 mm dia. pipes; plugged and screwed	23.26	0.42	5.41	24.68	m	30.09
extra for						
shoe	11.56	0.26	3.35	12.29	nr	15.64
bend	13.12	0.46	5.92	13.51	nr	19.43
single branch	15.68	0.69	8.89	16.56	nr	25.45
offset 228 projection	22.00	0.46	5.92	22.58	nr	28.50
offset 304 projection	24.42	0.46	5.92	25.06	nr	30.98
access pipe	27.22	0.46	5.92	27.93	nr	33.85
connection to clay pipes; cement and sand						
(1:2) joint	–	0.19	2.45	0.13	nr	2.58
Roof outlets; circular aluminium; with flat or domed						
grating; joint to pipe						
50 mm dia.	60.30	0.56	11.32	61.81	nr	73.13
75 mm dia.	79.03	0.60	12.13	81.01	nr	93.14
100 mm dia.	102.92	0.65	13.13	105.49	nr	118.62
150 mm dia.	131.85	0.69	13.94	135.15	nr	149.09
Roof outlets; d-shaped; balcony; with flat or domed						
grating; joint to pipe						
50 mm dia.	60.30	0.56	11.32	61.81	nr	73.13
75 mm dia.	80.02	0.60	12.13	82.02	nr	94.15
100 mm dia.	112.25	0.65	13.13	115.06	nr	128.19

33 DRAINAGE ABOVE GROUND

Item	PC £	Labour hours	Labour £	Material £	Unit	Total rate £
RAINWATER INSTALLATIONS – cont						
Aluminium – cont						
PVC balloon grating						
110 mm dia.	2.65	0.06	1.21	2.72	nr	**3.93**
63 mm dia.	1.66	0.06	1.21	1.70	nr	**2.91**
Aluminium gutters and fittings; BS EN 612; polyester powder coated finish						
100 mm half round gutters; on brackets; screwed to timber	14.43	0.32	5.71	18.48	m	**24.19**
extra for						
stop end	3.82	0.15	2.68	7.39	nr	**10.07**
running outlet	8.46	0.31	5.52	9.66	nr	**15.18**
stop end outlet	7.53	0.15	2.68	11.19	nr	**13.87**
angle	0.78	0.31	5.52	1.79	nr	**7.31**
113 mm half round gutters; on brackets; screwed to timber	15.12	0.32	5.71	19.21	m	**24.92**
extra for						
stop end	4.01	0.15	2.68	7.63	nr	**10.31**
running outlet	9.22	0.31	5.52	10.44	nr	**15.96**
stop end outlet	8.63	0.15	2.68	12.32	nr	**15.00**
angle	8.81	0.31	5.52	7.39	nr	**12.91**
125 mm half round gutters; on brackets; screwed to timber	16.97	0.37	6.60	23.07	m	**29.67**
extra for						
stop end	4.89	0.17	3.03	10.30	nr	**13.33**
running outlet	9.98	0.32	5.71	11.21	nr	**16.92**
stop end outlet	9.16	0.17	3.03	14.68	nr	**17.71**
angle	9.77	0.32	5.71	11.01	nr	**16.72**
100 mm ogee gutters; on brackets; screwed to timber	18.03	0.34	6.06	23.76	m	**29.82**
extra for						
stop end	4.03	0.16	2.85	4.88	nr	**7.73**
running outlet	9.94	0.32	5.71	10.53	nr	**16.24**
stop end outlet	7.70	0.16	2.85	12.77	nr	**15.62**
angle	8.38	0.32	5.71	5.80	nr	**11.51**
112 mm ogee gutters; on brackets; screwed to timber	20.05	0.39	6.95	26.16	m	**33.11**
extra for						
stop end	4.31	0.16	2.85	5.17	nr	**8.02**
running outlet	10.05	0.32	5.71	10.64	nr	**16.35**
stop end outlet	8.63	0.16	2.85	14.00	nr	**16.85**
angle	9.98	0.32	5.71	10.59	nr	**16.30**
125 mm ogee gutters; on brackets; screwed to timber	22.14	0.39	6.95	28.76	m	**35.71**
extra for						
stop end	4.72	0.18	3.21	5.59	nr	**8.80**
running outlet	10.99	0.34	6.06	11.60	nr	**17.66**
stop end outlet	9.79	0.18	3.21	15.62	nr	**18.83**
angle	11.64	0.34	6.06	8.44	nr	**14.50**

33 DRAINAGE ABOVE GROUND

Item	PC £	Labour hours	Labour £	Material £	Unit	Total rate £
Cast iron						
Pipes and fittings; BS 416; ears cast on; joints						
65 mm pipes; primed; nailed to masonry	27.21	0.48	6.18	28.33	m	**34.51**
extra for						
shoe	23.31	0.30	3.86	23.01	nr	**26.87**
bend	14.27	0.53	6.83	13.75	nr	**20.58**
single branch	28.05	0.67	8.63	27.95	nr	**36.58**
offset 225 mm projection	25.43	0.53	6.83	23.51	nr	**30.34**
offset 305 mm projection	29.78	0.53	6.83	27.42	nr	**34.25**
connection to clay pipes; cement and sand (1:2) joint	–	0.14	1.80	0.15	nr	**1.95**
75 mm pipes; primed; nailed to masonry	27.21	0.51	6.57	28.58	m	**35.15**
extra for						
shoe	23.31	0.32	4.12	23.44	nr	**27.56**
bend	17.32	1.11	17.27	17.30	nr	**34.57**
single branch	30.92	0.69	8.89	31.75	nr	**40.64**
offset 225 mm projection	25.43	0.56	7.22	23.94	nr	**31.16**
offset 305 mm projection	31.25	0.56	7.22	29.36	nr	**36.58**
connection to clay pipes; cement and sand (1:2) joint	–	0.16	2.06	0.15	nr	**2.21**
100 mm pipes; primed; nailed to masonry	36.53	0.56	7.22	38.42	m	**45.64**
extra for						
shoe	30.94	0.37	4.77	31.30	nr	**36.07**
bend	24.47	0.60	7.73	24.66	nr	**32.39**
single branch	36.03	0.74	9.53	37.45	nr	**46.98**
offset 225 mm projection	49.89	0.60	7.73	48.47	nr	**56.20**
offset 305 mm projection	50.88	0.60	7.73	48.74	nr	**56.47**
connection to clay pipes; cement and sand (1:2) joint	–	0.19	2.45	0.13	nr	**2.58**
100 mm × 75 mm rectangular pipes; primed; nailing to masonry	73.48	0.56	7.22	76.28	m	**83.50**
extra for						
shoe	87.31	0.37	4.77	86.05	nr	**90.82**
bend	83.13	0.60	7.73	81.76	nr	**89.49**
offset 225 mm projection	117.06	0.37	4.77	112.02	nr	**116.79**
offset 305 mm projection	125.12	0.37	4.77	118.78	nr	**123.55**
connection to clay pipes; cement and sand (1:2) joint	–	0.19	2.45	0.13	nr	**2.58**
Rainwater head; rectangular; for pipes						
65 mm dia.	71.64	0.53	6.83	74.77	nr	**81.60**
75 mm dia.	71.64	0.56	7.22	75.20	nr	**82.42**
100 mm dia.	98.91	0.60	7.73	103.97	nr	**111.70**
Rainwater head; octagonal; for pipes						
65 mm dia.	51.52	0.53	6.83	54.16	nr	**60.99**
75 mm dia.	51.52	0.56	7.22	54.59	nr	**61.81**
100 mm dia.	61.07	0.60	7.73	65.18	nr	**72.91**

33 DRAINAGE ABOVE GROUND

Item	PC £	Labour hours	Labour £	Material £	Unit	Total rate £
RAINWATER INSTALLATIONS – cont						
Cast iron – cont						
Gutters and fittings; BS EN 877						
100 mm half round gutters; primed; on brackets;						
screwed to timber	15.42	0.37	6.60	20.33	m	**26.93**
extra for						
stop end	3.69	0.16	2.85	6.11	nr	**8.96**
running outlet	10.74	0.32	5.71	10.37	nr	**16.08**
angle	11.02	0.32	5.71	12.61	nr	**18.32**
115 mm half round gutters; primed; on brackets;						
screwed to timber	16.07	0.37	6.60	21.09	m	**27.69**
extra for						
stop end	4.79	0.16	2.85	7.25	nr	**10.10**
running outlet	11.70	0.32	5.71	11.37	nr	**17.08**
angle	11.34	0.32	5.71	12.87	nr	**18.58**
125 mm half round gutters; primed; on brackets;						
screwed to timber	18.81	0.42	7.48	23.96	m	**31.44**
extra for						
stop end	4.79	0.19	3.38	7.37	nr	**10.75**
running outlet	13.36	0.37	6.60	12.90	nr	**19.50**
angle	13.36	0.37	6.60	14.53	nr	**21.13**
150 mm half round gutters; primed; on brackets;						
screwed to timber	30.61	0.46	8.20	37.14	m	**45.34**
extra for						
stop end	6.66	0.20	3.57	11.89	nr	**15.46**
running outlet	23.14	0.42	7.48	22.03	nr	**29.51**
angle	24.42	0.42	7.48	24.77	nr	**32.25**
100 mm ogee gutters; primed; on brackets;						
screwed to timber	17.19	0.39	6.95	22.20	m	**29.15**
extra for						
stop end	3.78	0.17	3.03	8.27	nr	**11.30**
running outlet	11.71	0.34	6.06	11.35	nr	**17.41**
angle	11.50	0.34	6.06	13.18	nr	**19.24**
115 mm ogee gutters; primed; on brackets;						
screwed to timber	18.92	0.39	6.95	24.02	m	**30.97**
extra for						
stop end	4.90	0.17	3.03	9.48	nr	**12.51**
running outlet	12.46	0.34	6.06	11.97	nr	**18.03**
angle	12.46	0.34	6.06	13.88	nr	**19.94**
125 mm ogee gutters; primed; on brackets;						
screwed to timber	19.84	0.43	7.67	25.49	m	**33.16**
extra for						
stop end	4.90	0.19	3.38	10.02	nr	**13.40**
running outlet	13.60	0.39	6.95	13.19	nr	**20.14**
angle	13.60	0.39	6.95	15.49	nr	**22.44**

33 DRAINAGE ABOVE GROUND

Item	PC £	Labour hours	Labour £	Material £	Unit	Total rate £
Steel						
3 mm thick zincalume coated pressed steel gutters and fittings; joggle joints; including bracket and stiffeners						
200 mm × 100 mm (400 mm girth) box gutter; screwed to timber	–	0.60	7.73	16.67	m	**24.40**
extra for						
stop end	–	0.32	4.12	4.42	nr	**8.54**
running outlet	–	0.65	8.37	25.93	nr	**34.30**
stop end outlet	–	0.32	4.12	33.27	nr	**37.39**
angle	–	0.65	8.37	28.31	nr	**36.68**
381 mm boundary wall gutters (900 mm girth); bent twice; screwed to timber	–	0.60	7.73	18.44	m	**26.17**
extra for						
stop end	–	0.37	4.77	4.89	nr	**9.66**
running outlet	–	0.65	8.37	34.02	nr	**42.39**
stop end outlet	–	0.32	4.12	60.85	nr	**64.97**
angle	–	0.65	8.37	39.28	nr	**47.65**
457 mm boundary wall gutters (1200 mm girth); bent twice; screwed to timber	–	0.69	8.89	70.26	m	**79.15**
extra for						
stop end	–	0.37	4.77	23.70	nr	**28.47**
running outlet	–	0.74	9.53	48.59	nr	**58.12**
stop end outlet	–	0.37	4.77	49.63	nr	**54.40**
angle	–	0.74	9.53	52.08	nr	**61.61**
750 mm girth; valley gutters; Kalzip Membrane lined composite gutter system	–	–	–	–	m	**88.44**
extra for						
stop end	–	–	–	–	nr	**66.34**
running outlet	–	–	–	–	nr	**84.02**
uPVC						
External rainwater pipes and fittings; BS EN 12200; slip-in joints						
50 mm pipes; fixing with pipe or socket brackets; plugged and screwed	9.32	0.28	3.61	12.81	m	**16.42**
extra for						
shoe	5.47	0.19	2.45	6.88	nr	**9.33**
bend	6.41	0.28	3.61	7.84	nr	**11.45**
two bends to form offset 229 mm projection	12.82	0.28	3.61	12.79	nr	**16.40**
connection to clay pipes; cement and sand (1:2) joint	–	0.12	1.55	0.15	nr	**1.70**
68 mm pipes; fixing with pipe or socket brackets; plugged and screwed	7.21	0.31	4.00	11.29	m	**15.29**
extra for						
shoe	5.47	0.20	2.57	7.47	nr	**10.04**
bend	7.02	0.31	4.00	9.06	nr	**13.06**
single branch	14.11	0.41	5.28	16.33	nr	**21.61**

33 DRAINAGE ABOVE GROUND

Item	PC £	Labour hours	Labour £	Material £	Unit	Total rate £
RAINWATER INSTALLATIONS – cont						
uPVC – cont						
External rainwater pipes and fittings – cont						
two bends to form offset 229 mm projection	14.04	0.31	4.00	15.00	nr	**19.00**
loose drain connector; cement and sand (1:2) joint	–	0.14	1.80	16.92	nr	**18.72**
110 mm pipes; fixing with pipe or socket brackets; plugged and screwed	13.93	0.33	4.25	22.37	m	**26.62**
extra for						
shoe	17.35	0.22	2.84	19.98	nr	**22.82**
bend	25.83	0.33	4.25	28.67	nr	**32.92**
single branch	38.07	0.44	5.67	41.22	nr	**46.89**
two bends to form offset 229 mm projection	51.66	0.33	4.25	53.15	nr	**57.40**
loose drain connector; cement and sand (1:2) joint	–	0.32	4.12	16.57	nr	**20.69**
65 mm square pipes; fixing with pipe or socket brackets; plugged and screwed	6.27	0.31	4.00	9.69	m	**13.69**
extra for						
shoe	4.58	0.20	2.57	6.26	nr	**8.83**
bend	7.02	0.31	4.00	8.75	nr	**12.75**
single branch	14.11	0.41	5.28	16.03	nr	**21.31**
two bends to form offset 229 mm projection	14.04	0.31	4.00	15.18	nr	**19.18**
drain connector; square to round; cement and sand (1:2) joint	–	0.32	4.12	7.19	nr	**11.31**
Rainwater head; rectangular; for pipes						
50 mm dia.	29.83	0.42	5.41	33.16	nr	**38.57**
68 mm dia.	24.09	0.43	5.55	28.45	nr	**34.00**
110 mm dia.	49.79	0.51	6.57	55.44	nr	**62.01**
65 mm square	20.15	0.43	5.55	23.81	nr	**29.36**
Gutters and fittings; BS EN 12200						
76 mm half round gutters; on brackets screwed to timber	7.08	0.28	4.99	10.53	m	**15.52**
extra for						
stop end	2.52	0.12	2.14	3.30	nr	**5.44**
running outlet	7.08	0.23	4.10	6.68	nr	**10.78**
stop end outlet	7.07	0.12	2.14	7.38	nr	**9.52**
angle	7.08	0.23	4.10	8.10	nr	**12.20**
112 mm half round gutters; on brackets screwed to timber	7.12	0.31	5.52	12.13	m	**17.65**
extra for						
stop end	3.94	0.12	2.14	5.26	nr	**7.40**
running outlet	7.73	0.26	4.63	7.34	nr	**11.97**
stop end outlet	7.73	0.12	2.14	8.56	nr	**10.70**
angle	7.22	0.26	4.63	9.26	nr	**13.89**
170 mm half round gutters; on brackets; screwed to timber	12.69	0.31	5.52	20.24	m	**25.76**
extra for						
stop end	5.56	0.15	2.68	7.68	nr	**10.36**
running outlet	12.41	0.29	5.17	11.69	nr	**16.86**
stop end outlet	11.80	0.15	2.68	13.03	nr	**15.71**
angle	16.17	0.29	5.17	19.49	nr	**24.66**

33 DRAINAGE ABOVE GROUND

Item	PC £	Labour hours	Labour £	Material £	Unit	Total rate £
114 mm rectangular gutters; on brackets; screwed to timber	7.30	0.31	5.52	13.49	m	**19.01**
extra for						
stop end	3.94	0.12	2.14	5.26	nr	**7.40**
running outlet	7.73	0.29	5.17	7.33	nr	**12.50**
stop end outlet	7.73	0.12	2.14	8.55	nr	**10.69**
angle	8.63	0.26	4.63	10.69	nr	**15.32**
FOUL DRAINAGE INSTALLATIONS						
Cast iron						
Cast iron Timesaver pipes and fittings or other equal and approved; BS 416						
50 mm pipes; primed; 3 m lengths; fixing with expanding bolts; to masonry	17.97	0.51	6.57	26.38	m	**32.95**
extra for						
fittings with two ends	–	0.51	6.58	22.95	nr	**29.53**
fittings with three ends	–	0.69	8.89	38.87	nr	**47.76**
bends; short radius	16.00	0.51	6.57	22.95	nr	**29.52**
access bends; short radius	39.42	0.51	6.57	46.96	nr	**53.53**
boss; 38 BSP	33.11	0.51	6.57	40.03	nr	**46.60**
single branch	24.06	0.69	8.89	39.70	nr	**48.59**
isolated Timesaver coupling joint	9.08	0.28	3.61	9.31	nr	**12.92**
connection to clay pipes; cement and sand (1:2) joint	–	0.12	1.55	0.13	nr	**1.68**
75 mm pipes; primed; 3 m lengths; fixing with standard brackets; plugged and screwed to masonry	20.09	0.51	6.57	32.60	m	**39.17**
extra for						
bends; short radius	18.10	0.55	7.08	25.74	nr	**32.82**
access bends; short radius	42.75	0.51	6.57	51.00	nr	**57.57**
boss; 38 BSP	33.11	0.55	7.08	41.12	nr	**48.20**
single branch	27.24	0.79	10.18	44.24	nr	**54.42**
double branch	40.46	1.02	13.14	68.07	nr	**81.21**
offset 115 mm projection	25.95	0.55	7.08	31.42	nr	**38.50**
offset 150 mm projection	30.50	0.55	7.08	35.46	nr	**42.54**
access pipe	38.49	0.55	7.08	43.95	nr	**51.03**
isolated Timesaver coupling joint	10.02	0.32	4.12	10.27	nr	**14.39**
connection to clay pipes; cement and sand (1:2) joint	–	0.14	1.80	0.13	nr	**1.93**
100 mm pipes; primed; 3 m lengths; fixing with standard brackets; plugged and screwed to masonry	24.29	0.55	7.08	45.35	m	**52.43**
extra for						
WC bent connector; 450 mm long tail	35.44	0.55	7.08	40.92	nr	**48.00**
bends; short radius	22.13	0.62	7.98	32.38	nr	**40.36**
access bends; short radius	46.85	0.62	7.98	57.71	nr	**65.69**
boss; 38 BSP	39.56	0.62	7.98	50.24	nr	**58.22**
single branch	34.22	0.93	11.98	56.07	nr	**68.05**
double branch	42.33	1.20	15.47	77.80	nr	**93.27**
offset 225 mm projection	33.32	0.62	7.98	40.60	nr	**48.58**
offset 300 mm projection	35.86	0.62	7.98	42.47	nr	**50.45**
access pipe	40.46	0.62	7.98	47.43	nr	**55.41**
roof connector; for asphalt	38.24	0.62	7.98	48.01	nr	**55.99**
isolated Timesaver coupling joint	13.09	0.39	5.02	13.42	nr	**18.44**
transitional clayware socket; cement and sand (1:2) joint	25.61	0.37	4.77	39.79	nr	**44.56**

33 DRAINAGE ABOVE GROUND

Item	PC £	Labour hours	Labour £	Material £	Unit	Total rate £
FOUL DRAINAGE INSTALLATIONS – cont						
Cast iron – cont						
Cast iron Timesaver pipes and fittings or other equal and approved – cont						
150 mm pipes; primed; 3 m lengths; fixing with standard brackets; plugged and screwed to						
masonry	50.72	0.69	8.89	91.77	m	**100.66**
extra for						
bends; short radius	39.56	0.77	9.92	59.54	nr	**69.46**
access bends; short radius	66.50	0.77	9.92	87.15	nr	**97.07**
boss; 38 BSP	64.54	0.77	9.92	83.83	nr	**93.75**
single branch	84.84	1.11	14.30	126.50	nr	**140.80**
double branch	119.21	1.48	19.08	186.17	nr	**205.25**
access pipe	67.29	0.77	9.92	77.36	nr	**87.28**
isolated Timesaver coupling joint	–	0.46	5.92	26.78	nr	**32.70**
transitional clayware socket; cement and sand (1:2) joint	44.85	0.48	6.18	72.88	nr	**79.06**
Cast iron Ensign lightweight pipes and fittings or other equal and approved; BS EN 877						
50 mm pipes; primed; 3 m lengths; fixing with standard brackets; plugged and screwed to						
masonry	12.40	0.31	4.22	18.81	m	**23.03**
extra for						
bends; short radius	9.65	0.27	3.68	15.61	nr	**19.29**
single branch	15.48	0.33	4.49	27.30	nr	**31.79**
access pipe	25.69	0.27	3.57	32.05	nr	**35.62**
70 mm pipes; primed; 3 m lengths; fixing with standard brackets; plugged and screwed to						
masonry	14.35	0.34	4.64	20.99	m	**25.63**
extra for						
bends; short radius	10.85	0.30	4.07	17.41	nr	**21.48**
single branch	16.33	0.37	5.02	29.33	nr	**34.35**
access pipe	27.17	0.30	4.07	34.14	nr	**38.21**
100 mm pipes; primed; 3 m lengths; fixing with standard brackets; plugged and screwed to						
masonry	17.07	0.37	5.02	24.96	m	**29.98**
extra for						
bends; short radius	12.85	0.32	4.37	21.35	nr	**25.72**
single branch	22.39	0.39	5.29	39.33	nr	**44.62**
double branch	29.93	0.46	6.25	55.24	nr	**61.49**
access pipe	29.88	0.32	4.37	38.81	nr	**43.18**
connector	27.20	0.21	2.84	36.06	nr	**38.90**
reducer	17.44	0.32	4.37	26.07	nr	**30.44**

33 DRAINAGE ABOVE GROUND

Item	PC £	Labour hours	Labour £	Material £	Unit	Total rate £
uPVC						
muPVC waste pipes and fittings; BS EN 1329; solvent welded joints						
32 mm pipes; fixing with pipe clips; plugged and screwed	1.36	0.23	2.96	2.04	m	**5.00**
extra for						
fittings with one end	–	0.16	2.06	1.05	nr	**3.11**
fittings with two ends	–	0.23	2.96	1.12	nr	**4.08**
fittings with three ends	–	0.31	4.00	1.48	nr	**5.48**
access plug	0.77	0.16	2.06	1.05	nr	**3.11**
straight coupling	0.82	0.16	2.06	1.10	nr	**3.16**
expansion coupling	1.46	0.23	2.96	1.75	nr	**4.71**
male iron to muPVC coupling	1.47	0.35	4.51	1.63	nr	**6.14**
sweep bend	0.84	0.23	2.96	1.12	nr	**4.08**
spigot/socket bend	–	0.23	2.96	1.67	nr	**4.63**
sweep tee	1.13	0.31	4.00	1.48	nr	**5.48**
40 mm pipes; fixing with pipe clips; plugged and screwed	1.68	0.28	3.61	2.40	m	**6.01**
extra for						
fittings with one end	–	0.18	2.32	1.05	nr	**3.37**
fittings with two ends	–	0.28	3.61	1.22	nr	**4.83**
fittings with three ends	–	0.37	4.77	1.78	nr	**6.55**
fittings with four ends	3.47	0.49	6.31	4.01	nr	**10.32**
access plug	0.77	0.18	2.32	1.05	nr	**3.37**
straight coupling	0.82	0.19	2.45	1.10	nr	**3.55**
expansion coupling	1.75	0.28	3.61	2.05	nr	**5.66**
male iron to muPVC coupling	1.47	0.35	4.51	1.63	nr	**6.14**
level invert taper	1.03	0.28	3.61	1.31	nr	**4.92**
sweep bend	0.94	0.28	3.61	1.22	nr	**4.83**
spigot/socket bend	1.58	0.28	3.61	1.88	nr	**5.49**
sweep tee	1.43	0.37	4.77	1.78	nr	**6.55**
sweep cross	3.47	0.49	6.31	4.01	nr	**10.32**
50 mm pipes; fixing with pipe clips; plugged and screwed	2.53	0.32	4.12	3.88	m	**8.00**
extra for						
fittings with one end	–	0.19	2.45	1.38	nr	**3.83**
fittings with two ends	–	0.32	4.12	1.94	nr	**6.06**
fittings with three ends	–	0.43	5.55	3.18	nr	**8.73**
fittings with four ends	–	0.57	7.35	4.18	nr	**11.53**
access plug	1.10	0.19	2.45	1.38	nr	**3.83**
straight coupling	1.50	0.21	2.71	1.80	nr	**4.51**
expansion coupling	2.37	0.32	4.12	2.69	nr	**6.81**
male iron to muPVC coupling	2.12	0.42	5.41	2.31	nr	**7.72**
level invert taper	1.28	0.32	4.12	1.57	nr	**5.69**
sweep bend	1.64	0.32	4.12	1.94	nr	**6.06**
spigot/socket bend	2.24	0.32	4.12	2.55	nr	**6.67**
sweep tee	1.43	0.37	4.77	1.78	nr	**6.55**
sweep cross	3.64	0.57	7.35	4.18	nr	**11.53**

33 DRAINAGE ABOVE GROUND

Item	PC £	Labour hours	Labour £	Material £	Unit	Total rate £
FOUL DRAINAGE INSTALLATIONS – cont						
uPVC – cont						
uPVC overflow pipes and fittings; solvent welded joints						
19 mm pipes; fixing with pipe clips; plugged and screwed	1.15	0.20	2.57	1.75	m	**4.32**
extra for						
splay cut end	–	0.01	0.14	–	nr	**0.14**
fittings with one end	–	0.16	2.06	1.03	nr	**3.09**
fittings with two ends	–	0.16	2.06	1.19	nr	**3.25**
fittings with three ends	–	0.20	2.57	1.34	nr	**3.91**
straight connector	0.88	0.16	2.06	1.03	nr	**3.09**
female iron to uPVC coupling	–	0.19	2.45	1.59	nr	**4.04**
bend	1.04	0.16	2.06	1.19	nr	**3.25**
bent tank connector	1.62	0.19	2.45	1.72	nr	**4.17**
uPVC pipes and fittings; with solvent welded joints (unless otherwise described)						
82 mm pipes; fixing with holderbats; plugged and screwed	6.44	0.37	4.77	9.26	m	**14.03**
extra for						
socket plug	4.74	0.19	2.45	5.56	nr	**8.01**
slip coupling; push fit	10.37	0.34	4.38	10.63	nr	**15.01**
expansion coupling	4.99	0.37	4.77	5.80	nr	**10.57**
sweep bend	8.37	0.37	4.77	9.28	nr	**14.05**
boss connector	4.58	0.25	3.22	5.39	nr	**8.61**
single branch	11.70	0.49	6.31	13.10	nr	**19.41**
access door	11.15	0.56	7.22	11.78	nr	**19.00**
110 mm pipes; fixing with holderbats; plugged and screwed	6.56	0.41	5.28	9.66	m	**14.94**
extra for						
socket plug	5.76	0.20	2.57	6.80	nr	**9.37**
slip coupling; push fit	12.98	0.37	4.77	13.30	nr	**18.07**
expansion coupling	5.10	0.41	5.28	6.13	nr	**11.41**
WC connector	9.27	0.27	3.47	9.98	nr	**13.45**
sweep bend	9.80	0.41	5.28	10.95	nr	**16.23**
WC connecting bend	15.21	0.27	3.47	16.07	nr	**19.54**
access bend	27.19	0.43	5.55	28.76	nr	**34.31**
boss connector	4.58	0.27	3.47	5.60	nr	**9.07**
single branch	12.96	0.54	6.96	14.67	nr	**21.63**
single branch with access	22.19	0.56	7.22	24.13	nr	**31.35**
double branch	32.04	0.68	8.76	34.71	nr	**43.47**
WC manifold	12.73	0.27	3.47	14.43	nr	**17.90**
access door	–	0.56	7.22	11.78	nr	**19.00**
access pipe connector	20.83	0.46	5.92	22.24	nr	**28.16**
connection to clay pipes; caulking ring and cement and sand (1:2) joint	–	0.39	5.02	8.63	nr	**13.65**
160 mm pipes; fixing with holderbats; plugged and screwed	17.01	0.46	5.92	24.97	m	**30.89**

33 DRAINAGE ABOVE GROUND

Item	PC £	Labour hours	Labour £	Material £	Unit	Total rate £
extra for						
socket plug	10.57	0.23	2.96	12.84	nr	**15.80**
slip coupling; push fit	33.23	0.42	5.41	34.06	nr	**39.47**
expansion coupling	15.35	0.46	5.92	17.74	nr	**23.66**
sweep bend	24.41	0.46	5.92	27.02	nr	**32.94**
boss connector	6.48	0.31	4.00	8.64	nr	**12.64**
single branch	27.52	0.61	7.86	31.18	nr	**39.04**
double branch	57.88	0.77	9.92	63.26	nr	**73.18**
access door	19.92	0.56	7.22	20.77	nr	**27.99**
access pipe connector	20.83	0.46	5.92	22.24	nr	**28.16**
Weathering apron; for pipe						
82 mm dia.	2.36	0.31	4.00	2.76	nr	**6.76**
110 mm dia.	2.71	0.35	4.51	3.26	nr	**7.77**
160 mm dia.	8.15	0.39	5.02	9.32	nr	**14.34**
Weathering slate; for pipe						
110 mm dia.	28.78	0.83	10.69	29.98	nr	**40.67**
Vent cowl; for pipe						
82 mm dia.	2.36	0.31	4.00	2.76	nr	**6.76**
110 mm dia.	2.39	0.31	4.00	2.93	nr	**6.93**
160 mm dia.	6.25	0.31	4.00	7.37	nr	**11.37**
Polypropylene						
Polypropylene (PP) waste pipes and fittings; BS EN 1451; push fit O-ring joints						
32 mm pipes; fixing with pipe clips; plugged and screwed	1.28	0.20	2.57	1.89	m	**4.46**
extra for						
fittings with one end	–	0.15	1.94	1.08	nr	**3.02**
fittings with two ends	–	0.20	2.57	1.09	nr	**3.66**
fittings with three ends	–	0.28	3.61	1.89	nr	**5.50**
access plug	1.05	0.15	1.94	1.08	nr	**3.02**
double socket	0.80	0.14	1.80	0.82	nr	**2.62**
male iron to PP coupling	2.21	0.26	3.35	2.27	nr	**5.62**
sweep bend	0.99	0.20	2.57	1.01	nr	**3.58**
spigot bend	1.45	0.23	2.96	1.49	nr	**4.45**
40 mm pipes; fixing with pipe clips; plugged and screwed	1.58	0.20	2.57	2.20	m	**4.77**
extra for						
fittings with one end	–	0.18	2.32	1.12	nr	**3.44**
fittings with two ends	–	0.28	3.61	1.28	nr	**4.89**
fittings with three ends	–	0.37	4.77	1.98	nr	**6.75**
access plug	1.09	0.18	2.32	1.12	nr	**3.44**
double socket	0.82	0.19	2.45	0.84	nr	**3.29**
universal connector	2.50	0.23	2.96	2.56	nr	**5.52**
sweep bend	1.12	0.28	3.61	1.15	nr	**4.76**
spigot bend	1.41	0.28	3.61	1.45	nr	**5.06**
reducer 40 mm–32 mm	0.99	0.28	3.61	1.01	nr	**4.62**

33 DRAINAGE ABOVE GROUND

Item	PC £	Labour hours	Labour £	Material £	Unit	Total rate £
FOUL DRAINAGE INSTALLATIONS – cont						
Polypropylene – cont						
Polypropylene (PP) waste pipes and fittings – cont						
50 mm pipes; fixing with pipe clips; plugged and screwed	2.03	0.32	4.12	3.23	m	**7.35**
extra for						
fittings with one end	–	0.19	2.45	1.99	nr	**4.44**
fittings with two ends	–	0.32	4.12	2.13	nr	**6.25**
fittings with three ends	–	0.43	5.55	2.96	nr	**8.51**
access plug	1.94	0.19	2.45	1.99	nr	**4.44**
double socket	1.64	0.21	2.71	1.68	nr	**4.39**
sweep bend	2.13	0.32	4.12	2.18	nr	**6.30**
spigot bend	3.34	0.32	4.12	3.42	nr	**7.54**
reducer 50 mm–40 mm	1.29	0.32	4.12	1.32	nr	**5.44**
Polypropylene ancillaries; screwed joint to waste fitting						
Tubular S trap; bath; shallow seal						
40 mm dia.	5.88	0.51	6.57	6.03	nr	**12.60**
Trap; P; two piece; 76 mm seal						
32 mm dia.	3.97	0.35	4.51	4.07	nr	**8.58**
40 mm dia.	4.59	0.42	5.41	4.70	nr	**10.11**
Trap; S; two piece; 76 mm seal						
32 mm dia.	5.03	0.35	4.51	5.16	nr	**9.67**
40 mm dia.	5.88	0.42	5.41	6.03	nr	**11.44**
Bottle trap; P; 76 mm seal						
32 dia.	4.43	0.35	4.51	4.54	nr	**9.05**
40 dia.	5.28	0.42	5.41	5.41	nr	**10.82**

34 DRAINAGE BELOW GROUND

Item	PC £	Labour hours	Labour £	Material £	Unit	Total rate £
TRENCHES						
NOTE: Prices for drain trenches are for excavation in firm soil and it has been assumed that earthwork support will only be required for trenches 1.00 m or more in depth.						
Excavating trenches; by machine; grading bottoms; earthwork support; filling with excavated material and compacting; disposal of surplus soil; spreading on site average 50 m from excavations						
Pipes not exceeding 200 mm nominal size						
average depth of trench 0.50 m	–	0.28	3.27	1.32	m	**4.59**
average depth of trench 0.75 m	–	0.37	4.32	1.96	m	**6.28**
average depth of trench 1.00 m	–	0.79	9.23	3.43	m	**12.66**
average depth of trench 1.25 m	–	1.16	13.54	3.82	m	**17.36**
average depth of trench 1.50 m	–	1.48	17.28	4.33	m	**21.61**
average depth of trench 1.75 m	–	1.85	21.60	4.71	m	**26.31**
average depth of trench 2.00 m	–	2.13	24.87	5.36	m	**30.23**
average depth of trench 2.25 m	–	2.64	30.82	6.67	m	**37.49**
average depth of trench 2.50 m	–	3.10	36.19	7.78	m	**43.97**
average depth of trench 2.75 m	–	3.42	39.92	8.71	m	**48.63**
average depth of trench 3.00 m	–	3.75	43.78	9.58	m	**53.36**
average depth of trench 3.25 m	–	4.07	47.52	10.17	m	**57.69**
average depth of trench 3.50 m	–	4.35	50.79	10.70	m	**61.49**
Pipes 225 mm nominal size						
average depth of trench 0.50 m	–	0.28	3.27	1.32	m	**4.59**
average depth of trench 0.75 m	–	0.37	4.32	1.96	m	**6.28**
average depth of trench 1.00 m	–	0.79	9.23	3.43	m	**12.66**
average depth of trench 1.25 m	–	1.16	13.54	3.82	m	**17.36**
average depth of trench 1.50 m	–	1.48	17.28	4.33	m	**21.61**
average depth of trench 1.75 m	–	1.85	21.60	4.71	m	**26.31**
average depth of trench 2.00 m	–	2.13	24.87	5.36	m	**30.23**
average depth of trench 2.25 m	–	2.64	30.82	6.67	m	**37.49**
average depth of trench 2.50 m	–	3.10	36.19	7.78	m	**43.97**
average depth of trench 2.75 m	–	3.42	39.92	8.71	m	**48.63**
average depth of trench 3.00 m	–	3.75	43.78	9.58	m	**53.36**
average depth of trench 3.25 m	–	4.07	47.52	10.17	m	**57.69**
average depth of trench 3.50 m	–	4.35	50.79	10.70	m	**61.49**
Pipes 300 mm nominal size						
average depth of trench 0.75 m	–	0.44	5.14	2.44	m	**7.58**
average depth of trench 1.00 m	–	0.93	10.85	3.43	m	**14.28**
average depth of trench 1.25 m	–	1.25	14.60	3.97	m	**18.57**
average depth of trench 1.50 m	–	1.62	18.91	4.47	m	**23.38**
average depth of trench 1.75 m	–	1.85	21.60	4.86	m	**26.46**
average depth of trench 2.00 m	–	2.13	24.87	5.85	m	**30.72**
average depth of trench 2.25 m	–	2.64	30.82	6.97	m	**37.79**
average depth of trench 2.50 m	–	3.10	36.19	7.97	m	**44.16**
average depth of trench 2.75 m	–	3.42	39.92	8.86	m	**48.78**
average depth of trench 3.00 m	–	3.75	43.78	9.73	m	**53.51**
average depth of trench 3.25 m	–	4.07	47.52	10.66	m	**58.18**
average depth of trench 3.50 m	–	4.35	50.79	11.05	m	**61.84**

34 DRAINAGE BELOW GROUND

Item	PC £	Labour hours	Labour £	Material £	Unit	Total rate £
TRENCHES – cont						
Excavating trenches; by machine – cont						
Pipes 375 mm nominal size						
average depth of trench 0.75 m	–	0.46	5.37	2.93	m	**8.30**
average depth of trench 1.00 m	–	0.97	11.33	3.92	m	**15.25**
average depth of trench 1.25 m	–	1.34	15.64	4.80	m	**20.44**
average depth of trench 1.50 m	–	1.71	19.97	5.10	m	**25.07**
average depth of trench 1.75 m	–	1.99	23.24	5.69	m	**28.93**
average depth of trench 2.00 m	–	2.27	26.50	6.00	m	**32.50**
average depth of trench 2.25 m	–	2.82	32.92	7.46	m	**40.38**
average depth of trench 2.50 m	–	3.38	39.46	8.61	m	**48.07**
average depth of trench 2.75 m	–	3.70	43.19	9.34	m	**52.53**
average depth of trench 3.00 m	–	4.02	46.93	10.08	m	**57.01**
average depth of trench 3.25 m	–	4.35	50.79	10.95	m	**61.74**
average depth of trench 3.50 m	–	4.67	54.52	11.69	m	**66.21**
Pipes 450 mm nominal size						
average depth of trench 0.75 m	–	0.51	5.96	2.93	m	**8.89**
average depth of trench 1.00 m	–	1.02	11.91	4.21	m	**16.12**
average depth of trench 1.25 m	–	1.48	17.28	5.14	m	**22.42**
average depth of trench 1.50 m	–	1.85	21.60	5.60	m	**27.20**
average depth of trench 1.75 m	–	2.13	24.87	6.03	m	**30.90**
average depth of trench 2.00 m	–	2.45	28.61	6.49	m	**35.10**
average depth of trench 2.25 m	–	3.05	35.61	7.80	m	**43.41**
average depth of trench 2.50 m	–	3.61	42.15	9.10	m	**51.25**
average depth of trench 2.75 m	–	3.98	46.46	9.97	m	**56.43**
average depth of trench 3.00 m	–	4.26	49.73	10.91	m	**60.64**
average depth of trench 3.25 m	–	4.63	54.05	11.93	m	**65.98**
average depth of trench 3.50 m	–	5.00	58.37	13.00	m	**71.37**
Pipes 600 mm nominal size						
average depth of trench 1.00 m	–	1.11	12.96	4.55	m	**17.51**
average depth of trench 1.25 m	–	1.57	18.33	5.43	m	**23.76**
average depth of trench 1.50 m	–	2.04	23.82	6.27	m	**30.09**
average depth of trench 1.75 m	–	2.31	26.97	6.52	m	**33.49**
average depth of trench 2.00 m	–	2.73	31.87	7.16	m	**39.03**
average depth of trench 2.25 m	–	3.28	38.29	8.77	m	**47.06**
average depth of trench 2.50 m	–	3.89	45.42	10.22	m	**55.64**
average depth of trench 2.75 m	–	4.30	50.20	11.44	m	**61.64**
average depth of trench 3.00 m	–	4.72	55.10	12.52	m	**67.62**
average depth of trench 3.25 m	–	5.09	59.42	13.40	m	**72.82**
average depth of trench 3.50 m	–	5.46	63.74	14.12	m	**77.86**
Pipes 900 mm nominal size						
average depth of trench 1.25 m	–	1.90	22.18	6.41	m	**28.59**
average depth of trench 1.50 m	–	2.41	28.14	7.26	m	**35.40**
average depth of trench 1.75 m	–	2.78	32.45	7.65	m	**40.10**
average depth of trench 2.00 m	–	3.10	36.19	8.78	m	**44.97**
average depth of trench 2.25 m	–	3.84	44.83	10.59	m	**55.42**
average depth of trench 2.50 m	–	4.53	52.89	12.18	m	**65.07**
average depth of trench 2.75 m	–	5.00	58.37	13.40	m	**71.77**
average depth of trench 3.00 m	–	5.46	63.74	14.62	m	**78.36**
average depth of trench 3.25 m	–	5.92	69.12	15.84	m	**84.96**
average depth of trench 3.50 m	–	6.38	74.49	16.91	m	**91.40**

34 DRAINAGE BELOW GROUND

Item	PC £	Labour hours	Labour £	Material £	Unit	Total rate £
Pipes 1200 mm nominal size						
average depth of trench 1.50 m	–	2.73	31.87	7.74	m	**39.61**
average depth of trench 1.75 m	–	3.19	37.24	8.96	m	**46.20**
average depth of trench 2.00 m	–	3.56	41.56	10.25	m	**51.81**
average depth of trench 2.25 m	–	4.35	50.79	12.34	m	**63.13**
average depth of trench 2.50 m	–	5.18	60.48	14.13	m	**74.61**
average depth of trench 2.75 m	–	5.69	66.43	15.69	m	**82.12**
average depth of trench 3.00 m	–	6.20	72.39	17.06	m	**89.45**
average depth of trench 3.25 m	–	6.75	78.80	18.47	m	**97.27**
average depth of trench 3.50 m	–	7.26	84.76	19.84	m	**104.60**
Extra over excavating trenches; irrespective of depth; breaking out existing materials						
brick	–	1.80	21.01	7.24	m³	**28.25**
concrete	–	2.54	29.65	9.99	m³	**39.64**
reinforced concrete	–	3.61	42.15	14.42	m³	**56.57**
Extra over excavating trenches; irrespective of depth; breaking out existing hard pavings; 75 mm thick						
tarmacadam	–	0.19	2.21	0.74	m²	**2.95**
Extra over excavating trenches; irrsepective of depth; breaking out existing hard pavings; 150 mm thick						
concrete	–	0.37	4.32	1.62	m²	**5.94**
tarmacadam and hardcore	–	0.28	3.27	0.90	m²	**4.17**
Excavating trenches; by hand; grading bottoms; earthwork support; filling with excavated material and compacting; disposal of surplus soil on site; spreading on site average 25 m from excavations						
Pipes not exceeding 200 mm nominal size						
average depth of trench 0.50 m	–	0.93	10.85	–	m	**10.85**
average depth of trench 0.75 m	–	1.39	16.23	–	m	**16.23**
average depth of trench 1.00 m	–	2.04	23.82	0.65	m	**24.47**
average depth of trench 1.25 m	–	2.87	33.51	0.89	m	**34.40**
average depth of trench 1.50 m	–	3.93	45.88	1.09	m	**46.97**
average depth of trench 1.75 m	–	5.18	60.48	1.29	m	**61.77**
average depth of trench 2.00 m	–	5.92	69.12	1.46	m	**70.58**
average depth of trench 2.25 m	–	7.40	86.39	1.94	m	**88.33**
average depth of trench 2.50 m	–	8.88	103.67	2.27	m	**105.94**
average depth of trench 2.75 m	–	9.76	113.94	2.50	m	**116.44**
average depth of trench 3.00 m	–	10.64	124.22	2.75	m	**126.97**
average depth of trench 3.25 m	–	11.52	134.49	2.98	m	**137.47**
average depth of trench 3.50 m	–	12.40	144.76	3.23	m	**147.99**
Pipes 225 mm nominal size						
average depth of trench 0.50 m	–	0.93	10.85	–	m	**10.85**
average depth of trench 0.75 m	–	1.39	16.23	–	m	**16.23**
average depth of trench 1.00 m	–	2.04	23.82	0.65	m	**24.47**
average depth of trench 1.25 m	–	2.87	33.51	0.89	m	**34.40**
average depth of trench 1.50 m	–	3.93	45.88	1.09	m	**46.97**
average depth of trench 1.75 m	–	5.18	60.48	1.29	m	**61.77**
average depth of trench 2.00 m	–	5.92	69.12	1.46	m	**70.58**

34 DRAINAGE BELOW GROUND

Item	PC £	Labour hours	Labour £	Material £	Unit	Total rate £
TRENCHES – cont						
Excavating trenches; by hand – cont						
Pipes 225 mm nominal size – cont						
average depth of trench 2.25 m	–	7.40	86.39	1.94	m	88.33
average depth of trench 2.50 m	–	8.88	103.67	2.27	m	105.94
average depth of trench 2.75 m	–	9.76	113.94	2.50	m	116.44
average depth of trench 3.00 m	–	10.64	124.22	2.75	m	126.97
average depth of trench 3.25 m	–	11.52	134.49	2.98	m	137.47
average depth of trench 3.50 m	–	12.40	144.76	3.23	m	147.99
Pipes 300 mm nominal size						
average depth of trench 0.75 m	–	1.62	18.91	–	m	18.91
average depth of trench 1.00 m	–	2.36	27.55	0.65	m	28.20
average depth of trench 1.25 m	–	3.33	38.88	0.89	m	39.77
average depth of trench 1.50 m	–	4.44	51.83	1.09	m	52.92
average depth of trench 1.75 m	–	5.18	60.48	1.29	m	61.77
average depth of trench 2.00 m	–	5.92	69.12	1.46	m	70.58
average depth of trench 2.25 m	–	7.40	86.39	1.94	m	88.33
average depth of trench 2.50 m	–	8.88	103.67	2.27	m	105.94
average depth of trench 2.75 m	–	9.76	113.94	2.50	m	116.44
average depth of trench 3.00 m	–	10.64	124.22	2.75	m	126.97
average depth of trench 3.25 m	–	11.52	134.49	2.98	m	137.47
average depth of trench 3.50 m	–	12.40	144.76	3.23	m	147.99
Pipes 375 mm nominal size						
average depth of trench 0.75 m	–	1.80	21.01	–	m	21.01
average depth of trench 1.00 m	–	2.64	30.82	0.65	m	31.47
average depth of trench 1.25 m	–	3.70	43.19	0.89	m	44.08
average depth of trench 1.50 m	–	4.93	57.55	1.09	m	58.64
average depth of trench 1.75 m	–	5.74	67.01	1.29	m	68.30
average depth of trench 2.00 m	–	6.57	76.70	1.46	m	78.16
average depth of trench 2.25 m	–	8.23	96.08	1.94	m	98.02
average depth of trench 2.50 m	–	9.90	115.58	2.27	m	117.85
average depth of trench 2.75 m	–	10.87	126.91	2.50	m	129.41
average depth of trench 3.00 m	–	11.84	138.23	2.75	m	140.98
average depth of trench 3.25 m	–	12.86	150.13	2.98	m	153.11
average depth of trench 3.50 m	–	13.88	162.04	3.23	m	165.27
Pipes 450 mm nominal size						
average depth of trench 0.75 m	–	2.04	23.82	–	m	23.82
average depth of trench 1.00 m	–	2.94	34.33	0.65	m	34.98
average depth of trench 1.25 m	–	4.13	48.22	0.89	m	49.11
average depth of trench 1.50 m	–	5.41	63.16	1.09	m	64.25
average depth of trench 1.75 m	–	6.31	73.67	1.29	m	74.96
average depth of trench 2.00 m	–	7.22	84.29	1.46	m	85.75
average depth of trench 2.25 m	–	9.05	105.66	1.94	m	107.60
average depth of trench 2.50 m	–	10.87	126.91	2.27	m	129.18
average depth of trench 2.75 m	–	11.96	139.63	2.50	m	142.13
average depth of trench 3.00 m	–	13.04	152.23	2.75	m	154.98
average depth of trench 3.25 m	–	14.11	164.73	2.98	m	167.71
average depth of trench 3.50 m	–	15.17	177.10	3.23	m	180.33

34 DRAINAGE BELOW GROUND

Item	PC £	Labour hours	Labour £	Material £	Unit	Total rate £
Pipes 600 mm nominal size						
average depth of trench 1.00 m	–	3.24	37.82	0.65	m	**38.47**
average depth of trench 1.25 m	–	4.63	54.05	0.89	m	**54.94**
average depth of trench 1.50 m	–	6.20	72.39	1.09	m	**73.48**
average depth of trench 1.75 m	–	7.17	83.70	1.29	m	**84.99**
average depth of trench 2.00 m	–	8.19	95.61	1.46	m	**97.07**
average depth of trench 2.25 m	–	9.20	107.41	1.94	m	**109.35**
average depth of trench 2.50 m	–	11.56	134.96	2.27	m	**137.23**
average depth of trench 2.75 m	–	12.35	144.18	2.50	m	**146.68**
average depth of trench 3.00 m	–	14.80	172.78	2.75	m	**175.53**
average depth of trench 3.25 m	–	16.03	187.14	2.98	m	**190.12**
average depth of trench 3.50 m	–	17.25	201.38	3.23	m	**204.61**
Pipes 900 mm nominal size						
average depth of trench 1.25 m	–	5.78	67.48	0.89	m	**68.37**
average depth of trench 1.50 m	–	7.63	89.07	1.09	m	**90.16**
average depth of trench 1.75 m	–	8.88	103.67	1.29	m	**104.96**
average depth of trench 2.00 m	–	10.13	118.26	1.46	m	**119.72**
average depth of trench 2.25 m	–	12.72	148.50	1.94	m	**150.44**
average depth of trench 2.50 m	–	15.31	178.74	2.27	m	**181.01**
average depth of trench 2.75 m	–	16.84	196.59	2.50	m	**199.09**
average depth of trench 3.00 m	–	18.32	213.88	2.75	m	**216.63**
average depth of trench 3.25 m	–	19.84	231.62	2.98	m	**234.60**
average depth of trench 3.50 m	–	21.37	249.48	3.23	m	**252.71**
Pipes 1200 mm nominal size						
average depth of trench 1.50 m	–	9.11	106.35	1.09	m	**107.44**
average depth of trench 1.75 m	–	10.59	123.64	1.29	m	**124.93**
average depth of trench 2.00 m	–	12.12	141.49	1.46	m	**142.95**
average depth of trench 2.25 m	–	15.20	177.46	1.94	m	**179.40**
average depth of trench 2.50 m	–	18.27	213.29	2.27	m	**215.56**
average depth of trench 2.75 m	–	20.07	234.30	2.50	m	**236.80**
average depth of trench 3.00 m	–	21.88	255.44	2.75	m	**258.19**
average depth of trench 3.25 m	–	23.66	276.22	2.98	m	**279.20**
average depth of trench 3.50 m	–	25.44	297.00	3.23	m	**300.23**
Extra over excavating trenches irrespective of depth; breaking out existing materials						
brick	–	2.78	32.45	5.92	m^3	**38.37**
concrete	–	4.16	48.56	9.88	m^3	**58.44**
reinforced concrete	–	5.55	64.79	13.84	m^3	**78.63**
concrete; 150 mm thick	–	0.65	7.59	1.38	m^2	**8.97**
tarmacadam and hardcore; 150 mm thick	–	0.46	5.37	0.98	m^2	**6.35**
Extra over excavating trenches irrespective of depth; breaking out existing hard pavings, 75 mm thick						
tarmacadam	–	0.37	4.32	0.80	m^2	**5.12**
Extra over excavating trenches irrespective of depth; breaking out existing hard pavings, 150 mm thick						
concrete	–	0.65	7.59	1.38	m^2	**8.97**
tarmacadam and hardcore	–	0.46	5.37	0.98	m^2	**6.35**

34 DRAINAGE BELOW GROUND

Item	PC £	Labour hours	Labour £	Material £	Unit	Total rate £
BEDS AND FILLINGS						
Sand filling						
Beds; to receive pitch fibre pipes						
600 mm × 50 mm thick	–	0.07	0.82	1.45	m	**2.27**
700 mm × 50 mm thick	–	0.09	1.06	1.62	m	**2.68**
800 mm × 50 mm thick	–	0.11	1.28	1.78	m	**3.06**
Granular (shingle) filling						
Beds; 100 mm thick; to pipes						
100 mm nominal size	–	0.09	1.06	2.50	m	**3.56**
150 mm nominal size	–	0.09	1.06	2.84	m	**3.90**
225 mm nominal size	–	0.11	1.28	3.19	m	**4.47**
300 mm nominal size	–	0.13	1.52	3.53	m	**5.05**
375 mm nominal size	–	0.15	1.75	3.86	m	**5.61**
450 mm nominal size	–	0.17	1.99	4.21	m	**6.20**
600 mm nominal size	–	0.19	2.21	4.55	m	**6.76**
Beds; 150 mm thick; to pipes						
100 mm nominal size	–	0.13	1.52	3.53	m	**5.05**
150 mm nominal size	–	0.15	1.75	3.86	m	**5.61**
225 mm nominal size	–	0.17	1.99	4.21	m	**6.20**
300 mm nominal size	–	0.19	2.21	4.55	m	**6.76**
375 mm nominal size	–	0.22	2.57	5.58	m	**8.15**
450 mm nominal size	–	0.24	2.80	5.91	m	**8.71**
600 mm nominal size	–	0.28	3.27	6.95	m	**10.22**
Beds and benchings; beds 100 mm thick; to pipes						
100 nominal size	–	0.21	2.45	4.21	m	**6.66**
150 nominal size	–	0.23	2.69	4.21	m	**6.90**
225 nominal size	–	0.28	3.27	5.58	m	**8.85**
300 nominal size	–	0.32	3.73	6.26	m	**9.99**
375 nominal size	–	0.42	4.90	8.31	m	**13.21**
450 nominal size	–	0.48	5.61	9.34	m	**14.95**
600 nominal size	–	0.62	7.24	12.07	m	**19.31**
Beds and benchings; beds 150 mm thick; to pipes						
100 nominal size	–	0.23	2.69	4.55	m	**7.24**
150 nominal size	–	0.26	3.03	4.89	m	**7.92**
225 nominal size	–	0.32	3.73	6.60	m	**10.33**
300 nominal size	–	0.42	4.90	7.96	m	**12.86**
375 nominal size	–	0.48	5.61	9.34	m	**14.95**
450 nominal size	–	0.57	6.65	11.05	m	**17.70**
600 nominal size	–	0.68	7.94	14.12	m	**22.06**
Beds and coverings; 100 mm thick; to pipes						
100 nominal size	–	0.33	3.85	5.58	m	**9.43**
150 nominal size	–	0.42	4.90	6.60	m	**11.50**
225 nominal size	–	0.56	6.54	8.99	m	**15.53**
300 nominal size	–	0.67	7.82	10.70	m	**18.52**
375 nominal size	–	0.80	9.34	12.75	m	**22.09**
450 nominal size	–	0.94	10.98	15.15	m	**26.13**
600 nominal size	–	1.22	14.25	19.25	m	**33.50**

34 DRAINAGE BELOW GROUND

Item	PC £	Labour hours	Labour £	Material £	Unit	Total rate £
Beds and coverings; 150 mm thick; to pipes						
100 nominal size	–	0.50	5.83	7.96	m	**13.79**
150 nominal size	–	0.56	6.54	8.99	m	**15.53**
225 nominal size	–	0.72	8.40	11.39	m	**19.79**
300 nominal size	–	0.86	10.05	13.44	m	**23.49**
375 nominal size	–	1.00	11.67	15.84	m	**27.51**
450 nominal size	–	1.19	13.89	18.91	m	**32.80**
600 nominal size	–	1.44	16.81	22.67	m	**39.48**
Plain in situ ready mixed designated concrete; C10 – 40 mm aggregate						
Beds; 100 mm thick; to pipes						
100 mm nominal size	–	0.17	2.33	4.52	m	**6.85**
150 mm nominal size	–	0.17	2.33	4.52	m	**6.85**
225 mm nominal size	–	0.20	2.74	5.32	m	**8.06**
300 mm nominal size	–	0.23	3.15	6.14	m	**9.29**
375 mm nominal size	–	0.27	3.70	6.95	m	**10.65**
450 mm nominal size	–	0.30	4.11	7.77	m	**11.88**
600 mm nominal size	–	0.33	4.52	8.58	m	**13.10**
900 mm nominal size	–	0.40	5.47	10.20	m	**15.67**
1200 mm nominal size	–	0.54	7.39	13.46	m	**20.85**
Beds; 150 mm thick; to pipes						
100 mm nominal size	–	0.23	3.15	6.14	m	**9.29**
150 mm nominal size	–	0.27	3.70	6.95	m	**10.65**
225 mm nominal size	–	0.30	4.11	7.77	m	**11.88**
300 mm nominal size	–	0.33	4.52	8.58	m	**13.10**
375 mm nominal size	–	0.40	5.47	10.20	m	**15.67**
450 mm nominal size	–	0.43	5.88	11.02	m	**16.90**
600 mm nominal size	–	0.50	6.85	12.65	m	**19.50**
900 mm nominal size	–	0.63	8.62	15.90	m	**24.52**
1200 mm nominal size	–	0.77	10.54	19.15	m	**29.69**
Beds and benchings; beds 100 mm thick; to pipes						
100 mm nominal size	–	0.33	4.52	7.77	m	**12.29**
150 mm nominal size	–	0.38	5.21	8.58	m	**13.79**
225 mm nominal size	–	0.45	6.16	10.20	m	**16.36**
300 mm nominal size	–	0.53	7.26	11.83	m	**19.09**
375 mm nominal size	–	0.68	9.31	15.08	m	**24.39**
450 mm nominal size	–	0.80	10.95	17.52	m	**28.47**
600 mm nominal size	–	1.02	13.96	22.41	m	**36.37**
900 mm nominal size	–	1.65	22.59	36.22	m	**58.81**
1200 mm nominal size	–	2.44	33.40	53.29	m	**86.69**
Beds and benchings; beds 150 mm thick; to pipes						
100 mm nominal size	–	0.38	5.21	8.58	m	**13.79**
150 mm nominal size	–	0.42	5.75	9.39	m	**15.14**
225 mm nominal size	–	0.53	7.26	11.83	m	**19.09**
300 mm nominal size	–	0.68	9.31	15.08	m	**24.39**
375 mm nominal size	–	0.80	10.95	17.52	m	**28.47**
450 mm nominal size	–	0.94	12.86	20.77	m	**33.63**
600 mm nominal size	–	1.20	16.43	26.47	m	**42.90**
900 mm nominal size	–	1.91	26.15	41.91	m	**68.06**
1200 mm nominal size	–	2.70	36.96	58.99	m	**95.95**

34 DRAINAGE BELOW GROUND

Item	PC £	Labour hours	Labour £	Material £	Unit	Total rate £
BEDS AND FILLINGS – cont						
Plain in situ ready mixed designated concrete – cont						
Beds and coverings; 100 mm thick; to pipes						
100 mm nominal size	–	0.50	6.85	10.20	m	**17.05**
150 mm nominal size	–	0.58	7.94	11.83	m	**19.77**
225 mm nominal size	–	0.83	11.37	16.71	m	**28.08**
300 mm nominal size	–	1.00	13.69	19.96	m	**33.65**
375 mm nominal size	–	1.21	16.56	24.03	m	**40.59**
450 mm nominal size	–	1.42	19.43	28.10	m	**47.53**
600 mm nominal size	–	1.83	25.05	36.22	m	**61.27**
900 mm nominal size	–	2.79	38.19	54.93	m	**93.12**
1200 mm nominal size	–	3.83	52.43	75.25	m	**127.68**
Beds and coverings; 150 mm thick; to pipes						
100 mm nominal size	–	0.75	10.27	15.08	m	**25.35**
150 mm nominal size	–	0.83	11.37	16.71	m	**28.08**
225 mm nominal size	–	1.08	14.78	21.59	m	**36.37**
300 mm nominal size	–	1.30	17.79	25.65	m	**43.44**
375 mm nominal size	–	1.50	20.53	29.71	m	**50.24**
450 mm nominal size	–	1.79	24.51	35.40	m	**59.91**
600 mm nominal size	–	2.16	29.57	42.72	m	**72.29**
900 mm nominal size	–	3.54	48.46	69.56	m	**118.02**
1200 mm nominal size	–	5.00	68.45	98.01	m	**166.46**
Plain in situ ready mixed designated concrete; C20 – 40 mm aggregate						
Beds; 100 mm thick; to pipes						
100 mm nominal size	–	0.17	2.33	4.60	m	**6.93**
150 mm nominal size	–	0.17	2.33	4.60	m	**6.93**
225 mm nominal size	–	0.20	2.74	5.43	m	**8.17**
300 mm nominal size	–	0.23	3.15	6.26	m	**9.41**
375 mm nominal size	–	0.27	3.70	7.09	m	**10.79**
450 mm nominal size	–	0.30	4.11	7.92	m	**12.03**
600 mm nominal size	–	0.33	4.52	8.75	m	**13.27**
900 mm nominal size	–	0.40	5.47	10.41	m	**15.88**
1200 mm nominal size	–	0.54	7.39	13.74	m	**21.13**
Beds; 150 mm thick; to pipes						
100 mm nominal size	–	0.23	3.15	6.26	m	**9.41**
150 mm nominal size	–	0.27	3.70	7.09	m	**10.79**
225 mm nominal size	–	0.30	4.11	7.92	m	**12.03**
300 mm nominal size	–	0.33	4.52	8.75	m	**13.27**
375 mm nominal size	–	0.40	5.47	10.41	m	**15.88**
450 mm nominal size	–	0.43	5.88	11.24	m	**17.12**
600 mm nominal size	–	0.50	6.85	12.90	m	**19.75**
900 mm nominal size	–	0.63	8.62	16.23	m	**24.85**
1200 mm nominal size	–	0.77	10.54	19.55	m	**30.09**

34 DRAINAGE BELOW GROUND

Item	PC £	Labour hours	Labour £	Material £	Unit	Total rate £
Beds and benchings; beds 100 mm thick; to pipes						
100 mm nominal size	–	0.33	4.52	7.92	m	**12.44**
150 mm nominal size	–	0.38	5.21	8.75	m	**13.96**
225 mm nominal size	–	0.45	6.16	10.41	m	**16.57**
300 mm nominal size	–	0.53	7.26	12.07	m	**19.33**
375 mm nominal size	–	0.68	9.31	15.40	m	**24.71**
450 mm nominal size	–	0.80	10.95	17.89	m	**28.84**
600 mm nominal size	–	1.02	13.96	22.88	m	**36.84**
900 mm nominal size	–	1.65	22.59	36.99	m	**59.58**
1200 mm nominal size	–	2.44	33.40	54.43	m	**87.83**
Beds and benchings; beds 150 mm thick; to pipes						
100 mm nominal size	–	0.38	5.21	8.75	m	**13.96**
150 mm nominal size	–	0.42	5.75	9.57	m	**15.32**
225 mm nominal size	–	0.53	7.26	12.07	m	**19.33**
300 mm nominal size	–	0.68	9.31	15.40	m	**24.71**
375 mm nominal size	–	0.80	10.95	17.89	m	**28.84**
450 mm nominal size	–	0.94	12.86	21.21	m	**34.07**
600 mm nominal size	–	1.20	16.43	27.02	m	**43.45**
900 mm nominal size	–	1.91	26.15	42.80	m	**68.95**
1200 mm nominal size	–	2.70	36.96	60.24	m	**97.20**
Beds and coverings; 100 mm thick; to pipes						
100 mm nominal size	–	0.50	6.85	10.41	m	**17.26**
150 mm nominal size	–	0.58	7.94	12.07	m	**20.01**
225 mm nominal size	–	0.83	11.37	17.06	m	**28.43**
300 mm nominal size	–	1.00	13.69	20.38	m	**34.07**
375 mm nominal size	–	1.21	16.56	24.53	m	**41.09**
450 mm nominal size	–	1.42	19.43	28.68	m	**48.11**
600 mm nominal size	–	1.83	25.05	36.99	m	**62.04**
900 mm nominal size	–	2.79	38.19	56.09	m	**94.28**
1200 mm nominal size	–	3.83	52.43	76.85	m	**129.28**
Beds and coverings; 150 mm thick; to pipes						
100 mm nominal size	–	0.75	10.27	15.40	m	**25.67**
150 mm nominal size	–	0.83	11.37	17.06	m	**28.43**
225 mm nominal size	–	1.08	14.78	22.04	m	**36.82**
300 mm nominal size	–	1.30	17.79	26.19	m	**43.98**
375 mm nominal size	–	1.50	20.53	30.34	m	**50.87**
450 mm nominal size	–	1.79	24.51	36.15	m	**60.66**
600 mm nominal size	–	2.16	29.57	43.63	m	**73.20**
900 mm nominal size	–	3.54	48.46	71.03	m	**119.49**
1200 mm nominal size	–	5.00	68.45	100.10	m	**168.55**

34 DRAINAGE BELOW GROUND

Item	PC £	Labour hours	Labour £	Material £	Unit	Total rate £
PIPES, FITTINGS AND ACCESSORIES						
Pipes and fittings						
NOTE: The following items unless otherwise described include for all appropriate joints/couplings in the running length. The prices for gullies and rainwater shoes, etc. include for appropriate joints to pipes and for setting on and surrounding accessory with site mixed in situ concrete 10.00 N/mm² – 40 mm aggregate (1:3:6).						
Cast Iron						
Timesaver drain pipes and fittings or other equal; BS 437; coated; with mechanical coupling joints						
100 mm dia. pipes						
laid straight	30.20	0.46	5.33	39.42	m	**44.75**
in runs not exceeding 3 m long	30.20	0.63	7.30	61.52	m	**68.82**
extra for						
bend; medium radius	37.11	0.56	6.49	53.51	nr	**60.00**
bend; medium radius with access	103.13	0.56	6.49	121.18	nr	**127.67**
bend; long radius	61.32	0.56	6.49	76.70	nr	**83.19**
rest bend	42.58	0.56	6.49	57.48	nr	**63.97**
single branch	49.24	0.69	8.00	83.20	nr	**91.20**
single branch; with access	113.57	0.79	9.15	149.14	nr	**158.29**
double branch	83.70	0.88	10.19	136.25	nr	**146.44**
isolated Timesaver joint	19.85	0.32	3.71	20.35	nr	**24.06**
transitional pipe; for WC	29.07	0.46	5.33	50.13	nr	**55.46**
150 mm dia. pipes						
laid straight	55.92	0.56	6.49	68.55	m	**75.04**
in runs not exceeding 3 m long	55.92	0.76	8.80	100.86	m	**109.66**
extra for						
bend; medium radius	85.39	0.65	7.52	103.14	nr	**110.66**
bend; medium radius with access	181.07	0.65	7.52	201.21	nr	**208.73**
bend; long radius	114.35	0.65	7.52	129.81	nr	**137.33**
diminishing pipe	48.38	0.65	7.52	62.19	nr	**69.71**
single branch	106.31	0.79	9.15	111.34	nr	**120.49**
isolated Timesaver joint	24.04	0.39	4.52	24.64	nr	**29.16**
Accessories in Timesaver cast iron or other equal and approved; with mechanical coupling joints						
Gully fittings; comprising low invert gully trap and round hopper						
100 mm outlet	47.91	0.88	10.19	74.84	nr	**85.03**
150 mm outlet	119.22	1.20	13.90	154.87	nr	**168.77**
Add to above for bellmouth 300 mm high; circular plain grating						
100 mm nominal size; 200 mm grating	49.90	0.42	4.87	79.20	nr	**84.07**
100 mm nominal size; 100 mm horizontal inlet; 200 mm grating	61.01	0.42	4.87	90.87	nr	**95.74**
100 mm nominal size; 100 mm vertical inlet; 200 mm grating	62.56	0.42	4.87	92.50	nr	**97.37**

34 DRAINAGE BELOW GROUND

Item	PC £	Labour hours	Labour £	Material £	Unit	Total rate £
Yard gully (Deans); trapped; galvanized sediment pan; 267 mm round heavy grating						
100 mm outlet	323.80	2.68	31.04	390.08	nr	**421.12**
Yard gully (garage); trapless; galvanized sediment pan; 267 mm round heavy grating						
100 mm outlet	330.30	2.50	28.96	372.35	nr	**401.31**
Yard gully (garage); trapped; with rodding eye, galvanized perforated sediment pan; stopper; 267 mm round heavy grating						
100 mm outlet	629.63	2.50	28.96	733.36	nr	**762.32**
Grease trap; internal access; galvanized perforated bucket; lid and frame						
100 mm outlet; 20 gallon capacity	685.13	3.70	42.86	755.40	nr	**798.26**
Ensign lightweight drain pipes and fittings or other equal and approved; BS EN 877; ductile iron couplings						
100 mm dia. pipes						
laid straight	24.21	0.19	2.57	28.94	m	**31.51**
extra for						
bend; long radius	37.48	0.19	2.57	50.79	nr	**53.36**
single branch	25.90	0.23	3.15	51.29	nr	**54.44**
150 mm dia. pipes						
laid straight	47.05	0.22	2.99	56.64	m	**59.63**
extra for						
bend; medium radius	112.34	0.22	2.99	140.40	nr	**143.39**
single branch	60.87	0.28	3.79	112.90	nr	**116.69**
Clay						
Extra strength vitrified clay pipes and fittings; Hepworth Supersleve or other equal; plain ends with push fit polypropylene flexible couplings						
100 mm dia. pipes						
laid straight	4.41	0.19	2.20	7.85	m	**10.05**
extra for						
bend	5.94	0.19	2.20	11.40	nr	**13.60**
rest bend	10.53	0.19	2.20	16.09	nr	**18.29**
rodding point	28.79	0.19	2.20	34.36	nr	**36.56**
socket adaptor	7.77	0.16	1.86	10.84	nr	**12.70**
saddle	12.61	0.69	8.00	16.26	nr	**24.26**
single junction	12.84	0.23	2.67	21.35	nr	**24.02**
single access junction	41.86	0.23	2.67	51.09	nr	**53.76**

34 DRAINAGE BELOW GROUND

Item	PC £	Labour hours	Labour £	Material £	Unit	Total rate £
PIPES, FITTINGS AND ACCESSORIES – cont						
Extra strength vitrified clay pipes and fittings – cont						
150 mm dia. pipes						
laid straight	8.91	0.23	2.67	15.18	m	**17.85**
extra for						
bend	12.24	0.22	2.55	21.91	nr	**24.46**
access bend	6.50	0.22	2.55	55.20	nr	**57.75**
90° bend	12.24	0.22	2.55	21.91	nr	**24.46**
taper pipe	15.16	0.22	2.55	22.19	nr	**24.74**
rodding point	44.16	0.22	2.55	53.72	nr	**56.27**
socket adaptor	12.42	0.19	2.20	17.88	nr	**20.08**
adaptor to HepSeal pipe	8.16	0.19	2.20	13.51	nr	**15.71**
saddle	18.75	0.83	9.61	25.28	nr	**34.89**
single junction	16.64	0.28	3.24	31.57	nr	**34.81**
single access junction	62.22	0.28	3.24	78.29	nr	**81.53**
225 mm dia. pipes						
laid straight	28.77	0.23	2.67	41.10	m	**43.77**
extra for						
bend	62.42	0.22	2.55	78.36	nr	**80.91**
90° bend	62.42	0.22	2.55	78.36	nr	**80.91**
taper pipe	15.16	0.22	2.55	18.31	nr	**20.86**
socket adaptor	26.69	0.19	2.20	36.02	nr	**38.22**
300 mm dia. pipes including epdm coupling						
laid straight	44.03	0.23	2.67	68.98	m	**71.65**
extra for						
bend	118.53	0.24	2.78	155.65	nr	**158.43**
90° bend	118.53	0.24	2.78	155.65	nr	**158.43**
taper pipe	15.16	0.24	2.78	25.84	nr	**28.62**
socket adaptor	42.37	0.20	2.32	62.76	nr	**65.08**
Extra strength vitrified clay pipes and fittings; Hepworth SuperSeal/Hepseal or equivalent; socketted; with push-fit flexible joints						
150 mm dia. pipes						
SuperSeal pipes; laid straight	14.25	0.30	3.47	14.61	m	**18.08**
extra for						
bend	27.39	0.23	2.67	23.69	nr	**26.36**
rest bend	14.71	0.20	2.32	10.69	nr	**13.01**
stopper	8.26	0.15	1.74	8.47	nr	**10.21**
taper reducer	14.18	0.23	2.67	10.16	nr	**12.83**
saddle	17.54	0.75	8.68	17.98	nr	**26.66**
single junction	35.80	0.30	3.47	30.85	nr	**34.32**
225 mm dia. pipes						
SuperSeal pipes; laid straight	29.57	0.38	4.40	30.31	m	**34.71**
extra for						
bend	64.19	0.30	3.47	56.69	nr	**60.16**
rest bend	78.41	0.30	3.47	71.28	nr	**74.75**
stopper	13.90	0.19	2.20	14.25	nr	**16.45**
taper reducer	44.22	0.30	3.47	36.22	nr	**39.69**
saddle	65.23	1.00	11.58	66.86	nr	**78.44**
single junction	114.01	0.38	4.40	104.74	nr	**109.14**

34 DRAINAGE BELOW GROUND

Item	PC £	Labour hours	Labour £	Material £	Unit	Total rate £
300 mm Superseal pipes						
SuperSeal pipes; laid straight	45.36	0.50	5.79	46.49	m	**52.28**
extra for						
bend	121.91	0.40	4.63	111.01	nr	**115.64**
rest bend	173.71	0.40	4.63	164.10	nr	**168.73**
stopper	29.69	0.25	2.90	30.43	nr	**33.33**
taper reducer	122.04	0.40	4.63	111.14	nr	**115.77**
saddle	113.58	1.33	15.41	116.42	nr	**131.83**
single junction	215.96	0.50	5.79	202.77	nr	**208.56**
400 mm dia. pipes						
Hepseal pipes; laid straight	111.56	0.67	7.76	114.35	m	**122.11**
extra for						
bend	419.23	0.54	6.25	395.40	nr	**401.65**
single unequal junction	392.82	0.67	7.76	356.89	nr	**364.65**
450 mm pipes						
Hepseal pipes; laid straight	144.92	0.83	9.61	148.54	m	**158.15**
extra for						
bend	552.06	0.67	7.76	521.29	nr	**529.05**
single unequal junction	469.53	0.83	9.61	421.85	nr	**431.46**
British Standard quality vitrified clay pipes and fittings; socketted; cement and sand (1:2) joints						
100 mm dia. pipes						
laid straight	9.01	0.37	4.28	9.36	m	**13.64**
extra for						
bend (short/medium/knuckle)	6.31	0.30	3.47	6.59	nr	**10.06**
bend (long/rest/elbow)	14.81	0.30	3.47	12.54	nr	**16.01**
single junction	16.55	0.37	4.28	13.42	nr	**17.70**
double collar	10.86	0.25	2.90	11.25	nr	**14.15**
150 mm dia. pipes						
laid straight	13.87	0.42	4.87	14.34	m	**19.21**
extra for						
bend (short/medium/knuckle)	13.73	0.33	3.82	9.93	nr	**13.75**
bend (long/rest/elbow)	24.80	0.33	3.82	21.28	nr	**25.10**
taper	32.83	0.33	3.82	29.08	nr	**32.90**
single junction	27.13	0.42	4.87	22.28	nr	**27.15**
double collar	18.09	0.28	3.24	18.67	nr	**21.91**
225 mm dia. pipes						
laid straight	27.46	0.51	5.90	28.44	m	**34.34**
extra for						
double collar	42.35	0.33	3.82	43.53	nr	**47.35**
300 mm pipes						
laid straight	46.04	0.69	8.00	47.48	m	**55.48**

34 DRAINAGE BELOW GROUND

Item	PC £	Labour hours	Labour £	Material £	Unit	Total rate £
PIPES, FITTINGS AND ACCESSORIES – cont						
Accessories in vitrified clay; set in concrete; with polypropylene coupling joints to pipes						
Rodding point; with square aluminium plate						
100 mm nominal size	28.79	0.46	5.33	34.46	nr	**39.79**
Gully fittings; comprising low back trap and square hopper; 150 mm × 150 mm square gully grid						
100 mm nominal size	28.52	0.79	9.15	37.52	nr	**46.67**
Gully fittings; comprising low back trap and square hopper with back inlet; 150 mm × 150 mm square gully grid						
100 mm nominal size	49.75	0.85	9.85	59.28	nr	**69.13**
Accessories in vitrified clay; set in concrete; with cement and sand (1:2) joints to pipes						
Yard gully; 225 mm dia.; including domestic duty grating and frame (up to 1 tonne) and combined filter and silk bucket						
100 mm outlet	126.87	2.50	28.96	130.54	nr	**159.50**
100 mm outlet; 100 mm back inlet	177.49	2.70	31.27	182.43	nr	**213.70**
150 mm outlet	126.87	3.50	40.54	130.54	nr	**171.08**
150 mm outlet; 150 mm back inlet	181.14	3.70	42.86	186.16	nr	**229.02**
Yard gully; 225 mm dia.; including medium duty grating and frame (up to 5 tonnes) and combined filter and silk bucket						
100 mm outlet	166.09	2.50	28.96	170.74	nr	**199.70**
100 mm outlet; 100 mm back inlet	220.27	2.70	31.27	226.28	nr	**257.55**
150 mm outlet	178.92	3.50	40.54	183.90	nr	**224.44**
150 mm outlet; 150 mm back inlet	223.94	3.70	42.86	230.04	nr	**272.90**
Road gully; trapped with rodding eye and stopper (grate not included)						
300 mm × 600 mm × 100 mm outlet	91.20	3.05	35.33	112.31	nr	**147.64**
300 mm × 600 mm × 150 mm outlet	93.39	3.05	35.33	114.55	nr	**149.88**
400 mm × 750 mm × 150 mm outlet	108.31	3.70	42.86	139.59	nr	**182.45**
450 mm × 900 mm × 150 mm outlet	146.54	4.65	53.86	184.47	nr	**238.33**
Grease trap; with internal access; galvanized perforated bucket; lid and frame						
600 mm × 450 mm × 600 mm deep; 100 mm outlet	746.63	3.89	45.06	793.41	nr	**838.47**
Interceptor; trapped with inspection arm; lever locking stopper; chain and staple; cement and sand (1:2) joints to pipes; building in, and cutting and fitting brickwork around						
100 mm outlet; 100 mm inlet	109.13	3.70	42.86	112.38	nr	**155.24**
150 mm outlet; 150 mm inlet	154.88	4.16	48.19	159.27	nr	**207.46**
225 mm outlet; 225 mm inlet	422.26	4.63	53.63	433.40	nr	**487.03**

34 DRAINAGE BELOW GROUND

Item	PC £	Labour hours	Labour £	Material £	Unit	Total rate £
Accessories in polypropylene; cover set in concrete; with coupling joints to pipes						
Inspection chamber; 5 nr 100 mm inlets; cast iron cover and frame						
475 mm dia. × 595 mm deep	207.48	2.13	24.67	219.99	nr	**244.66**
475 mm dia. × 940 mm deep	252.76	2.31	26.75	266.40	nr	**293.15**
Accessories; grates and covers						
Aluminium alloy gully grids; set in position						
120 mm × 120 mm	3.41	0.09	1.05	3.50	nr	**4.55**
150 mm × 150 mm	3.27	0.09	1.05	3.35	nr	**4.40**
225 mm × 225 mm	10.18	0.09	1.05	10.43	nr	**11.48**
100 mm dia.	3.41	0.09	1.05	3.50	nr	**4.55**
150 mm dia.	5.23	0.09	1.05	5.36	nr	**6.41**
225 mm dia.	11.39	0.09	1.05	11.67	nr	**12.72**
Aluminium alloy sealing plates and frames; set in cement and sand (1:3)						
150 mm × 150 mm	13.15	0.23	2.67	13.57	nr	**16.24**
225 mm × 225 mm	24.06	0.23	2.67	24.75	nr	**27.42**
140 mm dia. (for 100 mm)	10.71	0.23	2.67	11.08	nr	**13.75**
197 mm dia. (for 150 mm)	15.41	0.23	2.67	15.90	nr	**18.57**
273 mm dia. (for 225 mm)	24.67	0.23	2.67	25.39	nr	**28.06**
Polypropylene access covers and frames; supplied by Manhole Covers Ltd or other equal and approved; to suit PPIC inspection chambers; bedding and pointing in frame.						
450 mm dia.; class A15	31.98	1.30	15.06	34.39	nr	**49.45**
450 mm dia.; class B125; kite-marked	24.30	1.30	15.06	26.53	nr	**41.59**
Ductile iron heavy duty road gratings and frame; supplied by Manhole Covers Ltd or other equal and approved; bedding and pointing in cement and sand (1:3); one course half brick thick wall in semi-engineering bricks in cement: mortar (1:3)						
225 mm × 225 mm × 80 mm hinged and dished road grating and frame; class C250	21.33	2.25	26.07	24.67	nr	**50.74**
300 mm × 300 mm × 80 mm hinged and dished road grating and frame; class C250	35.24	2.25	26.07	38.93	nr	**65.00**
420 mm × 420 mm × 75 mm hinged road grating and frame; class C250; kite-marked	43.59	2.25	26.07	47.49	nr	**73.56**
445 mm × 445 mm × 75 mm double triangular road grating and frame; class C250; kite-marked	46.37	2.25	26.07	50.34	nr	**76.41**
435 mm × 435 mm × 100 mm pedestrian mesh road grating and frame; class D400	47.30	2.25	26.07	51.29	nr	**77.36**
440 mm × 400 mm × 150 mm hinged road grating and frame; class D400; kite-marked	61.21	2.25	26.07	65.55	nr	**91.62**

34 DRAINAGE BELOW GROUND

Item	PC £	Labour hours	Labour £	Material £	Unit	Total rate £
PIPES, FITTINGS AND ACCESSORIES – cont						
Concrete						
Vibrated concrete pipes and fittings; with flexible joints; BS 5911 Part 1; trench 2.00 m deep						
300 mm dia. pipes						
Class M; laid straight	11.91	0.65	7.52	18.07	m	25.59
extra for						
bend; <= 45°	–	0.65	7.52	122.06	nr	129.58
bend; > 45°	–	0.65	7.52	183.09	nr	190.61
junction; 300 mm × 100 mm	–	0.46	5.33	64.08	nr	69.41
450 mm dia. pipes						
Class H; laid straight	21.75	1.02	11.82	28.16	m	39.98
extra for						
bend; <= 45°	–	1.02	11.82	222.93	nr	234.75
bend; > 45°	–	1.02	11.82	334.40	nr	346.22
junction; 450 mm × 150 mm	–	0.65	7.52	117.03	nr	124.55
525 mm dia. pipes						
Class H; laid straight	28.51	1.48	17.14	35.09	m	52.23
extra for						
bend; <= 45°	–	1.48	17.14	298.09	nr	315.23
bend; > 45°	–	1.48	17.14	444.19	nr	461.33
junction; 600 mm × 150 mm	–	0.83	9.61	210.42	nr	220.03
900 mm dia. pipes						
Class H; laid straight	90.87	2.59	30.00	99.00	m	129.00
extra for						
bend; <= 45°	–	2.59	30.00	937.29	nr	967.29
bend; > 45°	–	2.59	30.00	1403.01	nr	1433.01
junction; 900 mm × 150 mm	–	1.02	11.82	425.01	nr	436.83
1200 mm dia. pipes						
Class H; laid straight	155.93	3.70	42.86	165.69	m	208.55
extra for						
bend; <= 45°	–	3.70	42.86	1604.09	nr	1646.95
bend; > 45°	–	3.70	42.86	2403.21	nr	2446.07
junction; 1200 mm × 150 mm	–	1.48	17.14	725.06	nr	742.20
Accessories in precast concrete; top set in with rodding eye and stopper; cement and sand (1:2) joint to pipe						
Concrete road gully; BS 5911; trapped with rodding eye and stopper; cement and sand (1:2) joint to pipe						
450 mm dia. × 900 mm deep; 100 mm or 150 mm outlet	35.46	4.39	50.85	58.11	nr	108.96
450 mm dia. × 1050 mm deep; 100 mm or 150 mm outlet	38.66	4.39	50.85	61.39	nr	112.24
Gully adapter type 1501	–	–	–	4.27	nr	4.27

34 DRAINAGE BELOW GROUND

Item	PC £	Labour hours	Labour £	Material £	Unit	Total rate £
uPVC						
Osmadrain uPVC pipes and fittings or other equal and approved; BS 4660; with ring seal joints						
82 mm dia. pipes						
laid straight	8.15	0.15	1.74	8.35	m	**10.09**
extra for						
bend; short radius	13.99	0.13	1.51	14.34	nr	**15.85**
spigot/socket bend	11.75	0.13	1.51	12.04	nr	**13.55**
adaptor	6.13	0.07	0.81	6.28	nr	**7.09**
single junction	18.19	0.18	2.08	18.64	nr	**20.72**
slip coupler	6.50	0.07	0.81	6.66	nr	**7.47**
100 mm dia. pipes						
laid straight	5.13	0.17	1.97	6.06	m	**8.03**
extra for						
bend; short radius	13.21	0.15	1.74	13.23	nr	**14.97**
bend; long radius	21.39	0.15	1.74	20.36	nr	**22.10**
spigot/socket bend	11.17	0.15	1.74	15.18	nr	**16.92**
socket plug	5.78	0.04	0.46	5.92	nr	**6.38**
adjustable double socket bend	15.81	0.15	1.74	19.99	nr	**21.73**
adaptor to clay	14.89	0.09	1.05	15.07	nr	**16.12**
single junction	15.76	0.21	2.43	14.58	nr	**17.01**
sealed access junction	40.77	0.19	2.20	40.21	nr	**42.41**
slip coupler	6.50	0.09	1.05	6.66	nr	**7.71**
160 mm dia. pipes						
laid straight	11.25	0.21	2.43	13.14	m	**15.57**
extra for						
bend; short radius	31.43	0.18	2.08	31.52	nr	**33.60**
spigot/socket bend	28.49	0.18	2.08	37.31	nr	**39.39**
socket plug	12.41	0.07	0.81	12.72	nr	**13.53**
adaptor to clay	32.39	0.12	1.39	32.63	nr	**34.02**
level invert taper	15.24	0.18	2.08	23.03	nr	**25.11**
single junction	51.46	0.24	2.78	52.75	nr	**55.53**
slip coupler	9.26	0.11	1.27	9.49	nr	**10.76**
uPVC Osma Ultra-Rib ribbed pipes and fittings or other equal and approved; WIS approval; with sealed ring push-fit joints						
150 mm dia. pipes						
laid straight	5.28	0.19	2.20	5.41	m	**7.61**
extra for						
bend; short radius	17.10	0.17	1.97	17.20	nr	**19.17**
adaptor to 160 mm dia. upvc	24.19	0.10	1.16	24.15	nr	**25.31**
adaptor to clay	49.66	0.10	1.16	50.58	nr	**51.74**
level invert taper	7.46	0.18	2.08	6.67	nr	**8.75**
single junction	30.75	0.22	2.55	29.89	nr	**32.44**

34 DRAINAGE BELOW GROUND

Item	PC £	Labour hours	Labour £	Material £	Unit	Total rate £
PIPES, FITTINGS AND ACCESSORIES – cont						
uPVC Osma Ultra-Rib ribbed pipes and fittings or other equal and approved – cont						
225 mm dia. pipes						
laid straight	13.23	0.22	2.55	13.56	m	**16.11**
extra for						
bend; short radius	68.74	0.20	2.32	69.65	nr	**71.97**
adaptor to clay	61.87	0.13	1.51	61.79	nr	**63.30**
level invert taper	11.96	0.20	2.32	9.82	nr	**12.14**
single junction	102.04	0.27	3.13	100.53	nr	**103.66**
300 mm dia. pipes						
laid straight	19.64	0.32	3.71	20.13	m	**23.84**
extra for						
bend; short radius	108.27	0.29	3.36	109.77	nr	**113.13**
adaptor to clay	162.75	0.14	1.62	164.40	nr	**166.02**
level invert taper	38.84	0.29	3.36	36.18	nr	**39.54**
single junction	235.79	0.37	4.28	235.65	nr	**239.93**
Accessories in uPVC; with ring seal joints to pipes (unless otherwise described)						
Rodding eye						
110 mm dia.	25.55	0.43	4.98	30.26	nr	**35.24**
Universal gulley fitting; comprising gulley trap, plain hopper						
150 mm × 150 mm grate	22.25	0.93	10.77	28.50	nr	**39.27**
Bottle gulley; comprising gulley with bosses closed; sealed access covers						
217 mm × 217 mm grate	44.15	0.78	9.03	50.95	nr	**59.98**
Shallow access pipe; light duty screw down access door assembly						
110 mm dia.	62.78	0.78	9.03	70.04	nr	**79.07**
Shallow access inspection junction; 3 nr 110 mm inlets; light duty screw down access door assembly						
110 mm dia.	97.26	1.11	12.85	102.14	nr	**114.99**
Shallow inspection chamber; 250 mm dia.; 600 mm deep; sealed cover and frame						
4 nr 110 mm outlets/inlets	80.22	1.28	14.82	102.55	nr	**117.37**
Universal inspection chamber; 450 mm dia.; single seal cast iron cover and frame; 4 nr 110 mm outlets/inlets						
500 mm deep	157.08	1.35	15.64	181.33	nr	**196.97**
730 mm deep	175.94	1.60	18.53	204.73	nr	**223.26**
960 mm deep	194.81	1.85	21.43	228.13	nr	**249.56**
Equal manhole base; 750 mm dia.						
6 nr 160 mm outlets/inlets	230.20	1.21	14.01	248.15	nr	**262.16**
Unequal manhole base; 750 mm dia.						
2 nr 160 mm, 4nr 110 mm outlets/inlets	177.98	1.21	14.01	194.63	nr	**208.64**
Kerb to gullies; class B engineering bricks on edge to three sides in cement: mortar (1:3) rendering in cement: mortar (1:3) to top and two sides and skirting to brickwork 230 mm high; dishing in cement: mortar (1:3) to gully; steel trowelled						
230 mm × 230 mm internally	–	1.39	16.10	1.24	nr	**17.34**

34 DRAINAGE BELOW GROUND

Item	PC £	Labour hours	Labour £	Material £	Unit	Total rate £
Slot and grate drainage						
ACO Multidrain M100D polymer concrete channel drainage system; galvanized steel edge trim; nominal bore 100 mm; type of fall constant; bedding and haunching in in situ concrete (not included)						
Slotted grating						
galvanized steel grating, load class A15 (pedestrian areas)	–	0.46	5.33	63.71	m	**69.04**
galvanized steel grating, load class C250 (cars and light vans)	–	0.46	5.33	84.05	m	**89.38**
ductile iron grating, load class D400 (driving lanes of roads)	–	0.46	5.33	82.54	m	**87.87**
Heelguard resin composite grating, load class C250 (cars and light vans)	–	0.46	5.33	90.46	m	**95.79**
extra for end caps	–	0.09	1.05	3.13	nr	**4.18**
extra for sump unit	–	1.39	16.10	93.59	nr	**109.69**
extra for ACO universal gully	–	1.50	17.37	401.19	nr	**418.56**
ACO S100 polymer concrete channel drainage system; bolted ductile iron grating, load class F900 (airfields); bedding and haunching in in situ concrete (not included)	–	1.00	11.58	137.49	m	**149.07**
extra for end caps	–	0.09	1.05	10.28	nr	**11.33**
extra for sump unit	–	1.50	17.37	158.21	nr	**175.58**
ACO Qmax large capacity slot drainage channel with MDPE body and hot dipped galvanized steel edge rail, up to load class F900; bedding and haunching in in situ concrete (not included)						
ACO Qmax 225	–	0.75	8.68	49.13	m	**57.81**
ACO Qmax 350	–	1.00	11.58	67.28	m	**78.86**
ACO Qmax 900	–	1.50	17.37	154.03	m	**171.40**
shallow access chamber	–	1.50	17.37	120.90	nr	**138.27**
deep access chamber	–	2.00	23.17	157.46	nr	**180.63**
ACO Kerbdrain one-piece polymer concrete combined drainage system, load class D400; bedding and haunching in in situ concrete (not included). Manufactured from recycled and recyclable material						
KerbDrain KD305	–	0.50	5.79	72.79	m	**78.58**
KerbDrain KD480	–	0.65	7.52	78.78	m	**86.30**
KerbDrain KD305 drop kerb (left drop, one centre stone and right drop) total length 2745 mm	–	2.00	23.17	148.98	nr	**172.15**
KerbDrain KD305 mitre unit	–	0.25	2.90	73.62	nr	**76.52**
KerbDrain KD end cap	–	0.09	1.05	35.17	nr	**36.22**
KerbDrain KD610 shallow gully assembly	–	1.50	17.37	567.91	nr	**585.28**

34 DRAINAGE BELOW GROUND

Item	PC £	Labour hours	Labour £	Material £	Unit	Total rate £
PIPES, FITTINGS AND ACCESSORIES – cont						
Interconnecting drainage channel; Birco-lite ref 8012 or other equal and approved; Marshalls Plc; galvanized steel grating ref 8041; bedding and haunching in in situ concrete (not included)						
100 mm wide						
laid level or to falls	–	0.46	5.33	39.21	m	**44.54**
100 mm dia. trapped outlet unit	–	1.39	16.10	86.28	nr	**102.38**
end caps	–	0.09	1.05	4.88	nr	**5.93**
LAND DRAINAGE						
Excavating; by hand; grading bottoms; earthwork support; filling to within 150 mm of surface with gravel rejects; remainder filled with excavated material and compacting; disposal of surplus soil on site; spreading on site average 50 m						
Pipes not exceeding 200 nominal size						
average depth of trench 0.75 m	–	1.57	18.33	9.61	m	**27.94**
average depth of trench 1.00 m	–	2.08	24.28	16.00	m	**40.28**
average depth of trench 1.25 m	–	2.91	33.97	20.32	m	**54.29**
average depth of trench 1.50 m	–	5.00	58.37	24.98	m	**83.35**
average depth of trench 1.75 m	–	5.92	69.12	29.28	m	**98.40**
average depth of trench 2.00 m	–	6.85	79.97	33.95	m	**113.92**
Disposal; load lorry by machine						
Excavated material						
off site; to tip not exceeding 13 km (using lorries); including Landfill Tax based on inactive waste	–	–	–	17.79	m³	**17.79**
Disposal; load lorry by hand						
Excavated material						
off site; to tip not exceeding 13 km (using lorries); including Landfill Tax based on inactive waste	–	0.75	8.75	16.75	m³	**25.50**
Vitrified clay perforated subsoil pipes; BS 65; Hepworth Hepline or other equal and approved						
Pipes; laid straight						
100 mm dia.	6.89	0.20	2.32	7.06	m	**9.38**
150 mm dia.	12.55	0.25	2.90	12.86	m	**15.76**
225 mm dia.	26.55	0.33	3.82	27.21	m	**31.03**
Concrete Canvas cement impregnated cloth lining to form ditch linings; holding ponds; slope protection and similar						
Type CC8, 8 mm thick and sprayed with water	–	0.10	1.34	32.41	m²	**33.75**

34 DRAINAGE BELOW GROUND

Item	PC £	Labour hours	Labour £	Material £	Unit	Total rate £
MANHOLES						
Excavating; by machine						
Manholes						
maximum depth not exceeding 1.00 m	–	0.19	2.21	3.42	m^3	**5.63**
maximum depth not exceeding 2.00 m	–	0.21	2.45	3.76	m^3	**6.21**
maximum depth not exceeding 4.00 m	–	0.25	2.92	4.40	m^3	**7.32**
Excavating; by hand						
Manholes						
maximum depth not exceeding 1.00 m	–	3.05	35.61	–	m^3	**35.61**
maximum depth not exceeding 2.00 m	–	3.61	42.15	–	m^3	**42.15**
maximum depth not exceeding 4.00 m	–	4.63	54.05	–	m^3	**54.05**
Earthwork support (average risk prices)						
Maximum depth not exceeding 1.00 m						
distance between opposing faces not exceeding 2.00 m	–	0.14	1.63	1.51	m^2	**3.14**
Maximum depth not exceeding 2.00 m						
distance between opposing faces not exceeding 2.00 m	–	0.18	2.10	2.86	m^2	**4.96**
Maximum depth not exceeding 4.00 m						
distance between opposing faces not exceeding 2.00 m	–	0.22	2.57	4.21	m^2	**6.78**
Disposal; by machine						
Excavated material						
off site; to tip not exceeding 13 km (using lorries) including Landfill Tax based on inactive waste	–	–	–	17.79	m^3	**17.79**
on site; depositing on site in spoil heaps; average 50 m distance	–	0.14	1.63	3.05	m^3	**4.68**
Disposal; by hand						
Excavated material						
off site; to tip not exceeding 13 km (using lorries) including Landfill Tax based on inactive waste	–	0.75	8.75	16.75	m^3	**25.50**
on site; depositing on site in spoil heaps; average 50 m distance	–	1.20	14.01	–	m^3	**14.01**
Filling to excavations; by machine						
Average thickness not exceeding 0.25 m						
arising excavations	–	0.14	1.63	1.61	m^3	**3.24**
Filling to excavations; by hand						
Average thickness not exceeding 0.25 m						
arising from excavations	–	0.93	10.85	–	m^3	**10.85**

34 DRAINAGE BELOW GROUND

Item	PC £	Labour hours	Labour £	Material £	Unit	Total rate £
MANHOLES – cont						
Plain in situ ready mixed designated concrete; C10 – 40 mm aggregate						
Beds						
thickness not exceeding 150 mm	83.29	2.78	38.06	85.37	m³	**123.43**
thickness 150 mm–450 mm	–	2.08	28.47	85.37	m³	**113.84**
thickness exceeding 450 mm	–	1.76	24.10	85.37	m³	**109.47**
Plain in situ ready mixed designated concrete; C20 – 20 mm aggregate						
Beds						
thickness not exceeding 150 mm	85.08	2.78	38.06	87.21	m³	**125.27**
thickness 150 mm–450 mm	–	2.08	28.47	87.21	m³	**115.68**
thickness exceeding 450 mm	–	1.76	24.10	87.21	m³	**111.31**
Plain in situ ready mixed designated concrete; C25 – 20 mm aggregate; (small quantities)						
Benching in bottoms						
150 mm–450 mm average thickness	83.28	8.33	121.26	85.36	m³	**206.62**
Reinforced in situ ready mixed designated concrete; C20 – 20 mm aggregate; (small quantities)						
Isolated cover slabs						
thickness not exceeding 150 mm	81.02	6.48	88.70	83.05	m³	**171.75**
Reinforcement; fabric to BS 4449; lapped; in beds or suspended slabs						
Ref D98 (1.54 kg/m²)						
400 mm minimum laps	1.09	0.11	1.78	1.12	m²	**2.90**
Ref A142 (2.22 kg/m²)						
400 mm minimum laps	1.39	0.11	1.78	1.42	m²	**3.20**
Ref A193 (3.02 kg/m²)						
400 mm minimum laps	1.88	0.11	1.78	1.93	m²	**3.71**
Formwork; basic finish						
Soffits of isolated cover slabs						
horizontal	–	2.64	42.77	3.35	m²	**46.12**
Edges of isolated cover slabs						
height not exceeding 250 mm	–	0.78	12.64	1.25	m	**13.89**
Precast concrete circular manhole rings; BS5911 Part 1; bedding, jointing and pointing in cement: mortar (1:3) on prepared bed						
Chamber or shaft rings; plain						
900 mm dia.	41.58	5.09	58.96	43.31	m	**102.27**
1050 mm dia.	43.88	6.01	69.61	46.35	m	**115.96**
1200 mm dia.	53.54	6.94	80.38	56.95	m	**137.33**

34 DRAINAGE BELOW GROUND

Item	PC £	Labour hours	Labour £	Material £	Unit	Total rate £
Chamber or shaft rings; reinforced						
1350 mm dia.	79.67	7.86	91.04	113.73	m	**204.77**
1500 mm dia.	89.17	8.79	101.81	124.86	m	**226.67**
1800 mm dia.	131.46	11.10	128.57	193.74	m	**322.31**
2100 mm dia.	260.18	13.88	160.76	333.61	m	**494.37**
extra for step irons built in	5.63	0.14	1.62	5.77	nr	**7.39**
extra for integrated ladder system 150 mm projection polypropylene encapusulated steps and rails	60.00	1.00	11.58	61.50	m	**73.08**
Reducing slabs						
1200 mm dia.	76.48	5.55	64.29	109.09	nr	**173.38**
1350 mm dia.	115.37	8.79	101.81	150.34	nr	**252.15**
1500 mm dia.	133.08	10.18	117.91	169.18	nr	**287.09**
1800 mm dia.	185.63	12.95	150.00	248.56	nr	**398.56**
Heavy duty cover slabs; to suit rings						
900 mm dia.	45.29	2.78	32.20	58.85	nr	**91.05**
1050 mm dia.	48.59	3.24	37.53	62.35	nr	**99.88**
1200 mm dia.	58.83	3.70	42.86	75.16	nr	**118.02**
1350 mm dia.	88.75	4.16	48.19	107.70	nr	**155.89**
1500 mm dia.	102.38	4.63	53.63	132.33	nr	**185.96**
1800 mm dia.	157.08	5.55	64.29	192.99	nr	**257.28**
2100 mm dia.	332.91	6.48	75.05	375.73	nr	**450.78**
Common bricks; in cement: mortar (1:3)						
Walls to manholes						
one brick thick	240.00	2.22	38.41	42.23	m²	**80.64**
one and a half brick thick	–	3.24	56.06	63.33	m²	**119.39**
Projections of footings						
two brick thick	–	4.53	78.37	84.45	m²	**162.82**
Class A engineering bricks; in cement: mortar (1:3)						
Walls to manholes						
one brick thick (PC £ per 1000)	450.00	2.50	43.26	56.74	m²	**100.00**
one and a half brick thick	–	3.61	62.45	58.81	m²	**121.26**
Projections of footings						
two brick thick	–	5.09	88.06	113.49	m²	**201.55**
Class B engineering bricks; in cement: mortar (1:3)						
Walls to manholes						
one brick thick (PC £ per 1000)	375.00	2.50	43.26	48.41	m²	**91.67**
one and a half brick thick	–	3.61	62.45	72.62	m²	**135.07**
Projections of footings						
two brick thick	–	5.09	88.06	96.83	m²	**184.89**

34 DRAINAGE BELOW GROUND

Item	PC £	Labour hours	Labour £	Material £	Unit	Total rate £
MANHOLES – cont						
Brickwork sundries						
Extra over for fair face; flush smooth pointing						
manhole walls	–	0.19	3.29	–	m²	**3.29**
Building ends of pipes into brickwork; making good fair face or rendering						
not exceeding 55 mm nominal size	–	0.09	1.56	–	nr	**1.56**
55 mm–110 mm nominal size	–	0.14	2.42	–	nr	**2.42**
over 110 mm nominal size	–	0.19	3.29	–	nr	**3.29**
Step irons; BS 1247; malleable; galvanized; building into joints						
general purpose pattern	–	0.14	2.42	3.08	nr	**5.50**
Cement and sand (1:3) in situ finishings; steel trowelled						
13 mm work to manhole walls; one coat; to						
brickwork base over 300 wide	–	0.65	11.24	1.38	m²	**12.62**
Cast iron inspection chambers; with bolted flat covers; BS 437; bedded in cement: mortar (1:3); with mechanical coupling joints						
100 mm × 100 mm						
one branch either side	232.58	1.40	16.22	239.08	nr	**255.30**
two branches either side	439.88	2.00	23.17	451.57	nr	**474.74**
150 mm × 100 mm						
one branch either side	287.93	1.55	17.96	295.83	nr	**313.79**
two branches either side	558.79	2.15	24.91	574.14	nr	**599.05**
150 mm × 150 mm						
one branch either side	356.62	1.80	20.85	366.92	nr	**387.77**
two branches either side	688.30	2.60	30.11	706.89	nr	**737.00**
Coated cast or ductile iron access covers and frames; to BS EN124; supplied by Manhole Covers Ltd or other equal; bedding frame in cement and sand (1:3); cover in grease and sand						
Light duty; cast iron; rectangular single seal solid top						
450 mm × 450 mm; class A15	33.83	1.50	17.37	36.29	nr	**53.66**
600 mm × 450 mm; class A15	33.83	1.50	17.37	36.44	nr	**53.81**
600 mm × 600 mm; class A15	52.27	1.50	17.37	55.50	nr	**72.87**
750 mm × 600 mm; class A15	102.70	1.50	17.37	107.19	nr	**124.56**
Light duty; cast iron; rectangular double seal solid top Medium duty; ductile iron; rectangular single seal solid top						
450 mm × 450 mm × 40 mm; class C250; kite-marked	62.73	2.00	23.17	66.21	nr	**89.38**
600 mm × 450 mm × 40 mm; slide-out; class C250; kite-marked	64.58	2.00	23.17	68.11	nr	**91.28**
600 mm × 600 mm × 40 mm; slide-out; class C250; kite-marked	78.11	2.00	23.17	81.98	nr	**105.15**
760 mm × 600 mm × 40 mm; slide-out; class C250; kite-marked	115.93	2.00	23.17	120.74	nr	**143.91**

34 DRAINAGE BELOW GROUND

Item	PC £	Labour hours	Labour £	Material £	Unit	Total rate £
Heavy duty; ductile iron; solid top						
450 mm × 450 mm × 75 mm; single seal; class C250; kite-marked	91.81	2.50	28.96	96.03	nr	**124.99**
600 mm × 450 mm × 75 mm; single seal; class C250; kite-marked	102.02	2.50	28.96	106.49	nr	**135.45**
600 mm × 600 mm × 75 mm; single seal; class C250; kite-marked	116.86	2.50	28.96	121.70	nr	**150.66**
450 mm × 450 mm × 100 mm; double triangular; class D400; kite-marked	83.47	2.50	28.96	87.47	nr	**116.43**
600 mm × 450 mm × 100 mm; double triangular; class D400; kite-marked	109.43	2.50	28.96	114.09	nr	**143.05**
600 mm × 600 mm × 100 mm; double triangular; class D400; kite-marked	66.78	2.50	28.96	70.37	nr	**99.33**
750 mm × 600 mm × 100 mm; double triangular; class D400; kite-marked	170.64	2.50	28.96	176.83	nr	**205.79**
1220 mm × 675 mm × 100 mm; double triangular; class D400; kite-marked	190.12	3.50	40.54	196.80	nr	**237.34**
British Standard best quality vitrified clay channels; bedding and jointing in cement and sand (1:2)						
Half section straight						
100 mm dia. × 1 m long	4.95	0.74	8.57	5.07	nr	**13.64**
150 mm dia. × 1 m long	8.71	0.93	10.77	8.93	nr	**19.70**
225 mm dia. × 1 m long	20.85	1.20	13.90	21.37	nr	**35.27**
300 mm dia. × 1 m long	43.92	1.48	17.14	45.02	nr	**62.16**
Half section bend						
100 mm dia.	5.02	0.56	6.49	5.15	nr	**11.64**
150 mm dia.	8.67	0.69	8.00	8.89	nr	**16.89**
225 mm dia.	33.62	0.93	10.77	34.46	nr	**45.23**
Taper straight						
150 mm–100 mm dia.	12.74	0.65	7.52	13.06	nr	**20.58**
300 mm–225 mm dia.	111.00	0.83	9.61	113.77	nr	**123.38**
Taper bend						
150 mm–100 mm dia.	39.69	0.83	9.61	40.68	nr	**50.29**
225 mm–150 mm dia.	113.59	1.06	12.28	116.43	nr	**128.71**
Three quarter section branch bend						
100 mm dia.	14.13	0.46	5.33	14.48	nr	**19.81**
150 mm dia.	23.72	0.69	8.00	24.31	nr	**32.31**
225 mm dia.	86.56	0.93	10.77	88.72	nr	**99.49**
uPVC channels; with solvent weld or lip seal coupling joints; bedding in cement and sand						
Half section cut away straight; with coupling either end						
110 mm dia.	36.25	0.28	3.24	47.56	nr	**50.80**
160 mm dia.	68.07	0.37	4.28	88.91	nr	**93.19**
Half section cut away long radius bend; with coupling either end						
110 mm dia.	59.43	0.28	3.24	71.33	nr	**74.57**
160 mm dia.	128.49	0.37	4.28	150.84	nr	**155.12**

34 DRAINAGE BELOW GROUND

Item	PC £	Labour hours	Labour £	Material £	Unit	Total rate £
MANHOLES – cont						
uPVC channels – cont						
Channel adaptor to clay; with one coupling						
110 mm dia.	13.91	0.23	2.67	19.46	nr	**22.13**
160 mm dia.	33.64	0.31	3.59	44.05	nr	**47.64**
Half section bend						
110 mm dia.	22.94	0.31	3.59	23.93	nr	**27.52**
160 mm dia.	39.41	0.46	5.33	41.37	nr	**46.70**
Half section channel connector						
110 mm dia.	6.28	0.07	0.81	7.27	nr	**8.08**
Half section channel junction						
110 mm dia.	17.80	0.46	5.33	18.65	nr	**23.98**
Polypropylene slipper bend						
110 mm dia.	15.44	0.37	4.28	16.25	nr	**20.53**
Glass fibre septic tank; Klargester or other equal and approved; fixing lockable manhole cover and frame; placing in position						
2800 litre capacity; depth to invert						
1000 mm deep	–	2.45	28.37	789.46	nr	**817.83**
1500 mm deep	–	2.73	31.62	836.50	nr	**868.12**
3800 litre capacity; depth to invert						
1000 mm deep	–	2.64	30.58	1008.70	nr	**1039.28**
1500 mm deep	–	2.91	33.70	1055.75	nr	**1089.45**
4600 litre capacity; depth to invert						
1000 mm deep	–	2.75	31.85	1156.43	nr	**1188.28**
1500 mm deep	–	3.20	37.06	1203.47	nr	**1240.53**

35 SITE WORKS

Item	PC £	Labour hours	Labour £	Material £	Unit	Total rate £
KERBS, EDGINGS AND CHANNELS						
Excavating; by machine						
Excavating trenches; to receive kerb foundations; average size						
300 mm × 100 mm	–	0.02	0.24	0.25	m	**0.49**
450 mm × 150 mm	–	0.02	0.24	0.49	m	**0.73**
600 mm × 200 mm	–	0.03	0.37	0.69	m	**1.06**
Excavating curved trenches; to receive kerb foundations; average size						
300 mm × 100 mm	–	0.01	0.12	0.39	m	**0.51**
450 mm × 150 mm	–	0.03	0.35	0.58	m	**0.93**
600 mm × 200 mm	–	0.04	0.47	0.73	m	**1.20**
Excavating; by hand						
Excavating trenches; to receive kerb foundations; average size						
150 mm × 50 mm	–	0.02	0.24	–	m	**0.24**
200 mm × 75 mm	–	0.06	0.70	–	m	**0.70**
250 mm × 100 mm	–	0.10	1.17	–	m	**1.17**
300 mm × 100 mm	–	0.13	1.52	–	m	**1.52**
Excavating curved trenches; to receive kerb foundations; average size						
150 mm × 50 mm	–	0.03	0.35	–	m	**0.35**
200 mm × 75 mm	–	0.07	0.82	–	m	**0.82**
250 mm × 100 mm	–	0.11	1.28	–	m	**1.28**
300 mm × 100 mm	–	0.14	1.63	–	m	**1.63**
Plain in situ ready mixed designated concrete; C7.5 – 40 mm aggregate; poured on or against earth or unblinded hardcore						
Foundations	82.94	1.16	15.88	85.01	m³	**100.89**
Blinding beds						
thickness not exceeding 150 mm	82.94	1.71	23.41	85.01	m³	**108.42**
Plain in situ ready mixed designated concrete; C10 – 40 mm aggregate; poured on or against earth or unblinded hardcore						
Foundations	83.29	1.16	15.88	85.37	m³	**101.25**
Blinding beds						
thickness not exceeding 150 mm	83.29	1.71	23.41	85.37	m³	**108.78**
Plain in situ ready mixed designated concrete; C20 – 20 mm aggregate; poured on or against earth or unblinded hardcore						
Foundations	85.08	1.16	15.88	87.21	m³	**103.09**
Blinding beds						
thickness not exceeding 150 mm	85.08	1.71	23.41	87.21	m³	**110.62**

35 SITE WORKS

Item	PC £	Labour hours	Labour £	Material £	Unit	Total rate £
KERBS, EDGINGS AND CHANNELS – cont						
Filling to make up levels; by machine						
Average thickness not exceeding 0.25 m						
obtained off site; hardcore	–	0.28	3.27	26.77	m³	**30.04**
obtained off site; granular fill type one	–	0.28	3.27	33.01	m³	**36.28**
obtained off site; granular fill type two	–	0.28	3.27	31.35	m³	**34.62**
Average thickness exceeding 0.25 m						
obtained off site; hardcore	–	0.24	2.80	23.00	m³	**25.80**
obtained off site; granular fill type one	–	0.24	2.80	32.93	m³	**35.73**
obtained off site; granular fill type two	–	0.24	2.80	31.28	m³	**34.08**
Filling to make up levels; by hand						
Average thickness not exceeding 0.25 m						
obtained off site; hardcore	–	0.61	7.12	27.37	m³	**34.49**
obtained off site; sand	–	0.71	8.29	43.91	m³	**52.20**
Average thickness exceeding 0.25 m						
obtained off site; hardcore	–	0.51	5.96	23.42	m³	**29.38**
obtained off site; sand	–	0.60	7.00	43.65	m³	**50.65**
Surface treatments						
Compacting						
filling; blinding with sand	–	0.04	0.47	2.15	m²	**2.62**
Precast concrete kerbs, channels, edgings, etc.;						
BS 340; bedded, jointed and pointed in cement:						
mortar (1:3); including haunching up one side						
with in situ ready mix designated concrete						
C10 – 40 mm aggregate; to concrete base						
Edgings; straight; square edge						
50 mm × 150 mm	–	0.23	4.64	2.95	m	**7.59**
50 mm × 200 mm	–	0.23	4.64	4.54	m	**9.18**
50 mm × 255 mm	–	0.23	4.64	5.78	m	**10.42**
Kerbs; straight						
125 mm × 255 mm; half battered	–	0.31	6.26	6.26	m	**12.52**
125 mm × 255 mm; half battered drop kerb	–	0.31	6.26	11.23	m	**17.49**
150 mm × 305 mm; half battered	–	0.31	6.26	14.05	m	**20.31**
150 mm × 305 mm; half battered drop kerb	–	0.31	6.26	39.11	m	**45.37**
Kerbs; curved						
125 mm × 255 mm; half battered	–	0.46	9.30	9.39	m	**18.69**
150 mm × 305 mm; half battered	–	0.46	9.30	15.42	m	**24.72**
Channels; 255 × 125 mm						
straight	–	0.31	6.26	10.08	m	**16.34**
Quadrants; half battered						
305 mm × 305 mm × 150 mm	–	0.32	6.47	11.11	nr	**17.58**
305 mm × 305 mm × 255 mm	–	0.32	6.47	12.15	nr	**18.62**
455 mm × 455 mm × 255 mm	–	0.37	7.47	15.77	nr	**23.24**

35 SITE WORKS

Item	PC £	Labour hours	Labour £	Material £	Unit	Total rate £
IN SITU CONCRETE ROADS AND PAVINGS						
Reinforced in situ ready mixed designated concrete; C10 – 40 mm aggregate						
Roads; to hardcore base						
thickness not exceeding 150 mm	79.32	1.85	25.33	81.30	m³	**106.63**
thickness 150 mm–450 mm	79.32	1.30	17.79	81.30	m³	**99.09**
Reinforced in situ ready mixed designated concrete; C20 – 20 mm aggregate						
Roads; to hardcore base						
thickness not exceeding 150 mm	81.02	1.85	25.33	83.05	m³	**108.38**
thickness 150 mm–450 mm	81.02	1.30	17.79	83.05	m³	**100.84**
Reinforced in situ ready mixed designated concrete; C25 – 20 mm aggregate						
Roads; to hardcore base						
thickness not exceeding 150 mm	83.28	1.85	25.33	85.36	m³	**110.69**
thickness 150 mm–450 mm	83.28	1.30	17.79	85.36	m³	**103.15**
Formwork; sides of foundations; basic finish						
Plain vertical						
height not exceeding 250 mm	–	0.39	6.31	1.48	m	**7.79**
height 250 mm–500 mm	–	0.57	9.24	2.40	m	**11.64**
height 500 mm–1.00 m	–	0.83	13.45	3.47	m	**16.92**
add to above for curved radius 6 m	–	0.03	0.48	0.15	m	**0.63**
Reinforcement; fabric; BS 4449; lapped; in roads, footpaths or pavings						
Ref A142 (2.22 kg/m²)						
400 mm minimum laps	1.39	0.14	2.27	1.42	m²	**3.69**
Ref A193 (3.02 kg/m²)						
400 mm minimum laps	–	0.14	2.27	1.93	m²	**4.20**
Formed joints; Fosroc Expandite Flexcell impregnated joint filler or other equal and approved						
Width not exceeding 150 mm						
12.50 mm thick	–	0.14	2.27	1.74	m	**4.01**
25 mm thick	–	0.19	3.08	4.91	m	**7.99**
Width 150–300 mm						
12.50 mm thick	–	0.19	3.08	2.58	m	**5.66**
25 mm thick	–	0.19	3.08	4.79	m	**7.87**
Width 300–450 mm						
12.50 mm thick	–	0.23	3.73	3.88	m	**7.61**
25 mm thick	–	0.23	3.73	7.17	m	**10.90**
Sealants; Fosroc Expandite Pliastic N2 hot poured rubberized bituminous compound or other equal and approved						
Width 25 mm						
25 mm depth	–	0.20	3.24	1.56	m	**4.80**

35 SITE WORKS

Item	PC £	Labour hours	Labour £	Material £	Unit	Total rate £
IN SITU CONCRETE ROADS AND PAVINGS – cont						
Concrete sundries						
Treating surfaces						
unset concrete; grading to cambers; tamping with a 75 mm thick steel shod tamper	–	0.23	3.15	–	m²	3.15
Sundries						
Line marking						
width not exceeding 300 mm; NB minimum charge usually applies	–	0.04	0.57	0.19	m	0.76
COATED MACADAM, ASPHALT ROADS AND PAVINGS						
NOTE: The prices for all bitumen macadam and hot rolled asphalt materials are for individual courses to roads and footpaths and need combining to arrive at complete specifications and costs for full construction. Costs include for work to falls, crossfalls or slopes not exceeding 15° from horizontal; for laying on prepared bases (prices not included) and for rolling with an appropriate roller. The following rates are based on black bitumen macadam. Red bitumen macadam rates are approximately 50% dearer. PSV is Polished Stone Value.						
Dense bitumen macadam base course; BS EN 13108; bitumen penetration 100/125						
Carriageway, hardshoulder and hardstrip						
100 mm thick; one coat; with 0/32 mm aggregate size	–	–	–	–	m²	18.54
200 mm thick; one coat; with 0/32 mm aggregate size	–	–	–	–	m²	32.64
extra for increase thickness in 10 mm increments	–	–	–	–	m²	1.33
Hot rolled asphalt base course; BS EN 13108						
Carriageway, hardshoulder and hardstrip						
150 mm thick; one coat; 60% 0/32 mm aggregate size; to column 2/5	–	–	–	–	m²	30.83
200 mm thick; one coat; 60% 0/32 mm aggregate size; to column 2/5	–	–	–	–	m²	41.01
extra for increase thickness in 10 mm increments	–	–	–	–	m²	1.70
Dense bitumen macadam binder course; BS EN 13108; bitumen penetration 100/125						
Carriageway, hardshoulder and hardstrip						
60 mm thick; one coat; with 0/32 mm aggregate size	–	–	–	–	m²	11.34
60 mm thick; one coat; with 0/32 mm aggregate size; to clause 6.5	–	–	–	–	m²	11.44
extra for increase thickness in 10 mm increments	–	–	–	–	m²	1.51

35 SITE WORKS

Item	PC £	Labour hours	Labour £	Material £	Unit	Total rate £
Hot rolled asphalt binder course; BS EN 13108						
Carriageway, hardshoulder and hardstrip						
40 mm thick; one coat; 50% 0/14 mm aggregate size; to column 2/2; 55 PSV	–	–	–	–	m²	**10.82**
60 mm thick; one coat; 50% 0/14 mm aggregate size; to column 2/2	–	–	–	–	m²	**12.46**
60 mm thick; one coat; 50% 0/20 mm aggregate size; to column 2/3	–	–	–	–	m²	**12.24**
60 mm thick; one coat; 60% 0/32 mm aggregate size; to column 2/5	–	–	–	–	m²	**11.62**
100 mm thick; one coat; 60% 0/32 mm aggregate size; to column 2/5	–	–	–	–	m²	**18.77**
extra for increase thickness in 10 mm increments	–	–	–	–	m²	**2.24**
Macadam surface course; BS EN 13108; bitumen penetration 100/125						
Carriageway, hardshoulder and hardstrip						
30 mm thick; one coat; medium graded with 0/6 mm nominal aggregate binder	–	–	–	–	m²	**8.76**
40 mm thick; one coat; close graded with 0/14 mm nominal aggregate binder; to clause 7.3	–	–	–	–	m²	**8.02**
40 mm thick; one coat; close graded with 0/10 mm nominal aggregate binder; to clause 7.4	–	–	–	–	m²	**8.76**
extra over above items for increase/reduction in 10 mm thick increments	–	–	–	–	m²	**1.58**
extra over above items for coarse aggregate 60–64 PSV	–	–	–	–	m²	**1.59**
extra over above items for coarse aggregate 65–67 PSV	–	–	–	–	m²	**1.74**
extra over above items for coarse aggregate 68 PSV	–	–	–	–	m²	**2.31**
Hot rolled asphalt surface course; BS EN 13108; bitumen penetration 40/60						
Carriageway, hardshoulder and hardstrip						
40 mm thick; one coat; 30% mix 0/10 mm aggregate size; to column 3/2; with 20 mm pre-coated chippings 60–64 PSV	–	–	–	–	m²	**10.81**
40 mm thick; one coat; 30% mix 0/10 mm aggregate size; to column 3/2; with 14 mm pre-coated chippings 60–64 PSV	–	–	–	–	m²	**10.90**
extra over above items for increase/reduction in 10 mm increments	–	–	–	–	m²	**1.80**
extra over above items for chippings with 65–67 PSV	–	–	–	–	m²	**0.10**
extra over above items for chippings with 68 PSV	–	–	–	–	m²	**0.16**
extra over above items for 6–10 KN High Traffic Flows	–	–	–	–	m²	**0.78**

35 SITE WORKS

Item	PC £	Labour hours	Labour £	Material £	Unit	Total rate £
COATED MACADAM, ASPHALT ROADS AND PAVINGS – cont						
Stone mastic asphalt surface course; BS EN 13108						
Carriageway, hardshoulder and hardstrip						
35 mm thick; one coat; with 0/14 mm nominal aggregate size; 55 PSV	–	–	–	–	m²	9.94
35 mm thick; one coat; with 0/10 mm nominal aggregate size; 55 PSV	–	–	–	–	m²	9.94
extra for increase thickness in 10 mm increments	–	–	–	–	m²	2.22
Thin surface course with 60 PSV						
Carriageway, hardshoulder and hardstrip						
35 mm thick; one coat; with 0/10 mm nominal aggregate size	–	–	–	–	m²	9.94
extra for increase thickness in 10 mm increments	–	–	–	–	m²	1.12
extra over above items for coarse aggregate 60–64 PSV	–	–	–	–	m²	0.27
extra over above items for coarse aggregate 65–67 PSV	–	–	–	–	m²	0.27
extra over above items for coarse aggregate 68 PSV	–	–	–	–	m²	0.51
Regulating courses						
Carriageway, hardshoulder and hardstrip						
Dense Bitumen Macadam; bitumen penetration 100/125; with 0/20 mm nominal aggregate regulating course	–	–	–	–	tonne	85.00
Hot rolled asphalt; 50% 0/20 mm aggregate size	–	–	–	–	tonne	94.00
Stone mastic asphalt; 0/6 mm aggregate	–	–	–	–	tonne	120.00
Bitumen Emulsion tack coats						
Carriageway, hardshoulder and hardstrip						
K1–40; applied 0.35–0.45 l/m²	–	–	–	–	m²	0.16
K1–70; applied 0.35–0.45 l/m²	–	–	–	–	m²	0.26
Sundries						
Line marking						
width not exceeding 300 mm; NB minimum charge usually applies	–	0.04	0.57	0.19	m	0.76

35 SITE WORKS

Item	PC £	Labour hours	Labour £	Material £	Unit	Total rate £
GRAVEL PAVINGS						
Two coat gravel paving; level and to falls; first layer course clinker aggregate and wearing layer fine gravel aggregate						
Pavings; over 300 mm wide						
50 mm thick	–	0.07	1.41	2.22	m²	**3.63**
63 mm thick	–	0.09	1.81	2.91	m²	**4.72**
Resin bonded gravel paving; level and to falls						
Pavings; over 300 mm wide						
50 mm thick	–	–	–	–	m²	**36.38**
SLAB, BRICK/BLOCK SETTS AND COBBLE PAVINGS						
Artificial stone paving; Charcon's Moordale Textured or other equal and approved; to falls or crossfalls; bedding 25 mm thick in cement: mortar (1:3); staggered joints; jointing in coloured cement: mortar (1:3), brushed in; to sand base						
Pavings; over 300 mm wide						
600 mm × 600 mm × 50 mm thick; natural	12.55	0.39	7.88	15.71	m²	**23.59**
Brick paviors; 215 mm × 103 mm × 65 mm rough stock bricks; to falls or crossfalls; bedding 10 mm thick in cement: mortar (1:3); jointing in cement: mortar (1:3); as work proceeds; to concrete base						
Pavings; over 300 mm wide; straight joints both ways						
bricks laid flat (PC £ per 1000)	400.00	0.74	12.80	19.70	m²	**32.50**
bricks laid on edge	–	1.04	17.99	29.21	m²	**47.20**
Pavings; over 300 mm wide; laid to herringbone pattern						
bricks laid flat bricks laid flat (PC £ per 1000)	400.00	0.93	16.09	19.70	m²	**35.79**
bricks laid on edge	–	1.30	22.49	29.21	m²	**51.70**
ADD or DEDUCT for variation of £10.00/1000 in PC of brick paviors						
bricks laid flat bricks laid flat (PC £ per 1000)	–	–	–	0.46	m²	**0.46**
bricks laid on edge	–	–	–	0.70	m²	**0.70**
River washed cobble paving; 50 mm–75 mm; to falls or crossfalls; bedding 13 mm thick in cement: mortar (1:3); jointing to a height of two thirds of cobbles in dry mortar (1:3); tightly butted, washed and brushed; to concrete						
Pavings; over 300 mm wide						
regular (PC £ per tonne)	84.98	3.70	74.75	20.26	m²	**95.01**
laid to pattern	–	4.63	93.54	20.26	m²	**113.80**

35 SITE WORKS

Item	PC £	Labour hours	Labour £	Material £	Unit	Total rate £
SLAB, BRICK/BLOCK SETTS AND COBBLE PAVINGS – cont						
Concrete paving flags; BS EN 1339; to falls or crossfalls; bedding 25 mm thick in cement and sand mortar (1:4); butt joints straight both ways; jointing in cement and sand (1:3); brushed in; to sand base						
Pavings; over 300 mm wide						
450 mm × 600 mm × 50 mm thick; grey	7.41	0.42	8.49	9.36	m²	**17.85**
450 mm × 600 mm × 60 mm thick; coloured	8.27	0.42	8.49	10.28	m²	**18.77**
600 mm × 600 mm × 50 mm thick; grey	5.75	0.39	7.88	7.59	m²	**15.47**
600 mm × 600 mm × 50 mm thick; coloured	6.89	0.39	7.88	8.82	m²	**16.70**
750 mm × 600 mm × 50 mm thick; grey	5.17	0.36	7.28	6.95	m²	**14.23**
750 mm × 600 mm × 50 mm thick; coloured	6.87	0.36	7.28	8.77	m²	**16.05**
900 mm × 600 mm × 50 mm thick; grey	4.60	0.33	6.66	6.33	m²	**12.99**
900 mm × 600 mm × 50 mm thick; coloured	6.31	0.33	6.66	8.17	m²	**14.83**
Blister Tactile paving flags; to falls or crossfalls; bedding 25 mm thick in cement and sand mortar (1:4); butt joints straight both ways; jointing in cement and sand (1:3); brushed in; to sand base						
Pavings; over 300 mm wide						
400 mm × 400 mm × 50 mm thick; buff	18.94	0.45	9.09	21.77	m²	**30.86**
400 mm × 400 mm × 60 mm thick; buff	20.58	0.45	9.09	23.52	m²	**32.61**
400 mm × 400 mm × 50 mm thick; buff	21.19	0.45	9.09	24.19	m²	**33.28**
400 mm × 400 mm × 70 mm thick; buff	23.56	0.45	9.09	26.74	m²	**35.83**
Concrete rectangular paving blocks; to falls or crossfalls; bedding 50 mm thick in dry sharp sand; filling joints with sharp sand brushed in; on earth base						
Pavings; Keyblock or other equal and approved; over 300 mm wide; straight joints both ways						
200 mm × 100 mm × 60 mm thick; grey	6.93	0.69	13.94	10.67	m²	**24.61**
200 mm × 100 mm × 60 mm thick; coloured	7.46	0.69	13.94	11.23	m²	**25.17**
200 mm × 100 mm × 60 mm thick; brindle	7.57	0.69	13.94	11.36	m²	**25.30**
200 mm × 100 mm × 80 mm thick; grey	7.72	0.74	14.95	11.79	m²	**26.74**
200 mm × 100 mm × 80 mm thick; coloured	8.71	0.74	14.95	12.85	m²	**27.80**
200 mm × 100 mm × 80 mm thick; brindle	8.91	0.74	14.95	13.07	m²	**28.02**
Pavings; Keyblock or other equal and approved; over 300 mm wide; laid to herringbone pattern						
200 mm × 100 mm × 60 mm thick; grey	6.93	0.88	17.77	10.67	m²	**28.44**
200 mm × 100 mm × 60 mm thick; coloured	7.46	0.88	17.77	11.23	m²	**29.00**
200 mm × 100 mm × 80 mm thick; grey	7.72	0.93	18.79	11.79	m²	**30.58**
200 mm × 100 mm × 80 mm thick; coloured	8.71	0.93	18.79	12.85	m²	**31.64**
Extra for two row boundary edging to herringbone pavings; 200 mm wide; including a 150 mm high in situ concrete mix C10 – 40 mm aggregate haunching to one side; blocks laid breaking joint						
200 mm × 100 mm × 60 mm; coloured	–	0.28	5.66	2.18	m	**7.84**
200 mm × 100 mm × 80 mm; coloured	–	0.28	5.66	2.29	m	**7.95**

35 SITE WORKS

Item	PC £	Labour hours	Labour £	Material £	Unit	Total rate £
Pavings; Europa or other equal; over 300 mm wide; straight joints both ways						
200 mm × 100 mm × 60 mm thick; grey	6.76	0.69	13.94	10.50	m²	**24.44**
200 mm × 100 mm × 60 mm thick; coloured	7.51	0.69	13.94	11.30	m²	**25.24**
200 mm × 100 mm × 80 mm thick; grey	8.05	0.74	14.95	12.15	m²	**27.10**
200 mm × 100 mm × 80 mm thick; coloured	8.83	0.74	14.95	12.98	m²	**27.93**
Pavings; Metropolitan or other equal; over 300 mm wide; straight joints both ways						
200 mm × 100 mm × 80 mm thick; grey	18.08	0.74	14.95	22.94	m²	**37.89**
200 mm × 100 mm × 80 mm thick; coloured	18.08	0.74	14.95	22.94	m²	**37.89**
Pavings; Intersett or other equal and approved; over 300 mm wide; straight joints both ways						
200 mm × 100 mm × 60 mm thick; grey	8.94	0.69	13.94	12.83	m²	**26.77**
200 mm × 100 mm × 60 mm thick; coloured	9.93	0.69	13.94	13.90	m²	**27.84**
200 mm × 100 mm × 80 mm thick; grey	10.70	0.74	14.95	15.00	m²	**29.95**
200 mm × 100 mm × 80 mm thick; coloured	11.88	0.74	14.95	16.27	m²	**31.22**
Concrete rectangular paving blocks; to falls or crossfalls; 6 mm wide joints; symmetrical layout; bedding in 15 mm semi-dry cement mortar (1:4); jointing and pointing in cement and sand (1:4); on concrete base						
Pavings; Trafica or other equal and approved; over 300 mm wide						
400 mm × 400 mm × 65 mm; Saxon textured; natural	37.76	0.44	8.89	21.02	m²	**29.91**
400 mm × 400 mm × 65 mm; Saxon textured; buff	21.43	0.44	8.89	24.31	m²	**33.20**
400 mm × 400 mm × 65 mm; Perfecta; natural	24.77	0.44	8.89	26.64	m²	**35.53**
400 mm × 400 mm × 65 mm; Perfecta; buff	27.24	0.44	8.89	30.57	m²	**39.46**
450 mm × 450 mm × 70 mm; Saxon textured; natural	20.15	0.43	8.69	22.94	m²	**31.63**
450 mm × 450 mm × 70 mm; Saxon textured; buff	23.16	0.43	8.69	26.17	m²	**34.86**
450 mm × 450 mm × 70 mm; Perfecta; natural	21.79	0.43	8.69	24.70	m²	**33.39**
450 mm × 450 mm × 70 mm; Perfecta; buff	25.50	0.43	8.69	28.69	m²	**37.38**
York stone slab pavings; to falls or crossfalls; bedding 25 mm thick in cement: sand mortar (1:4); 5 mm wide joints; jointing in coloured cement: mortar (1:3); brushed in; to sand base						
Pavings; over 300 mm wide						
50 mm thick; random rectangular pattern	64.74	0.69	11.94	69.69	m²	**81.63**
600 mm × 600 mm × 50 mm thick	61.65	0.39	6.74	66.45	m²	**73.19**
600 mm × 900 mm × 50 mm thick	61.65	0.33	5.71	66.45	m²	**72.16**

35 SITE WORKS

Item	PC £	Labour hours	Labour £	Material £	Unit	Total rate £
SLAB, BRICK/BLOCK SETTS AND COBBLE PAVINGS – cont						
Granite setts; BS EN 1342; 200 mm × 100 mm × 100 mm; standard 'C' dressing; tightly butted to falls or crossfalls; bedding 25 mm thick in cement: mortar (1:3); filling joints with dry mortar (1:6); washed and brushed; on concrete base						
Pavings; over 300 mm wide						
straight joints (PC £ per tonne)	141.66	1.48	29.90	43.81	m²	**73.71**
laid to pattern	141.66	1.85	37.37	43.81	m²	**81.18**
Boundary edging						
two rows of granite setts as boundary edging; 200 mm wide; including a 150 mm high ready mixed designated concrete C10 – 40 mm aggregate; haunching to one side; blocks laid						
breaking joint	–	0.65	13.13	10.36	m	**23.49**
EXTERNAL UNDERGROUND SERVICE RUNS						
Excavating trenches; by machine; grading bottoms; earthwork support; filling with excavated material and compacting; disposal of surplus soil on site; spreading on site average 50 m from excavations						
Services not exceeding 200 mm nominal size						
average depth of run not exceeding 0.50 m	–	0.28	3.27	0.97	m	**4.24**
average depth of run not exceeding 0.75 m	–	0.37	4.32	1.61	m	**5.93**
average depth of run not exceeding 1.00 m	–	0.79	9.23	2.70	m	**11.93**
average depth of run not exceeding 1.25 m	–	1.16	13.54	3.67	m	**17.21**
average depth of run not exceeding 1.50 m	–	1.48	17.28	4.85	m	**22.13**
average depth of run not exceeding 1.75 m	–	1.85	21.60	6.21	m	**27.81**
average depth of run not exceeding 2.00 m	–	2.13	24.87	7.16	m	**32.03**
Excavating trenches; by hand; grading bottoms; earthwork support; filling with excavated material and compacting; disposal; of surplus soil on site; spreading on site average 50 m from excavations						
Services not exceeding 200 mm nominal size						
average depth of run not exceeding 0.50 m	–	0.93	10.85	–	m	**10.85**
average depth of run not exceeding 0.75 m	–	1.39	16.23	–	m	**16.23**
average depth of run not exceeding 1.00 m	–	2.04	23.82	0.65	m	**24.47**
average depth of run not exceeding 1.25 m	–	2.87	33.51	0.89	m	**34.40**
average depth of run not exceeding 1.50 m	–	3.93	45.88	1.09	m	**46.97**
average depth of run not exceeding 1.75 m	–	5.18	60.48	1.31	m	**61.79**
average depth of run not exceeding 2.00 m	–	5.92	69.12	1.44	m	**70.56**

35 SITE WORKS

Item	PC £	Labour hours	Labour £	Material £	Unit	Total rate £
Stop cock pits, valve chambers and the like; excavating; half brick thick walls in common bricks in cement: mortar (1:3); on in situ concrete designated mix C20 – 20 mm aggregate bed; 100 mm thick						
Pits						
100 mm × 100 mm × 750 mm deep; internal holes for one small pipe; polypropylene hinged box cover; bedding in cement: mortar (1:3)	–	3.89	67.30	35.71	nr	**103.01**

36 FENCING

Item	PC £	Labour hours	Labour £	Material £	Unit	Total rate £
FENCING						
For more examples and depth of information please consult *Spon's External Works and Landscape Price Book*.						
NOTE: The prices for all fencing include for setting posts in position, to a depth of 0.60 m for fences not exceeding 1.40 m high and of 0.76 m for fences over 1.40 m high. The prices allow for excavating post holes; filling to within 150 mm of ground level with concrete and all necessary backfilling.						
Fencing; strained wire						
Strained wire fencing; BS 1722 Part 3; 4 mm dia. galvanized mild steel plain wire threaded through posts and strained with eye bolts						
900 mm high; three line; concrete posts at 2.75 m centres	–	–	–	–	m	16.19
end concrete straining post; one strut	–	–	–	–	nr	39.36
angle concrete straining post; two struts	–	–	–	–	nr	45.71
1.07 m high; six line; concrete posts at 2.75 m centres	–	–	–	–	m	16.84
end concrete straining post; one strut	–	–	–	–	nr	44.27
angle concrete straining post; two struts	–	–	–	–	nr	50.62
1.20 m high; six line; concrete posts at 2.75 m centres	–	–	–	–	m	16.94
end concrete straining post; one strut	–	–	–	–	nr	45.51
angle concrete straining post; two struts	–	–	–	–	nr	51.85
1.40 m high; eight line; concrete posts at 2.75 m centres	–	–	–	–	m	17.40
end concrete straining post; one strut	–	–	–	–	nr	46.50
angle concrete straining post; two struts	–	–	–	–	nr	52.84
Chain link fencing; BS 1722 Part 1; 3 mm dia. galvanized mild steel wire; 50 mm mesh; galvanized mild steel tying and line wire; three line wires threaded through posts and strained with eye bolts and winding brackets						
900 mm high; galvanized mild steel angle posts at 3.00 m centres	–	–	–	–	m	22.77
end steel straining post; one strut	–	–	–	–	nr	65.62
angle steel straining post; two struts	–	–	–	–	nr	75.68
900 mm high; concrete posts at 3.00 m centres	–	–	–	–	m	16.50
end concrete straining post; one strut	–	–	–	–	nr	35.24
angle concrete straining post; two struts	–	–	–	–	nr	41.60
1.20 m high; galvanized mild steel angle posts at 3.00 m centres	–	–	–	–	m	16.79
end steel straining post; one strut	–	–	–	–	nr	70.02
angle steel straining post; two struts	–	–	–	–	nr	89.67
1.20 m high; concrete posts at 3.00 m centres	–	–	–	–	m	16.08
end concrete straining post; one strut	–	–	–	–	nr	40.35
angle concrete straining post; two struts	–	–	–	–	nr	47.66

36 FENCING

Item	PC £	Labour hours	Labour £	Material £	Unit	Total rate £
1.80 m high; galvanized mild steel angle posts at 3.00 m centres	–	–	–	–	m	18.88
end steel straining post; one strut	–	–	–	–	nr	71.21
angle steel straining post; two struts	–	–	–	–	nr	88.59
1.80 m high; concrete posts at 3.00 m centres	–	–	–	–	m	21.80
end concrete straining post; one strut	–	–	–	–	nr	56.42
angle concrete straining post; two struts	–	–	–	–	nr	66.56
Gates						
Pair of gates and gate posts; gates to match galvanized chain link fencing, with angle framing, braces, etc., complete with hinges, locking bar, lock and bolts; two 100 mm × 100 mm angle section gate posts; each with one strut						
2.44 m × 0.90 m	–	–	–	–	nr	553.32
2.44 m × 1.20 m	–	–	–	–	nr	571.04
2.44 m × 1.80 m	–	–	–	–	nr	616.00
Fencing; chain link						
Chain link fencing; BS 1722 Part 1; 3 mm dia. plastic coated mild steel wire; 50 mm mesh; plastic coated mild steel tying and line wire; three line wires threaded through posts and strained with eye bolts and winding brackets						
900 mm high; galvanized mild steel angle posts at 3.00 m centres	–	–	–	–	m	20.88
end steel straining post; one strut	–	–	–	–	nr	57.85
angle steel straining post; two struts	–	–	–	–	nr	64.41
900 mm high; concrete posts at 3.00 m centres	–	–	–	–	m	15.57
end concrete straining post; one strut	–	–	–	–	nr	35.24
angle concrete straining post; two struts	–	–	–	–	nr	41.60
1.20 m high; galvanized mild steel angle posts at 3.00 m centres	–	–	–	–	m	15.42
end steel straining post; one strut	–	–	–	–	nr	60.70
angle steel straining post; two struts	–	–	–	–	nr	64.84
1.20 m high; concrete posts at 3.00 m centres	–	–	–	–	m	15.78
end concrete straining post; one strut	–	–	–	–	nr	40.35
angle concrete straining post; two struts	–	–	–	–	nr	47.66
1.80 m high; galvanized mild steel angle posts at 3.00 m centres	–	–	–	–	m	17.49
end steel straining post; one strut	–	–	–	–	nr	60.07
angle steel straining post; two struts	–	–	–	–	nr	71.74
1.80 m high; concrete posts at 3.00 mm centres	–	–	–	–	m	20.11
end concrete straining post; one strut	–	–	–	–	nr	56.42
angle concrete straining post; two struts	–	–	–	–	nr	66.56

36 FENCING

Item	PC £	Labour hours	Labour £	Material £	Unit	Total rate £
FENCING – cont						
Fencing; chain link – cont						
Chain link fencing for tennis courts; BS 1722 Part 13; 2.5 dia. galvanized mild wire; 45 mm mesh; line and tying wires threaded through 45 mm × 45 mm × 5 mm galvanized mild steel angle standards, posts and struts; 60 mm × 60 mm × 6 mm straining posts and gate posts;						
fencing to tennis court 36.00 m × 18.00 m × 2.745 m high; standards at 3.00 m centres	–	–	–	–	nr	109.20
fencing to tennis court 36.00 m × 18.00 m × 3.66 m high; standards at 2.50 m centres	–	–	–	–	nr	126.75
Gates						
Pair of gates and gate posts; gates to match plastic chain link fencing; with angle framing, braces, etc. complete with hinges, locking bar, lock and bolts; two 100 mm × 100 mm angle section gate posts; each with one strut						
2.44 m × 0.90 m	–	–	–	–	nr	483.83
2.44 m × 1.20 m	–	–	–	–	nr	496.44
2.44 m × 1.80 m	–	–	–	–	nr	534.68
Fencing; timber						
Cleft chestnut pale fencing; BS 1722 Part 4; pales spaced 51 mm apart; on two lines of galvanized wire; 64 mm dia. posts; 76 mm × 51 mm struts						
900 mm high; posts at 2.50 m centres	–	–	–	–	m	9.75
straining post; one strut	–	–	–	–	nr	25.84
corner straining post; two struts	–	–	–	–	nr	25.84
1.05 m high; posts at 2.50 m centres	–	–	–	–	m	10.92
straining post; one strut	–	–	–	–	nr	26.03
corner straining post; two struts	–	–	–	–	nr	26.03
Closeboarded fencing; BS 1722 Part 5; 76 mm × 38 mm softwood rails; 89 mm × 19 mm softwood pales lapped 13 mm; 152 mm × 25 mm softwood gravel boards; all softwood treated; posts at 3.00 m centres						
Fencing; two rail; concrete posts						
height 1.00 m	–	–	–	–	m	29.64
height 1.20 m	–	–	–	–	m	29.98

36 FENCING

Item	PC £	Labour hours	Labour £	Material £	Unit	Total rate £
Fencing; three rail; concrete posts						
height 1.40 m	–	–	–	–	m	**32.91**
height 1.60 m	–	–	–	–	m	**32.91**
height 1.80 m	–	–	–	–	m	**34.13**
Fencing; concrete						
Precast concrete slab fencing;						
305 mm × 38 mm × 1753 mm slabs; fitted into twice						
grooved concrete posts at 1.83 m centres						
height 1.50 m	–	–	–	–	m	**58.50**
height 1.80 m	–	–	–	–	m	**63.38**
Fencing; security						
Mild steel unclimbable fencing; in rivetted panels						
2440 mm long; 44 mm × 13 mm flat section top and						
bottom rails; two 44 mm × 19 mm flat section						
standards; one with foot plate; and 38 mm × 13 mm						
raking stay with foot plate; 20 mm dia. pointed						
verticals at 120 mm centres; two 44 mm × 19 mm						
supports 760 mm long with ragged ends to bottom						
rail; the whole bolted together; coated with red oxide						
primer; setting standards and stays in ground at						
2440 mm centres and supports at 815 mm centres						
height 1.67 m	–	–	–	–	m	**109.20**
height 2.13 m	–	–	–	–	m	**126.75**
Pair of gates and gate posts, to match mild steel						
unclimbable fencing; with flat section framing,						
braces, etc., complete with locking bar, lock,						
handles, drop bolt, gate stop and holding back						
catches; two 102 mm × 102 mm hollow section gate						
posts with cap and foot plates						
2.44 m × 1.67 m	–	–	–	–	nr	**940.88**
2.44 m × 2.13 m	–	–	–	–	nr	**1092.00**
4.88 m × 1.67 m	–	–	–	–	nr	**1462.50**
4.88 m × 2.13 m	–	–	–	–	nr	**1842.75**
PVC coated, galvanized mild steel high security						
fencing; Sentinal Sterling fencing or other equal and						
approved; 50 mm × 50 mm mesh; 3/3.50 mm gauge						
wire; barbed edge – 1; Sentinal Bi-steel colour						
coated posts or other equal and approved at 2.44 m						
centres						
1.80 m	–	0.93	10.85	29.94	m	**40.79**
2.10 m	–	1.16	13.54	33.07	m	**46.61**

37 SOFT LANDSCAPING

Item	PC £	Labour hours	Labour £	Material £	Unit	Total rate £
SEEDING AND TURFING						
For more examples and depth of information please consult *Spon's External Works and Landscape Price Book*.						
Top soil						
By machine						
Selected from spoil heaps not exceeding 50 m; grading; prepared for turfing or seeding; to general surfaces						
average 100 mm thick	–	–	0.06	0.99	m²	**1.05**
average 150 mm thick	–	0.01	0.07	1.32	m²	**1.39**
average 200 mm thick	–	0.01	0.09	2.31	m²	**2.40**
Selected from spoil heaps; grading; prepared for turfing or seeding; to cuttings or embankments						
average 100 mm thick	–	–	0.06	1.13	m²	**1.19**
average 150 mm thick	–	0.01	0.08	1.52	m²	**1.60**
average 200 mm thick	–	0.01	0.10	2.64	m²	**2.74**
By hand						
Selected from spoil heaps not exceeding 20 m; grading; prepared for turfing or seeding; to general surfaces						
average 100 mm thick	–	0.40	4.67	–	m²	**4.67**
average 150 mm thick	–	0.50	5.83	–	m²	**5.83**
average 200 mm thick	–	0.60	7.00	–	m²	**7.00**
Selected from spoil heaps; grading; prepared for turfing or seeding; to cuttings or embankments						
average 100 mm thick	–	0.51	5.90	–	m²	**5.90**
average 150 mm thick	–	0.63	7.38	–	m²	**7.38**
average 200 mm thick	–	0.76	8.86	–	m²	**8.86**
Imported top soil, planting quality						
By machine						
Grading; prepared for turfing or seeding; to general surfaces						
average 100 mm thick	–	–	0.06	3.76	m²	**3.82**
average 150 mm thick	–	0.01	0.07	5.47	m²	**5.54**
average 200 mm thick	–	0.01	0.09	7.38	m²	**7.47**
Grading; preparing for turfing or seeding; to cuttings or embankments						
average 100 mm thick	–	–	0.06	4.51	m²	**4.57**
average 150 mm thick	–	0.01	0.08	7.05	m²	**7.13**
average 200 mm thick	–	0.01	0.10	9.41	m²	**9.51**
By hand						
Grading; prepared for turfing or seeding; to general surfaces						
average 100 mm thick	–	0.51	5.90	3.38	m²	**9.28**
average 150 mm thick	–	0.63	7.38	5.54	m²	**12.92**
average 200 mm thick	–	0.76	8.86	6.76	m²	**15.62**

37 SOFT LANDSCAPING

Item	PC £	Labour hours	Labour £	Material £	Unit	Total rate £
Fertilizer						
Fertilizer 0.07 kg/m^2; raking in						
general surfaces (PC £ per 25 kg)	25.00	0.03	0.35	0.08	m^2	**0.43**
Selected grass seed						
Grass seed; sowing at a rate of 40 g/m^2 two applications; raking in						
general surfaces (PC £ per kg)	4.75	0.17	1.99	0.25	m^2	**2.24**
cuttings or embankments	–	0.20	2.34	0.25	m^2	**2.59**
Turfing						
Imported turf; cultivated						
general surfaces	2.39	0.10	1.17	2.45	m^2	**3.62**
cuttings or embankments; shallow	2.39	0.15	1.75	2.45	m^2	**4.20**
cuttings or embankments; steep; pegged	2.39	0.28	3.27	2.45	m^2	**5.72**
Preserved turf from stack on site; lay only						
general surfaces	–	0.19	2.21	–	m^2	**2.21**
cuttings or embankments; shallow	–	0.20	2.34	–	m^2	**2.34**
cuttings or embankments; steep; pegged	–	0.28	3.27	–	m^2	**3.27**
PLANTING						
Hedge plants						
height not exceeding 750 mm	–	0.23	2.69	–	nr	**2.69**
height 750 mm–1.50 m	–	0.56	6.54	–	nr	**6.54**
Saplings						
height not exceeding 3.00 m	–	1.57	18.33	–	nr	**18.33**

39 ELECTRICAL SERVICES

Item	PC £	Labour hours	Labour £	Material £	Unit	Total rate £
LIGHTNING PROTECTION						
For more examples and depth of information please consult *Spon's Mechanical and Electrical Services Price Book*.						
Lightning protection equipment						
Copper strip roof or down conductors fixed with bracket or saddle clips						
20 mm × 3 mm flat section	–	–	–	–	m	**20.05**
25 mm × 3 mm flat section	–	–	–	–	m	**23.39**
Aluminium strip roof or down conductors fixed with bracket or saddle clips						
20 mm × 3 mm flat section	–	–	–	–	m	**14.71**
25 mm × 3 mm flat section	–	–	–	–	m	**16.05**
Joints in tapes	–	–	–	–	nr	**11.38**
Bonding connections to roof and structural metalwork	–	–	–	–	nr	**66.85**
Testing points	–	–	–	–	nr	**54.93**
Earth electrodes						
16 mm dia. driven copper electrodes in 1220 mm long sectional lengths; 2440 mm long overall first 2440 mm length driven and tested	–	–	–	–	nr	**173.79**
25 mm × 3 mm copper strip electrode in 457 mm deep prepared trench	–	–	–	–	m	**13.37**

41 BUILDER'S WORK IN CONNECTION WITH SERVICES

Item	PC £	Labour hours	Labour £	Material £	Unit	Total rate £
HOLES, CHASES FOR SERVICES						
Electrical installations; cutting away for and making good after electrician; including cutting or leaving all holes, notches, mortices, sinkings and chases, in both the structure and its coverings, for the following electrical points						
Exposed installation						
lighting points	–	0.28	4.10	–	nr	**4.10**
socket outlet points	–	0.46	7.05	–	nr	**7.05**
fitting outlet points	–	0.46	7.05	–	nr	**7.05**
equipment points or control gear points	–	0.65	10.12	–	nr	**10.12**
Concealed installation						
lighting points	–	0.37	5.58	–	nr	**5.58**
socket outlet points	–	0.65	10.12	–	nr	**10.12**
fitting outlet points	–	0.65	10.12	–	nr	**10.12**
equipment points or control gear points	–	0.93	14.22	–	nr	**14.22**
Builder's work for other services installations						
Cutting chases in brickwork						
for one pipe; not exceeding 55 mm nominal size; vertical	–	0.37	4.28	–	m	**4.28**
for one pipe; 55 mm–110 mm nominal size; vertical	–	0.65	7.52	–	m	**7.52**
Cutting and pinning to brickwork or blockwork; ends of supports						
for pipes not exceeding 55 mm nominal size	–	0.19	3.29	–	nr	**3.29**
for cast iron pipes 55 mm–110 mm nominal size	–	0.31	5.36	–	nr	**5.36**
Cutting or forming holes for pipes or the like; not exceeding 55 mm nominal size; making good						
reinforced concrete; not exceeding 100 mm deep	–	0.75	10.27	0.56	nr	**10.83**
reinforced concrete; 100 mm–200 mm deep	–	1.15	15.74	0.86	nr	**16.60**
reinforced concrete; 200 mm–300 mm deep	–	1.50	20.53	1.13	nr	**21.66**
half brick thick	–	0.31	4.24	–	nr	**4.24**
one brick thick	–	0.51	6.98	–	nr	**6.98**
one and a half brick thick	–	0.83	11.37	–	nr	**11.37**
100 mm blockwork	–	0.28	3.83	–	nr	**3.83**
140 mm blockwork	–	0.37	5.06	–	nr	**5.06**
215 mm blockwork	–	0.46	6.29	–	nr	**6.29**
plasterboard partition or suspended ceiling	–	0.35	4.79	–	nr	**4.79**
Cutting or forming holes for pipes or the like; 55 mm–110 mm nominal size; making good						
reinforced concrete; not exceeding 100 mm deep	–	1.15	15.74	0.86	nr	**16.60**
reinforced concrete; 100 mm–200 mm deep	–	1.75	23.95	1.31	nr	**25.26**
reinforced concrete; 200 mm–300 mm deep	–	2.25	30.80	1.69	nr	**32.49**
half brick thick	–	0.37	5.06	–	nr	**5.06**
one brick thick	–	0.65	8.90	–	nr	**8.90**
one and a half brick thick	–	1.02	13.96	–	nr	**13.96**
100 mm blockwork	–	0.32	4.38	–	nr	**4.38**
140 mm blockwork	–	0.46	6.29	–	nr	**6.29**
215 mm blockwork	–	0.56	7.67	–	nr	**7.67**
plasterboard partition or suspended ceiling	–	0.40	5.47	–	nr	**5.47**

41 BUILDER'S WORK IN CONNECTION WITH SERVICES

Item	PC £	Labour hours	Labour £	Material £	Unit	Total rate £
HOLES, CHASES FOR SERVICES – cont						
Builder's work for other services installations – cont						
Cutting or forming holes for pipes or the like; over 110 mm nominal size; making good						
reinforced concrete; not exceeding 100 mm deep	–	1.15	15.74	0.86	nr	**16.60**
reinforced concrete; 100 mm–200 mm deep	–	1.75	23.95	1.31	nr	**25.26**
reinforced concrete; 200 mm–300 mm deep	–	2.25	30.80	1.69	nr	**32.49**
half brick thick	–	0.46	6.29	–	nr	**6.29**
one brick thick	–	0.79	10.81	–	nr	**10.81**
one and a half brick thick	–	1.25	17.11	–	nr	**17.11**
100 mm blockwork	–	0.42	5.75	–	nr	**5.75**
140 mm blockwork	–	0.56	7.67	–	nr	**7.67**
215 mm blockwork	–	0.69	9.45	–	nr	**9.45**
plasterboard partition or suspended ceiling	–	0.45	6.16	–	nr	**6.16**
Add for making good fair face or facings one side						
pipe; not exceeding 55 mm nominal size	–	0.07	1.21	–	nr	**1.21**
pipe; 55 mm–110 mm nominal size	–	0.09	1.56	–	nr	**1.56**
pipe; over 110 mm nominal size	–	0.11	1.91	–	nr	**1.91**
Add for fixing sleeve (supply not included)						
for pipe; small	–	0.14	2.42	–	nr	**2.42**
for pipe; large	–	0.19	3.29	–	nr	**3.29**
for pipe; extra large	–	0.28	4.85	–	nr	**4.85**
Add for supplying and fixing two hour intumescent sleeve						
for 55 mm uPVC pipe	–	0.25	3.42	6.36	nr	**9.78**
for 110 mm uPVC pipe	–	0.28	3.83	6.91	nr	**10.74**
for 200 mm uPVC pipe	–	0.30	4.11	45.20	nr	**49.31**
Cutting or forming holes for ducts; girth not exceeding 1.00 m; making good						
half brick thick	–	0.56	7.67	–	nr	**7.67**
one brick thick	–	0.93	12.73	–	nr	**12.73**
one and a half brick thick	–	1.48	20.26	–	nr	**20.26**
100 mm blockwork	–	0.46	6.29	–	nr	**6.29**
140 mm blockwork	–	0.65	8.90	–	nr	**8.90**
215 mm blockwork	–	0.83	11.37	–	nr	**11.37**
plasterboard partition or suspended ceiling	–	0.65	8.90	–	nr	**8.90**
Cutting or forming holes for ducts; girth 1.00 m–2.00 m; making good						
half brick thick	–	0.65	8.90	–	nr	**8.90**
one brick thick	–	1.11	15.19	–	nr	**15.19**
one and a half brick thick	–	1.76	24.10	–	nr	**24.10**
100 mm blockwork	–	0.56	7.67	–	nr	**7.67**
140 mm blockwork	–	0.74	10.13	–	nr	**10.13**
215 mm blockwork	–	0.93	12.73	–	nr	**12.73**
plasterboard partition or suspended ceiling	–	0.75	10.27	–	nr	**10.27**
Cutting or forming holes for ducts; girth 2.00 m–3.00 m; making good						
half brick thick	–	1.02	13.96	–	nr	**13.96**
one brick thick	–	1.76	24.10	–	nr	**24.10**
one and a half brick thick	–	2.78	38.06	–	nr	**38.06**

41 BUILDER'S WORK IN CONNECTION WITH SERVICES

Item	PC £	Labour hours	Labour £	Material £	Unit	Total rate £
100 mm blockwork	–	0.88	12.04	–	nr	**12.04**
140 mm blockwork	–	1.20	16.43	–	nr	**16.43**
215 mm blockwork	–	1.53	20.94	–	nr	**20.94**
plasterboard partition or suspended ceiling	–	1.00	13.69	–	nr	**13.69**
Cutting or forming holes for ducts; girth 3.00 m–4.00 m; making good						
half brick thick	–	1.39	19.02	–	nr	**19.02**
one brick thick	–	2.31	31.62	–	nr	**31.62**
one and a half brick thick	–	3.70	50.65	–	nr	**50.65**
100 mm blockwork	–	1.02	13.96	–	nr	**13.96**
140 mm blockwork	–	1.39	19.02	–	nr	**19.02**
215 mm blockwork	–	1.76	24.10	–	nr	**24.10**
plasterboard partition or suspended ceiling	–	1.25	17.11	–	nr	**17.11**
Mortices in brickwork						
for expansion bolt	–	0.19	2.60	–	nr	**2.60**
for 20 mm dia. bolt; 75 mm deep	–	0.14	1.92	–	nr	**1.92**
for 20 mm dia. bolt; 150 mm deep	–	0.23	3.15	–	nr	**3.15**
Mortices in brickwork; grouting with cement: mortar (1:1)						
75 mm × 75 mm × 200 mm deep	–	0.28	3.83	0.14	nr	**3.97**
75 mm × 75 mm × 300 mm deep	–	0.37	5.06	0.20	nr	**5.26**
Holes in softwood for pipes, bars, cables and the like						
12 mm thick	–	0.03	0.53	–	nr	**0.53**
25 mm thick	–	0.05	0.89	–	nr	**0.89**
50 mm thick	–	0.09	1.61	–	nr	**1.61**
100 mm thick	–	0.14	2.49	–	nr	**2.49**
Holes in hardwood for pipes, bars, cables and the like						
12 mm thick	–	0.05	0.89	–	nr	**0.89**
25 mm thick	–	0.08	1.42	–	nr	**1.42**
50 mm thick	–	0.14	2.49	–	nr	**2.49**
100 mm thick	–	0.20	3.57	–	nr	**3.57**
Diamond drilling for cutting holes and mortices in masonry or concrete						
Cutting holes and mortices in brickwork; per 25 mm depth						
25 mm dia.	–	–	–	–	nr	**1.60**
32 mm dia.	–	–	–	–	nr	**1.29**
52 mm dia.	–	–	–	–	nr	**1.55**
78 mm dia.	–	–	–	–	nr	**1.70**
107 mm dia.	–	–	–	–	nr	**1.79**
127 mm dia.	–	–	–	–	nr	**2.20**
152 mm dia.	–	–	–	–	nr	**2.60**
200 mm dia.	–	–	–	–	nr	**3.34**
250 mm dia.	–	–	–	–	nr	**5.04**
300 mm dia.	–	–	–	–	nr	**6.69**
Diamond chasing; per 25 × 25 mm section						
in facing or common brickwork	–	–	–	–	m	**3.04**
in semi-engineering brickwork	–	–	–	–	m	**6.09**
in engineering brickwork	–	–	–	–	m	**8.49**
in lightweight blockwork	–	–	–	–	m	**2.39**
in heavyweight blockwork	–	–	–	–	m	**4.79**
in render/screed	–	–	–	–	m	**9.43**

41 BUILDER'S WORK IN CONNECTION WITH SERVICES

Item	PC £	Labour hours	Labour £	Material £	Unit	Total rate £
HOLES, CHASES FOR SERVICES – cont						
Diamond drilling for cutting holes and mortices in masonry or concrete – cont						
Forming boxes; 100 × 100 mm; per 25 mm depth						
in facing or common brickwork	–	–	–	–	nr	**1.22**
in semi-engineering brickwork	–	–	–	–	nr	**2.43**
in engineering brickwork	–	–	–	–	nr	**3.39**
in lightweight blockwork	–	–	–	–	nr	**0.96**
in heavyweight blockwork	–	–	–	–	nr	**1.92**
in render/screed	–	–	–	–	nr	**3.77**
Other items						
diamond track mount or ring sawing brickwork	–	–	–	–	m	**5.99**
diamond floor sawing asphalte	–	–	–	–	m	**1.00**
stitch drilling 107 mm dia. hole in brickwork	–	–	–	–	nr	**1.29**
INTERNAL FLOOR DUCTS FOR SERVICES						
Screed Floor Ducting; with side flanges; laid within floor screed; galvanized mild steel						
Floor ducting						
100 mm wide × 50 mm deep	9.65	0.19	3.38	9.89	m	**13.27**
extra for						
bend	–	0.09	1.61	15.31	nr	**16.92**
tee section	–	0.09	1.61	15.31	nr	**16.92**
connector/stop end	–	0.09	1.61	1.76	nr	**3.37**
ply cover 15 mm/16 mm thick WBP exterior grade	–	0.09	1.61	2.19	m	**3.80**
100 mm wide × 70 mm deep	10.60	0.20	3.57	10.87	m	**14.44**
extra for						
bend	–	0.09	1.61	15.31	nr	**16.92**
tee section	–	0.09	1.61	15.31	nr	**16.92**
connector/stop end	–	0.09	1.61	1.76	nr	**3.37**
ply cover 15 mm/16 mm thick WBP exterior grade	–	0.09	1.61	2.19	m	**3.80**
200 mm wide × 50 mm deep	13.34	0.19	3.38	13.67	m	**17.05**
extra for						
bend	–	0.09	1.61	17.05	nr	**18.66**
tee section	–	0.09	1.61	17.05	nr	**18.66**
connector/stop end	–	0.09	1.61	1.76	nr	**3.37**
ply cover 15 mm/16 mm thick WBP exterior grade	–	0.09	1.61	3.98	m	**5.59**

PART 5

Fees for Professional Services

This part contains the following sections:

Design in Modular Construction

M. Lawson et al.

Modular construction can dramatically improve efficiency in construction, through factory production of pre-engineered building units and their delivery to the site either as entire buildings or as substantial elements. The required technology and application are developing rapidly, but design is still in its infancy. Good design requires a knowledge of modular production, installation and interface issues and also an understanding of the economics and client-related benefits which influence design decisions.

Looking at eight recent projects, along with background information, this guide gives you coverage of:

• generic types of module and their application

• vertical loading, stability and robustness

• dimensional and spacial planning

• hybrid construction

• cladding, services and building physics

• fire safety and thermal and acoustic performance

• logistical aspects – such as transport, tolerances and safe installation.

A valuable guide for professionals and a thorough introduction for advanced students.

March 2013: 267x203: 160pp
Hb: 978-0-415-55450-3: £60.00

To Order: Tel: +44 (0) 1235 400524 Fax: +44 (0) 1235 400525
or Post: Taylor and Francis Customer Services,
Bookpoint Ltd, Unit T1, 200 Milton Park, Abingdon, Oxon, OX14 4TA UK
Email: book.orders@tandf.co.uk

For a complete listing of all our titles visit:
www.tandf.co.uk

QUANTITY SURVEYORS' FEES

Guidance on basic quantity surveying services is set out in the RICS Standard Form of Consultant's Appointment for Quantity Surveyors. Services are separated into core services for a variety of contracts and supplementary services and it is advisable to refer to this guidance. Copies can be obtained from the Royal Institution of Chartered Surveyors (RICS) at www.ricsbooks.com.

Quantity surveying services

Preparation:

- liaising with clients and the professional team
- advice on cost
- preparation of initial budget/cost plan/cash flow forecasts

Design:

- prepare and maintain cost plan
- advise design team on impact of design development on cost

Pre-construction:

- liaise with professional team
- advise on procurement strategy
- liaise with client's legal advisors on contract matters
- prepare tender documents
- define prospective tenderers
- obtain tenders/check tenders/prepare recommendation for client
- maintain and develop cost plan

Construction:

- visit the site
- prepare interim valuations
- advise on the cost of variations
- agree the cost of claims
- advise on contractual matters

Supplementary services may include:

- preparation of mechanical and electrical tender documentation
- preparation of cost analyses
- advice on insurance claims
- facilitate value management exercises
- prepare life cycle calculations/sustainability
- capital allowance advice/VAT
- attend adjudication/mediation proceedings

Fee Guide

Quantity Surveying Services Benchmark	Mean	Lower Quartile	Upper Quartile
	2.02%	1.54%	2.32%

The level of fees above, are expressed as a percentage of the contract value of £3,500,000 for a new build project and do not include VAT.

Fee levels vary depending on many factors, including, but not limited to the following: type of project; complexity; procurement route.

ARCHITECTS' FEES

RIBA Agreements are designed to be:

- in line with current working practices, legislative changes and procurement methods
- attractive to clients, architects and other consultants, with robust but fair terms
- a flexible system of components that can be assembled and customized to create tailored and bespoke contracts
- suitable for a wide range of projects and services
- based upon the updated RIBA Outline Plan of Work 2013 (available for download free from www.ribabook-shops.com)
- available in electronic and printed formats

RIBA Agreements 2013	Suitable for
Standard Agreement	• a commission where detailed contract terms are necessary for a wide range of projects using most procurement methods • where the client is acting for business or commercial purposes • where the commission is for work to the client's home where the size or value of the Project merits use of the JCT Standard or Intermediate forms of building contract or similar and the terms have been negotiated with the client as a consumer
Concise Agreement	• for a commission where the concise contract terms are compatible with the complexity of the Project and the risks to each party • where the Client is acting for business or commercial purposes • where the commission is for work to the Client's home and the terms have been negotiated with the Client as a 'consumer'. A consumer is 'a natural person acting for purposes outside his trade, business or profession' • where the building works, including extensions and alterations, will be carried out using forms of building contract, such as JCT Agreement for Minor Works or JCT Intermediate Form of Building Contract
Domestic Project Agreement	• the commission relates to work to the client's home, provided that they have elected to use these conditions in their own name, i.e. not as a limited company or other legal entity • the contract terms are compatible with the complexity of the project and the risks to each party and have been negotiated with the client as a consumer • the building works, including extensions and alterations, will be carried out using forms of building contract, such as the JCT Building Contract for a homeowner/ occupier, JCT Agreement for Minor Works or JCT Intermediate Form of Building Contract>
Sub-consultant's Agreement	• a consultant wishes or perhaps is required by the client to appoint another consultant (thus, a sub-consultant) to perform part of the consultant's services • the contract terms are compatible with the (head) agreement between the consultant and the client, with the complexity of the project and the risks to each party
Electronic and print formats	• All the RIBA Agreements 2013 and their components are available as electronic files. A limited number of the conditions and core components are published in print. The electronic and printed versions may be used in combination

ARCHITECTS' FEES

RIBA Agreements 2013	Suitable for
Electronic components	• Conditions, notes and guides are available as locked PDFs and all other components, e.g. schedules and model letters, are available in Rich Text Format (RTF), which can be customized using most commonly used word processing software to meet project requirements or modified to match the house style of the practice

Each agreement comprises the selected Conditions of Appointment (i.e. Standard, Concise or Domestic), related components, and a schedule or schedules of Services.

Notes on use and completion and model letters for business clients and domestic clients are included with each pack.

For further information, readers are advised to log onto the RIBA Publications website at *www.ribabookshops.com/ agreements.*

RIBA Plan of Work 2013

The table below shows how the new Plan of Work maps to the old RIBA Plan of Work 2007 and the CIC's work stages

RIBA Plan of Work 2013		RIBA Plan of Work 2007		CIC/BIM Task Group Coordinated work stages	
0	Strategic definition	A	Appraisal	0	Strategy
1	Preparation and brief	B	Design Brief	1	Brief
2	Concept design	C	Concept	2	Concept
3	Developed design	D	Design development	3	Definition
4	Technical design	E	Technical design	4	Design
		F	Production information		
		G	Tender documentation		
		H	Tender Action		
5	Construction	J	Mobilization	5	Build and commission
		K	Construction to practical completion		
6	Handover and close out			6	Handover & close out
7	In use	L	Post practical completion	7	Operation and end of life

ARCHITECTS' FEES

Appointment guidance

A guide, 'A Client's Guide to Engaging an Architect', is available from RIBA Bookshops at www.ribabookshops.com.

This guide includes an introduction to the services an Architect can be expected to provide, advice on the forms to use, linking the RIBA Plan of Work Stages with fees (which are a matter of negotiation) and classifying buildings according to three levels of complexity.

Generally, the more complex the building the higher the level of fee.

Example categories include:

- Simple: for buildings such as car parks, warehouses, factories and speculative retail schemes
- Average: for buildings such as offices, most retail outlets, general housing, schools etc.
- Complex: for multi-purpose developments, specialist buildings e.g. hospitals, research laboratories etc.

Procurement option can also influence the architects' fees.

Fee Guide

Architectural Services Benchmark	Mean	Lower Quartile	Upper Quartile
	4.95%	3.97%	5.82%

The level of fees above, are expressed as a percentage of the contract value of £3,500,000 for a new build project and do not include VAT.

CONSULTING ENGINEERS' FEES

CONDITIONS OF APPOINTMENT

A scale of professional charges for consulting engineering services is published by the Association for Consultancy and Engineering (ACE)

Copies of the document can be obtained direct from:

Association for Consultancy and Engineering
Alliance House
12 Caxton Street
London SW1H OQL
Tel 020 7222 6557
Fax: 020 7222 0750

Comparisons

Instead of the previous arrangement of having different agreements designed for each major discipline of engineering, the current agreements have been developed primarily to suit the different roles that Consulting Engineers may be required to perform, with variants of some of them for different disciplines. The agreements have been standardized as far as possible whilst retaining essential differences.

Greater attention is required than with previous agreements to ensure the documents are completed properly. This is because of the perceived need to allow for a wider choice of arrangements, particularly of methods of payment.

The agreements are not intended to be used as unsigned reference material with the details of an engagement being covered in an exchange of letters, although much of their content could be used as a basis for drafting such correspondence.

For 2009 the ACE published a new suite of Agreements with a broader set of services, these are listed below.

Forms of Agreement

- ACE Agreement 1: Design
- ACE Agreement 2: Advise and Report
- ACE Agreement 3: Design and Construct
- ACE Agreement 4: Sub-Consultancy
- ACE Agreement 5: Homeowner
- ACE Agreement 6: Expert Witness (Sole Practitioner)
- ACE Agreement 7: Expert Witness (Firm)
- ACE Agreement 8: Adjudicator

To a number of the the above ACE Agreements Schedules of Services are appended and currently these are:

For use with ACE Agreement 1: Design

- ACE Schedule of Services – Part G(a):
  Civil and Structural Engineer – single consultant or non-lead consultant
- ACE Schedule of Services – Part G(b):
  Mechanical and Electrical Engineering (detailed design in buildings)
- ACE Schedule of Services – Part G(c):
  Mechanical and Electrical Engineering (performance design in buildings)
- ACE Schedule of Services – Part G(d):
  Civil and Structural Engineer – Lead consultant
- ACE Schedule of Services – Part G(e):
  Mechanical and Electrical Engineering Design in buildings – Lead consultant

CONSULTING ENGINEERS' FEES

For use with ACE Agreement 3: Design and Construct

- ACE Schedule of Services – Part G(f):
  Civil and Structural Engineer
- ACE Schedule of Services – Part G(g):
  Mechanical and Electrical Engineering (detailed design in buildings)
- ACE Schedule of Services – Part G(h):
  Mechanical and Electrical Engineering (performance design in buildings)

ACE Agreement 1: Design

Design for the appointment of a consultant by a client to undertake detailed design and/or specification of permanent works to be undertaken or installed by a contractor including any studies, appraisals, investigations, contract administration or construction monitoring leading to or resulting from such detailed design and/or specification.

ACE Agreement 2: Advise and Report

Advise and report for the appointment of a consultant by a client to provide any type of advisory, research, checking, reviewing, investigatory, monitoring, reporting or technical services in the built and natural environments where such services do not consist of detailed design or specification of permanent works to be constructed or installed by a contractor.

ACE Agreement 3: Design and Construct

For the appointment of a consultant by a contractor in circumstances where the contractor is to construct permanent works designed by the consultant.

ACE Agreement 4: Sub-consultancy

For the appointment of a sub-consultant by a consultant in circumstances where the consultant is appointed on the terms of an ACE Agreement by its client.

ACE Agreement 5: Homeowner

Model letter for the appointment of a consultant by a homeowner.

ACE Agreement 6: Expert Witness (Sole Practitioner)

For the appointment of an individual to act as an expert witness.

ACE Agreement 7: Expert Witness (Firm)

For the appointment of a firm to provide an expert witness.

ACE Agreement 8: Adjudicator

For the appointment of an adjudicator.

Fee guide

Profession

Structural Engineering Services Benchmark	Mean	Lower Quartile	Upper Quartile
	1.86%	1.31%	2.1%

Services Engineering Benchmark	Mean	Lower Quartile	Upper Quartile
	1.81%	1.45%	2.17%

THE TOWN AND COUNTRY PLANNING FEES AND BUILDING REGULATION FEES

THE TOWN AND COUNTRY PLANNING (APPPLICATION FEES) ORDER 2013

Author's note

This is only a small extract of typical fees chargeable. Users should always obtain actual fees from the local authority concerned with the particular planning application.

No.	Category of development	Fee
1.	**Application for approval in principle**	£135 for first 5000 m^2 (0.5 ha) of site area plus £135 for each additional 1000 m^2 (0.1 ha) (or part thereof) of site area up to a maximum of £7,710.
2.	**Application for approval of building, rebuilding, engineering, mining or other operations (other than approval in principle)**	
(a)	Estate layout (residential or industrial).	£415 for first 5000 m^2 (0.5 ha) of site area plus £135 for each additional 1000 m^2 (0.1 ha) (or part thereof) of site area up to a maximum of £7,710.
(b)	The erection of, or conversion of a building to, one or more dwellings.	£240 for each dwelling with a gross floor space up to 300 m^2 plus £240 for each additional 300 m^2 (or part thereof) of gross floor space for each dwelling created up to a maximum of £7,710.
(c)	The physical alteration of a dwelling house including the erection of: (i) an extension; (ii) a garage (whether attached or detached); or (iii) another building. Other works within the curtilage of a dwelling house including: (i) the erection of fences, walls (or other means of enclosure), or satellite dishes; or (ii) the laying of hard standing or landscaping works not covered by development permitted under the Permitted Development Order.	If no floor space is to be created by the development, £75 For development creating up to 15 m^2 of gross floor space, £135. For development exceeding 15 m^2 and up to 300 m^2 of gross floor space, £240 plus £135 for each additional 100 m^2 (or part thereof) of gross floor space up to a maximum of £7,710.

There is a very useful online calculator which will give prices for planning applications at
http: //www.planningportal.gov.uk/pins/FeeCalculatorStandalone

THE BUILDING (LOCAL AUTHORITY CHARGES) REGULATIONS

CHARGE SCHEDULES

Fees vary from one authority to another, so always check with your Local Authority if you decide to use their officers to provide building control services.

The Construction Industry Council (CIC) has been designated by the government as a body for approving inspectors (AI). Individual and Corporate Approved Inspectors registered with CIC are qualified to undertake building control work in accordance with section 49 of the Building Act 1984 and regulation 4 of the Building (Approved Inspectors etc.) Regulations 1985, and the Building (Approved Inspectors etc.) Regulations 2010.

Approved Inspectors provide building control services on all types of construction projects. The Construction Industry Council (CIC) maintains a list of approved inspectors, see www.cic.org.uk.

Fees vary. Each Local Authority publishes a schedule of rates applicable to project values commonly up to £250,000. The fee for a project of £250,000 in value is approximately £1,500.

For higher value projects the fees are individually determined but are typically approximately 0.1% of the estimated project cost.

Building Regulations in Brief
Seventh Edition

Ray Tricker & Sam Alford

The most popular and trusted guide to the building regulations, Building Regulations in Brief is updated regularly to reflect constant changes. Now in its seventh edition, it has sold over 28,000 copies since its first publication in 2003.

This new edition includes the latest on all the significant amendments to Building Regulations, Planning Permission and the Approved Documents that occurred in October 2010 and includes changes to Parts F and L, as well as Approved Documents A, C, and J. There are also changes reflecting the consolidation of the building regulations included.

The no-nonsense approach has made it a firm favourite with all involved in the building industry including designers, building surveyors and inspectors, students and architects. A ready reference giving practical information, it enables compliance in the simplest and most cost-effective manner possible. Building Regulations in Brief cuts through the confusion to explain the meaning of the regulations, their history, current status, requirements, associated documentation and how local authorities view their importance, as well as emphasizing the benefits and requirements of each regulation. It's an essential purchase for anyone needing to comply with the building regulations.

February 2012: 234 x156: 1,034 pp
Pb: 978-0-415-80969-6: £28.99

To Order: Tel: +44 (0) 1235 400524 Fax: +44 (0) 1235 400525
or Post: Taylor and Francis Customer Services,
Bookpoint Ltd, Unit T1, 200 Milton Park, Abingdon, Oxon, OX14 4TA UK
Email: book.orders@tandf.co.uk

For a complete listing of all our titles visit:
www.crcpress.com

PART 6

Daywork and Prime Cost

BUILDING INDUSTRY

When work is carried out which cannot be valued in any other way it is customary to assess the value on a cost basis with an allowance to cover overheads and profit. The basis of costing is a matter for agreement between the parties concerned, but definitions of prime cost for the building industry have been prepared and published jointly by the Royal Institution of Chartered Surveyors and the National Federation of Building Trades Employers (now the Construction Confederation) for the convenience of those who wish to use them. These documents are reproduced with the permission of the Royal Institution of Chartered Surveyors, which owns the copyright.

The daywork schedule published by the Civil Engineering Contractors Association is included in the *Architects' & Builders'* companion title, *Spon's Civil Engineering and Highway Works Price Book*.

For larger Prime Cost contracts the reader is referred to the form of contract issued by the Royal Institute of British Architects.

DEFINITION OF PRIME COST OF BUILDING WORKS OF A JOBBING OR MAINTENANCE CHARACTER (2010 EDITION)

This definition of Prime Cost is published by the Royal Institution of Chartered Surveyors and the National Federation of Building Trades Employers, for convenience and for use by people who choose to use it. Members of the National Federation of Building Trades Employers are not in any way debarred from defining Prime Cost and rendering their accounts for work carried out on that basis in any way they choose. Building owners are advised to reach agreement with contractors on the Definition of Prime Cost to be used prior to issuing instructions.

SECTION 1 – APPLICATION

1.1.   This definition provides a basis for the valuation of work of a jobbing or maintenance character executed under such building contracts as provide for its use.
1.2.   It is not applicable in any other circumstances, such as daywork executed under or incidental to a building contract.

SECTION 2 – COMPOSITION OF TOTAL CHARGES

2.1.   The prime cost of jobbing work comprises the sum of the following costs:
   (a)   Labour as defined in Section 3.
   (b)   Materials and goods as defined in Section 4.
   (c)   Plant, consumable stores and services as defined in Section 5.
   (d)   Subcontracts as defined in Section 6.
2.2.   Incidental costs, overhead and profit as defined in Section 7 and expressed as percentage adjustments are applicable to each of 2.1 (a)–(d).

SECTION 3 – LABOUR

3.1.   Labour costs comprise all payments made to or in respect of all persons directly engaged upon the work, whether on or off the site, except those included in Section 7.

BUILDING INDUSTRY

3.2.   Such payments are based upon the standard wage rates, emoluments and expenses as laid down for the time being in the rules or decisions of the National Joint Council for the Building Industry and the terms of the Building and Civil Engineering Annual and Public Holiday Agreements applying to the works, or the rules of decisions or agreements of such other body as may relate to the class of labour concerned, at the time when and in the area where the work is executed, together with the Contractor's statutory obligations, including:

(a)   Guaranteed minimum weekly earnings (e.g. Standard Basic Rate of Wages and Guaranteed Minimum Bonus Payment in the case of NJCBI rules).

(b)   All other guaranteed minimum payments (unless included in Section 7).

(c)   Payments in respect of incentive schemes or productivity agreements applicable to the works.

(d)   Payments in respect of overtime normally worked; or necessitated by the particular circumstances of the work; or as otherwise agreed between the parties.

(e)   Differential or extra payments in respect of skill, responsibility, discomfort or inconvenience.

(f)   Tool allowance.

(g)   Subsistence and periodic allowances.

(h)   Fares, travelling and lodging allowances.

(i)   Employer's contributions to annual holiday credits.

(j)   Employer's contributions to death benefit schemes.

(k)   Any amounts which may become payable by the Contractor to or in respect of operatives arising from the operation of the rules referred to in 3.2 which are not provided for in 3.2 (a)–(k) or in Section 7.

(l)   Employer's National Insurance contributions and any contribution, levy or tax imposed by statute, payable by the Contractor in his capacity as employer.

Note:

Any payments normally made by the Contractor which are of a similar character to those described in 3.2 (a)–(c) but which are not within the terms of the rules and decisions referred to above are applicable subject to the prior agreement of the parties, as an alternative to 3.2 (a)–(c).

3.3.   The wages or salaries of supervisory staff, timekeepers, storekeepers, and the like, employed on or regularly visiting site, where the standard wage rates, etc., are not applicable, are those normally paid by the Contractor together with any incidental payments of a similar character to 3.2 (c)–(k).

3.4.   Where principals are working manually their time is chargeable, in respect of the trades practised, in accordance with 3.2.

SECTION 4 – MATERIALS AND GOODS

4.1.   The prime cost of materials and goods obtained by the Contractor from stockists or manufacturers is the invoice cost after deduction of all trade discounts but including cash discounts not exceeding 5 per cent, and includes the cost of delivery to site.

4.2.   The prime cost of materials and goods supplied from the Contractor's stock is based upon the current market prices plus any appropriate handling charges.

4.3.   The prime cost under 4.1 and 4.2 also includes any costs of:

(a)   non-returnable crates or other packaging.

(b)   returning crates and other packaging less any credit obtainable.

4.4.   Any value added tax which is treated, or is capable of being treated, as input tax (as defined in the Finance Act, 1972 or any re-enactment thereof) by the Contractor is excluded.

SECTION 5 – PLANT, CONSUMABLE STORES AND SERVICES

5.1.   The prime cost of plant and consumable stores as listed below is the cost at hire rates agreed between the parties or in the absence of prior agreement at rates not exceeding those normally applied in the locality at the time when the works are carried out, or on a use and waste basis where applicable:

(a)   Machinery in workshops.

(b)   Mechanical plant and power-operated tools.

(c)   Scaffolding and scaffold boards.

BUILDING INDUSTRY

    (d)   Non-mechanical plant excluding hand tools.

    (e)   Transport including collection and disposal of rubbish.

    (f)   Tarpaulins and dust sheets.

    (g)   Temporary roadways, shoring, planking and strutting, hoarding, centring, formwork, temporary fans, partitions or the like.

    (h)   Fuel and consumable stores for plant and power-operated tools unless included in 5.1 (a), (b), (d) or (e) above.

    (i)   Fuel and equipment for drying out the works and fuel for testing mechanical services.

5.2.   The prime cost also includes the net cost incurred by the Contractor of the following services, excluding any such cost included under Sections 3, 4 or 7:

    (a)   Charges for temporary water supply including the use of temporary plumbing and storage.

    (b)   Charges for temporary electricity or other power and lighting including the use of temporary installations.

    (c)   Charges arising from work carried out by local authorities or public undertakings.

    (d)   Fees, royalties and similar charges.

    (e)   Testing of materials.

    (f)   The use of temporary buildings including rates and telephone and including heating and lighting not charged under (b) above.

    (g)   The use of canteens, sanitary accommodation, protective clothing and other provision for the welfare of persons engaged in the work in accordance with the current Working Rule Agreement and any Act of Parliament, statutory instrument, rule, order, regulation or bye-law.

    (h)   The provision of safety measures necessary to comply with any Act of Parliament.

    (i)   Premiums or charges for any performance bonds or insurances which are required by the Building Owner and which are not referred to elsewhere in this Definition.

SECTION 6 – SUBCONTRACTS

6.1.   The prime cost of work executed by subcontractors, whether nominated by the Building Owner or appointed by the Contractor, is the amount which is due from the Contractor to the subcontractors in accordance with the terms of the subcontracts after deduction of all discounts except any cash discount offered by any subcontractor to the Contractor not exceeding 2.5%.

SECTION 7 – INCIDENTAL COSTS, OVERHEADS AND PROFIT

7.1.   The percentage adjustments provided in the building contract, which are applicable to each of the totals of Sections 3–6, provide for the following:

    (a)   Head Office charges.

    (b)   Off-site staff including supervisory and other administrative staff in the Contractor's workshops and yard.

    (c)   Payments in respect of public holidays.

    (d)   Payments in respect of apprentices' study time.

    (e)   Sick pay or insurance in respect thereof.

    (f)   Third party employer's liability insurance.

    (g)   Liability in respect of redundancy payments made to employees.

    (h)   Use, repair and sharpening of non-mechanical hand tools.

    (i)   Any variations to basic rates required by the Contractor in cases where the building contract provides for the use of a specified schedule of basic plant charges (to the extent that no other provision is made for such variation).

    (j)   All other liabilities and obligations whatsoever not specifically referred to in this Section nor chargeable under any other section.

    (k)   Profit.

BUILDING INDUSTRY

SPECIMEN ACCOUNT FORMAT

If this Definition of Prime Cost is followed the Contractor's account could be in the following format:

	£
Labour (as defined in Section 3) Add _____ % (see Section 7)	
Materials and goods (as defined in Section 4) Add _____ % (see Section 7)	£
Plant, consumable stores and services (as defined in Section 5) Add _____ % (see Section 7)	£
Subcontracts (as defined in Section 6) Add _____ % (see Section 7)	£
Total	£

VAT to be added if applicable.

Example calculations of Prime Cost of Labour in Daywork

Example of calculation of typical standard hourly base rate (as defined in Section 3) for CIJC Building Craft operative and General Operative based upon rates applicable January 2013.

	Rate (£)	Craft Operative (£)	Rate (£)	General Operative (£)
Basic Wages	46.2 wks	416.13	19,225.21	313.17	14,468.45
Employer's National Insurance contribution (13.8% after the first £144 per week)			2046.64		1,057.14
Subtotal			21,271.85		15,525.49
Employer's Contributions to:					
CITB levy (0.5% of payroll)			108.19		81.42
Holiday Pay	226.2 hrs	10.67	2,413.55	8.03	1,816.39
Welfare Benefit – stamps	52 wks	11.39	592.28	11.39	592.28
Annual labour cost:			**24,385.87**		**18,015.68**
Hourly Base Rate:			**13.53**		**10.00**

Note:

(1)   Calculated following Definition of Prime Cost of Daywork carried out under a Building Contract, published by the Royal Institution of Chartered Surveyors and the Construction Confederation.

BUILDING INDUSTRY

(2)   Standard basic rates effective from 7 January 2013.
(3)   Standard working hours per annum calculated as follows:

52 weeks @ 39 hours =	2028
Less	
22 days holiday @ 39 hours = 163.8	
8 days public holidays @ 7.8 hours = 62.4	
= −226.2	
Standard working hours per year =	1801.8

(4)   All labour costs incurred by the contractor in his capacity as an employer, other than those contained in the hourly base rate, are to be taken into account under Section 6.

(5)   The above example is for guidance only and does not form part of the Definition; all the basic costs are subject to re-examination according to the time when and in the area where the daywork is executed.

(6)   N.I. payments are at not-contracted rates applicable from April 2013.

(7)   Basic rate and GMP number of weeks =
52 Weeks − 4.2 weeks annual holiday − 1.6 weeks public holiday = 46.2 weeks.

BUILDING INDUSTRY

SCHEDULE OF BASIC PLANT CHARGES (1st JULY 2010 ISSUE)

This Schedule is published by the Royal Institution of Chartered Surveyors, Parliament Square, London, and is for use in connection with Dayworks under a Building Contract.

EXPLANATORY NOTES

1. The rates in the Schedule are intended to apply solely to daywork carried out under and incidental to a Building Contract. They are NOT intended to apply to:
    (i)   Jobbing or any other work carried out as a main or separate contract; or
    (ii)   Work carried out after the date of commencement of the Defects Liability Period.
2. The rates apply only to plant and machinery already on site, whether hired or owned by the Contractor.
3. The rates, unless otherwise stated, include the cost of fuel and power of every description, lubricating oils, grease, maintenance, sharpening of tools, replacement of spare parts, all consumable stores and for licences and insurances applicable to items of plant.
4. The rates, unless otherwise stated, do not include the costs of drivers and attendants.
5. The rates do not include for any possible discounts which may be given to the Contractor if the plant or machinery was hired.
6. The rates are base costs and may be subject to the overall adjustment for price movement, overheads and profit, quoted by the Contractor prior to the placing of the Contract.
7. The rates should be applied to the time during which the plant is actually engaged in daywork.
8. Whether or not plant is chargeable on daywork depends on the daywork agreement in use and the inclusion of an item of plant in this schedule does not necessarily indicate that the item is chargeable.
9. Rates for plant not included in the Schedule or which is not already on site and is specifically provided or hired for daywork shall be settled at prices which are reasonably related to the rates in the Schedule having regard to any overall adjustment quoted by the Contractor in the Conditions of Contract.

Item of plant	Size/Rating	Unit	Rate per hour (£)
PUMPS			
Mobile Pumps			
Including pump hoses, values and strainers, etc.			
Diaphragm	50 mm dia.	Each	1.17
Diaphragm	76 mm dia.	Each	1.89
Diaphragm	102 mm dia.	Each	3.54
Submersible	50 mm dia.	Each	0.76
Submersible	76 mm dia.	Each	0.86
Submersible	102 mm dia.	Each	1.03
Induced Flow	50 mm dia.	Each	0.77
Induced Flow	76 mm dia.	Each	1.67
Centrifugal, self priming	25 mm dia.	Each	1.30
Centrifugal, self priming	50 mm dia.	Each	1.92
Centrifugal, self priming	75 mm dia.	Each	2.74
Centrifugal, self priming	102 mm dia.	Each	3.35
Centrifugal, self priming	152 mm dia.	Each	4.27

BUILDING INDUSTRY

Item of plant	Size/Rating	Unit	Rate per hour (£)
SCAFFOLDING, SHORING, FENCING			
Complete Scaffolding			
Mobile working towers, single width	2.0 m × 0.72 m base × 7.45 m high	Each	3.36
Mobile working towers, single width	2.0 m × 0.72 m base × 8.84 m high	Each	3.79
Mobile working towers, double width	2.0 m × 1.35 m × 7.45 m high	Each	3.79
Mobile working towers, double width	2.0 m × 1.35 m × 15.8 m high	Each	7.13
Chimney scaffold, single unit		Each	1.92
Chimney scaffold, twin unit		Each	3.59
Push along access platform	1.63–3.1 m	Each	5.00
Push along access platform	1.80 m × 0.70 m	Each	1.79
Trestles			
Trestle, adjustable	Any height	Pair	0.41
Trestle, painters	1.8 m high	Pair	0.31
Trestle, painters	2.4 m high	Pair	0.36
Shoring, Planking and Strutting			
'Acrow' adjustable prop	Sizes up to 4.9 m (open)	Each	0.06
'Strong boy' support attachment		Each	0.22
Adjustable trench strut	Sizes up to 1.67 m (open)	Each	0.16
Trench sheet		Metre	0.03
Backhoe trench box	Base unit	Each	1.23
Backhoe trench box	Top unit	Each	0.87
Temporary Fencing			
Including block and coupler			
Site fencing steel grid panel	3.5 m × 2.0 m	Each	0.05
Anti-climb site steel grid fence panel	3.5 m × 2.0 m	Each	0.08
Solid panel Heras	2.0 m × 2.0 m	Each	0.09
Pedestrian gate		Each	0.36
Roadway gate		Each	0.60

Item of plant	Size/Rating	Unit	Rate per hour (£)
LIFTING APPLIANCES AND CONVEYORS			
Cranes			
<u>Mobile Cranes</u>			
Rates are inclusive of drivers			
Lorry mounted, telescopic jib			
Two wheel drive	5 tonnes	Each	19.00
Two wheel drive	8 tonnes	Each	42.00
Two wheel drive	10 tonnes	Each	50.00

BUILDING INDUSTRY

Item of plant	Size/Rating	Unit	Rate per hour (£)
LIFTING APPLIANCES AND CONVEYORS – cont			
Two wheel drive	12 tonnes	Each	77.00
Two wheel drive	20 tonnes	Each	89.69
Four wheel drive	18 tonnes	Each	46.51
Four wheel drive	25 tonnes	Each	35.90
Four wheel drive	30 tonnes	Each	38.46
Four wheel drive	45 tonnes	Each	46.15
Four wheel drive	50 tonnes	Each	53.85
Four wheel drive	60 tonnes	Each	61.54
Four wheel drive	70 tonnes	Each	71.79
Static tower crane			
Rates inclusive of driver			

Note: Capacity equals maximum lift in tonnes times maximum radius at which it can be lifted

	Capacity (metre/tonnes) Up to	Height under hook above ground (m) Up to	Unit	Rate per hour (£)
Tower crane	30	22	Each	22.23
Tower crane	40	22	Each	26.62
Tower crane	40	30	Each	33.33
Tower crane	50	22	Each	29.16
Tower crane	60	22	Each	35.90
Tower crane	60	36	Each	35.90
Tower crane	70	22	Each	41.03
Tower crane	80	22	Each	39.12
Tower crane	90	42	Each	37.18
Tower crane	110	36	Each	47.62
Tower crane	140	36	Each	55.77
Tower crane	170	36	Each	64.11
Tower crane	200	36	Each	71.95
Tower crane	250	36	Each	84.77
Tower crane with luffing jig	30	25	Each	22.23
Tower crane with luffing jig	40	30	Each	26.62
Tower crane with luffing jig	50	30	Each	29.16
Tower crane with luffing jig	60	36	Each	41.03
Tower crane with luffing jig	65	30	Each	33.13
Tower crane with luffing jig	80	22	Each	48.72
Tower crane with luffing jig	100	45	Each	48.72

BUILDING INDUSTRY

Item of plant	Size/Rating		Unit	Rate per hour (£)
LIFTING APPLIANCES AND CONVEYORS				
	Capacity (metre/tonnes) Up to	Height under hook above ground (m) Up to		
Tower crane with luffing jig	125	30	Each	53.85
Tower crane with luffing jig	160	50	Each	53.85
Tower crane with luffing jig	200	50	Each	74.36
Tower crane with luffing jig	300	60	Each	100.00
Crane Equipment				
Muck tipping skip	Up to 200 litres		Each	0.67
Muck tipping skip	500 litres		Each	0.82
Muck tipping skip	750 litres		Each	1.08
Muck tipping skip	1000 litres		Each	1.28
Muck tipping skip	1500 litres		Each	1.41
Muck tipping skip	2000 litres		Each	1.67
Mortar skip	250 litres, plastic		Each	0.41
Mortar skip	350 litres steel		Each	0.77
Boat skip	250 litres		Each	0.92
Boat skip	500 litres		Each	1.08
Boat skip	750 litres		Each	1.23
Boat skip	1000 litres		Each	1.38
Boat skip	1500 litres		Each	1.64
Boat skip	2000 litres		Each	1.90
Boat skip	3000 litres		Each	2.82
Boat skip	4000 litres		Each	3.23
Master flow skip	250 litres		Each	0.77
Master flow skip	500 litres		Each	1.03
Master flow skip	750 litres		Each	1.28
Master flow skip	1000 litres		Each	1.44
Master flow skip	1500 litres		Each	1.69
Master flow skip	2000 litres		Each	1.85
Grand master flow skip	500 litres		Each	1.28
Grand master flow skip	750 litres		Each	1.64
Grand master flow skip	1000 litres		Each	1.69
Grand master flow skip	1500 litres		Each	1.95
Grand master flow skip	2000 litres		Each	2.21
Cone flow skip	500 litres		Each	1.33

BUILDING INDUSTRY

Item of plant	Size/Rating	Unit	Rate per hour (£)
LIFTING APPLIANCES AND CONVEYORS – cont			
Cone flow skip	1000 litres	Each	1.69
Geared rollover skip	500 litres	Each	1.28
Geared rollover skip	750 litres	Each	1.64
Geared rollover skip	1000 litres	Each	1.69
Geared rollover skip	1500 litres	Each	1.95
Geared rollover skip	2000 litres	Each	2.21
Multi skip, rope operated	200 mm outlet size, 500 litres	Each	1.49
Multi skip, rope operated	200 mm outlet size, 750 litres	Each	1.64
Multi skip, rope operated	200 mm outlet size, 1000 litres	Each	1.74
Multi skip, rope operated	200 mm outlet size, 1500 litres	Each	2.00
Multi skip, rope operated	200 mm outlet size, 2000 litres	Each	2.26
Multi skip, man riding	200 mm outlet size, 1000 litres	Each	2.00
Multi skip	4 point lifting frame	Each	0.90
Multi skip	Chain brothers	Set	0.87
Crane Accessories			
Multi-purpose crane forks	1.5 and 2 tonnes S.W.L.	Each	1.13
Self-levelling crane forks		Each	1.28
Man cage	1 man, 230 kg S.W.L.	Each	1.90
Man cage	2 man, 500 kg S.W.L.	Each	1.95
Man cage	4 man, 750 kg S.W.L.	Each	2.15
Man cage	8 man, 1000 kg S.W.L.	Each	3.33
Stretcher cage	500 kg, S.W.L.	Each	2.69
Goods carrying cage	1500 kg, S.W.L.	Each	1.33
Goods carrying cage	3000 kg, S.W.L.	Each	1.85
Builders' skip lifting cradle	12 tonnes, S.W.L.	Each	2.31
Board/pallet fork	1600 kg, S.W.L.	Each	1.90
Gas bottle carrier	500 kg, S.W.L.	Each	0.92
Hoists			
Scaffold hoist	200 kg	Each	2.46
Rack and pinion (goods only)	500 kg	Each	4.56
Rack and pinion (goods only)	1100 kg	Each	5.90
Rack and pinion (goods and passenger)	8 person, 80 kg	Each	7.44
Rack and pinion (goods and passenger)	14 person, 1400 kg	Each	8.72
Wheelbarrow chain sling		Each	1.67

BUILDING INDUSTRY

Item of plant	Size/Rating		Unit	Rate per hour (£)
LIFTING APPLIANCES AND CONVEYORS				
Conveyors				
Belt conveyors				
Conveyor	8 m long × 450 mm wide		Each	5.90
Miniveyor, control box and loading hopper	3 m unit		Each	4.49
Other Conveying Equipment				
Wheelbarrow			Each	0.62
Hydraulic superlift			Each	4.56
Pavac slab lifter (tile hoist)			Each	4.49
High lift pallet truck			Each	3.08
Lifting Trucks	Payload	Maximum lift		
Fork lift, two wheel drive	1100 kg	up to 3.0 m	Each	5.64
Fork lift, two wheel drive	2540 kg	up to 3.7 m	Each	5.64
Fork lift, four wheel drive	1524 kg	up to 6.0 m	Each	5.64
Fork lift, four wheel drive	2600 kg	up to 5.4 m	Each	7.44
Fork lift, four wheel drive	4000 kg	up to 17 m	Each	10.77
Lifting Platforms				
Hydraulic platform (Cherry picker)	9 m		Each	4.62
Hydraulic platform (Cherry picker)	12 m		Each	7.56
Hydraulic platform (Cherry picker)	15 m		Each	10.13
Hydraulic platform (Cherry picker)	17 m		Each	15.63
Hydraulic platform (Cherry picker)	20 m		Each	18.13
Hydraulic platform (Cherry picker)	25.6 m		Each	32.38
Scissor lift	7.6 m, electric		Each	3.85
Scissor lift	7.8 m, electric		Each	5.13
Scissor lift	9.7 m, electric		Each	4.23
Scissor lift	10 m, diesel		Each	6.41
Telescopic handler	7 m, 2 tonnes		Each	5.13
Telescopic handler	13 m, 3 tonnes		Each	7.18
Lifting and Jacking Gear				
Pipe winch including gantry	1 tonne		Set	1.92
Pipe winch including gantry	3 tonnes		Set	3.21
Chain block	1 tonne		Each	0.35
Chain block	2 tonnes		Each	0.58
Chain block	5 tonnes		Each	1.14
Pull lift (Tirfor winch)	1 tonne		Each	0.64
Pull lift (Tirfor winch)	1.6 tonnes		Each	0.90

BUILDING INDUSTRY

Item of plant	Size/Rating	Unit	Rate per hour (£)
LIFTING APPLIANCES AND CONVEYORS – cont			
Pull lift (Tirfor winch)	3.2 tonnes	Each	1.15
Brother or chain slings, two legs	not exceeding 3.1 tonnes	Set	0.21
Brother or chain slings, two legs	not exceeding 4.25 tonnes	Set	0.31
Brother or chain slings, four legs	not exceeding 11.2 tonnes	Set	1.09
CONSTRUCTION VEHICLES			
Lorries			
Plated lorries (Rates are inclusive of driver)			
Platform lorry	7.5 tonnes	Each	16.21
Platform lorry	17 tonnes	Each	22.90
Platform lorry	24 tonnes	Each	30.68
Extra for lorry with crane attachment	up to 2.5 tonnes	Each	3.25
Extra for lorry with crane attachment	up to 5 tonnes	Each	6.00
Extra for lorry with crane attachment	up to 7.5 tonnes	Each	9.10
Tipper Lorries			
(Rates are inclusive of driver)			
Tipper lorry	up to 11 tonnes	Each	15.78
Tipper lorry	up to 17 tonnes	Each	23.95
Tipper lorry	up to 25 tonnes	Each	31.35
Tipper lorry	up to 31 tonnes	Each	37.79
Dumpers			
Site use only (excl. tax, insurance and extra cost of DERV etc. when operating on highway)	Makers capacity		
Two wheel drive	1 tonne	Each	1.71
Four wheel drive	2 tonnes	Each	2.43
Four wheel drive	3 tonnes	Each	2.44
Four wheel drive	5 tonnes	Each	3.08
Four wheel drive	6 tonnes	Each	3.85
Four wheel drive	9 tonnes	Each	5.65
Tracked	0.5 tonnes	Each	3.33
Tracked	1.5 tonnes	Each	4.23
Tracked	3 tonnes	Each	8.33
Tracked	6 tonnes	Each	16.03

BUILDING INDUSTRY

Item of plant	Size/Rating	Unit	Rate per hour (£)
CONSTRUCTION VEHICLES			
Dumper Trucks (*Rates are inclusive of drivers*)			
Dumper truck	up to 15 tonnes	Each	28.56
Dumper truck	up to 17 tonnes	Each	32.82
Dumper truck	up to 23 tonnes	Each	54.64
Dumper truck	up to 30 tonnes	Each	63.50
Dumper truck	up to 35 tonnes	Each	73.02
Dumper truck	up to 40 tonnes	Each	87.84
Dumper truck	up to 50 tonnes	Each	133.44
Tractors			
Agricultural Type			
Wheeled, rubber-clad tyred	up to 40 kW	Each	8.63
Wheeled, rubber-clad tyred	up to 90 kW	Each	25.31
Wheeled, rubber-clad tyred	up to 140 kW	Each	36.49
Crawler Tractors			
With bull or angle dozer	up to 70 kW	Each	29.38
With bull or angle dozer	up to 85 kW	Each	38.63
With bull or angle dozer	up to 100 kW	Each	52.59
With bull or angle dozer	up to 115 kW	Each	55.85
With bull or angle dozer	up to 135 kW	Each	60.43
With bull or angle dozer	up to 185 kW	Each	76.44
With bull or angle dozer	up to 200 kW	Each	96.43
With bull or angle dozer	up to 250 kW	Each	117.68
With bull or angle dozer	up to 350 kW	Each	160.03
With bull or angle dozer	up to 450 kW	Each	219.86
With loading shovel	0.8 m³	Each	26.92
With loading shovel	1.0 m³	Each	32.59
With loading shovel	1.2 m³	Each	37.53
With loading shovel	1.4 m³	Each	42.89
With loading shovel	1.8 m³	Each	52.22
With loading shovel	2.0 m³	Each	57.22
With loading shovel	2.1 m³	Each	60.12
With loading shovel	3.5 m³	Each	87.26
Light vans			
VW Caddivan or the like		Each	5.26
VW Transport transit or the like	1 tonne	Each	6.03
Luton Box Van or the like	1.8 tonnes	Each	9.87

BUILDING INDUSTRY

Item of plant	Size/Rating	Unit	Rate per hour (£)
CONSTRUCTION VEHICLES – cont			
Water/Fuel Storage			
Mobile water container	110 litres	Each	0.62
Water bowser	1100 litres	Each	0.72
Water bowser	3000 litres	Each	0.87
Mobile fuel container	110 litres	Each	0.62
Fuel bowser	1100 litres	Each	1.23
Fuel bowser	3000 litres	Each	1.87
EXCAVATIONS AND LOADERS			
Excavators			
Wheeled, hydraulic	up to 11 tonnes	Each	25.86
Wheeled, hydraulic	up to 14 tonnes	Each	30.82
Wheeled, hydraulic	up to 16 tonnes	Each	34.50
Wheeled, hydraulic	up to 21 tonnes	Each	39.10
Wheeled, hydraulic	up to 25 tonnes	Each	43.81
Wheeled, hydraulic	up to 30 tonnes	Each	55.30
Crawler, hydraulic	up to 11 tonnes	Each	25.86
Crawler, hydraulic	up to 14 tonnes	Each	30.82
Crawler, hydraulic	up to 17 tonnes	Each	34.50
Crawler, hydraulic	up to 23 tonnes	Each	39.10
Crawler, hydraulic	up to 30 tonnes	Each	43.81
Crawler, hydraulic	up to 35 tonnes	Each	55.30
Crawler, hydraulic	up to 38 tonnes	Each	71.73
Crawler, hydraulic	up to 55 tonnes	Each	95.63
Mini excavator	1000/1500 kg	Each	4.87
Mini excavator	2150/2400 kg	Each	6.67
Mini excavator	2700/3500 kg	Each	7.31
Mini excavator	3500/4500 kg	Each	8.21
Mini excavator	4500/6000 kg	Each	9.23
Mini excavator	7000 kg	Each	14.10
Micro excavator	725 mm wide	Each	5.13
Loaders			
Shovel loader	0.4 m³	Each	7.69
Shovel loader	1.57 m³	Each	8.97
Shovel loader, four wheel drive	1.7 m³	Each	4.83
Shovel loader, four wheel drive	2.3 m³	Each	4.38

BUILDING INDUSTRY

Item of plant	Size/Rating	Unit	Rate per hour (£)
EXCAVATIONS AND LOADERS			
Shovel loader, four wheel drive	3.3 m³	Each	5.06
Skid steer loader wheeled	300/400 kg payload	Each	7.31
Skid steer loader wheeled	625 kg payload	Each	7.67
Tracked skip loader	650 kg	Each	4.42
Excavator Loaders			
Wheeled tractor type with black-hoe excavator			
Four wheel drive			
Four wheel drive, 2 wheel steer	6 tonnes	Each	6.41
Four wheel drive, 2 wheel steer	8 tonnes	Each	8.59
Attachments			
Breakers for excavator		Each	8.72
Breakers for mini excavator		Each	1.75
Breakers for back-hoe excavator/loader		Each	5.13
COMPACTION EQUIPMENT			
Rollers			
Vibrating roller	368–420 kg	Each	1.43
Single roller	533 kg	Each	1.94
Single roller	750 kg	Each	3.43
Twin roller	up to 650 kg	Each	6.03
Twin roller	up to 950 kg	Each	6.62
Twin roller with seat end steering wheel	up to 1400 kg	Each	7.68
Twin roller with seat end steering wheel	up to 2500 kg	Each	10.61
Pavement roller	3–4 tonnes dead weight	Each	6.00
Pavement roller	4–6 tonnes	Each	6.86
Pavement roller	6–10 tonnes	Each	7.17
Pavement roller	10–13 tonnes	Each	19.86
Rammers			
Tamper rammer 2 stroke-petrol	225 mm–275 mm	Each	1.52
Soil Compactors			
Plate compactor	75 mm–400 mm	Each	1.53
Plate compactor rubber pad	375 mm–1400 mm	Each	1.53
Plate compactor reversible plate-petrol	400 mm	Each	2.44

BUILDING INDUSTRY

Item of plant	Size/Rating	Unit	Rate per hour (£)
CONCRETE EQUIPEMENT			
Concrete/Mortar Mixers			
Open drum without hopper	0.09/0.06 m³	Each	0.61
Open drum without hopper	0.12/0.09 m³	Each	1.22
Open drum without hopper	0.15/0.10 m³	Each	0.72
Concrete/Mortar Transport Equipment			
Concrete pump incl. hose, valve and couplers			
Lorry mounted concrete pump	24 m max. distance	Each	50.00
Lorry mounted concrete pump	34 m max. distance	Each	66.00
Lorry mounted concrete pump	42 m max. distance	Each	91.50
Concrete Equipment			
Vibrator, poker, petrol type	up to 75 mm dia.	Each	0.69
Air vibrator (*excluding compressor and hose*)	up to 75 mm dia.	Each	0.64
Extra poker heads	25/36/60 mm dia.	Each	0.76
Vibrating screed unit with beam	5.00 m	Each	2.48
Vibrating screed unit with adjustable beam	3.00–5.00 m	Each	3.54
Power float	725 mm–900 mm	Each	2.56
Power float finishing pan		Each	0.62
Floor grinder	660 × 1016 mm, 110 V electric	Each	4.31
Floor plane	450 × 1100 mm	Each	4.31
TESTING EQUIPMENT			
Pipe Testing Equipment			
Pressure testing pump, electric		Set	2.19
Pressure test pump		Set	0.80
SITE ACCOMODATION AND TEMPORARY SERVICES			
Heating equipment			
Space heater – propane	80,000 Btu/hr	Each	1.03
Space heater – propane/electric	125,000 Btu/hr	Each	2.09
Space heater – propane/electric	250,000 Btu/hr	Each	2.33
Space heater – propane	125,000 Btu/hr	Each	1.54
Space heater – propane	260,000 Btu/hr	Each	1.88
Cabinet heater		Each	0.82
Cabinet heater, catalytic		Each	0.57

BUILDING INDUSTRY

Item of plant	Size/Rating	Unit	Rate per hour (£)
SITE ACCOMODATION AND TEMPORARY SERVICES			
Electric halogen heater		Each	1.27
Ceramic heater	3 kW	Each	0.99
Fan heater	3 kW	Each	0.66
Cooling fan		Each	1.92
Mobile cooling unit, small		Each	3.60
Mobile cooling unit, large		Each	4.98
Air conditioning unit		Each	2.81
Site Lighting and Equipment			
Tripod floodlight	500 W	Each	0.48
Tripod floodlight	1000 W	Each	0.62
Towable floodlight	4 × 100 W	Each	3.85
Hand held floodlight	500 W	Each	0.51
Rechargeable light		Each	0.41
Inspection light		Each	0.37
Plasterer's light		Each	0.65
Lighting mast		Each	2.87
Festoon light string	25 m	Each	0.55
Site Electrical Equipment			
Extension leads	240 V/14 m	Each	0.26
Extension leads	110 V/14 m	Each	0.36
Cable reel	25 m 110 V/240 V	Each	0.46
Cable reel	50 m 110 V240 V	Each	0.88
4 way junction box	110 V	Each	0.56
Power Generating Units			
Generator – petrol	2 kVA	Each	1.23
Generator – silenced petrol	2 kVA	Each	2.87
Generator – petrol	3 kVA	Each	1.47
Generator – diesel	5 kVA	Each	2.44
Generator – silenced diesel	10 kVA	Each	1.90
Generator – silenced diesel	15 kVA	Each	2.26
Generator – silenced diesel	30 kVA	Each	3.33
Generator – silenced diesel	50 kVA	Each	4.10
Generator – silenced diesel	75 kVA	Each	4.62
Generator – silenced diesel	100 kVA	Each	5.64
Generator – silenced diesel	150 kVA	Each	7.18
Generator – silenced diesel	200 kVA	Each	9.74

BUILDING INDUSTRY

Item of plant	Size/Rating	Unit	Rate per hour (£)
SITE ACCOMODATION AND TEMPORARY SERVICES – cont			
Generator – silenced diesel	250 kVA	Each	11.28
Generator – silenced diesel	350 kVA	Each	14.36
Generator – silenced diesel	500 kVA	Each	15.38
Tail adaptor	240 V	Each	0.10
Transformers			
Transformer	3 kVA	Each	0.32
Transformer	5 kVA	Each	1.23
Transformer	7.5 kVA	Each	0.59
Transformer	10 kVA	Each	2.00
Rubbish Collection and Disposal			
Equipment			
Rubbish Chutes			
Standard plastic module	1 m section	Each	0.15
Steel liner insert		Each	0.30
Steel top hopper		Each	0.22
Plastic side entry hopper		Each	0.22
Plastic side entry hopper liner		Each	0.22
Dust Extraction Plant			
Dust extraction unit, light duty		Each	2.97
Dust extraction unit, heavy duty		Each	2.97
SITE EQUIPMENT – Welding Equipment			
Arc-(Electric) Complete With Leads			
Welder generator – petrol	200 amp	Each	3.53
Welder generator – diesel	300/350 amp	Each	3.78
Welder generator – diesel	4000 amp	Each	7.92
Extra welding lead sets		Each	0.69
Gas-Oxy Welder			
Welding and cutting set (including oxygen and acetylene, excluding underwater equipment and thermic boring)			
Small		Each	2.24
Large		Each	3.75
Lead burning gun		Each	0.50
Mig welder		Each	1.38
Fume extractor		Each	2.46

BUILDING INDUSTRY

Item of plant	Size/Rating	Unit	Rate per hour (£)
SITE EQUIPMENT – Welding Equipment			
Road Works Equipment			
Traffic lights, mains/generator	2-way	Set	10.94
Traffic lights, mains/generator	3-way	Set	11.56
Traffic lights, mains/generator	4-way	Set	12.19
Flashing light		Each	0.10
Road safety cone	450 mm	Each	0.08
Safety cone	750 mm	Each	0.10
Safety barrier plank	1.25 m	Each	0.13
Safety barrier plank	2 m	Each	0.15
Safety barrier plank post		Each	0.13
Safety barrier plank post base		Each	0.10
Safety four gate barrier	1 m each gate	Set	0.77
Guard barrier	2 m	Each	0.19
Road sign	750 mm	Each	0.23
Road sign	900 mm	Each	0.31
Road sign	1200 mm	Each	0.42
Speed ramp/cable protection	500 mm section	Each	0.14
Hose ramp open top	3 m section	Each	0.07
DPC Equipment			
Damp-proofing injection machine		Each	2.56
Cleaning Equipment			
Vacuum cleaner (industrial wet) single motor		Each	1.08
Vacuum cleaner (industrial wet) twin motor	30 litre capacity	Each	1.79
Vacuum cleaner (industrial wet) twin motor	70 litre capacity	Each	2.21
Steam cleaner	Diesel/electric 1 phase	Each	3.33
Steam cleaner	Diesel/electric 3 phase	Each	3.85
Pressure washer, light duty electric	1450 PSI	Each	0.72
Pressure washer, heavy duty, diesel	2500 PSI	Each	1.33
Pressure washer, heavy duty, diesel	4000 PSI	Each	2.18
Cold pressure washer, electric		Each	2.39
Hot pressure washer, petrol		Each	4.19
Hot pressure washer, electric		Each	5.13
Cold pressure washer, petrol		Each	2.92

BUILDING INDUSTRY

Item of plant	Size/Rating	Unit	Rate per hour (£)
SITE EQUIPMENT – Welding Equipment – cont			
Sandblast attachment to last washer		Each	1.23
Drain cleaning attachment to last washer		Each	1.03
Surface Preparation Equipment			
Rotavator	5 h.p.	Each	2.46
Rotavator	9 h.p.	Each	5.00
Scabbler, up to three heads		Each	1.53
Scabbler, pole		Each	2.68
Scrabbler, multi-headed floor		Each	3.89
Floor preparation machine		Each	1.05
Compressors and Equipment			
Portable Compressors			
Compressor – electric	4 cfm	Each	1.36
Compressor – electric	8 cfm lightweight	Each	1.31
Compressor – electric	8 cfm	Each	1.36
Compressor – electric	14 cfm	Each	1.56
Compressor – petrol	24 cfm	Each	2.15
Compressor – electric	25 cfm	Each	2.10
Compressor – electric	30 cfm	Each	2.36
Compressor – diesel	100 cfm	Each	2.56
Compressor – diesel	250 cfm	Each	5.54
Compressor – diesel	400 cfm	Each	8.72
Mobile Compressors			
Lorry mounted compressor			
(machine plus lorry only)	up to 3 m³	Each	41.47
(machine plus lorry only)	up to 5 m³	Each	48.94
Tractor mounted compressor			
(machine plus rubber tyred tractor)	Up to 4 m³	Each	21.03
Accessories (Pneumatic Tools) *(with and including up to 15 m of air hose)*			
Demolition pick, medium duty		Each	0.90
Demolition pick, heavy duty		Each	1.03
Breakers (with six steels) light	up to 150 kg	Each	1.19
Breakers (with six steels) medium	295 kg	Each	1.24
Breakers (with six steels) heavy	386 kg	Each	1.44
Rock drill (for use with compressor) hand held		Each	1.18

BUILDING INDUSTRY

Item of plant	Size/Rating	Unit	Rate per hour (£)
SITE EQUIPMENT – Welding Equipment			
Additional hoses	15 m	Each	0.09
Breakers			
Demolition hammer drill, heavy duty, electric		Each	1.54
Road breaker, electric		Each	2.41
Road breaker, 2 stroke, petrol		Each	4.06
Hydraulic breaker unit, light duty, petrol		Each	3.06
Hydraulic breaker unit, heavy duty, petrol		Each	3.46
Hydraulic breaker unit, heavy duty, diesel		Each	4.62
Quarrying and Tooling Equipment			
Block and stone splitter, hydraulic	600 mm × 600 mm	Each	1.90
Block and stone splitter, manual		Each	1.64
Steel Reinforcement Equipment			
Bar bending machine – manual	up to 13 mm dia. rods	Each	1.03
Bar bending machine – manual	up to 20 mm dia. rods	Each	1.41
Bar shearing machine – electric	up to 38 mm dia. rods	Each	3.08
Bar shearing machine – electric	up to 40 mm dia. rods	Each	4.62
Bar cropper machine – electric	up to 13 mm dia. rods	Each	2.05
Bar cropper machine – electric	up to 20 mm dia. rods	Each	2.56
Bar cropper machine – electric	up to 40 mm dia. rods	Each	4.62
Bar cropper machine – 3 phase	up to 40 mm dia. rods	Each	4.62
Dehumidifiers			
110/240 V Water	68 litres extraction per 24 hours	Each	2.46
110/240 V Water	90 litres extraction per 24 hours	Each	3.38
SMALL TOOLS			
Saws			
Masonry bench saw	350 mm–500 mm dia.	Each	1.13
Floor saw	125 mm max. cut	Each	1.15
Floor saw	150 mm max. cut	Each	3.83
Floor saw, reversible	350 mm max. cut	Each	3.32
Wall saw, electric		Each	2.05
Chop/cut off saw, electric	350 mm dia.	Each	1.79
Circular saw, electric	230 mm dia.	Each	0.72
Tyrannosaw		Each	1.74
Reciprocating saw		Each	0.79

BUILDING INDUSTRY

Item of plant	Size/Rating	Unit	Rate per hour (£)
SMALL TOOLS – cont			
Door trimmer		Each	1.17
Stone saw	300 mm	Each	1.44
Chainsaw, petrol	500 mm	Each	3.92
Full chainsaw safety kit		Each	0.41
Worktop jig		Each	1.08
Pipe Work Equipment			
Pipe bender	15 mm–22 mm	Each	0.92
Pipe bender, hydraulic	50 mm	Each	1.76
Pipe bender, electric	50 mm–150 mm dia.	Each	2.19
Pipe cutter, hydraulic		Each	0.46
Tripod pipe vice		Set	0.75
Ratchet threader	12 mm–32 mm	Each	0.93
Pipe threading machine, electric	12 mm–75 mm	Each	3.07
Pipe threading machine, electric	12 mm–100 mm	Each	4.93
Impact wrench, electric		Each	1.33
Hand-held Drills and equipment			
Impact or hammer drill	up to 25 mm dia.	Each	1.03
Impact or hammer drill	35 mm dia.	Each	1.29
Dry diamond core cutter		Each	0.99
Angle head drill		Each	0.90
Stirrer, mixer drill		Each	1.13
Paint, Insulation Application Equipment			
Airless spray unit		Each	4.13
Portaspray unit		Each	1.16
HPVL turbine spray unit		Each	2.23
Compressor and spray gun		Each	1.91
Other Handtools			
Staple gun		Each	0.96
Air nail gun	110 V	Each	1.01
Cartridge hammer		Each	1.08
Tongue and groove nailer complete with mallet		Each	1.59
Diamond wall chasing machine		Each	2.63
Masonry chain saw	300 mm	Each	5.49
Floor grinder		Each	3.99
Floor plane		Each	1.79

BUILDING INDUSTRY

Item of plant	Size/Rating	Unit	Rate per hour (£)
SMALL TOOLS			
Diamond concrete planer		Each	1.93
Autofeed screwdriver, electric		Each	1.38
Laminate trimmer		Each	0.91
Biscuit jointer		Each	1.49
Random orbital sander		Each	0.97
Floor sander		Each	1.54
Palm, delta, flap or belt sander		Each	0.75
Disc cutter, electric	300 mm	Each	1.49
Disc cutter, 2 stroke petrol	300 mm	Each	1.24
Dust suppressor for petrol disc cutter		Each	0.51
Cutter cart for petrol disc cutter		Each	1.21
Grinder, angle or cutter	up to 225 mm	Each	0.50
Grinder, angle or cutter	300 mm	Each	1.41
Mortar raking tool attachment		Each	0.19
Floor/polisher scrubber	325 mm	Each	1.76
Floor tile stripper		Each	2.44
Wallpaper stripper, electric		Each	0.81
Hot air paint stripper		Each	0.50
Electric diamond tile cutter	all sizes	Each	2.42
Hand tile cutter		Each	0.82
Electric needle gun		Each	1.29
Needle chipping gun		Each	1.85
Pedestrian floor sweeper	250 mm dia.	Each	0.82
Pedestrian floor sweeper	Petrol	Each	2.20
Diamond tile saw		Each	1.84
Blow lamp equipment and glass		Set	0.50

Twenty Buildings Every Architect Should Understand

Simon Unwin

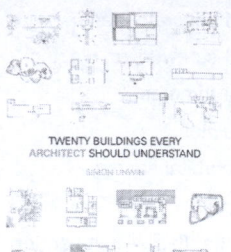

Have you ever wondered how the ideas behind the world's greatest architectural designs came about? What process does an architect go through to design buildings which become world-renowned for their excellence?

This book reveals the secrets behind these buildings. He asks you to 'read' the building and understand its starting point by analyzing its final form. Through the gradual revelations made by an understanding of the thinking behind the form, you learn a unique methodology which can be used every time you look at any building.

March 2010: 217x296: 230pp
Hb: 978-0-415-55251-6: £105.00
Pb: 978-0-415-55252-3: £18.99

To Order: Tel: +44 (0) 1235 400524 Fax: +44 (0) 1235 400525
or Post: Taylor and Francis Customer Services,
Bookpoint Ltd, Unit T1, 200 Milton Park, Abingdon, Oxon, OX14 4TA UK
Email: book.orders@tandf.co.uk

For a complete listing of all our titles visit:
www.tandf.co.uk

PART 7

Useful Addresses for Further Information

ACOUSTICAL INVESTIGATION & RESEARCH
ORGANISATION LTD (AIRO)
Duxon's Turn
Maylands Avenue
Hemel Hempstead
Hertfordshire
HP2 4SB
Tel: 01442 247 146
Fax: 01442 256 749
Website: www.airo.co.uk

ALUMINIUM FEDERATION LTD (ALFED)
National Metalforming Centre
47 Birmingham Road
West Bromwich
West Midlands
B70 6PY
Tel: 0121 601 6361
Fax: 0870 138 9714
Website: www.alfed.org.uk

AMERICAN HARDWOOD EXPORT COUNCIL (AHEC)
3 St. Michaels Alley
London
EC3V 9DS
Tel: 020 7626 4111
Fax: 020 7626 4222
Website: www.ahec-europe.org

ANCIENT MONUMENTS SOCIETY (AMS)
Saint Ann's Vestry Hall
2 Church Entry
London
EC4V 5HB
Tel: 020 7236 3934
Fax: 020 7329 3677
Website: www.ancientmonumentssociety.org.uk

APA – THE ENGINEERED WOOD ASSOCIATION
Claridge House
29 Barnes High Street
London
SW13 9LW
Tel: 0845 123 3721
Fax: 0208 282 1660
Website: http://www.apa-europe.org/

ARBORICULTURAL ASSOCIATION
The Moult House
Stround Green
Standish
Stonehouse
Gloustershire GL10 3DL
Tel: 01242 522152
Fax: 01242 577766
Website: www.trees.org.uk

ARCHITECTURAL ADVISORY SERVICE CENTRE
(POWDER/ANODIC METAL FINISHES)
Barn One
Barn Road
Longwick
Buckinghamshire
HP27 9RW
Tel: 01844 342 425
Fax: 01844 274 781
Website: http://www.aasc.org.uk/

ARCHITECTURAL ASSOCIATION (AA)
34–36 Bedford Square
London
WC1B 3ES
Tel: 020 7887 4000
Fax: 020 7414 0782
Website: http://www.aaschool.ac.uk/

ARCHITECTURAL CLADDING ASSOCIATION (ACA)
60 Charles Street
Leicester
Leicestershire
LE1 1FB
Tel: 0116 253 6161
Fax: 0116 251 4568
Website: http://www.architectural-cladding-
                association.org.uk/

ASBESTOS INFORMATION CENTRE (AIC)
ARCA House,
237 Branston Road
Burton upon Trent
Staffordshire
DE14 3BT
Tel: 01283 531 126
Fax: 01283 568 228
Website: http://www.aic.org.uk

ASSOCIATION FOR CONSULTANCY AND
ENGINEERING
Alliance House
12 Caxton Street
London
SW1H 0QL
Tel: 020 7222 6557
Fax: 020 7990 9202
Website: www.acenet.co.uk

ASSOCIATION OF INTERIOR SPECIALISTS
Olton Bridge
245 Warwick Road
Solihull
West Midlands
B92 7 AH
Tel: 0121 707 0077
Fax: 0121 706 1949
Website: www.ais-interiors.org.uk

ASSOCIATION OF LOADING AND ELEVATING
EQUIPMENT MANUFACTURERS
Airport House
Purley Way
Croydon
Surrey
CR0 0XY
Tel: 020 8253 4501
Fax: 020 8253 4510
Website: www.alem.org.uk

ASSOCIATION OF PROJECT MANAGEMENT
IBIS House
Summerleys Road
Princes Risborough
Bucks
HP27 9LE
Tel: 0845 458 1944
Fax: 01494 528 937
Website: www.apm.org.uk

BOX CULVERT ASSOCIATION (BCA)
60 Charles Street
Leicester
Leicestershire
LE1 1FB
Tel: 0116 253 6161
Fax: 0116 251 4568
Website: www.boxculvert.org.uk

BRITISH ADHESIVES AND SEALANTS
ASSOCIATION
5 Alderson Road
Worksop
Notts
S80 1UZ
Tel: 01909 480 888
Fax: 01909 473 834

BRITISH AGGREGATE CONSTRUCTION
MATERIALS INDUSTRIES LTD (BACMI)
156 Buckingham Palace Road
London
SW1W 9TR
Tel: 020 7730 8194

BRITISH APPROVALS FOR FIRE EQUIPMENT
(BAFE)
Thames House
29 Thames Street
Kingston upon Thames
Surrey
KT1 1PH
Tel: 020 8541 1950
Fax: 020 8547 1564
Website: www.bafe.org.uk

BRITISH APPROVALS SERVICE FOR CABLES
(BASEC)
23 Presley Way
Crownhill
Milton Keynes
Buckinghamshire
MK8 0ES
Tel: 01908 267 300
Fax: 01908 267 255
Website: www.basec.org.uk

BRITISH ARCHITECTURAL LIBRARY (BAL)
Royal Institute of British Architects
66 Portland Place
London
W1B 1AD
Tel: 020 7580 5533
Fax: 020 7631 1802
Website: www.architecture.com

BRITISH ASSOCIATION OF LANDSCAPE
INDUSTRIES (BALI)
Landscape House
National Agricultural Centre
Stoneleigh Park
Warwickshire
CV8 2LG
Tel: 024 7669 0333
Fax: 024 7669 0077
Website: www.bali.co.uk

BRITISH ASSOCIATION OF REINFORCEMENT
Riverside House
4 Meadows Business Park
Station Approach
Camberley
Surrey
GU17 9AB
Tel: 07802 747031
Website: www.uk-bar.org

BRITISH BATHROOM COUNCIL
(BATHROOM MANUFACTURERS ASSOCIATION)
Federation House
Station Road
Stoke-on-Trent
Staffordshire
ST4 2RT
Tel: 01782 747 123
Fax: 01782 747 161
Website: www.bathroom-association.org

BRITISH BOARD OF AGREMENT (BBA)
PO Box 195
Bucknalls Lane
Garston
Watford
Hertfordshire
WD25 9BA
Tel: 01923 665 300
Fax: 01923 665 301
Website: www.bbacerts.co.uk

BRITISH CABLES ASSOCIATION (BCA)
37a Walton Road
East Molesey
Surrey
KT8 0DH
Tel: 020 8941 4079
Fax: 020 8783 0104
Website: www.bcauk.org

BRITISH CARPET MANUFACTURERS
ASSOCIATION LTD (BCMA)
PO Box 1155
MCF Complex
60 New Road
Kidderminster
Worcestershire
DY10 1AQ
Tel: 01562 755 568
Fax: 01562 865 4055
Website: www.carpetfoundation.com

BRITISH CEMENT ASSOCIATION (BCA)
CENTRE FOR CONCRETE INFORMATION
Century House
Telford Avenue
Crowthorne
Berkshire
RG45 6YS
Tel: 01344 466 007
Fax: 01344 466 008
Website: www.cementindustry.co.uk

BRITISH CERAMIC CONFEDERATION (BCC)
Federation House
Station Road
Stoke-on-Trent
Staffordshire
ST4 2SA
Tel: 01782 744 631
Fax: 01782 744 102
Website: www.ceramfed.co.uk

BRITISH CERAMIC RESEARCH LTD (BCR)
Queens Road
Penkhull
Stoke-on-Trent
Staffordshire
ST4 7LQ
Tel: 01782 764 444
Fax: 01782 412 331
Website: www.ceram.co.uk

BRITISH CERAMIC TILE COUNCIL (BCTC TILE
ASSOCIATION)
Federation house
Station Road
Stoke On Trent
ST4 2RT
Tel: 01782 747 147
Fax: 01782 747 161
Website: www.tpb.org.uk/

BRITISH COMBUSTION EQUIPMENT
MANUFACTURERS ASSOCIATION (BCEMA)
58 London Road
Leicester
LE2 0QD
Tel: 0116 275 7111
Fax: 0116 275 7222
Website: bcema.co.uk

BRITISH CONSTRUCTIONAL STEELWORK
ASSOCIATION LTD (BCSA)
4 Whitehall Court
Westminster
London
SW1A 2ES
Tel: 0207 839 8566
Fax: 0207 976 1634
Website: www.steelconstruction.org

BRITISH CONTRACT FURNISHING ASSOCIATION
(BCFA)
Suite 2/4
The Business Design Centre
52 Upper Street
Islington Green
London
N1 0QH
Tel: 0207 226 6641
Fax: 0207 288 6190
Website: thebcfa.com

BRITISH ELECTROTECHNICAL APPROVALS
BOARD (BEAB)
1 Station View
Guildford
Surrey
GU1 4JY
Tel: 01483 455 466
Fax: 01483 455 477
Website: www.beab.co.uk

BRITISH FIRE PROTECTION SYSTEMS
ASSOCIATION LTD (BFPSA)
Thames House
29 Thames Street
Kingston-upon-Thames
Surrey
KT1 1PH
Tel: 0208 549 5855
Fax: 0208 547 1564
Website: www.bfpsa.org.uk

BRITISH FURNITURE MANUFACTURERS
FEDERATION LTD (BFM LTD)
30 Harcourt Street
London
W1H 2AA
Tel: 0207 724 0851
Fax: 0207 723 0622
Website: www.bfm.org.uk

BRITISH GEOLOGICAL SURVEY (BGS)
Keyworth Headquarters
Kingsley Drive
Dunham Centre
Nottingham
Nottinghamshire
NG12 5GG
Tel: 0115 936 3100
Fax: 0115 936 3200
Website: www.thebgs.co.uk

BRITISH INSTITUTE OF ARCHITECTURAL
TECHNOLOGISTS (BIAT)
397 City Road
London
EC1V 1NH
Tel: 0207 278 2206
Fax: 0207 837 3194
Website: www.biat.org.uk

BRITISH LAMINATED FABRICATORS ASSOCIATION
6 Bath Place
Rivington Street
London
EC2A 3JE
Tel: 020 7457 5000
Fax: 020 7457 5038
Website: www.bpf.co.uk

BRITISH LIBRARY BIBLIOGRAPHIC SERVICE AND
DOCUMENT SUPPLY
Boston Spa
Wetherby
West Yorkshire
LS23 7BQ
Tel: 01937 546 548
Fax: 01937 546 586
Website: www.bl.uk

BRITISH LIBRARY ENVIRONMENTAL INFORMATION
SERVICE
96 Euston Road
London
NW1 2DB
Tel: 020 7412 7000
Website: www.bl.uk/environment

BRITISH NON-FERROUS METALS FEDERATION
Broadway House
60 Calthorpe Road
Edgbaston
Birmingham
West Midlands
B15 1TN
Tel: 0121 456 6110
Fax: 0121 456 2274

BRITISH PLASTICS FEDERATION (BPF)
Plastics & Rubber Advisory Service
6 Bath Place
Rivington Street
London
EC2A 3JE
Tel: 020 7457 5000
Fax: 020 7457 5020
Website: www.bpf.co.uk

BRITISH PRECAST CONCRETE FEDERATION LTD
60 Charles Street
Leicester
Leicestershire
LE1 1FB
Tel: 0116 253 6161
Fax: 0116 251 4568
Website: www.britishprecast.org.uk

BRITISH PROPERTY FEDERATION (BPF)
1 Warwick Row
London
SW1E 5ER
Tel: 020 7828 0111
Fax: 020 7824 3442
Website: www.bpf.org.uk

BRITISH RUBBER MANUFACTURERS'
ASSOCIATION LTD (BRMA)
6 Bath Place
Rivington Street
London
EC2A 3JE
Tel: 020 7457 5040
Fax: 020 7972 9008
Website: www.brma.co.uk

BRITISH STAINLESS STEEL ASSOCIATION
Broomgrove
59 Clarkehouse Road
Sheffield
S10 2LE
Tel: 0114 267 1260
Fax: 0114 266 1252
Website: www.bssa.org.uk

BRITISH STANDARDS INSTITUTION (BSI)
389 Chiswick High Road
London
W4 4AL
Tel: 020 8996 9001
Fax: 020 8996 7001
Website: www.bsigroup.com

BRITISH WATER
1 Queen Anne's Gate
London
SW1H 9BT
Tel: 0207 957 4554
Fax: 0207 957 4565
Website: www.britishwater.co.uk

BRITISH WOOD PRESERVING & DAMP-PROOFING
ASSOCIATION (BWPDA)
6 Office Village
Romford Road
London
E15 4ED
Tel: 0208 519 2588
Fax: 0208 519 3444
Website: www.bwpda.co.uk

BRITISH WOODWORKING FEDERATION
55 Tufton Street
London
SW1 3QL
Tel: 0870 458 6939
Fax: 0870 458 6949
Website: www.bwf.org.uk

BUILDERS MERCHANTS FEDERATION
Soho Square
London
W1D 3HL
Tel: 020 7439 1753
Fax: 020 7734 2766
Website: www.bmf.org.uk

BUILDING & ENGINEERING SERVICES
ASSOCIATION
ESCA House
34 Palace Court
Bayswater
London
W2 4JG
Tel: 020 7313 4900
Fax: 020 7727 9268
Website: www.hvca.org.uk

BUILDING CENTRE
The Building Centre
26 Store Street
London
WC1E 7BT
Tel: 020 7692 4000
Fax: 020 7580 9641
Website: www.buildingcentre.co.uk

BUILDING COST INFORMATION SERVICE LTD
(BCIS)
Royal Institution of Chartered Surveyors
12 Great George Street
London
SW1P 3AD
Tel: 020 7695 1500
Fax: 020 7695 1501
Website: www.bcis.co.uk

BUILDING EMPLOYERS CONFEDERATION (BEC)
55 Tufton Street
Westminster
London
SW1P 3QL
Tel: 0870 898 9090
Fax: 0870 898 9095
Website: www.thecc.org.uk

BUILDING MAINTENANCE INFORMATION (BMI)
Royal Institution of Chartered Surveyors
12 Great George Street
London
SW1P 3 AD
Tel: 020 7695 1500
Fax: 020 7695 1501
Website: www.bcis.co.uk

BUILDING RESEARCH ESTABLISHMENT (BRE)
BRE Garston
Watford
WD5 9XX
Tel: 01923 664 000
Website: www.bre.co.uk

BUILDING RESEARCH ESTABLISHMENT:
SCOTLAND (BRE)
Kelvin Road
East Kilbride
Glasgow
G75 0RZ
Tel: 01355 576 200
Fax: 01355 241 895
Website: www.bre.co.uk

BUILDING SERVICES RESEARCH AND
INFORMATION ASSOCIATION LTD
Old Bracknell Lane West
Bracknell
Berkshire
RG12 7 AH
Tel: 01344 465 600
Fax: 01344 465 626
Website: www.bsria.co.uk

CASTINGS TECHNOLOGY INTERNATIONAL
7 East Bank Road
Sheffield
S2 3PT
Tel: 0114 272 8647
Fax: 0114 273 0854
Website: www.castingsdev.com

CATERING EQUIPMENT MANUFACTURERS
ASSOCIATION (CEMA)
Carlyle House
235 Vauxhall Bridge Road
London
SW1V 1EJ
Tel: 020 7233 7724
Fax: 020 7828 0667

CEMENT ADMIXTURES ASSOCIATION
38 Tilehouse
Green Lane
Knowle
West Midlands
B93 9EY
Tel + Fax: 01564 776 362
Website: www.admixtures.org.uk

CHARTERED INSTITUTE OF ARBITRATORS
12 Bloomsbury Square
London
WC1 A 2LP
Tel: 0207 421 7444
Fax: 0207 404 4023
Website: www.ciarb.org

CHARTERED INSTITUTE OF ARCHITECTURAL
TECHNOLOGISTS
397 City Road
London
EC1V 1NH
Tel: 0207 278 2206
Fax: 0207 837 3194
Website: www.ciat.org.uk

CHARTERED INSTITUTE OF BUILDING (CIOB)
Englemere
Kings Ride
Ascot
Berkshire
SL5 8BJ
Tel: 01344 630 700
Fax: 01344 630 777
Website: www.ciob.org.uk

CHARTERED INSTITUTION OF BUILDING
SERVICES ENGINEERS (CIBSE)
Delta House
222 Balham High Road
London
SW12 9BS
Tel: 020 8675 5211
Fax: 020 8675 5449
Website: www.cibse.org

CHARTERED INSTITUTE OF WASTES
MANAGEMENT
9 Saxon Court
St Peter's Gardens
Northampton
NN1 1SX
Tel: 01604 620 426
Fax: 01604 621 339
Website: www.iwm.co.uk

CIVIL ENGINEERING CONTRACTORS
ASSOCIATION
1 Birdcage Walk
London
SW1H 9JJ
Tel: 0207 340 0450
Website: www.ceca.co.uk

CLAY PIPE DEVELOPMENT ASSOCIATION (CPDA)
Copsham House
53 Broad Street
Chesham
HP5 3EA
Tel: 01494 791 456
Fax: 01494 792 378
Website: www.cpda.co.uk

CLAY ROOF TILE COUNCIL
Federation House
Station Road
Stoke-on-Trent
Staffordshire
ST4 2SA
Tel: 01782 744 631
Fax: 01782 744 102
Website: www.clayroof.co.uk

COLD ROLLED SECTIONS ASSOCIATION (CRSA)
National Metal Forming Centre
47 Birmingham Road
West Bromwich
West Midlands
B70 6PY
Tel: 0121 601 6350
Fax: 0121 601 6373
Website: www.crsauk.com

COMMONWEALTH ASSOCIATION OF ARCHITECTS
(CAA)
PO BOX 508
Edgware
HA8 9XZ
Tel: 020 8951 0550
Website: www.comarchitect.org

CONCRETE BRIDGE DEVELOPMENT GROUP
Riverside House
4 Meadows Business Park
Station Approach
Blackwater
Camberley
Surrey
GU17 9 AB
Tel: 01276 33777
Fax: 01276 38899
Website: www.concrete.org.uk

CONCRETE PIPE ASSOCIATION (CPA)
60 Charles Street
Leicester
Leicestershire
LE1 1FB
Tel: 0116 253 6161
Fax: 0116 251 4568
Website: www.britishprecast.org.uk

CONCRETE PIPELINE SYSTEMS ASSOCIATION
60 Charles St
Leicester
LE1 1FB
Tel: 0116 253 6161
Fax: 0116 251 4568
Website: www.concretepipes.co.uk

CONCRETE REPAIR ASSOCIATION (CRA)
Association House
235 Ash Road
Aldershot
Hampshire
GU12 4DD
Tel: 01252 321 302
Fax: 01252 333 901
Website: www.concreterepair.org.uk

CONCRETE SOCIETY ADVISORY SERVICE
Riverside House
4 Meadows Business Park
Station Approach
Blackwater
Camberley
Surrey
GU17 9AB
Tel: 01276 607 140
Fax: 01276 607 141
Website: www.concrete.org.uk

CONFEDERATION OF BRITISH INDUSTRY (CBI)
Centre Point
103 New Oxford Street
London
WC1A 1DU
Tel: 020 7379 7400
Fax: 020 7240 1578
Website: www.cbi.org.uk

CONSTRUCT – CONCRETE STRUCTURES
GROUP LTD
Riverside House
4 Meadows Business Park
Station Approach
Blackwater
Camberley
Surrey
GU17 9AB
Tel: 01276 38444
Fax: 01276 38899
Website: www.construct.org.uk

CONSTRUCTION EMPLOYERS FEDERATION LTD
(CEF)
143 Malone Road
Belfast
Northern Ireland
BT9 6SU
Tel: 028 9087 7143
Fax: 028 9087 7155
Website: www.cefni.co.uk

CONSTRUCTION INDUSTRY JOINT COUNCIL (CIJC)
55 Tufton Street
London
SW1P 3QL
Tel: 0870 898 9090
Fax: 0870 898 9095
Website: http://www.nscc.org.uk/Website

CONSTRUCTION INDUSTRY RESEARCH AND
INFORMATION ASSOCIATION
Classic House
174–180 Old Street
London
EC1V 9BP
Tel: 020 7549 3300
Fax: 020 7253 0523
Website: www.ciria.org.uk

CONSTRUCTION PLANT-HIRE ASSOCIATION (CPA)
27–28 Newbury Street
London
EC1A 7HU
Tel: 020 7796 3366
Website: www.cpa.uk.net

CONTRACT FLOORING ASSOCIATION (CFA)
4c Saint Mary's Place
The Lace Market
Nottingham
Nottinghamshire
NG1 1PH
Tel: 0115 941 1126
Fax: 0115 941 2238
Website: www.cfa.org.uk

CONTRACTORS MECHANICAL PLANT ENGINEERS
(CMPE)
43 Portsmouth Road
Horndeam
Waterlooville
Hampshire
PO8 9LN
Tel: 023 925 70011
Fax: 023 925 70022
Website: www.cmpe.co.uk/

COPPER DEVELOPMENT ASSOCIATION
Verulam Industrial Estate
224 London Road
Saint Albans
Hertfordshire
AL1 1AQ
Tel: 01727 731 200
Fax: 01727 731 216
Website: www.cda.org.uk

CORUS RESEARCH DEVELOPMENT AND
TECHNOLOGY
Swinden Technology Centre
Moorgate
Rotherham
South Yorkshire
S60 3 AR
Tel: 01709 825 335
Fax: 01709 825 464
Website: www.corusgroup.com/en/technology/
           research_and_development/

COUNCIL FOR ALUMINIUM IN BUILDING (CAB)
Bank House
Bond's Mill
Stonehouse
Gloucestershire
GL10 3RF
Tel: 01453 828851
Fax: 01453 828861
Website: www.c-a-b.org.uk

DEPARTMENT FOR BUSINESS, INNOVATION AND
SKILLS
1 Victoria Street
London
SW1H 0ET
Tel: 020 7215 5000
Website: www.gov.uk/government/organizations/
           department-for-business-innovation-skills

DEPARTMENT FOR TRANSPORT
Great Minister House
33 Horseferry Road,
London
SW1 P 4DR
Tel: 0300 330 3000
Website: www.dft.gov.uk

DOORS & HARDWARE FEDERATION
42 Heath Street
Tamworth
Staffordshire
B79 7JH
Tel: 01827 52337
Fax: 01827 310 827
Website: www.abhm.org.uk

DRY STONE WALLING ASSOCIATION OF GREAT
BRITAIN (DSWA)
Westmorland County Showground
Lane Fram
Crooklands, Milnthorpe
Cumbria
LA7 7NH
Tel: 01539 567 953
Website: www.dswa.org.uk

ELECTRICAL CONTRACTORS ASSOCIATION (ECA)
ESCA House
34 Palace Court
Bayswater
London
W2 4HY
Tel: 020 7313 4800
Fax: 020 7221 7344
Website: www.eca.co.uk

ELECTRICAL CONTRACTORS ASSOCIATION OF
SCOTLAND (SELECT)
The Walled Gardens
Bush Estate
Midlothian
Scotland
EH26 0SB
Tel: 0131 445 5577
Fax: 0131 445 5548
Website: www.select.org.uk

ELECTRICAL INSTALLATION EQUIPMENT
MANUFACTURERS ASSOCIATION LTD (EIEMA)
Beama Installation Ltd
Westminster Tower
3 Albert Embankment
London
SE1 7SL
Tel: 020 7793 3000
Fax: 020 7793 3003
Website: www.eiema.org.uk

EUROPEAN LIQUID ROOFING ASSOCIATION
(ELRA)
Fields House
Gower Road
Haywards Heath
West Sussex
RH16 4PL
Tel: 01444 417 458
Fax: 01444 415 616
Website: www.elra.org.uk

FEDERATION OF ENVIRONMENTAL TRADE
ASSOCIATIONS
2 Waltham Court
Milley Lane
Hare Hatch
Reading
RG10 9TH
Tel: 0118 940 3416
Fax: 0118 940 6258
Website: www.feta.co.uk

FEDERATION OF MANUFACTURERS OF
CONSTRUCTION EQUIPMENT & CRANES
Ambassador House
Brigstock Road
Thornton Heath
Surrey
CR7 7JG
Tel: 020 8665 5727
Fax: 020 8665 6447
Website: www.coneq.org.uk

FEDERATION OF MASTER BUILDERS
Gordon Fisher House
14–15 Great James Street
London
WC1N 3DP
Tel: 020 7242 7583
Fax: 020 7404 0296
Website: www.fmb.org.uk/

FEDERATION OF PILING SPECIALISTS
Forum Court
83 Coppers Cope Road
Beckenham
Kent
BR3 1NR
Tel: 020 8663 0947
Fax: 020 8663 0949
Website: www.fps.org.uk

FEDERATION OF PLASTERING & DRYWALL
CONTRACTORS
Construction House
56–64 Leonard Street
London
EC2A 4JX
Tel: 020 7608 5092
Fax: 020 7608 5081
Website: www.fpdc.org

FENCING CONTRACTORS ASSOCIATION
Warren Road
Trellech
Monmouthshire
NP5 4PQ
Tel: 07000 560 722
Fax: 01600 860 614
Website: www.fencingcontractors.org

FINNISH PLYWOOD INTERNATIONAL
PO BOX 99
Welwyn Garden City
Herts
AL6 0HS
Tel: 01438 798 746
Fax: 01438 798 305

FLAT ROOFING ALLIANCE
Fields House
Gower Road
Haywards Heath
West Sussex
RH16 4PL
Tel: 01444 440 027
Fax: 01444 415 616
Website: www.fra.org.uk

FURNITURE INDUSTRY RESEARCH ASSOCIATION
(FIRA INTERNATIONAL LTD)
Maxwell Road
Stevenage
Hertfordshire
SG1 2EW
Tel: 01438 777 700
Fax: 01438 777 800
Website: www.fira.co.uk

GLASS & GLAZING FEDERATION (GGF)
44–48 Borough High Street
London
SE1 1XB
Tel: 020 7403 7177
Website: www.ggf.org.uk

HOUSING CORPORATION HEADQUARTERS
Maple House
149 Tottenham Court Road
London
W1N 7BN
Tel: 0845 230 7000
Fax: 0207 393 2111
Website: www.housingcorp.gov.uk

ICOM Energy Association
Camden House
Warwick Road
Kenilworth
Warwickshire
CV8 1TH
Tel: 01926 513748
Fax: 01926 21 855017
Website: www.icome.org.uk

INSTITUTE OF ACOUSTICS
77 A Saint Peter' Street
Saint Albans
Hertfordshire
AL1 3BN
Tel: 01727 848 195
Fax: 01727 850 553
Website: www.ioa.org.uk

INSTITUTE OF ASPHALT TECHNOLOGY
Paper Mews Place
290 High Street
Dorking
Surrey
RH4 1QT
Tel: 01306 742 792
Fax: 01306 888 902
Website: www.instofasphalt.org

INSTITUTION OF CIVIL ENGINEERS
1 Great George Street
London
SW1P 3 AA
Tel: 0207 222 7722
Fax: 0207 222 7500
Website: www.ice.org.uk

INSTITUTE OF MAINTENANCE AND BUILDING
MANAGEMENT
Keets House
30 East Street
Farnham
Surrey
GU9 7SW
Tel: 01252 710 994
Fax: 01252 737 741
Website: www.imbm.org.uk

INSTITUTE OF MATERIALS
Headquarters
1 Carlton House Terrace
London
SW1Y 5 AF
Tel: 020 7451 7300
Fax: 020 7839 1702
Website: www.materials.org.uk

INSTITUTE OF PLUMBING
64 Station Lane
Hornchurch
Essex
RM12 6NB
Tel: 01708 472 791
Fax: 01708 448 987
Website: www.plumbers.org.uk

INSTITUTE OF WOOD SCIENCE
Stocking Lane
Hughenden Valley
High Wycombe
Buckinghamshire
HP14 4NU
Tel: 01494 565 374
Fax: 01494 565 395
Website: www.iwsc.org.uk

INSTITUTION OF CIVIL ENGINEERS (ICE)
1 Great George Street
London
SW1P 3 AA
Tel: 020 7222 7722
Fax: 020 7222 7500
Website: www.ice.org.uk

INSTITUTION OF ENGINEERING AND
TECHNOLOGY
The Institution of Engineering and Technology
Michael Faraday House
Stevenage
Herts
SG1 2AY
Tel: 01438 313 311
Fax: 01438 765 526
Website: www.theiet.org

INSTITUTION OF MECHANICAL ENGINEERS
1 Birdcage Walk
London
SW1H 9JJ
Tel: 0207 222 7899
Fax: 0207 222 4557
Website: www.imeche.org

INSTITUTION OF STRUCTURAL ENGINEERS (ISE)
11 Upper Belgrave Street
London
SW1X 8BH
Tel: 020 7235 4535
Fax: 020 7235 4294
Website: www.istructe.org

INSTITUTION OF WASTES MANAGEMENT
9 Saxon Court
St Peter's Gardens
Marefair
Northampton
NN1 1SX
Tel: 01604 620 426
Fax: 01604 621 339
Website: www.ciwm.co.uk

INTERNATIONAL LEAD ASSOCIATION
17 A Welbeck Way
London
W1G 9YJ
Tel: 0207 499 8422
Fax: 0207 493 1555
Website: www.ldaint.org

INTERPAVE (THE PRECAST CONCRETE PAVING
& KERB ASSOCIATION)
60 Charles Street
Leicester
Leicestershire
LE1 1FB
Tel: 0116 253 6161
Fax: 0116 251 4568
Website: www.paving.org.uk/

JOINT CONTRACTS TRIBUNAL LTD
9 Cavendish Place
London
W1G 0GD
Tel: 020 7630 8650
Fax: 020 7630 8670
Web site: www.jctltd.co.uk

KITCHEN SPECIALISTS ASSOCIATION
12 TopBarn Business Centre
Holt Heath
Worcester
Worcestershire
WR6 6NH
Tel: 01905 621 787
Fax: 01905 621 887
Website: www.kbsa.co.uk

LIGHTING ASSOCIATION LTD
Stafford Park 7
Telford
Shropshire
TF3 3BQ
Tel: 01952 290 905
Fax: 01952 290 906
Website: www.lightingassociation.com/

MASTIC ASPHALT COUNCIL LTD
PO BOX 77
Hastings
Kent
TN35 4WL
Tel: 01424 814 400
Fax: 01424 814 446
Website: www.masticasphaltcouncil.co.uk

MET OFFICE
Fitzroy Road
Exeter
Devon
EX1 3PB
Tel: 0870 900 0100
Fax: 0870 900 5050
Website: www.metoffice.gov.uk

METAL CLADDING & ROOFING MANUFACTURERS
ASSOCIATION
18 Mere Farm Road
Prenton
Wirral
Cheshire
CH43 9TT
Tel: 0151 652 3846
Fax: 0151 653 4080
Website: www.mcrma.co.uk

NATIONAL ASSOCIATION OF STEEL
STOCKHOLDERS
The Citadel
190 Corporation Street
Birmingham
B4 6QD
Tel: 0121 200 2288
Fax: 0121 236 7444
Website: www.nass.org.uk

NATIONAL HOUSE-BUILDING COUNCIL (NHBC)
NHBC
NHBC House
Davy Avenue
Knowlhill
Milton Keynes
MK5 8FP
Tel: 0844 633 1000
Website: www.nhbc.co.uk

NATURAL SLATE QUARRIES ASSOCIATION
26 Store Street
London
WC1E 7BT
Tel: 0207 323 3770
Fax: 0207 323 0307
Website: http://slateassociation.org

NHS ESTATES
Departments of Health
1 Trevelyan Square
Boar Lane
Leeds
West Yorkshire
LS1 6 AE
Tel: 0113 254 7000
Fax: 0113 254 7299
Website: http://www.buyingsolutions.gov.uk/
                   healthcms/

ORDNANCE SURVEY
Romsey Road
Southampton
SO16 4GU
Tel: 08456 050 504
Fax: 02380 792 615
Website: www.ordnancesurvey.co.uk

PAINTING AND DECORATING ASSOCIATION
32 Coton Road
Nuneaton
Warwickshire
CV11 5TW
Tel: 01203 353 776
Fax: 01203 354 4513
Website: www.paintingdecoratingassociation.co.uk

PIPELINE INDUSTRIES GUILD
14–15 Belgrave Square
London
SW1X 8PS
Tel: 020 7235 7938
Fax: 020 7235 0074
Website: www.pipeguild.co.uk

PLASTIC PIPES GROUP
c/o British Plastics Federation
6 Bath Place
Rivington Street
London
EC2 A 3JE
Tel: 0207 4575024
Website: www.plasticpipesgroup.com

PRECAST FLOORING FEDERATION
60 Charles Street
Leicester
Leicestershire
LE1 1FB
Tel: 0116 253 6161
Fax: 0116 251 4568
Website: www.pff.org.uk

PRESTRESSED CONCRETE ASSOCIATION
60 Charles Street
Leicester
Leicestershire
LE1 1FB
Tel: 0116 253 6161
Fax: 0116 251 4568
Website: www.britishprecast.org

PROPERTY CONSULTANTS SOCIETY LTD
Basement Office
1 Surrey Street
Arundel
West Sussex
BN18 9DT
Tel: 01903 883 787
Fax: 01903 889 590
Website: www.p-c-s.org.uk

QUARRY PRODUCTS ASSOCIATION
Gillingham House
38–44 Gillingham Street
London
SW1 V1HU
Tel: 0207 963 8000
Fax: 0207 963 8001
Website: www.qpa.org

READY-MIXED CONCRETE BUREAU
Century House
Telford Avenue
Crowthorne
Berkshire
RG45 6YS
Tel: 01344 725 732
Fax: 01344 774 976
Website: www.rcb.org.uk

REINFORCED CONCRETE COUNCIL
Riverside House
4 Meadows Business Park
Station Approach
Camberley
Surrey
GU17 9 AB
Tel: 01276 607140
Fax: 01276 607141
Website: www.rcc-info.org.uk

ROYAL INCORPORATION OF ARCHITECTS IN
SCOTLAND (RIAS)
15 Rutland Square
Edinburgh
Scotland
EH1 2BE
Tel: 0131 229 7545
Fax: 0131 228 2188
Website: www.rias.org.uk

ROYAL INSTITUTE OF BRITISH ARCHITECTS
(RIBA)
66 Portland Place
London
W1B 1 AD
Tel: 020 7580 5533
Fax: 020 7255 1541
Website: www.architecture.com

ROYAL INSTITUTION OF CHARTERED
SURVEYORS (RICS)
12 Great George Street
Parliament Square
London
SW1P 3 AD
Tel: 0207 222 7000
Fax: 0207 222 9430
Website: www.rics.org

ROYAL TOWN PLANNING INSTITUTE (RTPI)
41 Botolph Lane
London
EC3R 8DL
Tel: 020 7929 9494
Fax: 020 7323 1582
Website: www.rtpi.org.uk/

RURAL DESIGN AND BUILDING ASSOCIATION
ATSS House
Station Road East
Stowmarket
Suffolk
IP14 1RQ
Tel: 01449 676 049
Fax: 01449 770 028
Website: www.rdba.org.uk

SCOTTISH BUILDING EMPLOYERS FEDERATION
Carron Grange
Carron Grange Avenue
Stenhousemuir
Scotland
FK5 3BQ
Tel: 01324 555 550
Fax: 01324 555 551
Website: www.scottish-building.co.uk

SCOTTISH HOMES – Community Scotland
Thistle House
91 Haymarket Terrace
Edinburgh
Scotland
EH12 5HE
Tel: 0131 313 0044
Fax: 0131 313 2680
Website: www.scotland.gov.uk

SCOTTISH NATURAL HERITAGE
Communications Directorate
12 Hope Terrace
Edinburgh
EH9 2 AS
Tel: 0131 447 4784
Fax: 0131 446 2277
Website: www.snh.org.uk

SINGLE PLY ROOFING ASSOCIATION
177 Bagnall Road
Basford
Nottinghamshire
NG6 8SJ
Tel: 0115 914 4445
Fax: 0115 974 9827
Website: www.spra.co.uk/

SMOKE CONTROL ASSOCIATION
2 Waltham Court
Milley Lane
Hare Hatch
Reading
Berkshire
RG10 9TH
Tel: 0118 940 3416
Fax: 0118 940 6258
Website: www.feta.co.uk

SOCIETY FOR THE PROTECTION OF ANCIENT
BUILDINGS (SPAB)
37 Spital Square
London E1 6DY
Tel: 0207 377 1644
Fax: 0207 247 5296
Website: www.spab.org.uk

SOCIETY OF GLASS TECHNOLOGY
Don Valley House
Saville Street East
Sheffield
South Yorkshire
S4 7UQ
Tel: 0114 263 4455
Fax: 0114 263 4411
Website: www.sgt.org

SOIL SURVEY AND LAND RESEARCH INSTITUTE
Cranfield University
Silsoe Campus
Bedford
Bedfordshire
MK45 4DT
Tel: 01525 863 000
Fax: 01525 863 253
Website: www.cranfield.ac.uk/sslrc

SOLAR ENERGY SOCIETY
c/o School of Engineering
Oxford Brookes University
Gipsy Lane
Headington
Oxford
OX3 0BP
Tel: 01865 741 111
Fax: 01525 863 253
Website: www.uk-ises.org/

SPON'S PRICE BOOK EDITORS
Davis Langdon, An AECOM Company
MidCity Place
71 High Holborn
Holborn
London
WC1V 6QS
Tel: 020 7061 7000
Website: www.davislangdon.com

SPORT ENGLAND
3rd Floor, Victoria House
Bloomsbury Square
London
WC1B 4SE
Tel: 0845 850 8508
Fax: 020 7383 5740
Website: www.sportengland.org

SPORT SCOTLAND
Caledonia House
South Gyle
Edinburgh
EH12 9DQ
Tel: 0131 317 7200
Fax: 0131 317 7202
Website: www.sportscotland.org.uk

SPORTS COUNCIL FOR WALES
Welsh Institute of Sport
Sophia Gardens
Cardiff
CF11 9SW
Tel: 0845 045 0904
Fax: 0845 846 0014
Website: www.sports-council-wales.co.uk

SPORTS TURF RESEARCH INSTITUTE (STRI)
Saint Ives Estate
Bingley
West Yorkshire
BD16 1 AU
Tel: 01274 565 131
Fax: 01274 561 891
Website: www.stri.co.uk

SPRAYED CONCRETE ASSOCIATION
Association House
235 Ash Road
Aldershot
Hampshire
GU12 4DD
Tel: 01252 321 302
Fax: 01252 333 901
Website: www.sca.org.uk

STEEL CONSTRUCTION INSTITUTE
Silwood Park
Ascot
Berkshire
SL5 7QN
Tel: 01344 636 505
Fax: 01344 636 570
Website: www.steel-sci.org

STEEL WINDOW ASSOCIATION
The Building Centre
26 Store Street
London
WC1E 7BT
Tel: 020 7637 3571
Fax: 020 7637 3572
Website: www.steel-window-association.co.uk

STONE FEDERATION GREAT BRITAIN
Channel Business Centre
Ingles Manor
Castle Hill Avenue
Folkestone
Kent
CT20 2RD
Tel: 01303 856123
Fax: 01303 856117
Website: www.stone-federationgb.org.uk

SUSPENDED ACCESS EQUIPMENT
MANUFACTURERS ASSOCIATION
56–54 Leonard Street
London
EC2 A 4JX
Tel: 020 7608 5098
Fax: 020 7636 5984
Website: www.stone-federationgb.org.uk

SWIMMING POOL & ALLIED TRADES ASSOCIATION
(SPATA)
Spata House
1a Junction Road
Andover
Hampshire
SP10 3QT
Tel: 01264 356210
Fax: 01264 332628
Website: www.spata.co.uk

THE PIPELINE INDUSTRIES GUILD
F150 First Floor
Cherwell Business Village
Southam Road
Banbury, OX16 2SP
Tel: 0207 235 7938
Fax; 0207 235 0074
Website: www.pipeline.com

THERMAL INSULATION CONTRACTORS
ASSOCIATION
Tica House
Allington Way
Yarm Road Business Park
Darlington
County Durham
DL1 4QB
Tel: 01325 466 704
Fax: 01325 487 691
Website: www.tica-acad.co.uk

TIMBER RESEARCH & DEVELOPMENT
ASSOCIATION (TRADA)
Stocking Lane
Hughenden Valley
High Wycombe
Buckinghamshire
HP14 4ND
Tel: 01494 569 600
Fax: 01494 565 487
Website www.trada.co.uk

TIMBER TRADE FEDERATION
4th Floor
Clareville House
26–27 Oxenden Street
London
SW1Y 4EL
Tel: 020 7839 1891
Fax: 020 7930 0094
Website: www.ttf.co.uk

TOWN & COUNTRY PLANNING ASSOCIATION
(TCPA)
17 Carlton House Terrace
London
SW1Y 5 AS
Tel: 020 7930 8903
Fax: 020 7930 3280
Website: tcpa.org.uk

TREE COUNCIL
71 Newcomen Street
London
SE1 1WT
Tel: 020 7407 9992
Fax: 020 7407 9908
Website: www.treecouncil.org.uk

TRUSSED RAFTER ASSOCIATION
31 Station Road
Sutton
Retford
Nottinghamshire
DN22 8PZ
Tel: 01777 869 281
Fax: 01777 869 281
Website: www.tra.org.uk

TWI
Granta Park
Great Abington
Cambridge
Cambridgeshire
CB1 6 AL
Tel: 01223 899 000
Fax: 01223 892 588
Website: www.twi.co.uk

UNDERFLOOR HEATING MANUFACTURERS'
ASSOCIATION
Belhaven House
67 Walton Road
East Moseley
Surrey
KT8 0DB
Tel: 020 8941 7177
Fax: 020 8941 815
Website: www.uhma.org.uk

VERMICULITE INFORMATION SERVICE
1A Guildford Business Park
Guildford
Surrey
GU2 8XG
Tel: 01483 242 100
Fax: 01483 242 101

WALLCOVERING MANUFACTURERS ASSOCIATION
James House
Bridge Street
Leatherhead
Surrey
KT22 7EP
Tel: 01372 360 660
Fax: 01372 376 069

WATER RESEARCH CENTRE
Henley Road
Medmenham
Marlow
Buckinghamshire
SL7 2HD
Tel: 01491 636 500
Fax: 01491 636 501
Website: www.wrcplc.co.uk

WATER UK
1 Queen Anne's Gate
London
SW1H 9BT
Tel: 020 7344 1844
Fax: 020 7344 1866
Website: www.water.org.uk/

WATERHEATER MANUFACTURERS ASSOCIATION
C/O Andrews Waterheaters
Wednesbury One
Black Country New Road
Wednesbury
WS10 7NZ
Tel: 07775 754456
Fax: 0161 456 7106
Website: www.waterheating.fsnet.co.uk/wma.htm

WELDING MANUFACTURERS' ASSOCIATION
Westminster Tower
3 Albert Embankment
London
SE1 7SL
Tel: 020 7793 3041
Fax: 020 7582 8020
Website: www.wma.uk.com

WOOD PANEL INDUSTRIES FEDERATION
Grantham
Lincolnshire
NG31 6LR
Tel: 01476 563 707
Fax: 01476 579 314
Website: www.wpif.org.uk

WRAP
The Old Academy
21 Horse Fair
Banbury
OX16 0AH
Tel: 01295 819 900
Website: www.wrap.org.uk

ZINC DEVELOPMENT ASSOCIATION
42 Weymouth Street
London
W1N 3LQ
Tel: 0207 499 6636
Fax: 0207 493 135
Website: www.zincinfocentre.org

Carbon Management in the Built Environment

R Emmanuel & K Baker

CARBON MANAGEMENT IN THE BUILT ENVIRONMENT

ROHINTON EMMANUEL AND KEITH BAKER

Three broad sectors of the economy are generally recognised as key to a low carbon future: energy, construction and transportation. Of these, carbon management in the built environment remains the least well-studied.

This much-needed book brings together the latest developments in the field of climate change science, building design, materials science, energy and policy in a form readily accessible to both students of the built environment and practitioners. Although several books exist in the broad area of carbon management, this is the first to bring together carbon management technology, technique and policy as they apply to the building sector.

June 2012: 246x174: 240pp
Hb: 978-0-415-68406-4: £105.00
Pb: 978-0-415-68407-1: £29.99

To Order: Tel: +44 (0) 1235 400524 Fax: +44 (0) 1235 400525
or Post: Taylor and Francis Customer Services,
Bookpoint Ltd, Unit T1, 200 Milton Park, Abingdon, Oxon, OX14 4TA UK
Email: book.orders@tandf.co.uk

For a complete listing of all our titles visit:
www.tandf.co.uk

Taylor & Francis
Taylor & Francis Group

PART 8

Tables and Memoranda

This part of the book contains the following sections:

Estimator's Pocket Book

Duncan Cartlidge

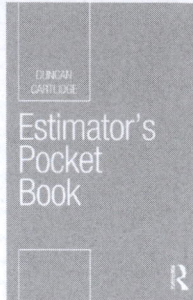

The Estimator's Pocket Book is a concise and practical reference covering the main pricing approaches, as well as useful information such as how to process sub-contractor quotations, tender settlement and adjudication. It is fully up-to-date with NRM2 throughout, features a look ahead to NRM3 and describes the implications of BIM for estimators.

It includes instructions on how to handle:

- the NRM order of cost estimate;
- unit-rate pricing for different trades;
- pro rata pricing and dayworks
- builders' quantities;
- approximate quantities.

Worked examples show how each of these techniques should be carried out in clear, easy-to-follow steps. This is the indispensible estimating reference for all quantity surveyors, cost managers, project managers and anybody else with estimating responsibilities. Particular attention is given to NRM2, but the overall focus is on the core estimating skills needed in practice.

May 2013 186x123: 310pp
Pb: 978-0-415-52711-8: £19.99

To Order: Tel: +44 (0) 1235 400524 Fax: +44 (0) 1235 400525
or Post: Taylor and Francis Customer Services,
Bookpoint Ltd, Unit T1, 200 Milton Park, Abingdon, Oxon, OX14 4TA UK
Email: book.orders@tandf.co.uk

For a complete listing of all our titles visit:
www.tandf.co.uk

CONVERSION TABLES

CONVERSION TABLES

Length	Unit	Conversion factors			
Millimetre	mm	1 in	= 25.4 mm	1 mm	= 0.0394 in
Centimetre	cm	1 in	= 2.54 cm	1 cm	= 0.3937 in
Metre	m	1 ft	= 0.3048 m	1 m	= 3.2808 ft
		1 yd	= 0.9144 m		= 1.0936 yd
Kilometre	km	1 mile	= 1.6093 km	1 km	= 0.6214 mile

Note:
1 cm	= 10 mm	1 ft	= 12 in	
1 m	= 1 000 mm	1 yd	= 3 ft	
1 km	= 1 000 m	1 mile	= 1 760 yd	

Area	Unit	Conversion factors			
Square Millimetre	mm^2	$1\ in^2$	$= 645.2\ mm^2$	$1\ mm^2$	$= 0.0016\ in^2$
Square Centimetre	cm^2	$1\ in^2$	$= 6.4516\ cm^2$	$1\ cm^2$	$= 1.1550\ in^2$
Square Metre	m^2	$1\ ft^2$	$= 0.0929\ m^2$	$1\ m^2$	$= 10.764\ ft^2$
		$1\ yd^2$	$= 0.8361\ m^2$	$1\ m^2$	$= 1.1960\ yd^2$
Square Kilometre	km^2	$1\ mile^2$	$= 2.590\ km^2$	$1\ km^2$	$= 0.3861\ mile^2$

Note:
$1\ cm^2$	$= 100\ mm^2$	$1\ ft^2$	$= 144\ in^2$
$1\ m^2$	$= 10\ 000\ cm^2$	$1\ yd^2$	$= 9\ ft^2$
$1\ km^2$	$= 100$ hectares	1 acre	$= 4\ 840\ yd^2$
		$1\ mile^2$	$= 640$ acres

Volume	Unit	Conversion factors			
Cubic Centimetre	cm^3	$1\ cm^3$	$= 0.0610\ in^3$	$1\ in^3$	$= 16.387\ cm^3$
Cubic Decimetre	dm^3	$1\ dm^3$	$= 0.0353\ ft^3$	$1\ ft^3$	$= 28.329\ dm^3$
Cubic Metre	m^3	$1\ m^3$	$= 35.3147\ ft^3$	$1\ ft^3$	$= 0.0283\ m^3$
		$1\ m^3$	$= 1.3080\ yd^3$	$1\ yd^3$	$= 0.7646\ m^3$
Litre	l	1 l	= 1.76 pint	1 pint	= 0.5683 l
			= 2.113 US pt		= 0.4733 US l

Note:
$1\ dm^3$	$= 1\ 000\ cm^3$	$1\ ft^3$	$= 1\ 728\ in^3$	1 pint	= 20 fl oz
$1\ m^3$	$= 1\ 000\ dm^3$	$1\ yd^3$	$= 27\ ft^3$	1 gal	= 8 pints
1 l	$= 1\ dm^3$				

Neither the Centimetre nor Decimetre are SI units, and as such their use, particularly that of the Decimetre, is not widespread outside educational circles.

Mass	Unit	Conversion factors			
Milligram	mg	1 mg	= 0.0154 grain	1 grain	= 64.935 mg
Gram	g	1 g	= 0.0353 oz	1 oz	= 28.35 g
Kilogram	kg	1 kg	= 2.2046 lb	1 lb	= 0.4536 kg
Tonne	t	1 t	= 0.9842 ton	1 ton	= 1.016 t

Note:
1 g	= 1000 mg	1 oz	= 437.5 grains	1 cwt	= 112 lb
1 kg	= 1000 g	1 lb	= 16 oz	1 ton	= 20 cwt
1 t	= 1000 kg	1 stone	= 14 lb		

Force	Unit	Conversion factors			
Newton	N	1 lbf	= 4.448 N	1 kgf	= 9.807 N
Kilonewton	kN	1 lbf	= 0.004448 kN	1 ton f	= 9.964 kN
Meganewton	MN	100 tonf	= 0.9964 MN		

CONVERSION TABLES

Pressure and stress	Unit	Conversion factors	
Kilonewton per square metre	kN/m^2	1 lbf/in^2	= 6.895 kN/m^2
		1 bar	= 100 kN/m^2
Meganewton per square metre	MN/m^2	1 tonf/ft^2	= 107.3 kN/m^2 = 0.1073 MN/m^2
		1 kgf/cm^2	= 98.07 kN/m^2
		1 lbf/ft^2	= 0.04788 kN/m^2

Coefficient of consolidation (Cv) or swelling	Unit	Conversion factors	
Square metre per year	m^2/year	1 cm^2/s	= 3 154 m^2/year
		1 ft^2/year	= 0.0929 m^2/year

Coefficient of permeability	Unit	Conversion factors	
Metre per second	m/s	1 cm/s	= 0.01 m/s
Metre per year	m/year	1 ft/year	= 0.3048 m/year
			= 0.9651 × (10)^8 m/s

Temperature	Unit	Conversion factors	
Degree Celsius	°C	°C = 5/9 × (°F − 32)	°F = (9 × °C)/5 + 32

CONVERSION TABLES

SPEED CONVERSION

km/h	m/min	mph	fpm
1	16.7	0.6	54.7
2	33.3	1.2	109.4
3	50.0	1.9	164.0
4	66.7	2.5	218.7
5	83.3	3.1	273.4
6	100.0	3.7	328.1
7	116.7	4.3	382.8
8	133.3	5.0	437.4
9	150.0	5.6	492.1
10	166.7	6.2	546.8
11	183.3	6.8	601.5
12	200.0	7.5	656.2
13	216.7	8.1	710.8
14	233.3	8.7	765.5
15	250.0	9.3	820.2
16	266.7	9.9	874.9
17	283.3	10.6	929.6
18	300.0	11.2	984.3
19	316.7	11.8	1038.9
20	333.3	12.4	1093.6
21	350.0	13.0	1148.3
22	366.7	13.7	1203.0
23	383.3	14.3	1257.7
24	400.0	14.9	1312.3
25	416.7	15.5	1367.0
26	433.3	16.2	1421.7
27	450.0	16.8	1476.4
28	466.7	17.4	1531.1
29	483.3	18.0	1585.7
30	500.0	18.6	1640.4
31	516.7	19.3	1695.1
32	533.3	19.9	1749.8
33	550.0	20.5	1804.5
34	566.7	21.1	1859.1
35	583.3	21.7	1913.8
36	600.0	22.4	1968.5
37	616.7	23.0	2023.2
38	633.3	23.6	2077.9
39	650.0	24.2	2132.5
40	666.7	24.9	2187.2

CONVERSION TABLES

km/h	m/min	mph	fpm
41	683.3	25.5	2241.9
42	700.0	26.1	2296.6
43	716.7	26.7	2351.3
44	733.3	27.3	2405.9
45	750.0	28.0	2460.6
46	766.7	28.6	2515.3
47	783.3	29.2	2570.0
48	800.0	29.8	2624.7
49	816.7	30.4	2679.4
50	833.3	31.1	2734.0

GEOMETRY

GEOMETRY

Two dimensional figures

Figure	Diagram of figure	Surface area	Perimeter
Square		a^2	$4a$
Rectangle		ab	$2(a+b)$
Triangle		$\frac{1}{2}ch$	$a+b+c$
Circle		πr^2 $\frac{1}{4}\pi d^2$ where $2r = d$	$2\pi r$ πd
Parallelogram		ah	$2(a+b)$
Trapezium		$\frac{1}{2}h(a+b)$	$a+b+c+d$
Ellipse		Approximately πab	$\pi(a+b)$
Hexagon		$2.6 \times a^2$	

GEOMETRY

Figure	Diagram of figure	Surface area	Perimeter
Octagon		$4.83 \times a^2$	6a
Sector of a circle		$\frac{1}{2}rb$ or $\frac{q}{360}\pi r^2$ note b = angle $\frac{q}{360} \times \pi 2r$	
Segment of a circle		S – T where S = area of sector, T = area of triangle	
Bellmouth		$\frac{3}{14} \times r^2$	

GEOMETRY

Three dimensional figures

Figure	Diagram of figure	Surface area	Volume
Cube		$6a^2$	a^3
Cuboid/ rectangular block		$2(ab + ac + bc)$	abc
Prism/ triangular block		$bd + hc + dc + ad$	$\frac{1}{2}hcd$
Cylinder		$2\pi r^2 + 2\pi h$	$\pi r^2 h$ $\frac{1}{4}\pi d^2 h$
Sphere		$4\pi r^2$	$\frac{4}{3}\pi r^3$
Segment of sphere		$2\pi Rh$	$\frac{1}{6}\pi h(3r^2 + h^2)$ $\frac{1}{3}\pi h^2(3R - H)$
Pyramid		$(a + b)l + ab$	$\frac{1}{3}abh$

GEOMETRY

Figure	Diagram of figure	Surface area	Volume
Frustum of a pyramid		$l(a + b + c + d) + \sqrt{(ab + cd)}$ [rectangular figure only]	$\frac{h}{3}(ab + cd + \sqrt{abcd})$
Cone		πrl (excluding base) $\pi rl + \pi r^2$ (including base)	$\frac{1}{3}\pi r^2 h$ $\frac{1}{12}\pi d^2 h$
Frustum of a cone		$\pi r^2 + \pi R^2 + \pi l(R + r)$	$\frac{1}{3}\pi(R^2 + Rr + r^2)$

FORMULAE

Formulae

Formula	Description
Pythagoras' Theorem	$A^2 = B^2 + C^2$ where A is the hypotenuse of a right-angled triangle and B and C are the two adjacent sides
Simpson's Rule	The Area is divided into an even number of strips of equal width, and therefore has an odd number of ordinates at the division points $$\text{area} = \frac{S(A + 2B + 4C)}{3}$$ where S = common interval (strip width) A = sum of first and last ordinates B = sum of remaining odd ordinates C = sum of the even ordinates The Volume can be calculated by the same formula, but by substituting the area of each coordinate rather than its length
Trapezoidal Rule	A given trench is divided into two equal sections, giving three ordinates, the first, the middle and the last $$\text{volume} = \frac{S \times (A + B + 2C)}{2}$$ where S = width of the strips A = area of the first section B = area of the last section C = area of the rest of the sections
Prismoidal Rule	A given trench is divided into two equal sections, giving three ordinates, the first, the middle and the last $$\text{volume} = \frac{L \times (A + 4B + C)}{6}$$ where L = total length of trench A = area of the first section B = area of the middle section C = area of the last section

TYPICAL THERMAL CONDUCTIVITY OF BUILDING MATERIALS

TYPICAL THERMAL CONDUCTIVITY OF BUILDING MATERIALS

(Always check manufacturer's details – variation will occur depending on product and nature of materials)

	Thermal conductivity (W/mK)		Thermal conductivity (W/mK)
Acoustic plasterboard	0.25	Oriented strand board	0.13
Aerated concrete slab (500 kg/m^3)	0.16	Outer leaf brick	0.77
Aluminium	237	Plasterboard	0.22
Asphalt (1700 kg/m^3)	0.5	Plaster dense (1300 kg/m^3)	0.5
Bitumen-impregnated fibreboard	0.05	Plaster lightweight (600 kg/m^3)	0.16
Blocks (standard grade 600 kg/m^3)	0.15	Plywood (950 kg/m^3)	0.16
Blocks (solar grade 460 kg/m^3)	0.11	Prefabricated timber wall panels (check manufacturer)	0.12
Brickwork (outer leaf 1700 kg/m^3)	0.84	Screed (1200 kg/m^3)	0.41
Brickwork (inner leaf 1700 kg/m^3)	0.62	Stone chippings (1800 kg/m^3)	0.96
Dense aggregate concrete block 1800 kg/m^3 (exposed)	1.21	Tile hanging (1900 kg/m^3)	0.84
Dense aggregate concrete block 1800 kg/m^3 (protected)	1.13	Timber (650 kg/m^3)	0.14
Calcium silicate board (600 kg/m^3)	0.17	Timber flooring (650 kg/m^3)	0.14
Concrete general	1.28	Timber rafters	0.13
Concrete (heavyweight 2300 kg/m^3)	1.63	Timber roof or floor joists	0.13
Concrete (dense 2100 kg/m^3 typical floor)	1.4	Roof tile (1900 kg/m^3)	0.84
Concrete (dense 2000 kg/m^3 typical floor)	1.13	Timber blocks (650 kg/m^3)	0.14
Concrete (medium 1400 kg/m^3)	0.51	Cellular glass	0.045
Concrete (lightweight 1200 kg/m^3)	0.38	Expanded polystyrene	0.034
Concrete (lightweight 600 kg/m^3)	0.19	Expanded polystyrene slab (25 kg/m^3)	0.035
Concrete slab (aerated 500 kg/m^3)	0.16	Extruded polystyrene	0.035
Copper	390	Glass mineral wool	0.04
External render sand/cement finish	1	Mineral quilt (12 kg/m^3)	0.04
External render (1300 kg/m^3)	0.5	Mineral wool slab (25 kg/m^3)	0.035
Felt – Bitumen layers (1700 kg/m^3)	0.5	Phenolic foam	0.022
Fibreboard (300 kg/m^3)	0.06	Polyisocyanurate	0.025
Glass	0.93	Polyurethane	0.025
Marble	3	Rigid polyurethane	0.025
Metal tray used in wriggly tin concrete floors (7800 kg/m^3)	50	Rock mineral wool	0.038
Mortar (1750 kg/m^3)	0.8		

EARTHWORK

EARTHWORK

Weights of Typical Materials Handled by Excavators

The weight of the material is that of the state in its natural bed and includes moisture
Adjustments should be made to allow for loose or compacted states

Material	Mass (kg/m³)	Mass (lb/cu yd)
Ashes, dry	610	1028
Ashes, wet	810	1365
Basalt, broken	1954	3293
Basalt, solid	2933	4943
Bauxite, crushed	1281	2159
Borax, fine	849	1431
Caliche	1440	2427
Cement, clinker	1415	2385
Chalk, fine	1221	2058
Chalk, solid	2406	4055
Cinders, coal, ash	641	1080
Cinders, furnace	913	1538
Clay, compacted	1746	2942
Clay, dry	1073	1808
Clay, wet	1602	2700
Coal, anthracite, solid	1506	2538
Coal, bituminous	1351	2277
Coke	610	1028
Dolomite, lumpy	1522	2565
Dolomite, solid	2886	4864
Earth, dense	2002	3374
Earth, dry, loam	1249	2105
Earth, Fullers, raw	673	1134
Earth, moist	1442	2430
Earth, wet	1602	2700
Felsite	2495	4205
Fieldspar, solid	2613	4404
Fluorite	3093	5213
Gabbro	3093	5213
Gneiss	2696	4544
Granite	2690	4534
Gravel, dry ¼ to 2 inch	1682	2835
Gravel, dry, loose	1522	2565
Gravel, wet ¼ to 2 inch	2002	3374
Gypsum, broken	1450	2444
Gypsum, solid	2787	4697
Hardcore (consolidated)	1928	3249
Lignite, dry	801	1350
Limestone, broken	1554	2619
Limestone, solid	2596	4375
Magnesite, magnesium ore	2993	5044
Marble	2679	4515
Marl, wet	2216	3735
Mica, broken	1602	2700
Mica, solid	2883	4859
Peat, dry	400	674
Peat, moist	700	1179
Peat, wet	1121	1889

EARTHWORK

Material	Mass (kg/m^3)	Mass (lb/cu yd)
Potash	1281	2159
Pumice, stone	640	1078
Quarry waste	1438	2423
Quartz sand	1201	2024
Quartz, solid	2584	4355
Rhyolite	2400	4045
Sand and gravel, dry	1650	2781
Sand and gravel, wet	2020	3404
Sand, dry	1602	2700
Sand, wet	1831	3086
Sandstone, solid	2412	4065
Shale, solid	2637	4444
Slag, broken	2114	3563
Slag, furnace granulated	961	1619
Slate, broken	1370	2309
Slate, solid	2667	4495
Snow, compacted	481	810
Snow, freshly fallen	160	269
Taconite	2803	4724
Trachyte	2400	4045
Trap rock, solid	2791	4704
Turf	400	674
Water	1000	1685

Transport Capacities

Type of vehicle	Capacity of vehicle	
	Payload	Heaped capacity
Wheelbarrow	150	0.10
1 tonne dumper	1250	1.00
2.5 tonne dumper	4000	2.50
Articulated dump truck (Volvo A20 6 × 4)	18500	11.00
Articulated dump truck (Volvo A35 6 × 6)	32000	19.00
Large capacity rear dumper (Euclid R35)	35000	22.00
Large capacity rear dumper (Euclid R85)	85000	50.00

EARTHWORK

Machine Volumes for Excavating and Filling

Machine type	Cycles per minute	Volume per minute (m³)
1.5 tonne excavator	1	0.04
	2	0.08
	3	0.12
3 tonne excavator	1	0.13
	2	0.26
	3	0.39
5 tonne excavator	1	0.28
	2	0.56
	3	0.84
7 tonne excavator	1	0.28
	2	0.56
	3	0.84
21 tonne excavator	1	1.21
	2	2.42
	3	3.63
Backhoe loader JCB3CX excavator Rear bucket capacity 0.28 m³	1	0.28
	2	0.56
	3	0.84
Backhoe loader JCB3CX loading Front bucket capacity 1.00 m³	1	1.00
	2	2.00

Machine Volumes for Excavating and Filling

Machine type	Loads per hour	Volume per hour (m³)
1 tonne high tip skip loader Volume 0.485 m³	5	2.43
	7	3.40
	10	4.85
3 tonne dumper Max volume 2.40 m³ Available volume 1.9 m³	4	7.60
	5	9.50
	7	13.30
	10	19.00
6 tonne dumper Max volume 3.40 m³ Available volume 3.77 m³	4	15.08
	5	18.85
	7	26.39
	10	37.70

EARTHWORK

Bulkage of Soils (after excavation)

Type of soil	Approximate bulking of 1 m³ after excavation
Vegetable soil and loam	25–30%
Soft clay	30–40%
Stiff clay	10–20%
Gravel	20–25%
Sand	40–50%
Chalk	40–50%
Rock, weathered	30–40%
Rock, unweathered	50–60%

Shrinkage of Materials (on being deposited)

Type of soil	Approximate bulking of 1 m³ after excavation
Clay	10%
Gravel	8%
Gravel and sand	9%
Loam and light sandy soils	12%
Loose vegetable soils	15%

Voids in Material Used as Subbases or Beddings

Material	m³ of voids/m³
Alluvium	0.37
River grit	0.29
Quarry sand	0.24
Shingle	0.37
Gravel	0.39
Broken stone	0.45
Broken bricks	0.42

Angles of Repose

Type of soil		Degrees
Clay	– dry	30
	– damp, well drained	45
	– wet	15–20
Earth	– dry	30
	– damp	45
Gravel	– moist	48
Sand	– dry or moist	35
	– wet	25
Loam		40

EARTHWORK

Slopes and Angles

Ratio of base to height	Angle in degrees
5:1	11
4:1	14
3:1	18
2:1	27
1½:1	34
1:1	45
1:1½	56
1:2	63
1:3	72
1:4	76
1:5	79

Grades (in Degrees and Percents)

Degrees	Percent	Degrees	Percent
1	1.8	24	44.5
2	3.5	25	46.6
3	5.2	26	48.8
4	7.0	27	51.0
5	8.8	28	53.2
6	10.5	29	55.4
7	12.3	30	57.7
8	14.0	31	60.0
9	15.8	32	62.5
10	17.6	33	64.9
11	19.4	34	67.4
12	21.3	35	70.0
13	23.1	36	72.7
14	24.9	37	75.4
15	26.8	38	78.1
16	28.7	39	81.0
17	30.6	40	83.9
18	32.5	41	86.9
19	34.4	42	90.0
20	36.4	43	93.3
21	38.4	44	96.6
22	40.4	45	100.0
23	42.4		

EARTHWORK

Bearing Powers

Ground conditions		Bearing power		
		kg/m²	lb/in²	Metric t/m²
Rock,	broken	483	70	50
	solid	2415	350	240
Clay,	dry or hard	380	55	40
	medium dry	190	27	20
	soft or wet	100	14	10
Gravel,	cemented	760	110	80
Sand,	compacted	380	55	40
	clean dry	190	27	20
Swamp and alluvial soils		48	7	5

Earthwork Support

Maximum depth of excavation in various soils without the use of earthwork support

Ground conditions	Feet (ft)	Metres (m)
Compact soil	12	3.66
Drained loam	6	1.83
Dry sand	1	0.3
Gravelly earth	2	0.61
Ordinary earth	3	0.91
Stiff clay	10	3.05

It is important to note that the above table should only be used as a guide. Each case must be taken on its merits and, as the limited distances given above are approached, careful watch must be kept for the slightest signs of caving in

CONCRETE WORK

CONCRETE WORK

Weights of Concrete and Concrete Elements

Type of material		kg/m³	lb/cu ft
Ordinary concrete (dense aggregates)			
Non-reinforced plain or mass concrete			
Nominal weight		2305	144
Aggregate	– limestone	2162 to 2407	135 to 150
	– gravel	2244 to 2407	140 to 150
	– broken brick	2000 (av)	125 (av)
	– other crushed stone	2326 to 2489	145 to 155
Reinforced concrete			
Nominal weight		2407	150
Reinforcement	– 1%	2305 to 2468	144 to 154
	– 2%	2356 to 2519	147 to 157
	– 4%	2448 to 2703	153 to 163
Special concretes			
Heavy concrete			
Aggregates	– barytes, magnetite	3210 (min)	200 (min)
	– steel shot, punchings	5280	330
Lean mixes			
Dry-lean (gravel aggregate)		2244	140
Soil-cement (normal mix)		1601	100

CONCRETE WORK

Type of material		kg/m² per mm thick	lb/sq ft per inch thick
Ordinary concrete (dense aggregates)			
Solid slabs (floors, walls etc.)			
Thickness:	75 mm or 3 in	184	37.5
	100 mm or 4 in	245	50
	150 mm or 6 in	378	75
	250 mm or 10 in	612	125
	300 mm or 12 in	734	150
Ribbed slabs			
Thickness:	125 mm or 5 in	204	42
	150 mm or 6 in	219	45
	225 mm or 9 in	281	57
	300 mm or 12 in	342	70
Special concretes			
Finishes etc.			
	Rendering, screed etc. Granolithic, terrazzo	1928 to 2401	10 to 12.5
	Glass-block (hollow) concrete	1734 (approx)	9 (approx)
Prestressed concrete		Weights as for reinforced concrete (upper limits)	
Air-entrained concrete		Weights as for plain or reinforced concrete	

CONCRETE WORK

Average Weight of Aggregates

Materials	Voids %	Weight kg/m³
Sand	39	1660
Gravel 10–20 mm	45	1440
Gravel 35–75 mm	42	1555
Crushed stone	50	1330
Crushed granite (over 15 mm)	50	1345
(n.e. 15 mm)	47	1440
'All-in' ballast	32	1800–2000

Material	kg/m³	lb/cu yd
Vermiculite (aggregate)	64–80	108–135
All-in aggregate	1999	125

Applications and Mix Design

Site mixed concrete

Recommended mix	Class of work suitable for	Cement (kg)	Sand (kg)	Coarse aggregate (kg)	Nr 25 kg bags cement per m³ of combined aggregate
1:3:6	Roughest type of mass concrete such as footings, road haunching over 300 mm thick	208	905	1509	8.30
1:2.5:5	Mass concrete of better class than 1:3:6 such as bases for machinery, walls below ground etc.	249	881	1474	10.00
1:2:4	Most ordinary uses of concrete, such as mass walls above ground, road slabs etc. and general reinforced concrete work	304	889	1431	12.20
1:1.5:3	Watertight floors, pavements and walls, tanks, pits, steps, paths, surface of 2 course roads, reinforced concrete where extra strength is required	371	801	1336	14.90
1:1:2	Works of thin section such as fence posts and small precast work	511	720	1206	20.40

CONCRETE WORK

Ready mixed concrete

Application	Designated concrete	Standardized prescribed concrete	Recommended consistence (nominal slump class)
Foundations			
Mass concrete fill or blinding	GEN 1	ST2	S3
Strip footings	GEN 1	ST2	S3
Mass concrete foundations			
Single storey buildings	GEN 1	ST2	S3
Double storey buildings	GEN 3	ST4	S3
Trench fill foundations			
Single storey buildings	GEN 1	ST2	S4
Double storey buildings	GEN 3	ST4	S4
General applications			
Kerb bedding and haunching	GEN 0	ST1	S1
Drainage works – immediate support	GEN 1	ST2	S1
Other drainage works	GEN 1	ST2	S3
Oversite below suspended slabs	GEN 1	ST2	S3
Floors			
Garage and house floors with no embedded steel	GEN 3	ST4	S2
Wearing surface: Light foot and trolley traffic	RC30	ST4	S2
Wearing surface: General industrial	RC40	N/A	S2
Wearing surface: Heavy industrial	RC50	N/A	S2
Paving			
House drives, domestic parking and external parking	PAV 1	N/A	S2
Heavy-duty external paving	PAV 2	N/A	S2

CONCRETE WORK

Prescribed Mixes for Ordinary Structural Concrete

Weights of cement and total dry aggregates in kg to produce approximately one cubic metre of fully compacted concrete together with the percentages by weight of fine aggregate in total dry aggregates

Conc. grade	Nominal max size of aggregate (mm)	40		20		14		10	
	Workability	Med.	High	Med.	High	Med.	High	Med.	High
	Limits to slump that may be expected (mm)	50–100	100–150	25–75	75–125	10–50	50–100	10–25	25–50
7	Cement (kg)	180	200	210	230	–	–	–	–
	Total aggregate (kg)	1950	1850	1900	1800	–	–	–	–
	Fine aggregate (%)	30–45	30–45	35–50	35–50	–	–	–	–
10	Cement (kg)	210	230	240	260	–	–	–	–
	Total aggregate (kg)	1900	1850	1850	1800	–	–	–	–
	Fine aggregate (%)	30–45	30–45	35–50	35–50	–	–	–	–
15	Cement (kg)	250	270	280	310	–	–	–	–
	Total aggregate (kg)	1850	1800	1800	1750	–	–	–	–
	Fine aggregate (%)	30–45	30–45	35–50	35–50	–	–	–	–
20	Cement (kg)	300	320	320	350	340	380	360	410
	Total aggregate (kg)	1850	1750	1800	1750	1750	1700	1750	1650
	Sand								
	Zone 1 (%)	35	40	40	45	45	50	50	55
	Zone 2 (%)	30	35	35	40	40	45	45	50
	Zone 3 (%)	30	30	30	35	35	40	40	45
25	Cement (kg)	340	360	360	390	380	420	400	450
	Total aggregate (kg)	1800	1750	1750	1700	1700	1650	1700	1600
	Sand								
	Zone 1 (%)	35	40	40	45	45	50	50	55
	Zone 2 (%)	30	35	35	40	40	45	45	50
	Zone 3 (%)	30	30	30	35	35	40	40	45
30	Cement (kg)	370	390	400	430	430	470	460	510
	Total aggregate (kg)	1750	1700	1700	1650	1700	1600	1650	1550
	Sand								
	Zone 1 (%)	35	40	40	45	45	50	50	55
	Zone 2 (%)	30	35	35	40	40	45	45	50
	Zone 3 (%)	30	30	30	35	35	40	40	45

REINFORCEMENT

REINFORCEMENT

Weights of Bar Reinforcement

Nominal sizes (mm)	Cross-sectional area (mm²)	Mass (kg/m)	Length of bar (m/tonne)
6	28.27	0.222	4505
8	50.27	0.395	2534
10	78.54	0.617	1622
12	113.10	0.888	1126
16	201.06	1.578	634
20	314.16	2.466	405
25	490.87	3.853	260
32	804.25	6.313	158
40	1265.64	9.865	101
50	1963.50	15.413	65

Weights of Bars (at specific spacings)

Weights of metric bars in kilogrammes per square metre

Size (mm)	Spacing of bars in millimetres									
	75	100	125	150	175	200	225	250	275	300
6	2.96	2.220	1.776	1.480	1.27	1.110	0.99	0.89	0.81	0.74
8	5.26	3.95	3.16	2.63	2.26	1.97	1.75	1.58	1.44	1.32
10	8.22	6.17	4.93	4.11	3.52	3.08	2.74	2.47	2.24	2.06
12	11.84	8.88	7.10	5.92	5.07	4.44	3.95	3.55	3.23	2.96
16	21.04	15.78	12.63	10.52	9.02	7.89	7.02	6.31	5.74	5.26
20	32.88	24.66	19.73	16.44	14.09	12.33	10.96	9.87	8.97	8.22
25	51.38	38.53	30.83	25.69	22.02	19.27	17.13	15.41	14.01	12.84
32	84.18	63.13	50.51	42.09	36.08	31.57	28.06	25.25	22.96	21.04
40	131.53	98.65	78.92	65.76	56.37	49.32	43.84	39.46	35.87	32.88
50	205.51	154.13	123.31	102.76	88.08	77.07	68.50	61.65	56.05	51.38

Basic weight of steelwork taken as 7850 kg/m³
Basic weight of bar reinforcement per metre run = 0.00785 kg/mm²
The value of π has been taken as 3.141592654

REINFORCEMENT

Fabric Reinforcement

Preferred range of designated fabric types and stock sheet sizes

Fabric reference	Longitudinal wires			Cross wires			
	Nominal wire size (mm)	Pitch (mm)	Area (mm^2/m)	Nominal wire size (mm)	Pitch (mm)	Area (mm^2/m)	Mass (kg/m^2)
Square mesh							
A393	10	200	393	10	200	393	6.16
A252	8	200	252	8	200	252	3.95
A193	7	200	193	7	200	193	3.02
A142	6	200	142	6	200	142	2.22
A98	5	200	98	5	200	98	1.54
Structural mesh							
B1131	12	100	1131	8	200	252	10.90
B785	10	100	785	8	200	252	8.14
B503	8	100	503	8	200	252	5.93
B385	7	100	385	7	200	193	4.53
B283	6	100	283	7	200	193	3.73
B196	5	100	196	7	200	193	3.05
Long mesh							
C785	10	100	785	6	400	70.8	6.72
C636	9	100	636	6	400	70.8	5.55
C503	8	100	503	5	400	49.0	4.34
C385	7	100	385	5	400	49.0	3.41
C283	6	100	283	5	400	49.0	2.61
Wrapping mesh							
D98	5	200	98	5	200	98	1.54
D49	2.5	100	49	2.5	100	49	0.77

Stock sheet size 4.8 m × 2.4 m, Area 11.52 m^2

Average weight kg/m^3 of steelwork reinforcement in concrete for various building elements

Substructure	kg/m^3 concrete	Substructure	kg/m^3 concrete
Pile caps	110–150	Plate slab	150–220
Tie beams	130–170	Cant slab	145–210
Ground beams	230–330	Ribbed floors	130–200
Bases	125–180	Topping to block floor	30–40
Footings	100–150	Columns	210–310
Retaining walls	150–210	Beams	250–350
Raft	60–70	Stairs	130–170
Slabs – one way	120–200	Walls – normal	40–100
Slabs – two way	110–220	Walls – wind	70–125

Note: For exposed elements add the following %:
Walls 50%, Beams 100%, Columns 15%

FORMWORK

FORMWORK

Formwork Stripping Times – Normal Curing Periods

Conditions under which concrete is maturing	Minimum periods of protection for different types of cement					
	Number of days (where the average surface temperature of the concrete exceeds 10°C during the whole period)			Equivalent maturity (degree hours) calculated as the age of the concrete in hours multiplied by the number of degrees Celsius by which the average surface temperature of the concrete exceeds 10°C		
	Other	SRPC	OPC or RHPC	Other	SRPC	OPC or RHPC
1. Hot weather or drying winds	7	4	3	3500	2000	1500
2. Conditions not covered by 1	4	3	2	2000	1500	1000

KEY
OPC – Ordinary Portland Cement
RHPC – Rapid-hardening Portland Cement
SRPC – Sulphate-resisting Portland Cement

Minimum Period before Striking Formwork

	Minimum period before striking		
	Surface temperature of concrete		
	16°C	17°C	t°C (0–25)
Vertical formwork to columns, walls and large beams	12 hours	18 hours	300 hours t+10
Soffit formwork to slabs	4 days	6 days	100 days t+10
Props to slabs	10 days	15 days	250 days t+10
Soffit formwork to beams	9 days	14 days	230 days t+10
Props to beams	14 days	21 days	360 days t+10

MASONRY

MASONRY

Number of Bricks Required for Various Types of Work per m² of Walling

Description	Brick size	
	215 × 102.5 × 50 mm	215 × 102.5 × 65 mm
Half brick thick		
Stretcher bond	74	59
English bond	108	86
English garden wall bond	90	72
Flemish bond	96	79
Flemish garden wall bond	83	66
One brick thick and cavity wall of two half brick skins		
Stretcher bond	148	119

Quantities of Bricks and Mortar required per m² of Walling

	Unit	No of bricks required	Mortar required (cubic metres)		
Standard bricks			No frogs	Single frogs	Double frogs
Brick size 215 × 102.5 × 50 mm					
half brick wall (103 mm)	m²	72	0.022	0.027	0.032
2 × half brick cavity wall (270 mm)	m²	144	0.044	0.054	0.064
one brick wall (215 mm)	m²	144	0.052	0.064	0.076
one and a half brick wall (322 mm)	m²	216	0.073	0.091	0.108
Mass brickwork	m³	576	0.347	0.413	0.480
Brick size 215 × 102.5 × 65 mm					
half brick wall (103 mm)	m²	58	0.019	0.022	0.026
2 × half brick cavity wall (270 mm)	m²	116	0.038	0.045	0.055
one brick wall (215 mm)	m²	116	0.046	0.055	0.064
one and a half brick wall (322 mm)	m²	174	0.063	0.074	0.088
Mass brickwork	m³	464	0.307	0.360	0.413
Metric modular bricks			Perforated		
Brick size 200 × 100 × 75 mm					
90 mm thick	m²	67	0.016	0.019	
190 mm thick	m²	133	0.042	0.048	
290 mm thick	m²	200	0.068	0.078	
Brick size 200 × 100 × 100 mm					
90 mm thick	m²	50	0.013	0.016	
190 mm thick	m²	100	0.036	0.041	
290 mm thick	m²	150	0.059	0.067	
Brick size 300 × 100 × 75 mm					
90 mm thick	m²	33	–	0.015	
Brick size 300 × 100 × 100 mm					
90 mm thick	m²	44	0.015	0.018	

Note: Assuming 10 mm thick joints

MASONRY

Mortar Required per m² Blockwork (9.88 blocks/m²)

Wall thickness	75	90	100	125	140	190	215
Mortar m³/m²	0.005	0.006	0.007	0.008	0.009	0.013	0.014

Mortar group	Cement: lime: sand	Masonry cement: sand	Cement: sand with plasticizer
1	1:0–0.25:3		
2	1:0.5:4–4.5	1:2.5-3.5	1:3–4
3	1:1:5–6	1:4–5	1:5–6
4	1:2:8–9	1:5.5–6.5	1:7–8
5	1:3:10–12	1:6.5–7	1:8

Group 1: strong inflexible mortar
Group 5: weak but flexible

All mixes within a group are of approximately similar strength
Frost resistance increases with the use of plasticizers
Cement: lime: sand mixes give the strongest bond and greatest resistance to rain penetration
Masonry cement equals ordinary Portland cement plus a fine neutral mineral filler and an air entraining agent

Calcium Silicate Bricks

Type	Strength	Location
Class 2 crushing strength	14.0 N/mm²	not suitable for walls
Class 3	20.5 N/mm²	walls above DPC
Class 4	27.5 N/mm²	cappings and copings
Class 5	34.5 N/mm²	retaining walls
Class 6	41.5 N/mm²	walls below ground
Class 7	48.5 N/mm²	walls below ground

The Class 7 calcium silicate bricks are therefore equal in strength to Class B bricks
Calcium silicate bricks are not suitable for DPCs

Durability of bricks	
FL	Frost resistant with low salt content
FN	Frost resistant with normal salt content
ML	Moderately frost resistant with low salt content
MN	Moderately frost resistant with normal salt content

MASONRY

Brickwork Dimensions

No. of horizontal bricks	Dimensions (mm)	No. of vertical courses	Height of vertical courses (mm)
½	112.5	1	75
1	225.0	2	150
1½	337.5	3	225
2	450.0	4	300
2½	562.5	5	375
3	675.0	6	450
3½	787.5	7	525
4	900.0	8	600
4½	1012.5	9	675
5	1125.0	10	750
5½	1237.5	11	825
6	1350.0	12	900
6½	1462.5	13	975
7	1575.0	14	1050
7½	1687.5	15	1125
8	1800.0	16	1200
8½	1912.5	17	1275
9	2025.0	18	1350
9½	2137.5	19	1425
10	2250.0	20	1500
20	4500.0	24	1575
40	9000.0	28	2100
50	11250.0	32	2400
60	13500.0	36	2700
75	16875.0	40	3000

TIMBER

TIMBER

Weights of Timber

Material	kg/m³	lb/cu ft
General	806 (avg)	50 (avg)
Douglas fir	479	30
Yellow pine, spruce	479	30
Pitch pine	673	42
Larch, elm	561	35
Oak (English)	724 to 959	45 to 60
Teak	643 to 877	40 to 55
Jarrah	959	60
Greenheart	1040 to 1204	65 to 75
Quebracho	1285	80
Material	**kg/m² per mm thickness**	**lb/sq ft per inch thickness**
Wooden boarding and blocks		
Softwood	0.48	2.5
Hardwood	0.76	4
Hardboard	1.06	5.5
Chipboard	0.76	4
Plywood	0.62	3.25
Blockboard	0.48	2.5
Fibreboard	0.29	1.5
Wood-wool	0.58	3
Plasterboard	0.96	5
Weather boarding	0.35	1.8

TIMBER

Conversion Tables (for timber only)

Inches	Millimetres	Feet	Metres
1	25	1	0.300
2	50	2	0.600
3	75	3	0.900
4	100	4	1.200
5	125	5	1.500
6	150	6	1.800
7	175	7	2.100
8	200	8	2.400
9	225	9	2.700
10	250	10	3.000
11	275	11	3.300
12	300	12	3.600
13	325	13	3.900
14	350	14	4.200
15	375	15	4.500
16	400	16	4.800
17	425	17	5.100
18	450	18	5.400
19	475	19	5.700
20	500	20	6.000
21	525	21	6.300
22	550	22	6.600
23	575	23	6.900
24	600	24	7.200

Planed Softwood

The finished end section size of planed timber is usually 3/16" less than the original size from which it is produced. This however varies slightly depending upon availability of material and origin of the species used.

Standards (timber) to cubic metres and cubic metres to standards (timber)

Cubic metres	Cubic metres standards	Standards
4.672	1	0.214
9.344	2	0.428
14.017	3	0.642
18.689	4	0.856
23.361	5	1.070
28.033	6	1.284
32.706	7	1.498
37.378	8	1.712
42.050	9	1.926
46.722	10	2.140
93.445	20	4.281
140.167	30	6.421
186.890	40	8.561
233.612	50	10.702
280.335	60	12.842
327.057	70	14.982
373.779	80	17.122

TIMBER

1 cu metre = 35.3148 cu ft = 0.21403 std

1 cu ft = 0.028317 cu metres

1 std = 4.67227 cu metres

Basic sizes of sawn softwood available (cross-sectional areas)

Thickness (mm)	Width (mm)								
	75	100	125	150	175	200	225	250	300
16	X	X	X	X					
19	X	X	X	X					
22	X	X	X	X					
25	X	X	X	X	X	X	X	X	X
32	X	X	X	X	X	X	X	X	X
36	X	X	X	X					
38	X	X	X	X	X	X	X		
44	X	X	X	X	X	X	X	X	X
47*	X	X	X	X	X	X	X	X	X
50	X	X	X	X	X	X	X	X	X
63	X	X	X	X	X	X	X		
75	X	X	X	X	X	X	X	X	
100		X		X		X		X	X
150				X		X			X
200						X			
250								X	
300									X

* This range of widths for 47 mm thickness will usually be found to be available in construction quality only

Note: The smaller sizes below 100 mm thick and 250 mm width are normally but not exclusively of European origin. Sizes beyond this are usually of North and South American origin

Basic lengths of sawn softwood available (metres)

1.80	2.10	3.00	4.20	5.10	6.00	7.20
	2.40	3.30	4.50	5.40	6.30	
	2.70	3.60	4.80	5.70	6.60	
		3.90			6.90	

Note: Lengths of 6.00 m and over will generally only be available from North American species and may have to be recut from larger sizes

TIMBER

Reductions from basic size to finished size by planning of two opposed faces

	Purpose	Reductions from basic sizes for timber			
		15–35 mm	36–100 mm	101–150 mm	over 150 mm
a)	Constructional timber	3 mm	3 mm	5 mm	6 mm
b)	Matching interlocking boards	4 mm	4 mm	6 mm	6 mm
c)	Wood trim not specified in BS 584	5 mm	7 mm	7 mm	9 mm
d)	Joinery and cabinet work	7 mm	9 mm	11 mm	13 mm

Note: The reduction of width or depth is overall the extreme size and is exclusive of any reduction of the face by the machining of a tongue or lap joints

Maximum Spans for Various Roof Trusses

Maximum permissible spans for rafters for Fink trussed rafters

Basic size	Actual size	Pitch (degrees)								
(mm)	(mm)	15 (m)	17.5 (m)	20 (m)	22.5 (m)	25 (m)	27.5 (m)	30 (m)	32.5 (m)	35 (m)
38 × 75	35 × 72	6.03	6.16	6.29	6.41	6.51	6.60	6.70	6.80	6.90
38 × 100	35 × 97	7.48	7.67	7.83	7.97	8.10	8.22	8.34	8.47	8.61
38 × 125	35 × 120	8.80	9.00	9.20	9.37	9.54	9.68	9.82	9.98	10.16
44 × 75	41 × 72	6.45	6.59	6.71	6.83	6.93	7.03	7.14	7.24	7.35
44 × 100	41 × 97	8.05	8.23	8.40	8.55	8.68	8.81	8.93	9.09	9.22
44 × 125	41 × 120	9.38	9.60	9.81	9.99	10.15	10.31	10.45	10.64	10.81
50 × 75	47 × 72	6.87	7.01	7.13	7.25	7.35	7.45	7.53	7.67	7.78
50 × 100	47 × 97	8.62	8.80	8.97	9.12	9.25	9.38	9.50	9.66	9.80
50 × 125	47 × 120	10.01	10.24	10.44	10.62	10.77	10.94	11.00	11.00	11.00

TIMBER

Sizes of Internal and External Doorsets

Description	Internal size (mm)	Permissible deviation	External size (mm)	Permissible deviation
Coordinating dimension: height of door leaf height sets	2100		2100	
Coordinating dimension: height of ceiling height set	2300 2350 2400 2700 3000		2300 2350 2400 2700 3000	
Coordinating dimension: width of all doorsets S = Single leaf set D = Double leaf set	600 S 700 S 800 S&D 900 S&D 1000 S&D 1200 D 1500 D 1800 D 2100 D		900 S 1000 S 1200 D 1800 D 2100 D	
Work size: height of door leaf height set	2090	± 2.0	2095	± 2.0
Work size: height of ceiling height set	2285 2335 2385 2685 2985	} ± 2.0	2295 2345 2395 2695 2995	} ± 2.0
Work size: width of all doorsets S = Single leaf set D = Double leaf set	590 S 690 S 790 S&D 890 S&D 990 S&D 1190 D 1490 D 1790 D 2090 D	} ± 2.0	895 S 995 S 1195 D 1495 D 1795 D 2095 D	} ± 2.0
Width of door leaf in single leaf sets F = Flush leaf P = Panel leaf	526 F 626 F 726 F&P 826 F&P 926 F&P	} ± 1.5	806 F&P 906 F&P	} ± 1.5
Width of door leaf in double leaf sets F = Flush leaf P = Panel leaf	362 F 412 F 426 F 562 F&P 712 F&P 826 F&P 1012 F&P	} ± 1.5	552 F&P 702 F&P 852 F&P 1002 F&P	} ± 1.5
Door leaf height for all doorsets	2040	± 1.5	1994	± 1.5

ROOFING

Total Roof Loadings for Various Types of Tiles/Slates

	Roof load (slope) kg/m^2		
	Slate/Tile	Roofing underlay and battens^2	Total dead load kg/m
Asbestos cement slate (600 × 300)	21.50	3.14	24.64
Clay tile interlocking	67.00	5.50	72.50
plain	43.50	2.87	46.37
Concrete tile interlocking	47.20	2.69	49.89
plain	78.20	5.50	83.70
Natural slate (18" × 10")	35.40	3.40	38.80
	Roof load (plan) kg/m^2		
Asbestos cement slate (600 × 300)	28.45	76.50	104.95
Clay tile interlocking	53.54	76.50	130.04
plain	83.71	76.50	60.21
Concrete tile interlocking	57.60	76.50	134.10
plain	96.64	76.50	173.14

ROOFING

Tiling Data

Product		Lap (mm)	Gauge of battens	No. slates per m²	Battens (m/m²)	Weight as laid (kg/m²)
CEMENT SLATES						
Eternit slates	600 × 300 mm	100	250	13.4	4.00	19.50
(Duracem)		90	255	13.1	3.92	19.20
		80	260	12.9	3.85	19.00
		70	265	12.7	3.77	18.60
	600 × 350 mm	100	250	11.5	4.00	19.50
		90	255	11.2	3.92	19.20
	500 × 250 mm	100	200	20.0	5.00	20.00
		90	205	19.5	4.88	19.50
		80	210	19.1	4.76	19.00
		70	215	18.6	4.65	18.60
	400 × 200 mm	90	155	32.3	6.45	20.80
		80	160	31.3	6.25	20.20
		70	165	30.3	6.06	19.60
CONCRETE TILES/SLATES						
Redland Roofing						
Stonewold slate	430 × 380 mm	75	355	8.2	2.82	51.20
Double Roman tile	418 × 330 mm	75	355	8.2	2.91	45.50
Grovebury pantile	418 × 332 mm	75	343	9.7	2.91	47.90
Norfolk pantile	381 × 227 mm	75	306	16.3	3.26	44.01
		100	281	17.8	3.56	48.06
Renown interlocking tile	418 × 330 mm	75	343	9.7	2.91	46.40
'49' tile	381 × 227 mm	75	306	16.3	3.26	44.80
		100	281	17.8	3.56	48.95
Plain, vertical tiling	265 × 165 mm	35	115	52.7	8.70	62.20
Marley Roofing						
Bold roll tile	420 × 330 mm	75	344	9.7	2.90	47.00
		100	–	10.5	3.20	51.00
Modern roof tile	420 × 330 mm	75	338	10.2	3.00	54.00
		100	–	11.0	3.20	58.00
Ludlow major	420 × 330 mm	75	338	10.2	3.00	45.00
		100	–	11.0	3.20	49.00
Ludlow plus	387 × 229 mm	75	305	16.1	3.30	47.00
		100	–	17.5	3.60	51.00
Mendip tile	420 × 330 mm	75	338	10.2	3.00	47.00
		100	–	11.0	3.20	51.00
Wessex	413 × 330 mm	75	338	10.2	3.00	54.00
		100	–	11.0	3.20	58.00
Plain tile	267 × 165 mm	65	100	60.0	10.00	76.00
		75	95	64.0	10.50	81.00
		85	90	68.0	11.30	86.00
Plain vertical tiles (feature)	267 × 165 mm	35	110	53.0	8.70	67.00
		34	115	56.0	9.10	71.00

ROOFING

Slate Nails, Quantity per Kilogram

Length	Type			
	Plain wire	Galvanized wire	Copper nail	Zinc nail
28.5 mm	325	305	325	415
34.4 mm	286	256	254	292
50.8 mm	242	224	194	200

Metal Sheet Coverings

Thicknesses and weights of sheet metal coverings								
Lead to BS 1178								
BS Code No	3	4	5	6	7	8		
Colour code	Green	Blue	Red	Black	White	Orange		
Thickness (mm)	1.25	1.80	2.24	2.50	3.15	3.55		
Density kg/m^2	14.18	20.41	25.40	30.05	35.72	40.26		
Copper to BS 2870								
Thickness (mm)			0.60	0.70				
Bay width								
Roll (mm)		500	650					
Seam (mm)		525	600					
Standard width to form bay	600	750						
Normal length of sheet	1.80	1.80						
Zinc to BS 849								
Zinc Gauge (Nr)	9	10	11	12	13	14	15	16
Thickness (mm)	0.43	0.48	0.56	0.64	0.71	0.79	0.91	1.04
Density (kg/m^2)	3.1	3.2	3.8	4.3	4.8	5.3	6.2	7.0
Aluminium to BS 4868								
Thickness (mm)	0.5	0.6	0.7	0.8	0.9	1.0	1.2	
Density (kg/m^2)	12.8	15.4	17.9	20.5	23.0	25.6	30.7	

Tables and Memoranda

ROOFING

Type of felt	Nominal mass per unit area (kg/10 m)	Nominal mass per unit area of fibre base (g/m²)	Nominal length of roll (m)
Class 1			
1B fine granule	14	220	10 or 20
surfaced bitumen	18	330	10 or 20
	25	470	10
1E mineral surfaced bitumen	38	470	10
1F reinforced bitumen	15	160 (fibre)	15
		110 (hessian)	
1F reinforced bitumen, aluminium faced	13	160 (fibre)	15
		110 (hessian)	
Class 2			
2B fine granule surfaced bitumen asbestos	18	500	10 or 20
2E mineral surfaced bitumen asbestos	38	600	10
Class 3			
3B fine granule surfaced bitumen glass fibre	18	60	20
3E mineral surfaced bitumen glass fibre	28	60	10
3E venting base layer bitumen glass fibre	32	60*	10
3H venting base layer bitumen glass fibre	17	60*	20

* Excluding effect of perforations

GLAZING

GLAZING

Nominal thickness (mm)	Tolerance on thickness (mm)	Approximate weight (kg/m²)	Normal maximum size (mm)
Float and polished plate glass			
3	+ 0.2	7.50	2140 × 1220
4	+ 0.2	10.00	2760 × 1220
5	+ 0.2	12.50	3180 × 2100
6	+ 0.2	15.00	4600 × 3180
10	+ 0.3	25.00)	6000 × 3300
12	+ 0.3	30.00)	
15	+ 0.5	37.50	3050 × 3000
19	+ 1.0	47.50)	3000 × 2900
25	+ 1.0	63.50)	
Clear sheet glass			
2 *	+ 0.2	5.00	1920 × 1220
3	+ 0.3	7.50	2130 × 1320
4	+ 0.3	10.00	2760 × 1220
5 *	+ 0.3	12.50)	2130 × 2400
6 *	+ 0.3	15.00)	
Cast glass			
3	+ 0.4 − 0.2	6.00)	2140 × 1280
4	+ 0.5	7.50)	
5	+ 0.5	9.50	2140 × 1320
6	+ 0.5	11.50)	3700 × 1280
10	+ 0.8	21.50)	
Wired glass (Cast wired glass)			
6	+ 0.3 − 0.7	−))	3700 × 1840
7	+ 0.7	−)	
(Polished wire glass)			
6	+ 1.0	−	330 × 1830

* The 5 mm and 6 mm thickness are known as *thick drawn sheet*. Although 2 mm sheet glass is available it is not recommended for general glazing purposes

METAL

METAL

Weights of Metals

Material	kg/m³	lb/cu ft
Metals, steel construction, etc.		
Iron		
– cast	7207	450
– wrought	7687	480
– ore – general	2407	150
– (crushed) Swedish	3682	230
Steel	7854	490
Copper		
– cast	8731	545
– wrought	8945	558
Brass	8497	530
Bronze	8945	558
Aluminium	2774	173
Lead	11322	707
Zinc (rolled)	7140	446
	g/mm² per metre	**lb/sq ft per foot**
Steel bars	7.85	3.4
Structural steelwork	Net weight of member @ 7854 kg/m³	
riveted	+ 10% for cleats, rivets, bolts, etc.	
welded	+ 1.25% to 2.5% for welds, etc.	
Rolled sections		
beams	+ 2.5%	
stanchions	+ 5% (extra for caps and bases)	
Plate		
web girders	+ 10% for rivets or welds, stiffeners, etc.	
	kg/m	**lb/ft**
Steel stairs: industrial type		
1 m or 3 ft wide	84	56
Steel tubes		
50 mm or 2 in bore	5 to 6	3 to 4
Gas piping		
20 mm or ¾ in	2	1¼

METAL

Universal Beams BS 4: Part 1: 2005

Designation	Mass (kg/m)	Depth of section (mm)	Width of section (mm)	Thickness		Surface area (m²/m)
				Web (mm)	Flange (mm)	
1016 × 305 × 487	487.0	1036.1	308.5	30.0	54.1	3.20
1016 × 305 × 438	438.0	1025.9	305.4	26.9	49.0	3.17
1016 × 305 × 393	393.0	1016.0	303.0	24.4	43.9	3.15
1016 × 305 × 349	349.0	1008.1	302.0	21.1	40.0	3.13
1016 × 305 × 314	314.0	1000.0	300.0	19.1	35.9	3.11
1016 × 305 × 272	272.0	990.1	300.0	16.5	31.0	3.10
1016 × 305 × 249	249.0	980.2	300.0	16.5	26.0	3.08
1016 × 305 × 222	222.0	970.3	300.0	16.0	21.1	3.06
914 × 419 × 388	388.0	921.0	420.5	21.4	36.6	3.44
914 × 419 × 343	343.3	911.8	418.5	19.4	32.0	3.42
914 × 305 × 289	289.1	926.6	307.7	19.5	32.0	3.01
914 × 305 × 253	253.4	918.4	305.5	17.3	27.9	2.99
914 × 305 × 224	224.2	910.4	304.1	15.9	23.9	2.97
914 × 305 × 201	200.9	903.0	303.3	15.1	20.2	2.96
838 × 292 × 226	226.5	850.9	293.8	16.1	26.8	2.81
838 × 292 × 194	193.8	840.7	292.4	14.7	21.7	2.79
838 × 292 × 176	175.9	834.9	291.7	14.0	18.8	2.78
762 × 267 × 197	196.8	769.8	268.0	15.6	25.4	2.55
762 × 267 × 173	173.0	762.2	266.7	14.3	21.6	2.53
762 × 267 × 147	146.9	754.0	265.2	12.8	17.5	2.51
762 × 267 × 134	133.9	750.0	264.4	12.0	15.5	2.51
686 × 254 × 170	170.2	692.9	255.8	14.5	23.7	2.35
686 × 254 × 152	152.4	687.5	254.5	13.2	21.0	2.34
686 × 254 × 140	140.1	383.5	253.7	12.4	19.0	2.33
686 × 254 × 125	125.2	677.9	253.0	11.7	16.2	2.32
610 × 305 × 238	238.1	635.8	311.4	18.4	31.4	2.45
610 × 305 × 179	179.0	620.2	307.1	14.1	23.6	2.41
610 × 305 × 149	149.1	612.4	304.8	11.8	19.7	2.39
610 × 229 × 140	139.9	617.2	230.2	13.1	22.1	2.11
610 × 229 × 125	125.1	612.2	229.0	11.9	19.6	2.09
610 × 229 × 113	113.0	607.6	228.2	11.1	17.3	2.08
610 × 229 × 101	101.2	602.6	227.6	10.5	14.8	2.07
533 × 210 × 122	122.0	544.5	211.9	12.7	21.3	1.89
533 × 210 × 109	109.0	539.5	210.8	11.6	18.8	1.88
533 × 210 × 101	101.0	536.7	210.0	10.8	17.4	1.87
533 × 210 × 92	92.1	533.1	209.3	10.1	15.6	1.86
533 × 210 × 82	82.2	528.3	208.8	9.6	13.2	1.85
457 × 191 × 98	98.3	467.2	192.8	11.4	19.6	1.67
457 × 191 × 89	89.3	463.4	191.9	10.5	17.7	1.66
457 × 191 × 82	82.0	460.0	191.3	9.9	16.0	1.65
457 × 191 × 74	74.3	457.0	190.4	9.0	14.5	1.64
457 × 191 × 67	67.1	453.4	189.9	8.5	12.7	1.63
457 × 152 × 82	82.1	465.8	155.3	10.5	18.9	1.51
457 × 152 × 74	74.2	462.0	154.4	9.6	17.0	1.50
457 × 152 × 67	67.2	458.0	153.8	9.0	15.0	1.50
457 × 152 × 60	59.8	454.6	152.9	8.1	13.3	1.50
457 × 152 × 52	52.3	449.8	152.4	7.6	10.9	1.48
406 × 178 × 74	74.2	412.8	179.5	9.5	16.0	1.51
406 × 178 × 67	67.1	409.4	178.8	8.8	14.3	1.50
406 × 178 × 60	60.1	406.4	177.9	7.9	12.8	1.49

METAL

Designation	Mass (kg/m)	Depth of section (mm)	Width of section (mm)	Thickness		Surface area (m²/m)
				Web (mm)	Flange (mm)	
406 × 178 × 50	54.1	402.6	177.7	7.7	10.9	1.48
406 × 140 × 46	46.0	403.2	142.2	6.8	11.2	1.34
406 × 140 × 39	39.0	398.0	141.8	6.4	8.6	1.33
356 × 171 × 67	67.1	363.4	173.2	9.1	15.7	1.38
356 × 171 × 57	57.0	358.0	172.2	8.1	13.0	1.37
356 × 171 × 51	51.0	355.0	171.5	7.4	11.5	1.36
356 × 171 × 45	45.0	351.4	171.1	7.0	9.7	1.36
356 × 127 × 39	39.1	353.4	126.0	6.6	10.7	1.18
356 × 127 × 33	33.1	349.0	125.4	6.0	8.5	1.17
305 × 165 × 54	54.0	310.4	166.9	7.9	13.7	1.26
305 × 165 × 46	46.1	306.6	165.7	6.7	11.8	1.25
305 × 165 × 40	40.3	303.4	165.0	6.0	10.2	1.24
305 × 127 × 48	48.1	311.0	125.3	9.0	14.0	1.09
305 × 127 × 42	41.9	307.2	124.3	8.0	12.1	1.08
305 × 127 × 37	37.0	304.4	123.3	7.1	10.7	1.07
305 × 102 × 33	32.8	312.7	102.4	6.6	10.8	1.01
305 × 102 × 28	28.2	308.7	101.8	6.0	8.8	1.00
305 × 102 × 25	24.8	305.1	101.6	5.8	7.0	0.992
254 × 146 × 43	43.0	259.6	147.3	7.2	12.7	1.08
254 × 146 × 37	37.0	256.0	146.4	6.3	10.9	1.07
254 × 146 × 31	31.1	251.4	146.1	6.0	8.6	1.06
254 × 102 × 28	28.3	260.4	102.2	6.3	10.0	0.904
254 × 102 × 25	25.2	257.2	101.9	6.0	8.4	0.897
254 × 102 × 22	22.0	254.0	101.6	5.7	6.8	0.890
203 × 133 × 30	30.0	206.8	133.9	6.4	9.6	0.923
203 × 133 × 25	25.1	203.2	133.2	5.7	7.8	0.915
203 × 102 × 23	23.1	203.2	101.8	5.4	9.3	0.790
178 × 102 × 19	19.0	177.8	101.2	4.8	7.9	0.738
152 × 89 × 16	16.0	152.4	88.7	4.5	7.7	0.638
127 × 76 × 13	13.0	127.0	76.0	4.0	7.6	0.537

METAL

Universal Columns BS 4: Part 1: 2005

Designation	Mass (kg/m)	Depth of section (mm)	Width of section (mm)	Thickness		Surface area (m²/m)
				Web (mm)	Flange (mm)	
356 × 406 × 634	633.9	474.7	424.0	47.6	77.0	2.52
356 × 406 × 551	551.0	455.6	418.5	42.1	67.5	2.47
356 × 406 × 467	467.0	436.6	412.2	35.8	58.0	2.42
356 × 406 × 393	393.0	419.0	407.0	30.6	49.2	2.38
356 × 406 × 340	339.9	406.4	403.0	26.6	42.9	2.35
356 × 406 × 287	287.1	393.6	399.0	22.6	36.5	2.31
356 × 406 × 235	235.1	381.0	384.8	18.4	30.2	2.28
356 × 368 × 202	201.9	374.6	374.7	16.5	27.0	2.19
356 × 368 × 177	177.0	368.2	372.6	14.4	23.8	2.17
356 × 368 × 153	152.9	362.0	370.5	12.3	20.7	2.16
356 × 368 × 129	129.0	355.6	368.6	10.4	17.5	2.14
305 × 305 × 283	282.9	365.3	322.2	26.8	44.1	1.94
305 × 305 × 240	240.0	352.5	318.4	23.0	37.7	1.91
305 × 305 × 198	198.1	339.9	314.5	19.1	31.4	1.87
305 × 305 × 158	158.1	327.1	311.2	15.8	25.0	1.84
305 × 305 × 137	136.9	320.5	309.2	13.8	21.7	1.82
305 × 305 × 118	117.9	314.5	307.4	12.0	18.7	1.81
305 × 305 × 97	96.9	307.9	305.3	9.9	15.4	1.79
254 × 254 × 167	167.1	289.1	265.2	19.2	31.7	1.58
254 × 254 × 132	132.0	276.3	261.3	15.3	25.3	1.55
254 × 254 × 107	107.1	266.7	258.8	12.8	20.5	1.52
254 × 254 × 89	88.9	260.3	256.3	10.3	17.3	1.50
254 × 254 × 73	73.1	254.1	254.6	8.6	14.2	1.49
203 × 203 × 86	86.1	222.2	209.1	12.7	20.5	1.24
203 × 203 × 71	71.0	215.8	206.4	10.0	17.3	1.22
203 × 203 × 60	60.0	209.6	205.8	9.4	14.2	1.21
203 × 203 × 52	52.0	206.2	204.3	7.9	12.5	1.20
203 × 203 × 46	46.1	203.2	203.6	7.2	11.0	1.19
152 × 152 × 37	37.0	161.8	154.4	8.0	11.5	0.912
152 × 152 × 30	30.0	157.6	152.9	6.5	9.4	0.901
152 × 152 × 23	23.0	152.4	152.2	5.8	6.8	0.889

METAL

Joists BS 4: Part 1: 2005 (retained for reference, Corus have ceased manufacture in UK)

Designation	Mass (kg/m)	Depth of section (mm)	Width of section (mm)	Thickness		Surface area (m²/m)
				Web (mm)	Flange (mm)	
254 × 203 × 82	82.0	254.0	203.2	10.2	19.9	1.210
203 × 152 × 52	52.3	203.2	152.4	8.9	16.5	0.932
152 × 127 × 37	37.3	152.4	127.0	10.4	13.2	0.737
127 × 114 × 29	29.3	127.0	114.3	10.2	11.5	0.646
127 × 114 × 27	26.9	127.0	114.3	7.4	11.4	0.650
102 × 102 × 23	23.0	101.6	101.6	9.5	10.3	0.549
102 × 44 × 7	7.5	101.6	44.5	4.3	6.1	0.350
89 × 89 × 19	19.5	88.9	88.9	9.5	9.9	0.476
76 × 76 × 13	12.8	76.2	76.2	5.1	8.4	0.411

Parallel Flange Channels

Designation	Mass (kg/m)	Depth of section (mm)	Width of section (mm)	Thickness		Surface area (m²/m)
				Web (mm)	Flange (mm)	
430 × 100 × 64	64.4	430	100	11.0	19.0	1.23
380 × 100 × 54	54.0	380	100	9.5	17.5	1.13
300 × 100 × 46	45.5	300	100	9.0	16.5	0.969
300 × 90 × 41	41.4	300	90	9.0	15.5	0.932
260 × 90 × 35	34.8	260	90	8.0	14.0	0.854
260 × 75 × 28	27.6	260	75	7.0	12.0	0.79
230 × 90 × 32	32.2	230	90	7.5	14.0	0.795
230 × 75 × 26	25.7	230	75	6.5	12.5	0.737
200 × 90 × 30	29.7	200	90	7.0	14.0	0.736
200 × 75 × 23	23.4	200	75	6.0	12.5	0.678
180 × 90 × 26	26.1	180	90	6.5	12.5	0.697
180 × 75 × 20	20.3	180	75	6.0	10.5	0.638
150 × 90 × 24	23.9	150	90	6.5	12.0	0.637
150 × 75 × 18	17.9	150	75	5.5	10.0	0.579
125 × 65 × 15	14.8	125	65	5.5	9.5	0.489
100 × 50 × 10	10.2	100	50	5.0	8.5	0.382

METAL

Equal Angles BS EN 10056-1

Designation	Mass (kg/m)	Surface area (m²/m)
200 × 200 × 24	71.1	0.790
200 × 200 × 20	59.9	0.790
200 × 200 × 18	54.2	0.790
200 × 200 × 16	48.5	0.790
150 × 150 × 18	40.1	0.59
150 × 150 × 15	33.8	0.59
150 × 150 × 12	27.3	0.59
150 × 150 × 10	23.0	0.59
120 × 120 × 15	26.6	0.47
120 × 120 × 12	21.6	0.47
120 × 120 × 10	18.2	0.47
120 × 120 × 8	14.7	0.47
100 × 100 × 15	21.9	0.39
100 × 100 × 12	17.8	0.39
100 × 100 × 10	15.0	0.39
100 × 100 × 8	12.2	0.39
90 × 90 × 12	15.9	0.35
90 × 90 × 10	13.4	0.35
90 × 90 × 8	10.9	0.35
90 × 90 × 7	9.61	0.35
90 × 90 × 6	8.30	0.35

Unequal Angles BS EN 10056-1

Designation	Mass (kg/m)	Surface area (m²/m)
200 × 150 × 18	47.1	0.69
200 × 150 × 15	39.6	0.69
200 × 150 × 12	32.0	0.69
200 × 100 × 15	33.7	0.59
200 × 100 × 12	27.3	0.59
200 × 100 × 10	23.0	0.59
150 × 90 × 15	26.6	0.47
150 × 90 × 12	21.6	0.47
150 × 90 × 10	18.2	0.47
150 × 75 × 15	24.8	0.44
150 × 75 × 12	20.2	0.44
150 × 75 × 10	17.0	0.44
125 × 75 × 12	17.8	0.40
125 × 75 × 10	15.0	0.40
125 × 75 × 8	12.2	0.40
100 × 75 × 12	15.4	0.34
100 × 75 × 10	13.0	0.34
100 × 75 × 8	10.6	0.34
100 × 65 × 10	12.3	0.32
100 × 65 × 8	9.94	0.32
100 × 65 × 7	8.77	0.32

METAL

Structural Tees Split from Universal Beams BS 4: Part 1: 2005

Designation	Mass (kg/m)	Surface area (m²/m)
305 × 305 × 90	89.5	1.22
305 × 305 × 75	74.6	1.22
254 × 343 × 63	62.6	1.19
229 × 305 × 70	69.9	1.07
229 × 305 × 63	62.5	1.07
229 × 305 × 57	56.5	1.07
229 × 305 × 51	50.6	1.07
210 × 267 × 61	61.0	0.95
210 × 267 × 55	54.5	0.95
210 × 267 × 51	50.5	0.95
210 × 267 × 46	46.1	0.95
210 × 267 × 41	41.1	0.95
191 × 229 × 49	49.2	0.84
191 × 229 × 45	44.6	0.84
191 × 229 × 41	41.0	0.84
191 × 229 × 37	37.1	0.84
191 × 229 × 34	33.6	0.84
152 × 229 × 41	41.0	0.76
152 × 229 × 37	37.1	0.76
152 × 229 × 34	33.6	0.76
152 × 229 × 30	29.9	0.76
152 × 229 × 26	26.2	0.76

Universal Bearing Piles BS 4: Part 1: 2005

Designation	Mass (kg/m)	Depth of Section (mm)	Width of Section (mm)	Thickness Web (mm)	Flange (mm)
356 × 368 × 174	173.9	361.4	378.5	20.3	20.4
356 × 368 × 152	152.0	356.4	376.0	17.8	17.9
356 × 368 × 133	133.0	352.0	373.8	15.6	15.7
356 × 368 × 109	108.9	346.4	371.0	12.8	12.9
305 × 305 × 223	222.9	337.9	325.7	30.3	30.4
305 × 305 × 186	186.0	328.3	320.9	25.5	25.6
305 × 305 × 149	149.1	318.5	316.0	20.6	20.7
305 × 305 × 126	126.1	312.3	312.9	17.5	17.6
305 × 305 × 110	110.0	307.9	310.7	15.3	15.4
305 × 305 × 95	94.9	303.7	308.7	13.3	13.3
305 × 305 × 88	88.0	301.7	307.8	12.4	12.3
305 × 305 × 79	78.9	299.3	306.4	11.0	11.1
254 × 254 × 85	85.1	254.3	260.4	14.4	14.3
254 × 254 × 71	71.0	249.7	258.0	12.0	12.0
254 × 254 × 63	63.0	247.1	256.6	10.6	10.7
203 × 203 × 54	53.9	204.0	207.7	11.3	11.4
203 × 203 × 45	44.9	200.2	205.9	9.5	9.5

METAL

Hot Formed Square Hollow Sections EN 10210 S275J2H & S355J2H

Size (mm)	Wall thickness (mm)	Mass (kg/m)	Superficial area (m²/m)
40 × 40	2.5	2.89	0.154
	3.0	3.41	0.152
	3.2	3.61	0.152
	3.6	4.01	0.151
	4.0	4.39	0.150
	5.0	5.28	0.147
50 × 50	2.5	3.68	0.194
	3.0	4.35	0.192
	3.2	4.62	0.192
	3.6	5.14	0.191
	4.0	5.64	0.190
	5.0	6.85	0.187
	6.0	7.99	0.185
	6.3	8.31	0.184
60 × 60	3.0	5.29	0.232
	3.2	5.62	0.232
	3.6	6.27	0.231
	4.0	6.90	0.230
	5.0	8.42	0.227
	6.0	9.87	0.225
	6.3	10.30	0.224
	8.0	12.50	0.219
70 × 70	3.0	6.24	0.272
	3.2	6.63	0.272
	3.6	7.40	0.271
	4.0	8.15	0.270
	5.0	9.99	0.267
	6.0	11.80	0.265
	6.3	12.30	0.264
	8.0	15.00	0.259
80 × 80	3.2	7.63	0.312
	3.6	8.53	0.311
	4.0	9.41	0.310
	5.0	11.60	0.307
	6.0	13.60	0.305
	6.3	14.20	0.304
	8.0	17.50	0.299
90 × 90	3.6	9.66	0.351
	4.0	10.70	0.350
	5.0	13.10	0.347
	6.0	15.50	0.345
	6.3	16.20	0.344
	8.0	20.10	0.339
100 × 100	3.6	10.80	0.391
	4.0	11.90	0.390
	5.0	14.70	0.387
	6.0	17.40	0.385
	6.3	18.20	0.384
	8.0	22.60	0.379
	10.0	27.40	0.374

Tables and Memoranda

METAL

Size (mm)	Wall thickness (mm)	Mass (kg/m)	Superficial area (m²/m)
120 × 120	4.0	14.40	0.470
	5.0	17.80	0.467
	6.0	21.20	0.465
	6.3	22.20	0.464
	8.0	27.60	0.459
	10.0	33.70	0.454
	12.0	39.50	0.449
	12.5	40.90	0.448
140 × 140	5.0	21.00	0.547
	6.0	24.90	0.545
	6.3	26.10	0.544
	8.0	32.60	0.539
	10.0	40.00	0.534
	12.0	47.00	0.529
	12.5	48.70	0.528
150 × 150	5.0	22.60	0.587
	6.0	26.80	0.585
	6.3	28.10	0.584
	8.0	35.10	0.579
	10.0	43.10	0.574
	12.0	50.80	0.569
	12.5	52.70	0.568
Hot formed from seamless hollow	16.0	65.2	0.559
160 × 160	5.0	24.10	0.627
	6.0	28.70	0.625
	6.3	30.10	0.624
	8.0	37.60	0.619
	10.0	46.30	0.614
	12.0	54.60	0.609
	12.5	56.60	0.608
	16.0	70.20	0.599
180 × 180	5.0	27.30	0.707
	6.0	32.50	0.705
	6.3	34.00	0.704
	8.0	42.70	0.699
	10.0	52.50	0.694
	12.0	62.10	0.689
	12.5	64.40	0.688
	16.0	80.20	0.679
200 × 200	5.0	30.40	0.787
	6.0	36.20	0.785
	6.3	38.00	0.784
	8.0	47.70	0.779
	10.0	58.80	0.774
	12.0	69.60	0.769
	12.5	72.30	0.768
	16.0	90.30	0.759
250 × 250	5.0	38.30	0.987
	6.0	45.70	0.985
	6.3	47.90	0.984
	8.0	60.30	0.979
	10.0	74.50	0.974
	12.0	88.50	0.969
	12.5	91.90	0.968
	16.0	115.00	0.959

METAL

Size (mm)	Wall thickness (mm)	Mass (kg/m)	Superficial area (m²/m)
300 × 300	6.0	55.10	1.18
	6.3	57.80	1.18
	8.0	72.80	1.18
	10.0	90.20	1.17
	12.0	107.00	1.17
	12.5	112.00	1.17
	16.0	141.00	1.16
350 × 350	8.0	85.40	1.38
	10.0	106.00	1.37
	12.0	126.00	1.37
	12.5	131.00	1.37
	16.0	166.00	1.36
400 × 400	8.0	97.90	1.58
	10.0	122.00	1.57
	12.0	145.00	1.57
	12.5	151.00	1.57
	16.0	191.00	1.56
(Grade S355J2H only)	20.00*	235.00	1.55

Note: * SAW process

METAL

Hot Formed Square Hollow Sections JUMBO RHS: JIS G3136

Size (mm)	Wall thickness (mm)	Mass (kg/m)	Superficial area (m²/m)
350 × 350	19.0	190.00	1.33
	22.0	217.00	1.32
	25.0	242.00	1.31
400 × 400	22.0	251.00	1.52
	25.0	282.00	1.51
450 × 450	12.0	162.00	1.76
	16.0	213.00	1.75
	19.0	250.00	1.73
	22.0	286.00	1.72
	25.0	321.00	1.71
	28.0 *	355.00	1.70
	32.0 *	399.00	1.69
500 × 500	12.0	181.00	1.96
	16.0	238.00	1.95
	19.0	280.00	1.93
	22.0	320.00	1.92
	25.0	360.00	1.91
	28.0 *	399.00	1.90
	32.0 *	450.00	1.89
	36.0 *	498.00	1.88
550 × 550	16.0	263.00	2.15
	19.0	309.00	2.13
	22.0	355.00	2.12
	25.0	399.00	2.11
	28.0 *	443.00	2.10
	32.0 *	500.00	2.09
	36.0 *	555.00	2.08
	40.0 *	608.00	2.06
600 × 600	25.0 *	439.00	2.31
	28.0 *	487.00	2.30
	32.0 *	550.00	2.29
	36.0 *	611.00	2.28
	40.0 *	671.00	2.26
700 × 700	25.0 *	517.00	2.71
	28.0 *	575.00	2.70
	32.0 *	651.00	2.69
	36.0 *	724.00	2.68
	40.0 *	797.00	2.68

Note: * SAW process

METAL

Hot Formed Rectangular Hollow Sections: EN10210 S275J2h & S355J2H

Size (mm)	Wall thickness (mm)	Mass (kg/m)	Superficial area (m²/m)
50 × 30	2.5	2.89	0.154
	3.0	3.41	0.152
	3.2	3.61	0.152
	3.6	4.01	0.151
	4.0	4.39	0.150
	5.0	5.28	0.147
60 × 40	2.5	3.68	0.194
	3.0	4.35	0.192
	3.2	4.62	0.192
	3.6	5.14	0.191
	4.0	5.64	0.190
	5.0	6.85	0.187
	6.0	7.99	0.185
	6.3	8.31	0.184
80 × 40	3.0	5.29	0.232
	3.2	5.62	0.232
	3.6	6.27	0.231
	4.0	6.90	0.230
	5.0	8.42	0.227
	6.0	9.87	0.225
	6.3	10.30	0.224
	8.0	12.50	0.219
76.2 × 50.8	3.0	5.62	0.246
	3.2	5.97	0.246
	3.6	6.66	0.245
	4.0	7.34	0.244
	5.0	8.97	0.241
	6.0	10.50	0.239
	6.3	11.00	0.238
	8.0	13.40	0.233
90 × 50	3.0	6.24	0.272
	3.2	6.63	0.272
	3.6	7.40	0.271
	4.0	8.15	0.270
	5.0	9.99	0.267
	6.0	11.80	0.265
	6.3	12.30	0.264
	8.0	15.00	0.259
100 × 50	3.0	6.71	0.292
	3.2	7.13	0.292
	3.6	7.96	0.291
	4.0	8.78	0.290
	5.0	10.80	0.287
	6.0	12.70	0.285
	6.3	13.30	0.284
	8.0	16.30	0.279

METAL

Size (mm)	Wall thickness (mm)	Mass (kg/m)	Superficial area (m²/m)
100 × 60	3.0	7.18	0.312
	3.2	7.63	0.312
	3.6	8.53	0.311
	4.0	9.41	0.310
	5.0	11.60	0.307
	6.0	13.60	0.305
	6.3	14.20	0.304
	8.0	17.50	0.299
120 × 60	3.6	9.70	0.351
	4.0	10.70	0.350
	5.0	13.10	0.347
	6.0	15.50	0.345
	6.3	16.20	0.344
	8.0	20.10	0.339
120 × 80	3.6	10.80	0.391
	4.0	11.90	0.390
	5.0	14.70	0.387
	6.0	17.40	0.385
	6.3	18.20	0.384
	8.0	22.60	0.379
	10.0	27.40	0.374
150 × 100	4.0	15.10	0.490
	5.0	18.60	0.487
	6.0	22.10	0.485
	6.3	23.10	0.484
	8.0	28.90	0.479
	10.0	35.30	0.474
	12.0	41.40	0.469
	12.5	42.80	0.468
160 × 80	4.0	14.40	0.470
	5.0	17.80	0.467
	6.0	21.20	0.465
	6.3	22.20	0.464
	8.0	27.60	0.459
	10.0	33.70	0.454
	12.0	39.50	0.449
	12.5	40.90	0.448
200 × 100	5.0	22.60	0.587
	6.0	26.80	0.585
	6.3	28.10	0.584
	8.0	35.10	0.579
	10.0	43.10	0.574
	12.0	50.80	0.569
	12.5	52.70	0.568
	16.0	65.20	0.559
250 × 150	5.0	30.40	0.787
	6.0	36.20	0.785
	6.3	38.00	0.784
	8.0	47.70	0.779
	10.0	58.80	0.774
	12.0	69.60	0.769
	12.5	72.30	0.768
	16.0	90.30	0.759

METAL

Size (mm)	Wall thickness (mm)	Mass (kg/m)	Superficial area (m²/m)
300 × 200	5.0	38.30	0.987
	6.0	45.70	0.985
	6.3	47.90	0.984
	8.0	60.30	0.979
	10.0	74.50	0.974
	12.0	88.50	0.969
	12.5	91.90	0.968
	16.0	115.00	0.959
400 × 200	6.0	55.10	1.18
	6.3	57.80	1.18
	8.0	72.80	1.18
	10.0	90.20	1.17
	12.0	107.00	1.17
	12.5	112.00	1.17
	16.0	141.00	1.16
450 × 250	8.0	85.40	1.38
	10.0	106.00	1.37
	12.0	126.00	1.37
	12.5	131.00	1.37
	16.0	166.00	1.36
500 × 300	8.0	98.00	1.58
	10.0	122.00	1.57
	12.0	145.00	1.57
	12.5	151.00	1.57
	16.0	191.00	1.56
	20.0	235.00	1.55

METAL

Hot Formed Circular Hollow Sections EN 10210 S275J2H & S355J2H

Outside diameter (mm)	Wall thickness (mm)	Mass (kg/m)	Superficial area (m²/m)
21.3	3.2	1.43	0.067
26.9	3.2	1.87	0.085
33.7	3.0	2.27	0.106
	3.2	2.41	0.106
	3.6	2.67	0.106
	4.0	2.93	0.106
42.4	3.0	2.91	0.133
	3.2	3.09	0.133
	3.6	3.44	0.133
	4.0	3.79	0.133
48.3	2.5	2.82	0.152
	3.0	3.35	0.152
	3.2	3.56	0.152
	3.6	3.97	0.152
	4.0	4.37	0.152
	5.0	5.34	0.152
60.3	2.5	3.56	0.189
	3.0	4.24	0.189
	3.2	4.51	0.189
	3.6	5.03	0.189
	4.0	5.55	0.189
	5.0	6.82	0.189
76.1	2.5	4.54	0.239
	3.0	5.41	0.239
	3.2	5.75	0.239
	3.6	6.44	0.239
	4.0	7.11	0.239
	5.0	8.77	0.239
	6.0	10.40	0.239
	6.3	10.80	0.239
88.9	2.5	5.33	0.279
	3.0	6.36	0.279
	3.2	6.76	0.27
	3.6	7.57	0.279
	4.0	8.38	0.279
	5.0	10.30	0.279
	6.0	12.30	0.279
	6.3	12.80	0.279
114.3	3.0	8.23	0.359
	3.2	8.77	0.359
	3.6	9.83	0.359
	4.0	10.09	0.359
	5.0	13.50	0.359
	6.0	16.00	0.359
	6.3	16.80	0.359

METAL

Outside diameter (mm)	Wall thickness (mm)	Mass (kg/m)	Superficial area (m²/m)
139.7	3.2	10.80	0.439
	3.6	12.10	0.439
	4.0	13.40	0.439
	5.0	16.60	0.439
	6.0	19.80	0.439
	6.3	20.70	0.439
	8.0	26.00	0.439
	10.0	32.00	0.439
168.3	3.2	13.00	0.529
	3.6	14.60	0.529
	4.0	16.20	0.529
	5.0	20.10	0.529
	6.0	24.00	0.529
	6.3	25.20	0.529
	8.0	31.60	0.529
	10.0	39.00	0.529
	12.0	46.30	0.529
	12.5	48.00	0.529
193.7	5.0	23.30	0.609
	6.0	27.80	0.609
	6.3	29.10	0.609
	8.0	36.60	0.609
	10.0	45.30	0.609
	12.0	53.80	0.609
	12.5	55.90	0.609
219.1	5.0	26.40	0.688
	6.0	31.50	0.688
	6.3	33.10	0.688
	8.0	41.60	0.688
	10.0	51.60	0.688
	12.0	61.30	0.688
	12.5	63.70	0.688
	16.0	80.10	0.688
244.5	5.0	29.50	0.768
	6.0	35.30	0.768
	6.3	37.00	0.768
	8.0	46.70	0.768
	10.0	57.80	0.768
	12.0	68.80	0.768
	12.5	71.50	0.768
	16.0	90.20	0.768
273.0	5.0	33.00	0.858
	6.0	39.50	0.858
	6.3	41.40	0.858
	8.0	52.30	0.858
	10.0	64.90	0.858
	12.0	77.20	0.858
	12.5	80.30	0.858
	16.0	101.00	0.858

METAL

Outside diameter (mm)	Wall thickness (mm)	Mass (kg/m)	Superficial area (m²/m)
323.9	5.0	39.30	1.02
	6.0	47.00	1.02
	6.3	49.30	1.02
	8.0	62.30	1.02
	10.0	77.40	1.02
	12.0	92.30	1.02
	12.5	96.00	1.02
	16.0	121.00	1.02
355.6	6.3	54.30	1.12
	8.0	68.60	1.12
	10.0	85.30	1.12
	12.0	102.00	1.12
	12.5	106.00	1.12
	16.0	134.00	1.12
406.4	6.3	62.20	1.28
	8.0	79.60	1.28
	10.0	97.80	1.28
	12.0	117.00	1.28
	12.5	121.00	1.28
	16.0	154.00	1.28
457.0	6.3	70.00	1.44
	8.0	88.60	1.44
	10.0	110.00	1.44
	12.0	132.00	1.44
	12.5	137.00	1.44
	16.0	174.00	1.44
508.0	6.3	77.90	1.60
	8.0	98.60	1.60
	10.0	123.00	1.60
	12.0	147.00	1.60
	12.5	153.00	1.60
	16.0	194.00	1.60

METAL

Spacing of Holes in Angles

Nominal leg length (mm)	Spacing of holes						Maximum diameter of bolt or rivet		
	A	B	C	D	E	F	A	B and C	D, E and F
200		75	75	55	55	55		30	20
150		55	55					20	
125		45	60					20	
120									
100	55						24		
90	50						24		
80	45						20		
75	45						20		
70	40						20		
65	35						20		
60	35						16		
50	28						12		
45	25								
40	23								
30	20								
25	15								

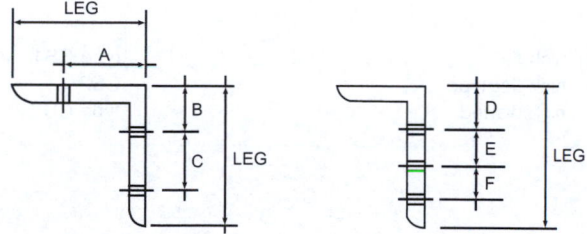

KERBS, PAVING, ETC.

KERBS/EDGINGS/CHANNELS

Precast Concrete Kerbs to BS 7263

Straight kerb units: length from 450 to 915 mm

150 mm high × 125 mm thick		
bullnosed	type BN	
half battered	type HB3	
255 mm high × 125 mm thick		
45° splayed	type SP	
half battered	type HB2	
305 mm high × 150 mm thick		
half battered	type HB1	
Quadrant kerb units		
150 mm high × 305 and 455 mm radius to match	type BN	type QBN
150 mm high × 305 and 455 mm radius to match	type HB2, HB3	type QHB
150 mm high × 305 and 455 mm radius to match	type SP	type QSP
255 mm high × 305 and 455 mm radius to match	type BN	type QBN
255 mm high × 305 and 455 mm radius to match	type HB2, HB3	type QHB
225 mm high × 305 and 455 mm radius to match	type SP	type QSP
Angle kerb units		
305 × 305 × 225 mm high × 125 mm thick		
bullnosed external angle	type XA	
splayed external angle to match type SP	type XA	
bullnosed internal angle	type IA	
splayed internal angle to match type SP	type IA	
Channels		
255 mm wide × 125 mm high flat	type CS1	
150 mm wide × 125 mm high flat type	CS2	
255 mm wide × 125 mm high dished	type CD	

KERBS, PAVING, ETC.

Transition kerb units			
from kerb type SP to HB	left handed	type TL	
	right handed	type TR	
from kerb type BN to HB	left handed	type DL1	
	right handed	type DR1	
from kerb type BN to SP	left handed	type DL2	
	right handed	type DR2	

Number of kerbs required per quarter circle (780 mm kerb lengths)

Radius (m)	Number in quarter circle
12	24
10	20
8	16
6	12
5	10
4	8
3	6
2	4
1	2

Precast Concrete Edgings

Round top type ER	Flat top type EF	Bullnosed top type EBN
150 × 50 mm	150 × 50 mm	150 × 50 mm
200 × 50 mm	200 × 50 mm	200 × 50 mm
250 × 50 mm	250 × 50 mm	250 × 50 mm

KERBS, PAVING, ETC.

BASES

Cement Bound Material for Bases and Subbases

CBM1:	very carefully graded aggregate from 37.5–75 mm, with a 7-day strength of 4.5 N/mm^2
CBM2:	same range of aggregate as CBM1 but with more tolerance in each size of aggregate with a 7-day strength of 7.0 N/mm^2
CBM3:	crushed natural aggregate or blast furnace slag, graded from 37.5 mm–150 mm for 40 mm aggregate, and from 20–75 mm for 20 mm aggregate, with a 7-day strength of 10 N/mm^2
CBM4:	crushed natural aggregate or blast furnace slag, graded from 37.5 mm–150 mm for 40 mm aggregate, and from 20–75 mm for 20 mm aggregate, with a 7-day strength of 15 N/mm^2

INTERLOCKING BRICK/BLOCK ROADS/PAVINGS

Sizes of Precast Concrete Paving Blocks

Type R blocks
200 × 100 × 60 mm
200 × 100 × 65 mm
200 × 100 × 80 mm
200 × 100 × 100 mm

Type S
Any shape within a 295 mm space

Sizes of clay brick pavers
200 × 100 × 50 mm
200 × 100 × 65 mm
210 × 105 × 50 mm
210 × 105 × 65 mm
215 × 102.5 × 50 mm
215 × 102.5 × 65 mm

Type PA: 3 kN
Footpaths and pedestrian areas, private driveways, car parks, light vehicle traffic and over-run

Type PB: 7 kN
Residential roads, lorry parks, factory yards, docks, petrol station forecourts, hardstandings, bus stations

KERBS, PAVING, ETC.

PAVING AND SURFACING

Weights and Sizes of Paving and Surfacing

Description of item	Size	Quantity per tonne
Paving 50 mm thick	900 × 600 mm	15
Paving 50 mm thick	750 × 600 mm	18
Paving 50 mm thick	600 × 600 mm	23
Paving 50 mm thick	450 × 600 mm	30
Paving 38 mm thick	600 × 600 mm	30
Path edging	914 × 50 × 150 mm	60
Kerb (including radius and tapers)	125 × 254 × 914 mm	15
Kerb (including radius and tapers)	125 × 150 × 914 mm	25
Square channel	125 × 254 × 914 mm	15
Dished channel	125 × 254 × 914 mm	15
Quadrants	300 × 300 × 254 mm	19
Quadrants	450 × 450 × 254 mm	12
Quadrants	300 × 300 × 150 mm	30
Internal angles	300 × 300 × 254 mm	30
Fluted pavement channel	255 × 75 × 914 mm	25
Corner stones	300 × 300 mm	80
Corner stones	360 × 360 mm	60
Cable covers	914 × 175 mm	55
Gulley kerbs	220 × 220 × 150 mm	60
Gulley kerbs	220 × 200 × 75 mm	120

KERBS, PAVING, ETC.

Weights and Sizes of Paving and Surfacing

Material	kg/m³	lb/cu yd
Tarmacadam	2306	3891
Macadam (waterbound)	2563	4325
Vermiculite (aggregate)	64–80	108–135
Terracotta	2114	3568
Cork – compressed	388	24
	kg/m²	**lb/sq ft**
Clay floor tiles, 12.7 mm	27.3	5.6
Pavement lights	122	25
Damp-proof course	5	1
	kg/m² per mm thickness	**lb/sq ft per inch thickness**
Paving slabs (stone)	2.3	12
Granite setts	2.88	15
Asphalt	2.30	12
Rubber flooring	1.68	9
Polyvinyl chloride	1.94 (avg)	10 (avg)

Coverage (m²) Per Cubic Metre of Materials Used as Subbases or Capping Layers

Consolidated thickness laid in (mm)	Square metre coverage		
	Gravel	Sand	Hardcore
50	15.80	16.50	–
75	10.50	11.00	–
100	7.92	8.20	7.42
125	6.34	6.60	5.90
150	5.28	5.50	4.95
175	–	–	4.23
200	–	–	3.71
225	–	–	3.30
300	–	–	2.47

KERBS, PAVING, ETC.

Approximate Rate of Spreads

Average thickness of course (mm)	Description	Approximate rate of spread			
		Open Textured		Dense, Medium & Fine Textured	
		(kg/m²)	(m²/t)	(kg/m²)	(m²/t)
35	14 mm open textured or dense wearing course	60–75	13–17	70–85	12–14
40	20 mm open textured or dense base course	70–85	12–14	80–100	10–12
45	20 mm open textured or dense base course	80–100	10–12	95–100	9–10
50	20 mm open textured or dense, or 28 mm dense base course	85–110	9–12	110–120	8–9
60	28 mm dense base course, 40 mm open textured or dense base course or 40 mm single course as base course	–	8–10	130–150	7–8
65	28 mm dense base course, 40 mm open textured or dense base course or 40 mm single course	100–135	7–10	140–160	6–7
75	40 mm single course, 40 mm open textured or dense base course, 40 mm dense roadbase	120–150	7–8	165–185	5–6
100	40 mm dense base course or roadbase	–	–	220–240	4–4.5

KERBS, PAVING, ETC.

Surface Dressing Roads: Coverage (m²) per Tonne of Material

Size in mm	Sand	Granite chips	Gravel	Limestone chips
Sand	168	–	–	–
3	–	148	152	165
6	–	130	133	144
9	–	111	114	123
13	–	85	87	95
19	–	68	71	78

Sizes of Flags

Reference	Nominal size (mm)	Thickness (mm)
A	600 × 450	50 and 63
B	600 × 600	50 and 63
C	600 × 750	50 and 63
D	600 × 900	50 and 63
E	450 × 450	50 and 70 chamfered top surface
F	400 × 400	50 and 65 chamfered top surface
G	300 × 300	50 and 60 chamfered top surface

Sizes of Natural Stone Setts

Width (mm)		Length (mm)		Depth (mm)
100	×	100	×	100
75	×	150 to 250	×	125
75	×	150 to 250	×	150
100	×	150 to 250	×	100
100	×	150 to 250	×	150

SEEDING/TURFING AND PLANTING

SEEDING/TURFING AND PLANTING

Topsoil Quality

Topsoil grade	Properties
Premium	Natural topsoil, high fertility, loamy texture, good soil structure, suitable for intensive cultivation.
General purpose	Natural or manufactured topsoil of lesser quality than Premium, suitable for agriculture or amenity landscape, may need fertilizer or soil structure improvement.
Economy	Selected subsoil, natural mineral deposit such as river silt or greensand. The grade comprises two subgrades; 'Low clay' and 'High clay' which is more liable to compaction in handling. This grade is suitable for low-production agricultural land and amenity woodland or conservation planting areas.

Forms of Trees

Standards:	Shall be clear with substantially straight stems. Grafted and budded trees shall have no more than a slight bend at the union. Standards shall be designated as Half, Extra light, Light, Standard, Selected standard, Heavy, and Extra heavy.
Sizes of Standards	
Heavy standard	12–14 cm girth × 3.50 to 5.00 m high
Extra heavy standard	14–16 cm girth × 4.25 to 5.00 m high
Extra heavy standard	16–18 cm girth × 4.25 to 6.00 m high
Extra heavy standard	18–20 cm girth × 5.00 to 6.00 m high
Semi-mature trees:	Between 6.0 m and 12.0 m tall with a girth of 20 to 75 cm at 1.0 m above ground.
Feathered trees:	Shall have a defined upright central leader, with stem furnished with evenly spread and balanced lateral shoots down to or near the ground.
Whips:	Shall be without significant feather growth as determined by visual inspection.
Multi-stemmed trees:	Shall have two or more main stems at, near, above or below ground.
Seedlings grown from seed and not transplanted shall be specified when ordered for sale as:	
1+0	one year old seedling
2+0	two year old seedling
1+1	one year seed bed, one year transplanted = two year old seedling
1+2	one year seed bed, two years transplanted = three year old seedling
2+1	two year seed bed, one year transplanted = three year old seedling
1u1	two years seed bed, undercut after one year = two year old seedling
2u2	four years seed bed, undercut after two years = four year old seedling

SEEDING/TURFING AND PLANTING

Cuttings

The age of cuttings (plants grown from shoots, stems, or roots of the mother plant) shall be specified when ordered for sale. The height of transplants and undercut seedlings/cuttings (which have been transplanted or undercut at least once) shall be stated in centimetres. The number of growing seasons before and after transplanting or undercutting shall be stated.

0 + 1	one year cutting
0 + 2	two year cutting
0 + 1 + 1	one year cutting bed, one year transplanted = two year old seedling
0 + 1 + 2	one year cutting bed, two years transplanted = three year old seedling

Grass Cutting Capacities in m² per hour

Speed mph	Width of cut in metres												
	0.5	0.7	1.0	1.2	1.5	1.7	2.0	2.0	2.1	2.5	2.8	3.0	3.4
1.0	724	1127	1529	1931	2334	2736	3138	3219	3380	4023	4506	4828	5472
1.5	1086	1690	2293	2897	3500	4104	4707	4828	5069	6035	6759	7242	8208
2.0	1448	2253	3058	3862	4667	5472	6276	6437	6759	8047	9012	9656	10944
2.5	1811	2816	3822	4828	5834	6840	7846	8047	8449	10058	11265	12070	13679
3.0	2173	3380	4587	5794	7001	8208	9415	9656	10139	12070	13518	14484	16415
3.5	2535	3943	5351	6759	8167	9576	10984	11265	11829	14082	15772	16898	19151
4.0	2897	4506	6115	7725	9334	10944	12553	12875	13518	16093	18025	19312	21887
4.5	3259	5069	6880	8690	10501	12311	14122	14484	15208	18105	20278	21726	24623
5.0	3621	5633	7644	9656	11668	13679	15691	16093	16898	20117	22531	24140	27359
5.5	3983	6196	8409	10622	12834	15047	17260	17703	18588	22128	24784	26554	30095
6.0	4345	6759	9173	11587	14001	16415	18829	19312	20278	24140	27037	28968	32831
6.5	4707	7322	9938	12553	15168	17783	20398	20921	21967	26152	29290	31382	35566
7.0	5069	7886	10702	13518	16335	19151	21967	22531	23657	28163	31543	33796	38302

Number of Plants per m²: For Plants Planted on an Evenly Spaced Grid

Planting distances

mm	0.10	0.15	0.20	0.25	0.35	0.40	0.45	0.50	0.60	0.75	0.90	1.00	1.20	1.50
0.10	100.00	66.67	50.00	40.00	28.57	25.00	22.22	20.00	16.67	13.33	11.11	10.00	8.33	6.67
0.15	66.67	44.44	33.33	26.67	19.05	16.67	14.81	13.33	11.11	8.89	7.41	6.67	5.56	4.44
0.20	50.00	33.33	25.00	20.00	14.29	12.50	11.11	10.00	8.33	6.67	5.56	5.00	4.17	3.33
0.25	40.00	26.67	20.00	16.00	11.43	10.00	8.89	8.00	6.67	5.33	4.44	4.00	3.33	2.67
0.35	28.57	19.05	14.29	11.43	8.16	7.14	6.35	5.71	4.76	3.81	3.17	2.86	2.38	1.90
0.40	25.00	16.67	12.50	10.00	7.14	6.25	5.56	5.00	4.17	3.33	2.78	2.50	2.08	1.67
0.45	22.22	14.81	11.11	8.89	6.35	5.56	4.94	4.44	3.70	2.96	2.47	2.22	1.85	1.48
0.50	20.00	13.33	10.00	8.00	5.71	5.00	4.44	4.00	3.33	2.67	2.22	2.00	1.67	1.33
0.60	16.67	11.11	8.33	6.67	4.76	4.17	3.70	3.33	2.78	2.22	1.85	1.67	1.39	1.11
0.75	13.33	8.89	6.67	5.33	3.81	3.33	2.96	2.67	2.22	1.78	1.48	1.33	1.11	0.89
0.90	11.11	7.41	5.56	4.44	3.17	2.78	2.47	2.22	1.85	1.48	1.23	1.11	0.93	0.74
1.00	10.00	6.67	5.00	4.00	2.86	2.50	2.22	2.00	1.67	1.33	1.11	1.00	0.83	0.67
1.20	8.33	5.56	4.17	3.33	2.38	2.08	1.85	1.67	1.39	1.11	0.93	0.83	0.69	0.56
1.50	6.67	4.44	3.33	2.67	1.90	1.67	1.48	1.33	1.11	0.89	0.74	0.67	0.56	0.44

SEEDING/TURFING AND PLANTING

Grass Clippings Wet: Based on 3.5 m³/tonne

Annual kg/100 m²	Average 20 cuts kg/100 m²	m²/tonne	m²/m³
32.0	1.6	61162.1	214067.3

Nr of cuts	22	20	18	16	12	4
kg/cut	1.45	1.60	1.78	2.00	2.67	8.00
Area capacity of 3 tonne vehicle per load						
m²	206250	187500	168750	150000	112500	37500
Load m³	**100 m² units/m³ of vehicle space**					
1	196.4	178.6	160.7	142.9	107.1	35.7
2	392.9	357.1	321.4	285.7	214.3	71.4
3	589.3	535.7	482.1	428.6	321.4	107.1
4	785.7	714.3	642.9	571.4	428.6	142.9
5	982.1	892.9	803.6	714.3	535.7	178.6

Transportation of Trees

To unload large trees a machine with the necessary lifting strength is required. The weight of the trees must therefore be known in advance. The following table gives a rough overview. The additional columns with root ball dimensions and the number of plants per trailer provide additional information for example about preparing planting holes and calculating unloading times.

Girth in cm	Rootball diameter in cm	Ball height in cm	Weight in kg	Numbers of trees per trailer
16–18	50–60	40	150	100–120
18–20	60–70	40–50	200	80–100
20–25	60–70	40–50	270	50–70
25–30	80	50–60	350	50
30–35	90–100	60–70	500	12–18
35–40	100–110	60–70	650	10–15
40–45	110–120	60–70	850	8–12
45–50	110–120	60–70	1100	5–7
50–60	130–140	60–70	1600	1–3
60–70	150–160	60–70	2500	1
70–80	180–200	70	4000	1
80–90	200–220	70–80	5500	1
90–100	230–250	80–90	7500	1
100–120	250–270	80–90	9500	1

Data supplied by Lorenz von Ehren GmbH
The information in the table is approximate; deviations depend on soil type, genus and weather

FENCING AND GATES

FENCING AND GATES

Types of Preservative

Creosote (tar oil) can be 'factory' applied	by pressure to BS 144: pts 1&2 by immersion to BS 144: pt 1 by hot and cold open tank to BS 144: pts 1&2
Copper/chromium/arsenic (CCA)	by full cell process to BS 4072 pts 1&2
Organic solvent (OS)	by double vacuum (vacvac) to BS 5707 pts 1&3 by immersion to BS 5057 pts 1&3
Pentachlorophenol (PCP)	by heavy oil double vacuum to BS 5705 pts 2&3
Boron diffusion process (treated with disodium octaborate to BWPA Manual 1986)	

Note: Boron is used on green timber at source and the timber is supplied dry

Cleft Chestnut Pale Fences

Pales	Pale spacing	Wire lines	
900 mm	75 mm	2	temporary protection
1050 mm	75 or 100 mm	2	light protective fences
1200 mm	75 mm	3	perimeter fences
1350 mm	75 mm	3	perimeter fences
1500 mm	50 mm	3	narrow perimeter fences
1800 mm	50 mm	3	light security fences

Close-Boarded Fences

Close-boarded fences 1.05 to 1.8 m high
Type BCR (recessed) or BCM (morticed) with concrete posts 140 × 115 mm tapered and Type BW with timber posts

Palisade Fences

Wooden palisade fences
Type WPC with concrete posts 140 × 115 mm tapered and Type WPW with timber posts

For both types of fence:
Height of fence 1050 mm: two rails
Height of fence 1200 mm: two rails
Height of fence 1500 mm: three rails
Height of fence 1650 mm: three rails
Height of fence 1800 mm: three rails

FENCING AND GATES

Post and Rail Fences

Wooden post and rail fences
Type MPR 11/3 morticed rails and Type SPR 11/3 nailed rails
Height to top of rail 1100 mm
Rails: three rails 87 mm, 38 mm

Type MPR 11/4 morticed rails and Type SPR 11/4 nailed rails
Height to top of rail 1100 mm
Rails: four rails 87 mm, 38 mm

Type MPR 13/4 morticed rails and Type SPR 13/4 nailed rails
Height to top of rail 1300 mm
Rail spacing 250 mm, 250 mm, and 225 mm from top
Rails: four rails 87 mm, 38 mm

Steel Posts

Rolled steel angle iron posts for chain link fencing

Posts	Fence height	Strut	Straining post
1500 × 40 × 40 × 5 mm	900 mm	1500 × 40 × 40 × 5 mm	1500 × 50 × 50 × 6 mm
1800 × 40 × 40 × 5 mm	1200 mm	1800 × 40 × 40 × 5 mm	1800 × 50 × 50 × 6 mm
2000 × 45 × 45 × 5 mm	1400 mm	2000 × 45 × 45 × 5 mm	2000 × 60 × 60 × 6 mm
2600 × 45 × 45 × 5 mm	1800 mm	2600 × 45 × 45 × 5 mm	2600 × 60 × 60 × 6 mm
3000 × 50 × 50 × 6 mm	1800 mm	2600 × 45 × 45 × 5 mm	3000 × 60 × 60 × 6 mm
with arms			

Concrete Posts

Concrete posts for chain link fencing

Posts and straining posts	Fence height	Strut
1570 mm 100 × 100 mm	900 mm	1500 mm × 75 × 75 mm
1870 mm 125 × 125 mm	1200 mm	1830 mm × 100 × 75 mm
2070 mm 125 × 125 mm	1400 mm	1980 mm × 100 × 75 mm
2620 mm 125 × 125 mm	1800 mm	2590 mm × 100 × 85 mm
3040 mm 125 × 125 mm	1800 mm	2590 mm × 100 × 85 mm (with arms)

FENCING AND GATES

Rolled Steel Angle Posts

Rolled steel angle posts for rectangular wire mesh (field) fencing

Posts	Fence height	Strut	Straining post
1200 × 40 × 40 × 5 mm	600 mm	1200 × 75 × 75 mm	1350 × 100 × 100 mm
1400 × 40 × 40 × 5 mm	800 mm	1400 × 75 × 75 mm	1550 × 100 × 100 mm
1500 × 40 × 40 × 5 mm	900 mm	1500 × 75 × 75 mm	1650 × 100 × 100 mm
1600 × 40 × 40 × 5 mm	1000 mm	1600 × 75 × 75 mm	1750 × 100 × 100 mm
1750 × 40 × 40 × 5 mm	1150 mm	1750 × 75 × 100 mm	1900 × 125 × 125 mm

Concrete Posts

Concrete posts for rectangular wire mesh (field) fencing

Posts	Fence height	Strut	Straining post
1270 × 100 × 100 mm	600 mm	1200 × 75 × 75 mm	1420 × 100 × 100 mm
1470 × 100 × 100 mm	800 mm	1350 × 75 × 75 mm	1620 × 100 × 100 mm
1570 × 100 × 100 mm	900 mm	1500 × 75 × 75 mm	1720 × 100 × 100 mm
1670 × 100 × 100 mm	600 mm	1650 × 75 × 75 mm	1820 × 100 × 100 mm
1820 × 125 × 125 mm	1150 mm	1830 × 75 × 100 mm	1970 × 125 × 125 mm

Cleft Chestnut Pale Fences

Timber Posts

Timber posts for wire mesh and hexagonal wire netting fences

Round timber for general fences

Posts	Fence height	Strut	Straining post
1300 × 65 mm dia.	600 mm	1200 × 80 mm dia.	1450 × 100 mm dia.
1500 × 65 mm dia.	800 mm	1400 × 80 mm dia.	1650 × 100 mm dia.
1600 × 65 mm dia.	900 mm	1500 × 80 mm dia.	1750 × 100 mm dia.
1700 × 65 mm dia.	1050 mm	1600 × 80 mm dia.	1850 × 100 mm dia.
1800 × 65 mm dia.	1150 mm	1750 × 80 mm dia.	2000 × 120 mm dia.

Squared timber for general fences

Posts	Fence height	Strut	Straining post
1300 × 75 × 75 mm	600 mm	1200 × 75 × 75 mm	1450 × 100 × 100 mm
1500 × 75 × 75 mm	800 mm	1400 × 75 × 75 mm	1650 × 100 × 100 mm
1600 × 75 × 75 mm	900 mm	1500 × 75 × 75 mm	1750 × 100 × 100 mm
1700 × 75 × 75 mm	1050 mm	1600 × 75 × 75 mm	1850 × 100 × 100 mm
1800 × 75 × 75 mm	1150 mm	1750 × 75 × 75 mm	2000 × 125 × 100 mm

FENCING AND GATES

Steel Fences to BS 1722: Part 9: 1992

	Fence height	Top/bottom rails and flat posts	Vertical bars
Light	1000 mm	40 × 10 mm 450 mm in ground	12 mm dia. at 115 mm cs
	1200 mm	40 × 10 mm 550 mm in ground	12 mm dia. at 115 mm cs
	1400 mm	40 × 10 mm 550 mm in ground	12 mm dia. at 115 mm cs
Light	1000 mm	40 × 10 mm 450 mm in ground	16 mm dia. at 120 mm cs
	1200 mm	40 × 10 mm 550 mm in ground	16 mm dia. at 120 mm cs
	1400 mm	40 × 10 mm 550 mm in ground	16 mm dia. at 120 mm cs
Medium	1200 mm	50 × 10 mm 550 mm in ground	20 mm dia. at 125 mm cs
	1400 mm	50 × 10 mm 550 mm in ground	20 mm dia. at 125 mm cs
	1600 mm	50 × 10 mm 600 mm in ground	22 mm dia. at 145 mm cs
	1800 mm	50 × 10 mm 600 mm in ground	22 mm dia. at 145 mm cs
Heavy	1600 mm	50 × 10 mm 600 mm in ground	22 mm dia. at 145 mm cs
	1800 mm	50 × 10 mm 600 mm in ground	22 mm dia. at 145 mm cs
	2000 mm	50 × 10 mm 600 mm in ground	22 mm dia. at 145 mm cs
	2200 mm	50 × 10 mm 600 mm in ground	22 mm dia. at 145 mm cs

Notes: Mild steel fences: round or square verticals; flat standards and horizontals. Tops of vertical bars may be bow-top, blunt, or pointed. Round or square bar railings

Timber Field Gates to BS 3470: 1975

Gates made to this standard are designed to open one way only
All timber gates are 1100 mm high
Width over stiles 2400, 2700, 3000, 3300, 3600 and 4200 mm
Gates over 4200 mm should be made in two leaves

Steel Field Gates to BS 3470: 1975

All steel gates are 1100 mm high
Heavy duty: width over stiles 2400, 3000, 3600 and 4500 mm
Light duty: width over stiles 2400, 3000 and 3600 mm

FENCING AND GATES

Domestic Front Entrance Gates to BS 4092: Part 1: 1966

Metal gates:	Single gates are 900 mm high minimum, 900 mm, 1000 mm and 1100 mm wide

Domestic Front Entrance Gates to BS 4092: Part 2: 1966

Wooden gates:	All rails shall be tenoned into the stiles
	Single gates are 840 mm high minimum, 801 mm and 1020 mm wide
	Double gates are 840 mm high minimum, 2130, 2340 and 2640 mm wide

Timber Bridle Gates to BS 5709:1979 (Horse or Hunting Gates)

Gates open one way only	
Minimum width between posts	1525 mm
Minimum height	1100 mm

Timber Kissing Gates to BS 5709:1979

Minimum width	700 mm
Minimum height	1000 mm
Minimum distance between shutting posts	600 mm
Minimum clearance at mid-point	600 mm

Metal Kissing Gates to BS 5709:1979

Sizes are the same as those for timber kissing gates
Maximum gaps between rails 120 mm

Categories of Pedestrian Guard Rail to BS 3049:1976

Class A for normal use
Class B where vandalism is expected
Class C where crowd pressure is likely

DRAINAGE

DRAINAGE

Width Required for Trenches for Various Diameters of Pipes

Pipe diameter (mm)	Trench n.e. 1.50 m deep	Trench over 1.50 m deep
n.e. 100 mm	450 mm	600 mm
100–150 mm	500 mm	650 mm
150–225 mm	600 mm	750 mm
225–300 mm	650 mm	800 mm
300–400 mm	750 mm	900 mm
400–450 mm	900 mm	1050 mm
450–600 mm	1100 mm	1300 mm

Weights and Dimensions – Vitrified Clay Pipes

Product	Nominal diameter (mm)	Effective length (mm)	BS 65 limits of tolerance min (mm)	max (mm)	Crushing strength (kN/m)	Weight (kg/pipe)	(kg/m)
Supersleve	100	1600	96	105	35.00	14.71	9.19
	150	1750	146	158	35.00	29.24	16.71
Hepsleve	225	1850	221	236	28.00	84.03	45.42
	300	2500	295	313	34.00	193.05	77.22
	150	1500	146	158	22.00	37.04	24.69
Hepseal	225	1750	221	236	28.00	85.47	48.84
	300	2500	295	313	34.00	204.08	81.63
	400	2500	394	414	44.00	357.14	142.86
	450	2500	444	464	44.00	454.55	181.63
	500	2500	494	514	48.00	555.56	222.22
	600	2500	591	615	57.00	796.23	307.69
	700	3000	689	719	67.00	1111.11	370.45
	800	3000	788	822	72.00	1351.35	450.45
Hepline	100	1600	95	107	22.00	14.71	9.19
	150	1750	145	160	22.00	29.24	16.71
	225	1850	219	239	28.00	84.03	45.42
	300	1850	292	317	34.00	142.86	77.22
Hepduct (conduit)	90	1500	–	–	28.00	12.05	8.03
	100	1600	–	–	28.00	14.71	9.19
	125	1750	–	–	28.00	20.73	11.84
	150	1750	–	–	28.00	29.24	16.71
	225	1850	–	–	28.00	84.03	45.42
	300	1850	–	–	34.00	142.86	77.22

DRAINAGE

Weights and Dimensions – Vitrified Clay Pipes

Nominal internal diameter (mm)	Nominal wall thickness (mm)	Approximate weight (kg/m)
150	25	45
225	29	71
300	32	122
375	35	162
450	38	191
600	48	317
750	54	454
900	60	616
1200	76	912
1500	89	1458
1800	102	1884
2100	127	2619

Wall thickness, weights and pipe lengths vary, depending on type of pipe required

The particulars shown above represent a selection of available diameters and are applicable to strength class 1 pipes with flexible rubber ring joints

Tubes with Ogee joints are also available

DRAINAGE

Weights and Dimensions – PVC-u Pipes

	Nominal size	Mean outside diameter (mm)		Wall thickness	Weight
		min	max	(mm)	(kg/m)
Standard pipes	82.4	82.4	82.7	3.2	1.2
	110.0	110.0	110.4	3.2	1.6
	160.0	160.0	160.6	4.1	3.0
	200.0	200.0	200.6	4.9	4.6
	250.0	250.0	250.7	6.1	7.2
Perforated pipes heavy grade	As above	As above	As above	As above	As above
thin wall	82.4	82.4	82.7	1.7	–
	110.0	110.0	110.4	2.2	–
	160.0	160.0	160.6	3.2	–

Width of Trenches Required for Various Diameters of Pipes

Pipe diameter (mm)	Trench n.e. 1.5 m deep (mm)	Trench over 1.5 m deep (mm)
n.e. 100	450	600
100–150	500	650
150–225	600	750
225–300	650	800
300–400	750	900
400–450	900	1050
450–600	1100	1300

DRAINAGE

DRAINAGE BELOW GROUND AND LAND DRAINAGE

Flow of Water Which Can Be Carried by Various Sizes of Pipe

Clay or concrete pipes

Pipe size	Gradient of pipeline							
	1:10	1:20	1:30	1:40	1:50	1:60	1:80	1:100
	Flow in litres per second							
DN 100 15.0	8.5	6.8	5.8	5.2	4.7	4.0	3.5	
DN 150 28.0	19.0	16.0	14.0	12.0	11.0	9.1	8.0	
DN 225 140.0	95.0	76.0	66.0	58.0	53.0	46.0	40.0	

Plastic pipes

Pipe size	Gradient of pipeline							
	1:10	1:20	1:30	1:40	1:50	1:60	1:80	1:100
	Flow in litres per second							
82.4 mm i/dia.	12.0	8.5	6.8	5.8	5.2	4.7	4.0	3.5
110 mm i/dia.	28.0	19.0	16.0	14.0	12.0	11.0	9.1	8.0
160 mm i/dia.	76.0	53.0	43.0	37.0	33.0	29.0	25.0	22.0
200 mm i/dia.	140.0	95.0	76.0	66.0	58.0	53.0	46.0	40.0

Vitrified (Perforated) Clay Pipes and Fittings to BS En 295-5 1994

Length not specified		
75 mm bore	250 mm bore	600 mm bore
100	300	700
125	350	800
150	400	1000
200	450	1200
225	500	

Precast Concrete Pipes: Prestressed Non-pressure Pipes and Fittings: Flexible Joints to BS 5911: Pt. 103: 1994

Rationalized metric nominal sizes: 450, 500	
Length:	500–1000 by 100 increments
	1000–2200 by 200 increments
	2200–2800 by 300 increments
Angles: length:	450–600 angles 45, 22.5,11.25°
	600 or more angles 22.5, 11.25°

DRAINAGE

Precast Concrete Pipes: Unreinforced and Circular Manholes and Soakaways to BS 5911: Pt. 200: 1994

Nominal sizes:	
Shafts:	675, 900 mm
Chambers:	900, 1050, 1200, 1350, 1500, 1800, 2100, 2400, 2700, 3000 mm
Large chambers:	To have either tapered reducing rings or a flat reducing slab in order to accept the standard cover
Ring depths:	1. 300–1200 mm by 300 mm increments except for bottom slab and rings below cover slab, these are by 150 mm increments
	2. 250–1000 mm by 250 mm increments except for bottom slab and rings below cover slab, these are by 125 mm increments
Access hole:	750 × 750 mm for DN 1050 chamber
	1200 × 675 mm for DN 1350 chamber

Calculation of Soakaway Depth

The following formula determines the depth of concrete ring soakaway that would be required for draining given amounts of water.

$$h = \frac{4ar}{3\pi D^2}$$

h = depth of the chamber below the invert pipe
a = the area to be drained
r = the hourly rate of rainfall (50 mm per hour)
π = pi
D = internal diameter of the soakaway

This table shows the depth of chambers in each ring size which would be required to contain the volume of water specified. These allow a recommended storage capacity of $\frac{1}{3}$ (one third of the hourly rainfall figure).

Table Showing Required Depth of Concrete Ring Chambers in Metres

Area m^2	50	100	150	200	300	400	500
Ring size							
0.9	1.31	2.62	3.93	5.24	7.86	10.48	13.10
1.1	0.96	1.92	2.89	3.85	5.77	7.70	9.62
1.2	0.74	1.47	2.21	2.95	4.42	5.89	7.37
1.4	0.58	1.16	1.75	2.33	3.49	4.66	5.82
1.5	0.47	0.94	1.41	1.89	2.83	3.77	4.72
1.8	0.33	0.65	0.98	1.31	1.96	2.62	3.27
2.1	0.24	0.48	0.72	0.96	1.44	1.92	2.41
2.4	0.18	0.37	0.55	0.74	1.11	1.47	1.84
2.7	0.15	0.29	0.44	0.58	0.87	1.16	1.46
3.0	0.12	0.24	0.35	0.47	0.71	0.94	1.18

DRAINAGE

Precast Concrete Inspection Chambers and Gullies to BS 5911: Part 230: 1994

Nominal sizes:	375 diameter, 750, 900 mm deep
	450 diameter, 750, 900, 1050, 1200 mm deep
Depths:	from the top for trapped or untrapped units:
	centre of outlet 300 mm
	invert (bottom) of the outlet pipe 400 mm
Depth of water seal for trapped gullies:	
	85 mm, rodding eye int. dia. 100 mm
Cover slab:	65 mm min

Bedding Flexible Pipes: PVC-u Or Ductile Iron

Type 1 =	100 mm fill below pipe, 300 mm above pipe: single size material
Type 2 =	100 mm fill below pipe, 300 mm above pipe: single size or graded material
Type 3 =	100 mm fill below pipe, 75 mm above pipe with concrete protective slab over
Type 4 =	100 mm fill below pipe, fill laid level with top of pipe
Type 5 =	200 mm fill below pipe, fill laid level with top of pipe
Concrete =	25 mm sand blinding to bottom of trench, pipe supported on chocks, 100 mm concrete under the pipe, 150 mm concrete over the pipe

DRAINAGE

Bedding Rigid Pipes: Clay or Concrete
(for vitrified clay pipes the manufacturer should be consulted)

Class D:	Pipe laid on natural ground with cut-outs for joints, soil screened to remove stones over 40 mm and returned over pipe to 150 m min depth. Suitable for firm ground with trenches trimmed by hand.
Class N:	Pipe laid on 50 mm granular material of graded aggregate to Table 4 of BS 882, or 10 mm aggregate to Table 6 of BS 882, or as dug light soil (not clay) screened to remove stones over 10 mm. Suitable for machine dug trenches.
Class B:	As Class N, but with granular bedding extending half way up the pipe diameter.
Class F:	Pipe laid on 100 mm granular fill to BS 882 below pipe, minimum 150 mm granular fill above pipe: single size material. Suitable for machine dug trenches.
Class A:	Concrete 100 mm thick under the pipe extending half way up the pipe, backfilled with the appropriate class of fill. Used where there is only a very shallow fall to the drain. Class A bedding allows the pipes to be laid to an exact gradient.
Concrete surround:	25 mm sand blinding to bottom of trench, pipe supported on chocks, 100 mm concrete under the pipe, 150 mm concrete over the pipe. It is preferable to bed pipes under slabs or wall in granular material.

PIPED SUPPLY SYSTEMS

Identification of Service Tubes From Utility to Dwellings

Utility	Colour	Size	Depth
British Telecom	grey	54 mm od	450 mm
Electricity	black	38 mm od	450 mm
Gas	yellow	42 mm od rigid 60 mm od convoluted	450 mm
Water	may be blue	(normally untubed)	750 mm

ELECTRICAL SUPPLY/POWER/LIGHTING SYSTEMS

ELECTRICAL SUPPLY/POWER/LIGHTING SYSTEMS

Electrical Insulation Class En 60.598 BS 4533

Class 1:	luminaires comply with class 1 (I) earthed electrical requirements
Class 2:	luminaires comply with class 2 (II) double insulated electrical requirements
Class 3:	luminaires comply with class 3 (III) electrical requirements

Protection to Light Fittings

BS EN 60529:1992 Classification for degrees of protection provided by enclosures.
(IP Code – International or Ingress Protection)

1st characteristic: against ingress of solid foreign objects		
The figure	2	indicates that fingers cannot enter
	3	that a 2.5 mm diameter probe cannot enter
	4	that a 1.0 mm diameter probe cannot enter
	5	the fitting is dust proof (no dust around live parts)
	6	the fitting is dust tight (no dust entry)
2nd characteristic: ingress of water with harmful effects		
The figure	0	indicates unprotected
	1	vertically dripping water cannot enter
	2	water dripping 15° (tilt) cannot enter
	3	spraying water cannot enter
	4	splashing water cannot enter
	5	jetting water cannot enter
	6	powerful jetting water cannot enter
	7	proof against temporary immersion
	8	proof against continuous immersion
Optional additional codes:		A–D protects against access to hazardous parts
	H	High voltage apparatus
	M	fitting was in motion during water test
	S	fitting was static during water test
	W	protects against weather
Marking code arrangement:		(example) IPX5S = IP (International or Ingress Protection)
		X (denotes omission of first characteristic)
		5 = jetting
		S = static during water test

RAIL TRACKS

RAIL TRACKS

	kg/m of track	lb/ft of track
Standard gauge Bull head rails, chairs, transverse timber (softwood) sleepers etc.	245	165
Main lines Flat-bottom rails, transverse prestressed concrete sleepers, etc.	418	280
Add for electric third rail	51	35
Add for crushed stone ballast	2600	1750
	kg/m²	**lb/sq ft**
Overall average weight – rails connections, sleepers, ballast, etc.	733	150
	kg/m of track	**lb/ft of track**
Bridge rails, longitudinal timber sleepers, etc.	112	75

RAIL TRACKS

Heavy Rails

British Standard Section No.	Rail height (mm)	Foot width (mm)	Head width (mm)	Min web thickness (mm)	Section weight (kg/m)
Flat Bottom Rails					
60 A	114.30	109.54	57.15	11.11	30.62
70 A	123.82	111.12	60.32	12.30	34.81
75 A	128.59	114.30	61.91	12.70	37.45
80 A	133.35	117.47	63.50	13.10	39.76
90 A	142.88	127.00	66.67	13.89	45.10
95 A	147.64	130.17	69.85	14.68	47.31
100 A	152.40	133.35	69.85	15.08	50.18
110 A	158.75	139.70	69.85	15.87	54.52
113 A	158.75	139.70	69.85	20.00	56.22
50 'O'	100.01	100.01	52.39	10.32	24.82
80 'O'	127.00	127.00	63.50	13.89	39.74
60R	114.30	109.54	57.15	11.11	29.85
75R	128.59	122.24	61.91	13.10	37.09
80R	133.35	127.00	63.50	13.49	39.72
90R	142.88	136.53	66.67	13.89	44.58
95R	147.64	141.29	68.26	14.29	47.21
100R	152.40	146.05	69.85	14.29	49.60
95N	147.64	139.70	69.85	13.89	47.27
Bull Head Rails					
95R BH	145.26	69.85	69.85	19.05	47.07

Light Rails

British Standard Section No.	Rail height (mm)	Foot width (mm)	Head width (mm)	Min web thickness (mm)	Section weight (kg/m)
Flat Bottom Rails					
20M	65.09	55.56	30.96	6.75	9.88
30M	75.41	69.85	38.10	9.13	14.79
35M	80.96	76.20	42.86	9.13	17.39
35R	85.73	82.55	44.45	8.33	17.40
40	88.11	80.57	45.64	12.3	19.89
Bridge Rails					
13	48.00	92	36.00	18.0	13.31
16	54.00	108	44.50	16.0	16.06
20	55.50	127	50.00	20.5	19.86
28	67.00	152	50.00	31.0	28.62
35	76.00	160	58.00	34.5	35.38
50	76.00	165	58.50	–	50.18
Crane Rails					
A65	75.00	175.00	65.00	38.0	43.10
A75	85.00	200.00	75.00	45.0	56.20
A100	95.00	200.00	100.00	60.0	74.30
A120	105.00	220.00	120.00	72.0	100.00
175CR	152.40	152.40	107.95	38.1	86.92

RAIL TRACKS

Fish Plates

British Standard Section No.	Overall plate length		Hole diameter	Finished weight per pair	
	4 Hole (mm)	6 Hole (mm)	(mm)	4 Hole (kg/pair)	6 Hole (kg/pair)
For British Standard Heavy Rails: Flat Bottom Rails					
60 A	406.40	609.60	20.64	9.87	14.76
70 A	406.40	609.60	22.22	11.15	16.65
75 A	406.40	–	23.81	11.82	17.73
80 A	406.40	609.60	23.81	13.15	19.72
90 A	457.20	685.80	25.40	17.49	26.23
100 A	508.00	–	pear	25.02	–
110 A (shallow)	507.00	–	27.00	30.11	54.64
113 A (heavy)	507.00	–	27.00	30.11	54.64
50 'O' (shallow)	406.40	–	–	6.68	10.14
80 'O' (shallow)	495.30	–	23.81	14.72	22.69
60R (shallow)	406.40	609.60	20.64	8.76	13.13
60R (angled)	406.40	609.60	20.64	11.27	16.90
75R (shallow)	406.40	–	23.81	10.94	16.42
75R (angled)	406.40	–	23.81	13.67	–
80R (shallow)	406.40	609.60	23.81	11.93	17.89
80R (angled)	406.40	609.60	23.81	14.90	22.33
For British Standard Heavy Rails: Bull Head Rails					
95R BH (shallow)	–	457.20	27.00	14.59	14.61
For British Standard Light Rails: Flat Bottom Rails					
30M	355.6	–	–	–	2.72
35M	355.6	–	–	–	2.83
40	355.6	–	–	3.76	–

FRACTIONS, DECIMALS AND MILLIMETRE EQUIVALENTS

Fractions	Decimals	(mm)		Fractions	Decimals	(mm)
1/64	0.015625	0.396875		33/64	0.515625	13.096875
1/32	0.03125	0.79375		17/32	0.53125	13.49375
3/64	0.046875	1.190625		35/64	0.546875	13.890625
1/16	0.0625	1.5875		9/16	0.5625	14.2875
5/64	0.078125	1.984375		37/64	0.578125	14.684375
3/32	0.09375	2.38125		19/32	0.59375	15.08125
7/64	0.109375	2.778125		39/64	0.609375	15.478125
1/8	0.125	3.175		5/8	0.625	15.875
9/64	0.140625	3.571875		41/64	0.640625	16.271875
5/32	0.15625	3.96875		21/32	0.65625	16.66875
11/64	0.171875	4.365625		43/64	0.671875	17.065625
3/16	0.1875	4.7625		11/16	0.6875	17.4625
13/64	0.203125	5.159375		45/64	0.703125	17.859375
7/32	0.21875	5.55625		23/32	0.71875	18.25625
15/64	0.234375	5.953125		47/64	0.734375	18.653125
1/4	0.25	6.35		3/4	0.75	19.05
17/64	0.265625	6.746875		49/64	0.765625	19.446875
9/32	0.28125	7.14375		25/32	0.78125	19.84375
19/64	0.296875	7.540625		51/64	0.796875	20.240625
5/16	0.3125	7.9375		13/16	0.8125	20.6375
21/64	0.328125	8.334375		53/64	0.828125	21.034375
11/32	0.34375	8.73125		27/32	0.84375	21.43125
23/64	0.359375	9.128125		55/64	0.859375	21.828125
3/8	0.375	9.525		7/8	0.875	22.225
25/64	0.390625	9.921875		57/64	0.890625	22.621875
13/32	0.40625	10.31875		29/32	0.90625	23.01875
27/64	0.421875	10.71563		59/64	0.921875	23.415625
7/16	0.4375	11.1125		15/16	0.9375	23.8125
29/64	0.453125	11.50938		61/64	0.953125	24.209375
15/32	0.46875	11.90625		31/32	0.96875	24.60625
31/64	0.484375	12.30313		63/64	0.984375	25.003125
1/2	0.5	12.7		1.0	1	25.4

IMPERIAL STANDARD WIRE GAUGE (SWG)

IMPERIAL STANDARD WIRE GAUGE (SWG)

SWG No.	Diameter (inches)	Diameter (mm)	SWG No.	Diameter (inches)	Diameter (mm)
7/0	0.5	12.7	23	0.024	0.61
6/0	0.464	11.79	24	0.022	0.559
5/0	0.432	10.97	25	0.02	0.508
4/0	0.4	10.16	26	0.018	0.457
3/0	0.372	9.45	27	0.0164	0.417
2/0	0.348	8.84	28	0.0148	0.376
1/0	0.324	8.23	29	0.0136	0.345
1	0.3	7.62	30	0.0124	0.315
2	0.276	7.01	31	0.0116	0.295
3	0.252	6.4	32	0.0108	0.274
4	0.232	5.89	33	0.01	0.254
5	0.212	5.38	34	0.009	0.234
6	0.192	4.88	35	0.008	0.213
7	0.176	4.47	36	0.008	0.193
8	0.16	4.06	37	0.007	0.173
9	0.144	3.66	38	0.006	0.152
10	0.128	3.25	39	0.005	0.132
11	0.116	2.95	40	0.005	0.122
12	0.104	2.64	41	0.004	0.112
13	0.092	2.34	42	0.004	0.102
14	0.08	2.03	43	0.004	0.091
15	0.072	1.83	44	0.003	0.081
16	0.064	1.63	45	0.003	0.071
17	0.056	1.42	46	0.002	0.061
18	0.048	1.22	47	0.002	0.051
19	0.04	1.016	48	0.002	0.041
20	0.036	0.914	49	0.001	0.031
21	0.032	0.813	50	0.001	0.025
22	0.028	0.711			

PIPES, WATER, STORAGE, INSULATION

WATER PRESSURE DUE TO HEIGHT

Imperial

Head (feet)	Pressure (lb/in²)		Head (feet)	Pressure (lb/in²)
1	0.43		70	30.35
5	2.17		75	32.51
10	4.34		80	34.68
15	6.5		85	36.85
20	8.67		90	39.02
25	10.84		95	41.18
30	13.01		100	43.35
35	15.17		105	45.52
40	17.34		110	47.69
45	19.51		120	52.02
50	21.68		130	56.36
55	23.84		140	60.69
60	26.01		150	65.03
65	28.18			

Metric

Head (m)	Pressure (bar)		Head (m)	Pressure (bar)
0.5	0.049		18.0	1.766
1.0	0.098		19.0	1.864
1.5	0.147		20.0	1.962
2.0	0.196		21.0	2.06
3.0	0.294		22.0	2.158
4.0	0.392		23.0	2.256
5.0	0.491		24.0	2.354
6.0	0.589		25.0	2.453
7.0	0.687		26.0	2.551
8.0	0.785		27.0	2.649
9.0	0.883		28.0	2.747
10.0	0.981		29.0	2.845
11.0	1.079		30.0	2.943
12.0	1.177		32.5	3.188
13.0	1.275		35.0	3.434
14.0	1.373		37.5	3.679
15.0	1.472		40.0	3.924
16.0	1.57		42.5	4.169
17.0	1.668		45.0	4.415

1 bar	=	14.5038 lbf/in²
1 lbf/in²	=	0.06895 bar
1 metre	=	3.2808 ft or 39.3701 in
1 foot	=	0.3048 metres
1 in wg	=	2.5 mbar (249.1 N/m²)

PIPES, WATER, STORAGE, INSULATION

Dimensions and Weights of Copper Pipes to BSEN 1057, BSEN 12499, BSEN 14251

Outside diameter (mm)	Internal diameter (mm)	Weight per metre (kg)	Internal diameter (mm)	Weight per metre (kg)	Internal diameter (mm)	Weight per metre (kg)
	Formerly Table X		Formerly Table Y		Formerly Table Z	
6	4.80	0.0911	4.40	0.1170	5.00	0.0774
8	6.80	0.1246	6.40	0.1617	7.00	0.1054
10	8.80	0.1580	8.40	0.2064	9.00	0.1334
12	10.80	0.1914	10.40	0.2511	11.00	0.1612
15	13.60	0.2796	13.00	0.3923	14.00	0.2031
18	16.40	0.3852	16.00	0.4760	16.80	0.2918
22	20.22	0.5308	19.62	0.6974	20.82	0.3589
28	26.22	0.6814	25.62	0.8985	26.82	0.4594
35	32.63	1.1334	32.03	1.4085	33.63	0.6701
42	39.63	1.3675	39.03	1.6996	40.43	0.9216
54	51.63	1.7691	50.03	2.9052	52.23	1.3343
76.1	73.22	3.1287	72.22	4.1437	73.82	2.5131
108	105.12	4.4666	103.12	7.3745	105.72	3.5834
133	130.38	5.5151	–	–	130.38	5.5151
159	155.38	8.7795	–	–	156.38	6.6056

Dimensions of Stainless Steel Pipes to BS 4127

Outside diameter (mm)	Maximum outside diameter (mm)	Minimum outside diameter (mm)	Wall thickness (mm)	Working pressure (bar)
6	6.045	5.940	0.6	330
8	8.045	7.940	0.6	260
10	10.045	9.940	0.6	210
12	12.045	11.940	0.6	170
15	15.045	14.940	0.6	140
18	18.045	17.940	0.7	135
22	22.055	21.950	0.7	110
28	28.055	27.950	0.8	121
35	35.070	34.965	1.0	100
42	42.070	41.965	1.1	91
54	54.090	53.940	1.2	77

PIPES, WATER, STORAGE, INSULATION

Dimensions of Steel Pipes to BS 1387

Nominal size	Approx. outside diameter	Outside diameter				Thickness		
		Light		Medium & heavy		Light	Medium	Heavy
		Max	Min	Max	Min			
(mm)	(mm)	(mm)	(mm)	(mm)	(mm)	(mm)	(mm)	(mm)
6	10.20	10.10	9.70	10.40	9.80	1.80	2.00	2.65
8	13.50	13.60	13.20	13.90	13.30	1.80	2.35	2.90
10	17.20	17.10	16.70	17.40	16.80	1.80	2.35	2.90
15	21.30	21.40	21.00	21.70	21.10	2.00	2.65	3.25
20	26.90	26.90	26.40	27.20	26.60	2.35	2.65	3.25
25	33.70	33.80	33.20	34.20	33.40	2.65	3.25	4.05
32	42.40	42.50	41.90	42.90	42.10	2.65	3.25	4.05
40	48.30	48.40	47.80	48.80	48.00	2.90	3.25	4.05
50	60.30	60.20	59.60	60.80	59.80	2.90	3.65	4.50
65	76.10	76.00	75.20	76.60	75.40	3.25	3.65	4.50
80	88.90	88.70	87.90	89.50	88.10	3.25	4.05	4.85
100	114.30	113.90	113.00	114.90	113.30	3.65	4.50	5.40
125	139.70	–	–	140.60	138.70	–	4.85	5.40
150	165.1*	–	–	166.10	164.10	–	4.85	5.40

* 165.1 mm (6.5 in) outside diameter is not generally recommended except where screwing to BS 21 is necessary
All dimensions are in accordance with ISO R65 except approximate outside diameters which are in accordance with ISO R64
Light quality is equivalent to ISO R65 Light Series II

Approximate Metres Per Tonne of Tubes to BS 1387

Nom. Size	BLACK						GALVANIZED					
	Plain/screwed ends			Screwed & socketed			Plain/screwed ends			Screwed & socketed		
	L	M	H	L	M	H	L	M	H	L	M	H
(mm)	(m)	(m)	(m)	(m)	(m)	(m)	(m)	(m)	(m)	(m)	(m)	(m)
6	2765	2461	2030	2743	2443	2018	2604	2333	1948	2584	2317	1937
8	1936	1538	1300	1920	1527	1292	1826	1467	1254	1811	1458	1247
10	1483	1173	979	1471	1165	974	1400	1120	944	1386	1113	939
15	1050	817	688	1040	811	684	996	785	665	987	779	661
20	712	634	529	704	628	525	679	609	512	673	603	508
25	498	410	336	494	407	334	478	396	327	474	394	325
32	388	319	260	384	316	259	373	308	254	369	305	252
40	307	277	226	303	273	223	296	268	220	292	264	217
50	244	196	162	239	194	160	235	191	158	231	188	157
65	172	153	127	169	151	125	167	149	124	163	146	122
80	147	118	99	143	116	98	142	115	97	139	113	96
100	101	82	69	98	81	68	98	81	68	95	79	67
125	–	62	56	–	60	55	–	60	55	–	59	54
150	–	52	47	–	50	46	–	51	46	–	49	45

The figures for 'plain or screwed ends' apply also to tubes to BS 1775 of equivalent size and thickness
Key:
L – Light
M – Medium
H – Heavy

PIPES, WATER, STORAGE, INSULATION

Flange Dimension Chart to BS 4504 & BS 10

Normal Pressure Rating (PN 6) 6 Bar

Nom. Size	Flange Outside Dia.	Table 6/2 Forged Welding Neck	Table 6/3 Plate Slip on	Table 6/4 Forged Bossed Screwed	Table 6/5 Forged Bossed Slip on	Table 6/8 Plate Blank	Raised Face Dia.	Raised Face T'ness	Nr. Bolt Hole	Size of Bolt
15	80	12	12	12	12	12	40	2	4	M10 × 40
20	90	14	14	14	14	14	50	2	4	M10 × 45
25	100	14	14	14	14	14	60	2	4	M10 × 45
32	120	14	16	14	14	14	70	2	4	M12 × 45
40	130	14	16	14	14	14	80	3	4	M12 × 45
50	140	14	16	14	14	14	90	3	4	M12 × 45
65	160	14	16	14	14	14	110	3	4	M12 × 45
80	190	16	18	16	16	16	128	3	4	M16 × 55
100	210	16	18	16	16	16	148	3	4	M16 × 55
125	240	18	20	18	18	18	178	3	8	M16 × 60
150	265	18	20	18	18	18	202	3	8	M16 × 60
200	320	20	22	–	20	20	258	3	8	M16 × 60
250	375	22	24	–	22	22	312	3	12	M16 × 65
300	440	22	24	–	22	22	365	4	12	M20 × 70

Normal Pressure Rating (PN 16) 16 Bar

Nom. Size	Flange Outside Dia.	Table 6/2 Forged Welding Neck	Table 6/3 Plate Slip on	Table 6/4 Forged Bossed Screwed	Table 6/5 Forged Bossed Slip on	Table 6/8 Plate Blank	Raised Face Dia.	Raised Face T'ness	Nr. Bolt Hole	Size of Bolt
15	95	14	14	14	14	14	45	2	4	M12 × 45
20	105	16	16	16	16	16	58	2	4	M12 × 50
25	115	16	16	16	16	16	68	2	4	M12 × 50
32	140	16	16	16	16	16	78	2	4	M16 × 55
40	150	16	16	16	16	16	88	3	4	M16 × 55
50	165	18	18	18	18	18	102	3	4	M16 × 60
65	185	18	18	18	18	18	122	3	4	M16 × 60
80	200	20	20	20	20	20	138	3	8	M16 × 60
100	220	20	20	20	20	20	158	3	8	M16 × 65
125	250	22	22	22	22	22	188	3	8	M16 × 70
150	285	22	22	22	22	22	212	3	8	M20 × 70
200	340	24	24	–	24	24	268	3	12	M20 × 75
250	405	26	26	–	26	26	320	3	12	M24 × 90
300	460	28	28	–	28	28	378	4	12	M24 × 90

PIPES, WATER, STORAGE, INSULATION

Minimum Distances Between Supports/Fixings

Material	BS Nominal pipe size		Pipes – Vertical	Pipes – Horizontal on to low gradients
	(inch)	(mm)	Support distance in metres	Support distance in metres
Copper	0.50	15.00	1.90	1.30
	0.75	22.00	2.50	1.90
	1.00	28.00	2.50	1.90
	1.25	35.00	2.80	2.50
	1.50	42.00	2.80	2.50
	2.00	54.00	3.90	2.50
	2.50	67.00	3.90	2.80
	3.00	76.10	3.90	2.80
	4.00	108.00	3.90	2.80
	5.00	133.00	3.90	2.80
	6.00	159.00	3.90	2.80
muPVC	1.25	32.00	1.20	0.50
	1.50	40.00	1.20	0.50
	2.00	50.00	1.20	0.60
Polypropylene	1.25	32.00	1.20	0.50
	1.50	40.00	1.20	0.50
uPVC	–	82.40	1.20	0.50
	–	110.00	1.80	0.90
	–	160.00	1.80	1.20
Steel	0.50	15.00	2.40	1.80
	0.75	20.00	3.00	2.40
	1.00	25.00	3.00	2.40
	1.25	32.00	3.00	2.40
	1.50	40.00	3.70	2.40
	2.00	50.00	3.70	2.40
	2.50	65.00	4.60	3.00
	3.00	80.40	4.60	3.00
	4.00	100.00	4.60	3.00
	5.00	125.00	5.50	3.70
	6.00	150.00	5.50	4.50
	8.00	200.00	8.50	6.00
	10.00	250.00	9.00	6.50
	12.00	300.00	10.00	7.00
	16.00	400.00	10.00	8.25

PIPES, WATER, STORAGE, INSULATION

Litres of Water Storage Required Per Person Per Building Type

Type of building	Storage (litres)
Houses and flats (up to 4 bedrooms)	120/bedroom
Houses and flats (more than 4 bedrooms)	100/bedroom
Hostels	90/bed
Hotels	200/bed
Nurses homes and medical quarters	120/bed
Offices with canteen	45/person
Offices without canteen	40/person
Restaurants	7/meal
Boarding schools	90/person
Day schools – Primary	15/person
Day schools – Secondary	20/person

Recommended Air Conditioning Design Loads

Building type	Design loading
Computer rooms	500 W/m² of floor area
Restaurants	150 W/m² of floor area
Banks (main area)	100 W/m² of floor area
Supermarkets	25 W/m² of floor area
Large office block (exterior zone)	100 W/m² of floor area
Large office block (interior zone)	80 W/m² of floor area
Small office block (interior zone)	80 W/m² of floor area

PIPES, WATER, STORAGE, INSULATION

Capacity and Dimensions of Galvanized Mild Steel Cisterns – BS 417

Capacity (litres)	BS type (SCM)	Dimensions		
		Length (mm)	Width (mm)	Depth (mm)
18	45	457	305	305
36	70	610	305	371
54	90	610	406	371
68	110	610	432	432
86	135	610	457	482
114	180	686	508	508
159	230	736	559	559
191	270	762	584	610
227	320	914	610	584
264	360	914	660	610
327	450/1	1220	610	610
336	450/2	965	686	686
423	570	965	762	787
491	680	1090	864	736
709	910	1070	889	889

Capacity of Cold Water Polypropylene Storage Cisterns – BS 4213

Capacity (litres)	BS type (PC)	Maximum height (mm)
18	4	310
36	8	380
68	15	430
91	20	510
114	25	530
182	40	610
227	50	660
273	60	660
318	70	660
455	100	760

PIPES, WATER, STORAGE, INSULATION

Minimum Insulation Thickness to Protect Against Freezing for Domestic Cold Water Systems (8 Hour Evaluation Period)

Pipe size (mm)	Insulation thickness (mm)					
	Condition 1			Condition 2		
	λ = 0.020	λ = 0.030	λ = 0.040	λ = 0.020	λ = 0.030	λ = 0.040
Copper pipes						
15	11	20	34	12	23	41
22	6	9	13	6	10	15
28	4	6	9	4	7	10
35	3	5	7	4	5	7
42	3	4	5	8	4	6
54	2	3	4	2	3	4
76	2	2	3	2	2	3
Steel pipes						
15	9	15	24	10	18	29
20	6	9	13	6	10	15
25	4	7	9	5	7	10
32	3	5	6	3	5	7
40	3	4	5	3	4	6
50	2	3	4	2	3	4
65	2	2	3	2	3	3

Condition 1: water temperature 7°C; ambient temperature −6°C; evaluation period 8 h; permitted ice formation 50%; normal installation, i.e. inside the building and inside the envelope of the structural insulation
Condition 2: water temperature 2°C; ambient temperature −6°C; evaluation period 8 h; permitted ice formation 50%; extreme installation, i.e. inside the building but outside the envelope of the structural insulation
λ = thermal conductivity [W/(mK)]

Insulation Thickness for Chilled And Cold Water Supplies to Prevent Condensation

On a Low Emissivity Outer Surface (0.05, i.e. Bright Reinforced Aluminium Foil) with an Ambient Temperature of +25°C and a Relative Humidity of 80%

Steel pipe size (mm)	t = +10			t = +5			t = 0		
	Insulation thickness (mm)			Insulation thickness (mm)			Insulation thickness (mm)		
	λ = 0.030	λ = 0.040	λ = 0.050	λ = 0.030	λ = 0.040	λ = 0.050	λ = 0.030	λ = 0.040	λ = 0.050
15	16	20	25	22	28	34	28	36	43
25	18	24	29	25	32	39	32	41	50
50	22	28	34	30	39	47	38	49	60
100	26	34	41	36	47	57	46	60	73
150	29	38	46	40	52	64	51	67	82
250	33	43	53	46	60	74	59	77	94
Flat surfaces	39	52	65	56	75	93	73	97	122

t = temperature of contents (°C)
λ = thermal conductivity at mean temperature of insulation [W/(mK)]

PIPES, WATER, STORAGE, INSULATION

Insulation Thickness for Non-domestic Heating Installations to Control Heat Loss

Steel pipe size (mm)	t = 75			t = 100			t = 150		
	Insulation thickness (mm)			Insulation thickness (mm)			Insulation thickness (mm)		
	λ = 0.030	λ = 0.040	λ = 0.050	λ = 0.030	λ = 0.040	λ = 0.050	λ = 0.030	λ = 0.040	λ = 0.050
10	18	32	55	20	36	62	23	44	77
15	19	34	56	21	38	64	26	47	80
20	21	36	57	23	40	65	28	50	83
25	23	38	58	26	43	68	31	53	85
32	24	39	59	28	45	69	33	55	87
40	25	40	60	29	47	70	35	57	88
50	27	42	61	31	49	72	37	59	90
65	29	43	62	33	51	74	40	63	92
80	30	44	62	35	52	75	42	65	94
100	31	46	63	37	54	76	45	68	96
150	33	48	64	40	57	77	50	73	100
200	35	49	65	42	59	79	53	76	103
250	36	50	66	43	61	80	55	78	105

t = hot face temperature (°C)
λ = thermal conductivity at mean temperature of insulation [W/(mK)]

Index

Understanding JCT Standard Building Contracts

Ninth Edition

David Chappell

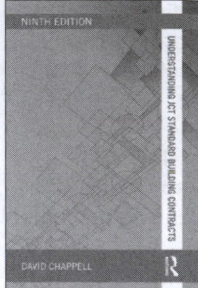

This ninth edition of David Chappell's bestselling guide has been revised to take into account changes made in 2011 to payment provisions, and elsewhere. This remains the most concise guide available to the most commonly used JCT building contracts: Standard Building Contract with quantities, 2011 (SBC11), Intermediate Building Contract 2011 (IC11), Intermediate Building Contract with contractor's design 2011 (ICD11), Minor Works Building Contract 2011 (MW11), Minor Works Building Contract with contractor's design 2011 (MWD11) and Design and Build Contract 2011 (DB11).

Chappell avoids legal jargon but writes with authority and precision. Architects, quantity surveyors, contractors and students of these professions will find this a practical and affordable reference tool arranged by topic.

April 2012: 234 x 156: 160 pp
Pb: 978-0-415-50890-2: £24.99

To Order: Tel: +44 (0) 1235 400524 Fax: +44 (0) 1235 400525
or Post: Taylor and Francis Customer Services,
Bookpoint Ltd, Unit T1, 200 Milton Park, Abingdon, Oxon, OX14 4TA UK
Email: book.orders@tandf.co.uk

For a complete listing of all our titles visit:
www.tandf.co.uk

Understanding the NEC3 ECC Contract

A Practical Handbook

Kelvin Hughes

As usage of the NEC (formerly the New Engineering Contract) family of contracts continues to grow worldwide, so does the importance of understanding its clauses and nuances to everyone working in the built environment. Currently in its third edition, this set of contracts is different to others in concept as well as format, so users may well find themselves needing a helping hand along the way.

Understanding the NEC3 ECC Contract uses plain English to lead the reader through the NEC3 Engineering and Construction Contract's key features, including:

*   main and secondary options
*   the use of early warnings
*   programme provisions
*   payment
*   compensation events
*   preparing and assessing tenders.

Common problems experienced when using the Engineering and Construction Contract are signalled to the reader throughout, and the correct way of reading each clause explained. The way the contract effects procurement processes, dispute resolution, project management, and risk management are all addressed in order to direct the user to best practice.

Written for construction professionals, by a practicing international construction contract consultant, this handbook is the most straightforward, balanced and practical guide to the NEC3 ECC available. An ideal companion for employers, contractors, project managers, supervisors, engineers, architects, quantity surveyors, subcontractors, and anyone else interested in working successfully with the NEC3 ECC.

October 2012: 234 x156: 272 pp
Pb: 978-0-415-61496-2: £29.99

To Order: Tel: +44 (0) 1235 400524 Fax: +44 (0) 1235 400525
or Post: Taylor and Francis Customer Services,
Bookpoint Ltd, Unit T1, 200 Milton Park, Abingdon, Oxon, OX14 4TA UK
Email: book.orders@tandf.co.uk

For a complete listing of all our titles visit:
www.crcpress.com

Whole Life Costing
A New Approach

Peter Caplehorn

Whole life costing has been a subject waiting to come of age for many years. What was previously of mainly academic interest is now becoming a key business tool in the procurement and construction of significant projects.

With the advent of PPP and in particular of PFI, details of the project life need to be assessed and tied in to funding and operation plans. Many of these projects run to millions of pounds and are of high political or social importance, so the implications of the life of materials is crucial. A fundamental requirement of these procurement routes has been that the whole enterprise should be included within the bid, so that a company takes on not only the construction but also the running and maintenance of any building.

Additionally as sustainability has emerged and grown in importance, so has the need for a whole life time costing approach, partly driven by governmental insistence. At the heart of sustainability is an understanding of what the specification means for the future of the building and how it will affect the environment. Whole life costing considers part of this and provides an understanding of how materials may perform and what allowances are needed at the end of their life.

This book sets out the practical issues involved in the selection of materials, their performance, and the issues that need to be taken into account. The emphasis, unlike in other publications, is not to formularise or to package the issues but to leave the reader with a clear understanding and a sensible practical way of arriving at conclusions in the future.

June 2012: 246x174: 160 pp
Hb: 978-0-415-43422-5 £105.00
Pb: 978-0-415-43423-2: £34.99

To Order: Tel: +44 (0) 1235 400524 Fax: +44 (0) 1235 400525
or Post: Taylor and Francis Customer Services,
Bookpoint Ltd, Unit T1, 200 Milton Park, Abingdon, Oxon, OX14 4TA UK
Email: book.orders@tandf.co.uk

For a complete listing of all our titles visit:
www.tandf.co.uk

Spon's Asia-Pacific Construction Costs Handbook

Fourth Edition

Davis Langdon

Spon's Asia Pacific Construction Costs Handbook includes construction cost data for twenty countries. This new edition has been extended to include Pakistan and Cambodia. Australia, UK and America are also included, to facilitate comparison with construction costs elsewhere.

Information is presented for each country in the same way, as follows:

- key data on the main economic and construction indicators.
- an outline of the national construction industry, covering structure, tendering and contract procedures, materials cost data, regulations and standards
- labour and materials cost data
- measured rates for a range of standard construction work items
- approximate estimating costs per unit area for a range of building types
- price index data and exchange rate movements against £ sterling, $US and Japanese Yen.

The book also includes a Comparative Data section to facilitate country-to-country comparisons. Figures from the national sections are grouped in tables according to national indicators, construction output, input costs and costs per square metre for factories, offices, warehouses, hospitals, schools, theatres, sports halls, hotels and housing.

This unique handbook will be an essential reference for all construction professionals involved in work outside their own country and for all developers or multinational companies assessing comparative development costs.

April 2010: 234x156: 480pp
Hb: 978-0-415-46565-6: £140.00

To Order: Tel: +44 (0) 1235 400524 Fax: +44 (0) 1235 400525
or Post: Taylor and Francis Customer Services,
Bookpoint Ltd, Unit T1, 200 Milton Park, Abingdon, Oxon, OX14 4TA UK
Email: book.orders@tandf.co.uk

For a complete listing of all our titles visit:
www.tandf.co.uk

Ebook Single-User Licence Agreement

We welcome you as a user of this Spon Price Book ebook and hope that you find it a useful and valuable tool. Please read this document carefully. **This is a legal agreement** between you (hereinafter referred to as the "Licensee") and Taylor and Francis Books Ltd. (the "Publisher"), which defines the terms under which you may use the Product. **By accessing and retrieving the access code on the label inside the front cover of this book you agree to these terms and conditions outlined herein. If you do not agree to these terms you must return the Product to your supplier intact, with the seal on the label unbroken and with the access code not accessed.**

1. **Definition of the Product**
    The product which is the subject of this Agreement, (the "Product") consists of online and offline access to the VitalSource ebook edition of *Spon's Architects' & Builders' Price Book 2014*.

2. **Commencement and licence**
    2.1 This Agreement commences upon the breaking open of the document containing the access code by the Licensee (the "Commencement Date").
    2.2 This is a licence agreement (the "Agreement") for the use of the Product by the Licensee, and not an agreement for sale.
    2.3 The Publisher licenses the Licensee on a non-exclusive and non-transferable basis to use the Product on condition that the Licensee complies with this Agreement. The Licensee acknowledges that it is only permitted to use the Product in accordance with this Agreement.

3. **Multiple use**
    For more than one user or for a wide area network or consortium, use of the Product is only permissible with the purchase from the Publisher of additional access codes..

4. **Installation and Use**
    4.1 The Licensee may provide access to the Product for individual study in the following manner: The Licensee may install the Product on a secure local area network on a single site for use by one user.
    4.2 The Licensee shall be responsible for installing the Product and for the effectiveness of such installation.
    4.3 Text from the Product may be incorporated in a coursepack. Such use is only permissible with the express permission of the Publisher in writing and requires the payment of the appropriate fee as specified by the Publisher and signature of a separate licence agreement.
    4.4 The Product is a free addition to the book and the Publisher is under no obligation to provide any technical support.

5. **Permitted Activities**
    5.1 The Licensee shall be entitled to use the Product for its own internal purposes;
    5.2 The Licensee acknowledges that its rights to use the Product are strictly set out in this Agreement, and all other uses (whether expressly mentioned in Clause 6 below or not) are prohibited.

6. **Prohibited Activities**
    The following are prohibited without the express permission of the Publisher:
    6.1 The commercial exploitation of any part of the Product.
    6.2 The rental, loan, (free or for money or money's worth) or hire purchase of this product, save with the express consent of the Publisher.
    6.3 Any activity which raises the reasonable prospect of impeding the Publisher's ability or opportunities to market the Product.
    6.4 Any networking, physical or electronic distribution or dissemination of the product save as expressly permitted by this Agreement.
    6.5 Any reverse engineering, decompilation, disassembly or other alteration of the Product save in accordance with applicable national laws.
    6.6 The right to create any derivative product or service from the Product save as expressly provided for in this Agreement.
    6.7 Any alteration, amendment, modification or deletion from the Product, whether for the purposes of error correction or otherwise.

7. **General Responsibilities of the License**

7.1 The Licensee will take all reasonable steps to ensure that the Product is used in accordance with the terms and conditions of this Agreement.

7.2 The Licensee acknowledges that damages may not be a sufficient remedy for the Publisher in the event of breach of this Agreement by the Licensee, and that an injunction may be appropriate.

7.3 The Licensee undertakes to keep the Product safe and to use its best endeavours to ensure that the product does not fall into the hands of third parties, whether as a result of theft or otherwise.

7.4 Where information of a confidential nature relating to the product of the business affairs of the Publisher comes into the possession of the Licensee pursuant to this Agreement (or otherwise), the Licensee agrees to use such information solely for the purposes of this Agreement, and under no circumstances to disclose any element of the information to any third party save strictly as permitted under this Agreement. For the avoidance of doubt, the Licensee's obligations under this sub-clause 7.4 shall survive the termination of this Agreement.

8. **Warrant and Liability**

8.1 The Publisher warrants that it has the authority to enter into this agreement and that it has secured all rights and permissions necessary to enable the Licensee to use the Product in accordance with this Agreement.

8.2 The Publisher warrants that the Product as supplied on the Commencement Date shall be free of defects in materials and workmanship, and undertakes to replace any defective Product within 28 days of notice of such defect being received provided such notice is received within 30 days of such supply. As an alternative to replacement, the Publisher agrees fully to refund the Licensee in such circumstances, if the Licensee so requests, provided that the Licensee returns this copy of *Spon's Architects' & Builders' Price Book 2014* to the Publisher. The provisions of this sub-clause 8.2 do not apply where the defect results from an accident or from misuse of the product by the Licensee.

8.3 Sub-clause 8.2 sets out the sole and exclusive remedy of the Licensee in relation to defects in the Product.

8.4 The Publisher and the Licensee acknowledge that the Publisher supplies the Product on an "as is" basis. The Publisher gives no warranties:

8.4.1 that the Product satisfies the individual requirements of the Licensee; or

8.4.2 that the Product is otherwise fit for the Licensee's purpose; or

8.4.3 that the Product is compatible with the Licensee's hardware equipment and software operating environment.

8.5 The Publisher hereby disclaims all warranties and conditions, express or implied, which are not stated above.

8.6 Nothing in this Clause 8 limits the Publisher's liability to the Licensee in the event of death or personal injury resulting from the Publisher's negligence.

8.7 The Publisher hereby excludes liability for loss of revenue, reputation, business, profits, or for indirect or consequential losses, irrespective of whether the Publisher was advised by the Licensee of the potential of such losses.

8.8 The Licensee acknowledges the merit of independently verifying the price book data prior to taking any decisions of material significance (commercial or otherwise) based on such data. It is agreed that the Publisher shall not be liable for any losses which result from the Licensee placing reliance on the data under any circumstances.

8.9 Subject to sub-clause 8.6 above, the Publisher's liability under this Agreement shall be limited to the purchase price.

9. **Intellectual Property Rights**

9.1 Nothing in this Agreement affects the ownership of copyright or other intellectual property rights in the Product.

9.2 The Licensee agrees to display the Publishers' copyright notice in the manner described in the Product.

9.3 The Licensee hereby agrees to abide by copyright and similar notice requirements required by the Publisher, details of which are as follows:
"© 2014 Taylor & Francis. All rights reserved. All materials in *Spon's Architects' & Builders' Price Book 2014* are copyright protected. All rights reserved. No such materials may be used, displayed, modified, adapted, distributed, transmitted, transferred, published or otherwise reproduced in any form or by any means now or hereafter developed other than strictly in accordance with the terms of the licence agreement enclosed with *Spon's Architects' & Builders' Price Book 2014*. However, text and images may be printed and copied for research and private study within the preset program limitations. Please note the copyright notice above, and that any text or images printed or copied must credit the source."

9.4 This Product contains material proprietary to and copyedited by the Publisher and others. Except for the licence granted herein, all rights, title and interest in the Product, in all languages, formats and media throughout the world, including copyrights therein, are and remain the property of the Publisher or other copyright holders identified in the Product.

10. Non-assignment

This Agreement and the licence contained within it may not be assigned to any other person or entity without the written consent of the Publisher.

11. Termination and Consequences of Termination.

11.1 The Publisher shall have the right to terminate this Agreement if:

11.1.1 the Licensee is in material breach of this Agreement and fails to remedy such breach (where capable of remedy) within 14 days of a written notice from the Publisher requiring it to do so; or

11.1.2 the Licensee becomes insolvent, becomes subject to receivership, liquidation or similar external administration; or

11.1.3 the Licensee ceases to operate in business.

11.2 The Licensee shall have the right to terminate this Agreement for any reason upon two month's written notice. The Licensee shall not be entitled to any refund for payments made under this Agreement prior to termination under this sub-clause 11.2.

11.3 Termination by either of the parties is without prejudice to any other rights or remedies under the general law to which they may be entitled, or which survive such termination (including rights of the Publisher under sub-clause 7.4 above).

11.4 Upon termination of this Agreement, or expiry of its terms, the Licensee must destroy all copies and any back up copies of the product or part thereof.

12. General

12.1 *Compliance with export provisions*

The Publisher hereby agrees to comply fully with all relevant export laws and regulations of the United Kingdom to ensure that the Product is not exported, directly or indirectly, in violation of English law.

12.2 *Force majeure*

The parties accept no responsibility for breaches of this Agreement occurring as a result of circumstances beyond their control.

12.3 *No waiver*

Any failure or delay by either party to exercise or enforce any right conferred by this Agreement shall not be deemed to be a waiver of such right.

12.4 *Entire agreement*

This Agreement represents the entire agreement between the Publisher and the Licensee concerning the Product. The terms of this Agreement supersede all prior purchase orders, written terms and conditions, written or verbal representations, advertising or statements relating in any way to the Product.

12.5 *Severability*

If any provision of this Agreement is found to be invalid or unenforceable by a court of law of competent jurisdiction, such a finding shall not affect the other provisions of this Agreement and all provisions of this Agreement unaffected by such a finding shall remain in full force and effect.

12.6 *Variations*

This agreement may only be varied in writing by means of variation signed in writing by both parties.

12.7 *Notices*

All notices to be delivered to: Spon's Price Books, Taylor & Francis Books Ltd., 3 Park Square, Milton Park, Abingdon, Oxfordshire, OX14 4RN, UK.

12.8 *Governing law*

This Agreement is governed by English law and the parties hereby agree that any dispute arising under this Agreement shall be subject to the jurisdiction of the English courts.

If you have any queries about the terms of this licence, please contact:

Spon's Price Books
Taylor & Francis Books Ltd.
3 Park Square, Milton Park, Abingdon, Oxfordshire, OX14 4RN
Tel: +44 (0) 20 7017 6000
www.sponpress.com

Multiple-user use of the Spon Press Software and eBook

For more than one user or for a wide area network or consortium, use of *Spon's Online* is only permissible with the purchase from the Publisher of a multiple-user licence and adherence to the terms and conditions of that licence.

To buy a licence and set up access to *Spon's Online* for a number of users of in your organisation please contact:

Spon's Price Books
eBooks & Online Sales
Taylor & Francis Books Ltd.
3 Park Square, Milton Park, Oxfordshire, OX14 4RN
Tel: +44 (0) 207 017 6000
www.pricebooks.co.uk

Number of users	Licence cost
2	£270
5	£620
10	£1200
other	Please contact Spon for details